Ad Hoc Wireless Networks

Prentice Hall Communications Engineering and Emerging Technologies Series

Theodore S. Rappaport, *Series Editor*

DOSTERT *Powerline Communications*

DURGIN *Space–Time Wireless Channels*

GANZ, GANZ, & WONGTHAVARAWAT *Multimedia Wireless Networks: Technologies, Standards, and QoS*

GARG *Wireless Network Evolution: 2G to 3G*

GARG *IS-95 CDMA and cdma2000: Cellular/PCS Systems Implementation*

GARG & WILKES *Principles and Applications of GSM*

KIM *Handbook of CDMA System Design, Engineering, and Optimization*

LIBERTI & RAPPAPORT *Smart Antennas for Wireless Communications: IS-95 and Third Generation CDMA Applications*

MURTHY & MANOJ *Ad Hoc Wireless Networks: Architectures and Protocols*

PAHLAVAN & KRISHNAMURTHY *Principles of Wireless Networks: A Unified Approach*

RAPPAPORT *Wireless Communications: Principles and Practice, Second Edition*

RAZAVI *RF Microelectronics*

REED *Software Radio: A Modern Approach to Radio Engineering*

STARR, CIOFFI, & SILVERMAN *Understanding Digital Subscriber Line Technology*

STARR, SORBARA, CIOFFI, & SILVERMAN *DSL Advances*

TRANTER, SHANMUGAN, RAPPAPORT, & KOSBAR *Principles of Communication Systems Simulation with Wireless Applications*

VANGHI, DAMNJANOVIC, & VOJCIC *The cdma2000 System for Mobile Communications*

WANG & POOR *Wireless Communication Systems: Advanced Techniques for Signal Reception*

Ad Hoc Wireless Networks

Architectures and Protocols

C. Siva Ram Murthy
B. S. Manoj

Pearson

This edition is published by arrangement with Pearson Education, Inc. and Dorling Kindersley Publishing Inc.

ISBN 978-81-317-0688-6

First Impression, 2007
Thirty Eighth Impression, 2023
Thirty Ninth Impression, 2024

Published by Pearson India Education Services Pvt. Ltd, CIN: U72200TN2005PTC057128.

Head Office: 1st Floor, Berger Tower, Plot No. C-001A/2, Sector 16B,
Noida – 201 301, Uttar Pradesh, India.
Registered Office: Featherlite, 'The Address' 5th Floor, Survey No 203/10B, 200 Ft
MMRD Road, Zamin Pallavaram, Chennai - 600044
Website: in.pearson.com, Email: companysecretary.india@pearson.com

Printed in India by Saurabh Printers Pvt. Ltd.

To my wife, Sharada,
my son, Chandrasekhar,
and my daughter, Sarita.

C. Siva Ram Murthy

To my wife, Swapna,
and my daughter, Gouri.

B.S. Manoj

CONTENTS

PREFACE

In the last few years, there has been a big interest in ad hoc wireless networks as they have tremendous military and commercial potential. An ad hoc wireless network is a wireless network, comprised of mobile computing devices that use wireless transmission for communication, having no fixed infrastructure (a central administration such as a base station in a cellular wireless network or an access point in a wireless local area network). The mobile devices also serve as routers due to the limited range of wireless transmission of these devices, that is, several devices may need to route or relay a packet before it reaches its final destination. Ad hoc wireless networks can be deployed quickly anywhere and anytime as they eliminate the complexity of infrastructure setup. These networks find applications in several areas. Some of these include: military communications (establishing communication among a group of soldiers for tactical operations when setting up a fixed wireless communication infrastructure in enemy territories or in inhospitable terrains may not be possible), emergency systems (for example, establishing communication among rescue personnel in disaster-affected areas) that need quick deployment of a network, collaborative and distributed computing, wireless mesh networks, wireless sensor networks, and hybrid (integrated cellular and ad hoc) wireless networks.

The purpose of this book is to provide students, researchers, network engineers, and network managers with an expert guide to the fundamental concepts, design issues, and solutions to the issues — architectures and protocols — and the state-of-the-art research developments in ad hoc wireless networking. A unique feature of the book is that it deals with the entire spectrum of issues that influence the design and performance of ad hoc wireless networks, and solutions to the issues, with easy-to-understand illustrative examples highlighting the intuition behind each of the solutions.

This book, organized into fourteen chapters, each covering a unique topic in detail, first presents (in Chapters 1-4) the fundamental topics involved with wireless networking such as wireless communications technology, wireless LANs and PANs, wireless WANs and MANs, and wireless Internet. It then covers all important design issues (in Chapters 5-11) — medium access control, routing, multicasting, transport layer, security, quality of service provisioning, energy management — in ad hoc wireless networking in considerable depth. Finally, some recent related important topics covered in this book (in Chapters 12-14) include wireless sensor networks, hybrid wireless architectures, pricing in multihop wireless networks, ultra wideband technology, Wi-Fi systems, optical wireless networks, and Multimode 802.11.

The book is intended as a textbook for senior undergraduate and graduate-level courses on ad hoc wireless networks. It can also be used as a supplementary textbook for undergraduate courses on wireless networks, wireless/mobile communications, mobile computing, and computer networks. The exercise problems provided at the end of each chapter add strength to the book. A solutions manual for instructors is available from Prentice Hall. The book is a useful resource for the students and researchers to learn all about ad hoc wireless networking and further their research work. In addition, the book will be valuable to professionals in the field of computer/wireless networking.

We owe our deepest gratitude to Karthigeyan, Jayashree, and Archana for reading line by line all the chapters and suggesting ways to correct technical and presentation problems. We wish to express our thanks to the following HPCN lab students who have contributed mightily to this book writing project: Archana, Bhaya Gaurav Ravindra, Bheemarjun, Jagadeesan, Jayashree, Karthigeyan, Rajendra Singh Sisodia, Srinivas, Subir Kumar Das, Vidhyashankar, and Vyas Sekar. Raj Kumar drew all the illustrations and we thank him for his excellent work. We appreciate the efforts of Steven M. Hirschman, Irving E. Hodnett, and Shivkumar Kalyanaraman in reviewing our draft manuscript and suggesting improvements. We would like to gratefully acknowledge the help rendered by the Indian Institute of Technology (IIT), Madras, especially for creating an excellent working environment, the Department of Science and Technology, New Delhi, and the Curriculum Development Cell of the Centre for Continuing Education, IIT Madras for providing the financial aid for writing this book. Infosys Technologies Ltd., Bangalore, provided financial support to the second author for wireless networking research over the last four years, and he is indebted to Infosys for the same. We are thankful to Bernard Goodwin and his colleagues at Prentice Hall for their excellent work in producing this book. Last though not least, we acknowledge the love and affection from our families. This project would never have been successfully completed but for their understanding and patience.

We have taken reasonable care in eliminating typographical or other errors that might have crept into the book. We encourage you to send your comments and suggestions to us via email. We appreciate your feedback and hope you enjoy reading the book.

C. Siva Ram Murthy, *murthy@iitm.ernet.in*
B. S. Manoj, *bsmanoj@cs.iitm.ernet.in*

Chapter 1

INTRODUCTION

A computer network is an interconnected collection of independent computers which aids communication in numerous ways. Apart from providing a good communication medium, sharing of available resources (programs and data on a computer are available to anyone on the network, regardless of the physical location of the computer and user), improved reliability of service (because of the presence of multiple computers), and cost-effectiveness (as small computers have a better price/performance ratio than larger ones) are some of the advantages of computer networking. In the early days, till even the early 1970s, computing and communication were considered to be two separate independent fields. The late 1970s and the early 1980s saw the merger of the fields of computing and communication. A modern-day computer network consists of two major components, namely, distributed applications and networking infrastructure. The distributed applications provide services to users/applications located on other computers. Common examples of distributed applications are Internet, electronic mail, air/train ticket reservation systems, and credit card authentication systems. The networking infrastructure provides support for data transfer among the interconnected computers where the applications reside. There are no restrictions on the physical media used for providing connectivity among the computers. Several physical media such as copper cable, optic fiber, and wireless radio waves are used in the present-day communication infrastructure. The main focus of this book is wireless networking in general, and ad hoc wireless networks in particular. The first half of this chapter consists of a broad discussion of the various concepts and principles involved in wireless communication. The second half of the chapter provides a brief introduction to computer networks, network reference models, and wireless networks.

1.1 FUNDAMENTALS OF WIRELESS COMMUNICATION TECHNOLOGY

As mentioned earlier, this book focuses on wireless networks, where electromagnetic radio waves are used for communication (exchange of information). In the following sections, the various characteristics of radio propagation are first discussed. Some of the important signal modulation mechanisms, multiple access techniques, and error control mechanisms are then described. A familiarity with all these fundamental aspects of wireless transmission is essential for understanding the issues involved in the design of wireless networks.

1.2 THE ELECTROMAGNETIC SPECTRUM

Wireless communication is based on the principle of broadcast and reception of electromagnetic waves. These waves can be characterized by their frequency (f) or their wavelength (λ). Frequency is the number of cycles (oscillations) per second of the wave and is measured in Hertz (Hz), in honor of Heinrich Hertz, the German physicist who discovered radio, and wavelength is the distance between two consecutive maxima or minima in the wave. The speed of propagation of these waves (c) varies from medium to medium, except in a vacuum where all electromagnetic waves travel at the same speed, the speed of light. The relation between the above parameters can be given as

$$c = \lambda \times f \tag{1.2.1}$$

where c is the speed of light (3×10^8 m/s), f is the frequency of the wave in Hz, and λ is its wavelength in meters. Table 1.1 shows the various frequency bands in the electromagnetic spectrum as defined by the International Telecommunications Union (ITU). ITU, located in Geneva and a sub-organization of the United Nations, coordinates wired and wireless telecommunication activities worldwide. There are no official names for the bands in which the very high-frequency X-rays and Gamma rays fall. A pictographic view of the electromagnetic spectrum is given in Figure 1.1.

The low-frequency bands comprised of the radio, microwave, infrared, and visible light portions of the spectrum can be used for information transmission by modulating the amplitude, frequency, or the phase of the waves (the different modulation techniques will be described later in this chapter). The high-frequency waves such as X-rays and Gamma rays, though theoretically better for information propagation, are not used due to practical concerns such as the difficulty to generate and modulate these waves, and the harm they could cause to living things. Also,

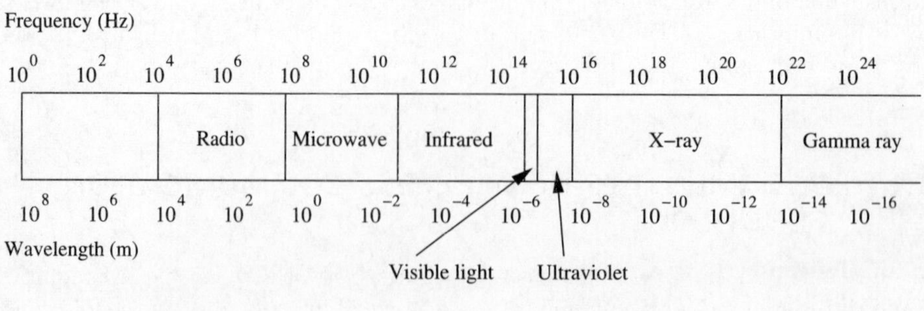

Figure 1.1. The electromagnetic spectrum.

Table 1.1. Frequency bands and their common uses

Band Name	Frequency	Wavelength	Applications
Extremely Low Frequency (ELF)	30 to 300 Hz	10,000 to 1,000 Km	Powerline frequencies
Voice Frequency (VF)	300 to 3,000 Hz	1,000 to 100 Km	Telephone communications
Very Low Frequency (VLF)	3 to 30 KHz	100 to 10 Km	Marine communications
Low Frequency (LF)	30 to 300 KHz	10 to 1 Km	Marine communications
Medium Frequency (MF)	300 to 3,000 KHz	1,000 to 100 m*	AM broadcasting
High Frequency (HF)	3 to 30 MHz	100 to 10 m	Long-distance aircraft/ship communications
Very High Frequency (VHF)	30 to 300 MHz	10 to 1 m	FM broadcasting
Ultra High Frequency (UHF)	300 to 3,000 MHz	100 to 10 cm	Cellular telephone
Super High Frequency (SHF)	3 to 30 GHz	10 to 1 cm	Satellite communications, microwave links
Extremely High Frequency (EHF)	30 to 300 GHz	10 to 1 mm	Wireless local loop
Infrared	300 GHz to 400 THz	1 mm to 770 nm	Consumer electronics
Visible Light	400 THz to 900 THz	770 nm to 330 nm	Optical communications

* Throughout this book, the unit *m* refers to meter(s).

such high-frequency waves do not propagate well through buildings. As we will see later, the amount of information that can be carried by an electromagnetic wave is determined by the width of the wavelength band (the corresponding frequency band) and the encoding technique (number of bits encoded/Hz) used.

Radio waves are easy to generate and are widely used for both indoor and outdoor communication due to properties such as their ability to pass through buildings and ability to travel long distances. Since radio transmission is omnidirectional (when radio waves are generated, they spread out from the transmitting antenna in all directions) in nature, the need to physically align the transmitter and receiver also does not arise. The frequency of the radio wave determines many of the characteristics of the transmission. At low frequencies the waves can pass through obstacles easily, but their power falls with an inverse-squared relation with respect to the distance. The higher frequency waves are more prone to absorption by rain drops, and they get reflected by obstacles. Due to the long transmission range of the radio waves, interference between transmissions is a problem that needs to be addressed.

In the VLF, LF, and MF bands the propagation of waves, also called as ground waves, follows the curvature of the Earth. The maximum transmission ranges of these waves are of the order of a few hundred kilometers. They are used for low bandwidth transmissions such as amplitude modulated (AM) radio broadcasting (amplitude modulation will be described in detail later in this chapter). The HF and VHF band transmissions are absorbed by the atmosphere near the Earth's surface. However, a portion of the radiation, called the sky wave, radiates outward and upward to the ionosphere in the upper atmosphere. The ionosphere contains ionized particles formed due to the sun's radiation. These ionized particles reflect the sky waves back to the Earth. A powerful sky wave may get reflected several times between the Earth and the ionosphere. Sky waves are used by amateur ham radio operators and for military communication.

Microwave transmissions (in the SHF band) tend to travel in straight lines and hence can be narrowly focused. Microwaves were widely used for long-distance telephony, before they got replaced by fiber optics. They are also widely used for mobile phones and television transmission. Since the energy is concentrated into a small beam, these transmissions have a higher signal-to-noise ratio (SNR). SNR is the ratio of the signal power to the noise power on a transmission medium, and is used to categorize the quality of a transmission. However, because of the higher frequency of operation they do not pass through buildings. Hence, proper line-of-sight alignment between the transmitting and the receiving antennas is required. Straight-line microwave transmission over long distances requires repeaters at regular distances as microwaves get attenuated by objects found in their path. The number of repeaters required is a function of the microwave transmission tower height.

Infrared waves and waves in the EHF band (also known as millimeter waves) are used for short-range communication. They are widely used in television, VCR, and stereo remote controls. They are relatively directional and inexpensive to build. They cannot travel through obstacles, which is a light-like (and not radio-like) property that is observed in very high-frequency transmissions.

The visible light part of the spectrum is just after the infrared portion. Unguided optical signaling using visible light provides very high bandwidth at a very low cost. Recent applications of light-wave transmissions involve the use of lasers to connect LANs on two buildings through roof-top antennas. But the main disadvantage here is that it is very difficult to focus a very narrow uni-directional laser beam, which limits the maximum distance between the transmitter and receiver. Also, such waves cannot penetrate through rain or thick fog.

1.2.1 Spectrum Allocation

Since the electromagnetic spectrum is a common resource which is open for access by anyone, several national and international agreements have been drawn regarding the usage of the different frequency bands within the spectrum. The individual national governments allocate spectrum for applications such as AM/FM radio broadcasting, television broadcasting, mobile telephony, military communication, and government usage. Worldwide, an agency of the International Telecommunications Union Radiocommunication (ITU-R) Bureau called World Administrative Radio Conference (WARC) tries to coordinate the spectrum allocation by the various national governments, so that communication devices that can work in multiple countries can be manufactured. However, the recommendations of ITU-R are not binding on any government. Regarding the national spectrum allocation, even when the government sets aside a portion of the spectrum for a particular application/use (*e.g.*, cellular telephony), another issue crops up – the issue of which carrier (company) is to use which set of frequencies. Many methods have been tried out for this frequency allocation among multiple competing carriers, some of which are described below.

The first and the oldest method, known as *comparative bidding* and often referred to as the *beauty contest* method, worked as follows. Each carrier would submit a proposal to the government explaining how it intends to use the frequency bands. The government, represented by officials, would go through the different proposals and come to a conclusion on which carrier would serve the public interest best, and accordingly allocate frequencies to that carrier. However, this method was often found to lead to bribery, corruption, and nepotism. Hence, this method was abandoned.

Governments then moved on to the lottery system, where a lottery would be held among the interested companies. But this method too had a serious problem: Companies which had no real interest in using the spectrum could participate in the lottery and win a segment of the spectrum. They would later resell the spectrum to genuine carriers and make huge profits. So this system also had to go.

The third method, which is still used today, is the auctioning method. The frequency bands would be auctioned off to the highest bidder. This method overcame all the problems associated with the previous two methods. Though governments were able to generate huge revenues by auctioning off frequencies, some companies that bought those frequencies at exorbitantly high rates through bidding are on the verge of going bankrupt.

The simplest method of allocating frequencies is not to allocate them at all. Internationally, the ITU has designated some frequency bands, called the ISM (industrial, scientific, medical) bands, for unlimited usage. These bands commonly used by wireless LANs and PANs are around the 2.4 GHz band. Parts of the 900 MHz and the 5 GHz bands are also available for unlicensed usage in countries such as the United States and Canada.

1.3 RADIO PROPAGATION MECHANISMS

Radio waves generally experience the following three propagation mechanisms:

- Reflection: When the propagating radio wave hits an object which is very large compared to its wavelength (such as the surface of the Earth, or tall buildings), the wave gets reflected by that object. Reflection causes a phase shift of 180 degrees between the incident and the reflected rays.

- Diffraction: This propagation effect is undergone by a wave when it hits an impenetrable object. The wave bends at the edges of the object, thereby propagating in different directions. This phenomenon is termed as diffraction. The dimensions of the object causing diffraction are comparable to the wavelength of the wave being diffracted. The bending causes the wave to reach places behind the object which generally cannot be reached by the line-of-sight transmission. The amount of diffraction is frequency-dependent, with the lower frequency waves diffracting more.

- Scattering: When the wave travels through a medium, which contains many objects with dimensions small when compared to its wavelength, scattering occurs. The wave gets scattered into several weaker outgoing signals. In practice, objects such as street signs, lamp posts, and foliage cause scattering.

Figure 1.2 depicts the various propagation mechanisms that a radio wave encounters. When the transmitted wave is received at the receiver, the received power of the signal is generally lower than the power at which it was transmitted. The loss in signal strength, known as attenuation, is due to several factors which are discussed in the next section.

1.4 CHARACTERISTICS OF THE WIRELESS CHANNEL

The wireless channel (transmission medium) is susceptible to a variety of transmission impediments such as path loss, interference, and blockage. These factors restrict the range, data rate, and the reliability of the wireless transmission. The extent to which these factors affect the transmission depends upon the environmental conditions and the mobility of the transmitter and receiver. Typically, the transmitted signal has a direct-path component between the transmitter and the receiver. Other components of the transmitted signal known as multipath components are reflected, diffracted, and scattered (explained in the previous section) by the environment, and arrive at the receiver shifted in amplitude, frequency, and phase with respect to the direct-path component. In what follows, the various characteristics of the wireless channel such as path loss, fading, interference, and Doppler shift are discussed. Also, two key constraints, Nyquist's and Shannon's theorems, that govern the ability to transmit information at different data rates, are presented.

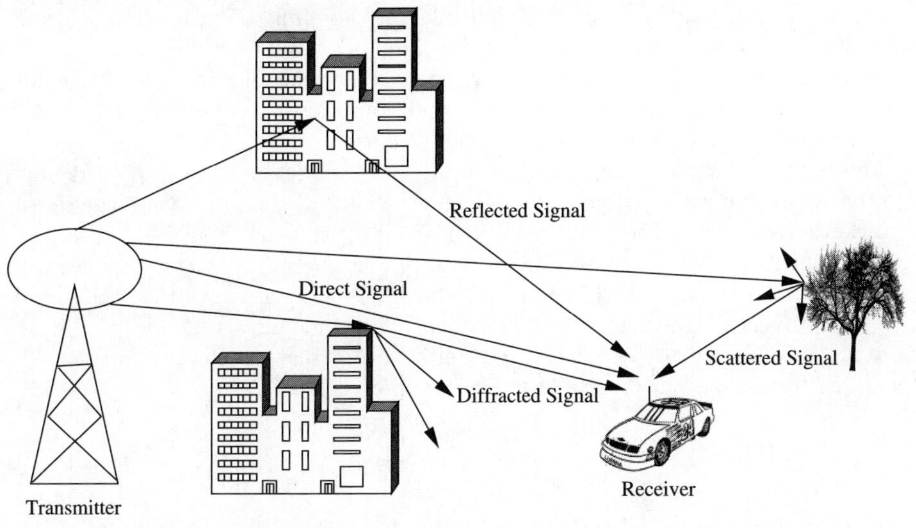

Figure 1.2. Propagation mechanisms.

1.4.1 Path Loss

Path loss can be expressed as the ratio of the power of the transmitted signal to the power of the same signal received by the receiver, on a given path. It is a function of the propagation distance. Estimation of path loss is very important for designing and deploying wireless communication networks. Path loss is dependent on a number of factors such as the radio frequency used and the nature of the terrain. Since several of these factors (in particular, the terrain) cannot be the same everywhere, a single model may not be enough, and hence cannot be used to describe the characteristics of every transmission. Therefore, in designing a network, several models are required to describe the variety of transmission environments.

The *free space* propagation model is the simplest path loss model in which there is a direct-path signal between the transmitter and the receiver, with no atmospheric attenuation or multipath components. In this model the relationship between the transmitted power P_t and the received power P_r is given by

$$P_r = P_t G_t G_r \left(\frac{\lambda}{4 \pi d} \right)^2 \tag{1.4.1}$$

where G_t and G_r are the transmitter and receiver antenna gains,[1] respectively, in the direction from the transmitter to the receiver, d is the distance between the transmitter and receiver, and $\lambda = c/f$ (from Equation 1.2.1) is the wavelength of

[1] Antenna gain is defined as the ratio of the power required at the input of a loss-free reference antenna to the power supplied to the input of the given antenna to produce, in a given direction, signals of the same strength at the same distance.

the signal. Realistic path loss models that take into account the propagation effects in specific environments can be obtained by solving Maxwell's equations (for details on Maxwell's equations, the reader can refer to [1]). Since obtaining solutions for such models involves complex algorithms and computation intensive operations, simpler models have been proposed to depict the loss.

Another popular path loss model is the *two-ray* model or the *two-path* model. The free space model described above assumes that there is only one single path from the transmitter to the receiver. But in reality, the signal reaches the receiver through multiple paths (because of reflection, refraction, and scattering). The two-path model tries to capture this phenomenon. The model assumes that the signal reaches the receiver through two paths, one a line-of-sight path, and the other the path through which the reflected (or refracted, or scattered) wave is received. According to the two-path model, the received power is given by

$$P_r = P_t G_t G_r \left(\frac{h_t h_r}{d^2}\right)^2 \tag{1.4.2}$$

where P_t is the transmitted power, G_t and G_r represent the antenna gains at the transmitter and the receiver, respectively, d is the distance between the transmitter and receiver, and h_t and h_r are the heights of the transmitter and the receiver, respectively.

In the general case, for isotropic antennas (antennas in which the power of the transmitted signal is the same in all directions), the received power of the signal is given by

$$P_r = P_t G_t G_r \left(\frac{\lambda}{4\pi}\right)^2 \frac{1}{d^\gamma} \tag{1.4.3}$$

where γ is the propagation coefficient that varies between 2 (free-space propagation) and 5 (strong attenuation). For more details on path loss models, the reader can refer to [2].

1.4.2 Fading

Fading refers to the fluctuations in signal strength when received at the receiver. Fading can be classified into two types: *fast fading/small-scale fading*, and *slow fading/large-scale fading*.

Fast fading refers to the rapid fluctuations in the amplitude, phase, or multipath delays of the received signal, due to the interference between multiple versions (copies) of the same transmitted signal arriving at the receiver at slightly different times. The time between the reception of the first version of the signal and the last echoed signal is called *delay spread*. The multipath propagation of the transmitted signal, which causes fast fading, is because of the three propagation mechanisms described previously, namely, reflection, diffraction, and scattering. The multiple signal paths may sometimes add constructively or sometimes destructively at the receiver, causing a variation in the power level of the received signal. The received signal envelope of a fast-fading signal is said to follow a Rayleigh distribution (refer to [3]) if there is no line-of-sight path between the transmitter and the receiver (applicable in outdoor environments), and a Ricean distribution (refer to [3]) if one such path is available (characterizes indoor settings).

Slow fading occurs when objects that partially absorb the transmissions lie between the transmitter and receiver. Slow fading is so called because the duration of the fade may last for multiple seconds or minutes. Slow fading may occur when the receiver is inside a building and the radio wave must pass through the walls of a building, or when the receiver is temporarily shielded from the transmitter by a building. The obstructing objects cause a random variation in the received signal power. Slow fading may cause the received signal power to vary, though the distance between the transmitter and receiver remains the same. Slow fading is also referred to as *shadow fading* since the objects that cause the fade, which may be large buildings or other structures, block the direct transmission path from the transmitter to the receiver.

Some of the common measures used for countering the effects of fading are diversity and adaptive modulation. Diversity mechanisms are based on the fact that independent paths between the same transmitter and receiver nodes experience independent fading effects, and therefore, by providing multiple logical channels between the transmitter and receiver, and sending parts of the signal over each channel, the error effects due to fading can be compensated. The independent paths can be distinct in space, time, frequency, or polarization. Time diversity mechanisms aim at spreading the data over time so that the effects of burst errors are minimized. Frequency diversity mechanisms spread the transmission over a wider frequency spectrum, or use multiple carriers for transmitting the information. The direct sequence spread spectrum and the frequency hopping spread spectrum techniques (which will be described later in this chapter) are some of the important frequency diversity mechanisms. Space diversity involves the use of different physical transmission paths. An antenna array could be used, where each antenna element receives an independent fading signal, each of the received signals following a different transmission path. The received signals are combined in some fashion such that the most likely transmitted signal could be reconstructed. In the adaptive modulation mechanisms, the channel characteristics are estimated at the receiver and the estimates are sent by the receiver to the transmitter through a feedback channel. The transmitter adapts its transmissions based on the received channel estimates in order to counter the errors that could occur due to the characteristics of the channel. Adaptive techniques are usually very complex to implement.

1.4.3 Interference

Wireless transmissions have to counter interference from a wide variety of sources. Two main forms of interference are adjacent channel interference and co-channel interference. In the adjacent channel interference case, signals in nearby frequencies have components outside their allocated ranges, and these components may interfere with on-going transmissions in the adjacent frequencies. It can be avoided by carefully introducing guard bands[2] between the allocated frequency ranges. Co-channel interference, sometimes also referred to as narrow-band interference, is due to other nearby systems (say, AM/FM broadcast) using the same transmission frequency. As we will see later in Chapter 3, narrow-band interference due to frequency reuse in cellular systems can be minimized with the use of multiuser detection mechanisms,[3] directional antennas, and dynamic channel allocation methods.

Inter-symbol interference is another type of interference, where distortion in the received signal is caused by the temporal spreading and the consequent overlapping of individual pulses in the signal. When this temporal spreading of individual pulses (delay spread) goes above a certain limit (symbol detection time), the receiver becomes unable to reliably distinguish between changes of state in the signal, that is, the bit pattern interpreted by the receiver is not the same as that sent by the sender. Adaptive equalization is a commonly used technique for combating inter-symbol interference. Adaptive equalization involves mechanisms for gathering the dispersed symbol energy into its original time interval. Complex digital processing algorithms are used in the equalization process. The main principle behind adaptive equalization is the estimation of the channel pulse response to periodically transmitted well-known bit patterns, known as training sequences. This would enable a receiver to determine the time dispersion of the channel and compensate accordingly.

1.4.4 Doppler Shift

The Doppler shift is defined as the change/shift in the frequency of the received signal when the transmitter and the receiver are mobile with respect to each other. If they are moving toward each other, then the frequency of the received signal will be higher than that of the transmitted signal, and if they are moving away from each other, the frequency of the signal at the receiver will be lower than that at the transmitter. The Doppler shift f_d is given by

$$f_d = \frac{v}{\lambda} \tag{1.4.4}$$

where v is the relative velocity between the transmitter and receiver, and λ is the wavelength of the signal.

1.4.5 Transmission Rate Constraints

Two important constraints that determine the maximum rate of data transmission on a channel are Nyquist's theorem and Shannon's theorem. The two theorems are presented below.

[2] A guard band is a small frequency band used to separate two adjacent frequency bands in order to avoid interference between them.

[3] Multiuser detection is an effective approach used to combat the multiuser interference problems inherent in CDMA systems.

Nyquist's Theorem

The signaling speed of a transmitted signal denotes the number of times per second the signal changes its value/voltage. The number of changes per second is measured in terms of *baud*. The baud rate is not the same as the bit rate/data rate of the signal since each signal value may be used to convey multiple bits. For example, if the voltage values used for transmission are 0, 1, 2, and 3, then each value can be used to convey two bits (00, 01, 10, and 11). Hence the bit rate here would be twice the baud rate. The *Nyquist theorem* gives the maximum data rate possible on a channel. If B is the bandwidth of the channel (in Hz) and L is the number of discrete signal levels/voltage values used, then the maximum channel capacity C according to the Nyquist theorem is given by

$$C = 2 \times B \times log_2 L \qquad \text{bits/sec} \tag{1.4.5}$$

The above condition is valid only for a noiseless channel.

Shannon's Theorem

Noise level in the channel is represented by the SNR. It is the ratio of signal power (S) to noise power (N), specified in decibels, that is, $SNR = 10\ log_{10}(S/N)$. One of the most important contributions of Shannon was his theorem on the maximum data rate possible on a noisy channel. According to *Shannon's theorem*, the maximum data rate C is given by

$$C = B \times log_2(1 + (S/N)) \qquad \text{bits/sec} \tag{1.4.6}$$

where B is the bandwidth of the channel (in Hz).

1.5 MODULATION TECHNIQUES

Having seen the various radio propagation mechanisms and wireless transmission impairments, the next step is to see how raw bits constituting the information are actually transmitted on the wireless medium. Data (whether in analog or in digital format) has to be converted into electromagnetic waves for transmission over a wireless channel. The techniques that are used to perform this conversion are called modulation techniques. The modulation process alters certain properties of a radio wave, called a carrier wave, whose frequency is the same as the frequency of the wireless channel being used for the transmission. The modulation scheme to be used must be chosen carefully so that the limited radio spectrum is used efficiently. Modulation schemes can be classified under two major categories: analog modulation schemes and digital modulation schemes. This classification is based on the nature of the data, analog or digital, to be transmitted. Some of the commonly used modulation techniques are discussed below.

1.5.1 Analog Modulation

As the name implies, analog modulation techniques are used for transmitting analog data. The analog data signal is superimposed on a carrier signal. This superimposition is aimed at altering a certain property (amplitude or frequency) of the carrier signal. Some of the commonly used analog modulation techniques are amplitude modulation, frequency modulation, and phase modulation. These techniques are described below.

Amplitude Modulation

Amplitude modulation (AM) is one of the simplest modulation schemes. It was the first method to be used for transmitting voice. The transmitter superimposes the information signal $x(t)$, also called the modulating signal, on the carrier signal $c(t)$. The result of AM of the carrier signal of Figure 1.3 (b), by the information/modulating signal of Figure 1.3 (a), is shown in Figure 1.3 (c). It can be seen that the frequency of the modulated wave remains constant, while its amplitude varies with that of the information signal. As the power of a signal depends on its amplitude, the power of the transmitted wave depends on the power of the modulating signal. When a cosine carrier wave is used, the AM wave can be mathematically represented as below:

$$s(t) = (1 + n_a x(t))\ cos(2\pi f_c t) \tag{1.5.1}$$

where n_a, known as the modulation index, is the ratio of the amplitude of the information signal to that of the carrier signal, f_c is the frequency of the carrier signal, $x(t)$ is the information signal, and $c(t) = cos(2\pi f_c t)$ is the carrier signal.

In modern communication systems, AM is not the preferred modulation mechanism. AM creates additional unwanted signals in frequencies on either side of the carrier during the superimposition process. These signals, called sidebands, lead to poor spectrum utilization, and they consume additional power. Hence, AM is considered to be an inefficient modulation mechanism. A variant of AM, where the sidebands are stripped away on one side, called single side band (SSB), is often used in place of plain AM. Broadcast radio is one of the applications which still uses AM.

Angle Modulation

Frequency modulation and phase modulation come under this category. The angle modulated signal can be mathematically represented as

$$s(t) = A_c\,[cos(2\pi f_c t) + \Phi(t)] \tag{1.5.2}$$

where A_c is the amplitude and f_c is the frequency of the carrier signal. Frequency modulation and phase modulation are special cases of angle modulation. Changing the frequency of a wave also changes its phase, and vice versa. The angle modulated signal has a constant amplitude, and as a consequence the transmitter operates at full power constantly, which also maximizes its range.

Frequency Modulation

In frequency modulation (FM), the amplitude of the modulated signal is kept constant, while the instantaneous frequency is altered to reflect the information signal

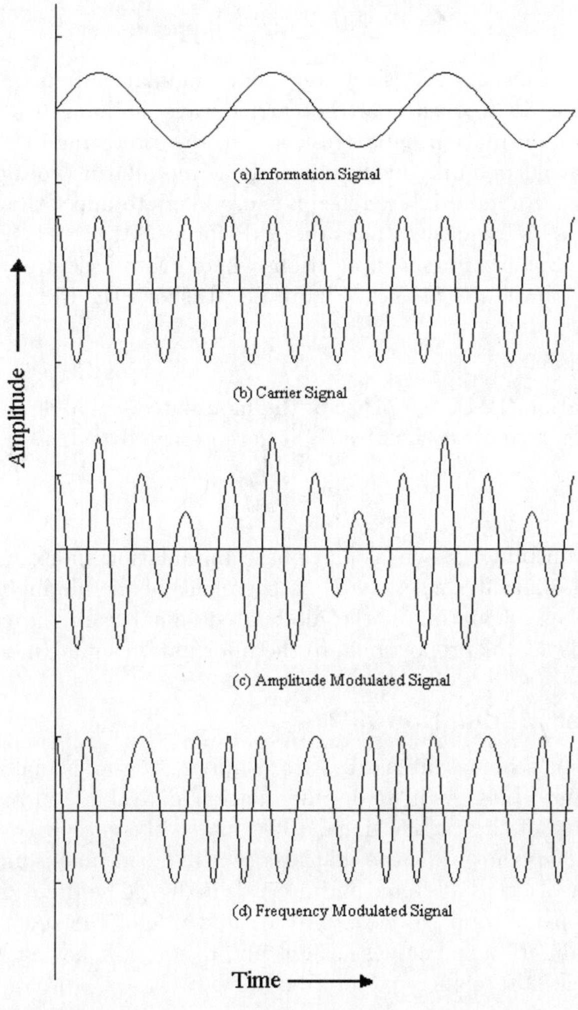

Figure 1.3. Analog modulation schemes.

that is being transmitted. Frequency modulation follows Equation 1.5.2, with the condition that the derivative of the phase, that is, $\Phi^{'}(t)$, is proportional to the information signal $x(t)$. Since frequency can also be defined as the rate of change of phase of the signal, here $\Phi^{'}(t)$ represents the deviation of the instantaneous frequency of the modulated signal from that of the carrier signal. In FM, this $\Phi^{'}(t)$ is directly proportional to the information signal/modulating signal. It can be represented as

$$\Phi'(t) = n_f x(t) \tag{1.5.3}$$

where n_f is a constant, known as the frequency modulation index.

The instantaneous frequency of the carrier wave is changed according to the amplitude of the information signal, resulting in the stretching or the compressing of the carrier wave depending on the value of the modulating voltage. FM is more resistant to noise. Random interference is more likely to affect the amplitude of a signal rather than its frequency. Thus the SNR for an FM wave is higher than that for an AM wave. Some common applications where FM is used are radio broadcasts and first-generation cellular phones. Figure 1.3 (d) shows the FM modulated signal.

Phase Modulation

In phase modulation (PM), the phase of the modulated signal $\Phi(t)$ is directly proportional to the information signal $x(t)$. It is represented as

$$\Phi(t) = n_p x(t) \tag{1.5.4}$$

where n_p is a constant, known as the phase modulation index. It can be seen from Equation 1.5.2 that $[2\pi f_c t + \Phi(t)]$ is the phase of the modulated signal. The instantaneous phase deviation of the modulated signal from the carrier signal is $\Phi(t)$. In PM, this $\Phi(t)$ is proportional to the information signal/modulating signal.

1.5.2 Digital Modulation

Digital modulation schemes are used for transmitting digital signals that consist of a sequence of 0 and 1 bits. As in analog modulation, digital modulation also changes a certain property of the carrier signal. The main difference between analog and digital modulation is that while the changes occur in a continuous manner in analog modulation, they occur at discrete time intervals in digital modulation. The number of such changes per second is known as the *baud rate* of the signal. Some of the basic digital modulation techniques such as amplitude shift keying, frequency shift keying, and phase shift keying are described below.

Amplitude Shift Keying

In amplitude shift keying (ASK), when a bit stream is transmitted, a binary 1 is represented by the presence of the carrier signal $c(t)$ for a specified interval of time, and a binary 0 is represented by the absence of the carrier signal for the same interval of time. Mathematically, ASK can be represented as

$$s(t) = \begin{cases} A_c\, cos(2\pi f_c t), & \text{for binary 1} \\ 0, & \text{for binary 0} \end{cases}$$

where A_c is the amplitude of the carrier signal and f_c is its frequency. The result of ASK when applied to the bit pattern shown in Figure 1.4 (a) is shown in Figure 1.4 (b).

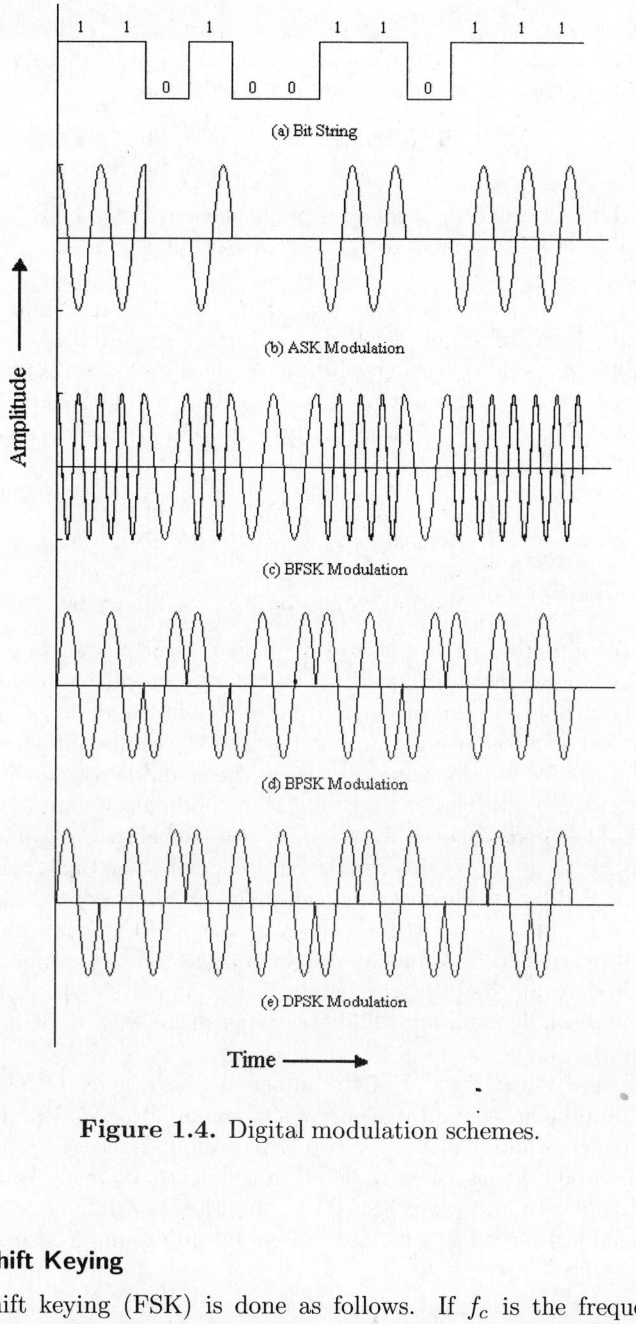

Figure 1.4. Digital modulation schemes.

Frequency Shift Keying

Frequency shift keying (FSK) is done as follows. If f_c is the frequency of the carrier signal and if k is a small frequency offset, then transmission of a binary 1 is represented by the presence of a carrier signal of frequency $f_c + k$ for a specified interval of time. Transmission of bit 0 is represented by the presence of a carrier signal of frequency $f_c - k$ for the same interval of time. Since two frequency levels are used in this technique, it is also known as two-level FSK or binary FSK (BFSK). FSK can be mathematically represented as follows:

$$s(t) = \begin{cases} A_c \, cos(2\pi(f_c + k)t), & \text{for binary 1} \\ A_c \, cos(2\pi(f_c - k)t), & \text{for binary 0} \end{cases}$$

where A_c and f_c are the amplitude and frequency, respectively, of the cosine carrier signal. FSK when applied to the bit pattern of Figure 1.4 (a) is shown in Figure 1.4 (c).

In the FSK mechanism described above, that is, BFSK, since only two frequency levels are used, each frequency level encodes just a single bit, 1 or 0. By using multiple frequency levels (by using multiple frequency offsets), a single frequency level can be used to encode multiple bits. An example is the four-level FSK. A four-level FSK can be used to transmit two bits per frequency shift. It can be mathematically represented as

$$s(t) = \begin{cases} A_c \, cos(2\pi(f_c + 3k)t), & \text{for binary 10} \\ A_c \, cos(2\pi(f_c + k)t), & \text{for binary 11} \\ A_c \, cos(2\pi(f_c - k)t), & \text{for binary 01} \\ A_c \, cos(2\pi(f_c - 3k)t), & \text{for binary 00} \end{cases}$$

where A_c is the amplitude and f_c is the frequency of the carrier signal.

The spacing between the frequencies used for representing bits 0 and 1 (also known as tone distance) is an important parameter in the design of FSK systems. The frequency spacing is directly proportional to the bandwidth occupied by the FSK signal. In order for a receiver to detect the 1 and 0 bits without any error, the tone distance must be chosen carefully. FSK modulation with a minimum tone distance of $1/2T$, where T is the duration of each transmitted bit, is referred to as minimum shift keying (MSK). In order to reduce side bands, the baseband signal to be transmitted is filtered before the frequency shift keying process. Gaussian filters are the most commonly used filters for this purpose, and the associated modulation mechanism is referred to as Gaussian MSK (GMSK). The filtering smoothens the rapid transmissions and reduces the bandwidth occupied by the signal. GMSK is widely used in second-generation cellular systems such as GSM, described in detail in Chapter 3.

Gaussian FSK (GFSK) is an FSK technique in which the data to be transmitted is first filtered in the baseband by means of a Gaussian filter, and is then modulated by simple frequency modulation. A two-level or four-level GFSK is used in the IEEE 802.11 standard (described in detail in the next chapter). An n-level GFSK technique is similar to an n-level FSK technique, the only difference being that the baseband signal is filtered using a Gaussian filter before being frequency modulated in the n-level GFSK.

Phase Shift Keying

In phase shift keying (PSK), change in phase of the carrier signal is used to represent the 0 and 1 bits. The transmission of bit 0 is represented by the presence of the carrier for a specific interval of time, while the transmission of bit 1 is represented by the presence of a carrier signal with a phase difference of π radians for the same interval of time. PSK using a cosine carrier wave with amplitude A_c and frequency f_c can be mathematically represented as

$$s(t) = \begin{cases} A_c\, cos(2\pi f_c t + \pi), & \text{for binary 1} \\ A_c\, cos(2\pi f_c t), & \text{for binary 0} \end{cases}$$

This technique is also known as binary PSK (BPSK) or two-level PSK since a single phase difference is used for representing 0 and 1 bits. Figure 1.4 (d) shows the BPSK modulation of the bit pattern of Figure 1.4 (a).

Just as multiple frequency levels are used in FSK, multiple phase deviations can be used in PSK. This enables encoding of multiple bits by each phase representation. Quadrature PSK (QPSK), for example, uses four different phases each separated by $\pi/2$ radians. This would enable transmission of two bits per phase shift. The mathematical representation of QPSK is given below.

$$s(t) = \begin{cases} A_c\, cos\left(2\pi f_c t + \frac{\pi}{4}\right), & \text{for binary 10} \\ A_c\, cos\left(2\pi f_c t + \frac{3\pi}{4}\right), & \text{for binary 11} \\ A_c\, cos\left(2\pi f_c t + \frac{5\pi}{4}\right), & \text{for binary 01} \\ A_c\, cos\left(2\pi f_c t + \frac{7\pi}{4}\right), & \text{for binary 00} \end{cases}$$

$\pi/4$ *shifted QPSK* ($\pi/4$-QPSK) is a QPSK mechanism where the maximum phase deviation is limited to $\pm$ 135 degrees. The main advantage of $\pi/4$ *shifted QPSK* is that it can be received non-coherently, that is, the receiver need not lock to the phase of the transmitted signal, which simplifies the receiver design. It provides the bandwidth efficiency of QPSK along with lesser fluctuations in amplitude. $\pi/4$ *shifted QPSK* is used in the North American digital cellular TDMA standard, IS-136, and also in the Japanese digital cellular (JDC) standard.

Differential PSK (DPSK) is a variation of the basic PSK mechanism. Here, binary 1 is represented by the presence of a carrier signal whose phase has been changed relative to the phase of the carrier used for representing the previous bit. A binary 0 is represented by the presence of a carrier wave whose phase is the same as that of the carrier used for transmitting the previous bit. Figure 1.4 (e) depicts the DPSK pattern when applied on the bit pattern of Figure 1.4 (a). DPSK comes with an added advantage. Since phase differences occur continuously for long runs of 1s, this can be used for self-clocking. If there are only two possible phase differences, used for representing bit 0 and bit 1, then the modulation technique is called *differential binary PSK* (DBPSK). If four phase differences are used, for representing the bit sequences 00, 01, 10, and 11, then the scheme is called *differential quadrature PSK* (DQPSK). If a greater number of phase differences, say, 8 or 16, is used, then the corresponding systems are called 8-DPSK and 16-DPSK, respectively. *$\pi/4$-DQPSK* is another modulation technique, which is a variant of the DQPSK technique. In $\pi/4$-DQPSK, an additional phase shift of $\pi/4$ radians is inserted in each symbol. In standard DQPSK, a long run of 0s at the data input would result in a signal with no phase shifts at all, which makes synchronization

Table 1.2. Phase changes used in $\pi/4$ shifted PSK

Pair of Bits	Phase Change
00	$\pi/4$
01	$5\pi/4$
10	$-\pi/4$
11	$-5\pi/4$

at the receiver very difficult. If $\pi/4$-DQPSK is used in such a situation, the phase shift of $\pi/4$ ensures that there is a phase transition for every symbol, which would enable the receiver to perform timing recovery and synchronization.

Another variant of PSK is the $\pi/4$ *shifted PSK* ($\pi/4$-PSK). This is a four-level PSK technique, hence each phase shift represents two bits. A pair of bits are represented by varying the phase of the carrier signal relative to the phase of the carrier signal used for representing the preceding pair of bits. Table 1.2 shows the phase changes used for various pairs of bits. This method also provides for self-clocking since there is always a phase shift between the transmissions of consecutive bits. Figure 1.5 illustrates the $\pi/4$ *shifted PSK* mechanism for a bit string 110001.

The next modulation mechanism to be discussed here is *quadrature amplitude modulation* (QAM). Both the amplitude and the phase are varied in QAM in order to represent bits. In QPSK, each shift in phase would be used to code a pair of bits. But if two amplitude values were used in combination with regular QPSK, it would be possible to obtain eight different combinations. And hence it would now be possible for each of those combinations to encode three bits. This effectively increases the data rate of the system. The different combinations possible are also known as constellation patterns. Figure 1.6 depicts the constellation pattern for a system using four phase changes with two amplitude values. By using a large number of

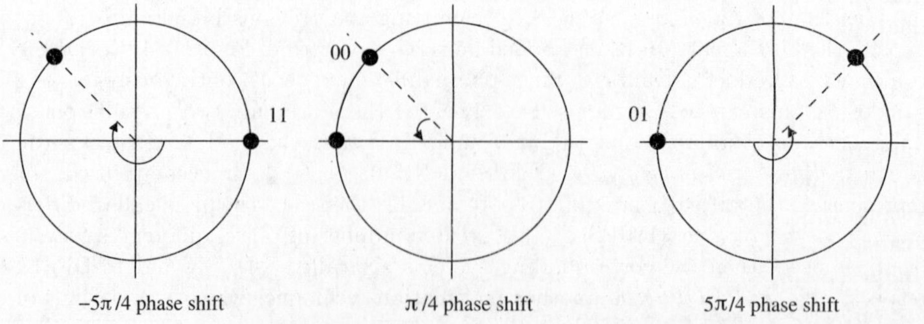

Figure 1.5. Operation of $\pi/4$ shifted PSK.

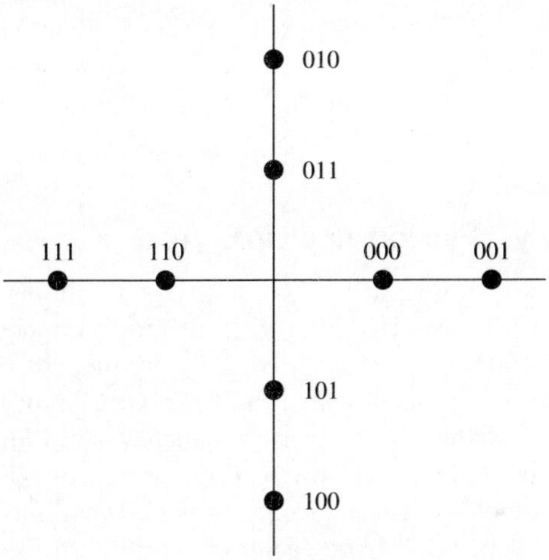

Figure 1.6. Constellation pattern in 8-QAM.

phase changes/amplitude combinations, it would be possible to encode multiple bits using each of those combinations. Hence, spectrum utilization is greatly improved in QAM. Several higher level QAM schemes which are more bandwidth-efficient are in use today. 16-QAM and 64-QAM are two such schemes where 16 and 64 different phase changes/amplitude combinations, respectively, are used. The main drawback of QAM systems is that, as the number of phase changes/amplitude combinations are increased, the system becomes more complex and is more susceptible to errors caused due to noise and distortion.

1.6 MULTIPLE ACCESS TECHNIQUES

The previous section discussed techniques used for the transmission of raw bits on the channel. Since the transmission medium in wireless networks is broadcast in nature, a node (wireless communication device) cannot transmit on the channel whenever it wants to. Multiple access techniques are used to control access to the shared channel. They determine the manner in which the channel is to be shared by the nodes. Some of the basic multiple access techniques used in wireless networks are presented in this section.

Multiple access techniques are based on the orthogonalization[4] of signals, each signal represented as a function of time, frequency, and code. Hence, multiplexing can be performed with respect to one of these three parameters; the respective techniques are termed frequency division multiple access, time division multiple access,

[4]Two vectors are said to be orthogonal if the normalized inner product of the two vectors, which is given by the cosine of the angle between the two vectors, is zero. The normalized inner product of two vectors is a measure of similarity between the two vectors. If it is zero, then it implies that the two vectors are totally dissimilar.

and code division multiple access. Apart from these techniques, the medium can be multiplexed with respect to space also, and this technique is called space division multiple access. A brief discussion of each of the above techniques is presented below.

1.6.1 Frequency Division Multiple Access

The frequency division multiple access (FDMA) mechanism operates as below. The available bandwidth is divided into multiple frequency channels/bands. A transmitter–receiver pair uses a single dedicated frequency channel for communication. Figure 1.7 depicts the principle behind the operation of FDMA. The frequency spectrum is in effect divided into several frequency sub-bands. Transmissions on the main band of a channel also result in the creation of additional signals on the side bands of the channel. Hence the frequency bands cannot be close to each other. Frequency bands are separated from each other by guard frequency bands in order to eliminate inter-channel interference. These guard bands result in the under-utilization of the frequency spectrum. This is the main disadvantage of FDMA. FDMA has been widely adopted in analog systems for portable telephones and automobile telephones.

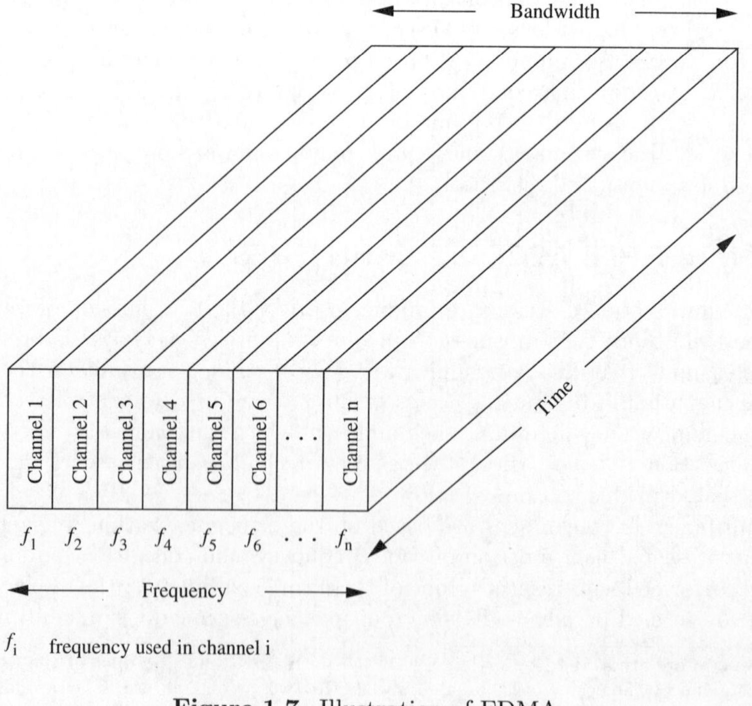

Figure 1.7. Illustration of FDMA.

In a cellular network, a central controller known as the base station (BS) dynamically allocates a different carrier frequency to each node, known as the mobile station (MS). Each node is in fact allocated a pair of frequencies for communication, one for the traffic from the MS to the BS (uplink frequency), and the other for carrying traffic from the BS to the MS (downlink frequency). This system, used for two-way communication between a pair of stations (MS and BS here), is called *frequency division duplexing* (FDD). Since high-frequency transmissions suffer greater attenuation when compared to low-frequency transmissions, high transmission power is required for high-frequency channels for compensating the transmission losses. Power available at an MS is limited. Hence, in order to conserve energy at the MS, the uplink frequency is always lower than the downlink frequency.

Orthogonal Frequency Division Multiplexing

Orthogonal frequency division multiplexing (OFDM) is a multi-carrier transmission mechanism. It resembles FDMA in that both OFDM and FDMA split the available bandwidth into a number of frequency channels. OFDM is based on the spreading of the data to be transmitted over multiple carriers, each of them being modulated at a low rate. The data signal is split into multiple smaller sub-signals that are then transmitted to the receiver simultaneously at different carrier frequencies (sub-carriers). Splitting the carrier into multiple smaller sub-carriers and then broadcasting the sub-carriers simultaneously reduce the signal distortion at the receiver caused due to multipath propagation of the transmitted signal. By appropriately choosing the frequency spacing between the sub-carriers, the sub-carriers are made orthogonal to each other. Though spectral overlapping among sub-carriers occurs, the orthogonality of the sub-carriers ensures error-free reception at the receiver, thereby providing better spectral efficiency. OFDM is sometimes also referred to as discrete multi-tone (DMT) modulation. OFDM is currently used in several applications such as wireless local area networks (WLANs) and digital broadcasting.

1.6.2 Time Division Multiple Access

Time division multiple access (TDMA) shares the available bandwidth in the time domain. Each frequency band is divided into several time slots (channels). A set of such periodically repeating time slots is known as the TDMA frame. Each node is assigned one or more time slots in each frame, and the node transmits only in those slots. Figure 1.8 depicts the concept behind TDMA. For two-way communication, the uplink and downlink time slots, used for transmitting and receiving data, respectively, can be on the same frequency band (TDMA frame) or on different frequency bands. The former is known as *time division duplex - TDMA* (TDD-TDMA), and the latter as *frequency division duplex - TDMA* (FDD-TDMA). Though TDMA is essentially a half-duplex mechanism, where only one of the two communicating nodes can transmit at a time, the small duration of time slots creates the illusion of a two-way simultaneous communication. Perfect synchronization is required between the sender and the receiver. To prevent synchronization errors and inter-symbol in-

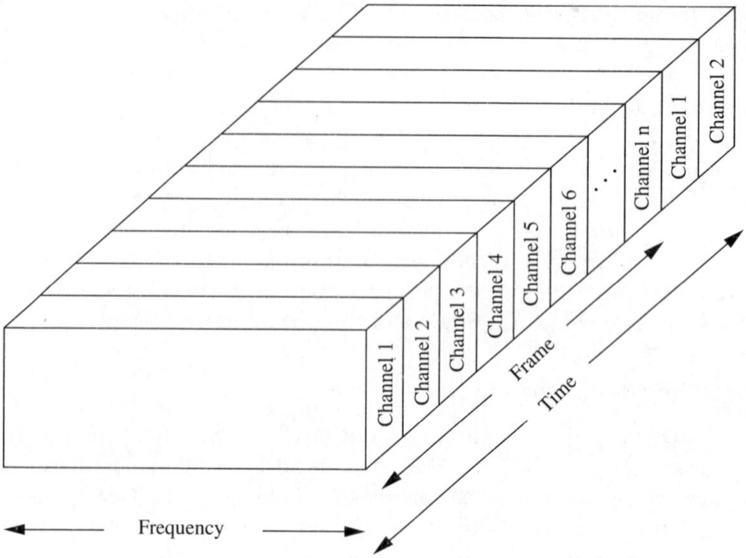

Figure 1.8. Illustration of TDMA.

terference due to signal propagation time differences, guard intervals are introduced between time slots. Since the sizes of slots are already small, the introduction of guard intervals results in a significant overhead for the system.

FDMA requires the device to have the capability of simultaneously receiving and transmitting signals, which leads to increased cost. But when TDMA is used, the device can switch between slots and hence use the same transmitter for receiving also. Hence, the equipment cost in TDMA is less. TDMA is widely used in second-generation cellular systems such as GSM, described in detail in Chapter 3.

1.6.3 Code Division Multiple Access

Unlike other systems such as TDMA and FDMA, code division multiple access (CDMA) does not assign a specific frequency to each user. Instead, every channel uses the entire spectrum. Individual conversations are encoded with a pseudo-random digital sequence. As the narrow-band transmission frequency is spread over the entire wideband spectrum, the technique is also called as spread spectrum. The transmissions are differentiated through a unique code that is independent of the data being transmitted, assigned to each user. The orthogonality of the codes enables simultaneous data transmissions from multiple users using the entire frequency spectrum, thus overcoming the frequency reuse limitations seen in FDMA and TDMA.

CDMA was first used during World War II by the English Allies to foil attempts by the German army to jam the transmissions of the Allies. The Allies decided to transmit signals over several frequencies in a specific pattern, instead of just one frequency, thereby making it very difficult for the Germans to get hold of the complete signal.

Two types of spread spectrum systems are widely in use today, namely, *frequency hopping spread spectrum* and *direct sequence spread spectrum*, which are described below.

Frequency Hopping Spread Spectrum

Frequency hopping spread spectrum (FHSS) is a simple technique in which the transmission switches across multiple narrow-band frequencies in a pseudo-random manner, that is, the sequence of transmission frequencies is known both at the transmitter and the receiver, but appears random to other nodes in the network. The process of switching from one channel to the other is termed frequency hopping. The radio frequency signal is de-hopped at the receiver by means of a frequency synthesizer controlled by a pseudo-random sequence generator. Figure 1.9 illustrates the concept behind FHSS. The figure shows two simultaneous transmissions. The first transmission (darker shade in figure) uses the hopping sequence $f_4\ f_7\ f_2\ f_1\ f_5\ f_3\ f_6\ f_2\ f_3$ and the second transmission uses the hopping sequence $f_1\ f_3\ f_6\ f_2\ f_4\ f_2\ f_7\ f_1\ f_5$. Frequency hopped systems are limited by the total number of frequencies available for hopping. FHSS can be classified into two types: fast FHSS and slow FHSS. In fast FHSS, the dwell time on each frequency is very small, that is, the rate of change of frequencies is much higher than the information

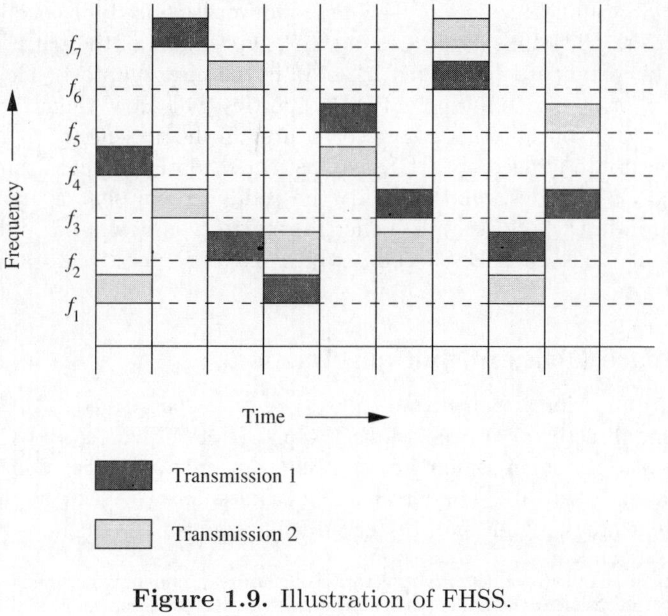

Figure 1.9. Illustration of FHSS.

bit rate, resulting in each bit being transmitted across multiple frequency hops. In slow FHSS, the dwell time on each frequency is high, hence multiple bits are transmitted on each frequency hop. FHSS is now used mainly for short-range radio signals, particularly those in the unlicensed bands.

Direct Sequence Spread Spectrum

The principle behind direct sequence spread spectrum (DSSS) can be explained easily by the following analogy. Suppose conversations in several languages are occurring in a room. People who understand only a particular language listen to and follow the conversation taking place in that language alone. They ignore and discard conversations in other languages as noise. The same principle applies to DSSS. In DSSS, each node is assigned a specific n-bit code, called a *chipping code.* n is known as the chipping rate of the system. These codes assigned to the nodes are orthogonal to each other, that is, the normalized inner product[5] of the vector representations of any two codes is zero. Each node transmits using its code. At the receiver, the transmission is received and information is extracted using the transmitter's code. For transmitting a binary 1, the sender transmits its code; for a binary 0, the one's complement of the code is transmitted. Hence, transmission of a signal using CDMA occupies n times the bandwidth that would be required for a narrow-band transmission of the same signal. Since the transmission is spread over a large bandwidth, resistance to multipath interference is provided for in CDMA. Figure 1.10 illustrates the DSSS transmission. DSSS is used in all CDMA cellular telephony systems.

Complementary code keying (CCK) is a modulation technique used in conjunction with DSSS. In CCK, a set of 64 8-bit code words is used for encoding data for the 5.5 Mbps and 11 Mbps data rates in the 2.4 GHz band of the IEEE 802.11 wireless networking standard (described in detail in the next chapter). The code words used in CCK have unique mathematical properties that allow them to be correctly distinguished from one another by a receiver even in the presence of significant noise and multipath interference. In CCK, sophisticated mathematical formulas are applied to the DSSS codes, which permit the codes to represent a greater amount of information in each clock cycle. The transmitter can now send multiple bits of information through each DSSS code, which makes it possible to achieve the 11 Mbps data rate.

1.6.4 Space Division Multiple Access

The fourth dimension in which multiplexing can be performed is space. Instead of using omnidirectional transmissions (as in FDMA, TDMA, and CDMA) that cover the entire circular region around the transmitter, space division multiple access (SDMA) uses directional transmitters/antennas to cover angular regions. Thus different areas/regions can be served using the same frequency channel. This

[5]The normalized inner product of two vectors is a measure of similarity between the two vectors. It is given by the cosine of the angle between the two vectors. If the normalized inner product is zero, then it implies that the two vectors are totally dissimilar.

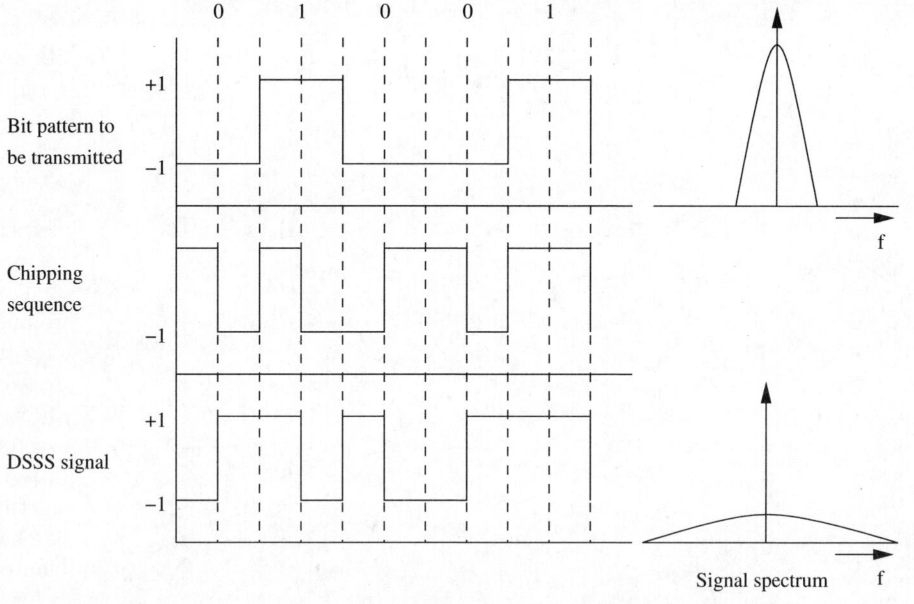

Figure 1.10. Illustration of DSSS.

method is best suited to satellite systems, which often need a narrowly focused beam to prevent the signal from spreading too widely and in the process becoming too weak. A satellite can thus reuse the same frequency to cover many different regions of the Earth's surface. Figure 1.11 shows how SDMA could be used for satellite communication.

1.7 VOICE CODING

The primary objective of most wireless networks in existence today is to support voice communication. If just a data signal consisting of binary digits is to be transmitted, the bits to be transmitted can be directly used to modulate the carrier signal to be used for the transmission. This process is not so direct in the case of analog voice signals. The voice coding process converts the analog signal into its equivalent digital representation. The analog speech information to be transmitted is first converted into a sequence of digital pulses. It should then be transmitted without any noticeable distortion. The devices that perform this analog to digital conversion (at the sender) and the reverse digital to analog signal conversion (at the receiver) are known as *codecs* (coder/decoder). The main goal of a codec is to convert the voice signal into a digital bit stream that has the lowest possible bit rate, while maintaining an acceptable level of quality of the signal, that is, it should be possible to reproduce the original analog speech signal at the receiver without any

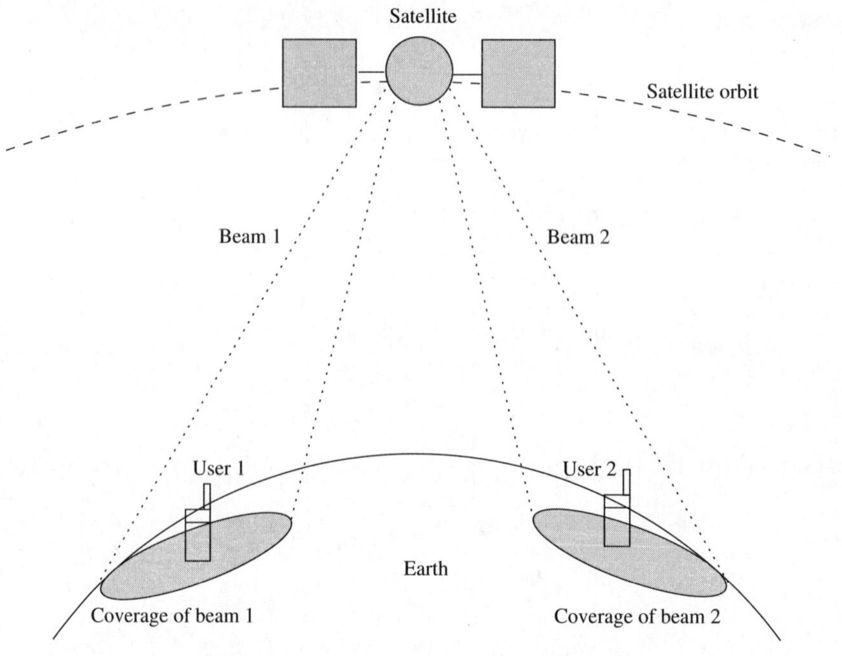

Figure 1.11. Illustration of SDMA.

perceivable distortion or disturbances. Pulse position modulation (PPM) is one of the techniques used for converting an analog signal into its digital representation. In PPM, the amplitude and width of the pulse are kept constant. The position of each pulse, with respect to the position of a recurrent reference pulse, is varied according to the instantaneous sampled amplitude of the information signal. PPM requires constant transmitter power since all pulses have the same constant amplitude and duration. PPM is used in the infrared physical layer specification of the IEEE 802.11 standard. The main disadvantage of PPM is that perfect synchronization is required between the transmitter and receiver. Pulse code modulation is another technique which is commonly used for conversion of an analog signal into its corresponding digital form.

1.7.1 Pulse Code Modulation

Pulse code modulation (PCM) essentially consists of three stages, namely, sampling of the analog signal, quantization, and binary encoding. Each of these stages is described below.

Sampling

The codec converts the analog speech signal to its digital representation by sampling the signal at regular intervals of time. The higher the sampling rate, that is, the shorter the time interval between successive samples, the better the description of the voice signal. The frequency of the samples to be taken, such that quality of the voice signal is preserved, is given by the *sampling theorem.* According to the sampling theorem, if the original voice signal has a limited bandwidth, that is, it does not contain any component with a frequency exceeding a given value B, then all information in the original signal will be present in the signal described by the samples, if the sampling frequency is greater than twice the highest frequency in the original signal, that is, it is greater than $2 \times B$. This theorem is also known as Nyquist's theorem. The series of pulses produced after sampling the analog signal, known as pulse amplitude modulation (PAM) pulses, have their amplitudes proportional to that of the original signal. Figure 1.12 (b) shows the PAM pulses generated for the analog waveform of Figure 1.12 (a).

Quantization

During the quantization phase, the amplitudes of the PAM pulses are measured and each pulse is assigned a numerical value. In order to avoid having to deal with an infinite number of values for the PAM pulses, a fixed number of amplitude levels are used. The amplitude of the analog signal at each sampling instant is rounded off to the nearest level. Figure 1.12 (c) depicts the output of quantization applied to the PAM pulses in Figure 1.12 (b). Since some of the information is lost due to the approximation of amplitude levels, quantization distorts the original analog signal. This distortion is also known as quantization error. But at the same time, since only a fixed number of amplitude levels are used, only a fixed number of numerical values need to be transmitted. This reduces the complexity of the equipment and also the probability of transmission errors.

Binary Encoding

In this final phase, the quantized PAM pulses are encoded into the binary format, which forms the final output of the analog-to-digital conversion process. The number of bits used to represent each amplitude level is based on the total number of amplitude levels that need to be represented. The quantized PAM pulses in Figure 1.12 (c) use seven quantization levels. Therefore, at least three bits would be required for representing each level. The encoded output of the quantized pulses in Figure 1.12 (c) would be 110 110 100 010 011 110 110 101 100 100 101. This bit pattern is then modulated and transmitted by the sender.

1.7.2 Vocoders

The PCM mechanism described above requires high bit rates. For example, if a sampling rate of 8,000 samples/second is used and each sample is encoded into eight bits, then the corresponding data rate would be 64 Kbps. Hence, PCM is

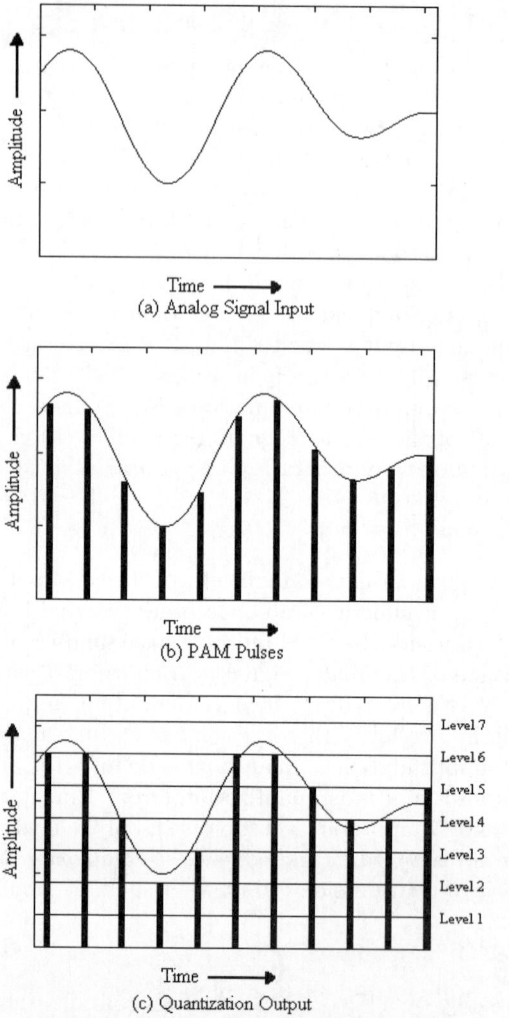

Figure 1.12. Pulse code modulation.

not very suitable for wireless networks where the limited bandwidth needs to be utilized efficiently. Vocoders are devices that operate making use of knowledge of the actual structure and operation of human speech production organs. They do not modulate the actual voice signal; instead they work by modeling the mechanics by which sound is produced (*e.g.*, lip movement, and the effect of the throat and nose on speech). This information is encoded and transmitted by the sender, and is received by the receiver, which then reconstructs the original signal. Vocoders achieve voice transfer at low bit rates. Hence they are more suitable compared to PCM for voice communication in wireless networks. But a drawback of using vocoders is that since the actual voice signals are not transmitted, the signals received and reconstructed at the receiver do not appear to be natural. The artificiality in the reconstructed signal makes it difficult to identify the actual speaker on the other side.

In order to overcome the above problem, certain hybrid codecs are used, which transmit both vocoding information and also the PCM signal information. Certain other codecs are also available, which vary the bit rate of the outgoing signal according to the characteristics of the speech signals fed to them. For more details on vocoders, the reader can refer to [4].

1.8 ERROR CONTROL

On any transmission channel, and especially on the wireless channel, it is never guaranteed that the transmitted data will be delivered as it is without any errors at the destination. There is always a non-zero probability that the bit stream gets altered while in transmission. The bit error rates (BERs – fraction of bits that are received in error) for the wired channel and the wireless channel can be as high as 10^{-9} and 10^{-4}, respectively. The unreliable nature of the wireless channel makes error control even more important in the wireless networks context.

Several coding techniques are available, which try to provide resistance against errors by adding redundant bits to the transmitted bits. These redundant bits help the receiver to either detect errors and request for a retransmission (error detection), or to identify and correct the faulty bits (error correction). We have different coding schemes for error detection and for error correction. The process of adding the redundant bits to the bit stream is known as *channel coding.* Some of the commonly used coding techniques are discussed below.

1.8.1 Parity Check

Parity check, which can detect single-bit errors, is the simplest of the error detection techniques. The parity check mechanism works as below. Initially, the transmitter and receiver come together in an agreement on whether the number of 1 bits in the messages they exchange would be odd or be even. If the number of 1s is odd, then the scheme is called odd parity; if it is even, the scheme is called even parity. Each message has an additional bit, called a parity bit, which is the last bit of the message. The parity bit is set to 1 or 0 based on the prior agreement on whether the message should contain an even or odd number of 1s. Consider the following example. If the message to be transmitted is 10111 and the odd parity scheme is used, the sender appends a parity bit (last bit) of 1 to make the number of 1s odd. Thus the message transmitted would be 101111. If no error occurs in transmission, the receiver would receive the message as it is, it would verify and find that the number of 1 bits is odd in number, and hence conclude that there was no error in transmission. On the other hand, if a single-bit error had occurred (a single bit would have been flipped from 0 to 1 or 1 to 0), the number of ones in the received message would not be odd. The receiver hence concludes that the message became corrupted in transmission and requests the sender for a retransmission.

The parity check mechanism cannot detect a multiple of two-bit errors in the same message. In the above example, if the received message had been 101001, the receiver would have wrongly concluded that the message had been received correctly, though actually two bits are in error. This scheme is guaranteed to detect only single-bit errors. Another disadvantage of this scheme is that the receiver can only detect the error and not correct it since it does not know the exact bit in error.

For correcting errors, the Hamming distance of the binary streams/binary words used in the communication becomes important. The Hamming distance of a set of binary streams is the minimum number of bits that need to be inverted for converting one stream from the set into another valid stream of the set. For example, if the set of binary streams used in the communication consists of only two patterns 000000 and 000111, the Hamming distance becomes three. Hence, up to two-bit errors can be detected successfully, and all one-bit errors in the received messages can be corrected. For example, if the message received is 100111, since the Hamming distance is three, the receiver would conclude that if a one-bit error had occurred, then the actual transmitted message should have been 000111. In general, to detect d errors, a Hamming distance of at least $d+1$ is required, and for correcting d errors a Hamming distance of at least $2d + 1$ is required. This would make the valid bit streams to be far enough from each other so that even when up to d bits go in error, the original transmitted bit stream is closer to one of the valid bit streams than any other.

1.8.2 Hamming Code

The Hamming code works as follows. Each valid binary stream of length n is said to consist of d data bits and r redundant/check bits ($n = d + r$). The bits of the bit stream are numbered consecutively from bit 1, the most significant bit (MSB), at the left end. Bits that are powers of 2 (1, 2, 4, 8, 16, etc.) are the check bits. The remaining positions (3, 5, 6, 7, 9, etc.) are taken up by the data bits. Each check bit is set to 1 or 0 such that the parity of the bits it represents, including itself, is even or odd (depending on whether even parity or odd parity is used). In order to determine to what check bits a data bit contributes to, the position of the data bit is written in powers of 2. The data bit is checked by the check bits occurring in the expansion. For example, 11 can be written as $1+2+8$. Hence it is checked by the check bits in positions 1, 2, and 8. When a bit stream arrives at the receiver, it initializes a counter to zero. It then examines each check bit in position i ($i = 1, 2, 4, 8, ...$) and checks whether it satisfies the parity condition (even or odd). If not, it adds the position of that check bit to the counter. After all check bits have been examined, if the counter value is zero, then it implies that the bit stream was received without any error. Otherwise, the counter contains the position of the incorrect bit. For example, if check bits 1 and 2 are in error, it implies that the third bit has been inverted. This is because it is only bit 3 that is checked by bits 1 and 2 alone. Hamming codes can correct only single-bit errors.

1.8.3 Cyclic Redundancy Check

Cyclic redundancy check (CRC) is one of the widely used error detection techniques. The CRC technique is used for protecting blocks of data called frames. The transmitter appends an additional n-bit sequence called frame check sequence (FCS) to each frame it transmits. The FCS holds redundant information regarding the frame, which helps the receiver detect errors in the frame. The CRC algorithm works as follows. It treats all bit streams as binary polynomials. The transmitter generates the FCS for a frame so that the resulting frame (FCS appended to the original frame) is exactly divisible by a certain pre-defined polynomial called the divisor or CRC polynomial. At the receiver, the received frame is divided by the same CRC polynomial. The frame is considered to be free of errors if the remainder of this division is zero; if not, it implies that the received frame is in error. CRC has the advantages of very good error detection capabilities, less overhead, and ease of implementation.

1.8.4 Convolutional Coding

Techniques such as the Hamming code and CRC work on fixed blocks of data. They would not be suitable for a continuous data stream. Convolutional coding is well-suited for long bit streams in noisy channels. Convolutional codes are referred to, based on their code rates and constraint length. The code rate r of a convolutional code gives the ratio of the number of bits k that were fed as input to the convolutional coder to the number of bits n at the output of the coder, that is, $r = k/n$. The constraint length K of a convolutional code denotes the length of the convolutional coder, that is, the number of stages used in the coder. The larger the constraint length, the lesser the probability of a bit suffering an error. Operation of the encoding process can be best explained using an example. Consider the convolutional coder shown in Figure 1.13. It consists of four stages, that is, $K = 4$, and a code rate $r = 1/3$. Initially, the first bit (MSB) is fed to the first stage of the coder, and three bits are obtained as the result of coding this bit. This process continues till the last bit of the input bit stream leaves the last stage of the coder. For a bit pattern of 1101, the output from the coder in Figure 1.13 would be 111 100 001 110 000 010 011 000. The operation of the coder depends on various factors such as the number of stages used, the number of XOR adders used, and the way connections are made from the stage outputs to the adders.

Two types of mechanisms are available for decoding the convolutional codes. The first mechanism called *sequential decoding* performs well with convolutional codes having a large K value, but it has a variable decoding time. The second type of decoding, called *Viterbi decoding*, has a fixed decoding time, but the computational complexity for this decoding is high. For more details on convolutional coding, the reader can refer to [5].

1.8.5 Turbo Codes

Turbo codes are very powerful error-correcting codes. Turbo codes were first proposed by C. Berrou *et al.* in 1993 [6]. Turbo codes are considered to be one of the

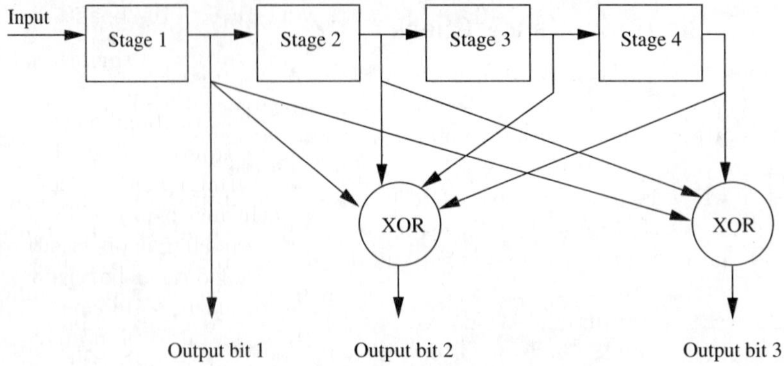

Figure 1.13. A convolutional coder.

very important breakthroughs in channel coding theory. The turbo codes exhibit performance closer to the theoretical Shannon limit than any other code. They have very good error-control properties, and at the same time the decoding process is computationally feasible. These two features make the turbo codes very powerful. Figure 1.14 shows a basic turbo code encoder. It uses two recursive[6] systematic[7] convolutional (RSC) codes with parallel concatenation. In Figure 1.14, the first RSC/convolutional encoder uses the bit stream as it arrives. The second convolutional encoder receives the bit stream through the *interleaver*. The interleaver

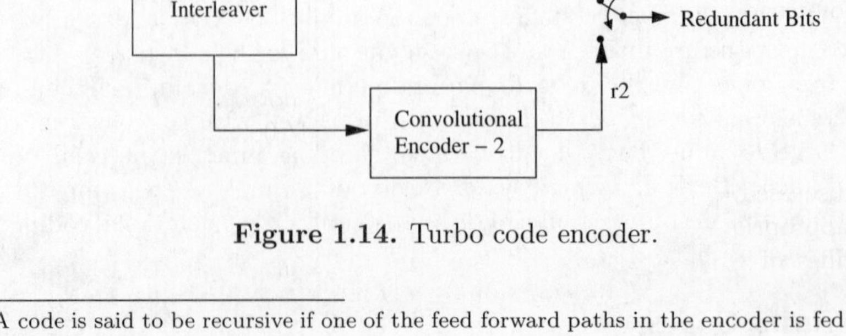

Figure 1.14. Turbo code encoder.

[6]A code is said to be recursive if one of the feed forward paths in the encoder is fed back into its input.

[7]A systematic code is one in which the original data bits are not modified; the redundant bits are just appended to the data bits.

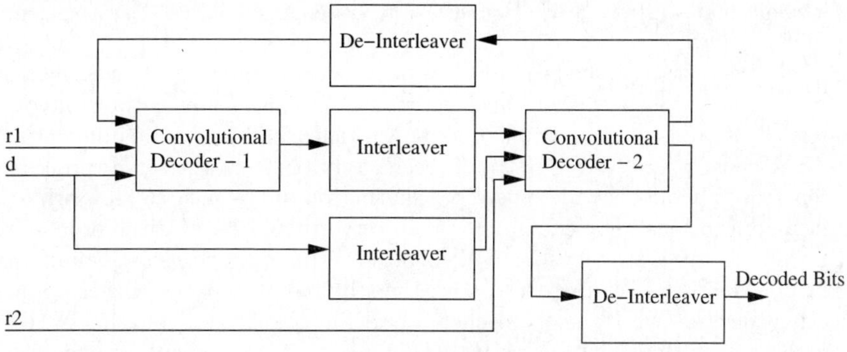

Figure 1.15. Turbo code decoder.

randomizes the input sequence of the second encoder, such that the inputs of the two encoders are uncorrelated. The two encoders introduce redundant bits for the same block of input data bits. However, because of the interleaver which alters the bit pattern input to the second encoder, the redundant bits generated by the two encoders are different. The received data would have two different coded sequences (generated by the two convolutional encoders). An iterative decoder is used at the receiver. A turbo decoder that uses a single iteration is shown in Figure 1.15. The first decoder decodes the input sequence to give a soft output. This soft output is a measure of the reliability of each decoded information bit (whether or not any errors had occurred). This reliability information is called extrinsic information, that is, it does not depend on the current inputs to the decoder. The extrinsic information from the first decoder, after interleaving, is passed to the second decoder. The second decoder uses this information to decode the interleaved input bit sequence. From the soft outputs of the second decoder the new extrinsic information, which again is a measure of the reliability of the decoded bits, is fed back to the first decoder, and the process continues. The final output is taken from the output of the second decoder. The performance of the turbo coding scheme improves as the number of decoder iterations is increased.

1.9 COMPUTER NETWORKS

Having discussed the fundamentals of wireless communication technology, it would now be appropriate to know how wireless networks operate. But before going into the specifics of wireless networks, it would do well to understand the principles behind the operation of a general computer network. A computer network interconnects autonomous computers/nodes. Computer networks are almost indispensable in today's world. They have become very essential for day-to-day life and are used by a wide gamut of applications. Some of the very common applications are electronic mail systems, ticket reservation systems, credit card systems, and video conferencing.

Computer networks can be broadly classified under two main categories, namely, client/server networks and peer-to-peer networks. A client/server network involves processes, client processes and server processes, which usually reside on different machines. The client processes request and receive services from the server processes. The client/server model is widely used for sharing resources such as files, printers, and other applications. For example, a Web server could be a server process receiving requests for Web pages from Web browsers (client processes) running on different machines, and sending them back the required Web pages. Other popular applications such as electronic mail clients and file transfer session clients (FTP clients) also follow the client/server model. A peer-to-peer network, on the other hand, does not require any dedicated server. Every computer on the network can share its resources (*e.g.*, files or printers) with any other computer on the network, provided sufficient access privileges are granted for those computers. In effect, every computer connected to the network is both a client and a server. The music file-sharing system, Napster, was once the most popular peer-to-peer application on the Internet.

Based on the transmission technology used, networks can be classified under two broad categories, namely, broadcast networks and point-to-point networks. A broadcast network uses a single shared communication channel. A message transmitted by a node can be heard by all other nodes in the network. Each message carries the address of its intended destination. The destination node, on finding its address on the message, accepts the message. The other nodes just ignore the message. A broadcast network is usually used to connect machines within a small geographic area. In a point-to-point network, the communication channel is not shared. Two nodes are connected through a dedicated link. Data transmitted by the source node can be received only by the intended next hop receiver node. Hence, the transmitted data might have to traverse through multiple hops in order to reach the final destination. Such point-to-point links are used to connect machines that are physically separated by large geographic distances. Based on the geographic span of the network, networks can be classified as LANs, MANs, or WANs.

Local area networks (LANs) are the most common networks one can find today. As the name suggests, a LAN is used to connect computers that are located within a small geographical area, such as within office buildings or homes. In a LAN, all the computers are attached to a common broadcast medium. Ethernet is the most common physical medium used for connecting computers in a LAN. A metropolitan area network (MAN) covers a much larger area compared to that covered by a LAN. A MAN may be used to connect computers that are located at different offices or those that are located in the same city or town. Some of the common technologies used in MANs are asynchronous transfer mode (ATM) (which will be described later in this chapter) and fiber distributed data interface (FDDI).[8] Networks that are used to connect computers located in a large geographical area,

[8]FDDI is a token-based LAN standard that uses a ring topology network in which nodes are interconnected by optical fiber links.

which may span different cities or even countries, are called wide area networks (WANs). WANs may be connected through phone lines, satellite links, or optical fibers. The transmission medium in a WAN may include several point-to-point links. A network that connects two or more of the above three types of networks, that is, LANs, WANs, or MANs, is called an internetwork.

Any network requires network hardware and software in order to operate. Network hardware may include devices such as network cards, switches, routers, and gateways. Network software was not given much importance in the early networks. But today, the network software plays an extremely important role in any type of network.

1.10 COMPUTER NETWORK SOFTWARE

In order to reduce its complexity, the computer network software is organized as a set of *layers*. Every layer has a specific function to perform. It offers its services to its next higher layer. In order for layers to access the services provided by their lower layers, each layer (except the bottom-most layer) is provided with a *service access point* (SAP). The SAP is an interface through which the higher layer can access the services offered by its immediately next lower layer. The lower ranking layer is referred to as the service provider, and the higher ranking layer as the service user. The information passed through the SAP is known as an *interface data unit* (IDU). The IDU typically consists of a *service data unit* (SDU), which is the actual data, and associated control information. For its convenience, a layer may split the received SDU into several units before sending it to the next lower layer. Each such unit is termed as a *protocol data unit* (PDU). The higher layer is in no way aware of the manner in which the services are actually implemented in the lower layer. Hence, effecting changes in a layer does not affect any of the other layers. Each layer at the sender machine is said to logically communicate with its peer layer on the destination machine. The rules used for this communication between peer layers, that is, layer n on the sender and the same layer n at the receiver, are collectively termed the *layer n protocol.* But the actual information flow between two connected machines occurs as shown in Figure 1.16. Figure 1.16 depicts a four-layer model. The collection of protocols used by the system, one protocol per network layer, is referred to as the *protocol stack.*

At the sender side, the top-most layer, on receiving data from the application, adds its own header to the data packet (a packet refers to the basic unit of data that is transmitted on the medium) and sends it to the next lower layer. Each intermediate layer, on receiving this data, appends its own layer-specific header and sends it to the next lower level layer. This process continues till the data reaches the lower-most layer. This lowest layer transmits the received data packet on the physical medium. When the packet transmitted by a node reaches its destination, each layer at the destination node strips off its corresponding peer layer header from the packet, and sends the packet to the next higher layer. This process continues till the data reaches the top-most layer, which then delivers it to the appropriate process. The layers, along with their respective protocols, constitute the *computer*

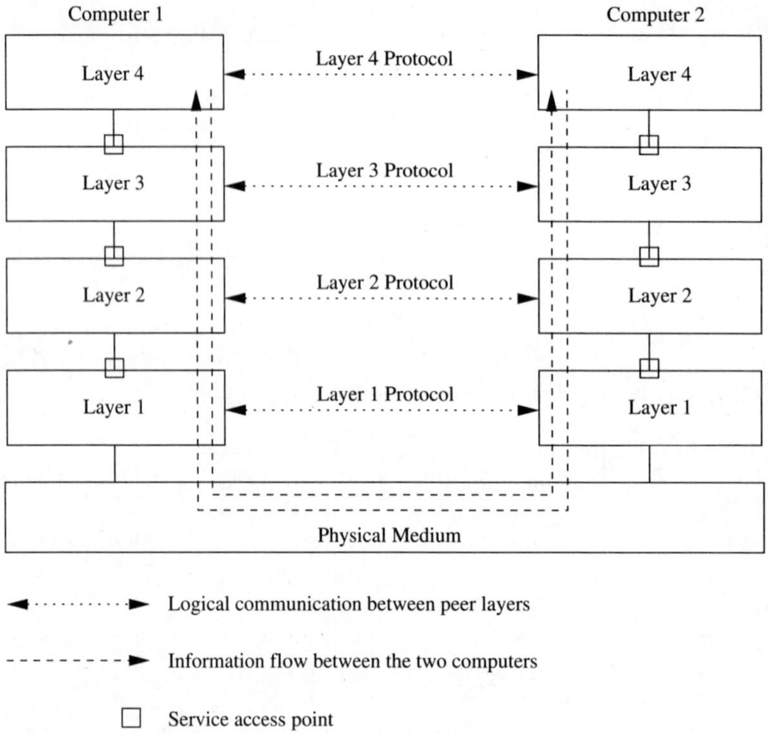

Figure 1.16. Different layers in computer network software.

network architecture. The computer network architecture does not pertain to any specific hardware or software. Using the network architecture, an implementer would be able to design his/her hardware and software for the network.

1.11 COMPUTER NETWORK ARCHITECTURE

As defined previously, the network layers, along with their respective protocols, constitute the computer network architecture. Several such computer network architectures/models are in existence today. Prominent models such as the OSI reference model, the TCP/IP reference model, and the ATM reference model will be discussed now.

1.11.1 The OSI Reference Model

This model was proposed by the International Organization for Standardization (called ISO in short), a worldwide federation of national standards bodies representing more than 140 countries. The objective of this standard was to standardize the protocols used in the various network layers. This model was named as the open systems interconnection (OSI) reference model, since the main objective of the model was to specify mechanisms for communication between systems (machines) in a telecommunication network that are open for communication with other systems. The OSI reference model is depicted in Figure 1.17. It consists of seven layers, namely, the physical layer, data link layer, network layer, transport layer, session layer, presentation layer, and the application layer. Each layer is associated with a unique set of functions and responsibilities. The following is a brief description of the seven layers.

The physical layer, which is the lowest layer, is responsible for the transmission of the bit stream over the physical medium. It deals with the mechanical and electrical specifications of the network hardware and the physical transmission medium to be used for the transmissions. The mechanical specifications refer to the physical dimensions of the devices such as cables, connectors, and pins used for interconnection. The electrical specifications include details such as voltage levels to be used for representing digital information (binary 0s and 1s) and the duration

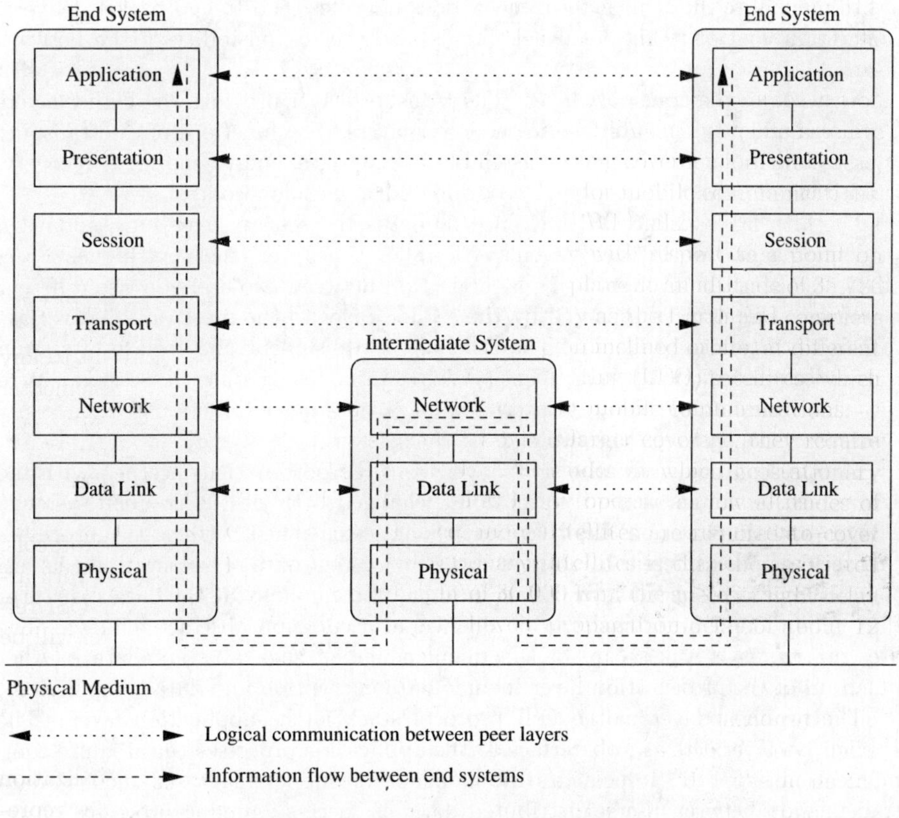

Figure 1.17. OSI reference model.

of each bit. The physical layer acts as a direct interface to the actual physical transmission medium.

The main objective of the next higher layer, the data link layer, is to ensure error-free transmission of data across a physical link. It receives data from the higher layer, splits the data into several frames/packets, and transmits the frames. It may have mechanisms for detecting and retransmitting damaged or lost frames. The data link layer is also responsible for implementing link-level congestion control mechanisms that ensure that a fast sender does not flood a slow receiver. In broadcast networks, the data link layer performs another important function, which is that of medium access control. It determines when, that is, at what point of time, the current node can access the channel for its transmissions so that collisions arising due to simultaneous transmissions by multiple nodes are minimized. A detailed description of the data link layer is provided in the next section.

The network layer is responsible for routing data packets from the source node to the destination node. The network layer is also responsible for node addressing. A packet may have to traverse across several networks in order to reach its destination. Each of these networks might follow different addressing schemes. The network layer has to make sure that this heterogeneity does not affect end-to-end packet delivery. Congestion control in the local network is also the responsibility of the network layer.

Next comes the transport layer. The transport layer provides the higher layers a network independent interface to the lower layers. The functions of the transport layer include segmentation and reassembly of messages, end-to-end error recovery, monitoring of quality of service (QoS), and end-to-end flow control.

The fifth layer in the OSI reference model is the session layer. A session can be defined as a connection between two presentation layer (the next higher layer) processes. It can be thought to be a connection between two users on two different machines. The services provided by the session layer to the presentation layer are with respect to these sessions. They include establishment and release of session connections, interaction management (which determines whether the session is two-way simultaneous, or two-way alternate, or a one-way interaction session), and synchronization between the two ends of the session.

The presentation layer, the sixth layer from the bottom, is concerned with the syntax and semantics of the information exchanged between the two end systems. The presentation layer ensures that the messages exchanged between two processes that reside on the different computers that use different data representations, have a common meaning. In order for such computers to communicate, the data structures that need to be transmitted can be defined in a common abstract manner. The presentation layer takes care of the management of such abstract data. Other functions of the presentation layer include data encryption and data compression.

The top-most layer in the OSI protocol stack is the application layer. The application layer acts as an interface to the application processes that require communication support. It provides mechanisms for supporting tasks such as data transmission between users, distributed database access, running processes on remote machines, controlling distributed systems, and electronic mail communication.

Some examples of protocols used in this layer are virtual terminal (TELNET), which provides remote login facility for users, file transfer protocol (FTP), used for transferring files between machines, simple mail transfer protocol (SMTP), that is used for electronic mail applications, and hypertext transfer protocol (HTTP), which is used for transferring Web pages over the Internet.

Shortcomings of the OSI Reference Model

Though the OSI model was elaborately designed, it never took off in reality. By the time the OSI model appeared, other protocols, in particular the TCP/IP protocol, were already in use widely. All products in the market were designed according to those protocols. The manufacturers were hesitant to again invest in products based on this new OSI model. Hence, the OSI model was never implemented.

Regarding the technical aspects, it was felt that the OSI model had too many layers. The session and presentation layers were not required for most applications. The model itself was thought to be very complex and difficult to implement efficiently. Several layers in the model were performing redundant operations. Tasks such as flow control and error control were being performed by many layers in the model. As such, the model was not very efficient. The few initial implementations of the model were bulky and slow. Since other better models such as the TCP/IP were already available, people preferred such models.

1.11.2 The TCP/IP Reference Model

The TCP/IP reference model is the network model that is used in today's Internet. The model is widely referred to as TCP/IP protocol stack since it was originally designed as a protocol stack for the Advanced Research Project Agency Network (ARPANET)[9] in the late sixties, rather than as a reference model. It is named so after two of its main protocols, transmission control protocol (TCP) and Internet protocol (IP). In contrast to the OSI reference model, the TCP/IP reference model has only four layers. This model was far better than the OSI model because, unlike the OSI model, the protocols that are part of this model were already in use before the model was actually designed. Hence, they fitted perfectly into the model. Figure 1.18 shows the layers present in the TCP/IP reference model. The figure also shows some of the protocols used in each layer. The four layers of the TCP/IP model are the host-to-network layer, internet/network layer,[10] transport layer, and the application layer. The session and presentation layers that found place in the OSI reference model were not included in the TCP/IP model since they were found to be not very significant.

The lowest layer, the host-to-network layer, is used to interface the TCP/IP protocol stack with the physical transmission medium. The operation of this layer

[9]The ARPANET, sponsored by the U.S. Department of Defense, was the first major effort at developing a network to interconnect computers over a wide geographical area.

[10]The term internet (i in lowercase) refers to the interconnection of individual networks using devices such as routers and gateways. The term Internet (I in uppercase) refers to a specific worldwide network.

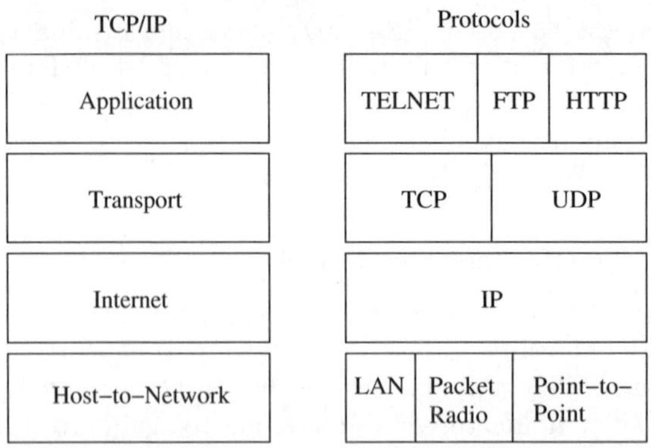

Figure 1.18. TCP/IP reference model.

is not very well-specified in the TCP/IP model. The host machine should connect to the network using some protocol in order to transmit the packets sent to it by the next higher layer, the internet layer.

The internet layer is similar in functionality to the network layer in the OSI reference model. The main functions of the internet layer are routing and congestion control. The internet layer should provide mechanisms for the host to inject packets into any network and have those packets travel independently to the destination. The internet layer defines a protocol called Internet protocol (IP). The packets sent by the internet layer to the lower layer follow this IP format. The routing of packets is done on an individual packet-by-packet basis. Hence the packets may arrive out of order at the destination. It is the responsibility of the next higher layer, the transport layer, to reorder the received packets.

The transport layer in the TCP/IP model performs similar operations as the transport layer in the OSI reference model. The fact is that the design of the transport layer in the OSI model was actually influenced by the operation and performance of TCP. Two end-to-end communication protocols, the transmission control protocol (TCP) and the user datagram protocol (UDP), have been defined in this layer. TCP is a reliable connection-oriented protocol. Here, establishment of an end-to-end connection that may traverse multiple hops is required before the actual data transfer can begin. All packets belonging to a session follow the same path, and are received at the destination in the same order as they were transmitted by the source node. The incoming byte stream at the source node is split into several messages and sent to the internet layer. The received messages are reassembled by the transport layer at the destination. In order that a fast sender does not swamp a slow receiver, TCP also performs flow control. UDP is an unreliable connectionless protocol. It does not have the flow control and sequencing mechanisms present in TCP. UDP is used by applications such as voice/video transmissions where fast delivery of information is more important than the accuracy of the information delivered.

The application layer is the topmost layer in the TCP/IP model. It consists of higher-level protocols, which include the FTP, TELNET, SMTP, and HTTP.

Shortcomings of the TCP/IP Reference Model

One of the main disadvantages of the TCP/IP model is that it is too specific to the TCP/IP protocol stack. It can be said that the model is not generic enough to describe other protocol stacks. It is difficult to design new networks with new technologies using the TCP/IP reference model. Another shortcoming of the TCP/IP model is the host-to-network layer. This layer is actually an interface connecting the machine to the network. An interface does not really require an exclusive layer for itself. Also, the data link layer which performs important functions such as framing, error control, link-level flow control, and medium access control (in broadcast networks) is absent in the TCP/IP model. Irrespective of all these shortcomings, the TCP/IP reference model still remains the most successful model and even forms the backbone of today's Internet.

1.11.3 The ATM Reference Model

The asynchronous transfer mode (ATM) model described below was developed by the ITU for broadband-integrated services digital network (B-ISDN). The ATM model is quite different from the OSI and TCP/IP reference models described previously. Here, all information is transmitted in the form of short fixed-sized packets called *cells*. The size of each cell is 53 bytes, a 5-byte header followed by a 48-byte payload. ATM uses a switching technology called cell switching. Before going into details of cell switching, the two other commonly used switching technologies, namely, packet switching and circuit switching, are described briefly below.

Circuit switching requires the setting up of an end-to-end path between the source and destination nodes before the actual data transmission begins. A request packet sent by the source node travels till the destination node. The destination responds with an acknowledgment. When the source node receives this acknowledgment, implying that the path has been set up, it starts transmitting packets. The request-response process also reserves the required resources (bandwidth) on the path. Once a path is set up, all packets belonging to the flow follow the same path. The packets are received at the destination in the same order they were transmitted at the source. Since resources are reserved *a priori*, the problem of link congestion rarely occurs in circuit switched networks. But if a reserved path is sparingly used or not used at all, then the reserved resources remain not utilized, resulting in under-utilization of the available network capacity. Circuit switching is widely used in telephone networks where once a connection is established between the two end users, the circuit (connection) continuously remains open from the start till completion of the call. Circuit switching may not be suitable for data or other non-conversational transmissions. Non-voice transmissions tend to occur in bursts, and if circuit switching is used for such transmissions, the transmission line often remains idle, resulting in under-utilization of the available network capacity.

Another switching technique, which is not very popular and is not used commonly, is message switching. Here the sender transmits data in blocks. There is no limit on the size of the blocks. An intermediate node receives the entire block of data from its uplink node (uplink node refers to the next hop neighbor node on the path from the current node to the source node of the data session), stores the block in its disk, checks for errors, and then transmits it to the next downlink node (downlink node refers to the next hop neighbor node of the current node on the path from the current node to the destination node). Since there is no limit on the size of blocks transmitted, a single large block of data may hog the channel for a long period of time. Hence, message switching is not suitable for interactive applications and time-sensitive applications such as voice and video transmissions. The third and most widely used switching technique, packet switching, solves the above problems.

In packet switching, a strict bound is set on the maximum size of the blocks transmitted. Hence the packets can now be held in the main memory of intermediate routers rather than in their disks. Since a single connection cannot hog a transmission channel indefinitely, it is suitable for non-conversational transmissions. A node can transmit each packet independently. Unlike message switching, it need not wait for a whole block of data (consisting of many packets) to be received before the transmission of the received block can be started. Hence, the end-to-end delay involved is less, and the throughput of the system is also high. Since resources are not reserved beforehand, the links may become congested when the network load increases, resulting in packet loss. In packet switching, each packet is routed (and not switched) independently at each intermediate node. Packets belonging to the same flow may hence take different paths to reach the destination. Due to this, it is highly probable that the packets are received out of order at the destination node.

Continuing with ATM and cell switching, cell switching is similar to packet switching in that it splits a data stream into multiple equal-sized cells which are then transmitted on a channel that is shared by other nodes also. The main difference between cell and packet switching techniques is that while cells have a fixed size, packets can have different sizes. One of the main reasons for having fixed-length cells is to facilitate the building of fast hardware switches. The design of fast and highly scalable hardware switches would be made simple if all packets were of fixed length. The processing of packets would be made simple if the size of the packets were known beforehand. Also, parallelism could be incorporated into the switching system if all packets were of the same length; multiple switching elements could be working in parallel performing the same operation on different packets. This improves the scalability of the switching system. Hence, the ATM design had fixed length cells.

Though an ATM network uses cell switching, it is in some ways similar to a circuit switched network. Every connection goes through a set-up phase, during which a *virtual circuit*[11] is established. The virtual circuit has several advantages

[11] A virtual circuit is a path between two points in the network that appears to be a dedicated physical path, but in reality is a managed pool of network resources from which specific flows/paths are allocated resources based on the need, such that the traffic requirements of the flows are met.

over a physical circuit. Reservations can make use of statistical multiplexing (multiplexing in which channels are established on a statistical basis, that is, connections are established on demand according to the probability of need, using the statistical information about the system, that is, information regarding the data arrival pattern at nodes). Also, once a virtual circuit is established for a connection, the switching time for packets belonging to that connection gets minimized. This increases the network throughput significantly. An ATM switch is never allowed to reorder the cells/packets. It may drop the packets, though, when the network becomes overloaded. Virtual circuits are unidirectional. The ATM model is depicted in Figure 1.19. Unlike the previous two models, the ATM model has been defined as a three-dimensional model consisting of layers and planes. The three main layers are the physical layer, the ATM layer, and the ATM adaptation layer. The three planes used in this model are the user plane, the control plane, and the management plane.

The user plane is responsible for operations such as cell transport, error control, and flow control. The control plane deals with connection management. Typical operations on this plane include call setup and call tear-down. The third plane, the

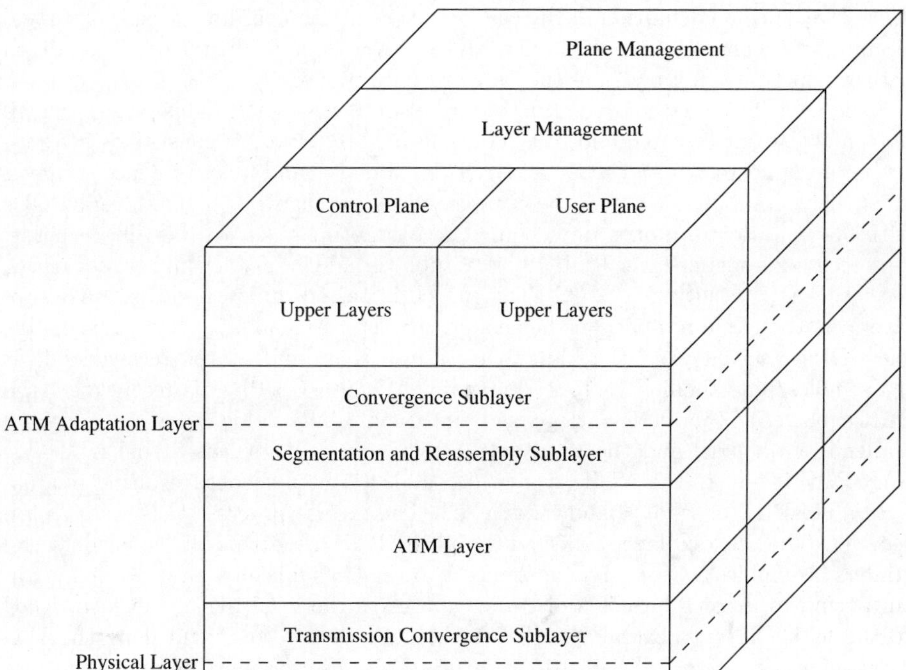

Figure 1.19. ATM reference model.

management plane, maintains the network and carries out operational functions. This plane is subdivided into layer management and plane management to manage the different layers and planes. The concept of user and control planes enables streamlining of the implementation in each plane, which improves the performance of the system.

As in the OSI reference model, the physical layer deals with the actual physical medium used, which includes the voltage representation to be used for bits, the duration of each bit, and other physical medium dependent issues. This layer is divided into two sublayers. The lower sublayer, called the physical medium dependent (PMD) sublayer, directly interfaces with the physical medium and takes care of the actual transmission of bits and the bit timing. The upper sublayer of the physical layer is called the transmission convergence (TC) sublayer. At the sender side, the TC sends the incoming cells as a string of bits to the PMD, which transmits the bit stream. At the receiver, the TC sublayer receives the bit stream from the PMD sublayer. The TC sublayer, which is aware of details such as where cells begin and end (cell boundaries), then converts the incoming bit stream into cells and delivers them to the next higher layer, the ATM layer.

The ATM layer is the core and the most important layer of the ATM reference model. It is responsible for routing cells across the network. Functions of the ATM layer include definition of the cell layout, establishment and release of virtual circuits, and congestion control. The ATM layer is equivalent to the combined network and data link layers in the OSI reference model.

The next higher layer is the ATM adaptation layer (AAL). This layer is where user information is created and received as 48-byte payloads. It consists of two sublayers, a segmentation and reassembly (SAR) sublayer and a convergence sublayer (CS). The CS sublayer which is the upper sublayer packages the higher layer's PDUs with any additional information required for adapting them to the ATM service. On receiving a higher-layer PDU, the CS sublayer adds a header and trailer to the received PDU, resulting in a CS-PDU, which is passed on to the lower SAR sublayer. As the name implies, on the sender side the SAR sublayer segments/breaks the PDUs received from the higher layers into cells, and at the receiver side it reassembles/puts them back together again. When a CS-PDU is received by the SAR sublayer at the sender, it breaks it into small units, adds its own header and trailer to each unit, and the resultant 48-byte unit is known as SAR-PDU. This SAR-PDU is sent to the ATM layer, which adds a 5-byte header, resulting in a 53-byte ATM-PDU or what is known as a cell. The reverse process takes place at the receiver. The SAR sublayer collects multiple SAR-PDUs, strips off the headers and trailers, assembles a CS-PDU, and sends it to the CS sublayer. The CS sublayer in turn removes its own header and trailer and gives the resulting higher-layer PDU to the higher layer. Several types of AALs have been recommended by the ITU Telecommunication Standardization Sector (ITU-T) for supporting different types of data. AAL1 and AAL2 provide real-time service guarantees to connections. AAL1 provides support for constant-bit-rate (CBR) connections, mostly used for voice and video transmissions. AAL2 supports variable-bit-rate (VBR) connections. AAL3/4 is used for supporting conventional packet switching services. AAL5, which is sometimes also known as simple and efficient adaptation layer (SEAL), is used by applications that can use ATM services directly.

The above described layers are the ones actually defined by the model. Additional upper layers may include optional protocol layers used for further encapsulating ATM services for use with other protocols such as TCP/IP. Figure 1.20 shows a comparison of the ATM reference model with the other two reference models described previously in this chapter, namely the OSI reference model and the TCP/IP reference model.

One of the main advantages of ATM is that it can provide differentiated services through the use of four service classes. The CBR class provides services similar to that of a dedicated communication line. It offers the best guarantees of packet delivery. The second service class is the VBR service class. This class is further divided into two subclasses, real-time (VBR-RT) and non-real-time (VBR-NRT). VBR-RT is commonly used for time-sensitive transmissions such as voice and real-time video. VBR-RT provides better congestion control and delay guarantees than VBR-NRT. The next service class, available-bit-rate (ABR) class, is sort of a hybrid between the CBR and the VBR classes. The ABR class delivers cells at a minimum guaranteed rate. If more bandwidth is available, the minimum rate can be exceeded. This class ideally suits bursty traffic. The fourth service class, called the unspecified-

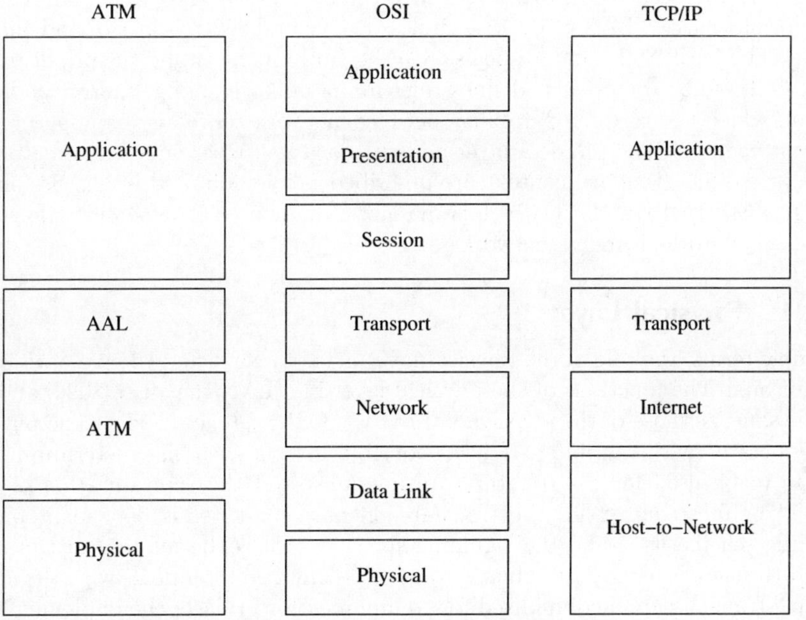

Figure 1.20. Comparison among ATM, OSI, and TCP/IP reference models.

bit-rate (UBR) class, supports best-effort delivery of cells. It provides no delay or bandwidth guarantees.

Shortcomings of the ATM Reference Model

The ATM reference model is not widely in use today. It has several shortcomings. Each 53-byte cell has a 5-byte header, which is nearly ten percent of the cell/frame. This constitutes a significant control overhead on the system. Complex mechanisms are required for ensuring fairness among connections and provisioning quality of service. The delay jitter due to the varying delays faced by the packets, requires complex packet scheduling algorithms at the ATM switches. The high cost and complexity of devices based on the ATM technology, and the lack of scalability are the other reasons that contributed to the failure of ATM.

1.12 IEEE 802 NETWORKING STANDARD

The Institute of Electrical and Electronics Engineers (IEEE) has defined several standards for LANs. Such standards collectively come under the IEEE 802 standard. The following are some of the original IEEE standards: 802.1 - internetworking; 802.2 - logical link control; 802.3 - Ethernet or CSMA/CD; 802.4 - token bus LANs; 802.5 - token ring LANs; 802.6 - MANs; 802.7 - broadband LANs; 802.8 - fiber optic LANs and MANs; 802.9 - integrated (voice/data) services LANs and MANs; 802.10 - security in LANs and MANs; 802.11 - wireless LANs; 802.12 - demand priority access LANs; 802.15 - wireless PANs; and 802.16 - broadband wireless MANs. The IEEE 802 standard deals with the data link layer and the physical layer of the OSI reference model. It defines rules for cabling, signaling, and media access control, which assure interoperability between network products manufactured by different vendors. The physical layer and the data link layer which assume prominence in the 802 standard context are described below, followed by discussions on the 802.3 Ethernet standard, which is the most widely used LAN standard, and the 802.11 standard for wireless networks.

1.12.1 Physical Layer

In simple terms, the role of the physical layer is to transmit and receive data in the form of bits. The functions of the physical layer in a LAN that uses 802 standards are the same as those of the physical layer in the OSI reference model. The physical layer operates with raw bits. It is responsible for bit encoding, determining the voltage to be used for the 0/1 bit transmissions, and the time duration of each bit. The time required for the transmission of a character is dependent on the encoding scheme used and the signaling speed, which is the number of times per second the signal changes its value/voltage. The physical layer deals with the actual physical transmission medium used for communication. Hence, the implementation of the physical layer varies depending on the physical medium used. Some of the commonly used physical transmission media are twisted pair, coaxial cable, optical fiber, and radio waves (wireless). The MAC sublayer of the data link layer uses the services of the physical layer to provide its own services to the LLC sublayer of the data link layer. The MAC and LLC sublayers are explained in the following section.

1.12.2 Data Link Layer

One of the most important layers in any network model is the data link layer (DLL). As mentioned earlier, the data link layer performs several important functions such as error control, flow control, addressing, framing, and medium access control. The data link layer consists of two sublayers, the logical link control sublayer (which takes care of error control and flow control) and the medium access control sublayer (which is responsible for addressing, framing, and medium access control). Figure 1.21 depicts the two sublayers. The following sections describe in detail the two sublayers.

Logical Link Control Sublayer

The logical link control sublayer (LLC) is defined by the IEEE 802.2 standard. It was originally designed by IBM as a sublayer in the IBM token ring architecture.[12] LLC provides physical medium independent data link layer services to the network layer. Below the LLC is the media access control sublayer. As mentioned previously, error control and flow control are some of the important services offered by the LLC. As defined in the IEEE 802.2 standard, the LLC adds the destination service access

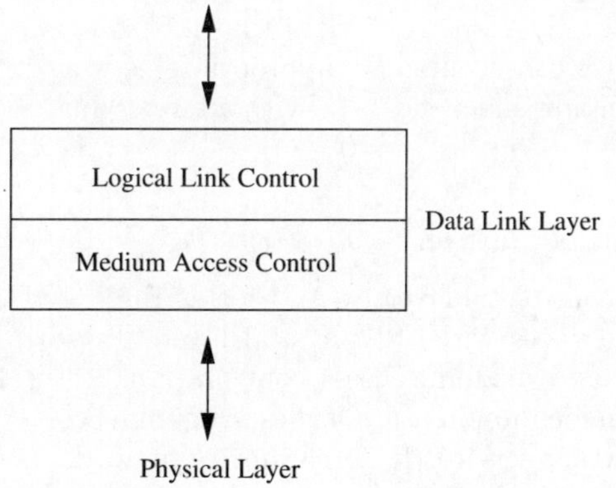

Figure 1.21. Data link layer.

[12]The IBM token ring architecture led to the popular IEEE 802.5 token ring standard. All nodes in a token ring network are connected in a ring. A special bit pattern, called a token, travels around the ring. In order to transmit a message, the sender node must be in receipt of the token. The token is attached to each message sent by a node. Once the sender completes its transmission, it releases the token to the next node that needs to transmit. The process continues and is used as the basis for resolving contention between multiple nodes waiting for transmission.

point (DSAP) and the source service access point (SSAP) labels to each packet it receives from the network layer. The SAPs identify the applications/users involved in the data transfer.

LLC provides three types of services, namely, unacknowledged connectionless service, connection-oriented service, and acknowledged connectionless service. The unacknowledged connectionless data transfer service is commonly known as LLC type 1, or LLC1 service. A connectionless service does not require the establishment of an end-to-end connection; each packet is routed independently to the destination. Once service access points (SAPs) are enabled at the hosts, data transfer can begin. The network layer at the source node sends its packets through the SAP provided by the LLC. When the packets are received at the destination node, the LLC of the destination delivers them to the network layer through its SAP. Connection-oriented data transfer service is referred to as LLC type 2, or LLC2 service. LLC2 guarantees delivery of data using sequence numbers on packets and acknowledgments. A connection-oriented service requires the establishment of an end-to-end path (that may traverse multiple hops) before the actual data transfer can begin. All packets belonging to a session are routed along the same path, and are received at the destination in the same order as they were transmitted by the source node. LLC2 is useful when the end nodes (source and destination) do not use a transport layer protocol for reliable packet delivery. LLC3 provides acknowledged connectionless service, which is nothing but connectionless transfer of individual packets, along with acknowledgments.

The LLC plays the role of a software bus. It enables different protocols of the higher layers of the protocol stack to access different types of physical networks. For example, a host may have multiple network interface cards, such as an Ethernet or a token ring card. It is the responsibility of the LLC to forward packets belonging to different higher-layer protocols to the appropriate network interfaces. Hence, the upper layers can function independently without worrying about the type of the physical network in use.

Medium Access Control Sublayer

The medium access control sublayer (MAC) forms the lower half of the data link layer. It directly interfaces with the physical layer. It provides services such as addressing, framing, and medium access control. Unlike the LLC, these services vary with the physical medium in use. Of these, medium access control is considered to be the most important service. It is relevant to networks (such as LANs) where a single broadcast transmission channel needs to be shared by multiple competing machines. Some of the important medium access control mechanisms are discussed below.

ALOHA

The ALOHA protocol [7] is one of the oldest multiple access mechanisms. ALOHA had its origins in the Hawaiian Islands. It was borne out of the need for interconnecting terminals at the campuses of the University of Hawaii, which were located on different islands. It was devised in the 1970s at the University of Hawaii. The original ALOHA system is referred to as pure ALOHA. Roberts extended this system and developed what is called the slotted ALOHA system [8], which doubled the throughput of the pure ALOHA system.

The pure ALOHA system is very simple. A radio transmitter is attached to each terminal. A user (terminal) transmits whenever data is ready for transmission. If more than one user transmits simultaneously, the transmitted packets end up getting collided and are lost. The sender would be able to detect collisions because of the feedback property of broadcast transmissions. If a sender detects that the frame transmitted by it got destroyed due to collisions, it waits for a random period of time and retransmits the frame. If the waiting time chosen is not random, the competing nodes would wait for the same period of time and so the frames transmitted by them will get collided time and again. In pure ALOHA the vulnerable period for a transmission by a node, which is the period during which no other nodes transmit so that a transmitted frame does not suffer a collision, is equal two twice the frame length/frame period. The throughput achieved in this scheme was found to be around 18 percent.

Slotted ALOHA reduces the probability of collisions by having the nodes transmit in a synchronized fashion. It requires the channel to be divided in time into discrete intervals/slots. The length of each slot is equal to the frame length used. The slot boundaries are the same for all users, that is, time is synchronized. Here a node does not transmit as soon a packet becomes ready for transmission. Instead, it waits till the beginning of the next slot interval and then transmits. If the transmitted frame suffers a collision, the node waits for a random period of time, and then retransmits the frame at the beginning of the next interval. It can be seen that the vulnerable period in the slotted ALOHA scheme is equal to the slot interval, which is equal to the frame length. This is half of that of the pure ALOHA scheme. Hence, the throughput achieved in slotted ALOHA is around 37 percent, which is almost double that of pure ALOHA.

Carrier Sense Multiple Access

The maximum achievable throughput in the ALOHA protocols is low because of wastage of bandwidth due to packet collisions. Packet collisions could be reduced by having the nodes sense for the carrier signal on the channel before they actually start transmitting. Carrier sense multiple access (CSMA) protocols are those in which nodes, before transmitting, first listen for a carrier (*i.e.*, transmission) on the channel, and make decisions on whether or not to transmit based on the absence or presence of the carrier. Several CSMA protocols have been proposed. Some of the important CSMA protocols are discussed below.

The first protocol discussed here is the *1-persistent* CSMA. Upon receiving a packet for transmission, a node first senses the channel to determine whether it is free. If free, the packet is transmitted immediately. Otherwise, it keeps sensing the channel till it becomes free. Once the channel becomes free, it immediately transmits the ready frame. This scheme is called 1-persistent CSMA, since the probability with which a ready node starts transmitting once it finds the channel to be idle, is 1. Propagation delay plays an important role here. Consider the case when the propagation delay is high. The sender has started transmitting the packet, but a ready node near the destination might sense the channel and find it to be idle since the already transmitted packet has not yet arrived at the destination. So this new ready node will start its own transmission which would finally result in a collision. The 1-persistent CSMA scheme performs well when the propagation delay is low.

In the *non-persistent* CSMA scheme, a ready node first senses the channel. If it finds the channel to be busy, it goes into a wait state. The wait period is randomly chosen. After waiting for the random time duration, it again senses the channel, and the algorithm is repeated. Consider the case when a transmission had already started, and two other nodes in the network receive packets for transmission from their higher layers. In the 1-persistent CSMA scheme, both of the ready nodes will wait till the channel becomes idle (current transmission ends) and immediately transmit. Hence, the packets suffer collisions. In the same situation, if the non-persistent scheme is followed, the two ready nodes will wait for different randomly chosen time periods before sensing the channel again. Hence, the probability of collisions is minimized.

The third scheme, *p-persistent* CSMA, combines the best features of the above two schemes. Here, the channel is assumed to be slotted as in the slotted ALOHA scheme. A ready node first senses the channel. If it finds the channel to be busy, it keeps sensing the channel until it can find an idle slot. Once it finds an idle slot, it transmits in the same slot with probability p, or defers the transmission to the next slot with probability $q = 1 - p$. If the next slot is again idle, then the node transmits in that slot or defers its transmission to the next slot with probabilities p and q, respectively. If a transmitted packet gets destroyed due to collisions, then the node waits for an idle slot and repeats the same algorithm.

The performance of all the CSMA protocols described above is greatly dependent on the end-to-end propagation delay of the medium, which constitutes the vulnerable period when collisions are most likely to occur.

Carrier Sense Multiple Access with Collision Detection

In CSMA with collision detection (CSMA/CD), the nodes, apart from sensing the channel, are also capable of detecting collisions in the channel. In the previous schemes, suppose two neighbor nodes start transmitting simultaneously; though the packets start getting collided and garbled at the very beginning, the two nodes still transmit the entire packets. In CSMA/CD, the moment a node detects a collision on the channel, it aborts its current transmission. Hence time, and therefore bandwidth, is saved. After aborting its current transmission, the node transmits a brief jamming signal. On hearing this jamming signal, all other nodes learn about the collision; any other simultaneously transmitting node stops its transmission once it hears the jamming signal. After transmitting the jamming signal, the node waits for a random period of time, and then restarts the process. A collision can be detected by a node by comparing the power or pulse width of the received signal with that of the transmitted signal, that is, if a node finds out that what it reads back from the channel is different from what it put onto the channel, it concludes that its current packet got collided. Assume that a node starts its transmission at time t_0 and let the propagation duration (delay) be t_d. Suppose the destination node starts transmitting at time $t_d - \delta$ (where δ is very small). It will immediately detect a collision and stop transmitting. But, the sender node would come to know of this collision only at time $2t_d - \delta$. In the worst case, for a sender node to determine whether its packet suffered any collision or not, it would take $2t_d$ units of time. Therefore, the channel here is slotted as in slotted ALOHA, with the slot length equal to $2t_d$. The IEEE 802.3 Ethernet standard follows CSMA/CD.

1.12.3 IEEE 802.3 Standard

IEEE 802.3 is a standard for CSMA/CD networks. It is commonly referred to as the Ethernet standard. Ethernet was so named based on the word *ether* so as to describe an essential feature of the system, which was the transfer of bits by the physical medium to all nodes in the LAN, in a manner similar to that of the old *luminiferous ether* which was once thought to propagate electromagnetic waves through space. Ethernet has proven itself for many years now as a reasonably fast and relatively inexpensive LAN technology, and hence is one of the most popular LAN technologies in use today. IEEE 802.3 defines the physical layer and the MAC sublayer for CSMA/CD LANs. The traditional Ethernet specification supports data transmission speeds of 10 Mbps. Additional specifications such as fast Ethernet (IEEE 802.3u) and gigabit Ethernet (IEEE 802.3z) have been published to extend the performance of traditional Ethernet up to 100 Mbps and 1,000 Mbps speeds, respectively. The physical layer and MAC sublayer of the IEEE 802.3 standard are described below.

Physical Layer

Four types of cabling have been specified in the 802.3 standard. They use the following physical transmission media: thick coaxial cable, thin coaxial cable, twisted pair, and optic fiber.

The first type of cabling called *10Base2* or *thin Ethernet* uses thin coaxial cable. Until recently, this was the most popular 802.3 cable because it was cheap and since standard BNC (Bayonet Neil-Concelman, or sometimes British Naval Connector) connectors, rather than vampire taps, were used to form T-junctions, the quality of the connection between the computer and the coaxial cable was good. As evident from its name, it operates at 10 Mbps, uses baseband signaling, and can support LAN segments (a LAN segment is a single section of network media that connects computers/nodes) of lengths up to 200 m. Each segment can have up to 30 nodes.

The second type of cabling, which uses thick coaxial cable, is called *10Base5* cabling. It is commonly referred to as *thick Ethernet*. Connections are made using *vampire taps*, that is, the connection is made by forcing a pin halfway into the coaxial cable's core. This results in a poor connection between the computer and the coaxial cable. This system is called 10Base5 as it operates at 10 Mbps, uses baseband signaling (*i.e.*, the signal is transmitted at its original frequency without using any modulation), and supports LAN segments of lengths up to 500 meters. It can support up to 100 nodes per segment. In the above two cabling techniques, detecting cable breaks, bad taps, or loose connections could be very tedious. These problems with coaxial cabling systems made people switch to the *10Base-T* cabling system.

In the 10Base-T system each node has a separate cable connecting it to a central *hub*. Unshielded twisted pair (UTP) is the most commonly used cable for this system. 10Base-T supports a maximum segment length of 100 m with a maximum of 1,024 nodes per segment.

The fourth cabling technique, called *10Base-F*, uses fiber optic cable. 10Base-F is costlier compared to the other three schemes. But it offers excellent noise immunity. Hence, it is the preferred cabling system when the distance between two connecting points is large. It can support segments of lengths up to 2,000 m, with a maximum of 1,024 nodes per segment.

IEEE came up with the 802.3u standard for supporting a faster LAN with speeds of 100 Mbps. It was backward compatible with the 802.3 standard. The 802.3u standard is commonly referred to as fast Ethernet. 802.3u uses three types of physical transmission media, namely category 3 UTP (*100Base-T4*), category 5 UTP (*100Base-TX*), and optic fiber (*100Base-F*). The IEEE 802.3z gigabit Ethernet standard extends fast Ethernet to provide speeds of 1,000 Mbps. It uses shielded twisted pair (STP) (*1000Base-CX*), UTP (*1000Base-T*), and optic fiber (*1000Base-SX* and *1000Base-LX*). Table 1.3 summarizes the above discussion. The reader can refer to [9] for more details on fast Ethernet and gigabit Ethernet cabling.

Table 1.3. Physical layer cabling

Name	Cable	Maximum Segment Length	Maximum Nodes/Segment
10Base2	Thin coaxial cable	200 m	30
10Base5	Thick coaxial cable	500 m	100
10Base-T	UTP	100 m	1,024
10Base-F	Optic fiber	2,000 m	1,024
100Base-T4	Category 3 UTP	100 m	*
100Base-TX	Category 5 UTP	100 m	*
100Base-F	Optic fiber	2,000 m	*
1000Base-CX	STP	25 m	*
1000Base-T	Category 5 UTP	100 m	*
1000Base-SX	Optic fiber	550 m	*
1000Base-LX	Optic fiber	5,000 m	*

* The maximum number of nodes/segment is not important here. What is more significant is the number of 10 Mbps or 100 Mbps LANs that can be connected through multiple hubs using the cable. The maximum distance between two hubs connected by the cable is more important, which is given by the maximum segment length field in the table.

MAC Sublayer

As mentioned previously, the MAC sublayer uses CSMA/CD technology. When a node wants to transmit a frame, it listens to the channel (cable) to check whether there are any on-going transmissions. If the channel is busy, the node waits for a random duration of time and attempts again. If the channel is idle, the node starts transmitting. It is possible that two nodes end up transmitting simultaneously. In order to minimize the duration of collisions in such circumstances, a transmitting node listens to what it is sending. If what it hears is different from what it is sending, then it implies that a collision has occurred. On detecting a collision, a transmitting node stops its data transmission and transmits a brief jamming signal. The jamming signal ensures that all other nodes get to know that there has been a collision. Any other simultaneously transmitting node, on hearing this jamming signal, stops its transmission. After transmitting the jamming signal, the node waits for a random period of time, and then attempts to transmit again. A receiver node operates as below. It listens to all frames on the cable. If the address of the frame is the same as that of its own address, or is the same as that of the group address of nodes of which the current node is a member, it copies the frame from the cable. Otherwise it just ignores the frame.

The 802.3 frame format is shown in Figure 1.22. The frame starts with a 7-byte *preamble*. It is used for synchronization purposes. Each byte in the preamble contains the bit pattern 10101010. 802.3 uses an encoding technique called Manchester encoding. Manchester encoding of the 10101010 produces a 10 MHz square wave for a duration of 5.6 μsec. This enables the receiver node's clock to synchronize with that of the sender. For more details on Manchester encoding and other encoding techniques, the reader can refer to [10]. The *preamble* field is followed by the *destination address* and *source address* fields. Each manufacturer of Ethernet network cards assigns a unique 48-bit Ethernet address to every Ethernet card manufactured. If all the bits are 1 in the destination address, the message is received

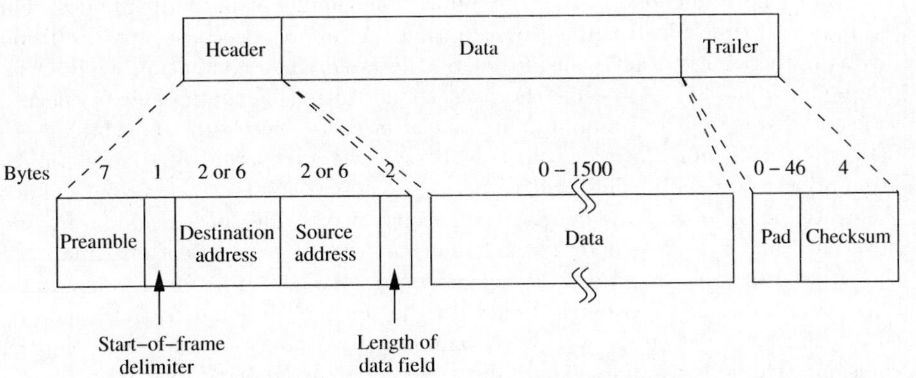

Figure 1.22. IEEE 802.3 frame format.

by all computers connected to the cable, that is, broadcast. If the first bit is 1, then the rest of the bits indicate a multicast group address; the frame would be delivered only to those nodes that are part of that particular multicast group. If the first bit is 0, the rest of the bits make up a normal node address, which is unique to that node on that LAN. The *length* field denotes the number of bytes present in the *data* field. It can vary from a minimum of 0 to a maximum of 1,500 bytes. Though a data field of length 0 bytes is perfectly legal, it would cause a problem. On detecting a collision, a transmitting node stops its transmission. Thus, collisions lead to the appearance of corrupted frames and stray bits on the cable all the time. In order to make it easier to distinguish valid frames from corrupted frames (frames that suffered collisions), 802.3 requires the length of valid frames to be at least 64 bytes from the destination address to the checksum. Another important reason for having a minimum frame length is to enable a node to detect whether the frame it transmitted was actually received without error at the destination or suffered collisions. Assume that the frame length is very small. The node would have completed transmitting the frame before initial bits of the frame reach the destination. If the destination node starts its own transmission just before the previously transmitted frame reaches it, it results in a collision. But since the source node has already completed its transmission, it does not get to know about the collision. Therefore, to avoid such situations, the transmission duration of a frame must be equal to at least twice the round-trip time of the cable. Hence, a minimum frame length is required. For a 10 Mbps LAN with a maximum length of 2,500 m and four repeaters (802.3 specification), the minimum frame duration must be 51.2 μsec. This corresponds to 64 bytes. If the frame length is less than 64 bytes, the *Pad* field is used to fill up the frame to 64 bytes. The *Checksum* field is used to detect if any data bits have been corrupted during transmission.

As described earlier, when a collision is detected, the transmitting node aborts its transmission and retries after a random duration of time. This randomization is achieved using an algorithm called the *binary exponential back-off* (BEB) algorithm. The BEB algorithm adjusts what is called the contention window size by indirectly estimating the traffic load in the communication channel at individual nodes. This estimation of traffic load is done by counting the number of consecutive collisions involving the same packet transmitted by the node. The contention window size of a node is doubled every time the packet transmitted by the node experiences a collision, and is reset to its minimum value after every successful transmission. On detecting a collision, the node must wait for a random period of time before reattempting to transmit. This waiting period is chosen based on the current value of the contention widow of the node. After k consecutive collisions, a random number between 0 and $2^k - 1$ (where 2^k is the current size of the contention window) is chosen, and the node remains idle, not attempting to transmit, for that number of slots (time is assumed to be slotted, with the slot length equal to the worst-case end-to-end propagation delay). For example, if two nodes collide, after the first collision, each node waits for 0 or 1 slot time before attempting to transmit again. If they collide again (second collision), each node picks 0, 1, 2, or 3 slots and waits for that number of slots. Thus the randomization interval grows exponentially after each collision. The randomization interval (0 to $2^k - 1$) is frozen once the maximum number of slots (1,023) is reached. If a node suffers repeated collisions, it retries for a certain maximum number of times (16), and then reports failure to the node. It is up to the higher layers to take recovery action.

1.12.4 IEEE 802.11 Standard

The IEEE 802.11 standard is one of the most popular standards for wireless LANs. Wireless LANs are used for providing network services in places where it may be very difficult or too expensive to lay cabling for a wireline network. The IEEE 802.11 standard comes under the IEEE 802.x LAN standards, and specifies the physical layer and the MAC layer, adapted to the specific requirements of wireless LANs. The objective of this standard is to provide wireless connectivity to wireless devices/nodes that require rapid deployment, which may be portable, or which may be mounted on moving vehicles within a local area. The IEEE 802.11 standard also aids the regulatory bodies in standardizing access to one or more radio frequency bands for the purpose of local area communication. The interfaces offered by 802.11 to the higher layers are the same as those offered in other 802.x standards. The MAC layer should be able to work with multiple physical layers catering to multiple transmission techniques such as infrared and spread spectrum.

Physical Layer

The basic 802.11 standard supports three different physical layers. Two of them, frequency hopping spread spectrum (FHSS) and direct sequence spread spectrum (DSSS) (which were described previously in Section 1.6.3) are based on radio transmissions, and the third is based on infrared. The physical layer provides mechanisms for sensing the wireless channel and determining whether or not it is idle. This mechanism is also called clear channel assessment (CCA).

MAC Sublayer

The 802.11 standard follows carrier sense multiple access with collision avoidance (CSMA/CA), which is a random access scheme with carrier sensing and collision avoidance through random back-off. Because of the nature of the radio environment, it is very difficult for a transmitting node to detect packet collisions in the network. Hence, CSMA/CD is not preferred in wireless LANs.

802.11 Task Groups

Several task groups have been working on different networking aspects of wireless LANs. The various 802.11 task groups are listed below.

- 802.11: This was the first 802.11 task group. The objective of this group was to develop MAC layer and physical layer specifications for wireless connectivity for fixed, portable, and mobile nodes within a local area. The 802.11 standard was first published in 1997.

- 802.11a: This group created a standard for wireless LAN operations in the 5 GHz frequency band, where data rates of up to 54 Mbps are possible. The 802.11a standard was ratified in 1999.
- 802.11b: This task group created a standard for wireless LAN operations in the 2.4 GHz Industrial, Scientific, and Medical (ISM) band, which is freely available for use throughout the world. This standard is popularly referred to as Wi-Fi, standing for Wireless-Fidelity. It can offer data rates of up to 11 Mbps. This standard was ratified in 1999.
- 802.11c: This group was constituted for devising standards for bridging operations. Manufacturers use this standard while developing bridges and access points. The standard was published in 1998.
- 802.11d: This group's main objective is publishing definitions and requirements for enabling the operation of the 802.11 standard in countries that are not currently served by the standard. This standard was published in 2001.
- 802.11e: The main objective of this group is to define an extension of the 802.11 standard for quality of service (QoS) provisioning and service differentiation in wireless LANs. Work is in progress in this regard.
- 802.11f: This group was created for developing specifications for implementing access points and distribution systems following the 802.11 standard, so that interoperability problems between devices manufactured by different vendors do not arise. This standard was published in 2003.
- 802.11g: This group was involved in extending the 802.11b standard to support high-speed transmissions of up to 54 Mbps in the 5 GHz frequency band, while maintaining backward compatibility with current 802.11b devices. The 802.11g standard was published in 2003.
- 802.11h: This is supplementary to the 802.11 standard. It was developed in order for the MAC layer to comply with European regulations for 5 GHz wireless LANs, which require products to have mechanisms for transmission power control and dynamic frequency selection. The standard was published in 2003.
- 802.11i: This group is working on mechanisms for enhancing security in the 802.11 standard.
- 802.11j: This task group is working on mechanisms for enhancing the current 802.11 MAC physical layer protocols to additionally operate in the newly available Japanese 4.9 GHz and 5 GHz bands.
- 802.11n: The objective of this group is to define standardized modifications to the 802.11 MAC and physical layers such that modes of operation that are capable of much higher throughputs at the MAC layer, with a maximum of at least 100 Mbps, can be enabled. Work on this is in progress.

A detailed description of the 802.11 standard is provided in the next chapter.

1.13 WIRELESS NETWORKS AND BOOK OVERVIEW

Wireless networks are computer networks that use radio frequency channels as their physical medium for communication. Each node in the network broadcasts information which can be received by all nodes within its direct transmission range. Since nodes transmit and receive over the air, they need not be physically connected to any network. Hence, such networks offer data connectivity along with user mobility.

The world's first wireless radio communication system was invented by Guglielmo Marconi in 1897. In 1901, he successfully demonstrated his wireless telegraph system to the world by transmitting radio signals across the Atlantic Ocean from England to America, covering more than 1,700 miles. Through his system, two end users could communicate by sending each other alphanumeric characters that were encoded in an analog signal. This signaled the beginning of the radio communications era.

Today, very advanced communication systems are available. Radio and television broadcasting are some of the very common applications that cannot do without wireless technologies. Voice communication has been the biggest beneficiary of the developments that took place, and are still taking place, in wireless communication technologies. Beginning in the 1960s, several communication satellites have been launched, and today a huge proportion of long-distance voice communication takes place through the numerous satellites orbiting around the Earth.

Wireless communications is one of the fastest growing industries in the world. The wireless communications industry has several segments such as cellular telephony, wireless LANs, and satellite-based communication networks. However, the major portion of the growth in wireless industry has been due to cellular networks. The early 1980s saw the commercial deployments of the world's first mobile cellular networks. Cellular networks provide two-way simultaneous voice communication. A fixed base station serves all mobile phones in its coverage area, also called a cell. The entire service area is divided into a number of non-overlapping cells (in actual deployments the cells overlap partially), communication within each cell being coordinated by a separate base station. The first-generation (1G) cellular networks used analog signal technology. They used frequency modulation for signal transmission. The mobile phones were bulky and not very compact. The coverage provided by the 1G networks was also not good. The 1G system deployed in the United States and Canada was known as the advanced mobile phone system (AMPS). Second-generation (2G) cellular systems used digital transmission mechanisms such as TDMA and CDMA. The global system for mobile communication (GSM) used in Europe, the IS-136 (Telecommunications Industry Association Interim Standard) system used in the United States, and the personal digital communication (PDC) system used in Japan are some of the popular 2G cellular systems. The 1G and 2G systems were designed primarily for voice communication. The present generation of wireless communication networks is often called 2.5G. 2.5G is usually associated with the general packet radio services (GPRS) system. GPRS has been commercially deployed in many countries. The third-generation (3G) systems are expected to provide services such as enhanced multimedia, bandwidth up to 2 Mbps, and roaming capability throughout the world. Commercial large scale deployments of 3G systems have not yet started. It is estimated that communication service providers would shift to 3G by 2005. However, the commercial viability of such a system which requires fresh investments is still in question. Wideband code division multiple access (W-CDMA) and universal mobile telecommunications system (UMTS) are some of the important 3G standards. Standardization efforts are on for the fourth-generation (4G) wireless networks. The 4G systems are expected to provide further improvements in the services provided by 3G, such as enhanced multimedia, universal access, and portability across all types of devices. 4G services are expected to be commercially introduced by 2010. The deployment of 4G services would transform the world truly into a *global village*.

Though the 2G and 2.5G systems were well-suited for voice communication, what fueled the development of newer technologies such as 3G and 4G, was the realization that wireless data services would also become very important in the next few years. Wireless data services are projected to be another huge market for the telecommunications companies. Wireless Internet forms the core of these data services. Wireless access to the Internet using mobile phones would require several modifications to the already existing Internet protocols, since such devices have limited display and input capabilities. Work is in progress for developing such protocols.

Wireless local area networks (WLANs) are another type of wireless networks that are increasingly popular today. WLANs are used for providing network services in places where it is very difficult or is too expensive to lay cabling for a wireline network. In a WLAN, a stationary node called an access point (AP) coordinates the communication taking place between nodes in the LAN. The two main standards for WLANs are the IEEE 802.11 standard and the European Telecommunications Standards Institute (ETSI) HIPERLAN standard.

Wireless personal area networks (WPANs) are short-distance wireless networks that have been specifically designed for interconnecting portable and mobile computing devices such as laptops, mobile phones, pagers, personal digital assistants (PDAs), and a number of other consumer electronic devices. Bluetooth is a popular WPAN specification and it defines how data is managed and physically carried over the WPAN. The typical range of a Bluetooth network is around 10 m. The Bluetooth Special Interest Group (SIG), comprised of several leading companies such as Ericsson, Intel, IBM, Nokia, and Toshiba, is driving the development of the Bluetooth technology and bringing it to the market. The IEEE 802.15 standard defines specifications for the media access control and physical layers for wireless devices in WPANs.

Another emerging type of wireless network is the ad hoc wireless network. An ad hoc wireless network is an autonomous system of mobile nodes connected through wireless links. It does not have any fixed infrastructure (such as base stations in cellular networks). The mobile stations/nodes in the network coordinate among themselves for communication. Hence, each node in the network, apart from being a source or destination, is also expected to route packets for other nodes in the network. Such networks find varied applications in real-life environments such as communication in battlefields, communication among rescue personnel in disaster-affected areas, law enforcement agencies, and wireless sensor networks.

The networks mentioned so far are either infrastructure-based where fixed base stations/access points are used, or infrastructure-less where no fixed infrastructure support is available. Several new hybrid networking architectures are emerging that are a combination of the above two types of networks, taking advantage of the best features of both. Some examples of such hybrid wireless network architectures are multi-hop cellular network (MCN), integrated cellular and ad hoc relaying system (iCAR), and multi-power architecture for cellular networks (MuPAC).

The rest of the book is organized into thirteen chapters. Chapters 2-4 address several issues concerning the design of wireless LANs and PANs, wireless WANs and MANs, and wireless Internet, respectively. Chapters 5-11 deal with all important design issues — medium access control, routing, multicasting, transport layer, security, quality of service provisioning, energy management — in ad hoc wireless networking in considerable depth. Chapters 12-14 cover some recent related important topics. Chapter 12 is devoted to wireless sensor networks. Chapter 13 describes several hybrid wireless architectures, and routing protocols and load balancing schemes for hybrid wireless networks. Finally, Chapter 14 provides recent advances in wireless networking such as ultra wideband technology, Wi-Fi systems, optical wireless networks, and multimode 802.11.

1.14 SUMMARY

The first half of this chapter provided a brief introduction to the fundamental aspects of wireless transmission, which include the characteristics of the wireless channel, the various modulation mechanisms, multiple access techniques, and coding and error control mechanisms. The second half of the chapter discussed the basic concepts and principles involved in computer networking. Some of the important computer network models such as the ISO OSI reference model, TCP/IP reference model, and the ATM reference model were described. This was followed by a detailed discussion of the IEEE 802.3 wired LAN standard and a brief introduction to the IEEE 802.11 wireless LAN standard. A brief discussion on the advent of wireless networks was also presented.

1.15 PROBLEMS

1. High-frequency X-rays and Gamma rays are not normally used for wireless communication. Explain why.

2. What is multipath propagation? Explain how it affects signal quality.

3. What is inter-symbol interference? Give a mechanism that is used for overcoming problems arising due to inter-symbol interference.

4. What is the maximum data rate that can be supported on a 10 MHz noise-less channel if the channel uses eight-level digital signals?

5. Calculate the maximum achievable data rate over a 9 KHz channel whose signal to noise ratio is 20 dB.

6. What is the main difference between analog modulation and digital modulation schemes?

7. Which modulation mechanism is better, amplitude modulation or frequency modulation? Give reasons to support your answer.

8. Explain the advantages and disadvantages of the quadrature amplitude modulation schemes.

9. Show the phase changes corresponding to the bit pattern 100001 when $\pi/4$ shifted PSK is used. Use Table 1.2.

10. With respect to pulse code modulation, what is quantization error? Explain how it is caused.

11. Consider the convolutional encoder shown in Figure 1.23. What would be the output of the encoder for an input of 1001?

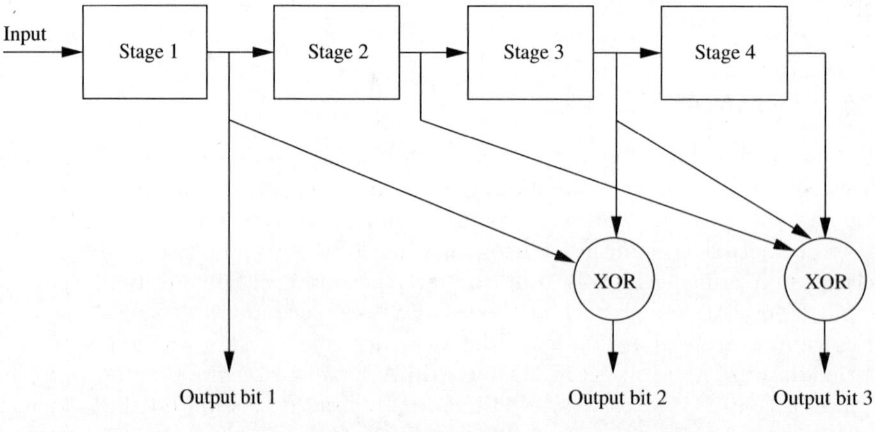

Figure 1.23. Convolutional coder.

12. Why does ATM use fixed sized cells?

13. What is meant by the vulnerable period of a transmission? What is the value of this vulnerable period for the ALOHA protocols?

14. Why is a minimum frame length required in CSMA/CD LANs that follow the IEEE 802.3 standard?

15. Why is CSMA/CD not generally used in wireless LANs?

BIBLIOGRAPHY

[1] D. J. Griffiths, *Introduction to Electrodynamics*, Prentice Hall PTR, New Jersey, 1998.

[2] K. Pahlavan and P. Krishnamurthy, *Principles of Wireless Networks*, Prentice Hall PTR, New Jersey, 2002.

[3] T. S. Rappaport, *Wireless Communications: Principles and Practice*, Prentice Hall PTR, New Jersey, 2001.

[4] N. Morgan, *Speech and Audio Signal Processing: Processing and Perception of Speech and Music*, John Wiley & Sons, 1999.

[5] R. Johannesson and K. S. Zigangirov, *Fundamentals of Convolutional Coding*, Wiley-IEEE Press, 2001.

[6] C. Berrou, A. Glavieux, and P. Thitimajshima, "Near Shannon Limit Error-Correcting Coding and Decoding: Turbo-Codes," *Proceedings of IEEE ICC 1993*, pp. 1064-1070, May 1993.

[7] N. Abramson, "The Aloha System – Another Alternative for Computer Communications," *Proceedings of AFIPS Fall Joint Computer Conference 1970*, vol. 37, pp. 281-285, November 1970.

[8] L. Roberts, "ALOHA Packet System with and without Slots and Capture," *Computer Communications Review*, vol. 5, pp. 28-42, April 1975.

[9] Andrew S. Tanenbaum, *Computer Networks*, Prentice Hall PTR, New Jersey, 2002.

[10] W. Stallings, *Data and Computer Communications*, Prentice Hall PTR, New Jersey, 1999.

Chapter 2

WIRELESS LANS AND PANS

2.1 INTRODUCTION

The field of computer networks has grown significantly in the last three decades. An interesting usage of computer networks is in offices and educational institutions, where tens (sometimes hundreds) of personal computers (PCs) are interconnected, to share resources (*e.g.*, printers) and exchange information, using a high-bandwidth communication medium (such as the Ethernet). These privately-owned networks are known as local area networks (LANs) which come under the category of small-scale networks (networks within a single building or campus with a size of a few kilometers). To do away with the wiring associated with the interconnection of PCs in LANs, researchers have explored the possible usage of radio waves and infrared light for interconnection [1]. This has resulted in the emergence of wireless LANs (WLANs), where wireless transmission is used at the physical layer of the network. Wireless personal area networks (WPANs) are the next step down from WLANs, covering smaller areas with low power transmission, for networking of portable and mobile computing devices such as PCs, personal digital assistants (PDAs), which are essentially very small computers designed to consume as little power as possible so as to increase the lifetime of their batteries, cell phones, printers, speakers, microphones, and other consumer electronics. This chapter highlights the issues involved in the design of WLANs and PANs. It consists of the following sections:

1. Fundamentals of WLANs: The technical issues in WLANs must be understood in order to appreciate the difference between wired networks and wireless networks. The use of WLANs and their design goals are then studied. The types of WLANs, their components, and their basic functionalities are also brought out in this section.

2. IEEE 802.11 Standard: This section introduces a prominent standard in WLANs, the IEEE 802.11 standard. The medium access control (MAC) layer and the physical layer mechanisms are explained here. This section also covers some of the optional functionalities, such as security and quality of service (QoS).

3. HIPERLAN Standard: This section describes another WLAN standard, HIPERLAN standard, which is a European standard based on radio access.

4. Bluetooth: This section deals with the Bluetooth standard, which enables personal devices to communicate with each other in the absence of infrastructure.

5. HomeRF: This section discusses the issues in home networking (HomeRF standard) and finally illustrates the technical differences between Bluetooth, HomeRF, and other technologies such as infrared [portable devices that use the infrared interface of the Infrared Data Association (IrDA) for transmission], which are the current technological alternatives in the PAN area.

2.2 FUNDAMENTALS OF WLANS

This section deals with the fundamental principles, concepts, and requirements of WLANs. This section also brings out WLAN types, their components, and some of their functionalities. In what follows, the terms "node," "station," and "terminal" are used interchangeably. While both portable terminals and mobile terminals can move from one place to another, portable terminals are accessed only when they are stationary. Mobile terminals (MTs), on the other hand, are more powerful, and can be accessed when they are in motion. WLANs aim to support truly mobile work stations.

2.2.1 Technical Issues

Here the technical issues that are encountered in the design and engineering of WLANs are discussed. In particular, the differences between wireless and wired networks, the use of WLANs, and the design goals for WLANs are studied.

Differences Between Wireless and Wired Transmission

- **Address is not equivalent to physical location:** In a wireless network, address refers to a particular station and this station need not be stationary. Therefore, address may not always refer to a particular geographical location.

- **Dynamic topology and restricted connectivity:** The mobile nodes may often go out of reach of each other. This means that network connectivity is partial at times.

- **Medium boundaries are not well-defined:** The exact reach of wireless signals cannot be determined accurately. It depends on various factors such as signal strength and noise levels. This means that the precise boundaries of the medium cannot be determined easily.

- **Error-prone medium:** Transmissions by a node in the wireless channel are affected by simultaneous transmissions by neighboring nodes that are located within the direct transmission range of the transmitting node. This means that the error rates are significantly higher in the wireless medium. Typical bit error rates (fractions of bits that are received in error) are of the order of 10^{-4} in a wireless channel as against 10^{-9} in fiber optic cables.

The above four factors imply that we need to build a reliable network on top of an inherently unreliable channel. This is realized in practice by having reliable protocols at the MAC layer, which hide the unreliability that is present in the physical layer.

Use of WLANs

Wireless computer networks are capable of offering versatile functionalities. WLANs are very flexible and can be configured in a variety of topologies based on the application. Some possible uses of WLANs are mentioned below.

- Users would be able to surf the Internet, check e-mail, and receive Instant Messages on the move.
- In areas affected by earthquakes or other such disasters, no suitable infrastructure may be available on the site. WLANs are handy in such locations to set up networks on the fly.
- There are many historic buildings where there has been a need to set up computer networks. In such places, wiring may not be permitted or the building design may not be conducive to efficient wiring. WLANs are very good solutions in such places.

Design Goals

The following are some of the goals which have to be achieved while designing WLANs:

- **Operational simplicity:** Design of wireless LANs must incorporate features to enable a mobile user to quickly set up and access network services in a simple and efficient manner.
- **Power-efficient operation:** The power-constrained nature of mobile computing devices such as laptops and PDAs necessitates the important requirement of WLANs operating with minimal power consumption. Therefore, the design of WLAN must incorporate power-saving features and use appropriate technologies and protocols to achieve this.
- **License-free operation:** One of the major factors that affects the cost of wireless access is the license fee for the spectrum in which a particular wireless access technology operates. Low cost of access is an important aspect for popularizing a WLAN technology. Hence the design of WLAN should consider the parts of the frequency spectrum (*e.g.,* ISM band) for its operation which do not require an explicit licensing.
- **Tolerance to interference:** The proliferation of different wireless networking technologies both for civilian and military applications and the use of the microwave frequency spectrum for non-communication purposes (*e.g.,* microwave ovens) have led to a significant increase in the interference level

across the radio spectrum. The WLAN design should account for this and take appropriate measures by way of selecting technologies and protocols to operate in the presence of interference.

- **Global usability:** The design of the WLAN, the choice of technology, and the selection of the operating frequency spectrum should take into account the prevailing spectrum restrictions in countries across the world. This ensures the acceptability of the technology across the world.
- **Security:** The inherent broadcast nature of wireless medium adds to the requirement of security features to be included in the design of WLAN technology.
- **Safety requirements:** The design of WLAN technology should follow the safety requirements that can be classified into the following: (i) interference to medical and other instrumentation devices and (ii) increased power level of transmitters that can lead to health hazards. A well-designed WLAN should follow the power emission restrictions that are applicable in the given frequency spectrum.
- **Quality of service requirements:** Quality of service (QoS) refers to the provisioning of designated levels of performance for multimedia traffic. The design of WLAN should take into consideration the possibility of supporting a wide variety of traffic, including multimedia traffic.
- **Compatibility with other technologies and applications:** The interoperability among the different LANs (wired or wireless) is important for efficient communication between hosts operating with different LAN technologies. In addition to this, interoperability with existing WAN protocols such as TCP/IP of the Internet is essential to provide a seamless communication across the WANs.

2.2.2 Network Architecture

This section lists the types of WLANs, the components of a typical WLAN, and the services offered by a WLAN.

Infrastructure Based Versus Ad Hoc LANs

WLANs can be broadly classified into two types, infrastructure networks and ad hoc LANs, based on the underlying architecture.

Infrastructure networks contain special nodes called *access points* (APs), which are connected via existing networks. APs are special in the sense that they can interact with wireless nodes as well as with the existing wired network. The other wireless nodes, also known as mobile stations (STAs), communicate via APs. The APs also act as bridges with other networks.

Ad hoc LANs do not need any fixed infrastructure. These networks can be set up on the fly at any place. Nodes communicate directly with each other or forward messages through other nodes that are directly accessible.

Components in a Typical IEEE 802.11 Network

IEEE 802.11 is the most popular WLAN standard that defines the specification for the physical and MAC layers. The success of this standard can be understood from the fact that the revenue from the products based on this standard touched \$730 million in the second quarter of the year 2003. The principles and mechanisms followed in this standard are explained later. In what follows, the basic components in a typical IEEE 802.11 WLAN [2] are listed.

The set of stations that can remain in contact (*i.e.,* are associated) with a given AP is called a basic service set (BSS). The coverage area of an AP within which member stations (STAs or MTs) may remain in communication is called the basic service area (BSA). The stations that are a part of a BSS need to be located within the BSA of the corresponding AP. A BSS is the basic building block of the network. BSSs are connected by means of a distribution system (DS) to form an extended network.

DS refers to an existing network infrastructure. The implementation of the DS is not specified by the IEEE 802.11 standard. The services of the DS, however, are specified rigidly. This gives a lot of flexibility in the design of the DS. The APs are connected by means of the DS.

Portals are logical points through which non-IEEE 802.11 packets (wired LAN packets) enter the system. They are necessary for integrating wireless networks with the existing wired networks. Just as an AP interacts with the DS as well as the wireless nodes, the portal interacts with the wired network as well as with the DS. The BSSs, DS, and the portals together with the stations they connect constitute the extended service set (ESS). An ad hoc LAN has only one BSS. Therefore, ad hoc LANs are also known as independent basic service sets (IBSSs). It may be noted that the ESS and IBSS appear identical to the logical link control (LLC). Figure 2.1 gives a schematic picture of what a typical ESS looks like.

Services Offered by a Typical IEEE 802.11 Network

The services offered by a typical IEEE 802.11 network can be broadly divided into two categories: AP services and STA services. The following are the AP services, which are provided by the DS:

- **Association:** The identity of an STA and its address should be known to the AP before the STA can transmit or receive frames on the WLAN. This is done during association, and the information is used by the AP to facilitate routing of frames.

- **Reassociation:** The established association is transferred from one AP to another using reassociation. This allows STAs to move from one BSS to another.

- **Disassociation:** When an existing association is terminated, a notification is issued by the STA or the AP. This is called disassociation, and is done when nodes leave the BSS or when nodes shut down.

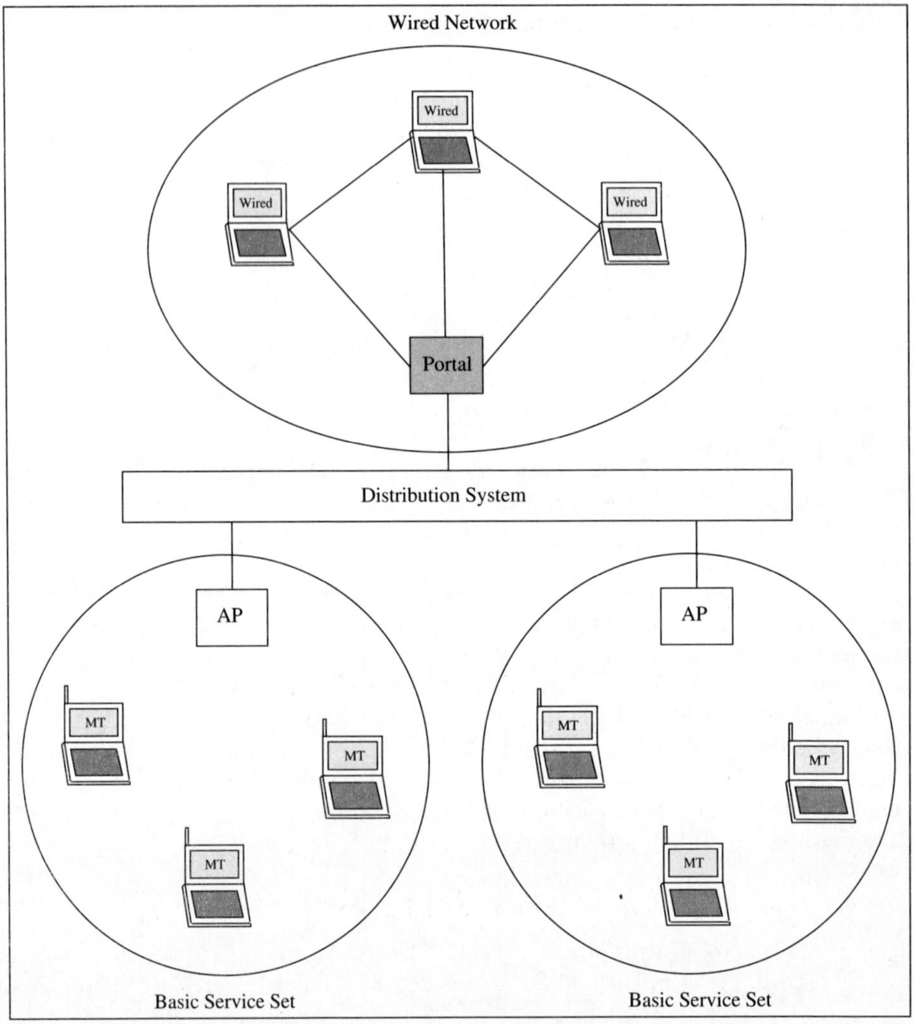

Figure 2.1. Extended Service Set.

- **Distribution:** Distribution takes care of routing frames. If the destination is in the same BSS, the frame is transmitted directly to the destination, otherwise the frame is sent via the DS.
- **Integration:** To send frames through non-IEEE 802.11 networks, which may have different addressing schemes or frame formats, the integration service is invoked.

The following are the STA services, which are provided by every station, including APs:

- **Authentication:** Authentication is done in order to establish the identity of stations to each other. The authentication schemes range from relatively insecure handshaking to public-key encryption schemes.
- **Deauthentication:** Deauthentication is invoked to terminate existing authentication.
- **Privacy:** The contents of messages may be encrypted (say, by using the WEP algorithm, which is explained later) to prevent eavesdroppers from reading the messages.
- **Data delivery:** IEEE 802.11 naturally provides a way to transmit and receive data. However, like Ethernet, the transmission is not guaranteed to be completely reliable.

2.3 IEEE 802.11 STANDARD

After the fundamental issues in WLANs are clearly understood, the reader is in a position to appreciate the *de facto* standards for WLANs. IEEE 802.11 is a prominent standard for WLANs, which is adopted by many vendors of WLAN products. A later version of this standard is the IEEE 802.11b [3], commercially known as *Wi-Fi* (wireless fidelity). The IEEE 802.11 standard, which deals with the physical and MAC layers in WLANs, was brought out in 1997. This standard is explained in this section.

It may be observed that IEEE 802.11 was the first WLAN standard that faced the challenge of organizing a systematic approach for defining a standard for wireless wideband local access (small-scale networks capable of transmitting data at high rates). As mentioned earlier, in contrast to other LAN standards, wireless standards need to have provisions to support mobility of nodes. The IEEE 802.11 working group had to examine connection management, link reliability management, and power management — none of which was a concern for other standards in IEEE 802. In addition, provision for security had to be introduced. For all these reasons and because of several competing proposals, it took nearly ten years for the development of IEEE 802.11, which was much longer compared to the time taken for the development of other 802 standards for the wired media. Once the overall picture and the ideas became clear, it took only a reasonable duration of time to develop the IEEE 802.11a and IEEE 802.11b enhancements. Under the IEEE 802.11 standard, MTs can operate in two modes: (i) *infrastructure mode*, in which MTs can communicate with one or more APs which are connected to a WLAN, and (ii) *ad hoc mode*, in which MTs can communicate directly with each other without using an AP.

2.3.1 Physical Layer

IEEE 802.11 supports three options for the medium to be used at the physical level — one is based on infrared [4] and the other two are based on radio transmission. The physical layer is subdivided conceptually into two parts — physical medium dependent sublayer (PMD) and physical layer convergence protocol (PLCP). PMD handles encoding, decoding, and modulation of signals and thus deals with the idiosyncrasies of the particular medium. The PLCP abstracts the functionality that the physical layer has to offer to the MAC layer. PLCP offers a service access point (SAP) that is independent of the transmission technology, and a clear channel assessment (CCA) carrier sense signal to the MAC layer. The SAP abstracts the channel which can offer up to 1 or 2 Mbps data transmission bandwidth. The CCA is used by the MAC layer to implement the CSMA/CA mechanism. The three choices for the physical layer in the original 802.11 standard are as follows: (i) frequency hopping spread spectrum (FHSS) operating in the license-free 2.4 GHz industrial, scientific, and medical (ISM) band, at data rates of 1 Mbps [using 2-level Gaussian frequency shift keying (GFSK) modulation scheme] and 2 Mbps (using 4-level GFSK); (ii) direct sequence spread spectrum (DSSS) operating in the 2.4 GHz ISM band, at data rates of 1 Mbps [using differential binary phase shift keying (DBPSK) modulation scheme] and 2 Mbps [using differential quadrature phase shift keying (DQPSK)]; (iii) infrared operating at wavelengths in 850-950 nm range, at data rates of 1 Mbps and 2 Mbps using pulse position modulation (PPM) scheme.

Carrier Sensing Mechanisms

In IEEE 802.3, sensing the channel is very simple. The receiver reads the peak voltage on the cable and compares it against a threshold. In contrast, the mechanism employed in IEEE 802.11 is relatively more complex. It is performed either physically or virtually. As mentioned earlier, the physical layer sensing is through the clear channel assessment (CCA) signal provided by the PLCP in the physical layer of the IEEE 802.11. The CCA is generated based on sensing of the air interface either by sensing the detected bits in the air or by checking the received signal strength (RSS) of the carrier against a threshold. Decisions based on the detected bits are made somewhat more slowly, but they are more reliable. Decisions based on the RSS can potentially create a false alarm caused by measuring the level of interference.

2.3.2 Basic MAC Layer Mechanisms

This section describes the MAC layer as specified by the IEEE 802.11 standard. The primary function of this layer is to arbitrate and statistically multiplex the transmission requests of various wireless stations that are operating in an area. This assumes importance because wireless transmissions are inherently broadcast in nature and contentions to access the shared channel need to be resolved prudently in order to avoid collisions, or at least to reduce the number of collisions. The MAC layer also supports many auxiliary functionalities such as offering support for roaming, authentication, and taking care of power conservation.

The basic services supported are the mandatory asynchronous data service and an optional real-time service. The asynchronous data service is supported for unicast packets as well as for multicast packets. The real-time service is supported only in infrastructure-based networks where APs control access to the shared medium.

Distributed Foundation Wireless Medium Access Control (DFWMAC)

The primary access method of IEEE 802.11 is by means of a distributed coordination function (DCF). This mandatory basic function is based on a version of carrier sense with multiple access and collision avoidance (CSMA/CA). To avoid the hidden terminal problem (which is explained later), an optional RTS-CTS mechanism is implemented. There is a second method called the point coordination function (PCF) that is implemented to provide real-time services. When the PCF is in operation, the AP controls medium access and avoids simultaneous transmissions by the nodes.

Inter-Frame Spacing (IFS)

Inter-frame spacing refers to the time interval between the transmission of two successive frames by any station. There are four types of IFS: SIFS, PIFS, DIFS, and EIFS, in order from shortest to longest. They denote priority levels of access to the medium. Shorter IFS denotes a higher priority to access the medium, because the wait time to access the medium is lower. The exact values of the IFS are obtained from the attributes specified in the physical layer management information base (PHYMIB) and are independent of the station bit rate.

- **Short inter-frame spacing (SIFS)** is the shortest of all the IFSs and denotes highest priority to access the medium. It is defined for short control messages such as acknowledgments for data packets and polling responses. The transmission of any packet should begin only after the channel is sensed to be idle for a minimum time period of at least SIFS.
- **PCF inter-frame spacing (PIFS)** is the waiting time whose value lies between SIFS and DIFS. This is used for real-time services.
- **DCF inter-frame spacing (DIFS)** is used by stations that are operating under the DCF mode to transmit packets. This is for asynchronous data transfer within the contention period.
- **Extended inter-frame spacing (EIFS)** is the longest of all the IFSs and denotes the least priority to access the medium. EIFS is used for resynchronization whenever physical layer detects incorrect MAC frame reception.

2.3.3 CSMA/CA Mechanism

Carrier sense with multiple access and collision avoidance (CSMA/CA) is the MAC layer mechanism used by IEEE 802.11 WLANs. Carrier sense with multiple access and collision detection (CSMA/CD) is a well-studied technique in IEEE 802.x wired LANs. This technique cannot be used in the context of WLANs effectively because the error rate in WLANs is much higher and allowing collisions will lead to a drastic reduction in throughput. Moreover, detecting collisions in the wireless medium is not always possible. The technique adopted here is therefore one of collision avoidance.

The Medium Access Mechanism

The basic channel access mechanism of IEEE 802.11 is shown in Figure 2.2 (a). If the medium is sensed to be idle for a duration of DIFS, the node accesses the medium for transmission. Thus the channel access delay at very light loads is equal to the DIFS. If the medium is busy, the node *backs off*, in which the station defers channel access by a random amount of time chosen within a *contention window* (CW). The value of CW can vary between CW_{min} and CW_{max}. The time intervals are all integral multiples of slot times, which are chosen judiciously using propagation delay, delay in the transmitter, and other physical layer dependent parameters. As soon as the back-off counter reaches zero and expires, the station can access the medium. During the back-off process, if a node detects a busy channel, it freezes the back-off counter and the process is resumed once the channel becomes idle for a period of DIFS. Each station executes the back-off procedure at least once between every successive transmission.

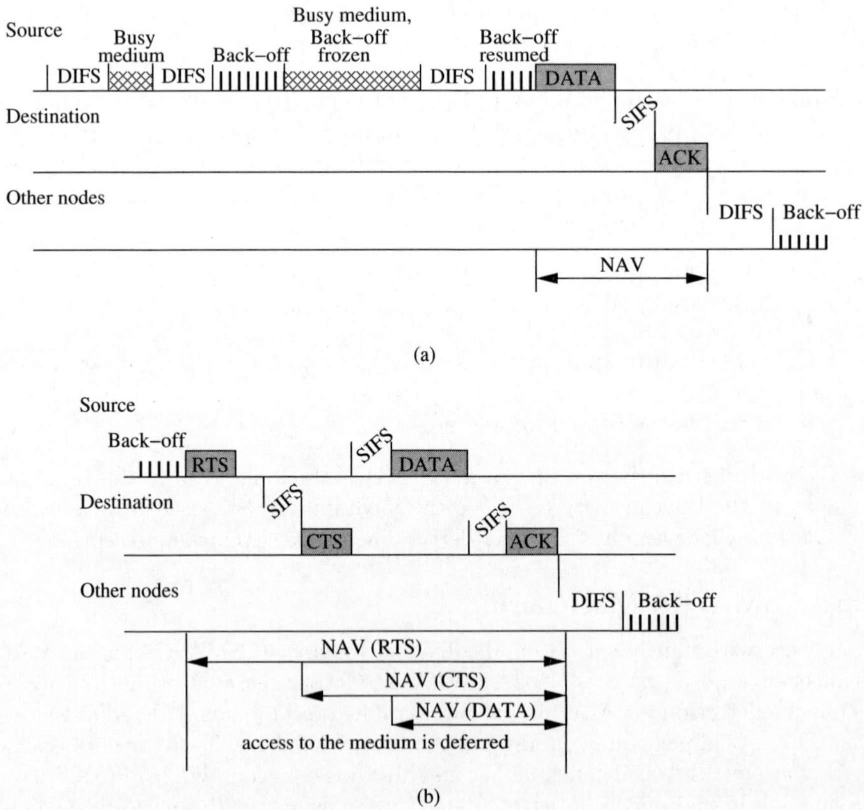

Figure 2.2. IEEE 802.11 DCF and RTS-CTS mechanism.

In the scheme discussed so far, each station has the same chances for transmitting data next time, independent of the overall waiting time for transmission. Such a system is clearly unfair. Ideally, one would like to give stations that wait longer a higher priority service in order to ensure that they are not starved. The back-off timer incorporated into the above mechanism tries to make it fair. Longer waiting stations, instead of choosing another random interval from the contention window, wait only for a residual amount of time that is specified by the back-off timer.

Contention Window Size

The size of the Contention Window (CW) is another important parameter. If the CW is small in size, then the random values will be close together and there is a high probability of packet collision. On the other hand, if the size of CW is very large, there will be some unnecessary delay because of large back-off values. Ideally, one would like the system to adapt to the current number of stations that are contending for channel access. To effect this, the truncated binary exponential back-off technique is used here, which is similar to the technique used in IEEE 802.3. The initial contention window is set to a random value between (0, CWmin) and each time a collision occurs, the CW doubles its size up to a maximum of CWmax. So at high load, the CW size is high and therefore the resolution power of the system is high. At low loads, small CW ensures low access delay. The specified values of CWmin and CWmax for different physical layer specifications are given in Table 2.1.

Table 2.1. IEEE 802.11 parameters

Parameter	802.11 (FHSS)	802.11 (DSSS)	802.11 (IR)	802.11b	802.11a
t_{slot}	50 μsec	20 μsec	8 μsec	20 μsec	9 μsec
SIFS	28 μsec	10 μsec	10 μsec	10 μsec	16 μsec
PIFS	SIFS $+t_{slot}$				
DIFS	SIFS+(2 $\times$ t_{slot})				
Operating Frequency	2.4 GHz	2.4 GHz	850-950 nm	2.4 GHz	5 GHz
Maximum Data Rate	2 Mbps	2 Mbps	2 Mbps	11 Mbps	54 Mbps
CWmin	15	31	63	31	15
CWmax	1,023	1,023	1,023	1,023	1,023

Acknowledgments

Acknowledgments (ACKs) must be sent for data packets in order to ensure their correct delivery. For unicast packets, the receiver accesses the medium after waiting for a SIFS and sends an ACK. Other stations have to wait for DIFS plus their back-off time. This reduces the probability of a collision. Thus higher priority is given for sending an ACK for the previously received data packet than for starting a new data packet transmission. ACK ensures the correct reception of the MAC layer frame by using cyclic redundancy checksum (CRC) technique. If no ACK is received by the sender, then a retransmission takes place. The number of retransmissions is limited, and failure is reported to the higher layer after the retransmission count exceeds this limit.

RTS-CTS Mechanism

The *hidden terminal problem* is a major problem that is observed in wireless networks. This is a classic example of problems arising due to incomplete topology information in wireless networks that was mentioned initially. It also highlights the non-transitive nature of wireless transmission. In some situations, one node can receive from two other nodes, which cannot hear each other. In such cases, the receiver may be bombarded by both the senders, resulting in collisions and reduced throughput. But the senders, unaware of this, may get the impression that the receiver can clearly listen to them without interference from anyone else. This is called the hidden terminal problem. To alleviate this problem, the RTS-CTS mechanism has been devised as shown in Figure 2.2 (b).

How RTS-CTS Works

The sender sends a request to send (RTS) packet to the receiver. The packet includes the receiver of the next data packet to be transmitted and the expected duration of the whole data transmission. This packet is received by all stations that can hear the sender. Every station that receives this packet will set its *network allocation vector* (NAV) accordingly. The NAV of a station specifies the earliest time when the station is permitted to attempt transmission. After waiting for SIFS, the intended receiver of the data packet answers with a clear to send (CTS) packet if it is ready to accept the data packet. The CTS packet contains the duration field, and all stations receiving the CTS packet also set their NAVs. These stations are within the transmission range of the receiver. The set of stations receiving the CTS packet may be different from the set of stations that received the RTS packet, which indicates the presence of some hidden terminals.

Once the RTS packet has been sent and CTS packet has been received successfully, all nodes within receiving distance from the sender and from the receiver are informed that the medium is reserved for one sender exclusively. The sender then starts data packet transmission after waiting for SIFS. The receiver, after receiving the packet, waits for another SIFS and sends the ACK. As soon as the transmission is over, the NAV in each node marks the medium as free (unless the node has meanwhile heard some other RTS/CTS) and the process can repeat again. The RTS packet is like any other packet and collisions can occur only at the beginning when RTS or CTS is being sent. Once the RTS and CTS packets are transmitted successfully, nodes that listen to the RTS or the CTS refrain from causing collision to the ensuing data transmission, because of their NAVs which will be set. The usage of RTS-CTS dialog before data packet transmission is a form of *virtual carrier sensing.*

Overhead Involved in RTS-CTS

It can be observed that the above mechanism is akin to reserving the medium prior to a particular data transfer sequence in order to avoid collisions during this transfer. But transmission of RTS-CTS can result in non-negligible overhead. Therefore, the RTS-CTS mechanism is used judiciously. An RTS threshold is used to determine whether to start the RTS-CTS mechanism or not. Typically, if the frame size is more than the RTS threshold, the RTS-CTS mechanism is activated and a four-way handshake (*i.e.*, RTS-CTS-DATA-ACK) follows. If the frame size is below the RTS threshold, the nodes resort to a two-way handshake (DATA-ACK).

MAC as a State Machine

Figure 2.3 diagrammatically shows what has been discussed so far. It models the MAC layer as a finite state-machine, and shows the permissible transitions. It

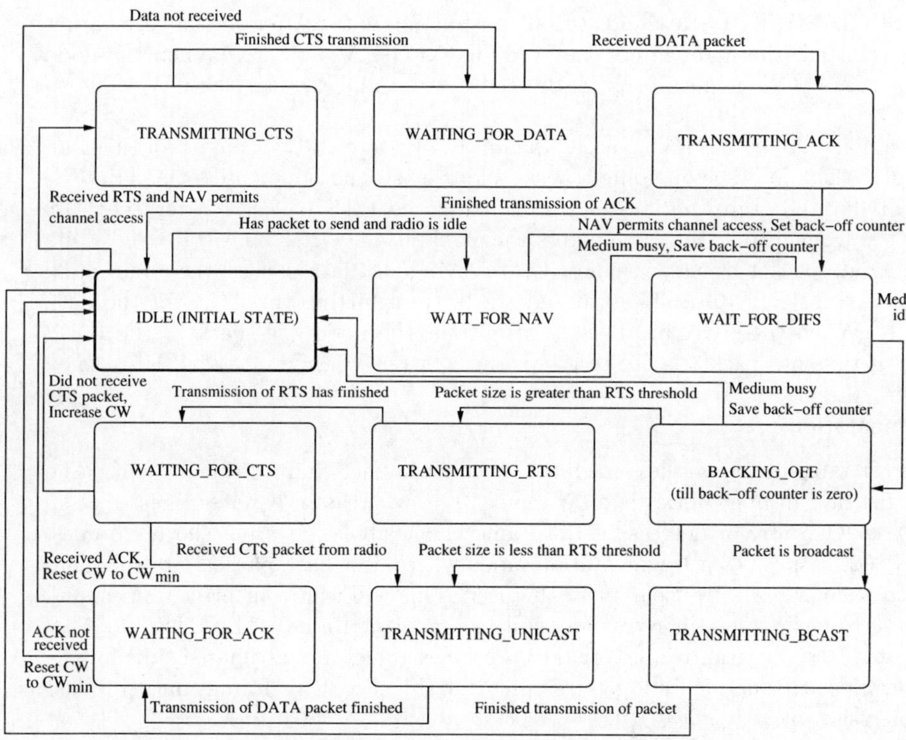

Figure 2.3. MAC state transition diagram.

must be noted that the state-machine is simplistic and is given only to ease the understanding of the fundamental mechanisms at the MAC layer. The functioning of the finite state-machine is explained in what follows.

If a node has a packet to send and is in the IDLE state, it goes into the WAIT_FOR_NAV state. After the on-going transmissions (if any) in the neighborhood are over, the node goes to the WAIT_FOR_DIFS state. After waiting for DIFS amount of time, if the medium continues to be idle, the station enters the BACKING_OFF state. Otherwise, the station sets its back-off counter (if the counter value is zero) and goes back to the IDLE state. During back-off, if the node senses a busy channel, the node saves the back-off counter and goes back to the IDLE state. Otherwise, it goes into one of three states. If the packet type is broadcast, the node enters the TRANSMITTING_BCAST state where it transmits the broadcast packet. If the packet type is unicast and the packet size is less than the RTS threshold, the node enters the TRANSMITTING_UNICAST state and starts transmitting data. If the packet size is greater than the RTS threshold, the node enters the TRANSMITTING_RTS state and starts transmitting the RTS packet. After the RTS transmission is over, the node enters the WAITING_FOR_CTS state. If the CTS packet is not received within a specified time, the node times out and goes back to the IDLE state, and increases the CW value exponentially up to a maximum of CW_{max}. If the CTS packet is received, the node enters the TRANSMITTING_UNICAST state and starts transmitting data. After the unicast packet is transmitted, the node enters the WAITING_FOR_ACK state. When the node receives the ACK, it goes back to the IDLE state and reduces the CW value to CW_{min}.

If a node receives an RTS packet when in IDLE state and if the NAV of the node indicates that no other on-going transmissions exist, the node enters the TRANSMITTING_CTS state and starts transmitting the CTS packet. After the CTS packet is transmitted, the node enters the WAITING_FOR_DATA state and waits for the data packet from the sender. On receiving the data packet, the node enters the TRANSMITTING_ACK state and starts transmitting the ACK for the data packet. When the ACK has been transmitted, the node goes back to the IDLE state. If the data packet is not received, the receiver returns to the IDLE state.

Fragmentation

Bit error rates in the wireless medium are much higher than in other media. The bit error rate in fiber optics is only about 10^{-9}, whereas in wireless, it is as large as 10^{-4}. One way of decreasing the frame error rate is by using shorter frames. IEEE 802.11 specifies a fragmentation mode where user data packets are split into several smaller parts transparent to the user. This will lead to shorter frames, and frame error will result in retransmission of a shorter frame. The RTS and CTS messages carry duration values for the current fragment and estimated time for the next fragment. The medium gets reserved for the successive frames until the last fragment is sent.

The length of each fragment is the same for all the fragments except the last fragment. The fragments contain information to allow the complete MAC protocol data unit (MPDU, informally referred to as packet) to be reassembled from the fragments that constitute it. The frame type, sender address, destination address, sequence control field, and indicator for more fragments to come are all present in the fragment header. The destination constructs the complete packet by reassembling the fragments in the order of the sequence number field. The receiving station ensures that all duplicate fragments are discarded and only one copy of each fragment is integrated. Acknowledgments for the duplicates may, however, be sent.

2.3.4 Other MAC Layer Functionalities

There are several other functionalities that the MAC layer provides in IEEE 802.11 WLANs. The functionalities described in this section are the point coordination function (PCF) which is used for QoS guarantees, timing synchronization, power management, and support for roaming.

Point Coordination Function

The objective of the point coordination function (PCF) is to provide guarantees on the maximum access delay, minimum transmission bandwidth, and other QoS parameters. Unlike the DCF, where the medium contention is resolved in a distributed manner, the PCF works by effecting a centralized contention resolution scheme, and is applicable only in networks where an AP polls the nodes in its BSS. A point coordinator (PC) at the AP splits the access time into super frame periods. The super frame period consists of alternating contention free periods (CFPs) and contention periods (CPs). The PC will determine which station has the right to transmit at any point of time. The PCF is essentially a polled service with the PC playing the role of the polling master. The operation of the PCF may require additional coordination to perform efficient operation in cases where multiple PCs are operating simultaneously such that their transmission ranges overlap. The IFS used by the PCF is smaller than the IFS of the frames transmitted by the DCF. This means that point-coordinated traffic will have higher priority access to the medium if DCF and PCF are concurrently in action. The PC controls frame transmissions so that contentions are eliminated over a limited period of time, that is, the CFP.

Synchronization

Synchronization of clocks of all the wireless stations is an important function to be performed by the MAC layer. Each node has an internal clock, and clocks are all synchronized by a timing synchronization function (TSF). Synchronized clocks are required for power management, PCF coordination, and frequency hopping spread spectrum (FHSS) hopping sequence synchronization. Without synchronization, clocks of the various wireless nodes in the network may not have a consistent view of the global time.

Within a BSS, quasi periodic beacon frames are transmitted by the AP, that is, one beacon frame is sent every target beacon transmission time (TBTT) and the transmission of a beacon is deferred if the medium is busy. A beacon contains a time-stamp that is used by the node to adjust its clock. The beacon also contains some management information for power optimization and roaming. Not all beacons need to be heard for achieving synchronization.

Power Management

Usage of power cords restricts the mobility that wireless nodes can potentially offer. The usage of battery-operated devices calls for power management because battery power is expensive. Stations that are always ready to receive data consume more power (the receiver current may be as high as 100 mA). The transceiver must be switched off whenever carrier sensing is not needed. But this has to be done in a manner that is transparent to the existing protocols. It is for this reason that power management is an important functionality in the MAC layer. Therefore, two states of the station are defined: sleep and awake. The sleep state refers to the state where the transceiver can not receive or send wireless signals. Longer periods in the sleep state mean that the average throughput will be low. On the other hand, shorter periods in the sleep state consume a lot of battery power and are likely to reduce battery life.

If a sender wants to communicate with a sleeping station, it has to buffer the data it wishes to send. It will have to wait until the sleeping station wakes up, and then send the data. Sleeping stations wake up periodically, when senders can announce the destinations of their buffered data frames. If any node is a destination, then that node has to stay awake until the corresponding transmission takes place.

Roaming

Each AP may have a range of up to a few hundred meters where its transmission will be heard well. The user may, however, walk around so that he goes from the BSS of one AP to the BSS of another AP. Roaming refers to providing uninterrupted service when the user walks around with a wireless station. When the station realizes that the quality of the current link is poor, it starts scanning for another AP. This scanning can be done in two ways: active scanning and passive scanning. Active scanning refers to sending a probe on each channel and waiting for a response. Passive scanning refers to listening into the medium to find other networks. The information necessary for joining the new BSS can be obtained from the beacon and probe frames.

2.3.5 Other Issues

Improvements in the IEEE 802.11 standard have been proposed to support higher data rates for voice and video traffic. Also, QoS provisioning and security issues have been addressed in extended versions of the standard. These will be discussed in the remainder of this section.

Newer Standards

The original standards for IEEE 802.11 came out in 1997 and promised a data rate of 1-2 Mbps in the license-free 2.4 GHz ISM band [5]. Since then, several improvements in technology have called for newer and better standards that offer higher data rates. This has manifested in the form of IEEE 802.11a and IEEE 802.11b standards, both of which came out in 1999. IEEE 802.11b, an extension of IEEE 802.11 DSSS scheme, defines operation in the 2.4 GHz ISM band at data rates of 5.5 Mbps and 11 Mbps, and is trademarked commercially by the Wireless Ethernet Compatibility Alliance (WECA) as Wi-Fi. It achieves high data rates due to the use of complimentary code keying (CCK). IEEE 802.11a operates in the 5 GHz band (unlicensed national information infrastructure band), and uses orthogonal frequency division multiplexing (OFDM) at the physical layer. IEEE 802.11a supports data rates up to 54 Mbps and is the fast Ethernet analogue to IEEE 802.11b.

Other IEEE 802.11 (c, d, and h) task groups are working on special regulatory and networking issues. IEEE 802.11e deals with the requirements of time-sensitive applications such as voice and video. IEEE 802.11f deals with inter-AP communication to handle roaming. IEEE 802.11g aims at providing the high speed of IEEE 802.11a in the ISM band. IEEE 802.11i deals with advanced encryption standards to support better privacy.

QoS for Voice and Video Packets

In order to offer QoS, delay-sensitive packets (such as voice and video packets) are to be given a higher priority to get ahead of less time-critical (*e.g.,* file transfer) traffic. Several mechanisms have been proposed to offer weighted priority. Hybrid coordination function (HCF) can be used where the AP polls the stations in a weighted way in order to offer QoS. Extended DCF is another mechanism which has been proposed where the higher priority stations will choose the random back-off interval from a smaller CW. Performance of WLANs where voice and data services are integrated is studied in [6] and [7].

Wired Equivalent Privacy

Security is a very important issue in the design of WLANs. In order to provide a modest level of physical security, the wired equivalent privacy (WEP) mechanism was devised. The name WEP implies that this mechanism is aimed at providing the level of privacy that is equivalent to that of a wired LAN. Data integrity, access control, and confidentiality are the three aims of WEP. It assumes the existence of an external key management service that distributes the key sequence used by the sender. This mechanism relies on the fact that the secret key cannot be determined by brute force. However, WEP has been proven to be vulnerable if more sophisticated mechanisms are used to crack the key. It uses the pseudo-random number key generated by RSA RC4 algorithm which has been efficiently implemented in hardware as well as in software. This mechanism makes use of the fact that if we take the plain text, XOR (bit-by-bit exclusive OR) it with a pseudo-random key sequence, and then XOR the result with the same key sequence, we get back the plain text.

2.4 HIPERLAN STANDARD

The European counterparts to the IEEE 802.11 standards are the high-performance radio LAN (HIPERLAN) standards defined by the European Telecommunications Standards Institute (ETSI). It is to be noted that while the IEEE 802.11 standards can use either radio access or infrared access, the HIPERLAN standards are based on radio access only. The standards have been defined as part of the ETSI broadband radio access networks (BRAN) project. In general, broadband systems are those in which user data rates are greater than 2 Mbps (and can go up to 100s of Mbps). Four standards have been defined for wireless networks by the ETSI.

- HIPERLAN/1 is a wireless radio LAN (RLAN) without a wired infrastructure, based on one-to-one and one-to-many broadcasts. It can be used as an extension to a wired infrastructure, thus making it suited to both ad hoc and infrastructure-based networks. It employs the 5.15 GHz and the 17.1 GHz frequency bands and provides a maximum data rate of 23.5 Mbps.

- The HIPERLAN/2 standard intends to provide short-range (up to 200 m) wireless access to Internet protocol (IP), asynchronous transfer mode (ATM[1]), and other infrastructure-based networks and, more importantly, to integrate WLANs into cellular systems. It employs the 5 GHz frequency band and offers a wide range of data rates from 6 Mbps to 54 Mbps. HIPERLAN/2 has been designed to meet the requirements of future wireless multimedia services.

- HIPERACCESS (originally called HIPERLAN/3) covers "the last mile" to the customer; it enables establishment of outdoor high-speed radio access networks, providing fixed radio connections to customer premises. HIPERACCESS provides a data rate of 25 Mbps. It can be used to connect HIPERLAN/2 deployments that are located far apart (up to 5 Km away). It offers point-to-multipoint communication.

- The HIPERLINK (originally called HIPERLAN/4) standard provides high-speed radio links for point-to-point static interconnections. This is used to connect different HIPERLAN access points or HIPERACCESS networks with high-speed links over short distances of up to 150 m. For example, the HIPERLINK can be employed to provide links between different rooms or floors within a large building. HIPERLINK operates on the 17 GHz frequency range.

Figure 2.4 shows a typical deployment of the ETSI standards. The standards excluding HIPERLAN/1 are grouped under the BRAN project. The scope of the BRAN has been to standardize the radio access network and the functions that serve as the interface to the infrastructural networks.

[1] ATM networks are connection-oriented and require a connection to set up prior to transfer of information from a source to a destination. All information to be transmitted — voice, data, image, and video — is first fragmented into small, fixed-size packets known as cells. These cells are then switched and routed using packet switching principles.

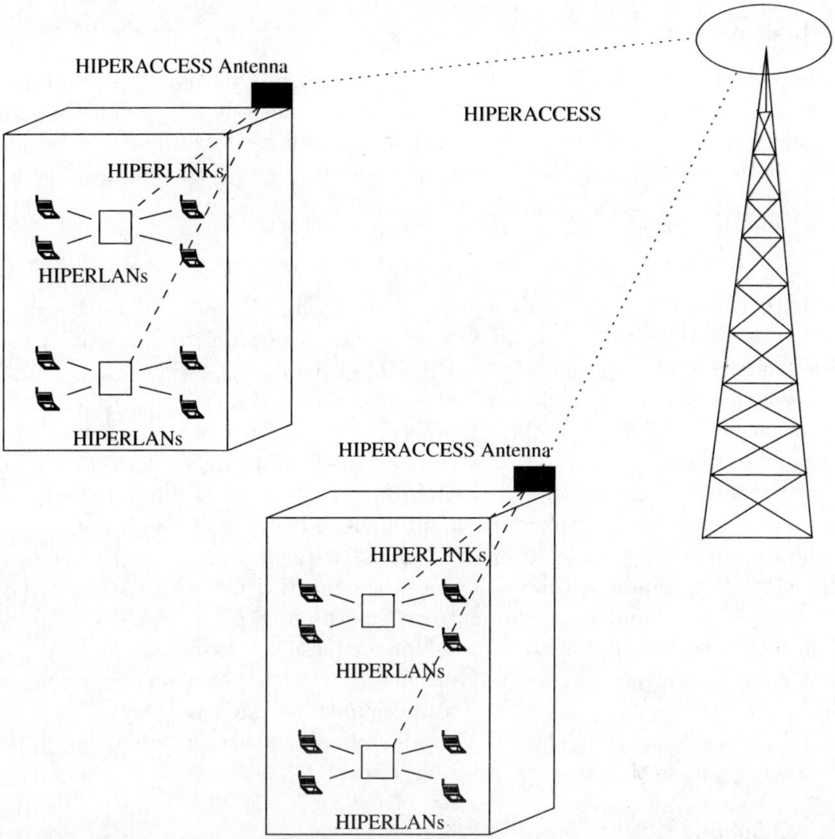

Figure 2.4. The ETSI-BRAN systems.

2.4.1 HIPERLAN/1

HIPERLAN/1 is a RLAN standard that was introduced by the ETSI in 1995. The standard allows nodes to be deployed either in a pre-arranged or in an ad hoc fashion. Apart from supporting node mobility, HIPERLAN/1 provides forwarding mechanisms (multi-hop routing). Thus, coverage is not limited to just the neighboring nodes. Using a clever framing scheme as explained later in this section, HIPERLAN/1 provides a data rate of around 23.5 Mbps without utilizing much power, thus having the capability to support multimedia data and asynchronous data effectively. This data rate is significantly higher than that provided by IEEE 802.11. The HIPERLAN/1 protocol stack is restricted to the two lower-most layers in the OSI reference model: the data link layer (DLL) and the physical layer. The DLL is further divided into the medium access control (MAC) sublayer and the channel access control (CAC) sublayer. The sections that follow describe the standard.

The Physical Layer

The tasks of the physical layer are modulation and demodulation of a radio carrier with a bit stream, forward error-correction mechanisms, signal strength measurement, and synchronization between the sender and the receiver. The standard uses the CCA scheme (similar to IEEE 802.11) to sense whether the channel is idle or busy.

The MAC Sublayer

The HIPERLAN/1 MAC (HM) sublayer is responsible for processing the packets from the higher layers and scheduling the packets according to the QoS requests from the higher layers specified by the HM QoS parameters. The MAC sublayer is also responsible for forwarding mechanisms, power conservation schemes, and communication confidentiality through encryption–decryption mechanisms.

Because of the absence of an infrastructure, the forwarding mechanism is needed to allow the physical extension of HIPERLAN/1 to go beyond the radio range of a single station. Topology-related data are exchanged between the nodes periodically with the help of special packets, for the purpose of forwarding.

In order to guarantee a time-bound service, the HM protocol data unit (HM-PDU) selected for channel access has to reflect the user priority and the residual lifetime of the packet (the time remaining for the packet to expire). The MAC layer computes the channel access priority for each of the PDUs following a mapping from the MAC priority to the channel access mechanism (CAM) priority. One among those PDUs which has the highest CAM priority and the least residual time will be selected for access to the channel.

The CAC Sublayer

The CAC sublayer offers a connectionless data service to the MAC sublayer. The MAC layer uses this service to specify a priority (called the CAM priority) which is the QoS parameter for the CAC layer. This is crucial in the resolution of contention in the CAM.

EY-NPMA

After a packet with an associated CAM priority has been chosen in the CAC sublayer for transmission, the next phase is to compete with packets of other nodes for channel access. The channel access mechanism is a dynamic, listen-and-then-talk protocol that is very similar to the CSMA/CA used in 802.11 and is called the elimination yield non-preemptive multiple access (EY-NPMA) mechanism. Figure 2.5 shows the operation of the EY-NPMA mechanism in which the nodes 1, 2, 3, and 4 have packets to be sent to the AP. The CAM priority for nodes 2 and 4 is higher with priority 2 followed by node 3 with priority 3, and node 1 with the least priority of 4. The prioritization phase will have k slots where k (can vary from 1 to 5 with $k-1$ having higher priority than k) refers to the number of priority levels.

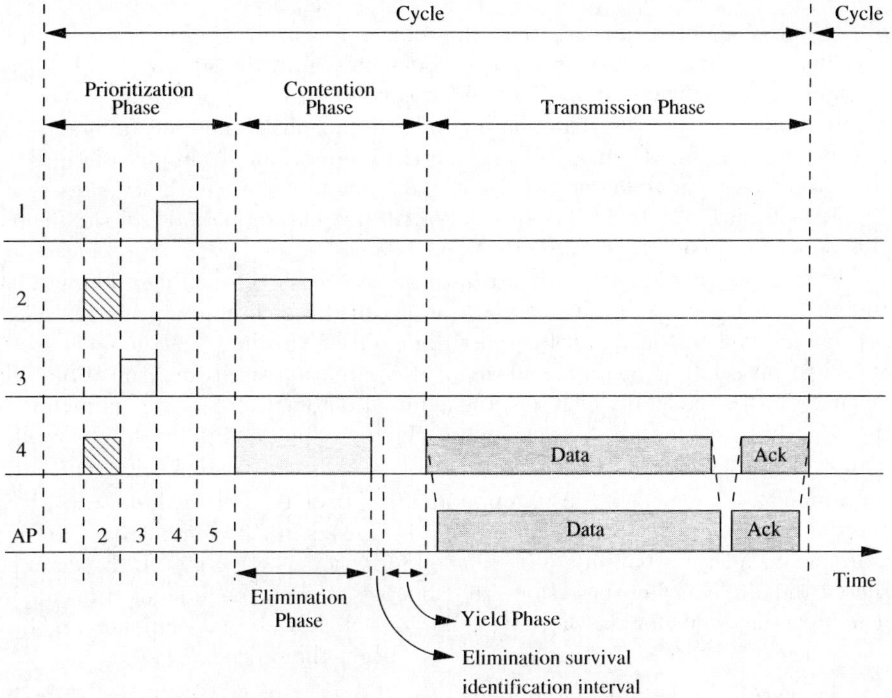

Figure 2.5. The operation of EY-NPMA.

The entire process of channel access occurs in the form of channel access cycles. A synchronization interval occurs after the end of every such cycle. This access cycle is comprised of three phases: prioritization, contention, and transmission.

(1) Prioritization: This phase culls out nodes with packets of the highest CAM priority and lets them participate in the next phase. The prioritization phase consists of two events, namely, priority detection and priority assertion. During the priority detection period, a node listens to the channel for a number of time slots proportional to the CAM priority assigned to the packet that the node wants to send. In Figure 2.5, the nodes 2 and 4 wait for one slot and assert their priority in the second slot as they hold packets with higher priority, and nodes 3 and 1 wait for slots equal to their priority level. By listening to the channel, nodes 3 and 1 detect the existence of other nodes with higher priority and hence leave the prioritization phase. If a low-priority node has succeeded in waiting up to this slot, it enters the priority assertion period during which it sends a burst, signaling its selection to the next stage. In this process, the node(s) with the highest CAM priority will finish the prioritization phase first and hence will be selected for the next phase.

(2) Contention: This phase is to eliminate as many nodes as possible, in order to minimize the collision rate during transmission. This phase extends to a maximum of 13 slots, each of the same width as that of the slots in the prioritization phase. In this phase, the nodes that transmitted a burst in the previous phase, resolve access to the channel by contention. This phase consists of two sub-phases, namely, the elimination phase and the yield phase. Nodes in this phase (nodes 2 and 4 in Figure 2.5) get to transmit a burst for a geometrically distributed number of time slots [the probability of a node's transmission extending to a slot length of k slots (where $k < 12$ slots) is 0.5^{k+1}] which is then followed by a sensing period of 1 slot. During this period, if a node detects another node's burst, it stops the contention process (node 2 in Figure 2.5). This period during which each contending node will have to listen to the channel for a slot duration is called the elimination survival identification interval. If the channel is sensed idle during this interval, the node reaches the yield phase. This period is also called elimination survival verification. This ensures that the node(s) which sent the elimination burst for the maximum number of slots will be chosen for the next phase. The next phase is the yield phase which complements the elimination phase; it involves each node listening to the channel for a number of time slots (up to a maximum of 15 slots, each with duration $\frac{1}{4}^{th}$ of the slot duration in the prioritization phase). This is in fact similar to the back-off state in which the probability of backing off for k slots is 0.1×0.9^k. If the channel is sensed to be idle during these slots, the node is said to be eligible for transmission. The node that waits for the shorter number of slots initiates transmission and other nodes defer their access to the next cycle to begin the process afresh.

(3) Transmission: This is the final stage in the channel access where the transmission of the selected packet takes place. During this phase, the successful delivery of a data packet is acknowledged with an ACK packet.

The performance of EY-NPMA protocol suffers from major factors such as packet length, number of nodes, and the presence of hidden terminals. The efficiency of this access scheme varies from 8% to 83% with variation of packet sizes from 50 bytes to 2 Kbytes.

The above-described channel access takes place during what is known as the channel synchronization condition. The other two conditions during which channel access can take place are (a) the channel free condition, when the node senses the channel free for some amount of time and then gains access, and (b) the hidden terminal condition, when a node is eliminated from contention, but still does not sense any data transmission, indicating the presence of a hidden node.

Power Conservation Issues

The HIPERLAN/1 standard has suggested power conservation schemes at both the MAC and the physical layers.

At the MAC level, the standard suggests awake/sleep modes similar to the DFWMAC in IEEE 802.11. Two roles defined for the nodes are the p-savers (nodes that want to implement the function) and the p-supporters (neighbors to the p-saver that are deputized to aid the latter's power conservation). The p-saver can receive packets only at predetermined time intervals and is active only during those intervals, in the process saving power.

At the physical level, a framing scheme has been adopted to conserve power. The physical burst is divided into high bit rate (HBR) and low bit rate (LBR) bursts. The difference between the two bursts lies in the keying mechanisms employed for them – the HBR burst is based on Gaussian minimum shift keying (GMSK) that yields a higher bit rate, but consumes more power than frequency shift keying (FSK) used for the LBR bursts. The LBR burst contains the destination address of the frame and precedes the HBR burst. Any node receiving a packet, first reads the LBR burst. The node will read the HBR burst only if it is the destination for that frame. Otherwise, the burst is simply ignored, thereby saving the power needed to read the HBR burst.

Failure of HIPERLAN/1

In spite of the high data rate that it promised, HIPERLAN/1 standard has always been considered unsuccessful. This is because IEEE Ethernet had been prevalent and hence, for its wireless counterpart too, everybody turned toward IEEE, which came out with its IEEE 802.11 standard. As a result, hardly any manufacturer adopted the HIPERLAN/1 standard for product development. However, the standard is still studied for the stability it provides and for the fact that many of the principles followed have been adopted in the other standards. For further details on the standard, readers are referred to [8] and [9].

2.4.2 HIPERLAN/2

As seen earlier, the IEEE 802.11 standard offers data rates of 1 Mbps while the newer standard IEEE 802.11a offers rates up to 54 Mbps. However, there was a necessity to support QoS, handoff (the process of transferring an MT from one channel/AP to another), and data integrity in order to satisfy the requirements of wireless LANs. This demand was the motivation behind the emergence of HIPERLAN/2. The standard has become very popular owing to the significant support it has received from cellular manufacturers such as Nokia and Ericsson. The HIPERLAN/2 tries to integrate WLANs into the next-generation cellular systems. It aims at converging IP and ATM type services at a high data rate of 54 Mbps for indoor and outdoor applications. The HIPERLAN/2, an ATM compatible WLAN, is a connection-oriented system, which uses fixed size packets and enables QoS applications easy to implement.

The HIPERLAN/2 network has a typical topology as shown in Figure 2.6. The figure shows MTs being centrally controlled by the APs which are in turn connected to the core network (infrastructure-based network). It is to be noted that, unlike the IEEE standards, the core network for HIPERLAN/2 is not just restricted to Ethernet. Also, the AP used in HIPERLAN/2 consists of one or many transceivers called access point transceivers (APTs) which are controlled by a single access point controller (APC).

There are two modes of communication in a HIPERLAN/2 network, which are described by the following two environments:

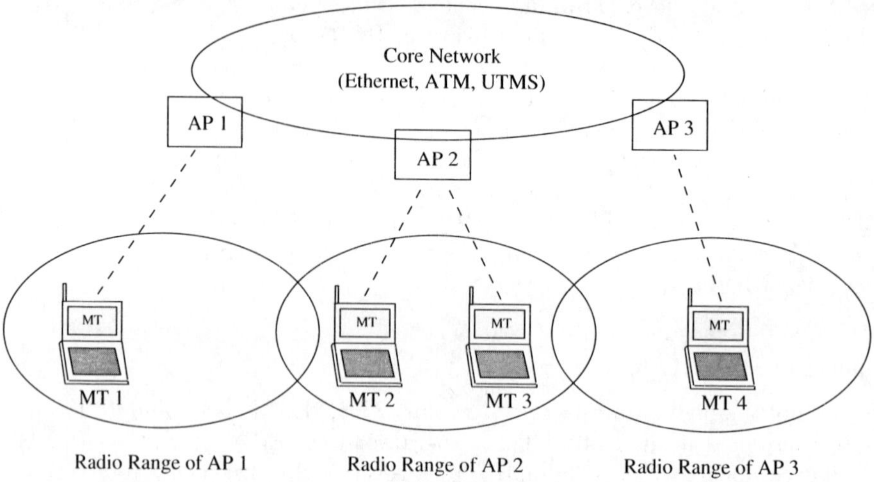

Figure 2.6. A typical deployment of HIPERLAN/2.

- **Business environment:** The ad hoc architecture of HIPERLAN/1 has been extended to support a centralized mode of communication using APs. This topology corresponds to business environments. Accordingly, each AP serves a number of MTs.

- **Home environment:** The home environment enables a direct mode of communication between the MTs. This corresponds to an ad hoc architecture that can be operated in a plug-and-play manner. The direct mode of communication is, however, managed by a central control entity elected from among the nodes called the central controller (CC).

There are several features of HIPERLAN/2 that have attracted many a cellular manufacturer. These features are part of the discussion on the protocol stack of HIPERLAN/2 below. The HIPERLAN/2 protocol stack consists of the physical layer, convergence layer (CL), and the data link control (DLC) layer.

The Physical Layer

The physical layer is responsible for the conversion of the PDU train from the DLC layer to physical bursts that are suitable for radio transmission. HIPERLAN/2, like IEEE 802.11a, uses OFDM for transmission. The HIPERLAN/2 allows bit rates from 6 Mbps to 54 Mbps using a scheme called link adaptation. This scheme allows the selection of a suitable modulation method for the required bit rate. This scheme is unique to HIPERLAN/2 and is not available in the IEEE standards and HIPERLAN/1. More details on the physical layer can be found in [14].

The CL

The topmost layer in the HIPERLAN/2 protocol stack is the CL. The functions of the layer are to adapt the requirements of the different higher layers of the core network with the services provided by the lower layers of HIPERLAN/2, and to convert the higher layer packets into ones of fixed size that can be used by the lower layers. A CL is defined for every type of core network supported. In short, this layer is responsible for the network-independent feature of HIPERLAN/2.

The CL is classified into two types, namely, the packet-based CL and the cell-based CL. The packet-based CL processes variable-length packets (such as IEEE 802.3, IP, and IEEE 1394). The cell-based CL processes fixed-sized ATM cells. The CL has two sublayers, namely, the common part (CP) and the service-specific convergence sublayer (SSCS). The CP is independent of the core network. It allows parallel segmentation and reassembly of packets. The CP comprises of two sublayers, namely, the common part convergence sublayer (CPCS) and the segmentation and reassembly (SAR) sublayer. The CPCS processes the packets from the higher layer and adds padding and additional information, so as to be segmented in the SAR. For further information on the CP, readers are referred to [10].

The SSCS consists of functions that are specific to the core network. For example, the Ethernet SSCS has been standardized in [11] for Ethernet core networks. The SSCS adapts the different data formats to the HIPERLAN/2 DLC format. It is also responsible for mapping the QoS requests of the higher layers to the QoS parameters of HIPERLAN/2 such as data rate, delay, and jitter.

The DLC Layer

The DLC layer constitutes the logical link between the AP and the MTs. This ensures a connection-oriented communication in a HIPERLAN/2 network, in contrast to the connectionless service offered by the IEEE standards. The DLC layer is organized into three functional units, namely, the radio link control (RLC) sublayer on the control plane, the error control (EC) sublayer on the user plane, and the MAC sublayer. The following discussion describes the features of the DLC layer. For further details, readers are referred to [12] and [13].

The RLC Sublayer

The RLC sublayer takes care of most of the control procedures on the DLC layer. The tasks of the RLC can be summarized as follows.

- **Association control function (ACF):** The ACF handles the registration and the authentication functions of an MT with an AP within a radio cell. Only after the ACF procedure has been carried out can the MT ever communicate with the AP.

- **DLC user connection control (DCC):** The DCC function is used to control DLC user connections. It can set up new connections, modify existing connections, and terminate connections.

- **Radio resource control (RRC):** The RRC is responsible for the surveillance and efficient utilization of the available frequency resources. It performs the following tasks:

  Dynamic frequency selection: This function is not available in IEEE 802.11, IEEE 802.11a, IEEE 802.11b, and HIPERLAN/1, and is thus unique to HIPERLAN/2. It allows the AP to select a channel (frequency) for communication with the MTs depending on the interferences in each channel, thereby aiding in the efficient utilization of the available frequencies.

  Handoff: HIPERLAN/2 supports three types of handoff, namely, sector handoff (moving to another sector of the same antenna of an APT), radio handoff (handoff between two APTs under the same APC), and network handoff (handoff between two APs in the same network).

  Power saving: Power-saving schemes much similar to those in HIPERLAN/1 and IEEE 802.11 have been implemented.

Error Control (EC)

Selective repeat (where only the specific damaged or lost frame is retransmitted) protocol is used for controlling the errors across the medium. To support QoS for stringent and delay-critical applications, a discard mechanism can be provided by specifying a maximum delay.

The MAC Sublayer

The MAC protocol is used for access to the medium, resulting in the transmission of data through that channel. However, unlike the IEEE standards and the HIPERLAN/1 in which channel access is made by sensing it, the MAC protocol follows a dynamic time division multiple access/time division duplexing (TDMA/TDD) scheme with centralized control. The protocol supports both AP-MT unicast and multicast transfer, and at the same time MT-MT peer-to-peer communication. The centralized AP scheduling provides QoS support and collision-free transmission. The MAC protocol provides a connection-oriented communication between the AP and the MT (or between MTs).

Security Issues

Elaborate security mechanisms exist in the HIPERLAN/2 system. The encryption procedure is optional and can be selected by the MT during association. Two strong encryption algorithms are offered, namely, the data encryption standard (DES) and the triple-DES algorithms.

2.5 BLUETOOTH

WLAN technology enables device connectivity to infrastructure-based services through a wireless carrier provider. However, the need for personal devices to communicate wirelessly with one another, without an established infrastructure, has led to the emergence of personal area networks (PANs). The first attempt to define a standard for PANs dates back to Ericsson's Bluetooth project[2] in 1994 to enable communication between mobile phones using low-power and low-cost radio interfaces. In May 1998, several companies such as Intel, IBM, Nokia, and Toshiba joined Ericsson to form the Bluetooth Special Interest Group (SIG) whose aim was to develop a *de facto* standard for PANs. Recently, IEEE has approved a Bluetooth-based standard (IEEE 802.15.1) for wireless personal area networks (WPANs). The standard covers only the MAC and the physical layers while the Bluetooth specification details the whole protocol stack. Bluetooth employs radio frequency (RF) technology for communication. It makes use of frequency modulation to generate radio waves in the ISM band.

Low power consumption of Bluetooth technology and an offered range of up to ten meters has paved the way for several usage models. One can have an interactive conference by establishing an ad hoc network of laptops. Cordless computer, instant postcard [sending digital photographs instantly (a camera is cordlessly connected to a mobile phone)], and three-in-one phone [the same phone functions as an intercom (at the office, no telephone charge), cordless phone (at home, a fixed-line charge), and mobile phone (on the move, a cellular charge)] are other indicative usage models.

2.5.1 Bluetooth Specifications

The Bluetooth specification consists of two parts: core and profiles. The core provides a common data link and physical layer to application protocols, and maximizes reusability of existing higher layer protocols. The profiles specifications classify Bluetooth applications into thirteen types. The protocol stack of Bluetooth performs the functions of locating devices, connecting other devices, and exchanging data. It is logically partitioned into three layers, namely, the transport protocol group, the middleware protocol group, and the application group.

The transport protocol group consists of the radio layer, baseband layer, link manager layer, logical link control and adaptation layer, and the host controller interface. The middleware protocol group comprises of RFCOMM, SDP, and IrDA (IrOBEX and IrMC). The application group consists of applications (profiles) using Bluetooth wireless links, such as the modem dialer and the Web-browsing client. The following sections discuss the concepts involved in the design of transport protocols in Bluetooth communications, and also provide an overview of the middleware and application layer protocols. Figure 2.7 shows the protocol stack of Bluetooth. The detailed specifications and explanation of the stack are available in [15], [16]. Readers may also refer to [17], [18] for more information.

2.5.2 Transport Protocol Group

This group is composed of the protocols designed to allow Bluetooth devices to locate each other and to create, configure, and manage the wireless links. Design

[2]The project was named after Danish King Harald Blatand (A.D. 940-981) (who was known as Bluetooth due to his fondness for blueberries), who unified the Scandinavians by introducing Christianity.

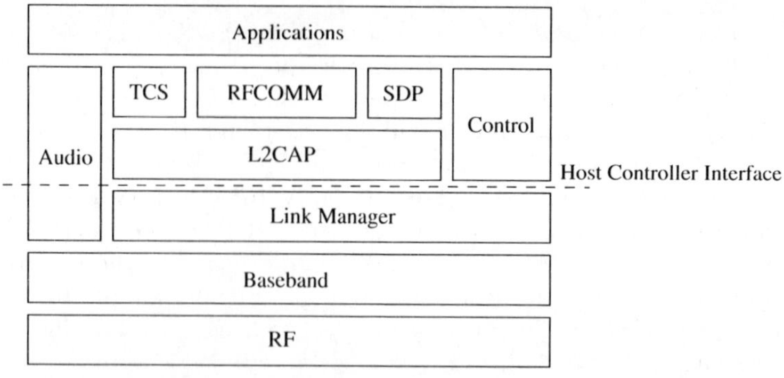

Figure 2.7. Bluetooth protocol stack.

of various protocols and techniques used in Bluetooth communications has been done with the target of low power consumption and ease of operation. This shall become evident in the design choice of FHSS and the master–slave architecture. The following sections study the various protocols in this group, their purpose, their modes of operation, and other specifications.

Radio (Physical) Layer

The radio part of the specification deals with the characteristics of the transceivers and design specifications such as frequency accuracy, channel interference, and modulation characteristics. The Bluetooth system operates in the globally available ISM frequency band and the frequency modulation is GFSK. It supports 64 Kbps voice channels and asynchronous data channels with a peak rate of 1 Mbps. The data channels are either asymmetric (in one direction) or symmetric (in both directions). The Bluetooth transceiver is a FHSS system operating over a set of m channels each of width 1 MHz. In most of the countries, the value of m is 79. Frequency hopping is used and hops are made at a rapid rate across the possible 79 hops in the band, starting at 2.4 GHz and stopping at 2.480 GHz. The choice of frequency hopping has been made to provide protection against interference.

The Bluetooth air interface is based on a nominal antenna power of 0 dBm (1 mW) with extensions for operating at up to 20 dBm (100 mW) worldwide. The nominal link range is from 10 centimeters to 10 meters, but can be extended to more than 100 meters by increasing the transmit power (using the 20 dBm option). It should be noted here that a WLAN cannot use an antenna power of less than 0 dBm (1 mW) and hence an 802.11 solution might not be apt for power-constrained devices as mentioned in [19].

Baseband Layer

The key functions of this layer are frequency hop selection, connection creation, and medium access control. Bluetooth communication takes place by ad hoc creation of a network called a *piconet*. The address and the clock associated with each Bluetooth device are the two fundamental elements governing the formation of a piconet.

Every device is assigned a single 48-bit address which is similar to the addresses of IEEE 802.xx LAN devices. The address field is partitioned into three parts and the lower address part (LAP) is used in several baseband operations such as piconet identification, error checking, and security checks. The remaining two parts are proprietary addresses of the manufacturing organizations. LAP is assigned internally by each organization. Every device also has a 28-bit clock (called the *native clock*) that ticks 3,200 times per second or once every 312.5 μs. It should be noted that this is twice the normal hopping rate of 1,600 hops per second.

Piconet

The initiator for the formation of the network assumes the role of the *master* (of the piconet). All the other members are termed as *slaves* of the piconet. A piconet can have up to seven active slaves at any instant. For the purpose of identification, each active slave of the piconet is assigned a locally unique active member address AM_ADDR. Other devices could also be part of the piconet by being in the parked mode (explained later). A Bluetooth device not associated with any piconet is said to be in standby mode. Figure 2.8 shows a piconet with several devices.

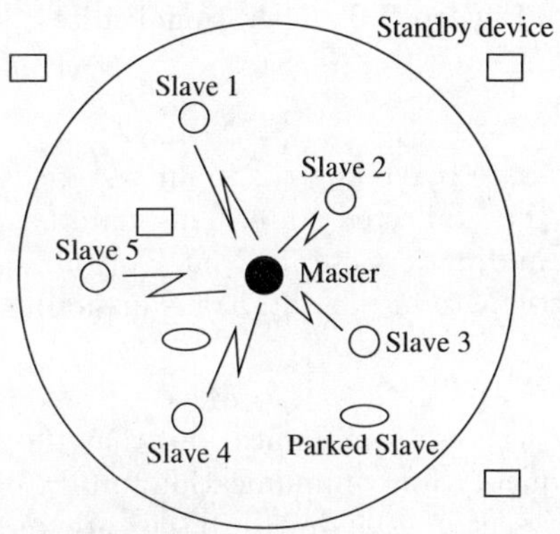

Figure 2.8. A typical piconet.

Operational States

Figure 2.9 shows the state diagram of Bluetooth communications. Initially, all the devices would be in the standby mode. Then some device (called the master) could begin the inquiry and get to know the nearby devices and, if needed, join them into its piconet. After the inquiry, the device could formally be joined by paging, which is a packet-exchange process between the master and a prospective slave to inform the slave of the master's clock. If the device was already inquired, the master

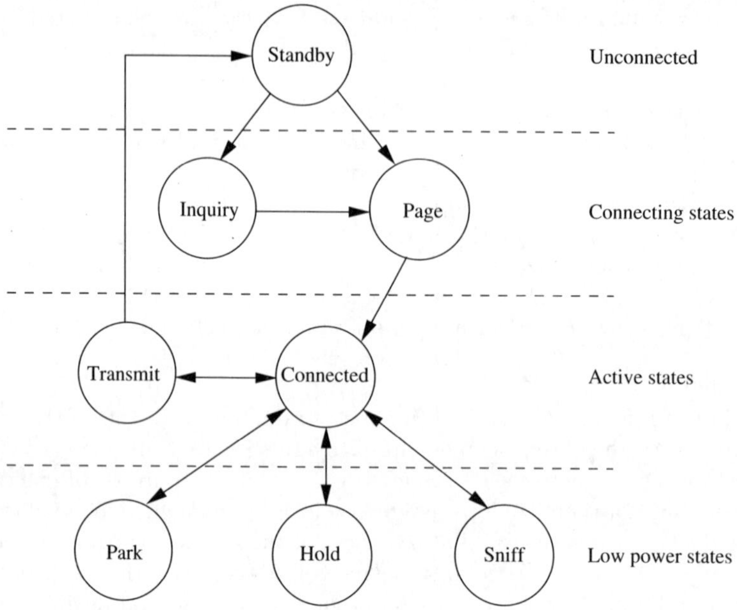

Figure 2.9. Operational states.

could get into the page state bypassing the inquiry state. Once the device finishes getting paged, it enters the connected state. This state has three power-conserving sub-states – hold, sniff, and park (described later in this section). A device in the connected state can participate in the data transmission.

Frequency Hopping Sequences

It is evident (in any wireless communication) that the sender and the receiver should use the same frequency for communication to take place. A frequency selection module (FSM) is present in each device to select the next frequency to be used under various circumstances. In the connected state, the clock and the address of the device (master) completely determine the hopping sequence. Different combination of inputs (clock, address) are used depending on the operational state. During the inquiry operation, the address input to FSM is a common inquiry address. This common address is needed because at the time of inquiry no device has information about the hopping sequence being followed. The address of the paged device is fed as input to the FSM for the paging state.

Communication Channel

The channel is divided into time slots, each 625 μs in length. The time slots are numbered according to the Bluetooth clock of the piconet master. A time division duplex (TDD) scheme is used where master and slave alternately transmit.

The master starts its transmission in even-numbered time slots only, and the slave starts its transmission in odd-numbered time slots only. This is clearly illustrated in Figure 2.10 (a). The packet start shall be aligned with the slot start. A Bluetooth device would determine slot parity by looking at the least significant bit (LSB) in the bit representation of its clock. If LSB is set to 1, it is the possible transmission slot for the slave. A slave in normal circumstances is allowed to transmit only if in the preceding slot it has received a packet from the master. A slave should know the master's clock and address to determine the next frequency (from the FSM). This information is exchanged during paging.

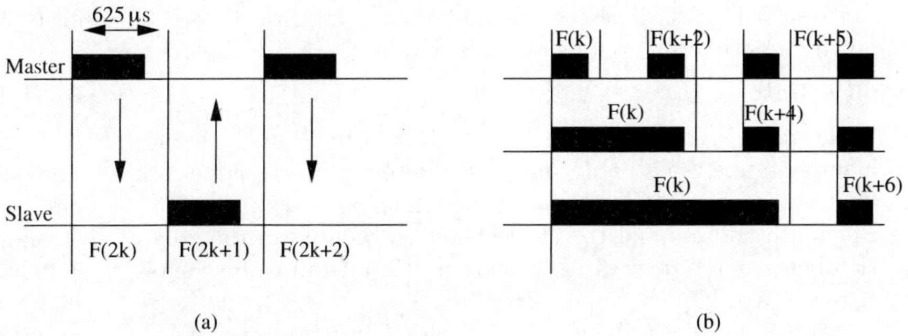

Figure 2.10. Transmission of packets over a channel.

Packet-Based Communication

Bluetooth uses packet-based communication where the data to be transmitted is fragmented into packets. Only a single packet can be transmitted in each slot. A typical packet used in these communications has three components: access code, header, and payload. The main component of the access code is the address of the piconet master. All packets exchanged on the channel are identified by the master's identity. The packet will be accepted by the recipient only if the access code matches the access code corresponding to the piconet master. This also helps in resolving conflicts in the case where two piconets are operating currently on the same frequency. A slave receiving two packets in the same slot can identify its packet by examining the access code.

The packet header contains many fields such as a three-bit active slave address, a one-bit ACK/NACK for ARQ scheme [Automatic Repeat reQuest — anytime an error is detected, a negative acknowledgment (NACK) is returned and the specified frames are retransmitted], a four-bit packet type to distinguish payload types, and an eight-bit header error check code to detect errors in the header. Depending on the payload size, one, three, or five slots may be used for the packet transmission. The hop frequency which is used for the first slot is used for the remainder of the packet. While transmitting packets in multiple slots, it is important that the frequencies used in the following time slots are those that are assigned to those slots, and that they do not follow the frequency sequence that should have normally applied. This is illustrated in Figure 2.10 (b). When a device uses five slots for packet transmission, the next packet transmission is allowed in F(k+6) and not in F(k+2). Also note that the receiving time slot becomes F(k+5) as opposed to F(k+1). On this slotted channel, both synchronous and asynchronous links are supported.

Between a master and a slave there is a single asynchronous connectionless link (ACL) supported. This is the default link that would exist once a link is established between a master and a slave. Whenever a master would like to communicate, it would, and then the slave would respond. Optionally, a piconet may also support synchronous connection oriented (SCO) links. SCO link is symmetric between master and slave with reserved bandwidth and regular periodic exchange of data in the form of reserved slots. These links are essential and useful for high-priority and time-bound information such as audio and video.

Inquiry State

As shown in Figure 2.9, a device which is initially in the standby state enters the inquiry state. As its name suggests, the sole purpose of this state is to collect information about other Bluetooth devices in its vicinity. This information includes the Bluetooth address and the clock value, as these form the crux of the communication between the devices. This state is classified into three sub-states: inquiry, inquiry scan, and inquiry response.

A potential master sends an inquiry packet in the inquiry state on the inquiry hop sequence of frequencies. This sequence is determined by feeding a common address as one of the inputs to the FSM. A device (slave) that wants to be discovered will periodically enter the inquiry scan state and listen for these inquiry packets. When an inquiry message is received in the inquiry scan state, a response packet called the frequency hopping sequence (FHS) containing the responding device address must be sent. Devices respond after a random jitter to reduce the chances of collisions.

Page State

A device enters this state to invite other devices to join its piconet. A device could invite only the devices known to itself. So normally the inquiry operation would precede this state. This state also is classified into three sub-states: page, page scan, and page response.

In the page mode, the master estimates the slave's clock based on the information received during the inquiry state, to determine where in the hop sequence the slave might be listening in the page scan mode. In order to account for inaccuracies in estimation, the master also transmits the page message through frequencies immediately preceding and succeeding the estimated one. On receiving the page message, the slave enters the slave page response sub-state. It sends back a page response consisting of its ID packet which contains its device access code (DAC). Finally, the master (after receiving the response from a slave) enters the page response state and informs the slave about its clock and address so that the slave can go ahead and participate in the piconet. The slave now calculates an offset to synchronize with the master clock, and uses that to determine the hopping sequence for communication in the piconet.

Scatternets and Issues

Piconets may overlap both spatially and temporally, that is, many piconets could operate in the same area at the same time. Each piconet is characterized by a unique master and hence the piconets hop independently, each with its own channel hopping sequence as determined by the respective master. In addition, the packets carried on the channels are preceded by different channel access codes as determined by the addresses of the master devices. As more piconets are added, the probability of collisions increases, and a degradation in performance results, as is common in FHSS systems.

In this scenario, a device can participate in two or more overlaying piconets by the process of time sharing. To participate on the proper channel, it should use the associated master device address and proper clock offset. A Bluetooth unit can act as a slave in several piconets, but as a master in only a single piconet. A group of piconets in which connections exist between different piconets is called a *scatternet* (Figure 2.11).

When a device changes its role and takes part in different piconets, it is bound to lead to a situation in which some slots remain unused (for synchronization). This implies that complete utilization of the available bandwidth is not achieved. An interesting proposition at this juncture would be to unite the timings of the whole of the scatternet as explained in [19]. But this may lead to an increase in the probability of packets colliding.

Another important issue is the timing that a device would be missing by participating in more than one piconet. A master that is missing from a piconet (by momentarily becoming a slave in another piconet) may miss polling slaves and must ensure that it does not miss beacons from its slaves. Similarly, a slave (by becoming a master or slave in another piconet) that is missing from a piconet could appear to its master to have gone out of range or to be connected through a poor-quality link.

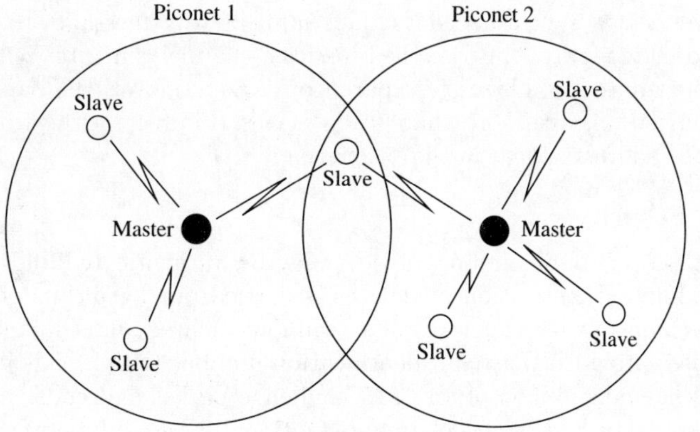

Figure 2.11. A typical scatternet.

Link Manager Protocol

Link manager protocol (LMP) is responsible for setting and maintaining the properties of the Bluetooth link. Currently, the major functionality of this layer is power management and security management. It also provides minimal QoS support by allowing control over parameters such as delay and delay jitter. Normally, a paging device is the default master of the piconet, but, depending on the usage scenario, the roles of the master and a slave could be switched and this is coordinated by exchange of LMP packets.

Power Management

The Bluetooth units can be in several modes of operation during the connection state, namely, active mode, sniff mode, hold mode, and park mode. These modes are now described.

- **Active mode:** In this mode, the Bluetooth unit actively participates in the piconet. Various optimizations are provided to save power. For instance, if the master informs the slave when it will be addressed, the slave may sleep until then. The active slaves are polled by the master for transmissions.

- **Sniff mode:** This is a low-power mode in which the listening activity of the slave is reduced. The LMP in the master issues a command to the slave to enter the sniff mode, giving it a sniff interval, and the slave listens for transmissions only at these fixed intervals.

- **Hold mode:** In this mode, the slave temporarily does not support ACL packets on the channel (possible SCO links will still be supported). In this mode, capacity is made available for performing other functions such as scanning, paging, inquiring, or attending another piconet.

- **Park mode:** This is a very low-power mode. The slave gives up its active member address and is given an eight-bit parked member address. The slave, however, stays synchronized to the channel. Any messages to be sent to a parked member are sent over the broadcast channel characterized by an active member address of all zeros. Apart from saving power, the park mode helps the master to have more than seven slaves (limited by the three-bit active member address space) in the piconet.

Bluetooth Security

In Bluetooth communications, devices may be authenticated and links may be encrypted. The authentication of devices is carried out by means of a challenge-response mechanism which is based on a commonly shared secret link key generated through a user-provided personal identification number (PIN). The authentication starts with the transmission of an LMP challenge packet and ends with the verification of result returned by the claimant. Optionally, the link between them could also be encrypted.

Logical Link Control and Adaptation Protocol (L2CAP)

This is the protocol with which most applications would interact unless a host controller is used. L2CAP supports protocol multiplexing to give the abstraction to each of the several applications running in the higher layers as if it alone is being run. Since the data packets defined by the baseband protocol are limited in size, L2CAP also segments large packets from higher layers such as RFCOMM or SDP into multiple smaller packets prior to their transmission over the channel. Similarly, multiple received baseband packets may be reassembled into a single larger L2CAP packet. This protocol provides QoS on certain parameters such as peak bandwidth, latency, and delay variation when the link is established between two Bluetooth units.

Host Controller Interface

This is the optional interface layer, provided between the higher (above LMP) and lower layers of the Bluetooth protocol stack, for accessing the Bluetooth hardware capabilities. Whenever the higher layers are implemented on the motherboard of a host device, this layer is needed. Such an approach could prove beneficial as the spare capacity of the host device (say, a personal computer) could be utilized. The specification defines details such as the different packet types as seen by this layer. Command packets that are used by the host to control the device, event packets that are used by the device to inform the host of the changes, and data packets come under this category.

2.5.3 Middleware Protocol Group

The basic functionality of the middleware protocol group is to present to the application layers a standard interface that may be used for communicating across the transport layer, that is, the applications need not know the transport layer's complexities, they can just use the application programming interfaces (APIs) or higher level functions provided by the middleware protocols. This group consists of the RFCOMM layer, service discovery protocol (SDP), IrDA interoperability protocols, telephony control specification (TCS), and audio.

The RFCOMM layer presents a virtual serial port to applications using the serial interface. Any application which is using the serial port can work seamlessly on Bluetooth devices. RFCOMM uses an L2CAP connection to establish a link between two devices. In the case of Bluetooth devices, there is no device which will be static and hence services offered by the other devices have to be discovered. This is achieved by using the service discovery protocol (SDP) of the Bluetooth protocol stack. Service discovery makes the device self-configured without manual intervention.

The IrDA interoperability protocol is not for communication between Bluetooth devices and Infrared devices. It is only for the existing IrDA applications to work on Bluetooth devices without any changes. The main protocols in the IrDA set are IrOBEX (IrDA object exchange) for exchanging objects between two devices and IrMC (infrared mobile communications) for synchronization.

Audio is the distinguishing part of Bluetooth. Audio is given the highest priority and is directly carried over the baseband at 64 Kbps so that a very good quality of voice is provided. Another important point to note here is that audio is actually not a layer of the protocol stack, but only a specific packet format that can be transmitted directly over the SCO links of the baseband layer.

Telephony control is implemented using the telephony control specification – binary (TCS-BIN) protocol. TCS defines three major functional areas: call control, group management, and connectionless TCS. Call control is used to set up calls which can be subsequently used to carry voice and data traffic. TCS operates in both point-to-point and point-to-multipoint configurations. One of the main concepts of TCS is that of the wireless user group (WUG). Group management enables multiple telephone extensions, call forwarding, and group calls. For example, consider multiple handsets and a single base set. When a call comes in to the base set, all the multiple handsets can receive this call. In a similar fashion, calls can also be forwarded.

The functionalities of TCS include *configuration distribution* and *fast intermember access*. Configuration distribution is the mechanism used to find the information about the other members in a group. Fast intermember access is a method for two slaves to create a new piconet. A WUG member uses the information from the configuration distribution and determines another member which it wants to contact. Then it sends the device's information to the master, which forwards it to this device. The contacted device then responds with its device address and clock information and places itself in a page scan state. Then the master contacts the device initiating the communication. This device now pages the contacted device and forms a new piconet. This explains how a new piconet is formed between two slaves with the help of the master.

In all the above cases, a connection-oriented channel is established. To exchange simple information such as adjusting volume or signaling information, establishing such a channel is overkill and hence connectionless TCS has been provided for having a connectionless channel.

2.5.4 Bluetooth Profiles

These profiles have been developed to promote interoperability among the many implementations of the Bluetooth protocol stack. Each Bluetooth profile specification has been defined to provide a clear and transparent standard that can be used to implement a specific user end function. Two Bluetooth devices can achieve a common functionality only if both devices support identical profiles. For example, a cellular phone and a headset both have to support the Bluetooth headset profile for the headset to work with the phone. The Bluetooth profiles spring up from the usage models. In all, 13 profiles have been listed and these can be broadly classified into the following four categories:

1. **Generic profiles:** The Generic access profile, which is not really an application, provides a way to establish and maintain secure links between the master and the slaves. The service discovery profile enables users to access

SDP to find out which applications (Bluetooth services) are supported by a specific device.

2. **Telephony profiles:** The cordless telephony profile is designed for three-in-one phones. The Intercom profile supports two-way voice communication between two Bluetooth devices within range of each other. The Headset profile specifies how Bluetooth can provide a wireless connection to a headset (with earphones/microphones) for use with a computer or a mobile phone.

3. **Networking profiles:** The LAN Access profile enables Bluetooth devices to either connect to a LAN through APs or form a small wireless LAN among themselves. The dial-up networking profile is designed to provide dial-up connections via Bluetooth-enabled mobile phones. The FAX profile, very similar to the dial-up networking profile, enables computers to send and receive faxes via a Bluetooth-enabled mobile phone.

4. **Serial and object exchange profiles:** The serial port profile emulates a serial line (RS232 and USB serial ports) for (legacy) applications that require a serial line. The other profiles, generic object exchange, object push, file transfer, and synchronization, are for exchanging objects between two wireless devices.

Bluetooth is the first wireless technology which has actually tried to attempt to make all the household consumer electronics devices follow one particular communication paradigm. It has been partially successful, but it does have its limitations. Bluetooth communication currently does not provide support for routing. It should be noted that some research efforts are under way to accommodate this in the Bluetooth specification. Once the routing provision is given, inter-piconet communication could be enhanced. The issues of handoffs also have not yet been dealt with till now. Although master–slave architecture has aided low cost, the master becomes the bottleneck for the whole piconet in terms of performance, fault tolerance, and bandwidth utilization. Most importantly, Bluetooth communication takes place in the same frequency band as that of WLAN and hence robust coexistence solutions need to be developed to avoid interference. The technology is still under development. Currently, there are nearly 1,800 adopter companies which are contributing toward the development of the technology.

2.6 HOMERF

Wireless home networking represents the use of the radio frequency (RF) spectrum to transmit voice and data in confined areas such as homes and small offices. One of the visionary concepts that home networking intends to achieve is the establishment of communication between home appliances such as computers, TVs, telephones, refrigerators, and air conditioners. Wireless home networks have an edge over their wired counterparts because features such as flexibility (enabling of file and drive sharing) and interoperability that exist in the wired networks are coupled with those in the wireless domain, namely, simplicity of installation and mobility.

The HIPERLAN/2, as mentioned earlier, has provisions for direct communication between the mobile terminals (the home environment). The home environment enables election of a central controller (CC) which coordinates the communication process. This environment is helpful in setting up home networks. Apart from this, an industry consortium known as the Home RF Working Group has developed a technology that is termed HomeRF. This technology intends to integrate devices used in homes into a single network and utilize RF links for communication. HomeRF is a strong competitor to Bluetooth as it operates in the ISM band.

Technical Features

The HomeRF provides data rates of 1.6 Mbps, a little higher than the Bluetooth rate, supporting both infrastructure-based and ad hoc communications. It provides a guaranteed QoS delivery to voice-only devices and best-effort delivery for data-only devices. The devices need to be plug-and-play enabled; this needs automatic device discovery and identification in the network. A typical HomeRF network consists of resource providers (through which communication to various resources such as the cable modem and phone lines is effected), and the devices connected to them (such as the cordless phone, printers, and file servers). The HomeRF technology follows a protocol called the shared wireless access protocol (SWAP). The protocol is used to set up a network that provides access to a public network telephone, the Internet (data), entertainment networks (cable television, digital audio, and video), transfer and sharing of data resources (such as disks and printers), and home control and automation.

The SWAP has been derived from the IEEE 802.11 and the European digitally enhanced cordless telephony (DECT) standards. It employs a hybrid TDMA/CSMA scheme for channel access. While TDMA handles isochronous transmission (similar to synchronous transmission, isochronous transmission is also used for multimedia communication where both the schemes have stringent timing constraints, but isochronous transmission is not as rigid as synchronous transmission in which data streams are delivered only at specific intervals), CSMA supports asynchronous transmission (in a manner similar to that of the IEEE 802.11 standard), thereby making the actual framing structure more complex. The SWAP, however, differs from the IEEE 802.11 specification by not having the RTS-CTS handshake since it is more economical to do away with the expensive handshake; moreover, the hidden terminal problem does not pose a serious threat in the case of small-scale networks such as the home networks.

The SWAP can support up to 127 devices, each identified uniquely by a 48-bit network identifier. The supported devices can fall into one (or more) of the following four basic types:

- Connection point that provides a gateway to the public switched telephone network (PSTN), hence supporting voice and data services.

- Asynchronous data node that uses the CSMA/CA mechanism to communicate with other nodes.

- Voice node that uses TDMA for communication.
- Voice and data node that can use both CSMA/CA and TDMA for channel access.

Home networking also needs strong security measures to safeguard against potential eavesdroppers. That is the reason why SWAP uses strong algorithms such as Blowfish encryption. HomeRF also includes support for optional packet compression which provides a trade-off between bandwidth and power consumption.

Because of its complex (hybrid) MAC and higher capability physical layer, the cost of HomeRF devices is higher than that of Bluetooth devices. HomeRF Version 2.0, released recently, offers higher data rates (up to 10 Mbps by using wider channels in the ISM band through FHSS).

Infrared

The infrared technology (IrDA) uses the infrared region of the light for communication [20]. Some of the characteristics of these communications are as follows:

- The infrared rays can be blocked by obstacles, such as walls and buildings.
- The effective range of infrared communications is about one meter. But when high power is used, it is possible to achieve better ranges.
- The power consumed by infrared devices is extremely low.
- Data rates of 4 Mbps are easily achievable using infrared communications.
- The cost of infrared devices is very low compared to that of Bluetooth devices.

Although the restriction of line of sight (LoS) is there on the infrared devices, they are extremely popular because they are cheap and consume less power. The infrared technology has been prevalent for a longer time than Bluetooth wireless communications. So it has more widespread usage than Bluetooth. Table 2.2 compares the technical features of Bluetooth, HomeRF, and IrDA technologies.

Table 2.2. Illustrative comparison among Bluetooth, HomeRF, and IrDA technologies

Feature	Bluetooth	HomeRF	IrDA
Peak Data Rate	1 Mbps	1.6 Mbps	4 Mbps
Data Network Support	via PPP*	TCP/IP	TCP/IP
Voice Network Support	via SCO	via IP & PSTN	via IP
Range	$<$ 10 meters	$>$ 50 meters	$<$ 10 meters
Power Consumption	0.25 - 100 mW	100 - 500 mW	10 mW (nominal)

*The point-to-point protocol is an Internet standard protocol for transporting IP datagrams over point-to-point links.

2.7 SUMMARY

This chapter has discussed networks of a small scale which use tetherless communication (ability to move without restriction due to wires). In most of these networks, communication has been using radio waves of appropriate wavelength. In certain scenarios, infrared light has been used for transmitting data. Efficient protocols are used at the physical and MAC layers in order to make the transition from wired to wireless networks appear seamless to the higher layers on the protocol stack. This chapter has dealt with two prominent standards for WLANs, IEEE 802.11 and ETSI HIPERLAN, and two technological alternatives, Bluetooth and HomeRF, in the PAN area. Table 2.3 compares the technical features of different WLAN and PAN standards discussed in this chapter.

The deployment considerations and choice of appropriate technology for a WLAN are network coverage, bandwidth requirement, expected traffic load, target users, security requirements, QoS requirements, scenario of deployment, and, finally, the cost of deployment.

2.8 PROBLEMS

1. Think of four scenarios where wireless networks can replace wired networks in order to improve the efficiency of people at their workplace. Briefly describe how in each case a wireless network will fit the role better than a wired network.

2. Compare and contrast infrastructure networks with ad hoc networks. Give example situations where one type of network is preferred to the other.

3. Does the IEEE 802.11 standard specify the implementation of the distribution system (DS)? If not, explain how DS is characterized.

4. Match the following pairs:

1. association	A. improves speed during roaming
2. authentication	B. is necessary for roaming
3. reassociation	C. is needed for STA-AP mapping
4. preauthentication	D. makes the transmission secure

5. Why is a back-off timer used in the CSMA/CA mechanism?

6. How can DCF and PCF coexist in the BSS?

7. How are fragments of one packet sent via a WLAN that uses 802.11 without getting interleaved with fragments of other packets?

8. What are the functions of beacon frames?

9. What design trade-offs have to be considered when designing stations capable of conserving power?

Table 2.3. A brief comparison among the different WLAN and PAN standards discussed in this chapter

Feature	802.11	802.11b	802.11a	802.11g	HIPERLAN/2	Bluetooth	HomeRF (V.2.0)
Frequency	2.4 GHz	2.4 GHz	5 GHz	2.4 GHz	5 GHz	2.4 GHz	2.4 GHz
Physical Layer	DSSS/ FHSS/IR (FHSS-2.5 hops/sec)	DSSS	OFDM	OFDM	OFDM	FHSS 1,600 hops/sec	FHSS 50-100 hops/sec
Maximum Transmission Rate	2 Mbps	11 Mbps	54 Mbps	54 Mbps	54 Mbps	1 Mbps	10 Mbps
Maximum Throughput	1.2 Mbps	5.5 Mbps	32 Mbps	24 Mbps*	32 Mbps	<700 Kbps	—
Frequency Management	None				Dynamic Selection	Dynamic Selection	Dynamic Selection
Medium Access	CSMA/CA				TDMA/TDD	Polling	CSMA/CA
Authentication	None				NAI/IEEE Add/X.509	E1 Challenge response scheme with 128-bit key	Shared key encryption same as DECT
Encryption	40-bit RC4				DES, 3DES	128-bit secret link key	128-bit key
QoS Support	PCF				ATM 802.1d/RSVP	Polling	TDMA
Wired Backbone	Ethernet				Ethernet/ATM/ UMTS/PPP	Ethernet/ PPP	Ethernet/ PPP/ WLL
Connectivity	Connectionless				Connection Oriented	Both	Both
Link Quality Control	None				Link Adaptation	None	None

*Observed throughput on early implementations and is likely to change with fine tuning of the parameters.

10. Match the following pairs:

1. PPM	A. IEEE 802.11a
2. OFDM	B. IEEE 802.11b
3. CCK	C. Infrared
4. RTS-CTS	D. MAC layer mechanism, mandatory
5. CSMA/CA	E. MAC layer mechanism, optional

11. Give two points for and against the use of infrared and radio as a physical layer medium in a WLAN.

12. Choose the correct alternative from the choices enclosed in the parentheses.

    (a) The power conservation problem in WLANs is that stations receive data (in bursts / constantly) but remain in an idle receive state (sporadically / constantly) which dominates the LAN adapter power consumption.

    (b) There are two types of authentication schemes in IEEE 802.11 — the default is (shared key authentication / open system authentication) whereas (shared key / open system) provides a greater amount of security.

    (c) The (DCF / PCF) mechanism is available only for infrastructure networks. In this mechanism, the AP organizes a periodical (CFP / CP) for the time-bounded information.

    (d) The (FHSS / DSSS) is easier for implementation because the sampling rate is of the order of the symbol rate of 1 Mbps. The (FHSS / DSSS) implementation provides a better coverage and a more stable signal because of its wider bandwidth.

    (e) The (IEEE 802.11 / HIPERLAN-2) camp is a connectionless WLAN camp that evolved from data-oriented computer communications. Its counterpart is the (HIPERLAN-2 / IEEE 802.11) camp that is more focused on connection-based WLANs addressing the needs of voice-oriented cellular telephony.

13. Why do we have four address fields in IEEE 802.11 MAC as against only two in IEEE 802.3 MAC frame?

14. Name the three MAC services provided by the IEEE 802.11 that are not provided in the traditional LANs, such as 802.3.

15. Determine the transfer time of a 22 KB file with a mobile data network (a) with a transmission rate of 10 Kbps and (b) repeat the same for 802.11 WLAN operating at 2 Mbps. (c) What is the length of the file that WLAN can carry in the time that mobile data service carried a 20 KB file? (d) What do you infer from the answers to the above questions?

16. What is normalized propagation delay? Determine the normalized propagation delay for the following: (a) IEEE 802.3 Ethernet (b) IEEE 802.11. Assume 802.11 LAN provides a coverage of 200 meters.

17. Discuss the deployment scenarios for various HIPERLAN standards in the ETSI BRAN system.

18. What is the probability that two HIPERLANs will have the same ID? Comment about the value that you have obtained.

19. What are the features of HIPERLAN/1 MAC sublayer that support QoS?

20. Compare the EY-NPMA and the CSMA/CA mechanisms.

21. Observe Table 2.4 and determine which node will get the chance of sending its packet.

Table 2.4. EY-NPMA scheme

Node	Priority	Elimination Burst Slots	Yield Slots
1	2	7	4
2	3	12	3
3	2	3	2
4	1	7	4
5	1	7	3
6	1	6	4
7	3	5	1
8	4	3	2
9	1	5	5

22. How is synchronization achieved in the LBR-HBR data burst in HIPERLAN/1?

23. Compare the handoff procedures of the HIPERLAN/2 and the IEEE 802.11 standards.

24. What do you think are the advantages of using a FHSS for the operation of Bluetooth devices?

25. Assume that in one slot in Bluetooth 256 bits of payload could be transmitted. How many slots are needed if the payload size is (a) 512 bits, (b) 728 bits, and (c) 1,024 bits. Assume that the non-payload portions do not change.

26. If a master leaves the piconet, what could possibly happen?

27. Why is a device constrained to act as a master in at most one piconet?

28. With reference to the discussion on scatternets, illustrate the trade-off of synchronization and bandwidth using a timing diagram. (Hint: Consider the case when a Bluetooth device acts as a master in one piconet and as a slave in another.)

29. In the serial communications, the baud rate of transmission is specified. Legacy applications on Bluetooth devices specify the baud rate. In the case of the old serial communications, the data rate is the specified baud rate, but in the case of Bluetooth serial communications, it is not so. Why?

30. Based on your understanding of the Bluetooth protocol stack, suggest a possible implementation of FTP over Bluetooth without using TCP/IP.

31. Why is the RTS-CTS handshake avoided in the HomeRF technology?

BIBLIOGRAPHY

[1] F. Gfeller, "INFRANET: Infrared Microbroadcasting Network for In-House Data Communication," *Proceedings of 7th European Conference on Optical Communication 1981*, pp. P27-1–P27-4, September 1981.

[2] B. P Crow, I. Widjaja, L. G. Kim, and P. T. Sakai, "IEEE 802.11 Wireless Local Area Networks," *IEEE Communications Magazine*, vol. 35, no. 9, pp. 116-126, September 1997.

[3] R. V. Nee, G. Awater, M. Morikura, H. Takanashi, M. Webster, and K. W. Halford, "New High Rate Wireless LAN Standards," *IEEE Communications Magazine*, vol. 37, no. 12, pp. 82-88, December 1999.

[4] R. T. Valadas, A. R. Tavares, A. M. D. Duarte, A. C. Moreira, and C. T. Lomba, "The Infrared Physical Layer of the IEEE 802.11 Standard for Wireless Local Area Networks," *IEEE Communications Magazine*, vol. 36, no. 12, pp. 106-114, December 1998.

[5] B. Tuch, "An ISM Band Spread Spectrum Local Area Network WaveLAN," *Proceedings of IEEE Workshop on WLANs 1991*, pp. 103-111, May 1991.

[6] A. Zahedi and K. Pahlavan, "Capacity of a WLAN with Voice and Data Services," *IEEE Transactions on Communications*, vol. 48, no. 7, pp. 1160-1170, July 2000.

[7] J. Feigin, K. Pahlavan, and M. Ylianttila, "Hardware Fitted Modeling and Simulation of VoIP over a Wireless LAN," *Proceedings of IEEE VTS Fall 2000*, vol. 3, pp. 1431-1438, September 2000.

[8] Radio Equipment and Systems (RES); HIgh PErformance Radio Local Area Networks (HIPERLAN) Type 1, Requirements and Architecture for Wireless ATM Access and Interconnection, 1998 (http://www.etsi.org).

[9] G. Anastasi, L. Lenzini, and E. Mingozzi, "Stability and Performance Analysis of HIPERLAN," *Proceedings of IEEE INFOCOM 1998*, vol. 1, pp. 134-141, March 1998.

[10] Broadband Radio Access Networks (BRAN); HIgh PErformance Radio Local Area Networks (HIPERLAN) Type 2; Packet Based Convergence Layer; Part 1: Common Part, 2000 (http://www.etsi.org).

[11] Broadband Radio Access Networks (BRAN); HIgh PErformance Radio Local Area Networks (HIPERLAN) Type 2; Packet Based Convergence Layer; Part 2: Ethernet Service Specific Sublayer (SSCS), 2001 (http://www.etsi.org).

[12] Broadband Radio Access Networks (BRAN); HIgh PErformance Radio Local Area Networks (HIPERLAN) Type 2; Data Link Control (DLC) Layer; Part 1: Basic Data Transport Functions, 2001 (http://www.etsi.org).

[13] Broadband Radio Access Networks (BRAN); HIgh PErformance Radio Local Area Networks (HIPERLAN) Type 2, Data Link Control (DLC) Layer, Part 2: RLC Sublayer, 2002 (http://www.etsi.org).

[14] Broadband Radio Access Networks (BRAN); HIgh PErformance Radio Local Area Networks (HIPERLAN) Type 2; Physical (PHY) Layer, 2001 (http://www.etsi.org).

[15] http://www.bluetooth.com/

[16] B. A. Miller and C. Bisdikian, *Bluetooth Revealed: The Insider's Guide to an Open Specification for Global Wireless Communications*, Prentice Hall PTR, New Jersey, November 2001.

[17] C. Bisdikian, "An Overview of Bluetooth Wireless Technology," *IEEE Communications Magazine*, vol. 39, no. 12, pp. 86-94, December 2001.

[18] http://www.palowireless.com

[19] J. Bray and C. F. Sturman, *Bluetooth: Connect Without Cables*, Prentice Hall PTR, New Jersey, December 2001.

[20] http://www.irda.org/

Chapter 3

WIRELESS WANS AND MANS

3.1 INTRODUCTION

This chapter first introduces the fundamental concepts of cellular (wide area) wireless networks. The cellular architecture is described and the evolution of different standards for cellular networks is traced. Then the wireless in local loop, known as WLL, which brings wireless access into the residence and office, is described. An overview of the major WLL standards adopted all over the world is presented and broadband access implementations of WLL are also briefly described. Wireless ATM, for supporting seamless interconnection with backbone ATM networks and QoS for mobile users, is then discussed. Finally, two standards for wireless broadband access (wireless metropolitan area networks), IEEE 802.16 and HIPER-ACCESS, are presented.

3.2 THE CELLULAR CONCEPT

The cellular concept is a novel way to ensure efficient utilization of the available radio spectrum. The area to be covered by a cellular network is divided into cells, which are usually considered to be hexagonal. This is because, of the shapes which can completely cover a two-dimensional region without overlaps, such as the triangle, square, and hexagon, the hexagon most closely resembles the nearly circular coverage area of a transmitter. An idealized model of the cellular radio system consists of an array of hexagonal cells with a base station (BS) located at the center of each cell. The available spectrum in a cell is used for *uplink* channels for mobile terminals (MTs) to communicate with the BS, and for *downlink* channels, for the BS to communicate with MTs.

The fundamental and elegant concept of cells relies on *frequency reuse*, that is, the usage of the same frequency by different users separated by a distance, without interfering with each other. Frequency reuse depends on the fact that the signal strength of an electromagnetic wave gets attenuated with distance. A *cluster* is a group of cells which uses the entire radio spectrum. The cluster size N is the number of cells in each cluster. No two cells within a cluster use channels of the same frequency. Clustering ensures that cells which use the same frequency are separated by a minimum distance, called the *reuse distance* D. Figure 3.1 depicts the concept of clustering. Cells that use different sets of frequencies are labeled differently, and the outlined set of seven cells forms a cluster. If the radius of a cell is R, then the distance between two adjacent cells is $\sqrt{3}R$. For a hexagonal cellular structure, the permissible values of the cluster size N are of the form

$$N = i^2 + j^2 + ij$$

where i and j are any non-negative integers. The lowest cluster size of three is obtained by setting $i = j = 1$. Figure 3.1 shows a cluster size of seven, given by $i = 1$ and $j = 2$. Let d be the distance between two adjacent cells. Applying the cosine law to the triangle ABC with sides a, b, and c,

$$cos(ABC) = \frac{a^2 + c^2 - b^2}{2ca}$$

Hence

$$cos(2\pi/3) = \frac{d^2 + 4d^2 - D^2}{2 \times d \times 2d}$$

which gives $D = \sqrt{21}R$. In general, it can be shown that the reuse factor, $D/R = \sqrt{3N}$ [1].

Two types of interference come into play when the cellular concept is used in the design of a network. The first one is the co-channel interference which results from the use of same frequencies in cells of different clusters. The reuse distance should

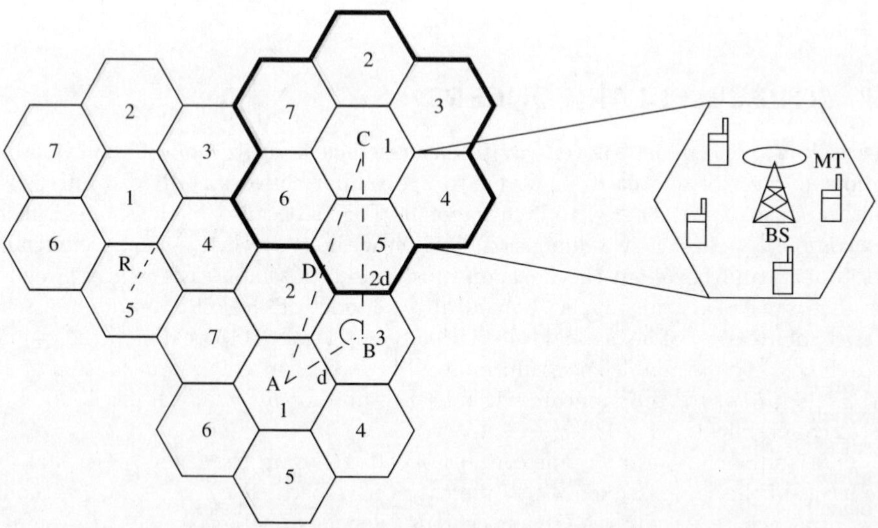

Figure 3.1. The cellular concept.

be such that co-channel interference does not adversely affect the signal strength. On the other hand, adjacent channel interference results due to usage of adjacent frequencies within a cluster. The channel assignment to different cells within the cluster must minimize adjacent channel interference by not assigning neighboring frequencies to the same cell.

The advantage of the cellular concept is that it can increase the number of users who can be supported in a given area, as illustrated in the following example. Consider a spectrum of bandwidth 25 MHz. Assume that every user requires 30 KHz bandwidth. Then a single city-wide cell can support only 25,000/30 = 833 users. However if the city is split into hexagonal cells with seven cells per cluster, only 1/7th of the spectrum is available in any cell. Hence each cell can support 119 users. If there are 20 cells in the city, then the system can support 2,380 users simultaneously.

In general, if S is the total available spectrum, W is the bandwidth needed per user, N is the cluster size, and k is the number of cells required to cover a given area, the number of users supported simultaneously (capacity) is

$$n = \frac{k(S/N)}{W}$$

In the previous example, with S = 25 MHz, W = 30 KHz, N = 7, and k = 20, n was calculated to be 2,380.

3.2.1 Capacity Enhancement

Each cellular service provider is allotted a certain band of frequencies. Considering the interference constraints, this restricts the number of channels that can be allotted to each cell. So, methods have been devised to enhance the capacity of cellular networks. It has been observed that the main reasons for reduction of cellular network capacity are off-center placement of antennas in the cell, limited frequency reuse imposed by a strict clustering scheme, and inhomogeneous propagation conditions. The nearest BS is not always the best for a mobile station, due to shadowing, reflections, and other propagation-based features. The simplistic model of cellular networks has to be modified to account for these variations. A few methods used to improve the capacity of cellular networks are discussed.

Cell-Splitting

Non-uniform traffic demand patterns create *hotspot* regions in cellular networks, which are small pockets with very high demand for channel access. In order to satisfy QoS constraints, the blocking probability in a hotspot region must not be allowed to shoot up. This gave rise to the concept of cell-splitting. A different layer of cells which are smaller in size, and support users with lower mobility rates, is overlaid on the existing (macro-cells) cellular network. These are called micro-cells. While macro-cells typically span across tens of kilometers, micro-cells are usually less than 1 Km in radius. Very small cells called pico-cells, of a few meters' radius,

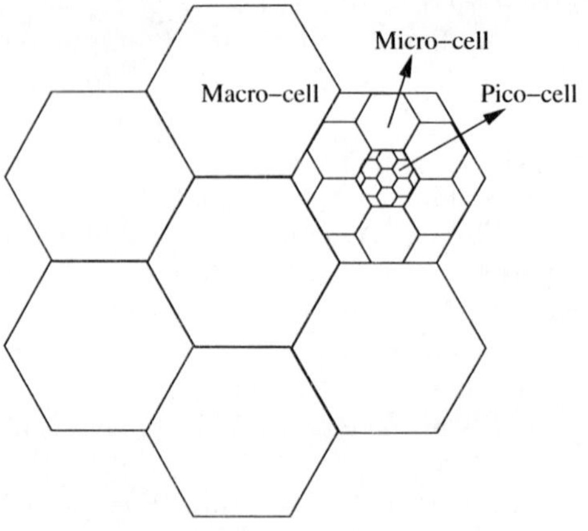

Figure 3.2. Cell-splitting.

are also in use to cover indoor areas. This is shown in Figure 3.2. Depending on the demands in traffic at a given point of time, channels can be allocated explicitly for the micro-cellular layer, or a sharing algorithm can be envisaged with the macro-cellular layer, by which channels can be allotted to the micro-cells as the demand arises. Users with lower rates of mobility are handled by the layer of smaller cell size. If a highly mobile user is handled by the micro-cellular layer, there is an overhead of too many handoffs. In fact, the handoffs may not be fast enough to let the call continue uninterrupted [2].

The introduction of a micro- or pico-cellular layer increases the available number of channels in hotspot regions and enables better frequency planning, so that the whole network does not have to be designed to handle the worst-case demand. This flexibility in frequency management leads to capacity enhancement.

Sectorization

This concept uses space division multiple access (SDMA) to let more channels be reused within a shorter distance. Antennas are modified from omnidirectional to sectorized, so that their signals are beamed only in a particular sector, instead of being transmitted symmetrically all around. This greatly reduces the downlink interference. A cell is normally partitioned into three 120-degree sectors or six 60-degree sectors. Further, the antennas can be down-tilted to reduce co-channel interference even more. Figure 3.3 shows the difference between omnidirectional and 60-degree sectored antennas. Reduction in interference allows shorter reuse distance, and hence increases capacity.

Cellular networks rely on trunking, the statistical multiplexing of a large number of users on limited number of channels to support many users on the wireless spectrum. Trunking efficiency increases when a larger channel pool is available,

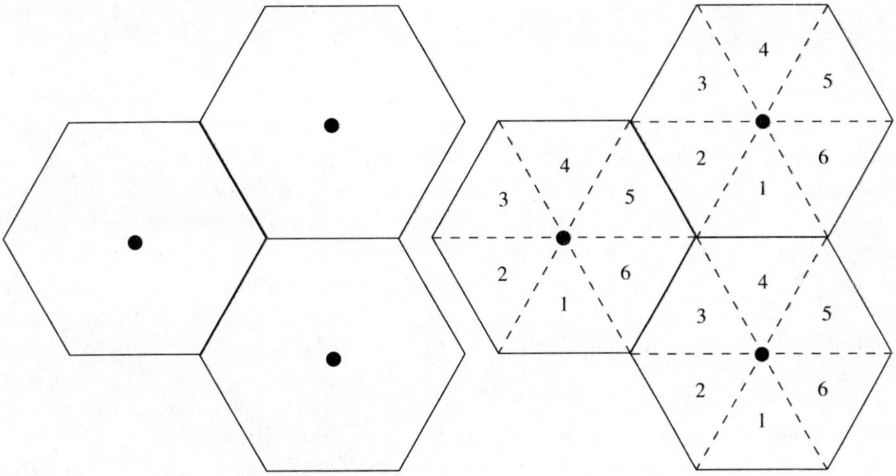

(a) Omnidirectional (b) Sectorized

Figure 3.3. Omnidirectional and sectorized antennas.

instead of many smaller subsets of channels. When sectoring is used, the available channels of the cell must be subdivided and allocated to the different sectors. This breaks up the available channel pool into smaller sub-groups, and reduces trunking efficiency, that is, blocking probability may increase. The number of handoffs is also increased due to inter-sector handoffs being introduced.

Power Control

Cellular networks face the "near-far" problem. An MT which is very close to the BS receives very strong signals from the BS, and its signals are also extremely strong at the BS. This can possibly drown out a weak signal of some far-away MT which is on an adjacent frequency. To avoid this problem, the BS must issue power control orders to the MTs to receive a fairly constant, equal power from all MTs, irrespective of their distance from the BS. MTs which are farther away from the BS transmit at higher power than nearby MTs, so that the received power at the BS is equal. This saves power for the MTs near the BS, and avoids excessive interference. Reduction in interference increases the capacity of the cellular network.

3.2.2 Channel Allocation Algorithms

Efficient allocation of channels to the different cells can greatly improve overall throughput of the network, in terms of the number of calls supported successfully. The channel allocation algorithms that can be used vary greatly in complexity and effectiveness. Fixed channel allocation algorithms allot a set of channels to each cell, and the number of channels per cell is determined *a priori*. Cells may be allowed to borrow some channels from their neighboring cells to tide over temporary increase in demand for channels. This should not violate the minimum reuse distance constraints. So, one borrowing may prevent some other cells from borrowing certain channels, which bring identical frequencies into use too close to each other. This is termed "channel locking." The main drawback of fixed channel allocation algorithms is that they assume a constant, or at least a predictable, distribution of load over the different cells. But, in reality, demands are very unpredictable, which may lead to a scarcity of channels in some cells and an excess in others.

The dynamic channel allocation algorithms do not have any local channels allotted to cells. The decision of which channel to lend is made dynamically. This makes dynamic algorithms extremely flexible and capable of dealing with large variations in demand, but they involve a lot of computation. Dynamic algorithms require a centralized arbitrator to allocate channels, which is a bottleneck. Hence, distributed channel allocation algorithms are preferred, especially for micro-cells. All BSs look at the local information about their possible interferers, and decide on a suitable allocation of channels to maximize bandwidth utilization.

The hybrid channel allocation schemes give two sets of channels to each cell; A is the set of local channels and B is the set of borrowable channels. Cells allow borrowing only from set B, and reallocate calls to their local channel set as soon as possible. The hybrid allocation algorithms introduce some flexibility into the channel allocation scheme. A subset of the channels can be borrowed to account for temporary variations in demand, while there is a minimum guarantee of channels for all cells at all times. These algorithms are intermediate in terms of complexity and efficiency of bandwidth utilization.

3.2.3 Handoffs

An important concept that is essential for the functioning of cellular networks is *handoffs*, also called *handovers*. When a user moves from the coverage area of one BS to the adjacent one, a handoff has to be executed to continue the call. There are two main parts to the handoff procedure: the first is to find an uplink-downlink channel pair from the new cell to carry on the call, and the second is to drop the link from the first BS.

Issues Involved in Handoffs

Certain issues that need to be addressed in a handoff algorithm are listed below.

- **Optimal BS selection:** The BS nearest to an MT may not necessarily be the best in terms of signal strength. Especially on the cell boundaries, it is very difficult to clearly decide to which BS the MT must be assigned.

- **Ping-pong effect:** If the handoff strategy had very strictly demarcated boundaries, there could be a series of handoffs between the two BSs whose cells touch each other. The call gets bounced back and forth between them like a ping-pong ball. This could totally be avoided, since signal strength is not significantly improved by such handoffs.

- **Data loss:** The interruption due to handoff may cause a loss in data. While the delay between relinquishing the channel in the old cell, and resuming the call in the new cell, may be acceptable for a voice call, it may cause a loss of few bits of data.

- **Detection of handoff requirement:** Handoffs may be mobile-initiated, in which case the MT monitors the signal strength received from the BS and requests a handoff when the signal strength drops below a threshold. In the case of network-initiated handoff, the BS forces a handoff if the signals from an MT weaken. The BS inquires from all its neighboring BSs about the signal strength they receive from the particular MT, and deduces to which BS the call should be handed over. The mobile-assisted scheme is a combination of the network and mobile-initiated schemes. It considers the evaluation of signal strength from the mobile, but the final handoff decision is made by the BS.

Handoff Quality

Handoff quality is measured by a number of parameters, and the performance of a handoff algorithm is judged in terms of how it improves these parameters. Some of the measures of handoff quality are as follows:

- **Handoff delay:** The signaling during a handoff causes a delay in the transfer of an on-going call from the current cell to the new cell. If the delay is too large, the signal may fall below the minimum carrier to interference ratio (C/I) required for continuation of the call, and the call may get dropped. The handoff protocol should aim to minimize this delay.

- **Duration of interruption:** In a conventional handoff algorithm, called *hard handoff*, the channel pair from the current BS is canceled, and then the channel pair from the next BS is used to continue the call. This can cause an interruption in the call. Though this is imperceptible to humans and the speech loss can be interpolated by the human ear, it adversely affects data transmission. So, handoff interruption must be minimized.

- **Handoff success:** The probability of a successful handoff, that is, continuation of the call when a mobile user crosses a cell boundary, is called the handoff success rate. This is influenced by the number of available channel pairs in the adjacent cells, and the capacity to switch before the signal falls below the acceptable C/I. The handoff strategies should maximize the handoff success rate.

- **Probability of unnecessary handoff:** Unnecessary handoffs, as in the ping-pong effect, increase the signaling overhead on the network and lead to unwanted delays and interruptions in calls. So, the probability of such unnecessary handoffs should be minimized.

Improved Handoff Strategies

Several improvements have been explored over the conventional hard handoff method, in order to tackle the issues related to handoffs.

- **Prioritization:** In order to reduce handoff failure, handoffs are given priority over new call requests. A certain number of channels may be reserved in each cell explicitly for handoffs. There is a trade-off here between probability of dropping a call due to handoff failure, and bandwidth utilization.

- **Relative signal strength:** There is a hysteresis margin required by which the signal strength received from the new BS has to be greater than the signal strength of the current BS signal, to ensure that the handoff leads to significant improvement in the quality of reception. This reduces the probability of unnecessary handoffs. There is a minimum time for which an MT must be in a cell before it can request a handoff, called the *dwell time*. The dwell timer keeps track of how long the MT has been in the cell. Dwell time has to be decided depending on the speed of the mobile users in the region. A very large dwell time may not allow a handoff when it really becomes necessary for a rapidly moving mobile. The concept of dwell time reduces the ping-pong effect as frequent handoffs will not occur on the border of two cells.

- **Soft handoffs:** Instead of a strict "handing over the baton" type of handoff, where one BS transfers the call to another, a period of time when more than one BS handles a call can be allowed, in the region of overlap between two or more adjacent cells. Fuzzy logic may be used to model the fading of one BS signal and the increasing intensity of the other. In the overlap region, the MT "belongs" to both cells, and listens to both BSs. This requires the MT to have the capability to tune in to two frequencies simultaneously.

- **Predictive handoffs:** The mobility patterns of users can be predicted by analysis of their regular movements, using the location tracking data. This can help in the prediction of the requirements of channels and handoffs in different cells.

- **Adaptive handoffs:** Users may have to be shifted across different layers, from micro- to macro-cellular or pico-cellular to micro-cellular, if their mobility differs during the call.

3.3 CELLULAR ARCHITECTURE

Every cell has a BS to which all the MTs in the cell transmit. The BS acts as an interface between the mobile subscriber and the cellular radio system. The BSs are themselves connected to a *mobile switching center* (MSC), as shown in Figure 3.4. The MSC acts as an interface between the cellular radio system and the public switched telephone network (PSTN). It performs overall supervision and control of the mobile communications, which include location update, call delivery, and user identification. The *authentication center* (AuC) validates the MTs by

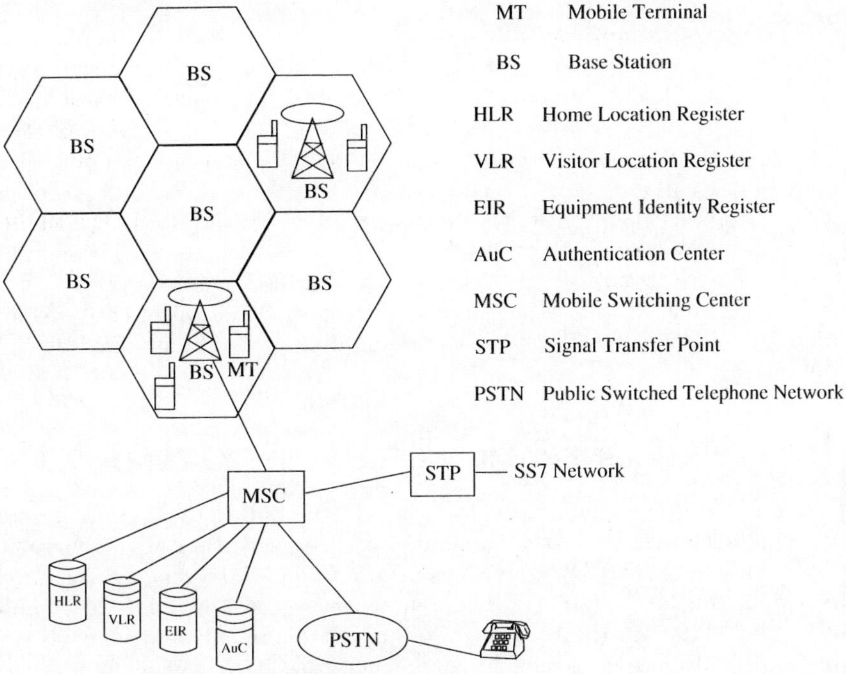

Figure 3.4. Cellular architecture.

verifying their identity with the *equipment identity register* (EIR). The MSCs are linked through a signaling system 7 (SS7) network, which controls the setting up, managing, and releasing of telephone calls. The SS7 protocol introduces certain nodes called signaling transfer points (STPs) which help in call routing. For a detailed description of mobile network architecture, refer to [3].

A subscriber in a mobile network should be efficiently traced to deliver calls to him/her, and he/she should be able to access the network through the mobile end-system, irrespective of his/her location. This involves management of different databases in the cellular network architecture, and is called location management. The method used currently for location management requires the MT, also called the mobile station (MS), to report its position to the network periodically. The network stores the location information of each MT in location databases and this information is retrieved during call delivery.

In the current location management scheme, each user is permanently associated with the *home location register* (HLR) in his/her subscribed cellular network. This HLR contains the *user profile* consisting of the services subscribed by the user (such as SMS and paging), billing information, and location information. The *visitor location register* (VLR) maintains the information regarding roaming users in the cell. VLRs download the information from the users' respective HLRs. The number and placement of the VLRs vary among networks. Registration of an MT in a new cell entails updates of its HLR, and the VLRs of its old and new cells. For ease of management, many cells may be grouped into one registration area (RA) and updates performed only on movement out of the RA. Delivery of a call requires tracing the current location of the handset by requesting the information from the HLR.

Recent improvements in location management include the use of distributed databases and replication of user profiles to enable faster access to user information. Partition-based architectures have been used to group MSCs into partitions, and update databases only if the MT moves out of the partition. This greatly reduces the number of updates, but the size of the database to be maintained at each partition increases.

3.4 THE FIRST-GENERATION CELLULAR SYSTEMS

After the basic concepts common to most cellular networks, the specific generations of cellular networks and the standards of each generation are discussed in the following sections. The first implementations of the cellular concept constitute the first-generation (1G) systems. These systems, such as the advanced mobile phone system (AMPS) in the United States and Nordic mobile telephony (NMT) deployed in several European countries, are analog-based (*i.e.*, they employ analog modulation and send information as a continuously varying waveform).

3.4.1 Advanced Mobile Phone System

AMPS divides the 800 MHz part of the frequency spectrum into several channels, each 30 KHz wide. The cellular structure uses a cluster size of seven, and each cell is roughly 10-20 Km across. The AMPS system uses 832 full-duplex channels, with FDM to separate the uplink and downlink channels. The channels are classified into four main categories: downlink *control channels* for system management, downlink *paging channels* for paging an MT (locating an MT in the network and alerting it when it receives a call), bidirectional *access channels* for call setup and channel assignment, and bidirectional *data channels* to carry user voice/data. AMPS provides a maximum data transmission rate of 10 Kbps.

When the MT is powered on, it scans for the most powerful control channel and broadcasts its 32-bit serial number and 10-digit telephone number on it. The BS which hears this broadcast registers the MT with the mobile switching office (MSO), and informs the HLR of the MT of its present location. The MT updates its position once every 15 minutes. To make a call, the MT sends the number to be called through an access channel. The BS sends this request to the MSO, which assigns a duplex channel for the call. MTs are alerted to incoming calls through the paging channel. The call is routed through the home to its current location. Once the MT is located, it takes up the call on the allotted voice channel.

In order to conserve the battery power of the handset, the MT goes into a "sleep state" when it is idle. The designed sleep time is 46.3 ms in the AMPS system.

The MT periodically wakes up to scan the paging channel and check if there is any call addressed to it.

The service of a mobile network has similar measures of performance like that of a wired network. The concept of *trunking*, used in all communication networks, exploits the statistical behavior of users to enable a fixed number of channels to be shared by a much larger number of users. The grade of service (GoS) of a trunked network is a measure of accessibility of the network during its peak traffic time. GoS is specified as the probability of a call being blocked, or experiencing a queuing delay greater than a threshold value. The AMPS system was designed for GoS of 2% blocking.

3.5 THE SECOND-GENERATION CELLULAR SYSTEMS

As mentioned earlier, 1G systems are analog which leads to the following problems: (i) No use of encryption (1G systems do not encrypt traffic to provide privacy and security, as analog signals do not permit efficient encryption schemes, and thus voice calls are subject to eavesdropping). (ii) Inferior call quality (This is due to analog traffic which is degraded by interference. In contrast to digital traffic stream, no coding or error correction is applied to combat interference). (iii) Spectrum inefficiency [This is because at any given time a channel is allocated to only one user, regardless of whether the user is active (speaking) or not. With digital traffic it is possible to share a channel by many users (which implies multiple conversations on a single channel) using TDMA or CDMA, and further, digital signals allow compression, which reduces the amount of capacity needed to send data by looking for repeated patterns].

The second-generation (2G) systems, the successors of 1G systems, are digital [*i.e.*, they convert speech into digital code (a series of pulses) which results in a clearer signal] and thus they overcome the deficiencies of 1G systems mentioned above. It may be noted that the user traffic (computer data which is inherently digital or digitized voice) must be converted into radio waves (analog signals) before it can be sent (between MT and BS). A 2G system is called personal communications services (PCS) in the marketing literature. There are several 2G standards followed in various parts of the world. Some of them are global system for mobile communications (GSM) in Europe, digital-AMPS (DAMPS) in United States, and personal digital cellular (PDC) in Japan.

3.5.1 Global System for Mobile Communications

GSM[1] is an extremely popular 2G system, which is fully digital, used across over 100 countries [4].

There are four variants of GSM:

- Operating around 900 MHz [This first variant reuses the spectrum intended for Europe's analog total access communication system (TACS)]

[1] GSM originally stood for Groupe Speciale Mobile, the name of the working group that designed it.

- Operating around 1,800 MHz [licensed in Europe specifically for GSM. This variant is sometimes called digital communications network (DCN)]
- Operating around 1,900 MHz (used in United States for several different digital networks)
- Operating around 450 MHz (latest variant for replacing aging analog networks based on NMT system)

Apart from voice service, GSM also offers a variety of data services.

The modulation scheme used in GSM is Gaussian minimum shift keying (GMSK). GSM uses frequency duplex communication, and each call is allotted a duplex channel. The duplex channels are separated by 45 MHz. Every channel is of 200 KHz bandwidth. Thus, GSM uses FDM to separate the channels. The downlink (BS-MT) channels are allotted 935-960 MHz, and the uplink (MT-BS) channels are on 890-915 MHz as shown in Figure 3.5. The uplink "frame" of eight slots is shifted by a delay of three slots from the downlink frame, so that the MT does not have to send and receive at the same time. A procedure called *adaptive frame alignment* is used to account for propagation delay. An MT which is far away from the BS starts its frame transmission slightly ahead of the actual slot commencement time, so that the reception at the BS is synchronized to the frame structure. The exact advancement is instructed by the BS to the MT, with MTs that are farther away from the BS requiring a greater advancement.

Each physical channel is divided into eight periodic time slots (each of which is 0.577 ms duration), which are time-shared between as many as eight users (logical channels) using TDMA. A complete cycle of eight time slots is called a (TDMA) frame. A slot comprises the following four parts: (i) header and footer (These are empty space at the beginning and end of the slot to separate a slot from its neighbors and to negate the effects of propagation delay for distances up to 35 Km from the BS), (ii) training sequence (This is to help a receiver lock on to the slot), (iii) stealing bits (These identify whether the slot carries data or control information), and (iv) traffic (This part carries user traffic (voice/data) as well as control information and error correction). Users cannot use all frames; for every 24 GSM frames that carry voice/data, one is "stolen" for signaling and another reserved for other types of traffic (such as short text messages or caller line identification). Thus eight slots make up a TDM frame and 26 TDM frames make up a multiframe. Multiframes are in turn grouped into superframes and hyperframes. Some of the slots here are used to hold several control channels for managing the GSM system.

The control channels in GSM are classified into three broad categories. The *broadcast control channel* (BCCH) is a downlink channel that contains the BS's identity and channel status. All MTs monitor the BCCH to detect if they have moved into a new cell. The *dedicated control channel* (DCCH) is used for call-setup, location updates, and all call-management related information exchange. Every call has its own allotted DCCH. The information obtained on the DCCH is used by the BS to keep a record of all the MTs currently in its footprint (coverage area). The third category, *common control channel* (CCCH), consists of the downlink paging

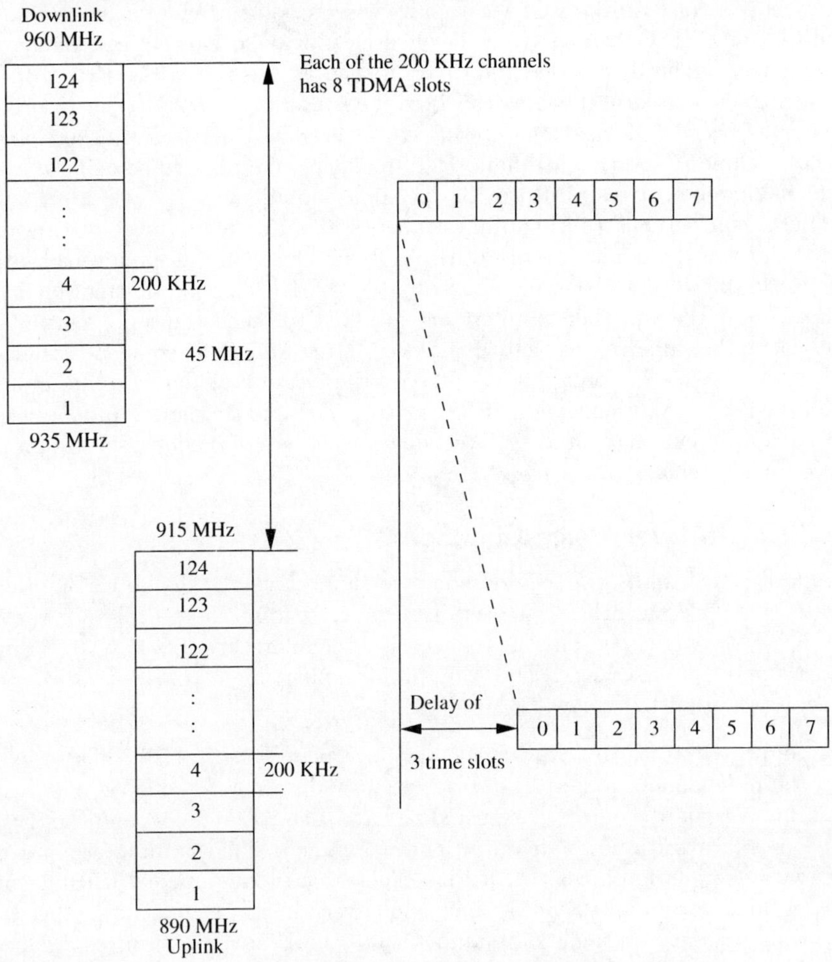

Figure 3.5. GSM frequency bands and TDMA frame.

channel, which is used to page any MT to alert it for an incoming call, the random access channel, which supports slotted ALOHA-based (random request without reservation of time slot) request from MT to BS for call-initiation, and the access grant channel, on which the BS informs the MT of an allotted duplex channel for a call.

As each MT is assigned only one slot within each frame, the maximum speed for data services is around 34 Kbps (1/8 of the 270.8 Kbps capacity of a 200 KHz GSM carrier). Forward error correction (FEC) and encryption reduce the data rate (by at least one-third; the precise amount depends on how the handset and network encode voice and data) to around 9.6 Kbps.

An important feature of GSM, from the user's point of view, is the subscriber identity module (SIM), a smart card which is pluggable into a GSM phone (mobile handset). The SIM stores information such as the subscriber's identification number, the networks and countries where the subscriber is entitled to service, and other user-specific information. By inserting the SIM card into another handset, the user is able to use the new handset to make/receive calls while using the same telephone number. Thus SIM provides personal mobility.

Due to conventional call-handling strategies, GSM requires the call to always be routed through the home network of the mobile. This causes a circuitous route for calls to be delivered to roaming subscribers. The HLR of the destination mobile is first contacted, and then pointers are followed to reach the MT's current cell, even if it is the same cell as that of the originating MT. This could be avoided by routing directly to the foreign network, once the location of the receiving mobile is established. On the other hand, such rerouting of traffic through a foreign network may be done intentionally if the call termination charges of the local service provider are very high. This is called *tromboning.*

3.5.2 Data Over Voice Channel

Cellular systems were primarily designed to support voice calls only. But the demand for supporting other data services, ranging from short messages to Web-browsing, slowly emerged. It was realized that cellular networks have to be modified to support data services. The main problems in using a voice network for data transmission are listed below [5].

- **Signal distortion:** Speech encoders are being used on all links to exploit the redundancy in speech and compress it, in order to conserve bandwidth. However, data cannot be expected to have such redundancy, and the data receivers cannot interpolate data the way a human listener can interpolate even with degradation in speech quality. This issue is compounded when encoders in tandem recompress data, that is, on every hop of the network, speech encoders compress the data, assuming it has the same amount of redundancy as speech.

- **Handoff error:** Handoffs introduce a certain delay in transfer of the call from one cell to another, which may lead to loss of data.

- **Interfacing with fixed network:** A PSTN modem expects a 2,100 Hz tone from the source when a data call is initiated. On the other hand, PSTN networks do not indicate non-voice service. So the cellular network should be able to differentiate between a data call and a voice call.

The main problem is in making the network recognize a data call, and handle it differently from a voice call, by probably disabling some optimizations made on voice calls. Some possible solutions that have been suggested are listed here.

- A control message could be transmitted all along the path of the call, to indicate a data call so that voice coding can be disabled.

- A two-stage dial-up operation can be used, first to the cellular carrier and then to the subscriber. The carrier has different numbers for each service it offers. For example, to set up a data call with an MT, the cellular carrier is first dialed, and it then informs the MT of a data call.
- A subscriber could be assigned separate subscriber numbers for each service he/she opts for.

3.5.3 GSM Evolution of Data Services

GSM and other 2G standards started providing data services overlaid on the cellular network. This was meant to be a stop-gap arrangement before the third-generation (3G) networks could be formally deployed, and these data services constituted the 2.5G of cellular networks. It started out with small messages through SMS, and today, Web-browsing is possible over the mobile network. Some of the popular data services provided over GSM are now briefly described.

Short Messaging Service (SMS)

SMS is a connectionless transfer of messages, each up to 160 alphanumeric characters in length. To send longer messages, small packets are either concatenated at the receiver, or the sender sends compressed messages. SMS can perform point-to-point data service as well as broadcast throughout the cell. The message transfer takes place over the control channel.

High-Speed Circuit-Switched Data (HSCSD)

HSCSD is a simple upgrade to GSM. As the name implies, it is a circuit-switched protocol for large file transfers and multimedia data transmissions. Contrary to GSM (one TDMA slot/user), it allots up to eight TDMA slots per user and hence the increased data rates. By using up to four consecutive time slots, HSCSD can provide a data rate of 57.6 Kbps to a user. Its disadvantage is that it increases blocking probability by letting the same user occupy more time slots that could have otherwise been used for many voice calls.

General Packet Radio Service (GPRS)

GPRS, which promises to give every user a high-capacity connection to the Internet, uses TCP/IP and X.25[2] and offers data rates up to 171.2 Kbps, when all eight time slots of a GSM radio channel are dedicated to GPRS. A variety of services has been provided on GPRS, ranging from bursty data transmission to large file downloads. Being a packetized service, GPRS does not need an end-to-end connection. It uses radio resources only during actual data transmission. So, GPRS is extremely well-suited for short bursts of data such as e-mail and faxes, and non-real-time Internet usage. Implementation of GPRS on the existing GSM network requires

[2] X.25 is an ITU-T (International Telecommunications Union–Telecommunication Standards Sector) protocol standard for packet-switched WAN communications that defines connection establishment and maintenance between user devices and network devices.

only the addition of packet data switching and gateway nodes, and has minimal impact on the already established network. The HLR is now enhanced to also record information about GPRS subscription.

Enhanced Data Rates for GSM Evolution (EDGE)

EDGE, also referred to as enhanced GPRS or EGPRS, inherits all the features from GSM and GPRS, including the eight-user TDMA slot structure and even the slot length of 0.577 ms. However, instead of the binary GMSK, it uses 8-PSK (octal PSK) modulation which triples the capacity compared to GSM. It provides cellular operators with a commercially viable and attractive method to support multimedia services. As 8-PSK is more susceptible to errors than GMSK, EDGE has nine different modulation and coding schemes (air interfaces), each designed for a different quality connection.

Cellular Digital Packet Data (CDPD)

CDPD is a packet-based data service that can be overlaid on AMPS and IS-136 systems. It supports a connectionless network service, where every packet is routed independently to the destination. The advantage of CDPD lies in being able to detect idle voice channels of the existing cellular network, and use them to transmit short data messages without affecting the voice network. The available channels are periodically scanned and a list of probable candidates for data transmission is prepared. The system uses *channel sniffing* to detect channels which have been idle and can carry data. The system must continuously hop channels so that it does not block a voice call on a channel it is currently using. To avoid this *channel stealing* from the voice network, channel hopping is performed very rapidly, at the rate of once every ten seconds. Usually, a timed hop is performed by using a *dwell timer*, which determines how long the CDPD can use a channel. If the channel has to be allocated in the meantime to a voice call, CDPD performs an *emergency hop*. So, it is essential to find alternate idle channels in real time to ensure that data is not lost.

3.5.4 Other 2G Standards

Digital-AMPS (D-AMPS) is the digital version of the first-generation AMPS and is designed to coexist with AMPS. It is a TDMA-based system known as IS-54 (Telecommunications Industry Association Interim Standard). It uses AMPS carriers to deploy digital channels, each of which can support three times the users that are supported by AMPS with the same carrier. Digital channels are organized into frames and there are six slots per frame. D-AMPS supports data rates of around 3 Kbps. D-AMPS+, an enhancement of D-AMPS for data (similar to HSCSD and GPRS in GSM), offers increased data rates in the range of 9.6-19.2 Kbps. While D-AMPS maintains the analog channels of AMPS, its successor known as IS-136 is a fully digital standard.

IS-95, another 2G system, also known as cdmaOne or IS-95a, was first deployed in South Korea and Hong Kong in 1993, and then in the United States in 1996. It is a fully digital standard that operates in the 800 MHz band (like AMPS, IS-136) and is the only 2G system based on CDMA. It is incompatible with IS-136. It supports data traffic at the rates of 4.8 and 14.4 Kbps. cdmaTwo or IS-95b, an extension of IS-95, offers support for 115.2 Kbps.

Personal digital cellular (PDC), deployed in Japan, is a system based on D-AMPS. The main reason for PDC's success is i-mode, which is a mobile Internet system developed by NTT DoCoMo (refer to Section 4.5.3). Using 9.6 Kbps or slower data rates, it provides access to thousands of Web sites specially adapted to fit into a mobile phone's low data rate and small screen.

3.6 THE THIRD-GENERATION CELLULAR SYSTEMS

The aim of the third-generation (3G) cellular system is to provide a *virtual home environment*, formally defined as a uniform and continuous presentation of services, independent of location and access. This means that the user must be able to access the services that he/she has subscribed to, from anywhere in the world, irrespective of his/her method of access, as if he/she were still at home. This requires very stringent QoS adherence, and highly effective and optimized architectures, algorithms, and operations of the network elements. The use of "intelligent network architecture" is foreseen here [6], [7], [8]. The networks should contain advanced algorithms to handle location information retrieval and update, handoffs, authentication, call routing, and pricing. From the user point of view, the 3G systems with very high-speed wireless communications (up to 2 Mbps) plan to offer Internet access (e-mail, Web surfing, including pages with audio and video), telephony (voice, video, fax, etc.), and multimedia (playing music, viewing videos, films, television, etc.) services through a single device. Such services cannot be realized over the present 2G systems — to effect a simple transfer of a 2 MB presentation, it would take approximately 28 minutes with a 9.6 Kbps GSM data rate.

3.6.1 3G Standards

An evolution plan has been charted out to upgrade 2G technologies into their 3G counterparts. Table 3.1 lists some existing 2G systems and their corresponding 3G versions [9].

Table 3.1. Evolution plan to 3G standards

Country	Existing 2G Standard	3G Standard
Europe	GSM	W-CDMA (UMTS)
Japan	PDC	W-CDMA (DoCoMo)
USA	IS-95/cdmaOne	cdma2000
USA	IS-136	UWC-136

In 1992, the International Telecommunications Union (ITU) initiated the standardization of 3G systems, and the outcome of this effort was called international mobile telecommunications-2000 (IMT-2000). The number 2000 stood for three things: (i) the system would become available in the year 2000, (ii) it has data rates of 2,000 Kbps, and (iii) it was supposed to operate in the 2,000 MHz region, which ITU wanted to make globally available for the new technology, so that users could roam seamlessly from country to country. None of these three things came to pass, but the name has remained. The service types that the IMT-2000 network is supposed to provide to its users are given in Table 3.2 [10]. As different parts of the world are dominated by different 2G standards, the hope for a single worldwide 3G standard has not materialized, as can be seen in Table 3.1.

Universal mobile telecommunications system (UMTS) is the European 3G standard which is also called wideband CDMA (W-CDMA) as the standard is based on CDMA (direct sequence spread spectrum) technology. The term "wideband" here denotes use of a wide carrier. W-CDMA uses a 5 MHz carrier (channel bandwidth) which is 25 times that of GSM and four times that of cdmaOne. It is designed to interwork with GSM networks, which means a caller can move from a W-CDMA cell to a GSM cell without losing the call. The first W-CDMA networks are being deployed in Japan by NTT DoCoMo. Of the existing 2G systems, cdmaOne is already based on CDMA. cdma2000, basically an extension of cdmaOne, is not designed to interwork with GSM, which means it cannot hand off calls to a GSM cell. Commercial versions of cdma2000 include 1Xtreme by Motorola and Nokia and high data rate (HDR) by Qualcomm. UWC-136 is a TDMA-based 3G standard developed by the Universal Wireless Communications Consortium, as an upgrade of IS-136.

Table 3.2. IMT-2000 service types

Service	Upstream	Downstream	Example	Switching
Interactive Multimedia	256 Kbps	256 Kbps	Video conference	Circuit
High Multimedia	20 Kbps	2 Mbps	TV	Packet
Medium Multimedia	19.2 Kbps	768 Kbps	Web surfing	Packet
Switched Data	43.2 Kbps	43.2 Kbps	Fax	Circuit
Simple Messaging	28.8 Kbps	28.8 Kbps	E-mail	Packet
Speech	28.8 Kbps	28.8 Kbps	Telephony	Circuit

3.6.2 The Problems with 3G Systems

3G systems fundamentally need greater bandwidth and lower interference to be able to meet QoS requirements for data and multimedia services. Field tests have indicated that CDMA is an attractive option for mobile cellular networks, as it can operate in the presence of interference, and can theoretically support very large bandwidth [11]. The performance of 3G systems is further improved by the use of smart antennas. This refers to the technology of controlling a directional antenna array with an advanced digital signal processing capability to optimize its radiation and/or reception pattern automatically in response to the signal environment [12].

3G systems promised to be the ultimate panacea to all problems of dropped calls, low data rates, and mobility restriction. The initial time-chart had scheduled complete deployment of 3G by the year 2000. But this has really not happened. The recommended spectrum allocation for 3G in the IMT-2000 could not be implemented as the requested frequencies by the ITU were partially or fully in use in many countries. And it is becoming increasingly difficult to find a common slice of spectrum to enable global roaming.

Another reason why 3G systems did not take off is the disappointing performance of CDMA in practice. CDMA was projected to be the ultimate solution to interference. Theoretically, an infinite number of users could use the same frequency band, since all of them hopped on all the frequencies in a pseudo-random sequence. But, implementations of CDMA have led critics to believe otherwise. Table 3.3 shows the stark difference between the theoretical and practical performances of CDMA [13].

It would be interesting to speculate where mobile telephony is heading at this point in time. The first event on the agenda, of course, would be to implement 3G networks completely [14]. The trend seems to be headed toward having cells of different sizes: on the one hand, large cells to handle global access, and at the other extreme, pico-cells to handle indoor communication.

Even though 3G networks are not yet fully deployed, some researchers started working on the the next generation of systems, called 4G systems, targeted for deployment in the year 2010, to provide seamless integration with wired networks and especially IP, high bandwidth (data rates of up to 100 Mbps), ubiquity (connectivity everywhere), high-quality multimedia services, adaptive resource management, and software-defined radio. [Efficient and relatively flexible handsets are required to combat the diverse range of cellular standards/air interfaces, without which roaming between different networks becomes difficult (without significant adjustment or replacement of handsets). Software-defined radio provides a solution to this problem,

Table 3.3. CDMA — The debate

Claims	Reality
Capacity of 20 times that of AMPS	Only 3-4 times that of AMPS
No more dropped calls	40 percent dropped calls when loaded
No problem of interference	Interference from existing AMPS
Quality of speech promised at 8 Kbps	Had to change to 13 Kbps

by implementing the radio functionality (such as modulation/demodulation, signal generation, coding, and link-layer protocols) of different standards as software modules running on a generic hardware platform. Also, the software modules which implement new features/services can be downloaded over the air onto the handsets.]

3.7 WIRELESS IN LOCAL LOOP

In cellular systems, mobility was the most important factor influencing the design of the networks. Wireless in local loop (WLL) technology is, on the other hand, in the scenario of limited mobility. The circuit (loop) that provides the last hop connectivity between the subscriber and PSTN is called the local loop. It has been conventionally implemented using common copper wiring. Optical fiber is not a popular local loop technology due to the high cost of deployment. WLL, also known as fixed wireless access (FWA) or radio in the local loop (RLL), is the use of wireless connectivity in the last hop between the subscriber and the telephone exchange (end office). It provides a cost-effective means of connecting far-flung areas. In urban areas where high capacity is required, the use of WLL can provide the means to extend the existing network [15].

WLL has many advantages for both the subscribers and the service providers. The deployment of WLL is much easier, and it can be extended to accommodate more customers as the demand increases. Lower investment, operations, and maintenance costs make WLL an attractive, cost-effective option for the customer. WLL offers a wide range of services from basic telephony to Internet surfing. With the introduction of broadband wireless systems, bandwidth-intensive applications such as video-on-demand can also be supported.

WLL must satisfy QoS requirements to compete with more efficient copper wire transmission technologies such as integrated services digital network (ISDN) and digital subscriber line (DSL). It must match the voice and data quality of regular telephone networks. This means it should adhere to 64 Kbps or 32 Kbps transmission channels, blocking probability of less than 10^{-2} and BER less than 10^{-3} for voice and 10^{-6} for data. The power consumption of subscriber units must be low since they run on batteries and are not powered by the exchange. It must support authentication and encryption schemes to ensure security and data privacy.

3.7.1 Generic WLL Architecture

The architecture of a WLL system is more or less similar to that of a mobile cellular system. The BS is implemented by the base transceiver station system (BTS) and the base station controller (BSC). The BTS, also called radio port (RP) or radio transceiver unit (RTU), performs channel coding, modulation/demodulation, and implements the radio interface for transmission and reception of radio signals. A BSC, alternatively called the radio port control unit (RPCU), controls one or more BTSs and provides them with an interface to the local exchange.

The fixed subscriber unit (FSU) or radio subscriber unit (RSU) is the interface between the subscriber's wired devices and the WLL network. The FSU performs

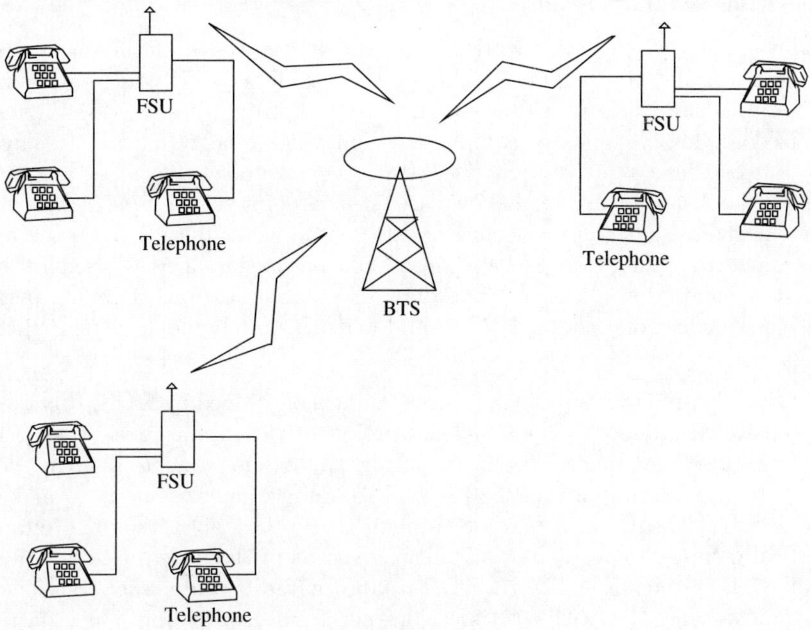

Figure 3.6. WLL architecture.

all physical and data-link layer functions from the subscriber end. The basic architecture of WLL networks is shown in Figure 3.6.

3.7.2 WLL Technologies

WLL technology is at the intersection between cellular and cordless telephony. Though there is no single global standard for WLL, many systems have been developed by suitable adaptation of cellular and cordless telephony standards.

Cellular-Based WLL Systems

WLL standards have been developed based on both TDMA (GSM-based) and CDMA. IS-95-based CDMA WLL supports channel rates of 9.6 Kbps and 14.4 Kbps. Since WLL FSUs are fixed, the BTS and FSU can be arranged in line-of-sight, which reduces interference. The cell can have six or more sectors, and hence increased capacity. For packetized data transfer, some dedicated channels are provided, besides statistical multiplexing with the voice channels. Up to 128 Kbps data services are available on CDMA-based WLL.

Cordless-Based WLL Systems

Cellular systems target maximization of bandwidth efficiency and frequency reuse, high coverage, and easy support for mobility in a high-speed fading environment. On the other hand, low-range (a few hundred meters) cordless-based systems are optimized for low complexity equipment, high voice quality, and low delays for data transfer, in a relatively confined static environment (with respect to user speeds). The range can also be increased using point-to-point microwave hops, where the radio signals are up-converted to microwave or optical frequencies, and down-converted again before the WLL terminal. The major TDMA-based low-range WLL systems are digital enhanced cordless telecommunications (DECT), personal access communication system (PACS), and personal handyphone system (PHS).

1. **DECT:** DECT[3] is a radio interface standard developed by ETSI. It is used extensively in indoor systems such as wireless private branch exchanges (PBXs) for intra-office connectivity and in cordless phones for residential areas. DECT system, operating in the 1,880-1,900 MHz range, supports users to make and receive calls within range of BSs around 100 m (indoor) and 500 m (outdoor). It also supports pedestrian mobility — speeds of the order of 10 Kmph. The DECT system uses TDMA/TDD mode for radio communications between handset and BS, with 24 time slots per frame, providing a net data rate of 32 Kbps. DECT provides 120 duplex channels with a bandwidth of 144 KHz/pair, unlike GSM, which offers only 50 KHz/duplex pair. DECT uses a dynamic channel allocation algorithm for channel assignment.

2. **PACS:** PACS system employs TDMA/TDM on the radio interface using $\pi/4$-QPSK modulation. It operates in TDD and FDD modes for the unlicensed PCS band and licensed PCS band, respectively. The radio frame is 2.5 ms in duration which is divided into eight time slots, each of which transports data at 32 Kbps. Channel assignment is performed using quasi-static autonomous frequency assignment (QSAFA), where an FSU listens to all the channel pairs and chooses the one with least interference. It then chooses a time-slot within the selected channel pair.

3. **PHS:** PHS, developed in Japan, supports very high-density pedestrian traffic. It uses the modulation format $\pi/4$-DQPSK, at a channel rate of 384 Kbps, in the RF band of 1,900 MHz (as DECT) with a bandwidth of 300 KHz per channel. In each of the 300 KHz wide RF carriers, there are four traffic channels, one of which is a dedicated control channel. PHS works on a TDMA/TDD mode for communication, whose frame duration is 5 ms, and uses a dynamic channel allocation scheme as in the case of DECT.

[3]It originally stood for digital European cordless telecommunications and today it stands for digital enhanced cordless telecommunications to underline its claim of being a worldwide standard for cordless telephony.

Proprietary Systems

Due to the absence of a universal standard specifically meant for WLL, a number of proprietary systems (independent technologies) have mushroomed. Cellular-based systems such as E-TDMA of Hughes Network Systems (HNS) and Lucent's Wireless Subscriber System attempt to maintain the desirable feature of cellular systems, namely, high coverage, while improving the capacity of the system. E-TDMA is an extension to the IS-136 cellular TDMA standard. It uses discontinuous transmission along with digital speech interpolation. This means both the FSU and BSC transmit only when speech is present, which is about 40% of the time. This leads to effective sharing of bandwidth. Proprietary systems Qualcomm's QCTel and Lucent's Airloop use CDMA for the MAC scheme, which helps in increasing transmission rates [15].

Satellite-Based Systems

In the absence of a common global standard for cellular systems, handheld devices which connect directly to a satellite present an attractive option to achieve worldwide communication. A satellite system consists of a satellite in space which links many Earth (ground) stations on the ground, and transponders, which receive, amplify, and retransmit signals to or from satellites. Most commercial communication satellites today use a 500 MHz uplink and downlink bandwidth. Mobile satellite systems complement existing cellular systems by providing coverage for rural areas, where it may not be economically viable to install BSs for mobile communications, and support for vehicular mobility rules out the use of WLL also.

Satellites may be geostationary, that is, stationary with respect to a point on Earth. These satellites have an orbit on the equatorial plane at an altitude of 35,786 Km such that they move with the same angular velocity as the Earth and complete a revolution in 24 hours. Satellites can also be placed on inclined orbits, at different altitudes and on non-equatorial planes. Low Earth orbiting (LEO) satellites, which orbit the Earth at lesser altitudes, play a major role in mobile communications.

While geostationary satellites can provide a much larger coverage, they require high subscriber unit power because of the high altitudes at which geostationary satellites operate. LEO satellites, on the other hand, operate at low altitudes of the order of hundreds of kilometers. Hence, more satellites are required to cover a given area. Another problem with geostationary satellites is that the associated propagation delay is significant. For a height of 36,000 Km, the speed of light delay is 0.24 s. On the other hand, LEO satellites have a propagation delay of about 12 ms [16].

The special aspect of satellite systems with respect to the MAC layer is that the time taken to detect collisions is very large. This is because collisions can be detected only when the transmission from the satellite down to the ground units arrives. Hence collision avoidance techniques are used instead of collision detection. A TDMA scheme with a centralized coordinator for allocation of time slots reduces contention. But, a fixed time-slot allocation leads to under-utilization of transponder capacity. The assignment should be made more flexible, with channels as well as frames pooled together, and allocated on demand to any Earth station. This method is termed demand assignment multiple access (DAMA), and can be used in conjunction with FDMA or TDMA. Two examples of satellite-based mobile telephony systems are studied below.

- **Iridium:** This was an ambitious project, initiated by Motorola in the early 1990s, with 66 LEO satellites[4] providing full global coverage (Earth's surface). The satellites operated at an altitude of 780 Km. The significant point about Iridium's architecture was that satellites did call routing and processing. There was line of sight (LoS) visibility between neighboring satellites and therefore the satellites communicated directly with one another to perform these call routing operations. Iridium ended up as a major commercial failure due to economic reasons in the year 2000.

- **Globalstar:** Iridium requires sophisticated switching equipment in the satellites because it relays calls from satellite to satellite. In Globalstar, which uses a traditional "bent-pipe" design, 48 LEO satellites are used as passive relays. All routing and processing of calls are done by ground stations (or gateways), so a satellite is useful only if there is a ground station underneath it. In this scheme, much of the complexity is on the ground, where it is easier to manage. Further, since typical Globalstar ground station antennas can have a range of many kilometers, a few such stations are needed to support the system. Globalstar can provide full global coverage except for the regions in the middle of oceans where ground station deployment is not possible or costs a lot, and those regions near the poles.

Global Positioning System (GPS)

Besides the use of satellite systems for WLL, another important use of satellites in the recent past has been the global positioning system (GPS)[17]. The GPS program is funded and controlled by the U.S. Department of Defense. The space segment of GPS consists of satellites which send encoded radio signals to the GPS receivers. The system has 24 satellites that orbit the Earth at about 18,000 Km in 12 hours, on six equally spaced orbital planes. The movement of the satellites is such that five to eight satellites are visible from any point on Earth. The control segment consists of five tracking stations located around the world which monitor the satellites and compute accurate orbital data and clock corrections for the satellites. The user segment consists of the GPS receivers, which compute the four dimensions – x, y, z coordinates of position, and the time – using the signals received from four GPS satellites.

GPS has been used primarily for navigation, on aircraft, ships, ground vehicles and by individuals. Relative positioning and time data has also been used in research areas such as plate tectonics, atmospheric parameter measurements, astronomical observatories, and telecommunications.

[4] The project initially called for the use of 77 LEO satellites and this fact gave it the name Iridium, as Iridium is the chemical element with an atomic number of 77.

3.7.3 Broadband Wireless Access

Fixed wireless systems are point-to-point (where a separate antenna transceiver is used for each user) or multipoint (where a single antenna transceiver is used to to provide links to many users). The latter one is the most popular and useful for WLL. A multipoint system is similar to a conventional cellular system. However, in a multipoint system (i) cells do not overlap, (ii) the same frequency is reused at each cell, and (iii) no handoff is provided as users are fixed. Two commonly used fixed wireless systems, which are typically used for high-speed Internet access, are local multipoint distribution service (LMDS) and multichannel multipoint distribution service (MMDS). These can be regarded as metropolitan area networks (MANs).

LMDS

LMDS, operating at around 28 GHz in the United States and at around 40 GHz in Europe, is a broadband wireless technology used to provide voice, data, Internet, and video services. Due to the high frequency at which LMDS systems operate, they are organized into smaller cells of 1-2 Km (against 45 Km in MMDS), which necessitates the use of a relatively large number of BSs in order to service a specific area. However, LMDS systems offer very high data rates (maximum cell capacity of 155 Mbps).

MMDS

MMDS operates at around 2.5 GHz in the licensed band and at 5.8 GHz in the unlicensed band in the United States, and LoS is required in most cases. The lower frequencies (or larger wavelengths of MMDS signals) facilitate longer ranges of around 45 Km. Due to the fact that equipment operating at low frequencies is less expensive, MMDS systems are also simpler and cheaper. However, they offer much less bandwidth (the maximum capacity of an MMDS cell is 36 Mbps). MMDS is also referred to as *wireless cable* as this service provides broadcast of TV channels in rural areas which are not reachable by broadcast TV or cable.

3.8 WIRELESS ATM

3.8.1 ATM — An Overview

ATM has been the preferred network mechanism for multimedia applications, LAN interconnections, imaging, video distribution, and other applications that require quality of service (QoS) guarantees. ATM uses an asynchronous mode of transfer [asynchronous time-division or statistical multiplexing (flexible bandwidth allocation)] to cope with the varying requirements of broadband services. ATM defines a fixed-size cell (which is the basic unit of data exchange) of length 53 bytes comprised of a 5-byte header and a 48-byte payload. The ATM layer provides the higher layers functions such as routing and generic flow control. First, a virtual connection is established between the source and the destination. This virtual connection is identified by the virtual path identifier (VPI) and virtual channel identifier (VCI).

After the connection is set up, all cells are sent over the same virtual connection using the VPI/VCI in the cell header. The VPI/VCI is used by the switching node to identify the virtual path and virtual channel on the link, and the routing table established at the connection setup time is used to route the cell to the correct output port. Cells are transmitted without consideration of their service requirements. The ATM adaptation layer (AAL) is responsible for providing the QoS requirements of the application. Each AAL layer will be supporting a subset of traffic and, by changing the AAL layer, different traffic requirements of the application can be transparently passed on to the ATM layer. The AAL's primary function is to provide the data flow to the upper layers taking into account the cell loss and incorrect delivery because of congestion or delay variation in the ATM layer.

In order to minimize the number of AAL protocols needed, services are classified on the basis of the following parameters:

- Timing relationship between source and destination
- Bit rate (constant or variable)
- Type of connection (connection-oriented or connection-less)

The service classes are further classified as constant bit rate (CBR), variable bit rate (VBR), available bit rate (ABR), and unspecified bit rate (UBR), on the basis of the data generation rate.

3.8.2 Motivation for WATM

The reasons for extending the ATM architecture to the wireless domain include supporting integrated multimedia services in next-generation wireless networks, seamless integration of wired ATM networks with wireless networks, the need for a scalable framework for QoS provisioning, and support for mobility. The introduction of ATM in the wireless domain creates new challenges as the original design did not consider varying conditions of the wireless domain. The issues involved in extending wired ATM to the wireless domain are location management, mobility management, maintaining end-to-end connection, support for QoS, and dealing with the characteristics of radio (wireless) links. Recall that HIPERLAN/2, a WLAN system, offers the capabilities of WATM.

3.8.3 Generic Reference Model

Figure 3.7 explains schematically the generic reference model that is used in WATM networks. Mobile WATM terminals (MTs) have radio links to base stations (BSs or radio transceivers) which are connected to a WATM access point (WATM-AP/AP) via wires. The APs are connected to enhanced mobility ATM switches entry (EMAS-E). The APs are traffic concentrators and lack routing capabilities. Rerouting is usually performed at the EMAS-E and the node that performs the rerouting is termed as the cross-over switch (COS) (Note that some APs may have routing capabilities, in which case they perform the functions of AP and EMAS-E). EMAS-Es are in turn connected to enhanced mobility aware ATM switches network

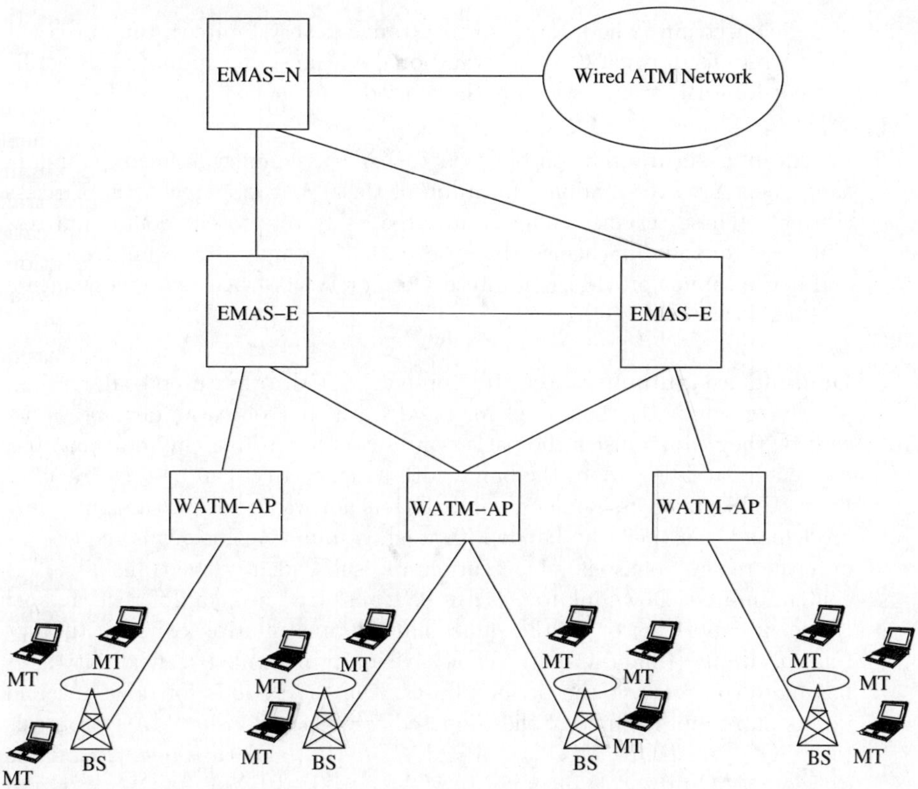

Figure 3.7. Generic reference model.

(EMAS-N). EMAS-E and EMAS-N are important entities that support mobility in WATM. The EMAS-N switches are finally connected to the ATM switches of the traditional ATM network, enabling wired and wireless networks to coexist.

3.8.4 MAC Layer for WATM

MAC protocols lay down the rules to control the access to the shared wireless medium among a number of users. The main challenge in the MAC protocol for the WATM is to provide the support for standard ATM services (including CBR, VBR, ABR and UBR) efficiently in the wireless environment. The three main categories of MAC protocols are:

1. **Fixed assignment:** Fixed assignment schemes essentially apportion the available resource (time or frequency) among the users in a definite manner in order to provide the QoS required. Fixed assignment schemes such as TDMA and FDMA suffer from inefficient bandwidth usage and, in the case

of radio spectrum (where the frequency range is very limited), this turns out to be a serious drawback. These traditional schemes are good for CBR traffic but not for VBR traffic, which is the most dominant traffic in WATM.

2. **Random assignment:** As opposed to fixed assignment schemes, random assignment involves random allocation of the resource, namely, the wireless channel. These schemes suffer from large delay due to the contention resolution algorithms. Schemes that use backoff techniques are unpredictable and hence cannot provide guaranteed QoS for WATM. An example of such a scheme is the CSMA/CA.

3. **Demand assignment:** Users (in contention with other users) address explicitly or implicitly their need for bandwidth, but once the demand is accepted, they can transfer the packets in a contention-free environment. If a user enters an idle period, the bandwidth assigned to it can be used by other users. Unlike fixed assignment, bandwidth is not wasted as it is assigned only on demand. Further, the bandwidth wastage due to collisions is reduced as only the request phase is in contention and subsequently the transmission is contention-free. Downlink frames can be transferred using a different channel (frequency division) or on the same channel time multiplexed with the uplink sub-frame (time division). Time division provides better flexibility as it has more control over the periods by varying the periods for downlink and uplink sub-frames. Some of the proposed frequency division MAC protocols for WATM are DQRUMA [18], PRMA/DA [19], and DSA++ [20]. Time division MAC protocols include MASCARA [21], PRMA/ATDD [22], and DTDMA/TDD [23].

3.8.5 Handoff Issues in WATM

Handoff is said to occur when a mobile terminal (MT) moves from the control of one BS to another, as discussed earlier. A handoff tends to disrupt existing connections of the MT, hence care needs to be taken during handoffs. This section deals with the issues related to handoffs and the proposed solutions to tackle the problems that arise. There are two levels of handoff, one at the radio layer and the other at the data link layer.

Types of Handoffs

Handoffs can be classified into two types: The first type of handoff occurs when the MT decides the possibility of handoff and the EMAS-E chooses the next BS from the prospective choices. This is called a backward handoff. During this time, there is a smooth transition in the power levels of the source and the destination BS. The next type of handoff occurs when the MT decides on the BS to which the handoff is to take place. This is called a forward handoff. There is an abrupt break in the radio connection during a forward handoff.

Different Situations of Backward Handoff

A backward handoff can occur in one of the following three situations:

1. **Intra AP:** In this case, the source and destination radio BSs belong to the same AP. The EMAS-E merely finds out the resource availability at the new BS and forwards the response to the MT. The issues involved in an Intra-AP handoff decision are similar to those in cellular networks.

2. **Inter AP/Intra EMAS-E:** In this case, the BSs belong to different APs connected to the same EMAS-E. The EMAS-E inquires about the availability of resources at the destination AP and forwards the response to the MT.

3. **Inter EMAS-E:** In this case, the BSs belong to different EMAS-Es. This is the most complicated of the handoffs. The MT asks the destination EMAS-E for the availability of resources. The destination EMAS-E in turn queries the corresponding AP and the response is sent back. The source EMAS-E now requests that the destination EMAS-E reserve resources and the handoff is performed.

Once the handoff is performed, the paths need to be reconfigured from the COS.

Different Situations of Forward Handoff

Similar to the backward handoff, the forward handoff takes place in the following three ways:

1. **Intra AP:** When a radio disruption occurs, the MT disassociates itself from the BS. The MT then tries to associate with another BS under the same AP. This is conveyed to the EMAS-E using a series of messages from MT and AP.

2. **Inter AP/Intra EMAS-E:** This case is similar to the Intra AP handoff except for the fact that source and destination APs are different.

3. **Inter EMAS-E:** The MT disassociates itself from the old BS and associates itself with the new BS. This is conveyed to the EMAS-E1 (source EMAS-E or the EMAS-E from which the handoff has taken place) and EMAS-E2 (destination EMAS-E or the EMAS-E to which the handoff has taken place) by MT. EMAS-E2 tries to reserve resources for the connection requested by the MT at the corresponding AP. EMAS-E2 establishes a connection to the COS. EMAS-E1 then releases the connection to the COS.

Protocols for Rerouting After Handoff

When an MT and its connections are handed over from one BS to another, the connections need to be reestablished for the data transfer to continue. In case of the intra AP handoff and inter AP/intra EMAS-E handoff, a change in the routing tables of the AP and EMAS-E, respectively, is enough to handle rerouting. However, the inter EMAS-E handoffs involve wide area network (WAN) rerouting, and hence are more challenging. This section describes the generic methods for rerouting. There are three generic methods for rerouting [24]:

1. **Partial path rerouting scheme:** This involves the tearing down of a portion of the current path and the establishing of a new sub-path. The new sub-path is formed from the COS to the destination switch. One way of implementing this scheme is to probe each switch on the path from the source switch to the end-point, to find the switch with which the current QoS requirements are best satisfied.

2. **Path splicing:** A probe packet is sent from the destination switch toward the source, searching for a switch that can act as the COS point. Once such a switch is found, the new path between the destination and the source passes through this COS, splicing the old path.

3. **Path extension:** This is the simplest of the rerouting schemes. Here the current path is extended from the source switch to the destination switch. This method usually results in a non-optimal path. Moreover, if the MT moves back to the source switch, a loop will be formed. This will result in needless delay and wastage of bandwidth. The protocols for implementing this scheme have to take the above aspects into consideration.

The important point to note in the above discussion is the trade-off between computational complexity and the optimality of the generated route. Some specific examples of rerouting are: Yuan-Biswas rerouting scheme, Bahama rerouting for WATM LAN, virtual connection tree rerouting scheme [25], source routing mobile circuit (SRMC) rerouting scheme [26], and nearest common node rerouting (NCNR) scheme [27].

Effect of Handoff on QoS

Handoffs have a major bearing on an WATM service QoS performance. The reasons for this are as follows [24]:

1. When a handoff occurs, there is a likelihood of incompatibility between the QoS requirements of the MT and the destination switch handling it.

2. There also exists a possibility of disruption of the active connection during the handoff.

The former is a post-handoff situation, whereas the latter occurs during the handoff.

When a network accepts a connection to a non-mobile end-point, it provides consistent QoS to the traffic on the connection for the lifetime of the connection. However, this cannot be guaranteed in case the destination is an MT. The failure may occur when the BS to which an MT is migrating is heavily loaded and hence cannot satisfy the QoS requirements of the MT. In such cases, one or more of the connections may have to be torn down. There are three ways in which this problem can be handled:

1. In case of a backward handoff, the EMAS-E chooses the best possible BS so that the number of connections affected by the handoff is minimized [28].

2. The probability of a switch not being able to match the migrating MT's QoS requirements is high; therefore, a method of preventing frequent complete tear-down of connections is to use soft parameters. In this paradigm, the parameters are actually a range of values which is acceptable instead of fixed numbers. As long as the destination can provide guaranteed QoS within this range, the MT continues with its connections. But if the range is violated, then the connection must be torn down.

3. In the worst case, when the network is forced to tear down a connection, one way of improving the situation is to provide some priority to the connections so that the MT can set some preferences for the higher priority connections. Low-priority connections are sacrificed to maintain QoS parameters of the higher priority ones.

QoS violations can also result from disruption of the connection during the course of the handoff. During a handoff, the connection gets temporarily disconnected for a small amount of time, and no data can be exchanged during this interval. Hence, the objective is to reduce this time as well as the disruption caused by it. This disruption can be of two kinds: loss of cells and corruption of their temporal sequence.

In order to prevent loss of cells, they are required to be buffered in the switches on the route between the source and the destination. A major problem associated with buffering is the storage required at the EMAS-Es. One way out is to buffer only when necessary and discard the cells belonging to real-time traffic whose value expires after a small amount of time. On the other hand, throughput-sensitive traffic cannot bear cell loss and hence all cells in transit during handoff have to be buffered.

3.8.6 Location Management

In a WATM network, the MTs are mobile and move from one BS to another over a period of time. Therefore, in order to enable communication with them some methods need to be developed to keep track of their current locations. This process comes under the purview of location management (LM). The following are some of the requirements of an LM system [29]:

1. **Transparency:** The LM system should be developed in such a manner that the user should be able to communicate irrespective of mobility.

2. **Security:** The system must guard against unauthorized access to the database of MT addresses and MT locations.

3. **Unambiguous identification:** The LM system should be capable of uniquely identifying MTs and their locations.

4. **Scalability:** The system must be scalable to various sizes of networks. For doing this efficiently, it should harness the advantages of the hierarchical nature of networks.

Location management needs to address three broad issues:

1. **Addressing:** This involves how the various entities such as MTs, switches, APs, and BSs are addressed. It specifies the location of the terminal in the network so that, given the address, a route between communicating terminals can be established.

2. **Location updating:** This involves how the LM system is notified about the change in an MT's location and how the LM system maintains this information.

3. **Location resolution:** This involves how the information maintained by the LM system is used to locate a specific MT.

Location Update

There are several entities that play a role in location update:

1. **Location server (LS):** The LS is responsible for maintaining the current location information for the MTs. Each LS is associated with an EMAS and maintains address for MTs.

2. **Authentication server (AuS):** The AuS is responsible for maintaining authentication information for each MT. Each AuS is associated with an EMAS. The AuS stores a table of permanent unique id of the MTs and the corresponding authentication key which the MT needs to supply during communication.

3. **End-user mobility-supporting ATM switch (EMAS):** An EMAS maintains the address of the LS and AuS associated with it. If it is on the edge of the network, then it also maintains ids of MTs currently associated with it.

Location Resolution

Location resolution deals with the methodology of obtaining the current location of an MT so the user can communicate with it. It is important to note that, if the current location maintained by the LS is the full address of the MT, that is, up to its current BS, then every time the MT migrates to a new BS the LS needs to be updated. This is not efficient, because messaging causes substantial bandwidth overhead. A better method would be to store the address of a switch under whose domain the MT is currently located. Hence as long as the MT moves within the domain of the switch, the LS need not be updated.

3.9 IEEE 802.16 STANDARD

Metropolitan area networks (MANs) are networks that span several kilometers and usually cover large parts of cities. These networks are much larger in size than LANs and their functionalities differ markedly from those of LANs. MANs often connect large buildings, each of which may contain several computers. Using fiber, coaxial cables, or twisted pair for interconnection is prohibitively expensive in such scenarios. The usage of broadband wireless access (BWA) for this purpose is an economical alternative which has been explored by researchers as well as by the industry. This led to the requirement of a standard for BWA to be used in wireless MANs (WMANs) and wireless local loops (WLLs). IEEE 802.16, which is officially called *air interface for fixed broadband wireless access systems*, was the result of the standardization efforts.

IEEE 802.16 is based on the OSI model, as shown in Figure 3.8 [30]. IEEE 802.16 specifies the air interface, including the data link layer (DLL) and the physical layer, of fixed point-to-multipoint BWA systems. The DLL is capable of supporting multiple physical layer specifications optimized for the frequency bands of the application. Base stations (BSs) are connected to public networks. A BS serves several subscriber stations (SSs), which in turn serve buildings. Thus the BS provides the SS with a last-mile (or first-mile) access to public networks. It may be noted that BSs and SSs are stationary. The challenge faced by this standard was to provide support for multiple services with different QoS requirements and priorities simultaneously.

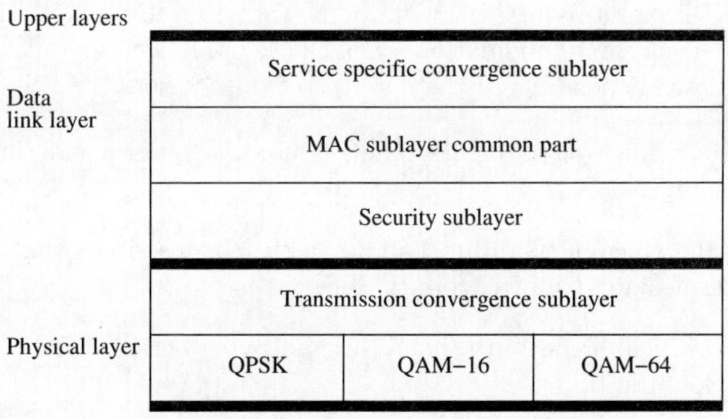

Figure 3.8. IEEE 802.16 protocol stack.

3.9.1 Differences Between IEEE 802.11 and IEEE 802.16

While IEEE 802.11 has been a successful standard for WLANs, it is not suited for use in BWA. This fact can be appreciated when the differences between IEEE 802.11 and IEEE 802.16, listed below, are studied.

- IEEE 802.11 has been designed for mobile terminals, which is irrelevant in the context of MANs. IEEE 802.16 has been designed for broadband data such as digital video and telephony.

- The number of users and bandwidth usage per user is much higher in a typical IEEE 802.16 network when compared to a typical IEEE 802.11 basic service set. This calls for usage of a larger frequency spectrum in IEEE 802.16 as against the ISM bands used by IEEE 802.11. BWA typically uses millimeter wave bands and microwave bands (above 1 GHz frequencies).

- IEEE 802.16 is completely connection-oriented and QoS guarantees are made for all transmissions. Though IEEE 802.11 provides some QoS support for real-time data (in the PCF mode), it has not been designed for QoS support for broadband usage.

3.9.2 Physical Layer

The physical layer uses traditional narrow-band radio (10-66 GHz) with conventional modulation schemes for transmission. Above the physical transmission layer, there is a convergence sublayer to hide the transmission technology from the DLL. Efforts are going on to add two new protocols, IEEE 802.16a and IEEE 802.16b, at the physical layer, which attempt to bring the IEEE 802.16 closer to IEEE 802.11. While IEEE 802.16a operates in the 2-11 GHz frequency range, IEEE 802.16b operates in the 5 GHz ISM band.

The signal strength of millimeter waves falls off sharply with distance from the BS, which results in a reduction in the signal to noise ratio (SNR). To account for this, the following three modulation schemes are used. The modulation scheme to be used is chosen depending on the distance of the SS from the BS.

1. QAM-64, which offers 6 bits/baud, is used by subscribers that are located near the BS.

2. QAM-16, which offers 4 bits/baud, is used by subscribers that are located at an intermediate distance from the BS.

3. QPSK, which offers 2 bits/baud, is used by subscribers that are located far away from the BS.

If we assume 30 MHz of spectrum, QAM-64 offers 180 Mbps, QAM-16 offers 120 Mbps, and QPSK offers 60 Mbps. It is apparent that subscribers farther away get lower data rates. It may also be noted that millimeter waves travel in straight lines, unlike the longer microwaves. This allows BSs to have multiple antennas, which point at different sectors of the surrounding terrain. The high error rates associated with millimeter waves have called for the usage of Hamming codes to do forward error correction in the physical layer. This is in contrast to most other networks where checksums detect errors and request retransmission of frames that are in error. The physical layer can pack multiple MAC frames in a single physical transmission to gain improved spectral efficiency, because of the reduction in the number of preambles and physical layer headers.

While voice traffic is generally symmetric, other applications such as Internet access have more downstream traffic than upstream traffic. To accommodate them, IEEE 802.16 provides a flexible way to accommodate bandwidth by using frequency division duplexing (FDD) and time division duplexing (TDD). The bandwidth devoted to each direction can be changed dynamically to match the traffic in the corresponding direction.

3.9.3 Data Link Layer

The DLL was designed with the wireless environment in mind, which demands an efficient usage of the spectrum. Broadband services call for very high uplink and downlink bit rates, and a range of QoS requirements. Security issues also assume importance in this scenario. It is preferred that the DLL be a protocol-independent engine, that is, the DLL should have convergence layers for all protocols including ATM, IP, and Ethernet. The DLL of IEEE 802.16 was designed to meet all these requirements. The DLL of IEEE 802.16 can be subdivided into three sublayers, whose functionalities are explained in this section. The sublayers are listed from the bottom up.

1. **Security sublayer:** This is the bottom-most sublayer, which deals with privacy and security. This is crucial for public outdoor networks where the transmission can be heard over a city. This layer manages encryption, decryption, and key management. It may be noted that only the payloads are encrypted, and the headers are kept intact. This means that the snooper can identify the participants in a transmission, but cannot read the data being transmitted.

2. **MAC sublayer common part:** This is the protocol-independent core, which deals with channel management and slot allocation to stations. Here the BS controls the system. MAC frames are integral multiples of physical layer time slots. Each frame contains the downstream (BS to SS) and upstream (SS to BS) maps, which indicate the traffic in the various time slots and other useful information. The MAC sublayer strikes a trade-off between the stability of contention-less operation and the efficiency of contention-based operation, using a TDM/TDMA mechanism. On the downlink, data to the SS is multiplexed using TDM, and on the uplink, the medium is shared by the SSs using TDMA.

   All services offered by IEEE 802.16 are connection-oriented, and each connection (uplink) is given one of the following four classes of service:

   (a) **Constant bit rate service:** This is intended for transmitting uncompressed voice, where a predetermined amount of data is generated at fixed time intervals. Certain time slots are dedicated to each connection of this type and they are available automatically without explicit request.

(b) **Real-time variable bit rate service:** This is intended for compressed multimedia and soft real-time[5] applications, where the amount of bandwidth required may vary with time. To accommodate such variances, the BS polls the SS periodically to query the bandwidth needed for the following period.

(c) **Non-real-time variable bit rate service:** This is intended for large file transfers and other such heavy transmissions that are not real-time. To accommodate them, the BS polls the SS often, but not periodically.

(d) **Best effort service:** All other applications contend for best effort service. Polling is absent and SSs contend by sending requests for bandwidth in time slots marked in the upstream map as available for contention. Collisions are reduced by using the binary exponential back-off algorithm.

3. **Service specific convergence sublayer:** This is the topmost sublayer in DLL and its function is to interface to the network layer, which is similar to the logical link sublayer in other 802 protocols. IEEE 802.16 has been designed to integrate seamlessly with both connection-less protocols such as PPP, IP, and Ethernet, and connection-oriented protocols such as ATM. While mapping ATM connections to IEEE 802.16 is quite straightforward, mapping packets to IEEE 802.16 is done in a judicious manner by this sublayer.

   A request/grant scheme is used for handling bandwidth allocation. Bandwidth requests are always per connection. Bandwidth grants may be per connection (GPC) or per SS (GPSS). Bandwidth GPSS is suitable if there are many connections per SS and this offloads the responsibilities of the BS. SS redistributes bandwidth among its connections maintaining QoS and service level agreements. While this method allows sophisticated QoS guarantees and a low overhead, it needs complex SSs. Bandwidth GPC is suitable if there are only a few users per SS. The BS grants bandwidth to each connection. This incurs a higher overhead, but allows a simpler SS.

3.10 HIPERACCESS

The HIPERACCESS standard of the ETSI BRAN (discussed in the previous chapter) pertains to broadband radio access systems [4] and is the European counterpart to the IEEE 802.16 standard. It uses fixed bidirectional radio connections to convey broadband services between users' premises and a broadband core network. HIPERACCESS systems are the means by which residential customers and small- to medium-sized enterprises can gain access to broadband communications delivered to their premises by radio. They provide support for a wide range of voice and data services and facilities. They use radio to connect the premises to other users and networks, and offer "bandwidth on demand" to deliver the appropriate

[5]This refers to the category of real-time traffic where the missing of deadlines results in non-catastrophic events such as degradation of the quality of communication.

data rate needed for various services. These systems intend to compete with other broadband wired access systems such as xDSL and cable modems.

HIPERACCESS network deployments can potentially cover large areas. A milliwave spectrum is employed in order to limit the transmission range to a few kilometers, owing to the large capacity requirements of the network. Hence, a typical network consists of a number of cells each operating in a point-to-multipoint (PMP) manner; each cell consists of an access point (AP) equipment device located approximately at the cell center and a number of access termination (AT) devices which are spread across the cell. The cell is divided into four sectors to increase the spectral efficiency by reusing the available radio frequency (RF) channels in a systematic manner within the deployment region. The protocol stack of the HIPERACCESS standard consists of the physical layer, the convergence layer, and the data link control (DLC) layer.

3.10.1 Physical Layer

The physical layer involves adaptive coding of the data obtained from the DLC layer, transmission of data, and support of the different duplex schemes, namely, frequency division duplexing (FDD), half-FDD (H-FDD), and time division duplexing (TDD). The AP equipment handles more than one RF channel and more than one user (AT), hence its architecture is different from that of the AT.

Modulation

Modulation techniques employed are based on QAM, much along the lines of IEEE 802.16 (with 2M points constellation, where M is the number of bits transmitted per modulated symbols). For the DL, QPSK (M = 2) and 16-QAM (M = 4) are mandatory and 64-QAM (M = 6) is optional. For the UL, QPSK is mandatory and 16-QAM is optional.

PHY-Modes

A PHYsical (PHY) mode includes a modulation and a coding scheme (FEC). Several sets of PHY-modes are specified for the DL. The reason for specifying different sets of PHY-modes is to offer a higher flexibility for the HA-standard deployment, where the adequate choice of a given set of PHY-modes will be determined by the parameters of the deployment scenario such as coverage, interference, and rain zone. The Reed Solomon coding scheme is generally employed for coding the data stream.

Duplex Schemes

As the communication channel between the AP and ATs is bidirectional, DL and UL paths must be established utilizing the spectrum resource available to the operator in an efficient manner. Two duplex schemes are available: one is frequency-domain-based and one is time-domain-based. FDD partitions the available spectrum into a DL block and an UL block. In HIPERACCESS, both DL and UL channels are equal in size, 28 MHz wide. In the H-FDD case, the AT radio equipment is limited to a half-duplex operation to reduce the cost. The AP recognizes in this case the fact that switching from transmission operation to reception operation (and vice versa) at the AT is not immediate. It is emphasized that the half-duplex operation is an AT feature only. The AP has a different impact on the deployment cost and on system capacity if a half-duplex operation is employed. In contrast to FDD, TDD uses the same RF channel for DL and UL communications. The DL and UL transmissions are established by time-sharing the radio channel where DL and UL transmission events never overlap. In HIPERACCESS, the channel is 28 MHz wide as in the FDD case. The AP establishes a frame-based transmission and allocates a portion of its frame for DL purposes and the remainder of the frame for UL purposes.

3.10.2 Convergence Layer

The task of the convergence layer is to adapt the service requirements of the applications of the higher layers to the services offered by the DLC layer. There are two types of convergence layers, namely, the cell-based convergence layer and the packet-based convergence layer. The classification is similar to the one discussed in the HIPERLAN/2 standard. The convergence layer is comprised of two parts, namely, the service-independent common part (CP) and the service-specific convergence sublayer (SSCS). This classification is similar to the discussion on the data link layer of IEEE 802.16.

3.10.3 DLC Layer

The basic features of the DLC layer are efficient use of the radio spectrum, high multiplex gain, and maintenance of QoS. Multiplexing schemes are employed to make a better use of the available frequency spectrum at a lower equipment cost. Unlike multiplexing, multiple access derives from the fact that every subscriber has access to every channel, instead of a fixed assignment as in most multiplex systems.

There are broadly two kinds of transmissions: uplink (UL) transmission (from AT to AP) and downlink (DL) transmission (from AP to AT). For the AP to control the access of ATs, TDMA is employed. UL transmission events of the ATs are scheduled by the AP that controls them. The DL data stream is multiplexed in the time domain (TDM). Each TDM region (part of the DL frame) is assigned a specific physical mode (consisting of coding and modulation schemes). The TDM regions are allocated in a robustness-descending order; for example, an AT with excellent link conditions, which is assigned to a spectrally efficient physical mode, starts its reception process at the beginning of the frame and continues through all TDM regions, ending the reception process with its associated TDM region. An AT with worse link conditions will be assigned to a more robust physical mode and its reception process will end before the AT of the previous example.

In addition to the DL TDM region, there could be TDMA transmissions present in a TDMA region on the DL. In this scheme, an AT may be assigned to receive DL transmissions either in a TDM region or in a TDMA region. With this option, the AT may seek DL reception opportunities immediately after it has stopped its UL transmission within the current DL frame.

The DLC layer is connection-oriented to guarantee QoS. Connections are set up over the air during the initialization of an AT, and additional new connections may be established when new services are required.

Radio Link Control (RLC) Sublayer

The RLC sublayer performs functions pertaining to radio resource control, initialization control, and connection control.

- **Radio resource control:** This comprises all mechanisms for load-leveling, power-leveling, and change of physical modes. These functions are specific to each AT.
- **Initialization control:** This includes functions for the initial access and release of a terminal to or from the network as well as the reinitialization process required in the case of link interruptions. These mechanisms are AT-specific.
- **DLC connection control:** This includes functions for the setup and release of connections and connection aggregates (groups of connections). These functions are usually connection-specific.

The ARQ protocol is implemented at the DLC level. It is based on a selective repeat approach, where only the erroneously received PDUs are retransmitted.

3.11 SUMMARY

Cellular systems offer city-wide or country-wide mobility, and may even provide compatibility across certain countries, but they are not truly global at present. Satellite-based systems offer mobility across the world. A WLL system aims at support for limited mobility, at pedestrian speeds only, with the main purpose being a cost-effective alternative to other wired local loop technologies. Therefore, it does not require elaborate diversity mechanisms, whereas frequency and spatial diversity are essential to cellular networks. The main purpose of broadband wireless networks is to offer huge bandwidth for applications such as multimedia and video-on-demand.

An increasing demand for mobile communications has led to efforts for capacity and QoS improvement. Superior algorithms are used for dynamic channel allocation, in cellular, WLL, and satellite networks. Location management techniques have been streamlined to involve minimal and quick database access. Strategies have been devised to minimize the handoff delay and maximize the probability of a successful handoff. The main standards and implementations have been discussed for the different kinds of wireless communication systems. It has been observed that

Table 3.4. A brief comparison among IEEE 802.11b WLANs, IEEE 802.16 WMANs, and GSM WWANs

Feature	**IEEE 802.11b WLANs**	**IEEE 802.16 WMANs**	**GSM WWANs**
Range	Few hundred meters	Several Km	Few tens of Km
Frequency	2.4 GHz ISM band	10-66 GHz	900 or 1,800 MHz
Physical Layer	CCK, BPSK, QPSK	QAM-64, QAM-16, QPSK	GMSK
Maximum Data Rates	11 Mbps	60-180 Mbps	9.6 Kbps/user
Medium Access	CSMA/CA	TDM/TDMA	FDD/TDMA
QoS Support	DCF - No PCF - Yes	Yes	Yes
Connectivity	DCF - Connectionless PCF - Connection Oriented	Connection Oriented	Connection Oriented
Typical Applications	Web browsing, e-mail	Multimedia, digital TV broadcasting	Voice

there are problems in reaching a global consensus on common standards for cellular networks and WLL, due to large investments already made in these networks employing different standards in different countries.

The mobile communication systems of the future aim at low cost, universal coverage, and better QoS in terms of higher bandwidth and lower delay. The ultimate goal is to provide seamless high bandwidth communication networks through a single globally compatible handset. This chapter also described the various issues in the design of a WATM network. As wireless technology and gadgets become increasingly popular, users will expect better QoS and reliability. WATM is one of the better ways of satisfying these needs. Special mechanisms need to be built to handle handoffs and to reroute the paths after the handoff and for location management. This chapter also described the IEEE 802.16 and HIPERACCESS standards, which show the state of the art in broadband wireless access standards. Table 3.4 compares the technical features of IEEE 802.11b WLANs, IEEE 802.16 WMANs, and GSM WWANs.

3.12 PROBLEMS

1. An alternative solution to the cellular system is the walkie-talkie system which provides a direct link between the mobile terminals. Compare the two systems.

2. How do you separate the different layers (macro, micro, and pico) of a cellular network in order to avoid co-channel interference across layers?

3. How does frequency reuse enhance cellular network capacity? Consider an area of 1,000 sq. Km to be covered by a cellular network. If each user requires 25 KHz for communication, and the total available spectrum is 50 MHz, how many users can be supported without frequency reuse? If cells of area 50 sq. Km are used, how many users can be supported with cluster sizes of 3, 4, and 7? Besides the number of users, what other major factor influences the decision on cluster size?

4. A particular cellular system has the following characteristics: cluster size = 7, uniform cell size (circular cells), user density = 100 users/sq. Km, allocated frequency spectrum = 900-949 MHz, bit rate required per user = 10 Kbps uplink and 10 Kbps downlink, and modulation code rate = 1 bps/Hz. Calculate the average cell radius for the above system if FDMA/FDD is used.

5. Using the same data as in the previous problem, if TDMA/TDD is adopted, explain how the wireless medium is shared.

6. Due to practical limitations, it is impossible to use TDMA over the whole 3.5 MHz spectrum calculated in Problem 4. Hence the channel is divided into 35 subchannels and TDMA is employed within each channel. Answer the following questions assuming the data and results of Problem 4.

   (a) How many time slots are needed in a TDMA frame to support the required number of users?

   (b) To have negligible delays, the frame is defined as 10 ms. How long is each time slot?

   (c) What is the data rate for each user? How many bits are transmitted in a time slot?

   (d) If one time slot of each frame is used for control and synchronization, and the same cell radius is maintained, how many users will the whole system support?

   (e) How long will it take for a signal from an MT farthest away from the BS to reach the BS?

   (f) The TDMA slots must be synchronized in time at the BS, but different MTs are at different distances and hence will have different delays to the BS. If all the nodes have a synchronized clock and synchronize transmission of slot at time $t = t0$, what guard time will we have to leave in each frame so that data will not overlap in time?

   (g) Suppose the clock synchronization error can be ± 10 ns. What guard time is needed to ensure uncorrupted data? What is the maximum data rate?

(h) Suppose the previous guard time is not acceptable for our system and that the clocks are not synchronized at all. The MTs have a mechanism to advance or delay transmission of their data, and the delay will be communicated to the MTs using the control channel. What can the MTs use as a reference to synchronize their data? How can a BS calculate the delay for each MT?

7. Briefly explain what happens to the cell size if each user needs to double the data rate.

8. What are the power-conserving strategies used in cellular systems?

9. A vehicle moves on a highway at an average speed of 60 Km/h. In the traffic of the city, the average speed drops down to 30 Km/h. The macro-cellular radius is 35 Km, and the micro-cellular radius is 3 Km. Assume the macro-cellular layer is used on the highways and the micro-cellular in the city.

   (a) How many handoffs are expected over a journey of six hours on a highway?

   (b) How many handoffs are there in a one-hour drive through the city?

   (c) What would happen if there was absolutely no traffic and the vehicle could move at 70 Km/h in the city?

   (d) What does this show about deciding which layer should handle a call?

10. The Internet is all set to take over the world as a very important form of communication. How does this affect the local loop?

11. Why are the ETSI recommended bit error rates lower for data than voice?

12. Consider a WLL system using FDMA and 64-QAM modulation. Each channel is 200 KHz and the total bandwidth is 10 MHz, used symmetrically in both directions. Assume that the BS employs 120-degree beam-width antennas. (Given spectral efficiency of 64-QAM is 5 bps/Hz.) Find:

    (a) The number of subscribers supported per cell.

    (b) The data capacity per channel.

    (c) The data capacity per cell.

    (d) The total data capacity, assuming 40 cells.

13. Frequency allocation is done automatically (dynamically) by DECT, PACS, and PHS. The alternative is manual assignment. How is manual (static/fixed) assignment of frequencies done?

14. Compare the use of satellite systems for last-mile connectivity to rural areas, with cellular and WLL systems.

15. GSM requires 48 bits for synchronization and 24 bits for guard time. Compute:

    (a) The size of a WATM packet.

    (b) The size of a WATM request packet if the request contains only MT ID of 1 byte.

    (c) Number of mini-slots per slot.

16. Suppose the frame in the case of TDD protocols is of fixed period, T, and the RTT for an MT and the BS is t. Calculate the time of waiting for an MT before it gets the ACK for its request for TDD and FDD schemes. Assume the number of packets that BS sends is only 70% of what MTs send to BS. Compute the percentage of used slots for FDD and an adaptive TDD.

17. In location management, it was said that the home base EMAS-E redirects the connection to the current location of the MT. Using a scheme similar to partial path rerouting, describe a way of doing this.

18. In case of the Inter EMAS-E forward handoff, unlike the backward handoff case, the destination EMAS-E has to notify the old EMAS-E about the handoff request from MT. Why is this so?

19. Why is the route generated by the path splicing scheme at least as optimal (if not more) as that generated by the partial path rerouting scheme?

BIBLIOGRAPHY

[1] B. H. Walke, *Mobile Radio Networks: Networking, Protocols, and Performance*, John Wiley & Sons, January 2002.

[2] J. Sarnecki, C. Vinodrai, A. Javed, P. O'Kelly, and K. Dick, "Microcell Design Principles," *IEEE Communications Magazine*, vol. 31, no. 4, pp. 76-82, April 1993.

[3] Y. B. Lin and I. Chlamtac, *Wireless and Mobile Network Architecture*, John Wiley & Sons, October 2000.

[4] P. Nicopolitidis, M. S. Obaidat, G. I. Papadimitriou, and A. S. Pomportsis, *Wireless Networks*, John Wiley & Sons, November 2002.

[5] D. Weissman, A. H. Levesque, and R. A. Dean, "Interoperable Wireless Data," *IEEE Communications Magazine*, vol. 31, no. 2, pp. 68-77, February 1993.

[6] M. Laitinen and J. Rantala, "Integration of Intelligent Network Services into Future GSM Networks," *IEEE Communications Magazine*, vol. 33, no. 6, pp. 76-86, June 1995.

[7] B. Jabbari, "Intelligent Network Concepts in Mobile Communications," *IEEE Communications Magazine*, vol. 30, no. 2, pp. 64-69, February 1992.

[8] J. Homa and S. Harris, "Intelligent Network Requirements for Personal Communications Services," *IEEE Communications Magazine*, vol. 30, no. 2, pp. 70-76, February 1992.

[9] V. K. Garg, *Wireless Network Evolution: 2G-3G*, Prentice Hall PTR, New Jersey, August 2001.

[10] A. Dornan, *Essential Guide to Wireless Communication*, Prentice Hall PTR, New Jersey, December 2000.

[11] A. J. Viterbi and R. Padovani, "Implications of Mobile Cellular CDMA," *IEEE Communications Magazine*, vol. 30, no. 12, pp. 38-41, December 1992.

[12] G. Tsoulos, M. Beach, and J. McGeehan, "Wireless Personal Communications for the 21st Century: European Technological Advances in Adaptive Antennas," *IEEE Communications Magazine*, vol. 35, no. 9, pp. 102-109, September 1997.

[13] J. D. Gibson, Ed., *The Mobile Communications Handbook*, IEEE Press/CRC Press, June 1996.

[14] N. Haardt and W. Mohr, "Complete Solution for 3G Wireless Communication: Two Modes on Air, One Winning Strategy," *IEEE Personal Communications Magazine*, vol. 7, no. 6, pp. 18-24, December 2000.

[15] P. Stavroulakis, Ed., *Wireless Local Loops: Theory and Applications*, John Wiley & Sons, 2001.

[16] U. D. Black, *Second Generation Mobile and Wireless Systems*, Prentice Hall, New Jersey, October 1998.

[17] GPS World-Home Page, http://www.gpsworld.com

[18] Z. Liu, M. J. Karol, and K. Y. Eng, "Distributed Queueing Request Update Multiple Access (DQRUMA) for Wireless Packet (ATM) Networks," *Proceedings of IEEE ICC 1995*, vol. 2, pp. 1224-1231, June 1995.

[19] J. G. Kim and I. Widjaja, "PRMA/DA: A New Media Access Control Protocol for Wireless ATM," *Proceedings of IEEE ICC 1996*, vol. 1, pp. 240-244, June 1996.

[20] D. Petras and A. Kramling, "MAC Protocols with Polling and Fast Collision Resolution Scheme for ATM Air Interface," *Proceedings of IEEE ATM Workshop 1996*, pp. 10-17, August 1996.

[21] F. Bauchot, "MASCARA, A Wireless ATM MAC Protocol," *Proceedings of IEEE ATM Workshop 1996*, pp. 647-651, August 1996.

[22] F. D. Priscoli, "Multiple Access Control for the MEDIAN System," *Proceedings of ACTS Mobile Summit 1996*, pp. 1-8, November 1996.

[23] D. Raychaudhuri *et al.*, "WATMnet: A Prototype for Wireless ATM System for Multimedia Applications," *IEEE Journal on Selected Areas in Communications*, vol. 15, no. 1, pp. 83-95, January 1997.

[24] A. Acharya, J. Li, B. Rajagopalan, and D. Raychaudhuri, "Mobility Management in Wireless ATM Networks," *IEEE Communications Magazine*, vol. 35, no. 11, pp. 100-109, November 1997.

[25] A. S. Acampora and M. Naghshineh, "An Architecture and Methodology for Mobile-Executed Handoff in Cellular ATM Networks," *IEEE Journal on Selected Areas in Communications*, vol. 12, no. 8, pp. 1365-1375, October 1994.

[26] O. T. W. Yu and V. C. M. Leung, "B-ISDN Architectures and Protocols to Support Wireless Personal Communications Internetworking," *Proceedings IEEE PIMRC 1995*, vol. 2, pp. 768-772, September 1995.

[27] B. A. Akyol and D. C. Cox, "Signaling Alternatives in a Wireless ATM Network," *IEEE Journal on Selected Areas in Communications*, vol. 15, no. 1, pp. 35-49, January 1997.

[28] J. Schiller, *Mobile Communications*, Addison-Wesley, January 2000.

[29] R. R. Bhat, "Wireless ATM Requirements Specification," *ATM Forum Draft:98-0196*, 1998.

[30] A. S. Tanenbaum, *Computer Networks*, Prentice Hall PTR, New Jersey, August 2002.

Chapter 4

WIRELESS INTERNET

4.1 INTRODUCTION

The advent of the Internet has caused a revolutionary change in the use of computers and the search for information. The Internet has affected the traditional way of information exchange and now almost every city, every town, and every street has access to the Internet. Some basic concepts about the Internet and some fundamental issues that are encountered when a transition is made from the wired domain to the wireless domain and the MobileIP framework are discussed. Some of the problems faced during a transition from the wired domain to the wireless domain arise due to the fact that the protocols that work very well in the former may perform poorly in the latter. TCP is a perfect example of this. The key issues involved in TCP for wireless networks and an analysis of the current set of proposals to enhance the performance of TCP in the wireless domain are also presented. The classical wired networks have given rise to a number of application protocols such as TELNET, FTP, and SMTP. The wireless application protocol (WAP) architecture aims at bridging the gap at the application level, between the wireless users and the services offered to them.

4.2 WHAT IS WIRELESS INTERNET?

Wireless Internet refers to the extension of the services offered by the Internet to mobile users, enabling them to access information and data irrespective of their location. The inherent problems associated with wireless domain, mobility of nodes, and the design of existing protocols used in the Internet, require several solutions for making the wireless Internet a reality. An illustration of wireless Internet with its layered protocol stack at wired and wireless parts is shown in Figure 4.1. The major issues that are to be considered for wireless Internet are the following.

- Address mobility
- Inefficiency of transport layer protocols
- Inefficiency of application layer protocols

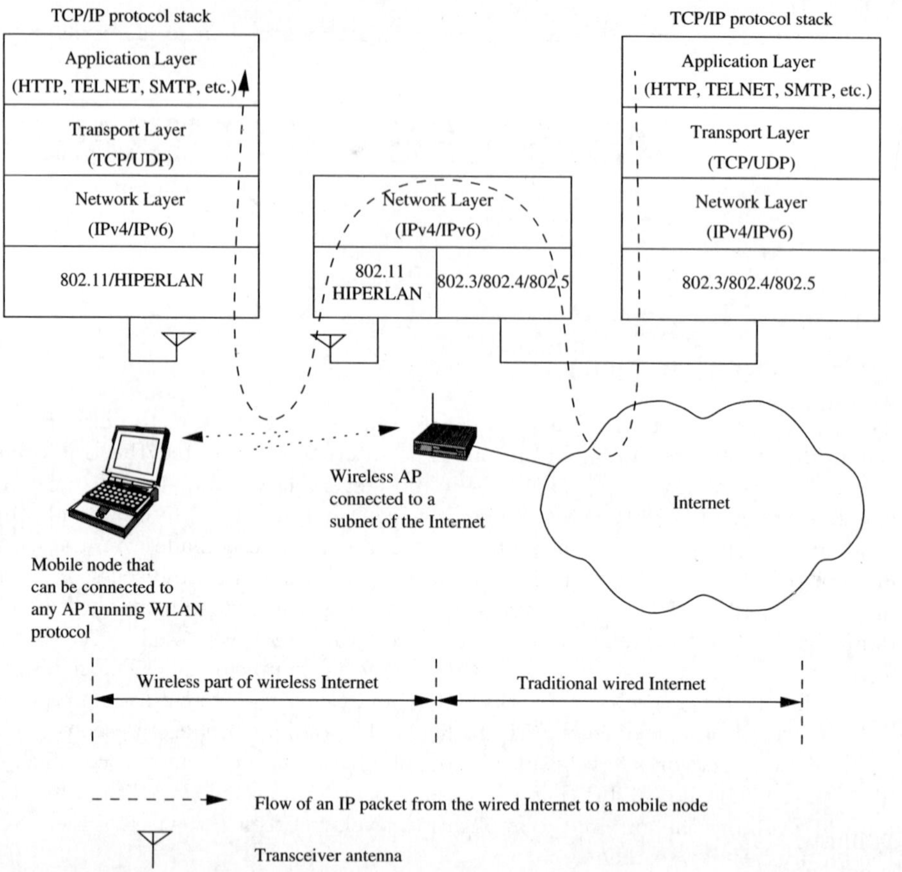

Figure 4.1. An illustration of wireless Internet.

4.2.1 Address Mobility

The network layer protocol used in the Internet is Internet protocol (IP) which was designed for wired networks with fixed nodes. IP employs a hierarchical addressing with a globally unique 32-bit address[1] which has two parts, network identifier and host identifier, as shown in Figure 4.2 (a). The network identifier refers to the subnet address to which the host is connected. This addressing scheme was used to reduce the routing table size in the core routers of the Internet, which uses only the network part of the IP address for making routing decisions. This addressing scheme may not work directly in the wireless extension of the Internet, as the mobile hosts may move from one subnet to another, but the packets addressed to the mobile host may be delivered to the old subnet to which the node was originally attached,

[1]The recently introduced IP Version 6 has a 128-bit address.

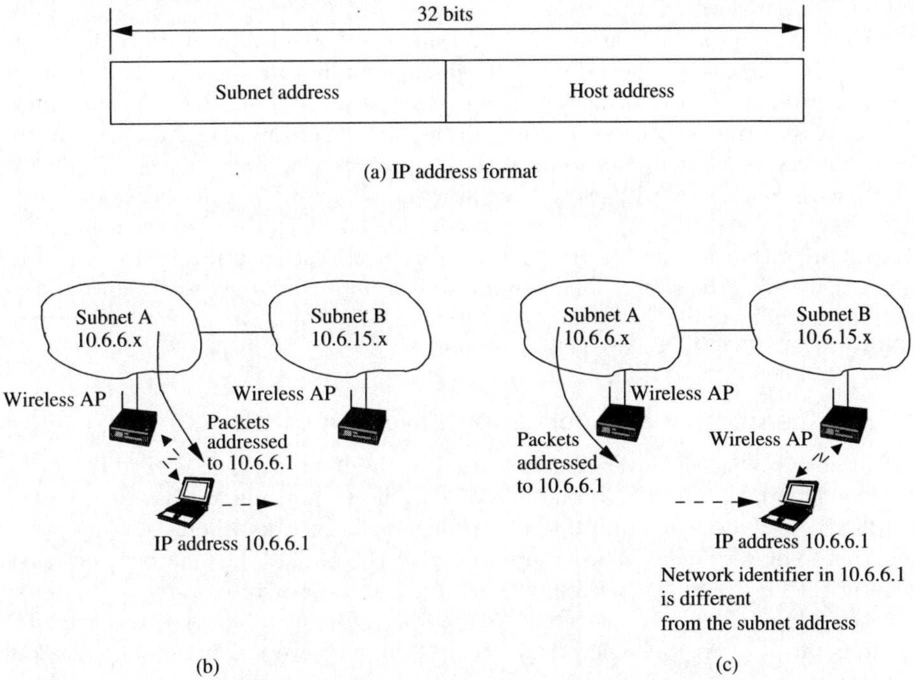

Figure 4.2. The address mobility problem.

as illustrated in Figures 4.2 (b) and 4.2 (c). Hence the traditional IP addressing is not supportive of address mobility which is essential in wireless Internet. Figure 4.2 shows the mobility of a node (with IP address 10.6.6.1) attached to subnet A (subnet address 10.6.6.x) moving over to another subnet B with address 10.6.15.x. In this case, the packets addressed to the node will be routed to the subnet A instead of the subnet B, as the network part in the mobile node's address is 10.6.6.x (see Figure 4.2 (c)). MobileIP[2] is a solution that uses an address redirection mechanism for this address mobility issue in wireless Internet.

4.2.2 Inefficiency of Transport Layer Protocols

The transport layer is very important in the Internet as it ensures setting up and maintaining end-to-end connections, reliable end-to-end delivery of data packets, flow control, and congestion control. TCP is the predominant transport layer protocol for wired networks, even though UDP, a connectionless unreliable transport layer protocol, is used by certain applications. Wireless Internet requires efficient operation of the transport layer protocols as the wireless medium is inherently

[2]Throughout this chapter, Mobile IP refers to the mobility aspect of IP address and MobileIP refers to one particular solution for Mobile IP.

unreliable due to its time-varying and environment-dependent characteristics. Traditional TCP invokes a congestion control algorithm in order to handle congestion in the networks. If a data packet or an ACK packet is lost, then TCP assumes that the loss is due to congestion and reduces the size of the congestion window by half. With every successive packet loss the congestion window is reduced, and hence TCP provides a degraded performance in wireless links. Even in situations where the packet loss is caused by link error or collision, the TCP invokes the congestion control algorithm leading to very low throughput. The identification of the real cause that led to the packet loss is important in improving the performance of the TCP over wireless links. Some of the solutions for the transport layer issues include indirect-TCP (ITCP), snoop TCP, and mobile TCP.

4.2.3 Inefficiency of Application Layer Protocols

Traditional application layer protocols used in the Internet such as HTTP,[3] TELNET, simple mail transfer protocol (SMTP), and several markup languages such as HTML were designed and optimized for wired networks. Many of these protocols are not very efficient when used with wireless links. The major issues that prevent HTTP from being used in wireless Internet are its stateless operation, high overhead due to character encoding, redundant information carried in the HTTP requests, and opening of a new TCP connection with every transaction. Wireless bandwidth is limited and much more expensive compared to wired networks. Also, the capabilities of the handheld devices are limited, making it difficult to handle computationally and bandwidth-wise expensive application protocols. Wireless application protocol (WAP) and optimizations over traditional HTTP are some of the solutions for the application layer issues.

4.3 MOBILE IP

Each computer connected to the Internet has a unique IP address, which helps not only in identifying the computer on the network but also routing the data to the computer. The problem of locating a mobile host in a mobile domain is now imminent as the IP address assigned can no longer be restricted to a region.

The first conceivable solution to the above problem would be to change the IP address when the host moves from one subnet to another. In this way, its address is consistent with the subnet it is currently in. The problems with changing the IP address as the host moves is that TCP identifies its connection with another terminal based on the IP address. Therefore, if the IP address itself changes, the TCP connection must be reestablished. Another method would be to continue to use the same IP address and add special routing entries for tracking the current location of the user. This solution is practical if the number of mobile users is small. The quick-fix solutions are inadequate, but they give valuable insight into the nature of the mobility problem and offer certain guidelines for the actual solution.

[3]Some documents mention HTTP as the session layer protocol. Since the TCP/IP stack does not have a session layer, in this, it is considered as part of the application layer.

Before providing the solution to the problem, some issues of utmost importance need to be enumerated. These are as follows:

- **Compatibility:** The existing wired Internet infrastructure is well-established today and it is economically impractical to try to alter the way it is working.
- **Scalability:** Wireless communication is the technology for the future, so the solution should be scalable to support a large number of users.
- **Transparency:** The mobility provided should be transparent in the sense that the user should not feel a difference when working in a wireless domain or in a wired one.

In Figure 4.3, mobile node (MN) is a mobile terminal system (end user) or a mobile router. It is the host for which the mobility support is to be provided. At the other end of the network is the system with which MN communicates. This is referred to as the correspondent node (CN), which may be a fixed or a mobile node. In this section, CN is considered to be a fixed, wired node. The node or router to which the MN is connected, which currently enjoys all the network facilities, is known as the foreign agent (FA). The subnet to which the MN's IP address belongs is the

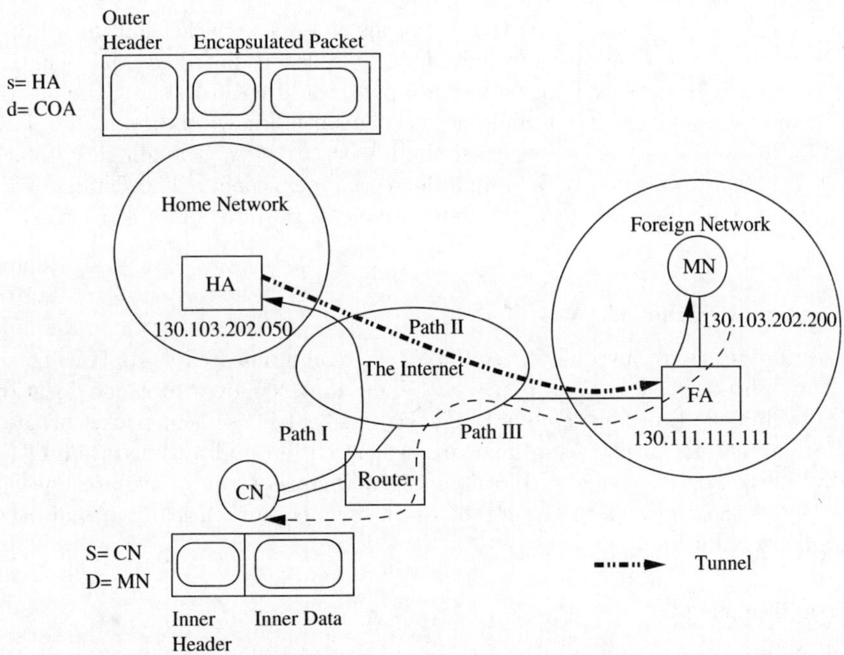

Figure 4.3. Routing in MobileIP.

home network, and the router or node under whose domain this IP address lies is the home agent (HA).

Suppose MN is currently in the subnet 130.111.*, hence as shown in the figure, 130.111.111.111 becomes the FA for MN. If CN sends a packet to MN, it reaches the HA of MN (130.103.202.050) along Path I. HA cannot find MN in the home network, but if it knows the location of MN, it can send the packet along Path II by creating a *tunnel*, as explained later.

4.3.1 MobileIP

The essence of the MobileIP scheme is the use of the old IP address but with a few additional mechanisms, to provide mobility support. MN is assigned another address, the care of address (COA). The COA can be one of the following types:

1. **Foreign agent-based COA:** The address of the FA to which the MN is connected can be used to locate the MN. The COA of the MN in this case is the address of its current FA.

2. **Colocated COA:** In this case MN acquires a topologically correct IP address. In effect, each MN now has two IP addresses assigned to it. In this case the CN sends data to the old IP address. The HA receives this packet and *tunnels* it to the MN using the new IP address.

In the case of FA-based COA, the FA decapsulates the packet and forwards it to MN, while in the case of colocated COA, it is decapsulated at MN. The HA encapsulates the data packet inside another packet addressed to the COA of MN. This is known as *encapsulation* and the mechanism is known as *tunneling*. Path II in Figure 4.3 shows the tunnel using the FA-based COA. Though the problem is solved, it has been done with a high degree of inefficiency. The details of the inefficiencies and the strategies adopted to avoid the overheads are discussed in Section 4.3.3.

Registration with the HA

This section discusses how the COA of an MN is communicated to its HA. This is done through the process of *registration*. When an MN moves to a new location, it tries to find the FA. This is done using the agent advertisement packet or agent solicitation packet. Registration involves authentication and authorization of the MN by the HA. In case of the colocated COA, there is no intermediate FA. MN simply sends the registration request to its HA, which authenticates it and sends back a registration reply.

Reverse Tunneling

It appears that there should not be any problem for the MN in sending a packet to the CN following path III. However, there are other practical constraints that play an important role here.

1. **Ingress filtering:** There are some routers which filter the packets going out of the network if the source IP address associated with them is not the subnet's IP address. This is known as ingress filtering where the MN's packet may get filtered in the foreign network if it uses its home IP address directly.

2. **Firewalls:** As a security measure, most firewalls will filter and drop packets that originate from outside the local network, but appear to have a source address of a node that belongs to the local network. Hence if MN uses its home IP address and if these packets are sent to the home network, then they will be filtered.

3. **Time to live (TTL):** The MN should be able to communicate transparently with all the CNs that it can communicate with while at home. Hence, in case of *triangular routing*, the TTL for the packets must be reduced only by one, up to the point where the packet is tunneled home.

Firewalls and ingress filtering have made a simple solution complicated. Therefore, to avoid these problems the idea of *reverse tunneling* is used, that is, MN encapsulates its packets using the source address of the encapsulated packet as its COA and destination as HA. The routing of packets from MN to CN takes place via the non-shortest path (as shown in Figure 4.3), that is, MN to HA to CN or vice versa is called *triangular routing*. This method, though not efficient, does work in practice.

4.3.2 Simultaneous Bindings

Simultaneous bindings is a feature of MobileIP that allows an MN to register more than one COA at the same time, that is, the HA allows MN to register more than one COA. MN can also deregister a specific COA. In such a situation, the HA must send multiple duplicated encapsulated data packets, one to each COA. The idea behind the use of simultaneous binding is to improve the reliability of data transmission.

4.3.3 Route Optimization

The packets sent to and from the HA are routed on non-optimal paths, hence the need for optimizations [1]. The CN is assumed to be mobility-aware, that is, it has the capability to deencapsulate the packets from the MN and send packets to the MN, bypassing the HA. The following are some of the concepts related to optimization strategies.

- **Binding cache:** The CN can keep the mapping of MN's IP address and COA in a cache. Such a cache is called a binding cache. Binding cache is used by the CN to find the COA of the MN in order to optimize the path length. Like any other cache, this may follow the update policies such as least recently used and first-in-first-out.

- **Binding request and binding update:** The CN can find the binding using a binding request message, to which the HA responds with a binding update message.

- **Binding warning:** In some cases, a handoff may occur, but CN may continue to use the old mapping. In such situations, the old FA sends a binding warning message to HA, which in turn informs the CN about the change, using a binding update message.

4.3.4 MobileIP Variations – The 4×4 Approach

As discussed in Section 4.3.1, MobileIP is a general-purpose solution to the mobility problem over IPv4. It uses encapsulation as a primary technique and thus introduces a huge overhead (approximately 20 bytes per packet). In the MobileIP scheme, the MN is dependent on the FA to provide a COA. The presence of the FA in all transactions prevents the MN from being able to perform any kind of optimization, and it is unable to forgo the MobileIP support even when it is not required. The key factors that affect any optimization scheme are the permissiveness of the network and the capabilities of the communicating nodes. In the following strategy presented, it is presumed that the MN does not depend on the FA for any support and it is able to acquire a COA from the subnet that it is present in.

Goals of Optimizations

Any optimization scheme should try to ensure guaranteed delivery, low latency, and low overhead. Deliverability is to be understood in terms of the traditional datagram network that provides only a best-effort service. The latency issue mainly deals with the route that is being followed by the packet from the source to the destination, either in terms of the hop count or the delay. The overhead in the MobileIP scheme is essentially the packet encapsulation overhead.

The 4×4 Approach

The strategy presented here provides four options for packets directed from the MN to the CN (OUT approaches) and four more options for packets directed from the CN to the MN (IN approaches). The set of options can be provided as a 4×4 [2] matrix to the hosts, which can decide on the appropriate combination depending on the situation. The IN and OUT strategies are summarized in Tables 4.1 and 4.2, respectively. **s** and **d** represent the outer source and destination in the encapsulated packet while **S** and **D** represent the inner source and destination of the packet (refer to Figure 4.3). Indirect transmission refers to the routing of packets between the CN and MN involving the HA, whereas direct transmission bypasses the HA. In Table 4.1 the four IN strategies are listed along with the respective source and destination fields, and the assumptions made and restrictions on usage of the strategies. For example, IN-IE uses the traditional MobileIP mechanism and works in all network environments irrespective of security considerations, while IN-DT is applicable for short-term communication wherein the mobility support

Table 4.1. The IN strategies in 4×4 approach

IN Strategy	s	d	S	D	Notes	Acceptable Combinations
Incoming Indirect Encapsulated (IN-IE)	IP address of HA	COA of the MN	IP address of the CN	Home IP address of the MN	1. Highest overhead 2. Guaranteed delivery 3. Uses tunneling 4. CN need not be mobility aware	OUT-IE OUT-DE OUT-DH
Incoming Direct Encapsulated (IN-DE)	IP address of CN	COA of the MN	IP address of the CN	Home IP address of the MN	1. CN is mobility aware 2. No tunneling	OUT-DE OUT-DH
Incoming Uses Home Address (IN-DH)	Not applicable	Not applicable	IP address of the CN	Home IP address of the MN	1. No encapsulation 2. Usable when there are no security constraints at intervening routers 3. MN and CN on same subnet	OUT-DH only
Incoming Direct Uses Temporary Address (IN-DT)	Not applicable	Not applicable	IP address of the CN	COA of MN	1. MN cannot receive packets addressed to its original IP address 2. Useful for short-term communication	OUT-DT only

Table 4.2. The OUT strategies in 4×4 approach

OUT Strategy	s	d	S	D	Notes	Acceptable Combinations
Outgoing Indirect Encapsulated (OUT-IE)	COA of the MN	IP address of HA	Home IP address of the MN	IP address of the CN	1.Highest overhead 2. Guaranteed delivery 3. No ingress filtering 4. CN need not be mobility aware	IN-IE only
Outgoing Direct Encapsulated (OUT-DE)	COA of the MN	IP address of CN	Home IP address of the MN	IP address of the CN	1. CN is mobility aware 2. No tunneling	IN-IE IN-DE
Outgoing Direct Home Address (OUT-DH)	Not applicable	Not applicable	Home IP address of the MN	IP address of the CN	1. No encapsulation 2. Usable when there are no security constraints at the intervening routers 3. MN and CN on the same subnet	IN-IE IN-DE IN-DH
Outgoing Direct uses Temporary Address (OUT-DT)	Not applicable	Not applicable	COA of MN	IP address of the CN	1. MN cannot receive packets addressed to its original IP address 2. Useful for short-term communication	IN-DT only

is compromised. In Table 4.2, the four OUT strategies are listed. For example, OUT-IE uses the traditional MobileIP reverse tunneling mechanism and works in all network scenarios, while OUT-DH avoids encapsulation overhead but can be used only when the MN and CN are in the same IP subnet.

Comparison and Evaluation of the Strategies

Having seen the four approaches for each of the two directions of packet transfer, different combinations of these strategies can be considered. Though there seem to be 16 combinations, some of them are inapplicable and some are redundant. There are also restrictions on when the approaches are valid; the characteristics of the situation will determine which approach to choose. The choice of a particular strategy can be made on a per session basis or on a packet-to-packet basis, as desired by the entities involved in the conversation. Tables 4.1 and 4.2 also show the acceptable combinations of the strategies.

4.3.5 Handoffs

A handoff is required when the MN is moving away from the FA it is connected to, and as a result the signals transmitted to and from the current FA become weak. If the MN can receive clearer signals from another FA, it breaks its connection with the current FA and establishes a connection with the new one. The typical phases involved in handoff are measuring the signal strength, decisions regarding where and when to hand off, and the establishment of a new connection breaking the old one.

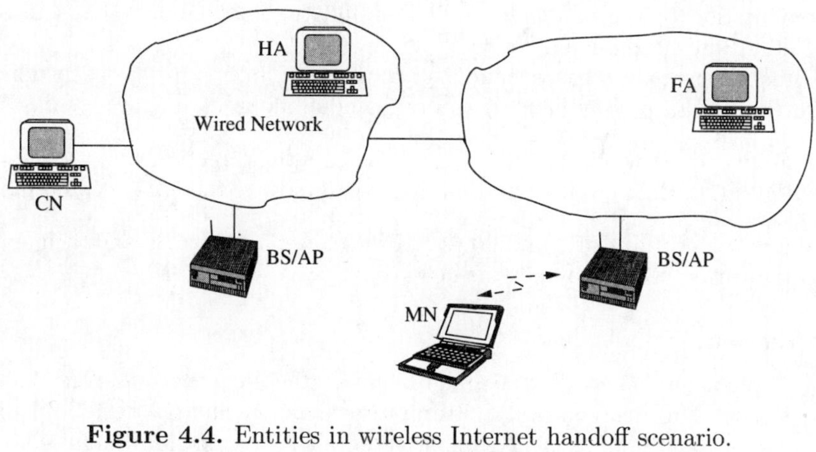

Figure 4.4. Entities in wireless Internet handoff scenario.

Classification of Handoffs

The issues in handoffs are on the same lines as those in cellular networks. Handoffs can be classified in three ways [3] based on functionalities of the entities involved, signaling procedure, and number of active connections. Function-based classification is based on the roles of the MN and FA during the handoff. Figure 4.4 shows the MN, BS, FA, and CN in the handoff scenario.

Here, handoffs can be classified into four categories as follows:

1. **Mobile initiated handoff:** In this case, the handoff is managed by the MN. The MN measures the signal strength, decides the target base station (BS), and triggers the handoff.

2. **Mobile evaluated handoff:** This is similar to the previous case except that the decision on the handoff lies within the network, perhaps with the BS.

3. **Network initiated handoff:** In this case, the network (BS) decides where the MN should be handed over. Also, only the network measures the signal strength of the uplink and the MN has very little role to play.

4. **Mobile assisted handoff:** The MN assists the network in the network initiated scenario by measuring the downlink signal strength. This is typically to avoid a *black hole* scenario. A black hole scenario occurs when the channel properties tend to be asymmetric. (Usually wireless channels are assumed to have the same properties in both uplink and downlink, but in certain circumstances the throughput on one of the directions may be significantly less than the other. This scenario is referred to as a black hole.)

The second kind of classification is based on the number of active connections, where the handoffs are classified into two types: the hard handoff (only one active connection to the new or the old FA) and the soft handoff (has two active connections during the handoff).

Signaling procedure-based handoffs are classified into two types depending on which FA (old FA or new FA) triggers the handoff along with MN.

- **Forward handoff:** In this case, MN decides the target BS and then requests the target BS to contact the current BS to initiate the handoff procedure.

- **Backward handoff:** In this case, MN decides the target BS and then requests the current BS to contact the new one.

Fast Handoffs

A typical handoff takes a few seconds to break the old connection and establish the new one. This delay may be split into three components [4]: delay in detection of a need for a handoff, layer2 handoff (a data link connection that needs to be established between the new FA and MN), and layer3 handoff or registration with HA. The first two components cannot be avoided; however, the delay due to the third can be reduced. Also, if the above operations are parallelized, the total delay will be reduced. Two techniques called pre- and post-registration handoffs are employed to perform the above operations. The difference lies in the order in which the operations are performed. In the case of the pre-registration handoff, the registration with the HA takes place before the handoff while the MN is still attached to the old FA, while in the case of the post-registration handoff, registration takes place after the MN is connected to the new FA. In this case, the MN continues to use the old FA, tunneling data via the new FA until the process of registration is completed.

4.3.6 IPv6 Advancements

The various optimizations provided over IPv4 (IP version 4) in order to avoid the inefficiencies in routing MN's data were discussed in Section 4.3.4. IPv6 (IP version 6) has a built-in support for mobility to a great extent. The features [5] are listed below:

- Route optimization is a built-in feature of IPv6.
- IPv6 has fields for specifying both new (COA) and home (IP) address. So problems that lead to reverse tunneling can be avoided.
- The problem of *ingress filtering* is also solved due to the above.
- Control packets such as those used in route optimization can be piggy-backed onto the data packets.
- *Detection of black holes:* Sometimes it might happen that the signals of one of the links (uplink or downlink) become weak while the other link has a good signal strength. Such a phenomenon is known as a *black hole* because data can go in one direction but cannot come out in the other. In such cases, a handoff may be required. IPv6 allows both MN and BS to detect the need for a handoff due to creation of black holes.
- IPv6 avoids overheads due to encapsulation because both the COA and the original IP address are included in the same packet in two different fields.

Apart from these, IPv6 allows 2^{128} addresses, thereby solving the IP address shortage problem, and includes advanced QoS features. It also supports encryption and decryption options to provide authentication and integrity.

4.3.7 IP for Wireless Domains

MobileIP is only a solution to the mobility of IP address problem, it is not a specific solution for wireless, especially cellular domains. The following discussion addresses certain protocols that are IP-based and suited for the wireless domain as well. In particular, we consider an approach which is terminal independent, that is, an approach aimed at giving a uniform service to both hosts that have the MobileIP capability as well as legacy hosts. The terminal independent mobility for IP (TIMIP) [6] strategy is based on two main protocols for the wireless networks, namely, HAWAII [7] and CellularIP [8]. Figure 4.5 gives the hierarchy of routers in the HAWAII, CellularIP, and TIMIP architectures. The access point (AP) is a router that is at the first level of the hierarchy and this is in direct communication

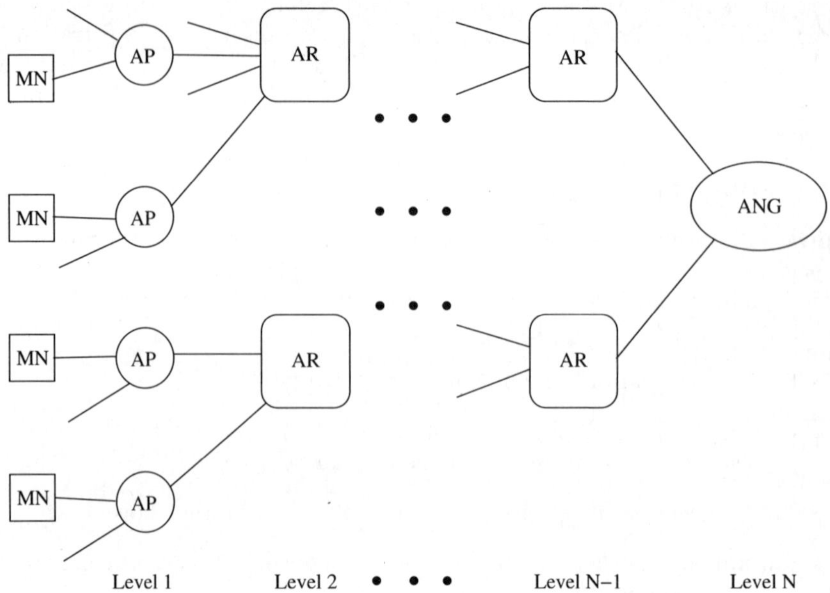

Figure 4.5. Hierarchical routers.

with the MN over the wireless interface. Access routers (AR) are interior routers in the tree. The access network gateway (ANG) is the router at the root of the tree that acts as the interface between the wireless (TIMIP) domain and the core wired IP network.

HAWAII

HAWAII stands for handoff aware wireless access Internet infrastructure. The infrastructure identifies two categories of mobility to be handled, micromobility (intra-domain) and macromobility (inter-domain), where domain refers to a part of the network under the control of a single authority, such as AR and ANG. The objective of the infrastructure is to solve the QoS and efficiency issues that are not addressed by MobileIP.

CellularIP

CellularIP offers an alternative to the handoff detection problem by using the MAC layer information based on the received signal strengths to detect handoffs, instead of using the network layer information. The routing nodes maintain both a paging cache and a routing cache; the routing cache is a mapping between an MN's IP address and its current location in the CellularIP domain. The paging cache is preferred for nodes that receive or send packets relatively infrequently, and it is maintained by paging update packets sent by the MN whenever it crosses between two APs. The routing cache will be updated whenever the MN has a packet to send. The MN will send the packet to the closest AP and this will update all routing caches all the way up to the ANG. It is to be noted that during a handoff or just after the handoff, packets meant for the MN will be routed to both the old as well as the current AP in charge of the MN for a time interval equal to the routing cache timeout.

TIMIP

In the terminal independent mobility for IP (TIMIP) approach, the emphasis is on providing a uniform service to both MobileIP-capable MNs as well as the legacy terminals. The MobileIP capability of the legacy terminals will be provided by the ANG. The ANG will keep track of the information regarding each of the MNs in its domain such as the MN's MAC and IP addresses, the MobileIP capabilities, and the authentication parameters.

Whenever an MN arrives in the TIMIP domain, a routing path has to be created in the domain so that all packets intended for this host can be efficiently routed. This will cause a trigger of updates to ensure route reconfiguration in the entire hierarchy. The ARs not involved in the route will be unaware of the new path to the MN. As a result, the default strategy for any packet in the TIMIP domain, that is, for any IP address that is unknown at a particular AP or AR, will be to route it to the ANG.

- **Micromobility:** Whenever the MN moves within the same TIMIP domain, it is referred to as micromobility. The route updates and the corresponding acknowledgments will propagate up the hierarchy until the crossover AR is reached. The old path needs to be deleted in all the routing tables of the nodes. Now the crossover AR will send a route update packet addressed to the MN, and this packet will propagate down the tree until the old AP in charge of the MN is reached.

- **Macromobility:** Similar to CellularIP and HAWAII, TIMIP relies purely on MobileIP to support macromobility. The ANG acts as the MobileIP proxy on behalf of the MN that does not have MobileIP capability, and does all the MobileIP signaling that the MN would have normally done. For the normal MobileIP capable MNs, however, the ANG performs the role of a FA.

The TIMIP approach also provides for seamless mobility through the context transfer framework. The context transfer essentially ensures that the data loss during handoff is minimized and this is transparent to the MN and the CN.

4.3.8 Security in MobileIP

The wireless domain is inherently insecure. Any data that needs to be transmitted has to be broadcast and anyone who can hear this can read it irrespective of the destination address.

Security Problems

The common security problems that may arise in wireless networks are as follows:

- **Registration request by a malicious node:** This is a problem because a malicious node can pose as a legitimate MN and use the MN's IP address for registration, thereby enjoying all the facilities meant for the MN.
- **Replay attacks:** Many times the above problem is solved by making the registration process encrypted. Though this appears to avoid the first problem, the malicious node may copy the MN's registration packet, which is encrypted when the MN tries to register with the FA. Though this packet cannot be decoded by this malicious node, it can certainly use this packet for registering itself as the MN at a later point of time, and hence enjoy all the facilities at the cost of the MN.
- **Tunnel hijacking:** In this case, the malicious node uses the tunnel built by the MN to break through the firewalls.
- **FA can itself be a malicious node.**

The MN and HA share the same security association and use the *message digest 5* (MD5) with 128-bit encryption. To circumvent the problem of replay attacks the MN and HA use a shared random number[4] (called Nonce) and this random number is sent along with the encrypted registration request. On registration, the HA verifies the random number and issues a new random number to be used for the next registration. Hence, even if the packet is copied by the malicious node, it becomes useless for a replay attack, as at the time of the next registration the random number would have changed anyway.

4.3.9 MRSVP – Resource Reservation

The following section describes a reservation protocol used to provide real-time services to mobile users. A major problem is that mobility affects the QoS adversely. Hence there is a need for advance reservations to be made on behalf of a mobile host at future locations that it is likely to visit. We notice that the current RSVP[5] structure is far from adequate and examine the proposed scheme.

Overview

The usual QoS parameters are delay, loss, throughput, and delay jitter. Whenever an MN moves across from one agent to another, there is obviously a change in the data flow path due to the handoff. The delay is likely to change due to the change in the data flow path and also due to the fact that the new location may vary from the

[4]Note: Time-stamps may also be used, but random number is a better option. Time-stamps may lead to synchronization problems.

[5]Resource reservation protocol (RSVP) is a resource reservation setup protocol designed for multicast, multimedia data streams or flows (RFC 2205). A flow is specified by attributes such as source-destination pair, average data rate, latency, and QoS (RFC 1363).

old location with respect to congestion characteristics. Again, if the new location is highly congested, the available bandwidth is less, hence the throughput guarantees that were provided earlier may be violated. In addition, under extreme cases there may be temporary disconnections immediately following a handoff, which causes significant data loss during the transit.

Requirements of a Mobility-Aware RSVP

A fundamental requirement is that an MN must be able to make advance reservations along data flow paths to and from locations that it is likely to visit in the lifetime of a particular connection or session. Such a protocol has to have information that we refer to as the MSPEC, which is the set of locations from which the MN requires reservations. The definition of the MSPEC may be either statically done or there may be additional options to update it dynamically while the flow is active. A hypothetical MRSVP [9] has two types of reservations: ACTIVE and PASSIVE. An ACTIVE reservation is a normal RSVP-like reservation that is on the data flow path from the current location of the MN. A PASSIVE reservation is made along all paths to and from other locations in the MSPEC of the MN. The path along which the reservation will be made is the path specified by the MobileIP protocol. Passive reservations become active reservations whenever there is a active sender or receiver involved in that data flow path. The paths along which passive reservations have been made can be used by other flows with weaker QoS guarantees, but appropriate action needs to be taken when the passive flow turns into an active one.

MRSVP – Implementation

In this section, we describe a basic implementation of the MRSVP framework. We have to identify proxy agents (PAs) that will make reservations on behalf of mobile senders and receivers. There are two types of PAs: remote and local. A local proxy agent (LPA) is that to which the MN is currently attached. Every other agent in the MSPEC will be a remote proxy agent (RPA).

The sender periodically generates ACTIVE PATH messages, and for a mobile sender the PAs will send PASSIVE PATH messages along the flow path to the destination. Similarly, the PAs for a mobile receiver send the PASSIVE RESV messages while the receiver itself sends the ACTIVE RESV message.

The framework also defines additional messages such as JoinGroup, RecvSpec, SenderSpec, and SenderMSpec [9]. The key issues in the implementation are as follows:

- The identification of proxy agents (local and remote) that will perform the reservations on behalf of an MN.

- The identification of flow anchors (proxy agents), a SenderAnchor when the MN is a sender and a ReceiverAnchor when the MN is a receiver, that will act as fixed points in the flow path.

- The establishment of both active and passive reservations (by the remote proxy agents) for the MN according to the MSPEC.

- The actual message sequences that lead to the reservation depends on the type of the flow and the strategy adopted. A detailed discussion of the protocol can be found in [9].

The MRSVP scheme is an initial approach to providing QoS guarantees within the MobileIP framework. The scheme considers both unicast as well as multicast traffic for all types of senders and receivers. The significant contribution of the approach is the notion of PASSIVE reservations that exist virtually on future routers that the MN's data flow is likely to use, but will turn into real flows when the MN moves into the new domain.

4.4 TCP IN WIRELESS DOMAIN

The topics discussed so far addressed the network layer modifications that are necessary to make an efficient transition from the wired to the wireless domain. The wireless domain is not only plagued by the mobility problem, but also by high error rates and low bandwidth. Obviously there needs to be a higher layer abstraction that would perform the error recovery and flow control. The traditional TCP, which guarantees in-order and reliable delivery, is the classical wired networks transmission protocol. Since the transition to the wireless domain should be compatible with the existing infrastructure, there is need for modifications of the existing protocols. This is the correct approach rather than resorting to a completely new set of protocols.

4.4.1 Traditional TCP

TCP provides a connection-oriented, reliable, and byte stream service. The term connection-oriented means the two applications using TCP must establish a TCP connection with each other before they can exchange data. It is a full duplex protocol, meaning that each TCP connection supports a pair of byte streams, one flowing in each direction. TCP includes a flow-control mechanism for each of these byte streams that allows the receiver to limit how much data the sender can transmit. TCP also implements a congestion-control mechanism.

TCP divides the data stream to be sent into smaller segments and assigns sequence numbers to them. The sequence number helps the receiver to provide the higher layers with in-order packet delivery, and also detect losses.

The sliding window mechanism employed by TCP guarantees the reliable delivery of data, ensures that the data is delivered in order, and enforces flow control between the sender and the receiver. In the sliding-window process, the sender sends several packets before awaiting acknowledgment of any of them, and the receiver acknowledges several packets at a time by sending to the transmitter the relative byte position of the last byte of the message that it has received successfully. The number of packets to be sent before the wait for acknowledgment (window size) is set dynamically, that is, it can change from time to time depending on network conditions.

Because the major cause of packet loss in the wired domain is congestion, TCP assumes that any loss is due to congestion. The TCP congestion control mechanism works as below. Initially, the TCP sender sets the congestion window to the size of one maximum TCP segment [also known as maximum segment size (MSS)]. The congestion window gets doubled for each successful transmission of the current window. This process continues until the size of the congestion window exceeds the size of the receiver window or the TCP sender notices a timeout for any TCP segment. The TCP sender interprets the timeout event as network congestion, initializes a parameter called *slow start threshold* to half the current congestion window size, and resets the congestion window size to one MSS. It then continues to double the congestion window on every successful transmission and repeats the process until the congestion window size reaches the slow start window threshold. Once the threshold is reached, the TCP sender increases the congestion window size by one MSS for each successful transmission of the window. This mechanism whereby the congestion window size is brought down to one MSS each time network congestion is detected and then is incremented as described above is referred to as *slow start*.

Another important characteristic of TCP is fast retransmit and recovery. If the receiver receives packets out of order, it continues to send the acknowledgment for the last packet received in sequence. This indicates to the sender that some intermediate packet was lost and the sender need not invoke the congestion control mechanism. The sender then reduces the window size by half and retransmits the missing packet. This avoids the slow start phase.

4.4.2 TCP Over Wireless

The adaptation of TCP to congestion causes a lot of problems in the wireless domain. The wireless domain has high packet loss and variable latency, which may cause TCP to respond with slow start. Bandwidth utilization is further reduced due to retransmission of lost packets.

One of the earliest suggested alternatives for improving the performance of TCP over wireless networks was to ensure that the link layer corrected all the errors itself over the wireless interface, thereby eliminating the need for error handling at the TCP layer. One of the suggestions in this category is the use of forward error correction (FEC) to correct small errors. FEC is a means of error control coding wherein redundancy is encoded into the sent message or binary stream to allow self-correction at the receiver. The main objective of these techniques is to hide errors from TCP as far as possible. However, FEC incurs overhead even when there are no errors as there must be the redundant parity bits to allow error detection and correction. The alternative is to use adaptive schemes, which are dynamic in the sense that when the error rate or error probability is found to be higher than usual, the redundancy introduced into the transmitted stream is also correspondingly increased. Under normal circumstances, the overhead is kept to a minimum. The other form of link layer recovery is to use retransmissions at the link layer. This incurs the overhead only on error. However, the link level recovery mechanism may cause head-of-the-line blocking, wherein the recovery mechanisms employed for one data stream consume the network resources and prevent others from being able to transmit packets. Some researchers have advocated the use of the retransmit when FEC capability is exceeded.

The most accepted role of the link layer strategy would be one in which the link layer helps TCP error recovery by providing "almost in order delivery" of packets. Not all connections can benefit from link level retransmission as it is dependent on the nature of the applications.

Several alternatives have been proposed to alter the existing TCP protocol to suit the wireless domain. The simplest idea would be to design a new TCP protocol for the wireless domain, but this will be incompatible with the wired domain. The following sections discuss various approaches to improve TCP performance in the wireless domain. Figure 4.6 provides a classification of the existing approaches.

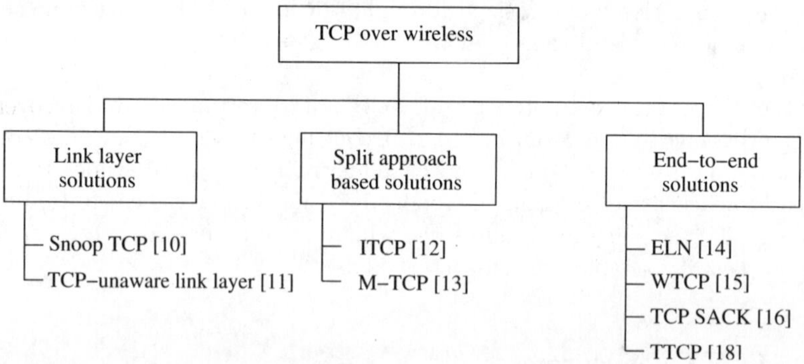

Figure 4.6. Classification of approaches for TCP over wireless.

4.4.3 Snoop TCP

The central idea used in snoop TCP [10] is to buffer the data as close to MN as possible in order to minimize the time for retransmission. The BS just snoops the packets being transmitted in both directions and recognizes the acknowledgments. The BS buffers the packets transmitted but does not acknowledge on behalf of MN. It simply removes the packet from the buffer when it sees an acknowledgment. If BS gets a duplicate acknowledgment (DUPACK) or no acknowledgment for quite some time, then it retransmits from the buffer after discarding the duplicate acknowledgment. This is to avoid unnecessary retransmissions from CN. The BS does not send acknowledgments to the CN on behalf of the MN, in order to retain the end-to-end semantics that traditional TCP provides. When the data transmission is from MN to CN, if the BS detects a gap in the sequence numbers acknowledged by the CN, it sends a NACK or negative acknowledgment to the MN to indicate loss over the wireless link.

4.4.4 TCP-Unaware Link Layer

This strategy particularly aims at simulating the behavior of the snoop-TCP protocol without requiring the link layer at the BS to be TCP-aware (hence the name TCP-unaware link layer even though TCP requires some information from the link layer). The usage of delayed DUPACKs [11] imitates snoop-TCP without requiring the link layer at BS to be TCP-aware. At the BS, as in snoop-TCP, link layer retransmission is used to perform local error recovery. But unlike snoop-TCP, where retransmissions are triggered by TCP DUPACKs, here retransmissions are triggered by link level ACKs. The MN reduces the interaction between the link layer and TCP using delayed DUPACKs. The advantages of this scheme are that the link layer need not be TCP-aware, it can be used even if headers are encrypted, which is not possible in snoop-TCP, which needs to look into the headers to see the sequence numbers, and it works well for small round trip times (RTTs) over the wireless link. The most significant disadvantage of this mechanism is that the optimum value of DUPACK delay is dependent on the wireless link, and this value is crucial in determining the performance.

4.4.5 Indirect TCP

This approach involves splitting of the TCP connection into two distinct connections, one TCP connection between the MN and BS[6] and another TCP connection between the BS and the CN.

Such a division splits the TCP connection based on the domain, the wireless domain, and the wired domain. The traditional TCP can be used in the wired part of the connection and some optimized version of TCP can be used in the wireless counterpart. In this case, the intermediate agent commonly known as the access point (AP) acts as a proxy for MN.

The indirect TCP (ITCP) mechanism [12] is shown in Figure 4.7. Loss of packets in the wireless domain, which would otherwise cause a retransmission in the wired domain, is now avoided by using a customized transport protocol between the AP and MN which accounts for the vagaries of the wireless medium. The AP acknowledges CN for the data sent to MN and buffers this data until it is successfully transmitted to MN. MN acknowledges the AP alone for the data received. Handoff

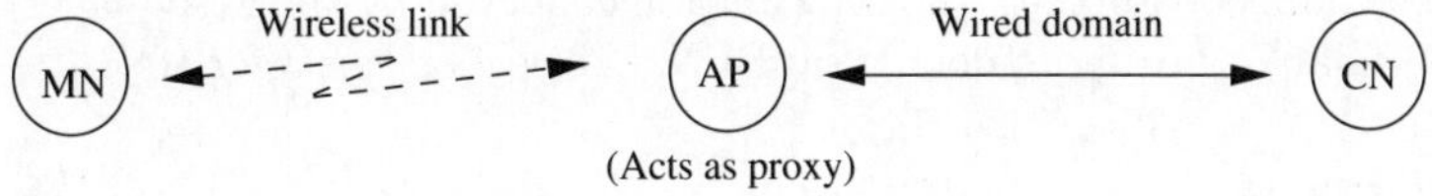

Figure 4.7. Indirect TCP.

[6]In this context of TCP over wireless, the terms BS and AP are used interchangeably.

may take a longer time as all the data acknowledged by AP and not transmitted to MN must be buffered at the new AP.

4.4.6 Mobile TCP

The most common problem associated with the wireless domain is that quite often the connection between MN and BS is lost for small intervals of time. This typically happens when MN moves behind a huge building or MN enters offices where the signals are filtered. In such cases, the sender will keep transmitting and times out eventually. In case of ITCP, the data buffered at AP may grow too large in size. It may also lead to slow start.

In such situations the sender needs to be informed. This situation is handled in mobile TCP (M-TCP) [13] by the supervisory host (the node in the wired network that controls a number of APs) which advertises the window size to be one, thus choking the sender and hence avoiding slow start. Connection may be resumed when MN can be contacted again. When the supervisory host receives a TCP packet, it forwards it to the M-TCP client. Upon reception of an ACK from M-TCP client, the supervisory host forwards the ACK to the TCP sender. Hence M-TCP maintains the end-to-end TCP semantics even though the TCP connection is split at the supervisory host. When the M-TCP client undergoes a temporary link break, the supervisory host avoids forwarding the ACK of the last byte to the sender and hence the sender TCP goes to the persist state by setting the window size to zero. This avoids retransmission, closing of the congestion window, and slow start at the sender. For more details on mobile TCP, the reader can refer to [13].

4.4.7 Explicit Loss Notification

Typically, the problem with TCP lies in the fact that it does not know the exact cause for packet loss, and hence has to invariably assume congestion loss. An ideal TCP simply retransmits the lost packets without any congestion control mechanism. The MAC layer, however, can identify the reason for the packet loss. Once the MAC layer detects that either a handoff is about to occur or realizes that the actual cause of the packet loss is not congestion, then it immediately informs the TCP layer of the possibility of a non-congestion loss. The crux of the strategy is to detect loss at MN and send an explicit loss notification (ELN) to the sender. The sender does not reduce window size on receiving the ELN as this message implies that there was an error and not congestion. This technique avoids slow start and can handle encrypted data. However, the protocol layer software at the MAC layer of MN needs to be changed. Further, the information conveyed by the MAC layer may not always be reliable. For more details on ELN, the reader can refer to [14].

4.4.8 WTCP

WTCP [15] aims at revamping the transport protocol for the wireless domain using (a) rate-based transmission at the source, (b) inter-packet separation at the receiver as the congestion metric, (c) mechanisms for detecting the reason for packet loss, and (d) bandwidth estimation, as some of the underlying principles. A unique characteristic of WTCP is the attempt to separate the congestion control and reliability mechanisms. WTCP uses separate sequence numbers for congestion control and reliability mechanisms in order to distinguish the two. The reliability mechanism involves a combination of selective and cumulative acknowledgments, and takes into account the reverse-path characteristics for determining the ACK frequency.

4.4.9 TCP SACK

The selective retransmission strategy [14] is more complex and requires more buffer space at the end-points. Hence TCP traditionally uses cumulative acknowledgments and the go-back-N strategy. Using selective retransmit reduces the overhead of retransmission on errors and therefore cannot be ruled out for use in wireless domains. The TCP with selective ACK scheme (TCP SACK) [16], [17] improves TCP performance by allowing the TCP sender to retransmit packets based on the selective ACKs provided by the receiver.

4.4.10 Transaction-Oriented TCP

The TCP connection setup and connection tear-down phases involve a huge overhead in terms of time and also in terms of the number of packets sent. This overhead is very costly, especially if the size of the data is small. An alternative for such transactions is transaction-oriented TCP (TTCP) [18]. The motivation behind this approach is to integrate the call setup, the call tear-down, and the actual data transfer into a single transaction, thereby avoiding separate packets for connecting and disconnecting. However, the flip side to the strategy is that changes must be made to TCP, which goes against some of the fundamental objectives that the changes to TCP must be transparent and must not affect the existing framework.

Table 4.3 shows a summary of the various approaches discussed so far. The next section briefly describes the impact of mobility on the performance of TCP.

4.4.11 Impact of Mobility

Handoffs occur in wireless domains when an MN moves into a new BS's domain (a cell in the cellular context). If the link layer ensures reliable delivery and guarantees zero loss during a handoff, then TCP will be totally unaware of the handoff and no measures need to be taken at the transport layer to support handoff. The only exception to this is when the handoff latency is too large and exceeds the TCP timeout; then the transparency of handoffs to TCP is lost.

Fast Retransmit/Recovery

The usual problem associated with handoffs is that the handoff may lead to packet loss during transit, either as a result of the intermediary routers' failure to allocate adequate buffers or their inability to forward the packets meant for the MN to the new BS. The result of the packet loss during handoff is slow start. The solution

Table 4.3. Summary of proposed protocols to improve the performance of TCP over wireless

Feature	**Snoop TCP**	**TCP-Unaware Link Layer**	**Mobile TCP**	**ITCP**	**ELN**	**WTCP**	**TCP SACK**	**TTCP**
Changes in:								
AP	Yes	Yes	Yes	Yes	No	No	No	No
CN	No	No	No	No	Yes	Yes	Yes	Yes
MN	Yes	No	Yes	Yes	Yes	Yes	No	No
Retransmitting Node	AP	AP	NA*	AP	NA	NA	NA	NA
Single Point Failure	No	No	No	Yes (AP)	No	No	No	No
Handoff Latency	Low	Low	Low	Low	High	High	High	High
Security	Breach at AP	No breach	NA	Breach at AP	No breach	No breach	No breach	No breach
End-to-End Semantics	Yes	Yes	Yes	No	Yes	Yes	Yes	Yes
Retransmissions by Intermediate Nodes	Yes	Yes	No	Yes	No	No	No	No
Slow Start	Yes	Yes	No	NA	No	No	No	Yes
Buffer at AP	Yes	Yes	No	Yes	No	No	No	No

*Not Applicable

involves artificially forcing the sender to go into fast retransmission mode immediately, by sending duplicate acknowledgments after the handoff, instead of going into slow start. The advantage of the strategy is its simplicity and the fact that it requires minimal changes to the existing TCP structure. However, the scheme does not consider the fact that there may be losses over the wireless links.

Using Multicast

Multicast has been suggested to improve the performance of TCP in the presence of handoffs [10]. The idea is similar to the one used in MRSVP [9], where the MN is required to define a group of BSs that it is likely to visit in the near future. These include the current cell (or the current BS) the MN is attached to and also the cells (BSs) likely to be visited by it. These BSs are then directed to join the multicast group, the address being the unique multicast address assigned to the MN. Packets destined for MN will have to be subsequently readdressed to the multicast group. In the implementation, only one BS is actually in contact with the MN and is responsible for transmitting the packets to it. If the rest of the BSs in the multicast group are able to buffer the packets addressed to the multicast address, then the loss of packets during the handoff can be significantly minimized. There is a trade-off between buffer allocation at the BSs and the loss during handoff. In practical situations, the number of buffers allocated can be minimized by buffering only when a handoff is likely to occur.

4.5 WAP

WAP stands for wireless application protocol. This name is a misnomer, because WAP represents a suite of protocols rather than a single protocol. WAP has today become the *de facto* standard for providing data and voice services to wireless handheld devices. WAP aims at integrating a simple lightweight browser also known as a micro-browser into handheld devices, thus requiring minimal amounts of resources such as memory and CPU at these devices. WAP tries to compensate for the shortfalls of the wireless handheld devices and the wireless link (low bandwidth, low processing capabilities, high bit-error rate, and low storage availability) by incorporating more intelligence into the network nodes such as the routers, Web servers, and BSs. The primary objectives of the WAP protocol suite are independence from the wireless network standards, interoperability among service providers, overcoming the shortfalls of the wireless medium (such as low bandwidth, high latency, low connection stability, and high transmission cost per bit), overcoming the drawbacks of handheld devices (small display, low memory, limited battery power, and limited CPU power), increasing efficiency and reliability, and providing security, scalability, and extensibility.

4.5.1 The WAP Model

WAP adopts a client-server approach. It specifies a proxy server that acts as an interface between the wireless domain and core wired network. This proxy server, also known as a WAP gateway, is responsible for a wide variety of functions such as protocol translation and optimizing data transfer over the wireless medium. Figure 4.8 illustrates the client-server model that WAP employs. The WAP-enabled handset communicates with a Web content server or an origin server [that may provide hypertext markup language (HTML)/common gateway interface (CGI) content] via a WAP gateway. It is at the WAP gateway that the convergence of the wireless and wired domains actually occurs. The gateway receives WAP requests from the handset, and these have to be converted into suitable HTTP requests to be sent to the origin server. If the origin server cannot provide the required information in wireless markup language (WML) form, then there must be an additional filter between the server and the gateway to convert the HTML content into WAP-compatible WML content. The gateway may additionally perform functions such as caching and user agent profiling as part of some optimization measures. This is also known as *capability and preference information.* By means of user agent profiling, the MN specifies its characteristics such as hardware characteristics, software capabilities, and user preferences, to the server so that the content can be formatted appropriately to be displayed correctly.

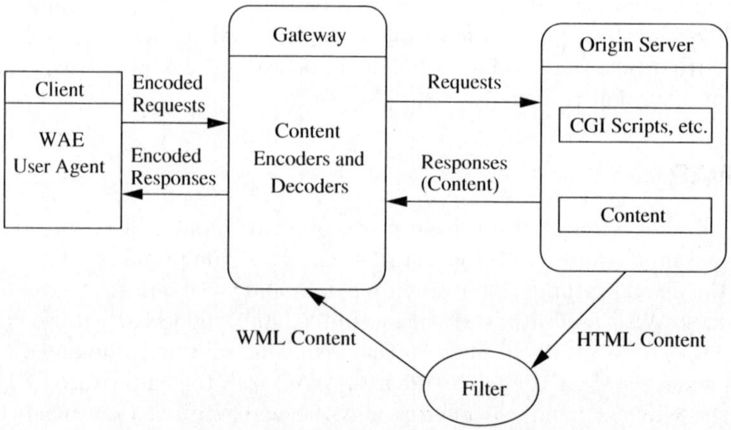

Figure 4.8. The WAP client-server model.

4.5.2 The WAP Protocol Stack

The WAP protocol stack is designed in a layered fashion that allows the architecture to provide an environment that is both extensible and scalable for application development. The WAP architecture allows other services to access the WAP stack at well-defined interfaces. Figure 4.9 gives an overview of the different layers in the WAP protocol suite and also their basic functionalities. This section provides a brief description of some of the important layers in the protocol stack.

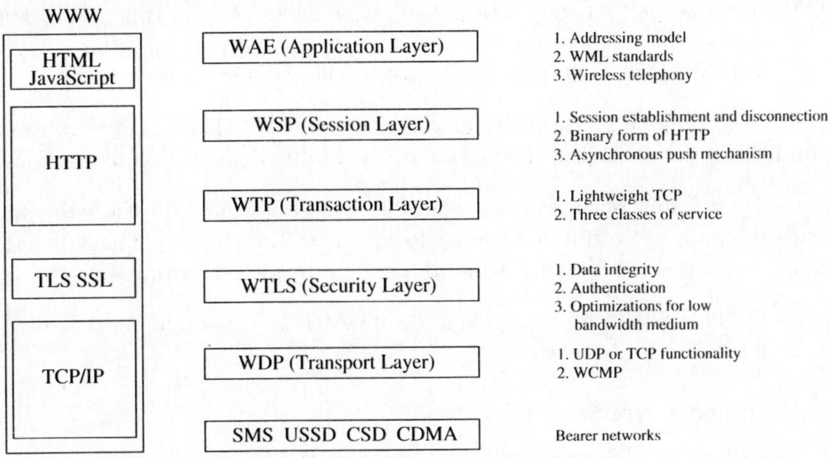

Figure 4.9. The WAP protocol stack.

The Wireless Application Environment

The wireless application environment (WAE) has a number of components that address specific issues in the application environment. The WAE provides for an addressing model for accessing both the WWW URLs and other resources specific to the wireless domain using uniform resource identifiers (URIs). The WAE uses WML as the standard markup language, which can be construed as an efficient binary encoded form of the traditional HTML. The WAE also provides a compact scripting language analogous to JavaScript. The WAE also provides for a set of telephony applications through the wireless telephony application interface (WTAI).

Wireless Session Protocol

The wireless session protocol (WSP) establishes a reliable session between the client and the server and also ensures that the session is released in an orderly manner. The push mechanism is a fundamental component of the WAP programming model aimed at reducing the number of requests made by the client to the server. A data server will asynchronously push the information to the registered client(s) efficiently using this mechanism. This is especially useful in multicast and broadcast applications. The WSP provides the equivalent of HTTP in the WWW domain. The core of the WSP design is a binary form of HTTP. A session may be suspended to save power at the clients, but the session reestablishment follows only a small procedure that avoids the overhead of starting a full-fledged session afresh.

Wireless Transaction Protocol

The wireless transaction protocol (WTP) can for all practical purposes be viewed as a lightweight version of TCP. A transaction is defined as a request/response cycle.

The WTP has no explicit setup and tear-down phases like TCP, as this would cause a tremendous overhead. There are no security options at the transaction layer in the WAP stack. WTP defines three categories or classes of service:

1. Class 0: Unreliable send (push model) with no ACK. There is no retransmission in case the message is lost. This is essentially a connection-less service.
2. Class 1: Reliable push service, where a request is sent and the responder sends the data as an implicit acknowledgment to the request. The responder maintains this state for some time to handle possible retransmissions.
3. Class 2: This is the classical request-data-ACK cycle providing a two-way reliable service.

Wireless Transport Layer Security

The objective of the wireless transport layer security (WTLS) is to provide transport layer security between the WAP client and a WAP server. WTLS is based on the industry standard transport layer security (TLS) protocol with certain features such as datagram support, optimized handshake, and dynamic key refreshing. The primary objectives of WTLS are data integrity, privacy, authentication, and denial of service (DoS) protection. WTLS has capabilities to detect and reject data that is not successfully verified; this protects servers from DoS attacks.

Wireless Datagram Protocol

The wireless datagram protocol (WDP) defines the WAP's transport layer in the protocol suite. The WDP has an adaptation layer that is bearer-specific that helps optimize the data transfer specific to a particular bearer service (such as SMS, USSD, CSD, and CDMA). If the underlying bearer service uses the IP standard user datagram protocol (UDP), then there is no necessity for a separate functionality at the WDP layer as UDP itself is used. The wireless control message protocol (WCMP) is responsible for providing the error-handling mechanisms analogous to Internet control message protocol (ICMP).

4.5.3 WAP 2.0 and i-mode

The i-mode (information-mode) system, developed in Japan and a major competitor to WAP, has three main components: a transmission system, a handset, and a language for designing Web pages. The transmission system consists of the existing mobile phone network (which is circuit-switched) and a new packet-switched network. Voice transmission uses the existing mobile phone network while data transmission uses the packet-switched network and is billed based on the number of packets transmitted as opposed to connection time. i-mode uses a subset of HTML called as cHTML (compact HTML). In contrast, WAP 2.0 was developed by the WAP Forum and is likely to use packet-switched network. WAP 2.0 has new features such as multimedia messaging, pull (request for data, then receive the data) as well as push model (asynchronous data transfer, without requiring explicit request messages, such as stock prices), integrated telephony, interoperability with WAP 1.0, and support for plug-ins in the browser. Unlike i-mode, WAP 2.0 charges the users based on connection time.

4.6 OPTIMIZING WEB OVER WIRELESS

The limitations of wireless networks that provide the motivation for such optimizations are low bandwidth, low reliability, high latency, and high cost per byte transferred. Integrating Web access over wireless devices would have to take into account the drawbacks of the wireless medium and the capabilities of the devices. Systems such as WebExpress [19] are aimed at optimizing routine repetitive browsing; many of the mechanisms suggested may not be suitable for random browsing (*i.e.*, there are no perceivable trends in the Web accesses). Web browsers must offer a good interface for the wireless devices, keeping in mind the network, memory, processing power, and power consumption constraints.

4.6.1 HTTP Drawbacks

The main protocol on which the Web operates today is the hypertext transfer protocol (HTTP), which is optimized mainly for the wired world. It has a lot of overhead, but it is acceptable when the network bandwidth is an inexpensive resource as in typical wired networks compared to wireless networks. HTTP has drawbacks such as high connection overhead (a new TCP socket is opened for every new HTML object), redundant capabilities transmission (information regarding the browser capabilities is included in every HTTP request), and verbosity (HTTP is ASCII-encoded and hence inherently verbose). The WebExpress system suggests that an Intercept model be applied for Web access over wireless interfaces. This allows the number of requests sent over the wireless channel to be optimized, and also avoids the connection setup overhead over the wireless interface. There are two main entities that are introduced into the system: the client side interface (CSI) and the server side interface (SSI). The CSI appears as a local Web proxy co-resident with the Web browser on the wireless rendering device, say, a mobile phone or a PDA. The communication between the CSI and the Web browser takes place through the loopback feature of the TCP/IP suite (wherein the host sends a packet to itself using an IP address like 127.0.0.1). The communication between the CSI and the SSI is the only interaction over the wireless network and this uses a reduced HTTP, as discussed later. The SSI communicates with the Web server over the wired network. The SSI could typically be resident at the network gateway or the FA in MobileIP. The intercept model (Figure 4.10) is transparent to browsers and servers, and is also insensitive to changes in HTTP/HTML technology.

4.6.2 Optimizations

Four main categories of optimizations that can improve the performance of Web access systems over wireless channels can be identified. These are:

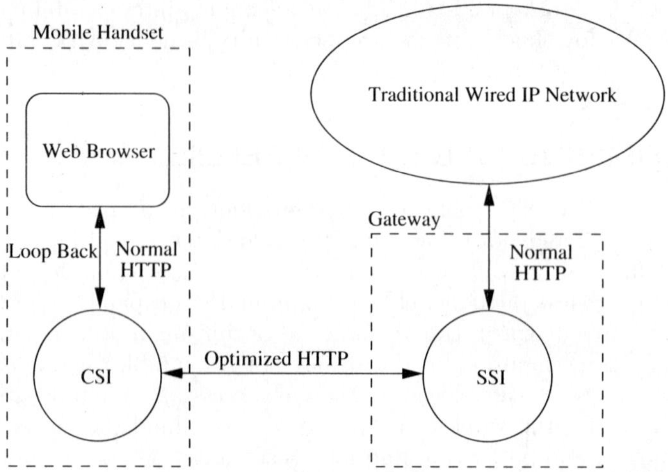

Figure 4.10. The intercept model.

- **Caching:** Current caching technologies are suited for wired applications. Cache objects are either purged at the end of the session or may persist across sessions. But it is advantageous to have cached data persist across browser sessions, as this increases cache hit ratios. Appropriate cache coherency methods are added to detect and change old information.

- **Differencing:** For transaction processing (involving forms) caching techniques do not help as different replies to the same application server are often different. Still, the fact that these replies tend to be similar can be exploited to reduce the network traffic over the wireless interface. A base object carries fundamental features that do not change across transactions and is created and maintained by both the client and server interfaces. Whenever a new transaction takes place, the server computes the difference stream and only the difference stream is transmitted.

- **Protocol reduction:** This approach aims at reducing the overhead of repeated setup and tear-down of TCP/IP connections for each Web-object to be transmitted. This can be eliminated by establishing a single TCP/IP connection between the CSI and the SSI that will persist for the entire session. The connection setup/tear-down overhead is on the local and wired connections only.

- **Header reduction:** HTTP requests are prefixed with headers that indicate to the origin server the rendering capabilities of the browser and also the various content formats handled by it. The alternative to this is that the CSI sends this information in the first request and SSI records this information.

For every subsequent request sent by the CSI, the SSI automatically inserts this capability list into each packet meant for the origin server.

4.7 SUMMARY

This chapter focused on the issues in wireless networks that are pertinent to the higher layers in the protocol stack, the network layer, the transport layer, and the application layer. The various aspects of the wireless Internet, that is, extension of the services offered by the Internet to the wireless domain, were discussed. Mobile IP aims at providing network connectivity to mobile hosts, and it is in a larger sense not restricted to wireless networks. The inefficiencies of MobileIP routing can be tackled in both generic techniques such as the optimizations incorporated in IPv6 and specific techniques as suggested in the 4×4 approach. The network layer also has to address the issues of security, accounting, and handoffs, as these are of great significance in wireless networks; some of the relevant issues were also discussed.

This chapter also discussed the issues in adaptation of TCP to the wireless domains, as it has been shown that the existing transport framework would perform miserably when used in its current form (optimized to work with high bandwidth, low-error wired networks). Most of the solutions involved some capability at the BS to buffer packets and also act on behalf of the MNs to send ACKs/NACKs. The strategies discussed in this chapter were broadly classified into various categories based on the role of the BS. The WAP architecture specified provides for an efficient, interoperable, and scalable framework for developing and using applications in the wireless domain. WAP 2.0 added more features to the previously existing WAP 1.0 protocol.

4.8 PROBLEMS

1. What are the major problems that arise in network and transport layers when an MN accesses the Internet from a different network?
2. Why should MN register with HA?
3. What is triangular routing?
4. Why is reverse tunneling necessary?
5. Why is it necessary to adopt different types of optimization strategies with regard to Mobile IP?
6. State the differences between network-initiated and mobile-initiated handoffs.
7. What are the various steps involved in a handoff?
8. What is post-registration handoff?
9. When do you think a forward handoff is needed as opposed to a backward handoff? What could be some advantages of a backward handoff as opposed to a forward handoff?

10. Can a soft handoff be a forward handoff? Can a forward handoff be a soft handoff?

11. What happens if an MN requests a handoff but then, after the resources are reserved, the MN does not hand off to that FA?

12. What are the two types of mobilities that are applicable in TIMIP domains?

13. What are the effects of mobility on the QoS provisioning for mobile nodes?

14. What do you think are the major drawbacks in the existing RSVP structure compared to MRSVP?

15. Briefly discuss the main goals of WAP.

16. Explain the WAP model and the WAP protocol stack.

17. What is the functionality of the session layer in the WAP stack?

18. What are the advantages of using the intercept model in WebExpress?

19. What are the four major categories of optimizations suggested in the WebExpress system?

BIBLIOGRAPHY

[1] C. E. Perkins, "Route Optimization in Mobile IP," *Internet draft (work in progress)*, draft-ietf-mobileip-optim-11.txt, September 2001.

[2] S. Cheshire and M. Stuart, "Internet Mobility 4×4," *Proceedings of ACM SIGCOMM 1996*, pp. 318-329, August 1996.

[3] P. Venkataram, R. Rajavelsamy, and S. Laxmaiah, "A Method of Data Transfer Control During Handoffs in MobileIP-Based Multimedia Networks," *ACM Mobile Computing and Communications Review*, vol. 5, no. 2, pp. 27-36, April 2001.

[4] K. E. Malki *et al.*, "Low Latency Handoffs in Mobile IPv4," *Internet draft (work in progress)*, draft-ietf-mobileip-lowlatency-handoffs-v4-03.txt, November 2001.

[5] D. B. Johnson and C. E. Perkins, "Mobility Support in IPv6," *Internet draft (work in progress)*, draft-ietf-mobileip-ipv6-15.txt, November 2001.

[6] A. Grilo, P. Estrela, and M. Nunes, "Terminal Independent Mobility for IP-TIMIP," *IEEE Communications Magazine*, vol. 39, no. 12, pp. 34-41, December 2001.

[7] R. Ramjee *et al.*, "IP-Based Access Network Infrastructure for Next-Generation Wireless Data Networks," *IEEE Personal Communications Magazine*, vol. 7, no. 4, pp. 34-41, August 2000.

[8] A. Campbell *et al.*, "Design Evolution and Implementation of Cellular IP," *IEEE Personal Communications Magazine*, vol. 7, no. 4, pp. 42-49, August 2000.

[9] A. K. Talukdar, B. R. Badrinath, and A. Acharya, "MRSVP: A Resource Reservation Protocol for an Integrated Service Network with Mobile Hosts," *ACM/Baltzer Wireless Networks Journal*, vol. 7, no. 1, pp. 5-19, January 2001.

[10] H. Balakrishnan, S. Seshan, and R. Katz, "Improving Reliable Transport and Handoff Performance in Cellular Wireless Networks," *ACM/Baltzer Wireless Networks Journal*, vol. 1, no. 4, pp. 469-481, December 1995.

[11] N. H. Vaidya, M. Mehta, C. Perkins, and G. Montenegro, "Delayed Duplicate Acknowledgments: A TCP-Unaware Approach to Improve Performance of TCP over Wireless," *Technical Report 99003*, Computer Science Department, Texas A&M University, February 1999.

[12] A. Bakre and B. R. Badrinath, "I-TCP: Indirect TCP for Mobile Hosts," *Proceedings of IEEE ICDCS 1995*, pp. 136-143, May 1995.

[13] K. Brown and S. Singh, "M-TCP: TCP for Mobile Cellular Networks," *ACM Computer Communication Review*, vol. 27, no. 5, pp. 19-43, October 1997.

[14] H. Balakrishnan, V. N. Padmanabhan, S. Seshan, and R. H. Katz, "A Comparison of Mechanisms for Improving TCP Performance over Wireless Links," *IEEE/ACM Transactions on Networking*, vol. 5, no. 6, pp. 756-769, December 1997.

[15] P. Sinha, N. Venkitaraman, R. Sivakumar, and V. Barghavan, "WTCP: A Reliable Transport Protocol for Wireless Wide-Area Networks," *Proceedings of ACM MOBICOM 1999*, pp. 231-241, August 1999.

[16] M. Mathis, J. Mahdavi, S. Floyd, and A. Romanow, "TCP Selective Acknowledgment Options," *IETF RFC 2018*, July 1997.

[17] M. Mathis, J. Semke, and J. Mahdavi, "The Macroscopic Behavior of the TCP Congestion Avoidance Algorithm," *ACM Computer Communications Review*, vol. 27, no. 3, pp. 67-82, July 1997.

[18] R. Braden, "T-TCP – TCP Extensions for Transactions Functional Specification," *IETF RFC 1644*, July 1994.

[19] B. C. Housel and D. B. Lindquist, "WebExpress: A System for Optimizing Web Browsing in a Wireless Environment," *Proceedings of ACM/IEEE MOBICOM 1996*, pp. 108-116, November 1996.

[20] C. E. Perkins, "Mobile Networking Terminology," *Internet draft (work in progress)*, draft-ietf-manet-term-01.txt, November 1998.

[21] C. E. Perkins, "Mobile IP," *IEEE Communications Magazine*, vol. 35, no. 5, pp. 84-99, May 1997.

[22] C. E. Perkins, "Mobile Networking Through Mobile IP," *IEEE Internet Computing*, vol. 2, no. 1, pp. 58-69, January-February 1998.

[23] C. E. Perkins, "Mobile IP Joins Forces with AAA," *IEEE Personal Communications Magazine*, vol. 7, no. 4, pp. 59-61, August 2000.

[24] S. Glass, T. Hiller, S. Jacobs, and C. Perkins, "Mobile IP Authentication, Authorization, and Accounting Requirements," *IETF RFC 2977*, October 2000.

[25] H. Chaskar, "Requirements of a QoS Solution for Mobile IP," *Internet draft (work in progress)*, draft-ietf-mobileip-qos-requirements-01.txt, August 2001.

[26] G. Dommety, "Fast Handovers for Mobile IPv6," *Internet draft (work in progress)*, draft-ietf-mobileip-fast-mipv6-03.txt, November 2001.

[27] B. Jabbari, R. Papneja, and E. Dinan, "Label Switched Packet Transfer for Wireless Cellular Networks," *Proceedings of IEEE WCNC 2000*, vol. 3, pp. 958-962, September 2000.

[28] H. Yumiba, K. Imai, and M. Yabusaki, "IP-Based IMT Network Platform," *IEEE Personal Communications Magazine*, vol. 8 no. 5, pp. 18-23, October 2001.

[29] V. Gupta and S. Gupta, "Securing the Wireless Internet," *IEEE Communications Magazine*, vol. 39, no. 12, pp. 68-74, December 2001.

[30] WAP 2.0 Technical White Paper, *WAP Forum Ltd., 2002.*

[31] J. Schiller, *Mobile Communications*, Addison-Wesley, January 2000.

[32] I. Stojmenovic, Ed., *Handbook of Wireless Networks and Mobile Computing*, Wiley Interscience, New York, February 2002.

Chapter 5

AD HOC WIRELESS NETWORKS

5.1 INTRODUCTION

The principle behind ad hoc networking is multi-hop relaying, which traces its roots back to 500 B.C. Darius I (522-486 B.C.), the king of Persia, devised an innovative communication system that was used to send messages and news from his capital to the remote provinces of his empire by means of a line of shouting men positioned on tall structures or heights. This system was more than 25 times faster than normal messengers available at that time. The use of ad hoc voice communication was used in many ancient/tribal societies with a string of repeaters of drums, trumpets, or horns. In 1970, Norman Abramson and his fellow researchers at the University of Hawaii invented the ALOHAnet, an innovative communication system for linking together the universities of the Hawaiian islands. ALOHAnet utilized single-hop wireless packet switching and a multiple access solution for sharing a single channel. Even though ALOHAnet was originally implemented for a fixed single-hop wireless network, the basic idea was compelling and applicable to any environment where access to a common resource had to be negotiated among a set of uncoordinated nodes. The success and novelty of ALOHAnet triggered widespread interest in different directions of computer communication, including the work that led to the development of Ethernet by Robert Metcalfe and the packet radio network (PRNET) project sponsored by the defense advanced research projects agency (DARPA) [1]. The PRNET project was aimed at developing a packet wireless network for military applications. Even though the initial attempt had a centralized control, it quickly evolved into a distributed multi-hop wireless communication system that could operate over a large geographical area. Each mobile node had a broadcast radio interface that provided many advantages such as the use of a single channel, simpler channel management techniques, and the ease of supporting mobility. PRNET used a combination of ALOHA and carrier sense multiple access (CSMA) for access to the shared radio channel. The radio interface employed the direct-sequence (DS) spread spectrum scheme. The system was designed to self-organize, self-configure, and detect radio connectivity for the dynamic operation of a routing protocol without any support from fixed infrastructure. The major issues that the PRNET project faced include those of obtaining, maintaining, and utilizing the topology information, error and flow control over the wireless links, reconfiguration of paths to handle path breaks arising due to the mobility of nodes and routers, processing and storage capability of nodes, and distributed channel sharing. The successful demonstrations of the PRNET proved the feasibility and efficiency of infrastructure-less networks and their applications for civilian and military purposes. DARPA extended the work on multi-hop wireless networks through the survivable radio networks (SURAN) project that aimed at providing ad hoc networking with small, low-cost, low-power devices with efficient protocols and improved scalability and survivability (the ability of a network to survive the failure of network nodes and links). During the 1980s, research on military applications was extensively funded across the globe. Realizing the necessity of open standards in this emerging area of computer communication, a working group within the Internet Engineering Task Force (IETF), termed the mobile ad hoc networks (MANET) working group [2], was formed to standardize the protocols and functional specifications of ad hoc wireless networks. The vision of the IETF effort in the MANET working group is to provide improved standardized routing functionality to support self-organizing mobile networking infrastructure.

In 1994, the Swedish communication equipment maker Ericsson proposed to develop a short-range, low-power, low-complexity, and inexpensive radio interface and associated communication protocols referred to as *Bluetooth* for ubiquitous connectivity among heterogeneous devices, as discussed in Section 2.5. This effort was later taken over by a Special Interest Group (SIG) formed by several major computer and telecommunication vendors such as 3Com, Ericsson, IBM, Intel, Lucent, Microsoft, Motorola, Nokia, and Toshiba. The Bluetooth SIG aims at delivering a universal solution for connectivity among heterogeneous devices. This is one of the first commercial realizations of ad hoc wireless networking. Bluetooth standardizes the single-hop point-to-point wireless link that helps in exchanging voice or data, and formation of *piconets* that are formed by a group of nodes in a smaller geographical region where every node can reach every other node in the group within a single-hop. Multiple piconets can form a *scatternet*, which necessitates the use of multi-hop routing protocols.

Even though ad hoc wireless networks are expected to work in the absence of any fixed infrastructure, recent advances in wireless network architectures reveal interesting solutions that enable the mobile ad hoc nodes to function in the presence of infrastructure. Multi-hop cellular networks (MCNs) [3] and self-organizing packet radio ad hoc networks with overlay (SOPRANO) [4] are examples of such types of networks. These hybrid architectures (which combine the benefits of cellular and ad hoc wireless networks) improve the capacity of the system significantly. Even with all the promises that are offered by ad hoc wireless networks, successful commercial deployment requires realistic solutions to different problems, including support for QoS provisioning and real-time applications, pricing, cooperative functioning, energy-efficient relaying, load balancing, and support for multicast traffic.

5.1.1 Cellular and Ad Hoc Wireless Networks

Figure 5.1 shows a representation of different wireless networks. The current cellular wireless networks (depicted in Figure 5.2) are classified as the infrastructure dependent networks. The path setup for a call between two nodes, say, node C to node E, is completed through the base station as illustrated in Figure 5.2.

Ad hoc wireless networks are defined as the category of wireless networks that utilize multi-hop radio relaying and are capable of operating without the support of any fixed infrastructure (hence they are also called infrastructureless networks). The absence of any central coordinator or base station makes the routing a complex one compared to cellular networks. Ad hoc wireless network topology for the cellular network shown in Figure 5.2 is illustrated in Figure 5.3. Note that in Figure 5.3 the cell boundaries are shown purely for comparison with the cellular network in Figure 5.2 and do not carry any special significance. The path setup for a call between two nodes, say, node C to node E, is completed through the intermediate mobile node F, as illustrated in Figure 5.3. Wireless mesh networks and wireless sensor networks are specific examples of ad hoc wireless networks.

The major differences between cellular networks and ad hoc wireless networks are summarized in Table 5.1. The presence of base stations simplifies routing and resource management in a cellular network as the routing decisions are made in a centralized manner with more information about the destination node. But in an ad hoc wireless network, the routing and resource management are done in a distributed manner in which all nodes coordinate to enable communication among themselves. This requires each node to be more intelligent so that it can function both as a network host for transmitting and receiving data and as a network router for routing packets from other nodes. Hence the mobile nodes in ad hoc wireless networks are more complex than their counterparts in cellular networks.

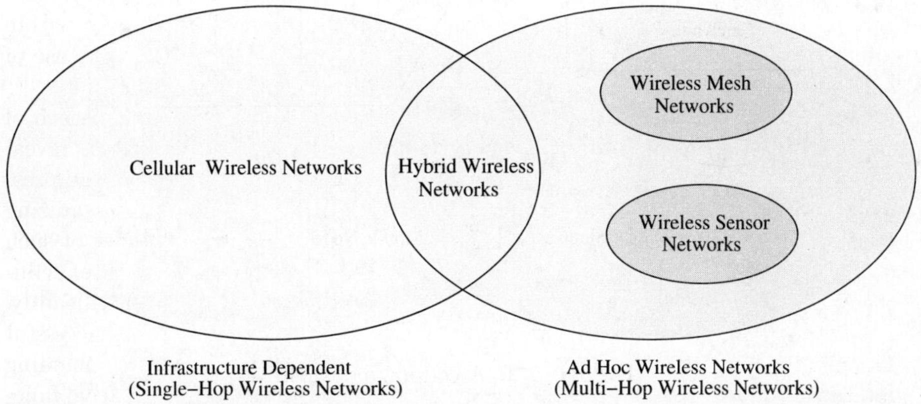

Figure 5.1. Cellular and ad hoc wireless networks.

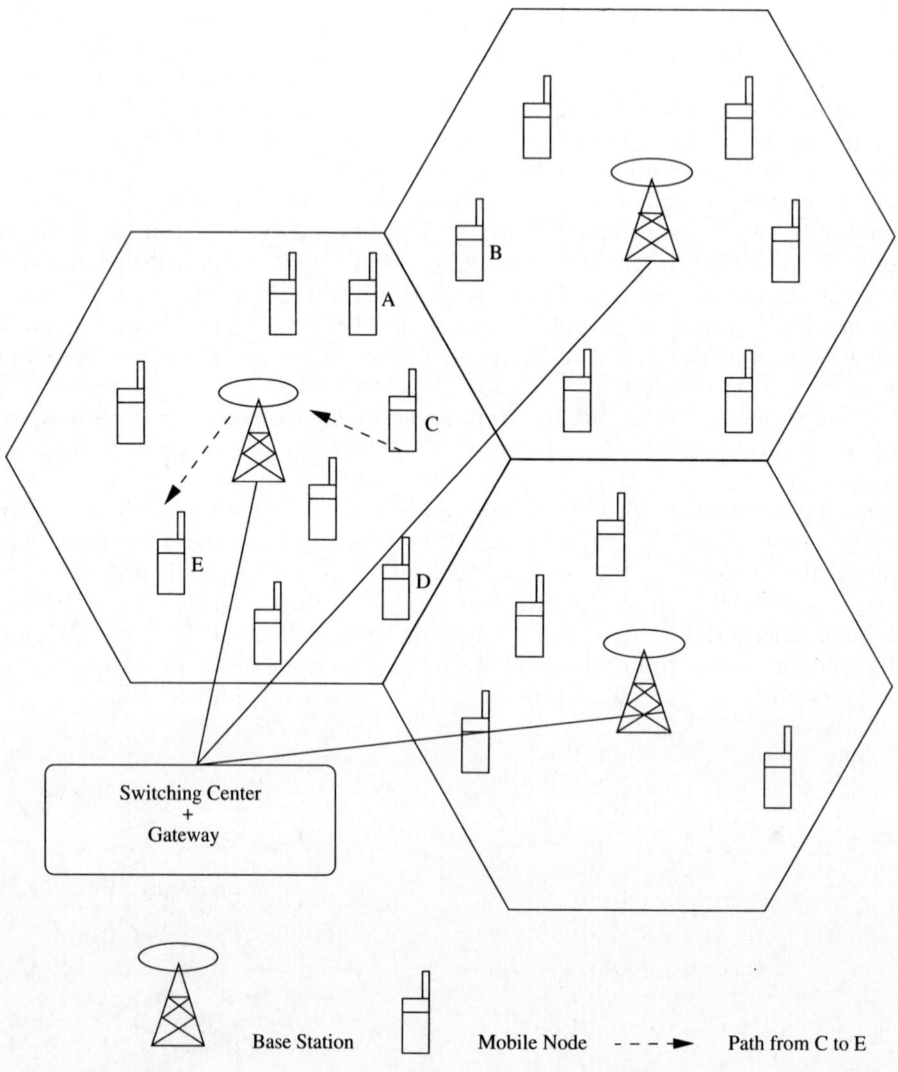

Figure 5.2. A cellular network.

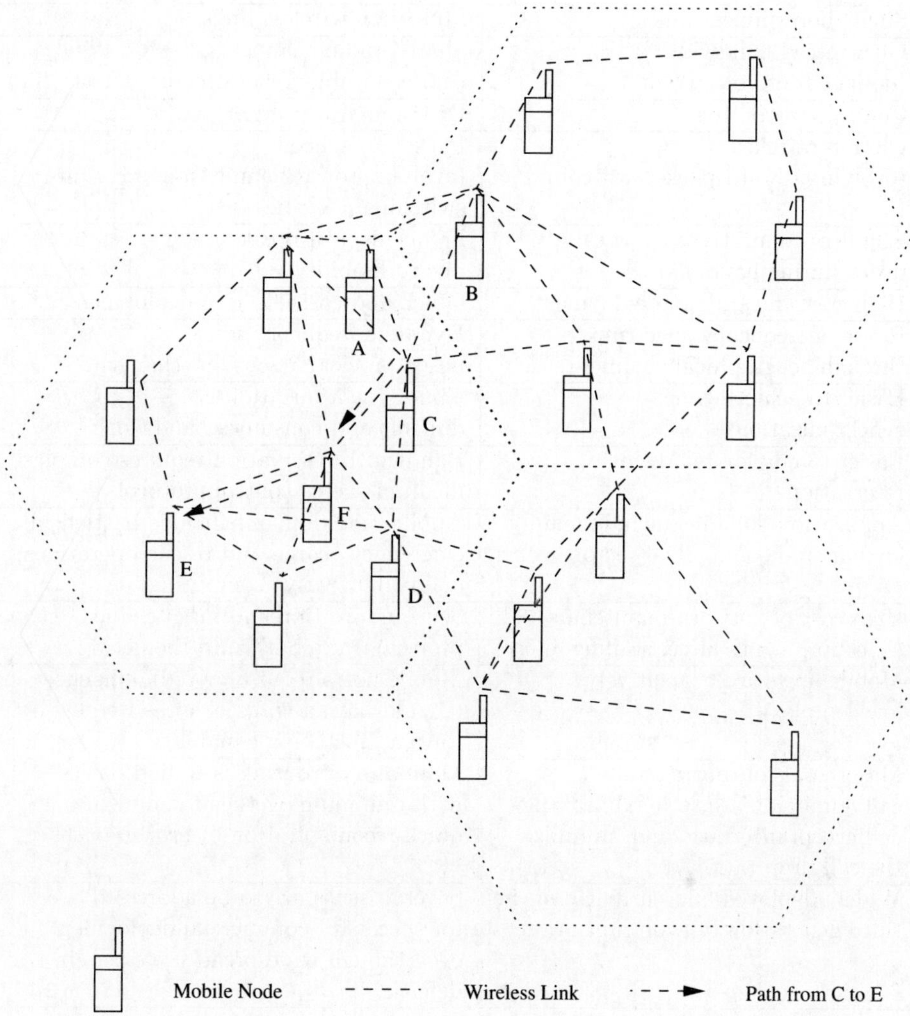

Figure 5.3. An ad hoc wireless network.

Table 5.1. Differences between cellular networks and ad hoc wireless networks

Cellular Networks	Ad Hoc Wireless Networks
Fixed infrastructure-based	Infrastructure-less
Single-hop wireless links	Multi-hop wireless links
Guaranteed bandwidth (designed for voice traffic)	Shared radio channel (more suitable for best-effort data traffic)
Centralized routing	Distributed routing
Circuit-switched (evolving toward packet switching)	Packet-switched (evolving toward emulation of circuit switching)
Seamless connectivity (low call drops during handoffs)	Frequent path breaks due to mobility
High cost and time of deployment	Quick and cost-effective deployment
Reuse of frequency spectrum through geographical channel reuse	Dynamic frequency reuse based on carrier sense mechanism
Easier to achieve time synchronization	Time synchronization is difficult and consumes bandwidth
Easier to employ bandwidth reservation	Bandwidth reservation requires complex medium access control protocols
Application domains include mainly civilian and commercial sectors	Application domains include battlefields, emergency search and rescue operations, and collaborative computing
High cost of network maintenance (backup power source, staffing, etc.)	Self-organization and maintenance properties are built into the network
Mobile hosts are of relatively low complexity	Mobile hosts require more intelligence (should have a transceiver as well as routing/switching capability)
Major goals of routing and call admission are to maximize the call acceptance ratio and minimize the call drop ratio	Main aim of routing is to find paths with minimum overhead and also quick reconfiguration of broken paths
Widely deployed and currently in the third generation of evolution	Several issues are to be addressed for successful commercial deployment even though widespread use exists in defense

5.1.2 Applications of Ad Hoc Wireless Networks

Ad hoc wireless networks, due to their quick and economically less demanding deployment, find applications in several areas. Some of these include: military applications, collaborative and distributed computing, emergency operations, wireless mesh networks, wireless sensor networks, and hybrid wireless network architectures.

Military Applications

Ad hoc wireless networks can be very useful in establishing communication among a group of soldiers for tactical operations. Setting up a fixed infrastructure for communication among a group of soldiers in enemy territories or in inhospitable terrains may not be possible. In such environments, ad hoc wireless networks provide the required communication mechanism quickly. Another application in this area can be the coordination of military objects moving at high speeds such as fleets of airplanes or warships. Such applications require quick and reliable communication. Secure communication is of prime importance as eavesdropping or other security threats can compromise the purpose of communication or the safety of personnel involved in these tactical operations. They also require the support of reliable and secure multimedia multicasting. For example, the leader of a group of soldiers may want to give an order to all the soldiers or to a set of selected personnel involved in the operation. Hence, the routing protocol in these applications should be able to provide quick, secure, and reliable multicast communication with support for real-time traffic.

As the military applications require very secure communication at any cost, the vehicle-mounted nodes can be assumed to be very sophisticated and powerful. They can have multiple high-power transceivers, each with the ability to hop between different frequencies for security reasons. Such communication systems can be assumed to be equipped with long-life batteries that might not be economically viable for normal usage. They can even use other services such as location tracking [using the global positioning system (GPS)] or other satellite-based services for efficient communication and coordination. Resource constraints such as battery life and transmitting power may not exist in certain types of applications of ad hoc wireless networks. For example, the ad hoc wireless network formed by a fleet of military tanks may not suffer from the power source constraints present in the ad hoc network formed by a set of wearable devices used by the foot soldiers.

In short, the primary nature of the communication required in a military environment enforces certain important requirements on ad hoc wireless networks, namely, reliability, efficiency, secure communication, and support for multicast routing.

Collaborative and Distributed Computing

Another domain in which the ad hoc wireless networks find applications is collaborative computing. The requirement of a temporary communication infrastructure for quick communication with minimal configuration among a group of people in a conference or gathering necessitates the formation of an ad hoc wireless network. For example, consider a group of researchers who want to share their research findings or presentation materials during a conference, or a lecturer distributing notes to the class on the fly. In such cases, the formation of an ad hoc wireless network with the necessary support for reliable multicast routing can serve the purpose. The distributed file sharing applications utilized in such situations do not require the level of security expected in a military environment. But the reliability of data transfer is of high importance. Consider the example where a node that is part of an ad hoc wireless network has to distribute a file to other nodes in the network. Though this application does not demand the communication to be interruption-free, the goal of the transmission is that all the desired receivers must have the replica of the transmitted file. Other applications such as streaming of multimedia objects among the participating nodes in an ad hoc wireless network may require support for soft real-time communication. The users of such applications prefer economical and portable devices, usually powered by battery sources. Hence, a mobile node may drain its battery and can have varying transmission power, which may result in unidirectional links with its neighbors. Devices used for such applications could typically be laptops with add-on wireless interface cards, enhanced personal digital assistants (PDAs), or mobile devices with high processing power. In the presence of such heterogeneity, interoperability is an important issue.

Emergency Operations

Ad hoc wireless networks are very useful in emergency operations such as search and rescue, crowd control, and commando operations. The major factors that favor ad hoc wireless networks for such tasks are self-configuration of the system with minimal overhead, independent of fixed or centralized infrastructure, the nature of the terrain of such applications, the freedom and flexibility of mobility, and the unavailability of conventional communication infrastructure. In environments where the conventional infrastructure-based communication facilities are destroyed due to a war or due to natural calamities such as earthquakes, immediate deployment of ad hoc wireless networks would be a good solution for coordinating rescue activities. Since the ad hoc wireless networks require minimum initial network configuration for their functioning, very little or no delay is involved in making the network fully operational. The above-mentioned scenarios are unexpected, in most cases unavoidable, and can affect a large number of people. Ad hoc wireless networks employed in such circumstances should be distributed and scalable to a large number of nodes. They should also be able to provide fault-tolerant communication paths. Real-time communication capability is also important since voice communication predominates data communication in such situations.

Wireless Mesh Networks

Wireless mesh networks are ad hoc wireless networks that are formed to provide an alternate communication infrastructure for mobile or fixed nodes/users, without the spectrum reuse constraints and the requirements of network planning of cellular networks. The mesh topology of wireless mesh networks provides many alternate paths for a data transfer session between a source and destination, resulting in quick reconfiguration of the path when the existing path fails due to node failures. Wireless mesh networks provide the most economical data transfer capability coupled with the freedom of mobility. Since the infrastructure built is in the form of small radio relaying devices fixed on the rooftops of the houses in a residential zone as shown in Figure 5.4, or similar devices fitted on the lamp posts as depicted in Figure 5.5, the investment required in wireless mesh networks is much less than what is

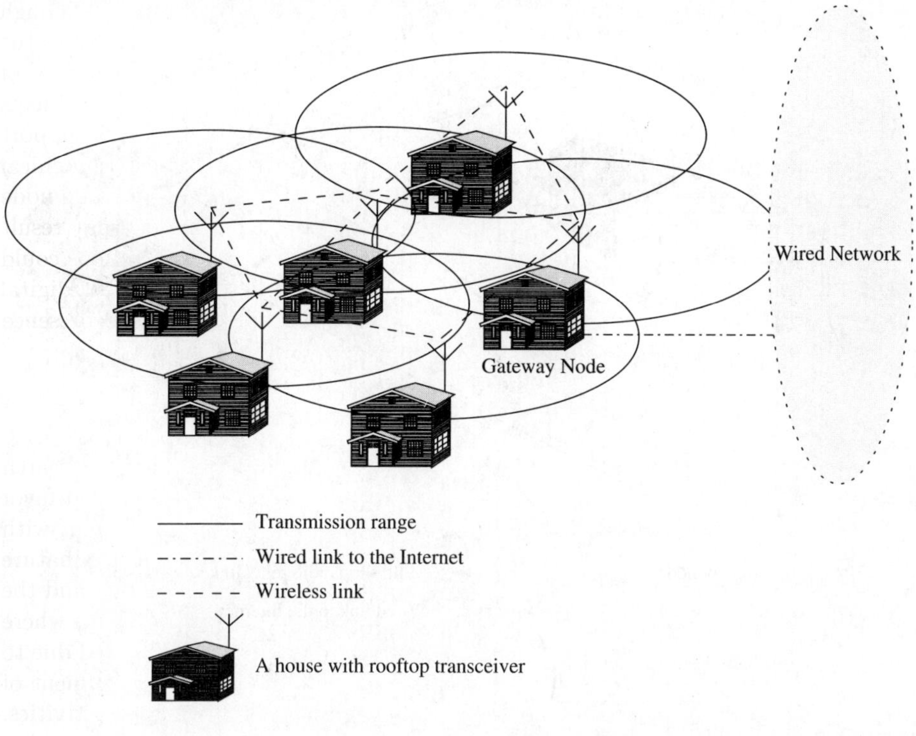

Figure 5.4. Wireless mesh network operating in a residential zone.

required for the cellular network counterparts. Such networks are formed by placing wireless relaying equipment spread across the area to be covered by the network. The possible deployment scenarios of wireless mesh networks include: residential zones (where broadband Internet connectivity is required), highways (where a communication facility for moving automobiles is required), business zones (where an alternate communication system to cellular networks is required), important civilian regions (where a high degree of service availability is required), and university campuses (where inexpensive campus-wide network coverage can be provided). Wireless mesh networks should be capable of self-organization and maintenance. The ability of the network to overcome single or multiple node failures resulting from disasters makes it convenient for providing the communication infrastructure for strategic applications. The major advantages of wireless mesh networks are support for a high data rate, quick and low cost of deployment, enhanced services, high scalability, easy extendability, high availability, and low cost per bit. Wireless mesh networks operate at the license-free ISM bands around 2.4 GHz and 5 GHz. Depending on the technology used for the physical layer and MAC layer communication, data rates of 2 Mbps to 60 Mbps can be supported. For example, if IEEE 802.11a is

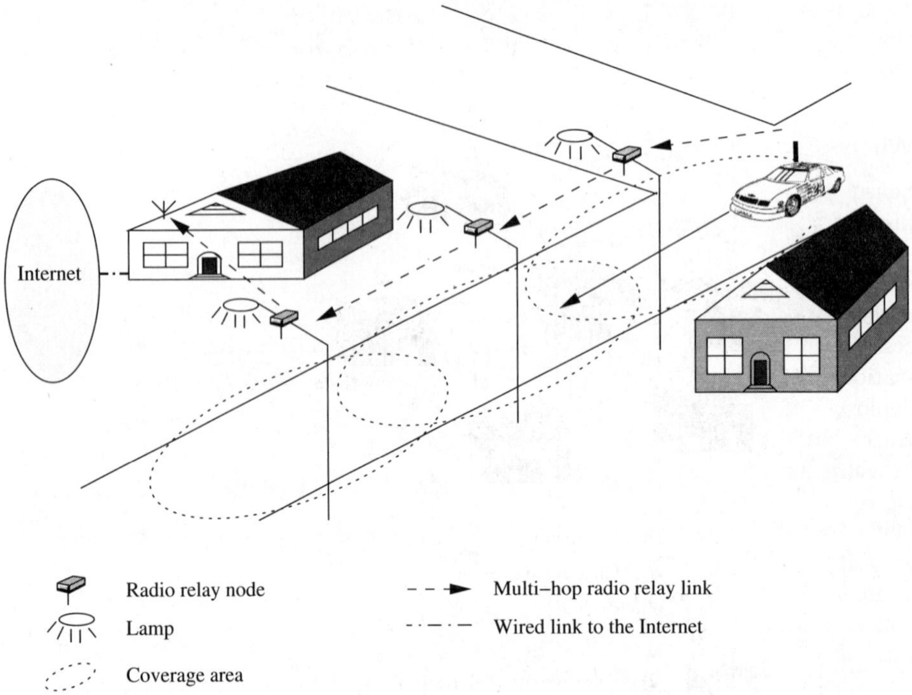

Figure 5.5. Wireless mesh network covering a highway.

used, a maximum data rate of 54 Mbps can be supported. The deployment time required for this network is much less than that provided by other infrastructure-based networks. Incremental deployment or partial batch deployment can also be done. Wireless mesh networks provide a very economical communication infrastructure in terms of both deployment and data transfer costs. Services such as smart environments that update information about the environment or locality to the visiting nodes are also possible in such an environment. A truck driver can utilize enhanced location discovery services, and hence spotting his location on an updated digital map is possible. Mesh networks scale well to provide support to a large number of nodes. Even at a very high density of mobile nodes, by employing power control at the mobile nodes and relay nodes, better system throughput and support for a large number of users can be achieved. But in the case of cellular networks, improving scalability requires additional infrastructural nodes, which in turn involves high cost. As mentioned earlier, mesh networks provide expandability of service in a cost-effective manner. Partial roll out and commissioning of the network and extending the service in a seamless manner without affecting the existing installation are the benefits from the viewpoint of service providers. Wireless mesh networks provide very high availability compared to the existing cellular architecture, where the presence of a fixed base station that covers a much larger area involves the risk of a single point of failure.

Wireless Sensor Networks

Sensor networks are a special category of ad hoc wireless networks that are used to provide a wireless communication infrastructure among the sensors deployed in a specific application domain. Recent advances in wireless communication technology and research in ad hoc wireless networks have made smart sensing a reality. Sensor nodes are tiny devices that have the capability of sensing physical parameters, processing the data gathered, and communicating over the network to the monitoring station. A sensor network is a collection of a large number of sensor nodes that are deployed in a particular region. The activity of sensing can be periodic or sporadic. An example for the periodic type is the sensing of environmental factors for the measurement of parameters such as temperature, humidity, and nuclear radiation. Detecting border intrusion, sensing the temperature of a furnace to prevent it rising beyond a threshold, and measuring the stress on critical structures or machinery are examples of the sensing activities that belong to the sporadic type. Some of the domains of application for sensor networks are military, health care, home security, and environmental monitoring. The issues that make sensor networks a distinct category of ad hoc wireless networks are the following:

- **Mobility of nodes:** Mobility of nodes is not a mandatory requirement in sensor networks. For example, the nodes deployed for periodic monitoring of soil properties are not required to be mobile. However, the sensor nodes that are fitted on the bodies of patients in a post-surgery ward of a hospital may be designed to support limited or partial mobility. In general, sensor networks need not in all cases be designed to support mobility of sensor nodes.

- **Size of the network:** The number of nodes in the sensor network can be much larger than that in a typical ad hoc wireless network.

- **Density of deployment:** The density of nodes in a sensor network varies with the domain of application. For example, military applications require high availability of the network, making redundancy a high priority.

- **Power constraints:** The power constraints in sensor networks are much more stringent than those in ad hoc wireless networks. This is mainly because the sensor nodes are expected to operate in harsh environmental or geographical conditions, with minimum or no human supervision and maintenance. In certain cases, the recharging of the energy source is impossible. Running such a network, with nodes powered by a battery source with limited energy, demands very efficient protocols at network, data link, and physical layer. The power sources used in sensor networks can be classified into the following three categories:

- **Replenishable power source:** In certain applications of sensor networks, the power source can be replaced when the existing source is fully drained (*e.g.*, wearable sensors that are used to sense body parameters).
- **Non-replenishable power source:** In some specific applications of sensor networks, the power source cannot be replenished once the network has been deployed. The replacement of the sensor node is the only solution to it (*e.g.*, deployment of sensor nodes in a remote, hazardous terrain).
- **Regenerative power source:** Power sources employed in sensor networks that belong to this category have the capability of regenerating power from the physical parameter under measurement. For example, the sensor employed for sensing temperature at a power plant can use power sources that can generate power by using appropriate transducers.

- **Data/information fusion:** The limited bandwidth and power constraints demand aggregation of bits and information at the intermediate relay nodes that are responsible for relaying. Data fusion refers to the aggregation of multiple packets into one before relaying it. This mainly aims at reducing the bandwidth consumed by redundant headers of the packets and reducing the media access delay involved in transmitting multiple packets. Information fusion aims at processing the sensed data at the intermediate nodes and relaying the outcome to the monitor node.

- **Traffic distribution:** The communication traffic pattern varies with the domain of application in sensor networks. For example, the environmental sensing application generates short periodic packets indicating the status of the environmental parameter under observation to a central monitoring station. This kind of traffic demands low bandwidth. The sensor network employed in detecting border intrusions in a military application generates traffic on detection of certain events; in most cases these events might have time constraints for delivery. In contrast, ad hoc wireless networks generally carry user traffic such as digitized and packetized voice stream or data traffic, which demands higher bandwidth.

Hybrid Wireless Networks

One of the major application areas of ad hoc wireless networks is in hybrid wireless architectures such as multi-hop cellular networks (MCNs) [5] and [3] and integrated cellular ad hoc relay (iCAR) networks [6]. The tremendous growth in the subscriber base of existing cellular networks has shrunk the cell size up to the pico-cell level. The primary concept behind cellular networks is geographical channel reuse. Several techniques such as cell sectoring, cell resizing, and multi-tier cells have been proposed to increase the capacity of cellular networks. Most of these schemes also increase the equipment cost. The capacity (maximum throughput) of a cellular network can be increased if the network incorporates the properties of multi-hop relaying along with the support of existing fixed infrastructure. MCNs combine the reliability and support of fixed base stations of cellular networks with flexibility and multi-hop relaying of ad hoc wireless networks.

The MCN architecture is depicted in Figure 5.6. In this architecture, when two nodes (which are not in direct transmission range) in the same cell want to communicate with each other, the connection is routed through multiple wireless hops over the intermediate nodes. The base station maintains the information about the topology of the network for efficient routing. The base station may or may not be involved in this multi-hop path. Suppose node A wants to communicate with node B. If all nodes are capable of operating in MCN mode, node A can reach node B directly if the node B is within node A's transmission range. When node C wants to communicate with node E and both are in the same cell, node C can reach node E through node D, which acts as an intermediate relay node. Such hybrid wireless

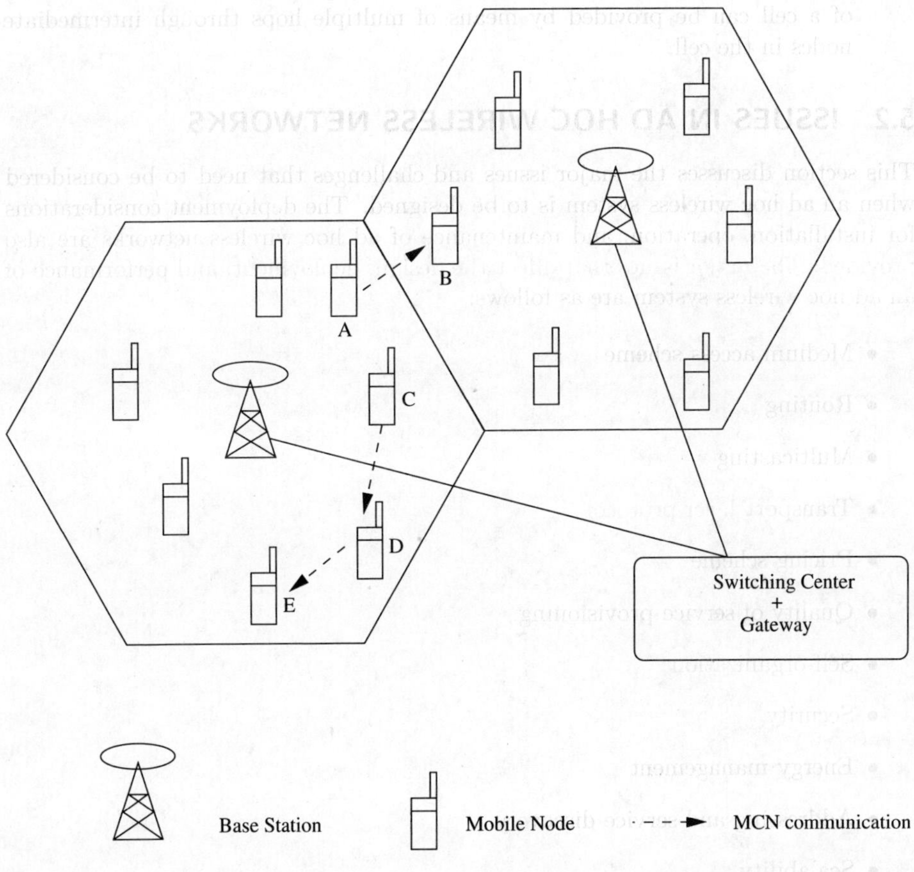

Figure 5.6. MCN architecture.

networks can provide high capacity resulting in lowering the cost of communication to less than that in single-hop cellular networks. The major advantages of hybrid wireless networks are as follows:

- Higher capacity than cellular networks obtained due to the better channel reuse provided by reduction of transmission power, as mobile nodes use a power range that is a fraction of the cell radius.
- Increased flexibility and reliability in routing. The flexibility is in terms of selecting the best suitable nodes for routing, which is done through multiple mobile nodes or through base stations, or by a combination of both. The increased reliability is in terms of resilience to failure of base stations, in which case a node can reach other nearby base stations using multi-hop paths.
- Better coverage and connectivity in holes (areas that are not covered due to transmission difficulties such as antenna coverage or the direction of antenna) of a cell can be provided by means of multiple hops through intermediate nodes in the cell.

5.2 ISSUES IN AD HOC WIRELESS NETWORKS

This section discusses the major issues and challenges that need to be considered when an ad hoc wireless system is to be designed. The deployment considerations for installation, operation, and maintenance of ad hoc wireless networks are also provided. The major issues that affect the design, deployment, and performance of an ad hoc wireless system are as follows:

- Medium access scheme
- Routing
- Multicasting
- Transport layer protocol
- Pricing scheme
- Quality of service provisioning
- Self-organization
- Security
- Energy management
- Addressing and service discovery
- Scalability
- Deployment considerations

5.2.1 Medium Access Scheme

The primary responsibility of a medium access control (MAC) protocol in ad hoc wireless networks is the distributed arbitration for the shared channel for transmission of packets. The performance of any wireless network hinges on the MAC protocol, more so for ad hoc wireless networks. The major issues to be considered in designing a MAC protocol for ad hoc wireless networks are as follows:

- **Distributed operation:** The ad hoc wireless networks need to operate in environments where no centralized coordination is possible. The MAC protocol design should be fully distributed involving minimum control overhead. In the case of polling-based MAC protocols, partial coordination is required.

- **Synchronization:** The MAC protocol design should take into account the requirement of time synchronization. Synchronization is mandatory for TDMA-based systems for management of transmission and reception slots. Synchronization involves usage of scarce resources such as bandwidth and battery power. The control packets used for synchronization can also increase collisions in the network.

- **Hidden terminals:** Hidden terminals are nodes that are hidden (or not reachable) from the sender of a data transmission session, but are reachable to the receiver of the session. In such cases, the hidden terminal can cause collisions at the receiver node. The presence of hidden terminals can significantly reduce the throughput of a MAC protocol used in ad hoc wireless networks. Hence the MAC protocol should be able to alleviate the effects of hidden terminals.

- **Exposed terminals:** Exposed terminals, the nodes that are in the transmission range of the sender of an on-going session, are prevented from making a transmission. In order to improve the efficiency of the MAC protocol, the exposed nodes should be allowed to transmit in a controlled fashion without causing collision to the on-going data transfer.

- **Throughput:** The MAC protocol employed in ad hoc wireless networks should attempt to maximize the throughput of the system. The important considerations for throughput enhancement are minimizing the occurrence of collisions, maximizing channel utilization, and minimizing control overhead.

- **Access delay:** The access delay refers to the average delay that any packet experiences to get transmitted. The MAC protocol should attempt to minimize the delay.

- **Fairness:** Fairness refers to the ability of the MAC protocol to provide an equal share or weighted share of the bandwidth to all competing nodes. Fairness can be either node-based or flow-based. The former attempts to provide an equal bandwidth share for competing nodes whereas the latter provides an equal share for competing data transfer sessions. In ad hoc wireless networks, fairness is important due to the multi-hop relaying done by the nodes. An unfair relaying load for a node results in draining the resources of that node much faster than that of other nodes.

- **Real-time traffic support:** In a contention-based channel access environment, without any central coordination, with limited bandwidth, and with location-dependent contention, supporting time-sensitive traffic such as voice, video, and real-time data requires explicit support from the MAC protocol.

- **Resource reservation:** The provisioning of QoS defined by parameters such as bandwidth, delay, and jitter requires reservation of resources such as bandwidth, buffer space, and processing power. The inherent mobility of nodes in ad hoc wireless networks makes such reservation of resources a difficult task. A MAC protocol should be able to provide mechanisms for supporting resource reservation and QoS provisioning.

- **Ability to measure resource availability:** In order to handle the resources such as bandwidth efficiently and perform call admission control based on their availability, the MAC protocol should be able to provide an estimation of resource availability at every node. This can also be used for making congestion-control decisions.

- **Capability for power control:** The transmission power control reduces the energy consumption at the nodes, causes a decrease in interference at neighboring nodes, and increases frequency reuse. Support for power control at the MAC layer is very important in the ad hoc wireless environment.

- **Adaptive rate control:** This refers to the variation in the data bit rate achieved over a channel. A MAC protocol that has adaptive rate control can make use of a high data rate when the sender and receiver are nearby and adaptively reduce the data rate as they move away from each other.

- **Use of directional antennas:** This has many advantages that include increased spectrum reuse, reduction in interference, and reduced power consumption. Most of the existing MAC protocols that use omnidirectional antennas do not work with directional antennas.

5.2.2 Routing

The responsibilities of a routing protocol include exchanging the route information; finding a feasible path to a destination based on criteria such as hop length, minimum power required, and lifetime of the wireless link; gathering information about the path breaks; mending the broken paths expending minimum processing power and bandwidth; and utilizing minimum bandwidth. The major challenges that a routing protocol faces are as follows:

- **Mobility:** One of the most important properties of ad hoc wireless networks is the mobility associated with the nodes. The mobility of nodes results in frequent path breaks, packet collisions, transient loops, stale routing information, and difficulty in resource reservation. A good routing protocol should be able to efficiently solve all the above issues.

- **Bandwidth constraint:** Since the channel is shared by all nodes in the broadcast region (any region in which all nodes can hear all other nodes), the bandwidth available per wireless link depends on the number of nodes and the traffic they handle. Thus only a fraction of the total bandwidth is available for every node.

- **Error-prone and shared channel:** The bit error rate (BER) in a wireless channel is very high (of the order of 10^{-5} to 10^{-3}) compared to that in its wired counterparts (of the order of 10^{-12} to 10^{-9}). Routing protocols designed for ad hoc wireless networks should take this into account. Consideration of the state of the wireless link, signal-to-noise ratio, and path loss for routing in ad hoc wireless networks can improve the efficiency of the routing protocol.

- **Location-dependent contention:** The load on the wireless channel varies with the number of nodes present in a given geographical region. This makes the contention for the channel high when the number of nodes increases. The high contention for the channel results in a high number of collisions and a subsequent wastage of bandwidth. A good routing protocol should have built-in mechanisms for distributing the network load uniformly across the network so that the formation of regions where channel contention is high can be avoided.

- **Other resource constraints:** The constraints on resources such as computing power, battery power, and buffer storage also limit the capability of a routing protocol.

The major requirements of a routing protocol in ad hoc wireless networks are the following:

- **Minimum route acquisition delay:** The route acquisition delay for a node that does not have a route to a particular destination node should be as minimal as possible. This delay may vary with the size of the network and the network load.

- **Quick route reconfiguration:** The unpredictable changes in the topology of the network require that the routing protocol be able to quickly perform route reconfiguration in order to handle path breaks and subsequent packet losses.

- **Loop-free routing:** This is a fundamental requirement of any routing protocol to avoid unnecessary wastage of network bandwidth. In ad hoc wireless networks, due to the random movement of nodes, transient loops may form in the route thus established. A routing protocol should detect such transient routing loops and take corrective actions.

- **Distributed routing approach:** An ad hoc wireless network is a fully distributed wireless network and the use of centralized routing approaches in such a network may consume a large amount of bandwidth.

- **Minimum control overhead:** The control packets exchanged for finding a new route and maintaining existing routes should be kept as minimal as possible. The control packets consume precious bandwidth and can cause collisions with data packets, thereby reducing network throughput.

- **Scalability:** Scalability is the ability of the routing protocol to scale well (*i.e.*, perform efficiently) in a network with a large number of nodes. This requires minimization of control overhead and adaptation of the routing protocol to the network size.

- **Provisioning of QoS:** The routing protocol should be able to provide a certain level of QoS as demanded by the nodes or the category of calls. The QoS parameters can be bandwidth, delay, jitter, packet delivery ratio, and throughput. Supporting differentiated classes of service may be of importance in tactical operations.

- **Support for time-sensitive traffic:** Tactical communications and similar applications require support for time-sensitive traffic. The routing protocol should be able to support both hard real-time and soft real-time traffic.

- **Security and privacy:** The routing protocol in ad hoc wireless networks must be resilient to threats and vulnerabilities. It must have inbuilt capability to avoid resource consumption, denial-of-service, impersonation, and similar attacks possible against an ad hoc wireless network.

5.2.3 Multicasting

Multicasting plays an important role in the typical applications of ad hoc wireless networks, namely, emergency search-and-rescue operations and military communication. In such an environment, nodes form groups to carry out certain tasks that require point-to-multipoint and multipoint-to-multipoint voice and data communication. The arbitrary movement of nodes changes the topology dynamically in an unpredictable manner. The mobility of nodes, with the constraints of power source and bandwidth, makes multicast routing very challenging. Traditional wired network multicast protocols such as core based trees (CBT), protocol independent multicast (PIM), and distance vector multicast routing protocol (DVMRP) do not perform well in ad hoc wireless networks because a tree-based multicast structure is highly unstable and needs to be frequently readjusted to include broken links. Use of any global routing structure such as the link-state table results in high control overhead. The use of single-link connectivity among the nodes in a multicast group results in a tree-shaped multicast routing topology. Such a tree-shaped topology provides high multicast efficiency, with low packet delivery ratio due to the frequent tree breaks. Provisioning of multiple links among the nodes in an ad hoc wireless network results in a mesh-shaped structure. The mesh-based multicast routing structure may work well in a high-mobility environment. The major issues in designing multicast routing protocols are as follows:

- **Robustness:** The multicast routing protocol must be able to recover and reconfigure quickly from potential mobility-induced link breaks thus making it suitable for use in highly dynamic environments.

- **Efficiency:** A multicast protocol should make a minimum number of transmissions to deliver a data packet to all the group members.

- **Control overhead:** The scarce bandwidth availability in ad hoc wireless networks demands minimal control overhead for the multicast session.

- **Quality of service:** QoS support is essential in multicast routing because, in most cases, the data transferred in a multicast session is time-sensitive.

- **Efficient group management:** Group management refers to the process of accepting multicast session members and maintaining the connectivity among them until the session expires. This process of group management needs to be performed with minimal exchange of control messages.

- **Scalability:** The multicast routing protocol should be able to scale for a network with a large number of nodes.

- **Security:** Authentication of session members and prevention of non-members from gaining unauthorized information play a major role in military communications.

5.2.4 Transport Layer Protocols

The main objectives of the transport layer protocols include setting up and maintaining end-to-end connections, reliable end-to-end delivery of data packets, flow control, and congestion control. There exist simple connectionless transport layer protocols (*e.g.*, UDP) which neither perform flow control and congestion control nor provide reliable data transfer. Such unreliable connectionless transport layer protocols do not take into account the current network status such as congestion at the intermediate links, the rate of collision, or other similar factors affecting the network throughput. This behavior of the transport layer protocols increases the contention of the already-choked wireless links. For example, in an ad hoc wireless network that employs a contention-based MAC protocol, nodes in a high-contention region experience several backoff states, resulting in an increased number of collisions and a high latency. Connectionless transport layer protocols, unaware of this situation, increase the load in the network, degrading the network performance.

The major performance degradation faced by a reliable connection-oriented transport layer protocol such as transmission control protocol (TCP) in an ad hoc wireless network arises due to frequent path breaks, presence of stale routing information, high channel error rate, and frequent network partitions.

Further discussion of each of the above properties and their effect on the performance of the transport layer protocol assumes TCP as the transport layer protocol. Due to the mobility of nodes and limited transmission range, an existing path to a destination node experiences frequent path breaks. Each path break results in route reconfiguration that depends on the routing protocol employed. The process of finding an alternate path or reconfiguring the broken path might take longer than the retransmission timeout of the transport layer at the sender, resulting in retransmission of packets and execution of the congestion control algorithm. The congestion control algorithm decreases the size of the congestion window, resulting in low throughput. In an environment where path breaks are frequent, the execution of congestion control algorithms on every path break affects the throughput drastically.

The latency associated with the reconfiguration of a broken path and the use of route caches result in stale route information at the nodes. Hence the packets will be forwarded through multiple paths to a destination, causing an increase in the number of out-of-order packets. Also, multipath routing protocols such as temporally-ordered routing algorithm (TORA) [7] and split multipath routing (SMR) protocols ([8] and [9]) employ multiple paths between a source-destination pair. Out-of-order packet arrivals force the receiver of the TCP connection to generate duplicate acknowledgments (ACKs). On receiving duplicate ACKs, the sender invokes the congestion control algorithm.

Wireless channels are inherently unreliable due to the high probability of errors caused by interference. In addition to the error due to the channel noise, the presence of hidden terminals also contributes to the increased loss of TCP data packets or ACKs. When the TCP ACK is delayed more than the round-trip timeout, the congestion control algorithm is invoked.

Due to the mobility of the nodes, ad hoc wireless networks frequently experience isolation of nodes from the rest of the network or occurrence of partitions in the network. If a TCP connection spans across multiple partitions, that is, the sender and receiver of the connection are in two different partitions, all the packets get dropped. This tends to be more serious when the partitions exist for a long duration, resulting in multiple retransmissions of the TCP packets and subsequent increase in the retransmission timers. Such a behavior causes long periods of inactivity even when a transient partition in the network lasts for a short while.

Adaptation of the existing transport layer protocols should attempt to handle the above issues for performing efficiently in ad hoc wireless networks.

5.2.5 Pricing Scheme

An ad hoc wireless network's functioning depends on the presence of relaying nodes and their willingness to relay other nodes' traffic. Even if the node density is sufficient enough to ensure a fully connected network, a relaying neighbor node may not be interested in relaying a call and may just decide to power down. Assume that an optimal route from node A to node B passes through node C, and node C is not powered on. Then node A will have to set up a costlier and non-optimal route to B. The non-optimal path consumes more resources and affects the throughput of the system. As the intermediate nodes in a path that relay the data packets expend their resources such as battery charge and computing power, they should be properly compensated. Hence pricing schemes that incorporate service compensation or service reimbursement are required. Ad hoc wireless networks employed for special tasks such as military missions, rescue operations, and law enforcement do not require such pricing schemes, whereas the successful commercial deployment of ad hoc wireless networks requires billing and pricing. The obvious solution to provide participation guarantee is to provide incentives to forwarding nodes.

5.2.6 Quality of Service Provisioning

Quality of service (QoS) is the performance level of services offered by a service provider or a network to the user. QoS provisioning often requires negotiation between the host and the network, resource reservation schemes, priority scheduling, and call admission control. Rendering QoS in ad hoc wireless networks can be on a per flow, per link, or per node basis. In ad hoc wireless networks, the boundary between the service provider (network) and the user (host) is blurred, thus making it essential to have better coordination among the hosts to achieve QoS. The lack of central coordination and limited resources exacerbate the problem. In this section, a brief discussion of QoS parameters, QoS-aware routing, and QoS frameworks in ad hoc wireless networks is provided.

- **QoS parameters:** As different applications have different requirements, their level of QoS and the associated QoS parameters also differ from application to application. For example, for multimedia applications, the bandwidth and delay are the key parameters, whereas military applications have the additional requirements of security and reliability. For defense applications, finding trustworthy intermediate hosts and routing through them can be a QoS parameter. For applications such as emergency search-and-rescue operations, availability is the key QoS parameter. Multiple link disjoint paths can be the major requirement for such applications. Applications for hybrid wireless networks can have maximum available link life, delay, channel utilization, and bandwidth as the key parameters for QoS. Finally, applications such as communication among the nodes in a sensor network require that the transmission among them results in minimum energy consumption, hence battery life and energy conservation can be the prime QoS parameters here.

- **QoS-aware routing:** The first step toward a QoS-aware routing protocol is to have the routing use QoS parameters for finding a path. The parameters that can be considered for routing decisions are network throughput, packet delivery ratio, reliability, delay, delay jitter, packet loss rate, bit error rate, and path loss. Decisions on the level of QoS and the related parameters for such services in ad hoc wireless networks are application-specific and are to be met by the underlying network. For example, in the case where the QoS parameter is bandwidth, the routing protocol utilizes the available bandwidth at every link to select a path with necessary bandwidth. This also demands the capability to reserve the required amount of bandwidth for that particular connection.

- **QoS framework:** A framework for QoS is a complete system that attempts to provide the promised services to each user or application. All the components within this subsystem should cooperate in providing the required services. The key component of QoS framework is a QoS service model which defines the way user requirements are served. The key design issue is whether to serve the user on a per-session basis or a per-class basis. Each class represents an aggregation of users based on certain criteria. The other key components of this framework are QoS routing to find all or some feasible paths in the network that can satisfy user requirements, QoS signaling for resource reservation required by the user or application, QoS medium access control, connection admission control, and scheduling schemes pertaining to that service model. The QoS modules such as routing protocol, signaling protocol, and resource management should react promptly according to changes in the network state (topology change in ad hoc wireless networks) and flow state (change in end-to-end view of service delivered).

5.2.7 Self-Organization

One very important property that an ad hoc wireless network should exhibit is organizing and maintaining the network by itself. The major activities that an ad hoc wireless network is required to perform for self-organization are neighbor discovery, topology organization, and topology reorganization. During the neighbor discovery phase, every node in the network gathers information about its neighbors and maintains that information in appropriate data structures. This may require periodic transmission of short packets named *beacons*, or promiscuous snooping on the channel for detecting activities of neighbors. Certain MAC protocols permit varying the transmission power to improve upon spectrum reusability. In the topology organization phase, every node in the network gathers information about the entire network or a part of the network in order to maintain topological information.

During the topology reorganization phase, the ad hoc wireless networks require updating the topology information by incorporating the topological changes occurred in the network due to the mobility of nodes, failure of nodes, or complete depletion of power sources of the nodes. The reorganization consists of two major activities. First is the periodic or aperiodic exchange of topological information. Second is the adaptability (recovery from major topological changes in the network).

Similarly, network partitioning and merging of two existing partitions require major topological reorganization. Ad hoc wireless networks should be able to perform self-organization quickly and efficiently in a way transparent to the user and the application.

5.2.8 Security

The security of communication in ad hoc wireless networks is very important, especially in military applications. The lack of any central coordination and shared wireless medium makes them more vulnerable to attacks than wired networks. The attacks against ad hoc wireless networks are generally classified into two types: passive and active attacks. Passive attacks refer to the attempts made by malicious nodes to perceive the nature of activities and to obtain information transacted in the network without disrupting the operation. Active attacks disrupt the operation of the network. Those active attacks that are executed by nodes outside the network are called external attacks, and those that are performed by nodes belonging to the same network are called internal attacks. Nodes that perform internal attacks are compromised nodes. The major security threats that exist in ad hoc wireless networks are as follows:

- **Denial of service:** The attack effected by making the network resource unavailable for service to other nodes, either by consuming the bandwidth or by overloading the system, is known as denial of service (DoS). A simple scenario in which a DoS attack interrupts the operation of ad hoc wireless networks is by keeping a target node busy by making it process unnecessary packets.

- **Resource consumption:** The scarce availability of resources in ad hoc wireless network makes it an easy target for internal attacks, particularly aiming at consuming resources available in the network. The major types of resource-consumption attacks are the following:

  - **Energy depletion:** Since the nodes in ad hoc wireless networks are highly constrained by the energy source, this type of attack is basically aimed at depleting the battery power of critical nodes by directing unnecessary traffic through them.
  - **Buffer overflow:** The buffer overflow attack is carried out either by filling the routing table with unwanted routing entries or by consuming the data packet buffer space with unwanted data. Such attacks can lead to a large number of data packets being dropped, leading to the loss of critical information. Routing table attacks can lead to many problems, such as preventing a node from updating route information for important destinations and filling the routing table with routes for nonexistent destinations.

- **Host impersonation:** A compromised internal node can act as another node and respond with appropriate control packets to create wrong route entries, and can terminate the traffic meant for the intended destination node.

- **Information disclosure:** A compromised node can act as an informer by deliberate disclosure of confidential information to unauthorized nodes. Information such as the amount and the periodicity of traffic between a selected pair of nodes and pattern of traffic changes can be very valuable for military applications. The use of filler traffic (traffic generated for the sole purpose of changing the traffic pattern) may not be suitable in resource-constrained ad hoc wireless networks.

- **Interference:** A common attack in defense applications is to jam the wireless communication by creating a wide-spectrum noise. This can be done by using a single wide-band jammer, sweeping across the spectrum. The MAC and the physical layer technologies should be able to handle such external threats.

5.2.9 Addressing and Service Discovery

Addressing and service discovery assume significance in ad hoc wireless networks due to the absence of any centralized coordinator. An address that is globally unique in the connected part of the ad hoc wireless network is required for a node in order to participate in communication. Auto-configuration of addresses is required to allocate non-duplicate addresses to the nodes. In networks where the topology is highly dynamic, frequent partitioning and merging of network components require duplicate address-detection mechanisms in order to maintain unique addressing throughout the connected parts of the network. Nodes in the network should be able to locate services that other nodes provide. Hence efficient service advertisement mechanisms are necessary. Topological changes force a change in the location of the service provider as well, hence fixed positioning of a server providing a particular service is ruled out. Rather, identifying the current location of the service provider gathers importance. The integration of service discovery with the route-acquisition mechanism, though it violates the traditional design objectives of the routing protocol, is a viable alternative. However, provisioning of certain kinds of services demands authentication, billing, and privacy that in turn require the service discovery protocols to be separated from the network layer protocols.

5.2.10 Energy Management

Energy management is defined as the process of managing the sources and consumers of energy in a node or in the network as a whole for enhancing the lifetime of the network. Shaping the energy discharge pattern of a node's battery to enhance the battery life; finding routes that result in minimum total energy consumption in the network; using distributed scheduling schemes to improve battery life; and handling the processor and interface devices to minimize power consumption are some of the functions of energy management. Energy management can be classified into the following categories:

- **Transmission power management:** The power consumed by the radio frequency (RF) module of a mobile node is determined by several factors such as the state of operation, the transmission power, and the technology used for the RF circuitry. The state of operation refers to the transmit, receive, and sleep modes of the operation. The transmission power is determined by the reachability requirement of the network, the routing protocol, and the MAC protocol employed.

The RF hardware design should ensure minimum power consumption in all the three states of operation. Going to the sleep mode when not transmitting or receiving can be done by additional hardware that can wake up on reception of a control signal. Power conservation responsibility lies across the data link, network, transport, and application layers. By designing a data link layer protocol that reduces unnecessary retransmissions, by preventing collisions, by switching to standby mode or sleep mode whenever possible, and by reducing the transmit/receive switching, power management can be performed at the data link layer.

The use of a variable power MAC protocol can lead to several advantages that include energy-saving at the nodes, increase in bandwidth reuse, and reduction in interference. Also, MAC protocols for directional antennas are at their infancy. The network layer routing protocols can consider battery life and relaying load of the intermediate nodes while selecting a path so that the load can be balanced across the network, in addition to optimizing and reducing the size and frequency of control packets. At the transport layer, reducing the number of retransmissions, and recognizing and handling the reason behind the packet losses locally, can be incorporated into the protocols. At the application layer, the power consumption varies with applications. In a mobile computer, the image/video processing/playback software and 3D gaming software consume higher power than other applications. Hence application software developed for mobile computers should take into account the aspect of power consumption as well.

- **Battery energy management:** The battery management is aimed at extending the battery life of a node by taking advantage of its chemical properties, discharge patterns, and by the selection of a battery from a set of batteries that is available for redundancy. Recent studies showed that pulsed discharge of a battery gives longer life than continuous discharge. Controlling the charging rate and discharging rate of the battery is important in avoiding early charging to the maximum charge or full discharge below the minimum threshold. This can be achieved by means of embedded charge controllers in the battery pack. Also, the protocols at the data link layer and network layer can be designed to make use of the discharge models. Monitoring of the battery for voltage levels, remaining capacity, and temperature so that proactive actions (such as incremental powering off of certain devices, or shutting down of the mobile node when the voltage crosses a threshold) can be taken is required.

- **Processor power management:** The clock speed and the number of instructions executed per unit time are some of the processor parameters that affect power consumption. The CPU can be put into different power saving modes during low processing load conditions. The CPU power can be completely turned off if the machine is idle for a long time. In such cases, interrupts can be used to turn on the CPU upon detection of user interaction or other events.

- **Devices power management:** Intelligent device management can reduce power consumption of a mobile node significantly. This can be done by the operating system (OS) by selectively powering down interface devices that are not used or by putting devices into different power-saving modes, depending on their usage. Advanced power management features built into the operating system and application softwares for managing devices effectively are required.

5.2.11 Scalability

Even though the number of nodes in an ad hoc wireless network does not grow in the same magnitude as today's Internet, the operation of a large number of nodes in the ad hoc mode is not far away. Traditional applications such as military, emergency operations, and crowd control may not lead to such a big ad hoc wireless network. Commercial deployments of ad hoc wireless networks that include wireless mesh networks show early trends for a widespread installation of ad hoc wireless networks for mainstream wireless communication.

For example, the latency of path-finding involved with an on-demand routing protocol in a large ad hoc wireless network may be unacceptably high. Similarly, the periodic routing overhead involved in a table-driven routing protocol may consume a significant amount of bandwidth in such large networks.

Also a large ad hoc wireless network cannot be expected to be formed by homogeneous nodes, raising issues such as widely varying resource capabilities across the nodes. A hierarchical topology-based system and addressing may be more suitable for large ad hoc wireless networks. Hybrid architectures that combine the multi-hop radio relaying in the presence of infrastructure may improve scalability.

5.2.12 Deployment Considerations

The deployment of ad hoc wireless networks involves actions different from those of wired networks. It requires a good amount of planning and estimation of future traffic growth over any link in the network. The time-consuming planning stage is followed by the actual deployment of the network. The cost and time required for laying copper cables or fiber cables make it difficult to reconfigure any partial deployment that has already been done. The deployment of a commercial ad hoc wireless network has the following benefits when compared to wired networks:

- **Low cost of deployment:** The use of multi-hop wireless relaying essentially eliminates the requirement of laying cables and maintenance in a commercial deployment of communication infrastructure. Hence the cost involved is much lower than that of wired networks.

- **Incremental deployment:** In commercial wireless WANs based on ad hoc wireless networks, deployment can be performed incrementally over geographical regions of the city. The deployed part of the network starts functioning immediately after the minimum configuration is done. For example, during the deployment process for covering a highway, whenever each radio relaying equipment is installed on the highway side, it can be commissioned.

- **Short deployment time:** Compared to wired networks, the deployment time is considerably less due to the absence of any wired links. Also, wiring a dense urban region is extremely difficult and time-consuming in addition to the inconvenience caused.

- **Reconfigurability:** The cost involved in reconfiguring a wired network covering a metropolitan area network (MAN) is very high compared to that of an ad hoc wireless network covering the same service area. Also, the incremental deployment of ad hoc wireless networks might demand changes in the topology of the fixed part (*e.g.,* the relaying devices fixed on lamp posts or rooftops) of the network at a later stage.

The issues and solutions for deployment of ad hoc wireless networks vary with the type of applications and the environment in which the networks are to be deployed. The following are the major issues to be considered in deploying an ad hoc wireless network:

- **Scenario of deployment:** The scenario of deployment assumes significance because the capability required for a mobile node varies with the environment in which it is used. The capabilities required for the mobile nodes that form an ad hoc wireless network among a fleet of ships are not the same as those required for forming an ad hoc wireless network among a set of notebook computers at a conference. The following are some of the different scenarios in which the deployment issues vary widely.

  - **Military deployment:** The military deployment of an ad hoc wireless network may be data-centric (*e.g.,* a wireless sensor network) or user-centric (*e.g.,* soldiers or armored vehicles carrying soldiers equipped with wireless communication devices). The data-centric networks handle a different pattern of data traffic and can be partially comprised of static nodes, whereas the user-centric network consists of highly mobile nodes with or without any support from any infrastructure (*e.g.,* military satellite constellations). The vehicle-mounted nodes have at their disposal better power sources and computational resources, whereas the hand-held devices are constrained by energy and computational resources. Thus the resource availability demands appropriate changes in the protocols employed. Also, the military environment requires secure communication. Routing should involve as few nodes as possible to avoid possible leakage of information. Flat addressing schemes are preferred to hierarchical addressing since the latter addressing requires paths to be set up through the hierarchy, and hence the chances of unreliable nodes forwarding the packets are high.

   - **Emergency operations deployment:** This kind of application scenario demands a quick deployment of rescue personnel equipped with hand-held communication equipment. Essentially, the network should provide support for time-sensitive traffic such as voice and video. Short data messaging can be used in case the resource constraints do not permit voice communication. Also in this scenario, a flat fixed addressing scheme with a static configuration is preferred. Typically, the size of the network for such applications is not more than 100 nodes. The nodes are fully mobile without expecting support from any fixed infrastructure.
   - **Commercial wide-area deployment:** One example of this deployment scenario is the wireless mesh networks. The aim of the deployment is to provide an alternate communication infrastructure for wireless communication in urban areas and areas where a traditional cellular base station cannot handle the traffic volume. This scenario assumes significance as it provides very low cost per bit transferred compared to the wide-area cellular network infrastructure. Another major advantage of this application is the resilience to failure of a certain number of nodes. Addressing, configuration, positioning of relaying nodes, redundancy of nodes, and power sources are the major issues in deployment. Billing, provisioning of QoS, security, and handling mobility are major issues that the service providers need to address.
   - **Home network deployment:** The deployment of a home area network needs to consider the limited range of the devices that are to be connected by the network. Given the short transmission ranges of a few meters, it is essential to avoid network partitions. Positioning of relay nodes at certain key locations of a home area network can solve this. Also, network topology should be decided so that every node is connected through multiple neighbors for availability.

- **Required longevity of network:** The deployment of ad hoc wireless networks should also consider the required longevity of the network. If the network is required for a short while (*e.g.,* the connectivity among a set of researchers at a conference and the connectivity required for coordination of a crowd control team), battery-powered mobile nodes can be used. When the connectivity is required for a longer duration of time, fixed radio relaying equipment with regenerative power sources can be deployed. A wireless mesh network with roof-top antennas deployed at a residential zone requires weather-proof packages so that the internal circuitry remains unaffected by the environmental conditions. In such an environment, the mesh connectivity is planned in such a way that the harsh atmospheric factors do not create network partitions.

- **Area of coverage:** In most cases, the area of coverage of an ad hoc wireless network is determined by the nature of application for which the network is set up. For example, the home area network is limited to the surroundings of a home. The mobile nodes' capabilities such as the transmission range and associated hardware, software, and power source should match the area of coverage required. In some cases where some nodes can be fixed and the network topology is partially or fully fixed, the coverage can be enhanced by means of directional antennas.

- **Service availability:** The availability of network service is defined as the ability of an ad hoc wireless network to provide service even with the failure of certain nodes. Availability assumes significance both in a fully mobile ad hoc wireless network used for tactical communication and in partially fixed ad hoc wireless networks used in commercial communication infrastructure such as wireless mesh networks. In the case of wireless mesh networks, the fixed nodes need to be placed in such a way that the failure of multiple nodes does not lead to lack of service in that area. In such cases, redundant inactive radio relaying devices can be placed in such a way that on the event of failure of an active relaying node, the redundant relaying device can take over its responsibilities.

- **Operational integration with other infrastructure:** Operational integration of ad hoc wireless networks with other infrastructures can be considered for improving the performance or gathering additional information, or for providing better QoS. In the military environment, integration of ad hoc wireless networks with satellite networks or unmanned aerial vehicles (UAVs) improves the capability of the ad hoc wireless networks. Several routing protocols assume the availability of the global positioning system (GPS), which is a satellite-based infrastructure by which the geographical location information can be obtained as a resource for network synchronization and geographical positioning. In the commercial world, the wireless mesh networks that service a given urban region can interoperate with the wide-area cellular infrastructure in order to provide better QoS and smooth handoffs across the networks. Handoff to a different network can be done in order to avoid call drops when a mobile node with an active call moves into a region where service is not provided by the current network.

- **Choice of protocols:** The choice of protocols at different layers of the protocol stack is to be done taking into consideration the deployment scenario. A TDMA-based and insecure MAC protocol may not be the best suited compared to a CDMA-based MAC protocol for a military application. The MAC protocol should ensure provisioning of security at the link level. At the network layer, the routing protocol has to be selected with care. A routing protocol that uses geographical information (GPS information) may not work well in situations where such information is not available. For example, the search-and-rescue operation teams that work in extreme terrains or underground or inside a building may not be able to use such a routing protocol. An ad hoc wireless network with nodes that cannot have their power sources replenished should use a routing protocol that does not employ periodic *bea-*

cons for routing. The periodic beacons, or routing updates, drain the battery with time. In situations of high mobility, for example, an ad hoc wireless network formed by devices connected to military vehicles, the power consumption may not be very important and hence one can employ beacon-based routing protocols for them. The updated information about connectivity leads to improved performance. In the case of deployment of wireless mesh networks, the protocols should make use of the fixed nodes to avoid unstable paths due to the mobility of the relaying nodes.

At the transport layer, the connection-oriented or connectionless protocol should be adapted to work in the environment in which the ad hoc wireless network is deployed. In a high-mobility environment, path breaks, network partitions, and remerging of partitions are to be considered, and appropriate actions should be taken at the higher layers. This can be extended to connectionless transport protocols to avoid congestion. Also, packet loss arising out of congestion, channel error, link break, and network partition is to be handled differently in different applications. The timer values at different layers of the protocol stack should be adapted to the deployment scenarios.

5.3 AD HOC WIRELESS INTERNET

Similar to the wireless Internet discussed in Chapter 4, the ad hoc wireless Internet extends the services of the Internet to the end users over an ad hoc wireless network. Some of the applications of the ad hoc wireless Internet are wireless mesh networks, provisioning of temporary Internet services to major conference venues, sports venues, temporary military settlements, battlefields, and broadband Internet services in rural regions. A schematic diagram of the ad hoc wireless Internet is shown in Figure 5.7.

The major issues to be considered for a successful ad hoc wireless Internet are the following:

- **Gateways:** Gateway nodes in the ad hoc wireless Internet are the entry points to the wired Internet. The major part of the service provisioning lies with the gateway nodes. Generally owned and operated by a service provider, gateways perform the following tasks: keeping track of the end users, bandwidth management, load balancing, traffic shaping, packet filtering, bandwidth fairness, and address, service, and location discovery.

- **Address mobility:** Similar to the Mobile IP discussed in Chapter 4, the ad hoc wireless Internet also faces the challenge of address mobility. This problem is worse here as the nodes operate over multiple wireless hops. Solutions such as Mobile IP can provide temporary alternatives for this.

- **Routing:** Routing is a major problem in the ad hoc wireless Internet, due to the dynamic topological changes, the presence of gateways, multi-hop relaying, and the hybrid character of the network. The possible solution for this is the use of a separate routing protocol, which is discussed in Chapter 7, for the

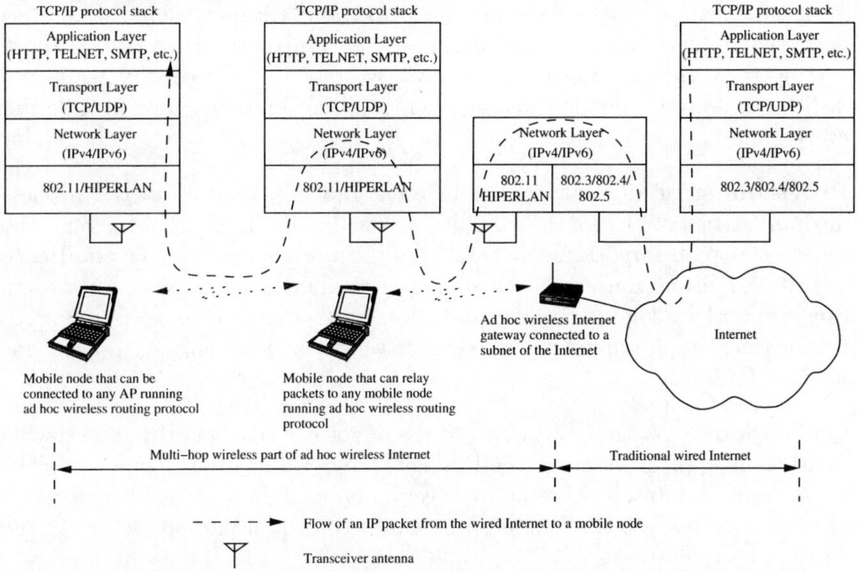

Figure 5.7. A schematic diagram of the ad hoc wireless Internet.

wireless part of the ad hoc wireless Internet. Routing protocols discussed in Chapter 13 are more suitable as they exploit the presence of gateway nodes.

- **Transport layer protocol:** Even though several solutions for transport layer protocols exist for ad hoc wireless networks, unlike other layers, the choice lies in favor of TCP's extensions proposed for ad hoc wireless networks. Split approaches that use traditional wired TCP for the wired part and a specialized transport layer protocol for the ad hoc wireless network part can also be considered where the gateways act as the intermediate nodes at which the connections are split. Several factors are to be considered here, the major one being the state maintenance overhead at the gateway nodes.

- **Load balancing:** It is likely that the ad hoc wireless Internet gateways experience heavy traffic. Hence the gateways can be saturated much earlier than other nodes in the network. Load balancing techniques are essential to distribute the load so as to avoid the situation where the gateway nodes become bottleneck nodes. Gateway selection strategies and load balancing schemes discussed in Chapter 13 can be used for this purpose.

- **Pricing/billing:** Since Internet bandwidth is expensive, it becomes very important to introduce pricing/billing strategies for the ad hoc wireless Internet. Gateway is the preferred choice for charging the traffic to and from the Internet. Pricing schemes discussed in Chapter 13 can be used for this purpose. A much more complex case is pricing the local traffic (traffic within the wireless part, that is, it originated and terminated within the wireless part without passing through the gateway nodes), where it becomes necessary to have a dedicated, secure, and lightweight pricing/billing infrastructure installed at every node.

- **Provisioning of security:** The inherent broadcast nature of the wireless medium attracts not just the mobility seekers but also potential hackers who love to snoop on important information sent unprotected over the air. Hence security is a prime concern in the ad hoc wireless Internet. Since the end users can utilize the ad hoc wireless Internet infrastructure to make e-commerce transactions, it is important to include security mechanisms in the ad hoc wireless Internet.

- **QoS support:** With the widespread use of voice over IP (VoIP) and growing multimedia applications over the Internet, provisioning of QoS support in the ad hoc wireless Internet becomes a very important issue. As discussed in Chapter 10, this is a challenging problem in the wired part as well as in the wireless part.

- **Service, address, and location discovery:** Service discovery in any network refers to the activity of discovering or identifying the party which provides a particular service or resource. In wired networks, service location protocols exist to do the same, and similar systems need to be extended to operate in the ad hoc wireless Internet as well. Address discovery refers to the services such as those provided by address resolution protocol (ARP) or domain name service (DNS) operating within the wireless domain. Location discovery refers to different activities such as detecting the location of a particular mobile node in the network or detecting the geographical location of nodes. Location discovery services can provide enhanced services such as routing of packets, location-based services, and selective region-wide broadcasts.

Figure 5.8 shows a wireless mesh network that connects several houses to the Internet through a gateway node. Such networks can provide highly reliable broadband wireless networks for the urban as well as the rural population in a cost-effective manner with fast deployment and reconfiguration. This wireless mesh network is a special case of the ad hoc wireless Internet where mobility of nodes is not a major concern as most relay stations and end users use fixed transceivers. Figure 5.8 shows that house A is connected to the Internet over multiple paths (path 1 and path 2).

5.4 SUMMARY

In this chapter, the major issues and applications of ad hoc wireless networks were described. The design issues for every layer of protocol stack and deployment scenarios were discussed. The applications of ad hoc wireless networks include military

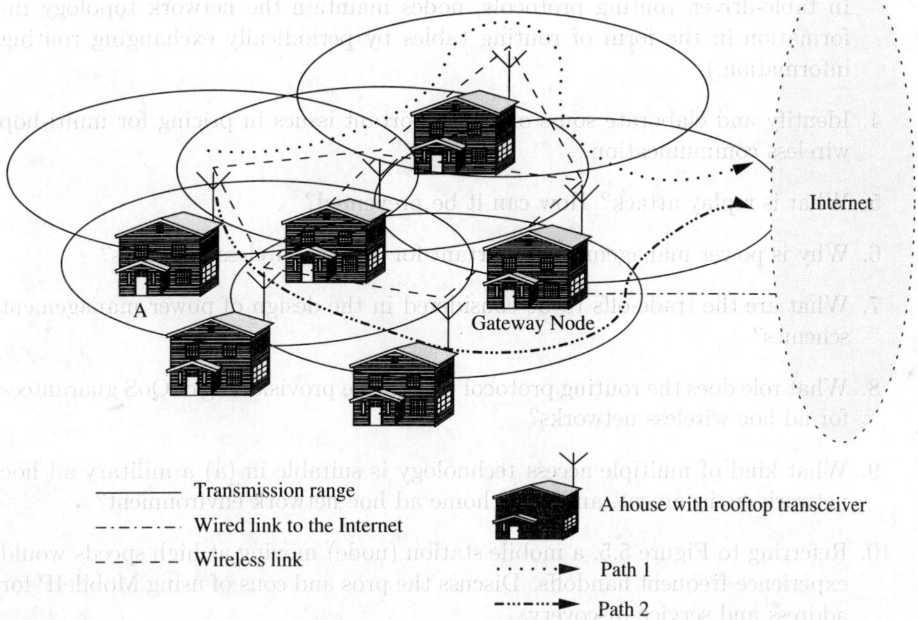

Figure 5.8. An illustration of the ad hoc wireless Internet implemented by a wireless mesh network.

applications, collaborative and distributed computing, emergency operations, and hybrid wireless architectures. The important deployment issues for ad hoc wireless networks are scenario of deployment, required longevity of network, area of coverage, service availability, operational integration with other infrastructure, and choice of protocols. Each of the deployment issues has been discussed in detail. The ad hoc wireless Internet was another important topic discussed in this chapter, in which the objectives, challenges, and the application scenarios were discussed.

5.5 PROBLEMS

1. Describe a common method used in alleviating the hidden terminal problem at the MAC layer.

2. How is the loop-free property ensured in on-demand routing protocols? (Hint: On-demand routing protocols do not maintain network topology information and obtain the necessary path as and when required by using a connection-establishment process.)

3. How is the loop-free property ensured in table-driven routing protocols? (Hint: In table-driven routing protocols, nodes maintain the network topology information in the form of routing tables by periodically exchanging routing information.)

4. Identify and elaborate some of the important issues in pricing for multi-hop wireless communication.

5. What is replay attack? How can it be prevented?

6. Why is power management important for ad hoc wireless networks?

7. What are the trade-offs to be considered in the design of power management schemes?

8. What role does the routing protocol play in the provisioning of QoS guarantees for ad hoc wireless networks?

9. What kind of multiple access technology is suitable in (a) a military ad hoc network environment, and (b) a home ad hoc network environment?

10. Referring to Figure 5.5, a mobile station (node) moving at high speeds would experience frequent handoffs. Discuss the pros and cons of using MobileIP for address and service discovery.

11. Referring to Figure 5.5, what is the rate of handoffs that a mobile node experiences when it moves at a speed of 144 Km/hour, assuming that the relay node's transmission range (radius) is 500 m.

12. Discuss the pros and cons of a routing protocol that uses GPS information for an ad hoc wireless network for search-and-rescue operations.

13. List the major advantages of the ad hoc wireless Internet.

BIBLIOGRAPHY

[1] "DARPA Home Page," http://www.darpa.mil

[2] "IETF MANET Working Group Information," http://www.ietf.org/html.charters/manet-charter.html

[3] Y. D. Lin and Y. C. Hsu, "Multi-Hop Cellular: A New Architecture for Wireless Communications," *Proceedings of IEEE INFOCOM 2000*, pp. 1273-1282, March 2000.

[4] A. N. Zadeh, B. Jabbari, R. Pickholtz, and B. Vojcic, "Self-Organizing Packet Radio Ad Hoc Networks with Overlay," *IEEE Communications Magazine*, vol. 40, no. 6, pp. 140-157, June 2002.

[5] R. Ananthapadmanabha, B. S. Manoj, and C. Siva Ram Murthy, "Multi-Hop Cellular Networks: The Architecture and Routing Protocol," *Proceedings of IEEE PIMRC 2001*, vol. 2, pp. 78-82, October 2001.

[6] H. Wu, C. Qiao, S. De, and O. Tonguz, "Integrated Cellular and Ad Hoc Relaying Systems: iCAR," *IEEE Journal on Selected Areas in Communications*, vol. 19, no. 10, pp. 2105-2115, October 2001.

[7] V. D. Park and M. S. Corson, "A Highly Adaptive Distributed Routing Algorithm for Mobile Wireless Networks," *Proceedings of IEEE INFOCOM 1997*, pp. 1405-1413, April 1997.

[8] S. J. Lee and M. Gerla, "SMR: Split Multipath Routing with Maximally Disjoint Paths in Ad Hoc Networks," *Proceedings of IEEE ICC 2001*, vol. 10, pp. 3201-3205, June 2001.

[9] J. Raju and J. J. Garcia-Luna-Aceves, "A New Approach to On-Demand Loop-Free Multipath Routing," *Proceedings of IEEE ICCCN 1999*, pp. 522-527, October 1999.

[10] I. F. Akyildiz, W. Su, Y. Sankarasubramaniam, and E. Cayirci, "A Survey on Sensor Networks," *IEEE Communications Magazine*, vol. 40, no. 8, pp 102-114, August 2002.

[11] D. Christo Frank, B. S. Manoj, and C. Siva Ram Murthy, "Throughput Enhanced Wireless in Local Loop (TWiLL) — The Architecture, Protocols, and Pricing Schemes," *Proceedings of IEEE LCN 2002*, pp. 177-186, November 2002.

[12] H. Deng, W. Li, and D. P. Agrawal, "Routing Security in Wireless Ad Hoc Networks," *IEEE Communications Magazine*, vol. 40, no. 10, pp. 70-75, October 2002.

[13] C. L. Fullmer and J. J. Garcia-Luna-Aceves, "Solutions to Hidden Terminal Problems in Wireless Networks," *Proceedings of ACM SIGCOMM 1997*, pp. 39-49, September 1997.

[14] Z. J. Haas, "The Routing Algorithm for the Reconfigurable Wireless Networks," *Proceedings of IEEE ICUPC 1997*, vol. 2, pp. 562-566, October 1997.

[15] S. B. Lee, A. Gahng-seop, X. Zhang, and A. T. Campbell, "INSIGNIA: An IP-Based Quality of Service Framework for Mobile Ad Hoc Networks," *Journal of Parallel and Distributed Computing*, vol. 60, no. 4, pp. 374-406, April 2000.

[16] C. R. Lin and M. Gerla, "Asynchronous Multimedia Multi-Hop Wireless Networks," *Proceedings of IEEE INFOCOM 1997*, vol. 1, pp. 118-125, April 1997.

[17] J. Liu and S. Singh, "ATCP: TCP for Mobile Ad Hoc Networks," *IEEE Journal on Selected Areas in Communications*, vol. 19, no. 7, pp. 1300-1315, July 2001.

[18] R. Sivakumar, P. Sinha, and V. Bharghavan, "CEDAR: A Core-Extraction Distributed Ad Hoc Routing Algorithm," *IEEE Journal on Selected Areas in Communications*, vol. 17, no. 8, pp. 1454-1465, August 1999.

[19] L. Subramanian and R. H. Katz, "An Architecture for Building Self-Configurable Systems," *Proceedings of ACM MOBIHOC 2000*, pp. 63-73, August 2000.

[20] N. H. Vaidya, "Weak Duplicate Address Detection in Mobile Ad Hoc Networks," *Proceedings of ACM MOBIHOC 2002*, pp. 206-216, June 2002.

[21] B. S. Manoj and C. Siva Ram Murthy, "Ad Hoc Wireless Networks: Issues and Challenges," *Technical Report*, Department of Computer Science and Engineering, Indian Institute of Technology, Madras, India, November 2003.

Chapter 6

MAC PROTOCOLS FOR AD HOC WIRELESS NETWORKS

6.1 INTRODUCTION

Nodes in an ad hoc wireless network share a common broadcast radio channel. Since the radio spectrum is limited, the bandwidth available for communication in such networks is also limited. Access to this shared medium should be controlled in such a manner that all nodes receive a fair share of the available bandwidth, and that the bandwidth is utilized efficiently. Since the characteristics of the wireless medium are completely different from those of the wired medium, and since ad hoc wireless networks need to address unique issues (such as node mobility, limited bandwidth availability, error-prone broadcast channel, hidden and exposed terminal problems, and power constraints) that are not applicable to wired networks, a different set of protocols is required for controlling access to the shared medium in such networks. This chapter focuses on media access protocols for ad hoc wireless networks. First, the issues involved in designing a medium access control (MAC) protocol for ad hoc wireless networks are presented, followed by several classifications of the currently existing MAC protocols. This chapter then provides detailed descriptions of several existing MAC protocols.

6.2 ISSUES IN DESIGNING A MAC PROTOCOL FOR AD HOC WIRELESS NETWORKS

The following are the main issues that need to be addressed while designing a MAC protocol for ad hoc wireless networks.

6.2.1 Bandwidth Efficiency

As mentioned earlier, since the radio spectrum is limited, the bandwidth available for communication is also very limited. The MAC protocol must be designed in such a way that the scarce bandwidth is utilized in an efficient manner. The control overhead involved must be kept as minimal as possible. Bandwidth efficiency can be defined as the ratio of the bandwidth used for actual data transmission to the total available bandwidth. The MAC protocol must try to maximize this bandwidth efficiency.

6.2.2 Quality of Service Support

Due to the inherent nature of the ad hoc wireless network, where nodes are usually mobile most of the time, providing quality of service (QoS) support to data sessions in such networks is very difficult. Bandwidth reservation made at one point of time may become invalid once the node moves out of the region where the reservation was made. QoS support is essential for supporting time-critical traffic sessions such as in military communications. The MAC protocol for ad hoc wireless networks that are to be used in such real-time applications must have some kind of a resource reservation mechanism that takes into consideration the nature of the wireless channel and the mobility of nodes.

6.2.3 Synchronization

The MAC protocol must take into consideration the synchronization between nodes in the network. Synchronization is very important for bandwidth (time slot) reservations by nodes. Exchange of control packets may be required for achieving time synchronization among nodes. The control packets must not consume too much of network bandwidth.

6.2.4 Hidden and Exposed Terminal Problems

The hidden and exposed terminal problems are unique to wireless networks. The hidden terminal problem refers to the collision of packets at a receiving node due to the simultaneous transmission of those nodes that are not within the direct transmission range of the sender, but are within the transmission range of the receiver. Collision occurs when both nodes transmit packets at the same time without knowing about the transmission of each other. For example, consider Figure 6.1. Here, if both node S1 and node S2 transmit to node R1 at the same time, their packets collide at node R1. This is because both nodes S1 and S2 are hidden from each other as they are not within the direct transmission range of each other and hence do not know about the presence of each other.

The exposed terminal problem refers to the inability of a node, which is blocked due to transmission by a nearby transmitting node, to transmit to another node. Consider the example in Figure 6.1. Here, if a transmission from node S1 to another node R1 is already in progress, node S3 cannot transmit to node R2, as it concludes that its neighbor node S1 is in transmitting mode and hence it should not interfere with the on-going transmission.

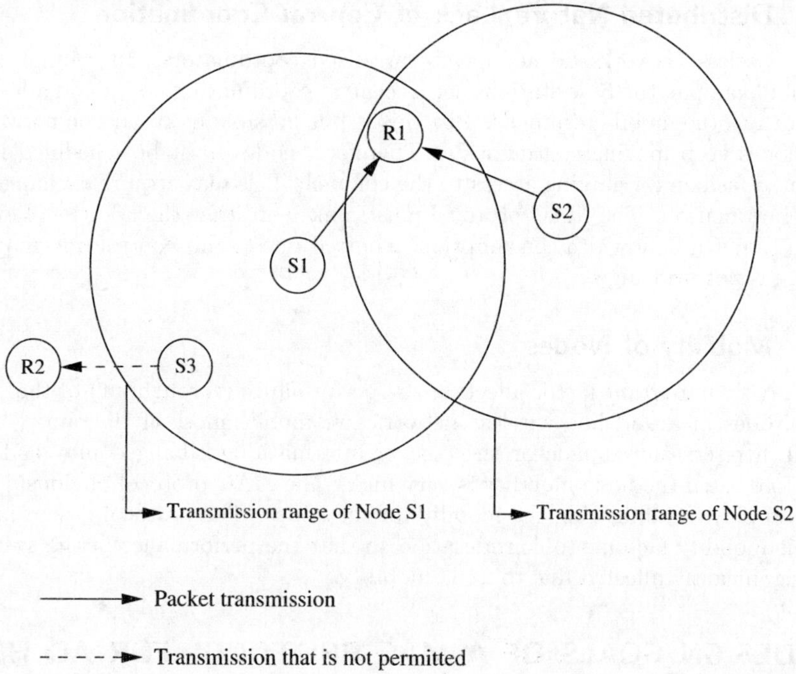

Figure 6.1. Hidden and exposed terminal problems.

The hidden and exposed terminal problems significantly reduce the throughput of a network when the traffic load is high. It is therefore desirable that the MAC protocol be free from the hidden and exposed terminal problems.

6.2.5 Error-Prone Shared Broadcast Channel

Another important factor in the design of a MAC protocol is the broadcast nature of the radio channel, that is, transmissions made by a node are received by all nodes within its direct transmission range. When a node is receiving data, no other node in its neighborhood, apart from the sender, should transmit. A node should get access to the shared medium only when its transmissions do not affect any ongoing session. Since multiple nodes may contend for the channel simultaneously, the possibility of packet collisions is quite high in wireless networks. A MAC protocol should grant channel access to nodes in such a manner that collisions are minimized. Also, the protocol should ensure that all nodes are treated fairly with respect to bandwidth allocation.

6.2.6 Distributed Nature/Lack of Central Coordination

Ad hoc wireless networks do not have centralized coordinators. In cellular networks, for example, the base stations act as central coordinating nodes and allocate bandwidth to the mobile terminals. But this is not possible in an ad hoc network, where nodes keep moving continuously. Therefore, nodes must be scheduled in a distributed fashion for gaining access to the channel. This may require exchange of control information. The MAC protocol must make sure that the additional overhead, in terms of bandwidth consumption, incurred due to this control information exchange is not very high.

6.2.7 Mobility of Nodes

This is a very important factor affecting the performance (throughput) of the protocol. Nodes in an ad hoc wireless network are mobile most of the time. The bandwidth reservations made or the control information exchanged may end up being of no use if the node mobility is very high. The MAC protocol obviously has no role to play in influencing the mobility of the nodes. The protocol design must take this mobility factor into consideration so that the performance of the system is not significantly affected due to node mobility.

6.3 DESIGN GOALS OF A MAC PROTOCOL FOR AD HOC WIRELESS NETWORKS

The following are the important goals to be met while designing a medium access control (MAC) protocol for ad hoc wireless networks:

- The operation of the protocol should be distributed.
- The protocol should provide QoS support for real-time traffic.
- The access delay, which refers to the average delay experienced by any packet to get transmitted, must be kept low.
- The available bandwidth must be utilized efficiently.
- The protocol should ensure fair allocation (either equal allocation or weighted allocation) of bandwidth to nodes.
- Control overhead must be kept as low as possible.
- The protocol should minimize the effects of hidden and exposed terminal problems.
- The protocol must be scalable to large networks.
- It should have power control mechanisms in order to efficiently manage energy consumption of the nodes.

- The protocol should have mechanisms for adaptive data rate control (adaptive rate control refers to the ability to control the rate of outgoing traffic from a node after taking into consideration such factors as load in the network and the status of neighbor nodes).

- It should try to use directional antennas which can provide advantages such as reduced interference, increased spectrum reuse, and reduced power consumption.

- Since synchronization among nodes is very important for bandwidth reservations, the protocol should provide time synchronization among nodes.

6.4 CLASSIFICATIONS OF MAC PROTOCOLS

MAC protocols for ad hoc wireless networks can be classified into several categories based on various criteria such as initiation approach, time synchronization, and reservation approaches. Figure 6.2 provides a detailed classification tree. In this section, some of the classifications of MAC protocols are briefly discussed. Ad hoc network MAC protocols can be classified into three basic types:

- Contention-based protocols

- Contention-based protocols with reservation mechanisms

- Contention-based protocols with scheduling mechanisms

Apart from these three major types, there exist other MAC protocols that cannot be classified clearly under any one of the above three types of protocols.

6.4.1 Contention-Based Protocols

These protocols follow a contention-based channel access policy. A node does not make any resource reservation *a priori*. Whenever it receives a packet to be transmitted, it contends with its neighbor nodes for access to the shared channel. Contention-based protocols cannot provide QoS guarantees to sessions since nodes are not guaranteed regular access to the channel. Random access protocols can be further divided into two types:

- Sender-initiated protocols: Packet transmissions are initiated by the sender node.

- Receiver-initiated protocols: The receiver node initiates the contention resolution protocol.

Sender-initiated protocols can be further divided into two types:

- Single-channel sender-initiated protocols: In these protocols, the total available bandwidth is used as it is, without being divided. A node that wins the contention to the channel can make use of the entire bandwidth.

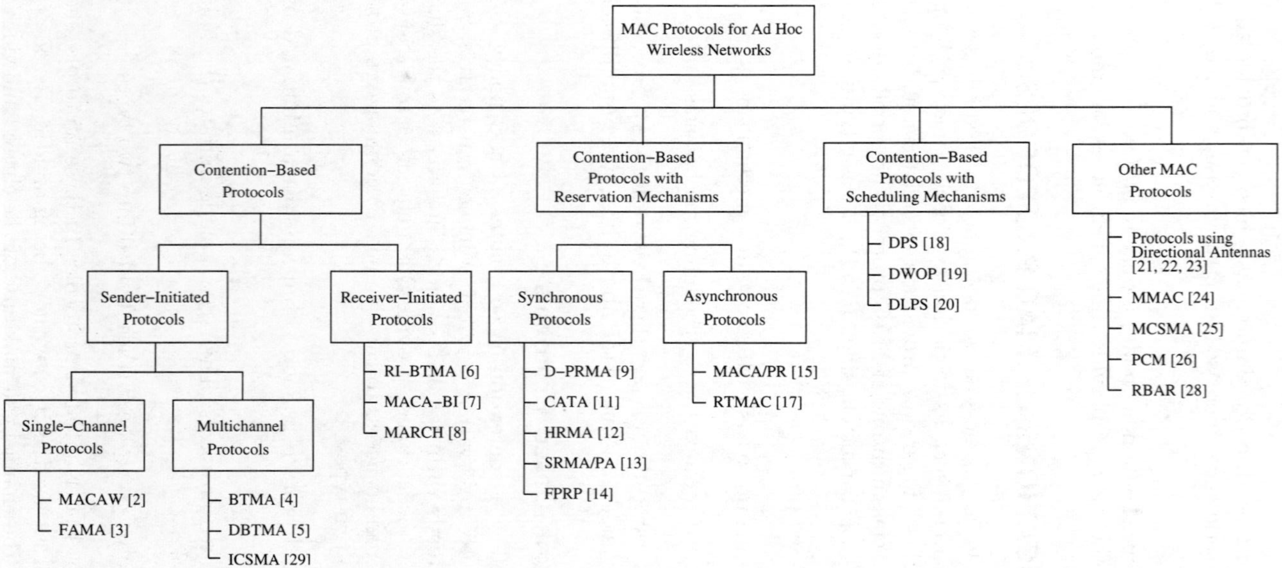

Figure 6.2. Classifications of MAC protocols.

- Multichannel sender-initiated protocols: In multichannel protocols, the available bandwidth is divided into multiple channels. This enables several nodes to simultaneously transmit data, each using a separate channel. Some protocols dedicate a frequency channel exclusively for transmitting control information.

Several contention-based MAC protocols are discussed in detail in Section 6.5.

6.4.2 Contention-Based Protocols with Reservation Mechanisms

Ad hoc wireless networks sometimes may need to support real-time traffic, which requires QoS guarantees to be provided. In contention-based protocols, nodes are not guaranteed periodic access to the channel. Hence they cannot support real-time traffic. In order to support such traffic, certain protocols have mechanisms for reserving bandwidth *a priori*. Such protocols can provide QoS support to time-sensitive traffic sessions. These protocols can be further classified into two types:

- Synchronous protocols: Synchronous protocols require time synchronization among all nodes in the network, so that reservations made by a node are known to other nodes in its neighborhood. Global time synchronization is generally difficult to achieve.

- Asynchronous protocols: They do not require any global synchronization among nodes in the network. These protocols usually use relative time information for effecting reservations.

Various reservation-based random access protocols are discussed in detail in Section 6.6.

6.4.3 Contention-Based Protocols with Scheduling Mechanisms

As mentioned earlier, these protocols focus on packet scheduling at nodes, and also scheduling nodes for access to the channel. Node scheduling is done in a manner so that all nodes are treated fairly and no node is starved of bandwidth. Scheduling-based schemes are also used for enforcing priorities among flows whose packets are queued at nodes. Some scheduling schemes also take into consideration battery characteristics, such as remaining battery power, while scheduling nodes for access to the channel. Some of the existing scheduling-based protocols are discussed in Section 6.7.

6.4.4 Other Protocols

There are several other MAC protocols that do not strictly fall under the above categories. Some of these protocols are discussed in Section 6.9.

6.5 CONTENTION-BASED PROTOCOLS

As mentioned earlier, contention-based protocols do not have any bandwidth reservation mechanisms. All ready nodes contend for the channel simultaneously, and the winning node gains access to the channel. Since nodes are not guaranteed bandwidth, these protocols cannot be used for transmitting real-time traffic, which requires QoS guarantees from the system. In this section, several contention-based MAC protocols are described in detail.

6.5.1 MACAW: A Media Access Protocol for Wireless LANs

This protocol is based on the multiple access collision avoidance protocol (MACA) proposed by Karn [1]. MACA was proposed due to the shortcomings of CSMA protocols when used for wireless networks. In what follows, a brief description on why CSMA protocols fail in wireless networks is given. This is followed by detailed descriptions of the MACA protocol and the MACAW protocol.

MACA Protocol

The MACA protocol was proposed as an alternative to the traditional carrier sense multiple access (CSMA) protocols used in wired networks. In CSMA protocols, the sender first senses the channel for the carrier signal. If the carrier is present, it retries after a random period of time. Otherwise, it transmits the packet. CSMA senses the state of the channel only at the transmitter. This protocol does not overcome the hidden terminal problem. In a typical ad hoc wireless network, the transmitter and receiver may not be near each other at all times. In such situations, the packets transmitted by a node are prone to collisions at the receiver due to simultaneous transmissions by the hidden terminals. Also, the bandwidth utilization in CSMA protocols is less because of the exposed terminal problem.

MACA does not make use of carrier-sensing for channel access. It uses two additional signaling packets: the request-to-send (RTS) packet and the clear-to-send (CTS) packet. When a node wants to transmit a data packet, it first transmits an RTS packet. The receiver node, on receiving the RTS packet, if it is ready to receive the data packet, transmits a CTS packet. Once the sender receives the CTS packet without any error, it starts transmitting the data packet. This data transmission mechanism is depicted in Figure 6.3. If a packet transmitted by a node is lost, the node uses the binary exponential back-off (BEB) algorithm to back off for a random interval of time before retrying. In the binary exponential back-off mechanism, each time a collision is detected, the node doubles its maximum back-off window. Neighbor nodes near the sender that hear the RTS packet do not transmit for a long enough period of time so that the sender could receive the CTS packet. Both the RTS and the CTS packets carry the expected duration of the data packet transmission. A node near the receiver, upon hearing the CTS packet, defers its transmission till the receiver receives the data packet. Thus, MACA overcomes the hidden node problem. Similarly, a node receiving an RTS defers only for a short period of time till the sender could receive the CTS. If no CTS is heard by the node during its waiting period, it is free to transmit packets once the waiting interval is over. Thus, a node that hears only the RTS packet is free to transmit simultaneously when the sender of the RTS is transmitting data packets. Hence, the exposed terminal problem is also overcome in MACA. But MACA still has certain problems, which was why MACAW, described below, was proposed.

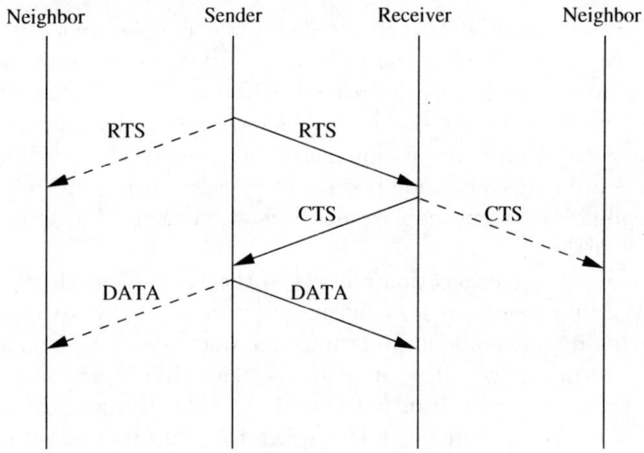

Figure 6.3. Packet transmission in MACA.

MACAW Protocol

The binary exponential back-off mechanism used in MACA at times starves flows. For example, consider Figure 6.4. Here both nodes S1 and S2 keep generating a high volume of traffic. The node that first captures the channel (say, node S1) starts transmitting packets. The packets transmitted by the other node S2 get collided, and the node keeps incrementing its back-off window according to the BEB algorithm. As a result, the probability of node S2 acquiring the channel keeps decreasing, and over a period of time it gets completely blocked. To overcome this problem, the back-off algorithm has been modified in MACAW [2]. The packet header now has an additional field carrying the current back-off counter value of the transmitting node. A node receiving the packet copies this value into its own back-off counter. This mechanism allocates bandwidth in a fair manner. Another problem with BEB algorithm is that it adjusts the back-off counter value very rapidly, both when a node successfully transmits a packet and when a collision is detected by the node. The back-off counter is reset to the minimum value after every successful transmission. In the modified back-off process, this would require a period

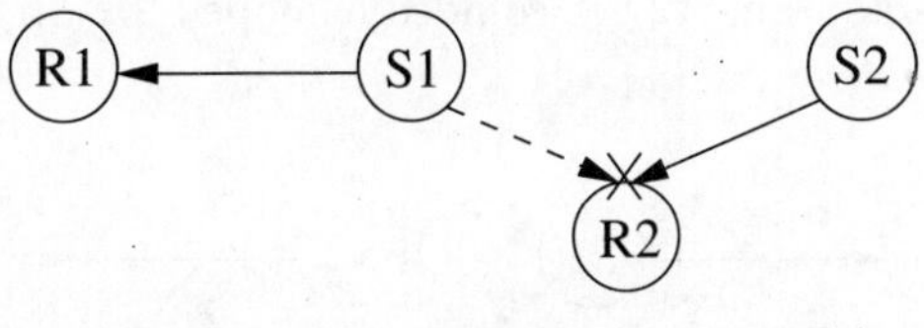

Figure 6.4. Example topology.

of contention to be repeated after each successful transmission in order to build up the back-off timer values. To prevent such large variations in the back-off values, a multiplicative increase and linear decrease (MILD) back-off mechanism is used in MACAW. Here, upon a collision, the back-off is increased by a multiplicative factor (1.5), and upon a successful transmission, it is decremented by one. This eliminates contention and hence long contention periods after every successful transmission, at the same time providing a reasonably quick escalation in the back-off values when the contention is high.

In MACAW another modification related to the back-off mechanism has been made. MACAW implements per flow fairness as opposed to the per node fairness in MACA. This is done by maintaining multiple queues at every node, one each for each data stream, and running the back-off algorithm independently for each queue. A node that is ready to transmit packets first determines how long it needs to wait before it could transmit an RTS packet to each of the destination nodes corresponding to the top-most packets in the node's queues. It then selects the packet for which the waiting time is minimal.

In addition to the RTS and CTS control packets used in MACA, MACAW uses another new control packet called acknowledgment (ACK) packet. The need for using this additional packet arises because of the following reason. In MACA, the responsibility of recovering from transmission errors lies with the transport layer. As many TCP implementations have a minimum timeout period of about 0.5 sec, significant delay is involved while recovering from errors. But in MACAW, the error recovery responsibility is given to the data link layer (DLL). In DLL, the recovery process can be made quicker as the timeout periods can be modified in order to suit the physical media being employed. In MACAW, after successful reception of each data packet, the receiver node transmits an ACK packet. If the sender does not receive the ACK packet, it reschedules the same data packet for transmission. The back-off counter is incremented if the ACK packet is not received by the sender. If the ACK packet got lost in transmission, the sender would retry by transmitting an RTS for the same packet. But now the receiver, instead of sending back a CTS, sends an ACK for the packet received, and the sender moves on to transmit the next data packet.

In MACA, an exposed node (which received only the RTS and not the CTS packet) is free to transmit simultaneously when the source node is transmitting packets. For example, in Figure 6.5, when a transmission is going on between nodes S1 and R1, node S2 is free to transmit. RTS transmissions by node S2 are of no use, as it can proceed further only if it can receive a CTS from R2. But this is not possible as CTS packets from R2 get collided at node S2 with packets transmitted

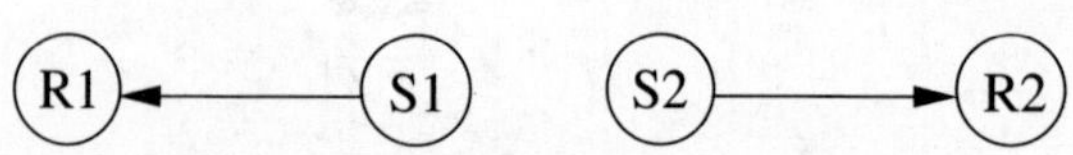

Figure 6.5. Example topology.

by node S1. As a result, the back-off counter at node S2 builds up unnecessarily. So an exposed node should not be allowed to transmit. But an exposed node, since it can hear only the RTS sent by the source node and not the CTS sent by the receiver, does not know for sure whether the RTS-CTS exchange was successful. To overcome this problem, MACAW uses another small (30-bytes) control packet called the data-sending (DS) packet. Before transmitting the actual data packet, the source node transmits this DS packet. The DS packet carries information such as the duration of the data packet transmission, which could be used by the exposed nodes for updating information they hold regarding the duration of the data packet transmission. An exposed node, overhearing the DS packet, understands that the previous RTS-CTS exchange was successful, and so defers its transmissions until the expected duration of the DATA-ACK exchange. If the DS packet was not used, the exposed node (node S2) would retransmit after waiting for random intervals of time, and with a high probability the data transmission (between nodes S1 and R1) would be still going on when the exposed node retransmits. This would result in a collision and the back-off period being further incremented, which affects the node even more.

The MACAW protocol uses one more control packet called the request-for-request-to-send (RRTS) packet. The following example shows how this RRTS packet proves to be useful. Consider Figure 6.6. Here assume transmission is going on between nodes S1 and R1. Now node S2 wants to transmit to node R2. But since R2 is a neighbor of R1, it receives CTS packets from node R1, and therefore it defers its own transmissions. Node S2 has no way to learn about the contention periods during which it can contend for the channel, and so it keeps on trying, incrementing its back-off counter after each failed attempt. Hence the main reason for this problem is the lack of synchronization information at source S2. MACAW overcomes this problem by using the RRTS packet. In the same example shown in Figure 6.6, receiver node R2 contends for the channel on behalf of source S2. If node R2 had received an RTS previously for which it was not able to respond immediately because of the on-going transmission between nodes S1 and R1, then node R2 waits for the next contention period and transmits the RRTS packet. Neighbor nodes that hear the RRTS packet (including node R1) are made to wait for two successive slots (for the RTS-CTS exchange to take place). The source node S2, on receiving the RRTS from node R2, transmits the regular RTS packet to node R2, and the normal packet exchange (RTS-CTS-Data-ACK) continues from here. Figure 6.7 shows the operation of the MACAW protocol. In the figure, S is the source node and R denotes the receiver node. N1 and N2 are neighbor nodes. When RTS transmitted by node S is overheard by node N1, it refrains from transmitting until

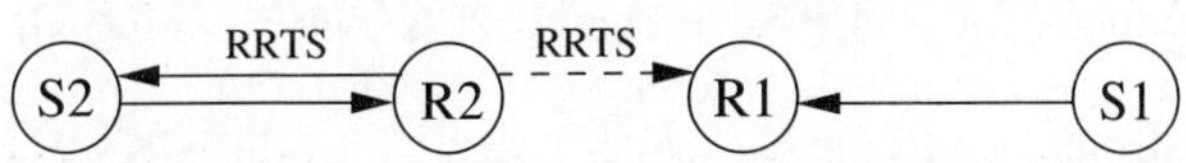

Figure 6.6. Example topology.

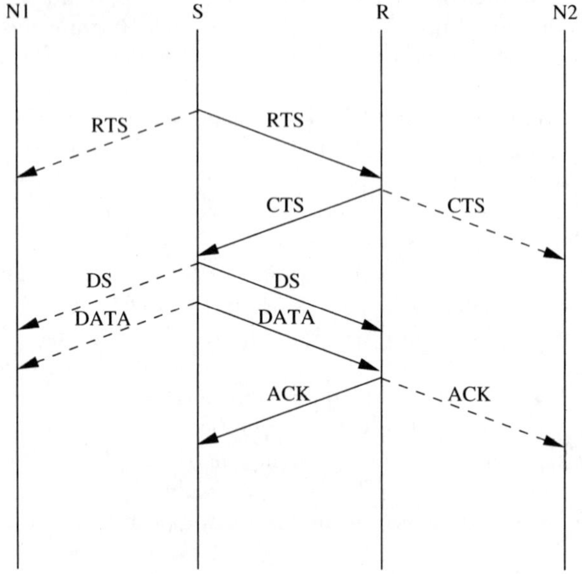

Figure 6.7. Packet exchange in MACAW.

node S receives the CTS. Similarly, when the CTS transmitted by node R is heard by neighbor node N2, it defers its transmissions until the data packet is received by receiver R. On receiving this CTS packet, node S immediately transmits the DS message carrying the expected duration of the data packet transmission. On hearing this packet, node N1 backs off until the data packet is transmitted. Finally, after receiving the data packet, node R acknowledges the reception by sending node S an ACK packet.

To summarize, the MACAW protocol has been designed based on four main observations. The first is that the relevant congestion occurs at the receiver node and not at the sender. This realization makes CSMA protocols unsuitable for ad hoc wireless networks, and therefore the RTS-CTS-DATA exchange mechanism of MACA becomes necessary. MACAW further improves upon this scheme using the RTS-CTS-DS-DATA-ACK exchange mechanism. The second observation is that congestion is dependent on the location of the receiver. Therefore, instead of characterizing back-off by a single back-off parameter, separate back-off parameters have been introduced for each flow. The third is that learning about congestion at various nodes must be a collective enterprise. Therefore, the notion of copying back-off values from overheard packets has been introduced in MACA. And the final observation is that in order that nodes contend effectively for the channel, the synchronization information needs to be propagated to the concerned nodes at appropriate times. This is done in MACAW through the DS and RRTS packets.

Because of the various changes described above, the performance of MACAW is significantly improved when compared to the MACA protocol.

6.5.2 Floor Acquisition Multiple Access Protocols

The floor acquisition multiple access (FAMA) protocols [3] are based on a channel access discipline which consists of a carrier-sensing operation and a collision-avoidance dialog between the sender and the intended receiver of a packet. Floor acquisition refers to the process of gaining control of the channel. At any given point of time, the control of the channel is assigned to only one node, and this node is guaranteed to transmit one or more data packets to different destinations without suffering from packet collisions. Carrier-sensing by the sender, followed by the RTS-CTS control packet exchange, enables the protocol to perform as efficiently as MACA [1] in the presence of hidden terminals, and as efficiently as CSMA otherwise. FAMA requires a node that wishes to transmit packets to first acquire the floor (channel) before starting to transmit the packets. The floor is acquired by means of exchanging control packets. Though the control packets themselves may collide with other control packets, it is ensured that data packets sent by the node that has acquired the channel are always transmitted without any collisions. Any single-channel MAC protocol that does not require a transmitting node to sense the channel can be adapted for performing floor acquisition tasks. Floor acquisition using the RTS-CTS exchange is advantageous as the mechanism also tries to provide a solution for the hidden terminal problem.

Two FAMA protocol variants are discussed in this section: RTS-CTS exchange with no carrier-sensing, and RTS-CTS exchange with non-persistent carrier-sensing. The first variant uses the ALOHA protocol for transmitting RTS packets, while the second variant uses non-persistent CSMA for the same purpose.

Multiple Access Collision Avoidance

Multiple access collision avoidance (MACA) [1], which was discussed earlier in this chapter, belongs to the category of FAMA protocols. In MACA, a ready node transmits an RTS packet. A neighbor node receiving the RTS defers its transmissions for the period specified in the RTS. On receiving the RTS, the receiver node responds by sending back a CTS packet, and waits for a long enough period of time in order to receive a data packet. Neighbor nodes of the receiver which hear this CTS packet defer their transmissions for the time duration of the impending data transfer. In MACA, nodes do not sense the channel. A node defers its transmissions only if it receives an RTS or CTS packet. In MACA, data packets are prone to collisions with RTS packets.

According to the FAMA principle, in order for data transmissions to be collision-free, the duration of an RTS must be at least twice the maximum channel propagation delay. Transmission of bursts of packets is not possible in MACA. In FAMA-NTR (discussed below) the MACA protocol is modified to permit transmission of packet bursts by enforcing waiting periods on nodes, which are proportional to the channel propagation time.

FAMA – Non-Persistent Transmit Request

This variant of FAMA, called FAMA – non-persistent transmit request (FAMA-NTR), combines non-persistent carrier-sensing along with the RTS-CTS control packet exchange mechanism. Before sending a packet, the sender node senses the channel. If the channel is found to be busy, then the node backs off for a random time period and retries later. If the channel is found to be free, it transmits the RTS packet. After transmitting the RTS, the sender listens to the channel for one round-trip time in addition to the time required by the receiver node to transmit a CTS. If it does not receive the CTS within this time period or if the CTS received is found to be corrupted, then the node takes a random back-off and retries later. Once the sender node receives the CTS packet without any error, it can start transmitting its data packet burst. The burst is limited to a maximum number of data packets, after which the node releases the channel, and contends with other nodes to again acquire the channel.

In order to allow the sender node to send a burst of packets once it acquires the floor, the receiver node is made to wait for a time duration of τ seconds after processing each data packet received. Here, τ denotes the maximum channel propagation time. A waiting period of 2τ seconds is enforced on a transmitting node after transmitting any control packet. This is done to allow the RTS-CTS exchange to take place without any error. A node transmitting an RTS is required to wait for 2τ seconds after transmitting the RTS in order to enable the receiver node to receive the RTS and transmit the corresponding CTS packet. After sending the final data packet, a sender node is made to wait for τ seconds in order to allow the destination to receive the data packet and to account for the enforced waiting time at the destination node.

6.5.3 Busy Tone Multiple Access Protocols

Busy Tone Multiple Access

The busy tone multiple access (BTMA) protocol [4] is one of the earliest protocols proposed for overcoming the hidden terminal problem faced in wireless environments. The transmission channel is split into two: a data channel and a control channel. The data channel is used for data packet transmissions, while the control channel is used to transmit the busy tone signal.

When a node is ready for transmission, it senses the channel to check whether the busy tone is active. If not, it turns on the busy tone signal and starts data transmission; otherwise, it reschedules the packet for transmission after some random rescheduling delay. Any other node which senses the carrier on the incoming data channel also transmits the busy tone signal on the control channel. Thus, when a node is transmitting, no other node in the two-hop neighborhood of the transmitting node is permitted to simultaneously transmit. Though the probability of collisions is very low in BTMA, the bandwidth utilization is very poor. Figure 6.8 shows the worst-case scenario where the node density is very high; the dotted circle shows the region in which nodes are blocked from simultaneously transmitting when node N1 is transmitting packets.

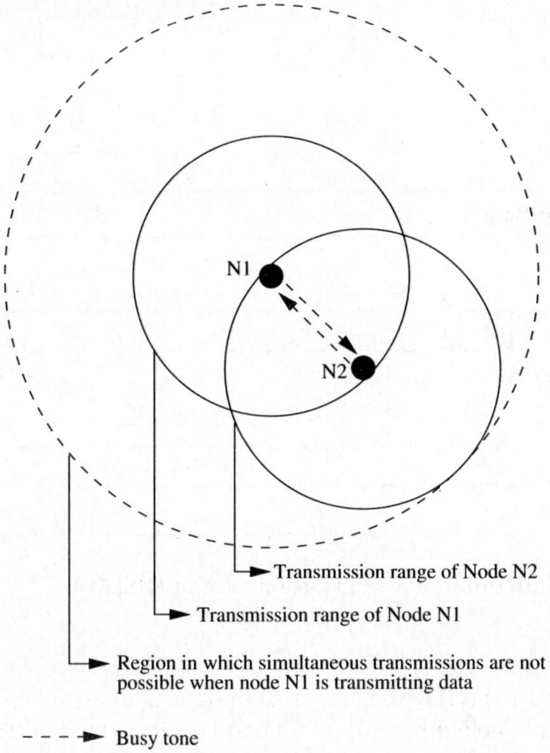

Figure 6.8. Transmission in BTMA.

Dual Busy Tone Multiple Access Protocol

The dual busy tone multiple access protocol (DBTMA) [5] is an extension of the BTMA scheme. Here again, the transmission channel is divided into two: the data channel and the control channel.

As in BTMA, the data channel is used for data packet transmissions. The control channel is used for control packet transmissions (RTS and CTS packets) and also for transmitting the busy tones. DBTMA uses two busy tones on the control channel, BT_t and BT_r. The BT_t tone is used by the node to indicate that it is transmitting on the data channel. The BT_r tone is turned on by a node when it is receiving data on the data channel. The two busy tone signals are two sine waves at different well-separated frequencies.

When a node is ready to transmit a data packet, it first senses the channel to determine whether the BT_r signal is active. An active BT_r signal indicates that a node in the neighborhood of the ready node is currently receiving packets. If the ready node finds that there is no BT_r signal, it transmits the RTS packet on the control channel. On receiving the RTS packets, the node to which the RTS was

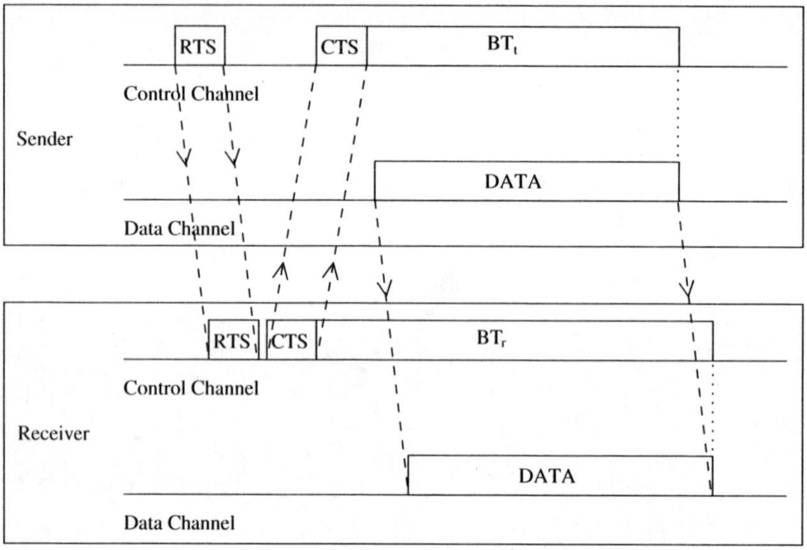

Figure 6.9. Packet transmission in DBTMA.

destined checks whether the BT_t tone is active in its neighborhood. An active BT_t implies that some other node in its neighborhood is transmitting packets and so it cannot receive packets for the moment. If the node finds no BT_t signal, it responds by sending a CTS packet and then turns on the BT_r signal (which informs other nodes in its neighborhood that it is receiving). The sender node, on receiving this CTS packet, turns on the BT_t signal (to inform nodes in its neighborhood that it is transmitting) and starts transmitting data packets. After completing transmission, the sender node turns off the BT_t signal. The receiver node, after receiving all data packets, turns off the BT_r signal. The above process is depicted in Figure 6.9.

When compared to other RTS/CTS-based medium access control schemes (such as MACA [1] and MACAW [2]), DBTMA exhibits better network utilization. This is because the other schemes block both the forward and reverse transmissions on the data channel when they reserve the channel through their RTS or CTS packets. But in DBTMA, when a node is transmitting or receiving, only the reverse (receive) or forward (transmit) channels, respectively, are blocked. Hence the bandwidth utilization of DBTMA is nearly twice that of other RTS/CTS-based schemes.

Receiver-Initiated Busy Tone Multiple Access Protocol

In the receiver-initiated busy tone multiple access protocol (RI-BTMA) [6], similar to BTMA, the available bandwidth is divided into two channels: a data channel for transmitting data packets and a control channel. The control channel is used by a node to transmit the busy tone signal. A node can transmit on the data channel only if it finds the busy tone to be absent on the control channel.

The data packet is divided into two portions: a preamble and the actual data packet. The preamble carries the identification of the intended destination node. Both the data channel and the control channel are slotted, with each slot equal to the length of the preamble. Data transmission consists of two steps. First, the preamble needs to be transmitted by the sender. Once the receiver node acknowledges the reception of this preamble by transmitting the busy tone signal on the control channel, the actual data packet is transmitted. A sender node that needs to transmit a data packet first waits for a free slot, that is, a slot in which the busy tone signal is absent on the control channel. Once it finds such a slot, it transmits the preamble packet on the data channel. If the destination node receives this preamble packet correctly without any error, it transmits the busy tone on the control channel. It continues transmitting the busy tone signal as long as it is receiving data from the sender. If preamble transmission fails, the receiver does not acknowledge with the busy tone, and the sender node waits for the next free slot and tries again. The operation of the RI-BTMA protocol is shown in Figure 6.10. The busy tone serves two purposes. First, it acknowledges the sender about the successful reception of the preamble. Second, it informs the nearby hidden nodes about the impending transmission so that they do not transmit at the same time.

There are two types of RI-BTMA protocols: the basic protocol and the controlled protocol. The basic packet transmission mechanism is the same in both

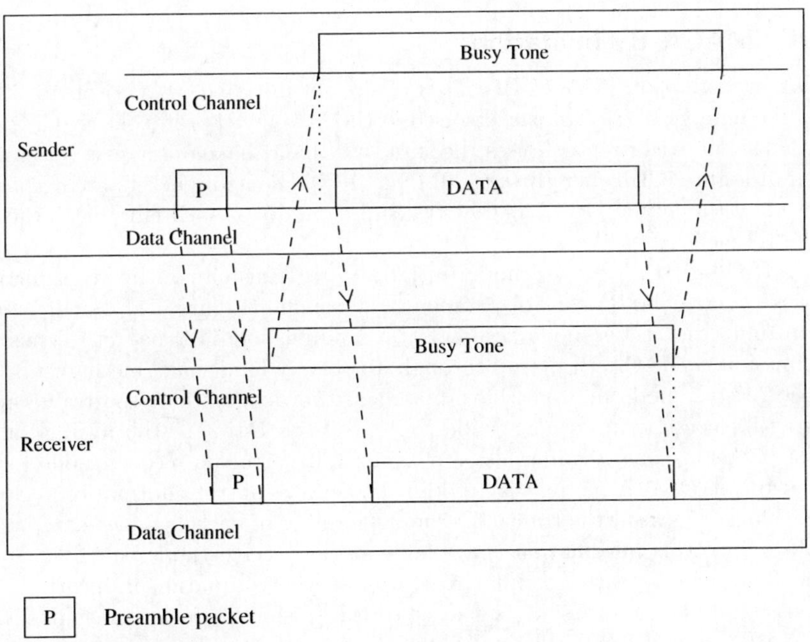

Figure 6.10. Packet transmission in RI-BTMA.

protocols. In the basic protocol, nodes do not have backlog buffers to store data packets. Hence packets that suffer collisions cannot be retransmitted. Also, when the network load increases, packets cannot be queued at the nodes. This protocol would work only when the network load is not high; when network load starts increasing, the protocol becomes unstable.

The controlled protocol overcomes this problem. This protocol is the same as the basic protocol, the only difference being the availability of backlog buffers at nodes. Therefore, packets that suffer collisions, and those that are generated during busy slots, can be queued at nodes. A node is said to be in the backlogged mode if its backlog buffer is non-empty. When a node in the backlogged mode receives a packet from its higher layers, the packet is put into the buffer and transmitted later. Suppose the packet arrives at a node when it is not in the backlogged mode, then if the current slot is free, the preamble for the packet is transmitted with probability p in the current slot itself (not transmitted in the same slot with probability $(1-p)$). If the packet was received during a busy slot, the packet is just put into the backlog buffer, where it waits until the next free slot. A backlogged node transmits a backlogged packet in the next idle slot with a probability q. All other packets in the backlog buffer just keep waiting until this transmission succeeds.

This protocol can work for multi-hop radio networks as well as for single-hop fully connected networks.

6.5.4 MACA-By Invitation

MACA-by invitation (MACA-BI) [7] is a receiver-initiated MAC protocol. It reduces the number of control packets used in the MACA [1] protocol. MACA, which is a sender-initiated protocol, uses the three-way handshake mechanism (which was shown in Figure 6.3), where first the RTS and CTS control packets are exchanged, followed by the actual DATA packet transmission. MACA-BI eliminates the need for the RTS packet.

In MACA-BI the receiver node initiates data transmission by transmitting a ready to receive (RTR) control packet to the sender (Figure 6.11). If it is ready to transmit, the sender node responds by sending a DATA packet. Thus data transmission in MACA-BI occurs through a two-way handshake mechanism.

The receiver node may not have an exact knowledge about the traffic arrival rates at its neighboring sender nodes. It needs to estimate the average arrival rate of packets. For providing necessary information to the receiver node for this estimation, the DATA packets are modified to carry control information regarding the backlogged flows at the transmitter node, number of packets queued, and packet lengths. Once this information is available at the receiver node, the average rate of the flows can be easily estimated. Suppose the estimation is incorrect or is not possible (when the first data packet of the session is to be transmitted), the MACA-BI protocol can be extended by allowing the sender node to declare its backlog through an RTS control packet, if an RTR packet is not received within a given timeout period.

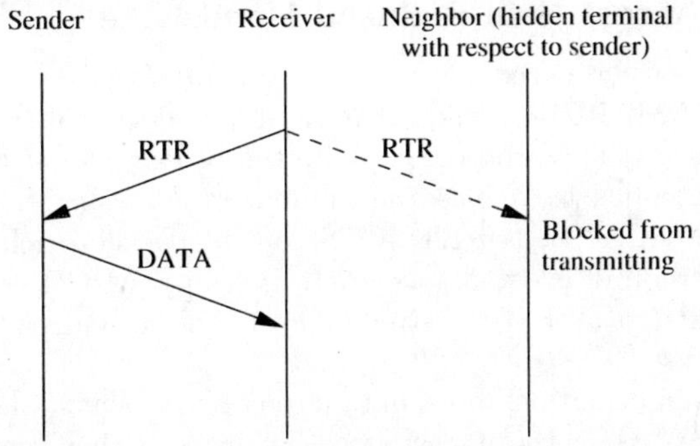

Figure 6.11. Packet transmission in MACA-BI.

In MACA, the CTS packet was used to inform the hidden terminals (nodes) about the impending DATA packet transmission, so that they do not transmit at the same time and disrupt the session. This role is played in MACA-BI by the RTR packets. An RTR packet carries information about the time interval during which the DATA packet would be transmitted. When a node hears RTR packets transmitted by its neighbors, it can obtain information about the duration of DATA packet transmissions by nodes that may be either its direct one-hop neighbors or its two-hop neighbors, that is, hidden terminals. Since it has information about transmissions by the hidden terminals, it refrains from transmitting during those periods (Figure 6.11). Hence the hidden terminal problem is overcome in MACA-BI. Collision among DATA packets is impossible.

However, the hidden terminal problem still affects the control packet transmissions. This leads to protocol failure, as in certain cases the RTR packets can collide with DATA packets. One such scenario is depicted in Figure 6.12. Here, RTR packets transmitted by receiver nodes R1 and R2 collide at node A. So node A is not aware of the transmissions from nodes S1 and S2. When node A transmits RTR packets, they collide with DATA packets at receiver nodes R1 and R2.

The efficiency of the MACA-BI scheme is mainly dependent on the ability of the receiver node to predict accurately the arrival rates of traffic at the sender nodes.

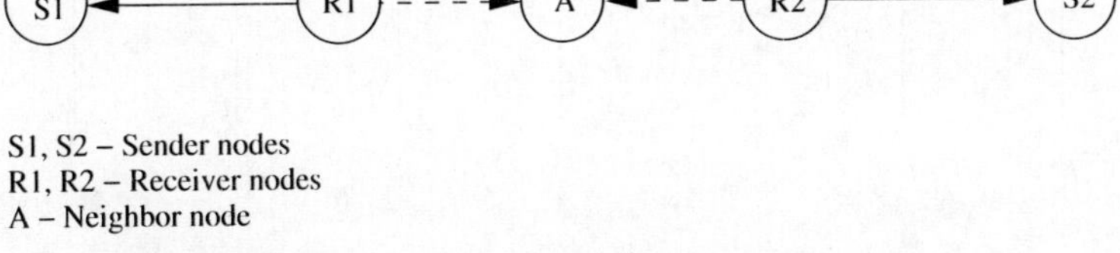

Figure 6.12. Hidden terminal problem in MACA-BI.

6.5.5 Media Access with Reduced Handshake

The media access with reduced handshake protocol (MARCH) [8] is a receiver-initiated protocol. MARCH, unlike MACA-BI [7], does not require any traffic prediction mechanism. The protocol exploits the broadcast nature of traffic from omnidirectional antennas to reduce the number of handshakes involved in data transmission. In MACA [1], the RTS-CTS control packets exchange takes place before the transmission of every data packet. But in MARCH, the RTS packet is used only for the first packet of the stream. From the second packet onward, only the CTS packet is used.

A node obtains information about data packet arrivals at its neighboring nodes by overhearing the CTS packets transmitted by them. It then sends a CTS packet to the concerned neighbor node for relaying data from that node. This mechanism is illustrated in Figure 6.13. Figure 6.13 (a) depicts the packet exchange mechanism of MACA. Here two control packets RTS and CTS need to be exchanged before each data packet is transmitted. It can be seen from this figure that node C, for example, can hear both the CTS and the RTS packets transmitted by node B. MARCH uses this property of the broadcast channel to reduce the two-way handshake into a single CTS-only handshake. Figure 6.13 (b) shows the handshake mechanism of MARCH. Here, when node B transmits the CTS_1 packet, this packet is also heard by node C. A CTS packet carries information regarding the duration of the next data packet. Node C therefore determines the time at which the next data packet would be available at node B. It sends the CTS_2 packet at that point of time. On receiving the CTS_2 packet, node B sends the data packet directly to node C. It can

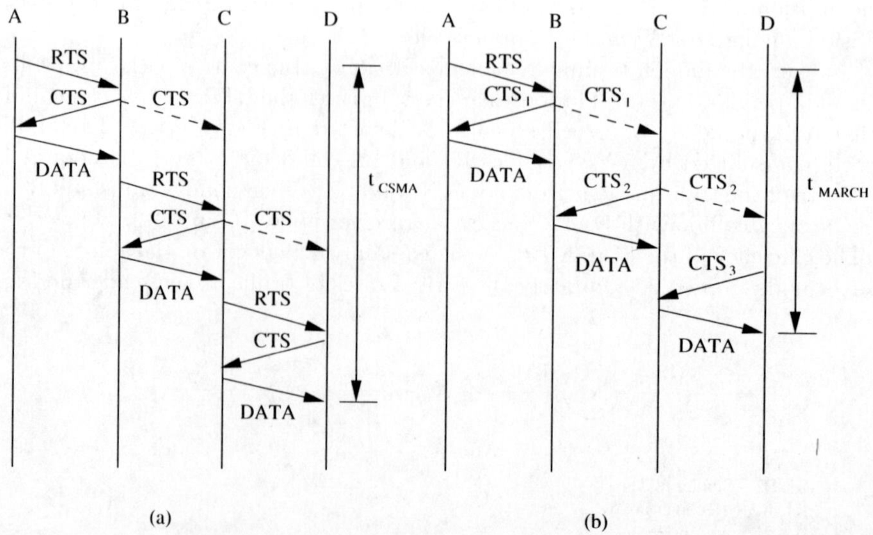

Figure 6.13. Handshake mechanism in (a) MACA and (b) MARCH.

be observed from the figure that the time taken for a packet transmitted by node A to reach node D in MARCH, that is, t_{MARCH}, is less compared to the time taken in MACA, t_{MACA}.

The CTS packet carries the MAC addresses of the sender and the receiver node, and the route identification number (RT_{id}) for that flow. The RT_{id} is used by nodes in order to avoid misinterpretation of CTS packets and initiation of false CTS-only handshakes. Consider Figure 6.14. Here there are two routes – Route 1: A-B-C-D-E and Route 2: X-C-Y. When node C hears a CTS packet transmitted by node B, by means of the RT_{id} field on the packet, it understands that the CTS was transmitted by its upstream node (upstream node refers to the next hop neighbor node on the path from the current node to the source node of the data session) on Route 1. It invokes a timer T which is set to expire after a certain period of time, long enough for node B to receive a packet from node A. A CTS packet is transmitted by node C once the timer expires. This CTS is overheard by node Y also, but since the RT_{id} carried on the CTS is different from the RT_{id} corresponding to Route 2, node Y does not respond. In MARCH, the MAC layer has access to tables that maintain routing information (such as RT_{id}), but the protocol as such does not get involved in routing.

The throughput of MARCH is significantly high when compared to MACA, while the control overhead is much less. When the network is heavily loaded, the average end-to-end delay in packet delivery for MARCH is very low compared to that of MACA. All the above advantages are mainly due to the fact that MARCH has a lower number of control packet handshakes compared to MACA. The lower number of control packets transmitted reduces the control overhead while improving the throughput, since less bandwidth is being consumed for control traffic.

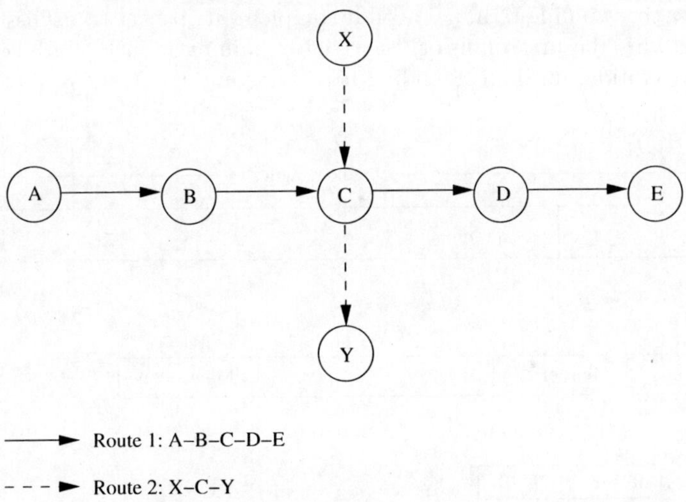

Figure 6.14. Example topology.

6.6 CONTENTION-BASED PROTOCOLS WITH RESERVATION MECHANISMS

Protocols described in this section have certain mechanisms that aid the nodes in effecting bandwidth reservations. Though these protocols are contention-based, contention occurs only during the resource (bandwidth) reservation phase. Once the bandwidth is reserved, the node gets exclusive access to the reserved bandwidth. Hence, QoS support can be provided for real-time traffic.

6.6.1 Distributed Packet Reservation Multiple Access Protocol

The distributed packet reservation multiple access protocol (D-PRMA) [9] extends the earlier centralized packet reservation multiple access (PRMA) [10] scheme into a distributed scheme that can be used in ad hoc wireless networks. PRMA was proposed for voice support in a wireless LAN with a base station, where the base station serves as the fixed entity for the MAC operation. D-PRMA extends this protocol for providing voice support in ad hoc wireless networks.

D-PRMA is a TDMA-based scheme. The channel is divided into fixed- and equal-sized frames along the time axis (Figure 6.15). Each frame is composed of s slots, and each slot consists of m minislots. Each minislot can be further divided into two control fields, RTS/BI and CTS/BI (BI stands for busy indication), as shown in the figure. These control fields are used for slot reservation and for overcoming the hidden terminal problem. Details on how this is done will be explained later in this section. All nodes having packets ready for transmission contend for the first minislot of each slot. The remaining $(m-1)$ minislots are granted to the node that wins the contention. Also, the same slot in each subsequent frame can be reserved for this winning terminal until it completes its packet transmission session. If no node wins the first minislot, then the remaining minislots are continuously used for contention, until a contending node wins any minislot. Within a reserved

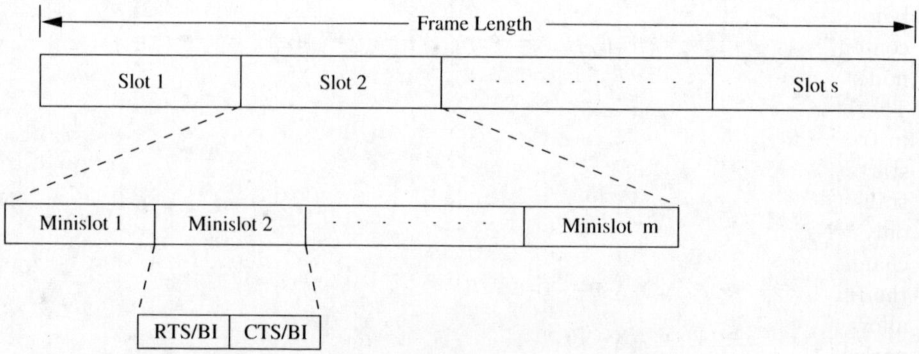

Figure 6.15. Frame structure in D-PRMA.

slot, communication between the source and receiver nodes takes place by means of either time division duplexing (TDD) or frequency division duplexing (FDD). Any node that wants to transmit packets has to first reserve slots, if they have not been reserved already. A certain period at the beginning of each minislot is reserved for carrier-sensing. If a sender node detects the channel to be idle at the beginning of a slot (minislot 1), it transmits an RTS packet (slot reservation request) to the intended destination through the RTS/BI part of the current minislot. On successfully receiving this RTS packet, the receiver node responds by sending a CTS packet through the CTS/BI of the same minislot. If the sender node receives this CTS successfully, then it gets the reservation for the current slot and can use the remaining minislots, that is, minislots 2 to m. Otherwise, it continues the contention process through the subsequent minislots of the same slot.

In order to prioritize nodes transmitting voice traffic (voice nodes) over nodes transmitting normal data traffic (data nodes), two rules are followed in D-PRMA. According to the first rule, the voice nodes are allowed to start contending from minislot 1 with probability $p = 1$; data nodes can start contending only with probability $p < 1$. For the remaining $(m - 1)$ minislots, both the voice nodes and the data nodes are allowed to contend with probability $p < 1$. This is because the reservation process for a voice node is triggered only after the arrival of voice traffic at the node; this avoids unnecessary reservation of slots. According to the second rule, only if the node winning the minislot contention is a voice node, is it permitted to reserve the same slot in each subsequent frame until the end of the session. If a data node wins the contention, then it is allowed to use only one slot, that is, the current slot, and it has to make fresh reservations for each subsequent slot.

Nodes that are located within the radio coverage of the receiver should not be permitted to transmit simultaneously when the receiver is receiving packets. If permitted, packets transmitted by them may collide with the packets of the on-going traffic being received at the receiver. Though a node which is located outside the range of a receiver is able to hear packets transmitted by the sender, it should still be allowed to transmit simultaneously. The above requirements, in essence, mean that the protocol must be free of the hidden terminal and exposed terminal problems. In D-PRMA, when a node wins the contention in minislot 1, other terminals must be prevented from using any of the remaining $(m-1)$ minislots in the same slot for contention (*requirement 1*). Also, when a slot is reserved in subsequent frames, other nodes should be prevented from contending for those reserved slots (*requirement 2*). The RTS-CTS exchange mechanism taking place in the reservation process helps in trying to satisfy *requirement 1*. A node that wins the contention in minislot 1 starts transmitting immediately from minislot 2. Any other node that wants to transmit will find the channel to be busy from minislot 2. Since an RTS can be sent only when the channel is idle, other neighboring nodes would not contend for the channel until the on-going transmission gets completed. A node sends an RTS in the RTS/BI part of a minislot. Only a node that receives an RTS destined to it is allowed to use the CTS/BI part of the slot for transmitting the CTS. So the CTS packet does not suffer any collision due to simultaneous RTS packet transmissions. This improves the probability for a successful reservation. In order to avoid the hidden terminal problem, all nodes hearing the CTS sent by the receiver are not allowed to transmit during the remaining period of that same slot. In order to avoid the exposed terminal problem, a node hearing the RTS but not the CTS (sender node's neighbor) is still allowed to transmit. But, if the communication is duplex in nature, where a node may transmit and receive simultaneously, even such exposed nodes (that hear RTS alone) should not be allowed to transmit. Therefore, D-PRMA makes such a node defer its transmissions for the remaining time period of the same slot. If an RTS or CTS packet collides, and a successful reservation cannot be made in the first minislot, then the subsequent $(m - 1)$ minislots of the same slot are used for contention.

For satisfying *requirement 2*, the following is done. The receiver of the reserved slot transmits a busy indication (BI) signal through the RTS/BI part of minislot 1 of the same slot in each of the subsequent frames, without performing a carrier-sense. The sender also performs a similar function, transmitting the BI through the CTS/BI part of minislot 1 of the same slot in each subsequent frame. When any node hears a BI signal, it does not further contend for that slot in the current frame. Because of this, the reserved slot in each subsequent frame is made free of contention. Also, making the receiver transmit the BI signal helps in eliminating the hidden terminal problem, since not all neighbors of the receiver can hear from the sender. Finally, after a node that had made the reservation completes its data transmission and does not anymore require a reserved slot, it just stops transmitting the BI signal. D-PRMA is more suited for voice traffic than for data traffic applications.

6.6.2 Collision Avoidance Time Allocation Protocol

The collision avoidance time allocation protocol (CATA) [11] is based on dynamic topology-dependent transmission scheduling. Nodes contend for and reserve time slots by means of a distributed reservation and handshake mechanism. CATA supports broadcast, unicast, and multicast transmissions simultaneously. The operation of CATA is based on two basic principles:

- The receiver(s) of a flow must inform the potential source nodes about the reserved slot on which it is currently receiving packets. Similarly, the source node must inform the potential destination node(s) about interferences in the slot.

- Usage of negative acknowledgments for reservation requests, and control packet transmissions at the beginning of each slot, for distributing slot reservation information to senders of broadcast or multicast sessions.

Time is divided into equal-sized frames, and each frame consists of S slots (Figure 6.16). Each slot is further divided into five minislots. The first four minislots are used for transmitting control packets and are called control minislots (CMS1, CMS2, CMS3, and CMS4). The fifth and last minislot, called data minislot (DMS), is meant for data transmission. The data minislot is much longer than the control minislots as the control packets are much smaller in size compared to data packets.

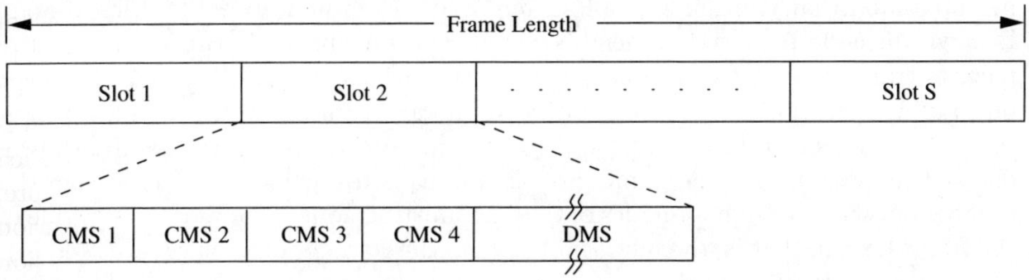

Figure 6.16. Frame format in CATA.

Each node that receives data during the DMS of the current slot transmits a slot reservation (SR) packet during the CMS1 of the slot. This serves to inform other neighboring potential sender nodes about the currently active reservation. The SR packet is either received without error at the neighbor nodes or causes noise at those nodes, in both cases preventing such neighbor nodes from attempting to reserve the current slot. Every node that transmits data during the DMS of the current slot transmits a request-to-send (RTS) packet during CMS2 of the slot. This RTS packet, when received by other neighbor nodes or when it collides with other RTS packets at the neighbor nodes, causes the neighbor nodes to understand that the source node is scheduled to transmit during the DMS of the current slot. Hence they defer their transmissions during the current slot.

The control minislots CMS3 and CMS4 are used as follows. The sender of an intended reservation, if it senses the channel to be idle during CMS1, transmits an RTS packet during CMS2. The receiver node of a unicast session transmits a clear-to-send (CTS) packet during CMS3. On receiving this packet, the source node understands that the reservation was successful and transmits data during the DMS of that slot, and during the same slot in subsequent frames, until the unicast flow gets terminated. Once the reservation has been made successfully in a slot, from the next slot onward, both the sender and receiver do not transmit anything during CMS3, and during CMS4 the sender node alone transmits a not-to-send (NTS) packet. The purpose of the NTS packet is explained below. If a node receives an RTS packet for broadcast or multicast during CMS2, or if it finds the channel to be free during CMS2, it remains idle and does not transmit anything during CMS3 and CMS4. Otherwise, it sends a not-to-send (NTS) packet during CMS4. The NTS packet serves as a negative acknowledgment; a potential multicast or broadcast source node that receives the NTS packet during CMS4, or that detects noise during CMS4, understands that its reservation request had failed, and it does not transmit during the DMS of the current slot. If it finds the channel to be free during CMS4, which implies that its reservation request was successful, it starts transmitting the multicast or broadcast packets during the DMS of the slot.

The length of the frame is very important in CATA. For any node (say, node A) to broadcast successfully, there must be no other node (say, node B) in its two-hop neighborhood that transmits simultaneously. If such a node B exists, then if node B is within node A's one-hop neighborhood, node A and node B cannot hear the packets transmitted by each other. If node B is within the two-hop neighborhood of node A, then the packets transmitted by nodes A and B would collide at their common neighbor nodes. Therefore, for any node to transmit successfully during one slot in every frame, the number of slots in each frame must be larger than the number of two-hop neighbor nodes of the transmitting node. The worst-case value of the frame length, that is, the number of slots in the frame, would be $Min(d^2+1, N)$, where d is the maximum degree (degree of a node refers to the count of one-hop neighbors of the node) of a node in the network, and N is the total number of nodes in the network.

CATA works well with simple single-channel half-duplex radios. It is simple and provides support for collision-free broadcast and multicast traffic.

6.6.3 Hop Reservation Multiple Access Protocol

The hop reservation multiple access protocol (HRMA) [12] is a multichannel MAC protocol which is based on simple half-duplex, very slow frequency-hopping spread spectrum (FHSS) radios. It uses a reservation and handshake mechanism to enable a pair of communicating nodes to reserve a frequency hop, thereby guaranteeing collision-free data transmission even in the presence of hidden terminals. HRMA can be viewed as a time slot reservation protocol where each time slot is assigned a separate frequency channel.

Out of the available L frequency channels, HRMA uses one frequency channel, denoted by f_0, as a dedicated synchronizing channel. The nodes exchange synchronization information on f_0. The remaining $L-1$ frequencies are divided into $M = \lfloor\frac{(L-1)}{2}\rfloor$ frequency pairs (denoted by (f_i, $f_i{}^*$), $i = 1, 2, 3, .., M$), thereby restricting the length of the hopping sequence to M. f_i is used for transmitting and receiving hop-reservation (HR) packets, request-to-send (RTS) packets, clear-to-send (CTS) packets, and data packets. $f_i{}^*$ is used for sending and receiving acknowledgment (ACK) packets for the data packets received or transmitted on frequency f_i. In HRMA, time is slotted, and each slot is assigned a separate frequency hop, which is one among the M frequency hops in the hopping sequence. Each time slot is divided into four periods, namely, synchronizing period, HR period, RTS period, and CTS period, each period meant for transmitting or receiving the synchronizing packet, FR packet, RTS packet, and CTS packet, respectively. All idle nodes, that is, nodes that do not transmit or receive packets currently, hop together. During the synchronizing period of each slot, all idle nodes hop to the synchronizing frequency f_0 and exchange synchronization information. During the HR, RTS, and CTS periods, they just stay idle, dwelling on the common frequency hop assigned to each slot. In addition to the synchronization period used for synchronization purposes, an exclusive synchronization slot is also defined at the beginning of each HRMA frame (Figure 6.17). This slot is of the same size as that of the other normal slots. All idle nodes dwell on the synchronizing frequency f_0 during the synchronizing slot and exchange synchronization information that may be used to identify

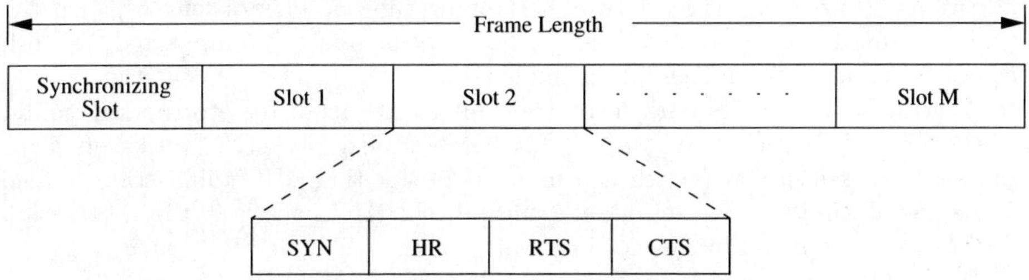

Figure 6.17. Frame format in HRMA.

the beginning of a frequency hop in the common hopping sequence, and also the frequency to be used in the immediately following hop. Thus the HRMA frame, as depicted in Figure 6.17, is composed of the single synchronizing slot, followed by M consecutive normal slots.

When a new node enters the network, it remains on the synchronizing frequency f_0 for a long enough period of time so as to gather synchronization information such as the hopping pattern and the timing of the system. If it receives no synchronization information, it assumes that it is the only node in the network, broadcasts its own synchronization information, and forms a one-node system. Since synchronization information is exchanged during every synchronization slot, new nodes entering the system can easily join the network. If μ is the length of each slot and μ_s the length of the synchronization period on each slot, then the dwell time of f_o at the beginning of each frame would be $\mu + \mu_s$. Consider the case where nodes from two different disconnected network partitions come nearby. Figure 6.18 depicts the worst-case frequency overlap scenario. In the figure, the maximum number of frequency hops $M = 5$. It is evident from the figure that within any time period equal to the duration of a HRMA frame, any two nodes from the two disconnected

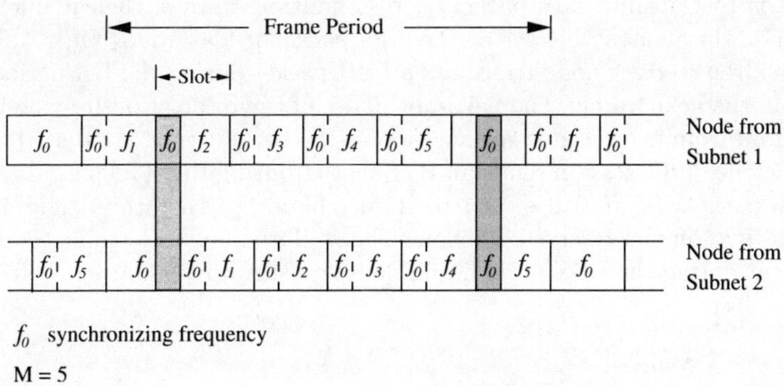

Figure 6.18. Merging of subnets.

partitions always have at least two overlapping time periods of length μ_s on the synchronizing frequency f_0. Therefore, nodes belonging to disconnected network components can easily merge into a single network.

When a node receives data to be transmitted, it first listens to the HR period of the immediately following slot. If it hears an HR packet, it backs off for a randomly chosen period (which is a multiple of slot time). If it finds the channel to be free during the SR period, it transmits an RTS packet to the destination during the RTS period of the slot and waits for the CTS packet. On receiving the RTS, the destination node transmits the CTS packet during the CTS period of the same slot, stays on the same frequency currently being used, and waits for the data packet. If the source node receives the CTS packet correctly, it implies that the source and receiver nodes have successfully reserved the current hop. In case the source node does not receive any CTS packet, it backs off for a random number of time slots and repeats the entire process again. The source and receiver nodes dwell on the same reserved frequency throughout the data transmission process, which starts immediately after the CTS period. As mentioned earlier, a separate frequency $(f_i{}^*, i = 1, 2, ..., M)$ is used for transmitting acknowledgments. After transmitting each data packet, the source node hops onto this acknowledgment frequency. The receiver sends an acknowledgment (ACK) packet back to the source on this acknowledgment frequency. Once the ACK packet transmission/reception is over, both the source and receiver hop back to the reserved frequency to continue with the data transmission.

After the CTS period of a slot, the idle nodes that do not transmit or receive packets hop onto the synchronization frequency f_0 and exchange synchronization information. They dwell on f_0 for a time period of μ_s and then hop onto the next frequency hop in the common hopping sequence.

The data packets transmitted can be of any size. Data transmission can take place through a single packet or through a train of packets. A maximum dwell period has been defined in order to prevent nodes from hogging onto a particular frequency channel. Therefore, the transmission time for the data packet, or the train of data packets, should not exceed this maximum dwell time. Suppose the sender needs to transmit data packets across multiple frames, then it informs the receiver node through the header of the data packet it transmits. On reading this information, the receiver node transmits an HR packet during the HR period of the same slot in the next frame. The neighbor nodes of the receiver on hearing this HR packet refrain from using the frequency hop reserved. On receiving the HR packet, the source node of the session sends an RTS packet during the RTS period and jams other RTS packets (if any) destined to its neighbors, so that the neighbor nodes do not interfere on the reserved frequency hop. Both the sender and the receiver remain silent during the CTS period, and data transmission resumes once this CTS period gets over.

6.6.4 Soft Reservation Multiple Access with Priority Assignment

Soft reservation multiple access protocol with priority assignment (SRMA/PA) [13] was developed with the main objective of supporting integrated services of real-time and non-real-time applications in ad hoc wireless networks, at the same time maximizing the statistical multiplexing gain. Nodes use a *collision-avoidance* handshake mechanism and a *soft reservation* mechanism in order to contend for and effect reservation of time slots. The soft reservation mechanism allows any urgent node, transmitting packets generated by a real-time application, to take over the radio resource from another node of a non-real-time application on an on-demand basis. SRMA/PA is a TDMA-based protocol in which nodes are allocated different time slots so that the transmissions are collision-free. The main features of SRMA/PA are a unique frame structure and soft reservation capability for distributed and dynamic slot scheduling, dynamic and distributed access priority assignment and update policies, and a time-constrained back-off algorithm.

Time is divided into frames, with each frame consisting of a fixed number (N) of time slots. The frame structure is shown in Figure 6.19. Each slot is further divided into six different fields, SYNC, soft reservation (SR), reservation request (RR), reservation confirm (RC), data sending (DS), and acknowledgment (ACK). The SYNC field is used for synchronization purposes. The SR, RR, RC, and ACK fields are used for transmitting and receiving the corresponding control packets. The DS field is used for data transmission. The SR packet serves as a busy tone. It informs the nodes in the neighborhood of the transmitting node about the reservation of the slot. The SR packet also carries the access priority value assigned to the node that has reserved the slot. When an idle node receives a data packet for transmission, the node waits for a free slot and transmits the RR packet in the RR field of that slot. A node determines whether or not a slot is free through the SR field of that slot. In case of a voice terminal node, the node tries to take control of the slot already reserved by a data terminal if it finds its priority level to be higher than that of the data terminal. This process is called *soft reservation*. This makes the SRMA/PA different from other protocols where even if a node has lower access priority compared to other ready nodes, it proceeds to complete the transmission of the entire data burst once it has reserved the channel.

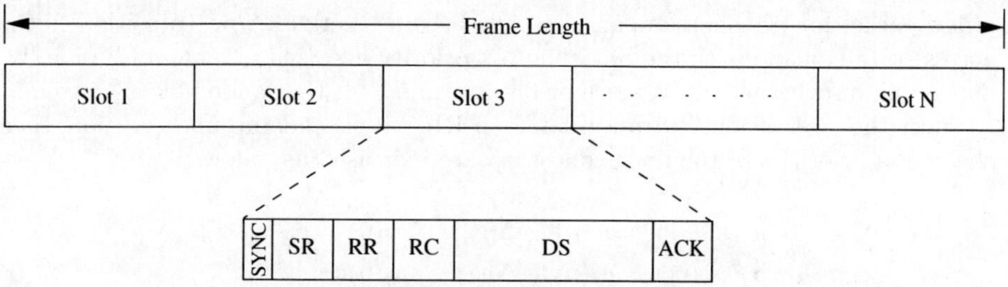

Figure 6.19. Frame structure in SRMA/PA.

Priority levels are initially assigned to nodes based on the service classes (real-time or non-real-time) in a static manner. Once the node acquires the channel, the corresponding slot stays reserved for the node until the node completes transmitting the entire data burst. The node is assigned a prespecified priority, ${p_v}^{(R)}$ or ${p_d}^{(R)}$, respectively, for voice and data terminals. R denotes that the node is a reserved node, that is, a node that has successfully reserved the slot. It is required that ${p_v}^{(R)} > {p_d}^{(R)}$, such that delay-sensitive voice applications get preference over normal data applications. Whenever the reservation attempt fails due to collision, the access priority of the node is updated based on the urgency of its packets.

A node that is currently transmitting is said to be in the active state. A node is said to be in the idle state if it does not have any packet to be transmitted. In the active state itself, nodes can be in one of the two states: access state and reserved state. Access state is one in which the node is backlogged and is trying to reserve a slot for transmission. The node is said to be in the reserved state if it has already reserved the slot for transmission. Whenever the access priority level of a voice terminal in the access state becomes greater than that of a data terminal in the reserved state, which could be known from the SR field, the corresponding slot is taken over by the prioritized voice terminal. In order to effect this mechanism, the values of priority levels must be such that ${p_v}^{(R)} > p_v(n) > {p_d}^{(R)} > p_d(n)$, where $p_v(n)$ and $p_d(n)$ are the access priority values of a voice terminal and data terminal, respectively, after its nth reservation attempt results in a collision. This soft reservation feature of SRMA/PA, where a voice terminal can take over the slots reserved by a data terminal whenever, due to the urgent nature of its traffic, its access priority becomes higher than that of the data terminal, helps in maximizing the statistical multiplexing gain for voice-data integrated services.

The RR-RC-DS-ACK exchange mechanism of SRMA/PA is similar to the RTS-CTS-DATA-ACK exchange mechanism of MACAW [2]. The RR and RC packets help in eliminating the hidden terminal problem. The major difference between SRMA/PA and CATA protocol [11] is that, while in CATA the slot reservation (SR) packet is transmitted by the receiver of the session, in SRMA/PA it is sent by the source node. Also, the soft reservation feature of SRMA/PA is absent in CATA.

The access priorities are assigned to nodes and updated in a distributed and dynamic manner. This allows dynamic sharing of the shared channel. On receiving a new packet for transmission, an idle node becomes active. Now, transition to the access state is made with the initial access priority assigned a value ${p_v}^{(0)}$ or ${p_d}^{(0)}$, depending on whether it is a voice or data terminal. If the random access attempt to effect the reservation by transmitting an RR packet ends up in a collision, then the access priorities of the node concerned are increased as follows:

$$p_v(n+1) = p_v(n) + \Delta p_v, \qquad p_v(0) = {p_v}^{(0)}$$
$$p_d(n+1) = p_d(n) + \Delta p_d, \qquad p_d(0) = {p_d}^{(0)}$$

where Δp_v and Δp_d are the incremental access priorities for voice and data services, respectively. They reflect the urgency of the traffic queued at the two types of terminals, and are given as below:

$$\Delta p_v = \Delta p_v{}^{(R)} \times \tau_S/\tau_r, \qquad \Delta p_d = \alpha \times l_Q$$

where τ_S is the slot duration, τ_r is the residual lifetime for the voice service, l_Q is the queue length , and α is a scaling coefficient. In order that the access priority of a voice terminal is always higher than that of a data terminal, the following constraint is followed:

$$p_d{}^{(0)} < p_d(n) < p_d{}^{(R)} = p_d{}^{(max)} < p_v{}^{(0)} < p_v(n) < p_v{}^{(R)} = p_v{}^{(max)}$$

Though dynamic assignment and update of access priority values are followed in SRMA/PA, collisions among nodes with the same priority and carrying traffic of the same service types cannot be avoided completely. Collisions occur during the RR field of the slot. In order to avoid collisions, a binary exponential back-off algorithm is used for non-real-time connections, and a modified binary exponential back-off algorithm is used for real-time connections. The modified algorithm implements a priority access policy in order to meet the delay requirements of real-time sessions. Here the back-off window is divided into two different regions, each region having a length of N_{B1} and N_{B2}, respectively, for real-time and non-real-time traffic. Each node checks the laxity of its head-of-line packet (laxity is the difference between the maximum access delay allowed and the residual lifetime of the packet). If the laxity exceeds the threshold T_{limit} slots, one slot out of the N_{B1} slots is selected randomly. Otherwise, one out of the N_{B2} slots is chosen randomly, that is, if the node is unable to make a reservation within the given time (T_{limit} slots), it is given a higher priority compared to other non-real-time nodes by choosing a slot from N_{B1}; otherwise, the node selects a slot from N_{B2}. The RR packet is transmitted on this chosen slot. Here again, if more than one node selects the same random slot and their RR packets collide again, a new back-off window starts immediately after the current slot. This is shown in Figure 6.20. The above back-off mechanism, which gives high preference to nodes transmitting delay-sensitive traffic, helps in guaranteeing the QoS requirements of real-time services in the network. The parameters N_{B1}, N_{B2}, and T_{limit} significantly affect the performance of the protocol, and must be chosen carefully based on the traffic load expected on the network.

6.6.5 Five-Phase Reservation Protocol

The five-phase reservation protocol (FPRP) [14] is a single-channel time division multiple access (TDMA)-based broadcast scheduling protocol. Nodes use a contention mechanism in order to acquire time slots. The protocol is fully distributed, that is, multiple reservations can be simultaneously made throughout the network. No ordering among nodes is followed; nodes need not wait for making time slot reservations. The slot reservations are made using a five-phase reservation process. The reservation process is localized; it involves only the nodes located within the two-hop radius of the node concerned. Because of this, the protocol is insensitive to the network size, that is, it is scalable. FPRP also ensures that no collisions occur due to the hidden terminal problem.

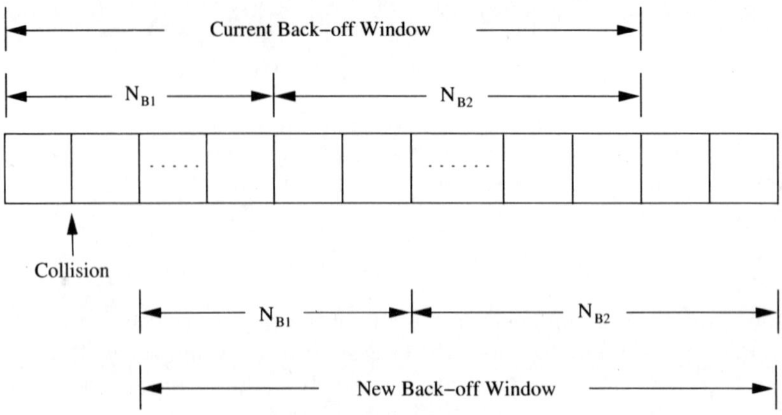

Figure 6.20. Back-off windows.

Time is divided into frames. There are two types of frames: reservation frame (RF) and information frame (IF). Each RF is followed by a sequence of IFs. Each RF has N reservation slots (RS), and each IF has N information slots (IS). In order to reserve an IS, a node needs to contend during the corresponding RS. Based on these contentions, a TDMA schedule is generated in the RF and is used in the subsequent IFs until the next RF. The structure of the frames is shown in Figure 6.21. Each RS is composed of M reservation cycles (RC). Within each RC, a five-phase dialog takes place, using which a node reserves slots. If a node wins the contention in an RC, it is said to have reserved the IS corresponding to the current RS in the subsequent IFs of the current frame. Otherwise, the node contends during

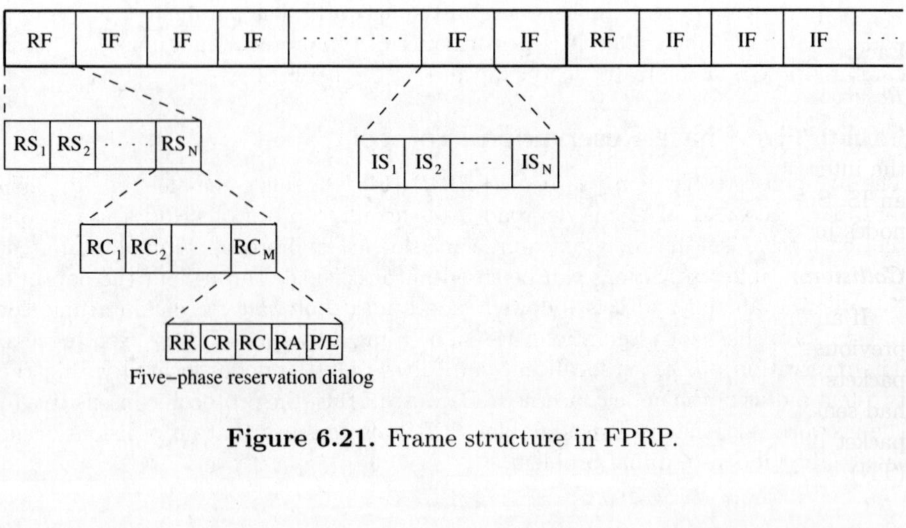

Figure 6.21. Frame structure in FPRP.

the subsequent RCs of the current RS until itself or any other node (1-hop or 2-hop neighbor) succeeds. During the corresponding IS, a node would be in one of the following three states: transmit (T), receive (R), or blocked (B). The five-phase dialog ensures that the protocol is free from the hidden node problem, and also ensures that once a reservation is made by a node with a high probability, it gets sole access to the slot within its neighborhood.

The protocol assumes the availability of global time at all nodes. Each node therefore knows when a five-phase cycle would start. The five phases of the reservation process are as follows:

1. Reservation request phase: Nodes that need to transmit packets send reservation request (RR) packets to their destination nodes.

2. Collision report phase: If a collision is detected by any node during the reservation request phase, then that node broadcasts a collision report (CR) packet. The corresponding source nodes, upon receiving the CR packet, take necessary action.

3. Reservation confirmation phase: A source node is said to have won the contention for a slot if it does not receive any CR messages in the previous phase. In order to confirm the reservation request made in the reservation request phase, it sends a reservation confirmation (RC) message to the destination node in this phase.

4. Reservation acknowledgment phase: In this phase, the destination node acknowledges reception of the RC by sending back a reservation acknowledgment (RA) message to the source. The hidden nodes that receive this message defer their transmissions during the reserved slot.

5. Packing and elimination (P/E) phase: Two types of packets are transmitted during this phase: packing packet and elimination packet. The details regarding the use of these packets will be described later in this section.

Each of the above five phases is described below.

Reservation request phase:

In this phase, each node that needs to transmit packets sends an RR packet to the intended destination node with a contention probability p, in order to reserve an IS. Such nodes that send RR packets are called requesting nodes (RN). Other nodes just keep listening during this phase.

Collision report phase:

If any of the listening nodes detects collision of RR packets transmitted in the previous phase, it broadcasts a collision report (CR) packet. By listening for CR packets in this phase, an RN comes to know about collision of the RR packet it had sent. If no CR is heard by the RN in this phase, then it assumes that the RR packet did not collide in its neighborhood. It then becomes a transmitting node (TN). Once it becomes a transmitting node, the node proceeds to the next phase, the reservation confirmation phase. On the other hand, if it hears a CR packet in this phase, it waits until the next reservation request phase, and then tries again. Thus, if two RNs are hidden from each other, their RR packets collide, both receive CR packets, and no reservation is made, thereby eliminating the hidden terminal problem.

Reservation confirmation phase:

An RN that does not receive any CR packet in the previous phase, that is, a TN, sends an RC packet to the destination node. Each neighbor node that receives this packet understands that the slot has been reserved, and defers its transmission during the corresponding information slots in the subsequent information frames until the next reservation frame.

Reservation acknowledgment phase:

On receiving the RC packet, the intended receiver node responds by sending an RA packet back to the TN. This is used to inform the TN that the reservation has been established. In case the TN is isolated and is not connected to any other node in the network, then it would not receive the RA packet, and thus becomes aware of the fact that it is isolated. Thus the RC packet prevents such isolated nodes from transmitting further. The reservation acknowledgment phase also serves another purpose. Other two-hop neighbor nodes that receive this RA packet get blocked from transmitting. Therefore, they do not disturb the transmission that is to take place in the reserved slots.

When more than two TNs are located nearby, it results in a deadlock condition. Such situations may occur when there is no common neighbor node present when the RNs transmit RR packets. Collisions are not reported in the next phase, and so each node claims success and becomes a TN. Deadlocks are of two types: isolated and non-isolated. An isolated deadlock is a condition where none of the deadlocked nodes is connected to any non-deadlocked node. In the non-isolated deadlock situation, at least one deadlocked node is connected to a non-deadlocked neighbor node. The RA phase can resolve isolated deadlocks. None of the nodes transmits RA, and hence the TNs abort their transmissions.

Packing/elimination (P/E) phase:

In this phase, a packing packet (PP) is sent by each node that is located within two hops from a TN, and that had made a reservation since the previous P/E phase. A node receiving a PP understands that there has been a recent success in slot reservation three hops away from it, and because of this some of its neighbors would have been blocked during this slot. The node can take advantage of this and adjust its contention probability p, so that convergence is faster.

In an attempt to resolve a non-isolated deadlock, each TN is required to transmit an elimination packet (EP) in this phase, with a probability 0.5. A deadlocked TN, on receiving an EP before transmitting its own EP, gets to know about the deadlock. It backs off by marking the slot as reserved and does not transmit further during the slot.

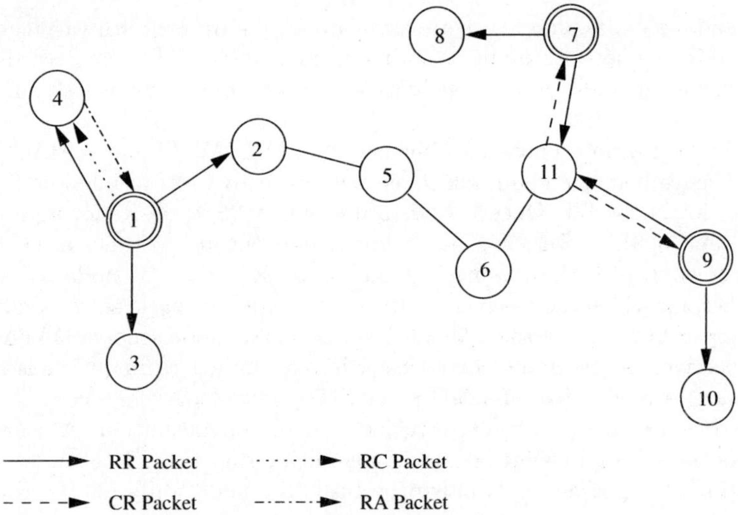

Figure 6.22. FPRP - Example.

Consider Figure 6.22. Here nodes 1, 7, and 9 have packets ready to be transmitted to nodes 4, 8, and 10, respectively. During the reservation request phase, all three nodes transmit RR packets. Since no other node in the two-hop neighborhood of node 1 transmits simultaneously, node 1 does not receive any CR message in the collision report phase. So node 1 transmits an RC message in the next phase, for which node 4 sends back an RA message, and the reservation is established. Node 7 and node 9 both transmit the RR packet in the reservation request phase. Here node 9 is within two hops from node 7. So if both nodes 7 and 9 transmit simultaneously, their RR packets collide at common neighbor node 11. Node 11 sends a CR packet which is heard by nodes 7 and 9. On receiving the CR packet, nodes 7 and 9 stop contending for the current slot.

The reservation process in FPRP is simple. No information needs to be distributed to nodes other than the one-hop neighbor nodes before the reservation becomes successful.

6.6.6 MACA with Piggy-Backed Reservation

MACA with piggy-backed reservation (MACA/PR) [15] is a protocol used to provide real-time traffic support in multi-hop wireless networks. The MAC protocol used is based on the MACAW protocol [2], with the provisioning of non-persistent CSMA (as in FAMA [3]). The main components of MACA/PR are: a MAC protocol, a reservation protocol, and a QoS routing protocol. MACA/PR differentiates real-time packets from the best-effort packets. While providing guaranteed bandwidth support for real-time packets, at the same time it provides reliable transmission of best-effort packets. Time is divided into slots. The slots are defined by the reservations made at nodes, and hence are asynchronous in nature with varying lengths. Each node in the network maintains a reservation table (RT) that records all the reserved transmit and receive slots/windows of all nodes within its transmission range.

In order to transmit a non-real-time packet, a MACAW [2]-based MAC protocol is used. The ready node (a node which has packets ready for transmission) first waits for a free slot in the RT. Once it finds a free slot, it again waits for an additional random time of the order of a single-hop round-trip delay time, after which it senses the channel. If the channel is found to be still free, the node transmits an RTS packet, for which the receiver, if it is ready to receive packets, responds with a CTS packet. On receiving the CTS packet, the source node sends a DATA packet, and the receiver, on receiving the packet without any error, finally sends an ACK packet back to the source. The RTS and CTS control packets contain, in them, the time duration in which the DATA packet is to be transmitted. A nearby node that hears these packets avoids transmission during that time. If, after the random waiting time, the channel is found to be busy, the node waits for the channel to become idle again, and then repeats the same procedure.

For real-time traffic, the reservation protocol of MACA/PR functions as follows. The sender is assumed to transmit real-time packets at certain regular intervals, say, every CYCLE time period. The first data packet of the session is transmitted in the usual manner just as a best-effort packet would be transmitted. The source node first sends an RTS packet, for which the receiver node responds with a CTS packet. Now the source node sends the first DATA packet of the real-time session. Reservation information for the next DATA packet to be transmitted (which is scheduled to be transmitted after CYCLE time period) is piggy-backed on this current DATA packet. On receiving this DATA packet, the receiver node updates its reservation table with the piggy-backed reservation information. It then sends an ACK packet back to the source. The receiver node uses the ACK packet to confirm the reservation request that was piggy-backed on the previous DATA packet. It piggy-backs the reservation confirmation information on the ACK packet. Neighbor nodes that hear the DATA and ACK packets update their reservation tables with the reservation information carried by them, and refrain from transmitting when the next packet is to be transmitted. Unlike MACAW [2], MACA/PR does not make use of RTS/CTS packets for transmission of the subsequent DATA packets. After receiving the ACK, the source node directly transmits the next DATA packet at its scheduled transmission time in the next CYCLE. This DATA packet in turn would carry reservation information for the next DATA packet. Real-time data transmission, hence, occurs as a series of DATA-ACK packet exchanges. The real-time packets (except for the first packet of the session that is used to initiate the reservation process) are transmitted only once. If an ACK packet is not received for a DATA packet, the source node just drops the packet. The ACK packet therefore serves the purpose of renewing the reservation, in addition to recovering from packet loss. If the source node fails to receive ACK packets for a certain number of consecutive DATA packets, it then assumes the reservation to have been lost. It restarts the real-time session again with an RTS-CTS control packet exchange, either on a different slot on the same link, or on a different link in case of a path break. In order to transmit an RTS to the receiver node, the source needs to find a slot that is free at both the nodes. For maintaining consistent information regarding free slots at all nodes, MACA/PR uses periodic exchange of reservation tables. This periodic table exchange automatically overcomes the hidden terminal problem. When a hidden terminal receives a reservation table from a node, it refrains from transmitting in the reserved slots of that node. Slot reservation information maintained in the reservation tables is refreshed every cycle. If the reservation is not refreshed for a certain number of consecutive cycles, it is then dropped. The transmission of packets in MACA/PR is depicted in Figure 6.23. It can be seen from the figure that the RTS-CTS exchange is used only for transmitting the first packet of the session. Since each DATA packet carries reservation information for the next DATA packet that would be transmitted after CYCLE time period, RTS-CTS exchange is not required for the subsequent DATA packets. Neighbor nodes

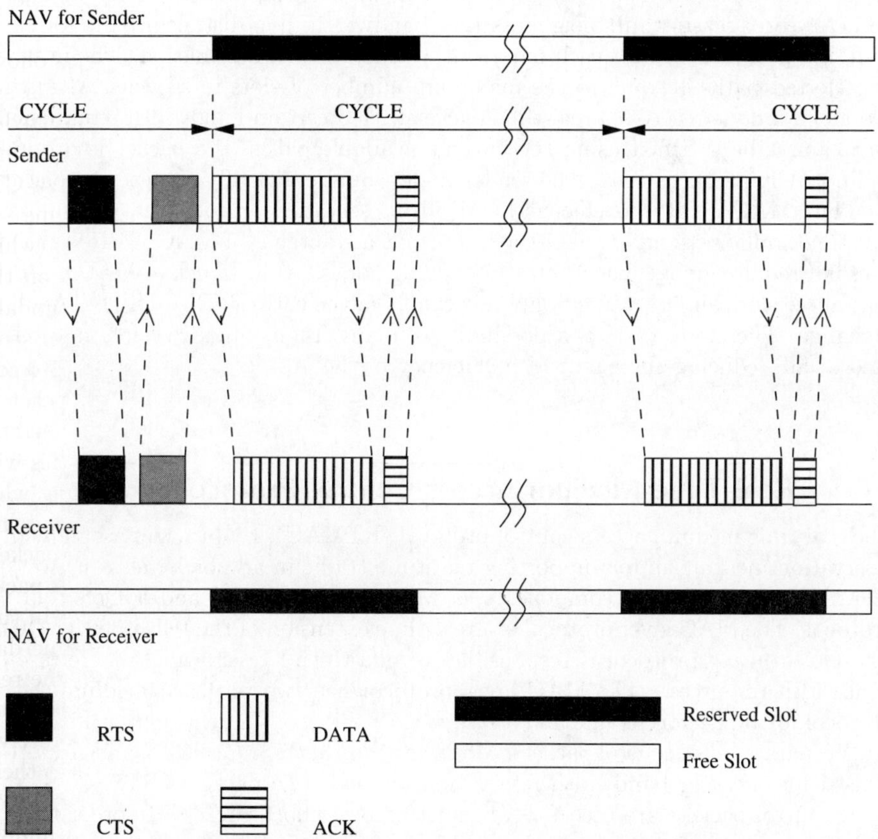

Figure 6.23. Packet transmission in MACA/PR.

that receive DATA packets update their reservation tables accordingly and do not contend for the channel during the reserved slots. The network allocation vector (NAV) at each node reflects the current and future state of the channel as perceived by the node.

Best-effort packet transmissions and real-time packet transmissions can be interleaved at nodes, with higher priority being given to real-time packets. For real-time packets, MACA/PR effectively works as a TDM system, with a superframe time of CYCLE. The best-effort packets are transmitted in the empty slots (which have not been reserved) of the cycle.

When a new node joins the network, it initially remains in the listening mode during which it receives reservation tables from each of its neighbors and learns about the reservations made in the network. After this initial period, the node shifts to its normal mode of operation.

The QoS routing protocol used with MACA/PR is the destination sequenced distance vector (DSDV) routing protocol [16] (described in detail in Chapter 7). Bandwidth constraint has been introduced in the routing process. Each node periodically broadcasts to its neighbors the (bandwidth, hop distance) pairs for each preferred path, that is, for each bandwidth value, to each destination. The number of preferred paths is equal to the maximum number of slots in a cycle. After this is done, if a node receives a real-time packet with a certain bandwidth requirement that cannot be satisfied using the current available paths, the packet is dropped and no ACK packet is sent. The sender node would eventually reroute the packet.

Thus, MACA/PR is an efficient bandwidth reservation protocol that can support real-time traffic sessions. One of the important advantages of MACA/PR is that it does not require global synchronization among nodes. A drawback of MACA/PR is that a free slot can be reserved only if it can fit in the entire RTS-CTS-DATA-ACK exchange. Therefore, there is a possibility of many fragmented free slots not being used at all, reducing the bandwidth efficiency of the protocol.

6.6.7 Real-Time Medium Access Control Protocol

The real-time medium access control protocol (RTMAC) [17] provides a bandwidth reservation mechanism for supporting real-time traffic in ad hoc wireless networks. RTMAC consists of two components, a MAC layer protocol and a QoS routing protocol. The MAC layer protocol is a real-time extension of the IEEE 802.11 DCF. The QoS routing protocol is responsible for end-to-end reservation and release of bandwidth resources. The MAC layer protocol has two parts: a medium-access protocol for best-effort traffic and a reservation protocol for real-time traffic.

A separate set of control packets, consisting of *ResvRTS*, *ResvCTS*, and *ResvACK*, is used for effecting bandwidth reservation for real-time packets. RTS, CTS, and ACK control packets are used for transmitting best-effort packets. In order to give higher priority for real-time packets, the wait time for transmitting a *ResvRTS* packet is reduced to half of DCF inter-frame space (DIFS), which is the wait time used for best-effort packets.

Time is divided into superframes. As can be seen from Figure 6.24, the superframe for each node may not strictly align with the other nodes. Bandwidth reservations can be made by a node by reserving variable-length time slots on superframes, which are sufficient enough to carry the traffic generated by the node. The core concept of RTMAC is the flexibility of slot placement in the superframe. Each superframe consists of a number of reservation-slots (resv-slots). The time duration of each resv-slot is twice the maximum propagation delay. Data transmission normally requires a block of resv-slots. A node that needs to transmit real-time packets first reserves a set of resv-slots. The set of resv-slots reserved by a node for a connection on a superframe is called a connection-slot. A node that has made reservations on the current superframe makes use of the same connection-slot in the successive superframes for transmitting packets. Each node maintains a reservation table containing information such as the sender id, receiver id, and starting and ending times of reservations that are currently active within its direct transmission range.

In RTMAC, no time synchronization is assumed. The protocol uses relative time for all reservation purposes. When a node receives this relative-time-based

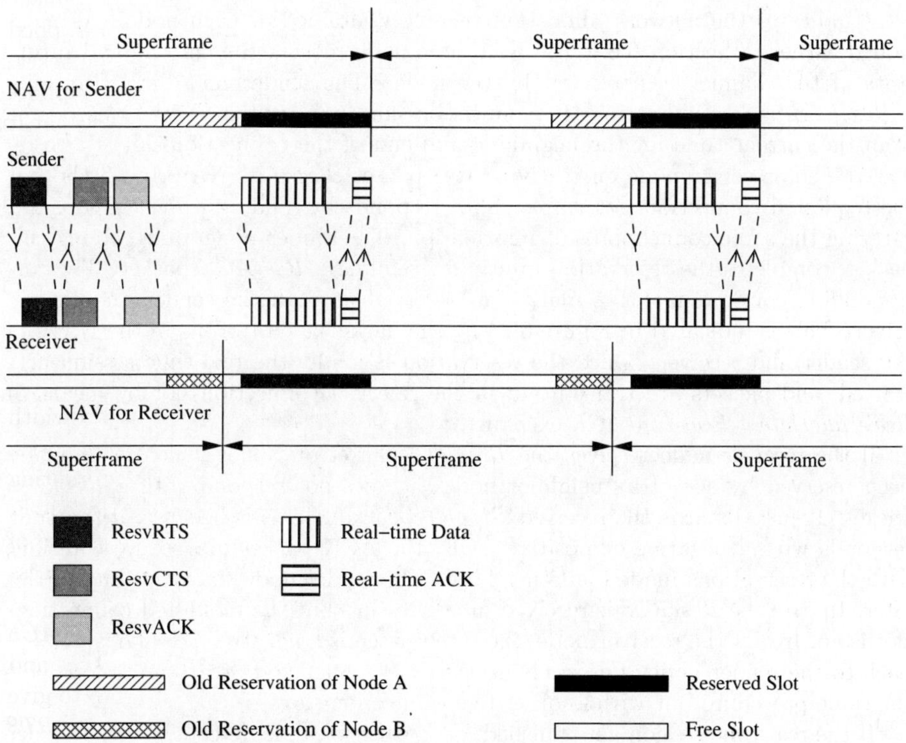

Figure 6.24. Reservation mechanism in RTMAC.

information, it converts the relative time to absolute time by adding its current time maintained in its clock. A three-way handshake protocol is used for effecting the reservation. For example, node A, which wants to reserve a slot with node B, sends a *ResvRTS* packet which contains the relative time information of starting and ending of the connection-slot (a number of resv-slots) to be reserved. Node B, on receiving this packet, first checks its reservation table to see whether it can receive on those resv-slots. If it can, it replies by sending a *ResvCTS* packet containing the relative time information of the same resv-slots to be reserved. Neighbor nodes of the receiver, on receiving the *ResvCTS*, update their reservation tables accordingly. Source node A, on receiving the *ResvCTS* packet, responds by sending a *ResvACK* packet. This packet also carries relative time information regarding the reserved slots. The *ResvACK* packet is meant for the neighbor nodes of the source node (node A) which are not aware of the reservation as they may not receive the *ResvCTS* packet. Such nodes update their reservation tables on receiving the *ResvACK* packet. Transmission of the *ResvACK* packet completes the reservation process. Once the reservation is made, real-time packets are transmitted in these reserved slots. Transmission of each real-time packet is followed by the transmission of a real-time ACK (RTACK) packet by the receiver.

The bandwidth reservation process is illustrated in Figure 6.24. In the figure, NAV indicates the network allocation vector maintained at each node. As mentioned earlier in Section 6.6.6, the NAV at a node reflects the current and future state of the channel as perceived by the node. The sender node first transmits a *ResvRTS* packet indicating the connection-slot (represented by the offset time from the current time for the beginning and end of the connection-slot) to be reserved. The receiver node on receiving this packet checks its NAV and finds that the requested connection-slot is free. So it responds by sending a *ResvCTS* packet carrying the same connection-slot information. The sender node, on receiving this packet, completes the reservation process by sending a *ResvACK* packet. The corresponding connection-slot is marked as reserved at both the sender and the receiver. This is indicated in Figure 6.24 by the dark-shaded regions in the NAVs of the sender and receiver. Once the reservation is made, the real-time session gets started, and packets are transmitted in the reserved connection-slot by means of *Real-timeData* – *Real-timeACK* exchanges.

If the receiver node receives the *ResvRTS* packet on a slot which has already been reserved by one of its neighbor nodes, it does not respond with a *ResvCTS* packet. It just discards the received *ResvRTS* packet. This is because, if the node responds with a negative or positive ACK, the ACK packet may cause collisions with the reservations made by its neighbor. The sender node times out and retries later. In case the *ResvRTS* is received on a free slot, but the requested connection-slot is not free at the receiver node, the receiver sends a negative CTS (*ResvNCTS*) back to the sender node. On receiving this, the sender node reattempts following the same procedure but with another free connection-slot.

If the real-time session gets finished, or a route break is detected by the sender node, the node releases the resources reserved for that session by sending a reservation release RTS (*ResvRelRTS*) packet. The *ResvRelRTS* packet is a broadcast packet. Nodes hearing this packet update their reservation tables in order to free the corresponding connection slots. In case the receiver node receives this (*ResvRelRTS*) packet, it responds by broadcasting a (*ResvRelCTS*) packet. The receiver's neighbor nodes, on receiving this (*ResvRelCTS*) packet, free up the corresponding reservation slots. In case the downstream node of the broken link does not receive the *ResvRelRTS* packet, since it also does not receive any DATA packet belonging to the corresponding connection, it times out and releases the reservations made.

A QoS routing protocol is used with RTMAC to find an end-to-end path that matches the QoS requirements (bandwidth requirements) of the user. The QoS routing protocol used here is an extension of the destination sequenced distance vector (DSDV) routing protocol. When a node receives a data packet for a new connection, the node reserves bandwidth on the forward link and forwards the packet to the next node on the path to the destination. In order to maintain a consistent view of reservation tables of neighboring nodes at each node, each node transmits its reservation information along with the route update packet, which is defined as part of the DSDV protocol. The routing protocol can specify a specific connection-slot to be reserved for a particular connection; this gives flexibility for the routing protocol to decide on the positioning of the connection-slot. But generally, the first available connection-slot is used.

One of the main advantages of RTMAC is its bandwidth efficiency. Since nodes operate in the asynchronous mode, successive reservation slots may not strictly align with each other. Hence small fragments of free slots may occur in between reservation slots. If the free slot is just enough to accommodate a DATA and ACK packet, then RTMAC can make use of the free slot by transmitting *ResvRTS-ResvCTS-ResvACK* in some other free slot. Such small free slots cannot be made use of in other protocols such as MACA/PR, which require the free slot to accommodate the entire RTS-CTS-DATA-ACK exchange. Another advantage of RTMAC is its asynchronous mode of operation where nodes do not require any global time synchronization.

6.7 CONTENTION-BASED MAC PROTOCOLS WITH SCHEDULING MECHANISMS

Protocols that fall under this category focus on packet scheduling at the nodes and transmission scheduling of the nodes. Scheduling decisions may take into consideration various factors such as delay targets of packets, laxities of packets, traffic load at nodes, and remaining battery power at nodes. In this section, some of the scheduling-based MAC protocols are described.

6.7.1 Distributed Priority Scheduling and Medium Access in Ad Hoc Networks

This work, proposed in [18], presents two mechanisms for providing quality of service (QoS) support for connections in ad hoc wireless networks. The first technique, called distributed priority scheduling (DPS), piggy-backs the priority tag of a node's current and head-of-line packets on the control and data packets. By retrieving information from such packets transmitted in its neighborhood, a node builds a scheduling table from which it determines its rank (information regarding its position as per the priority of the packet to be transmitted next) compared to other nodes in its neighborhood. This rank is incorporated into the back-off calculation mechanism in order to provide an approximate schedule based on the ranks of the nodes. The second scheme, called multi-hop coordination, extends the DPS scheme to carry out scheduling over multi-hop paths. The downstream nodes in the path to the destination increase the relative priority of a packet in order to compensate for the excessive delays incurred by the packet at the upstream nodes.

Distributed Priority Scheduling

The distributed priority scheduling scheme (DPS) is based on the IEEE 802.11 distributed coordination function. DPS uses the same basic RTS-CTS-DATA-ACK packet exchange mechanism. The RTS packet transmitted by a ready node carries the priority tag/priority index for the current DATA packet to be transmitted. The priority tag can be the delay target for the DATA packet. On receiving the RTS packet, the intended receiver node responds with a CTS packet. The receiver node copies the priority tag from the received RTS packet and piggy-backs it along with the source node id, on the CTS packet. Neighbor nodes receiving the RTS or CTS packets (including the hidden nodes) retrieve the piggy-backed priority tag information and make a corresponding entry for the packet to be transmitted, in their scheduling tables (STs). Each node maintains an ST holding information about packets, which were originally piggy-backed on control and data packets. The entries in the ST are ordered according to their priority tag values. When the source node transmits a DATA packet, its head-of-line packet information (consisting of the destination and source ids along with the priority tag) is piggy-backed on the DATA packet (head-of-line packet of a node refers to the packet to be transmitted next by the node). This information is copied by the receiver onto the ACK packet it sends in response to the received DATA packet. Neighbor nodes receiving the DATA or ACK packets retrieve the piggy-backed information and update their STs accordingly. When a node hears an ACK packet, it removes from its ST any entry made earlier for the corresponding DATA packet.

Figure 6.25 illustrates the piggy-backing and table update mechanism. Node 1 needs to transmit a DATA packet (with priority index value 9) to node 2. It first transmits an RTS packet carrying piggy-backed information about this DATA packet. The initial state of the ST of node 4 which is a neighbor of nodes 1 and 2 is shown in ST (a). Node 4, on hearing this RTS packet, retrieves the piggy-backed priority information and makes a corresponding entry in its ST, as shown in ST (b). The destination node 2 responds by sending a CTS packet. The actual DATA packet is sent by the source node once it receives the CTS packet. This DATA packet carries piggy-backed priority information regarding the head-of-line packet at node 1. On hearing this DATA packet, neighbor node 4 makes a corresponding entry for the head-of-line packet of node 1, in its ST. ST (c) shows the new updated status of the ST at node 4. Finally, the receiver node sends an ACK packet to

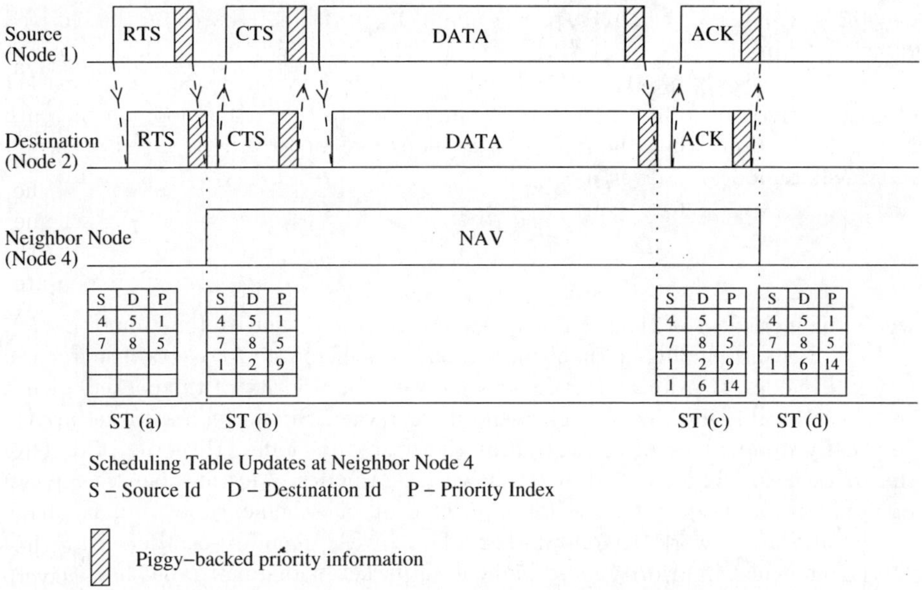

Figure 6.25. Piggy-backing and scheduling table update mechanism in DPS.

node 1. When this packet is heard by node 4, it removes the entry made for the corresponding DATA packet from its ST. The state of the scheduling table at the end of this data transfer session is depicted in ST (d).

In essence, each node's scheduling table gives the rank of the node with respect to other nodes in its neighborhood. This rank information is used to determine the back-off period to be taken by the node. The back-off distribution is given by

$$\text{back}-\text{off} = \begin{cases} Uniform[0,(2^t CW_{min})-1] & r=1,\, n<nmax \\ \alpha \times CW_{min} + Uniform[0,\gamma CW_{min}-1] & r>1,\, n=0 \\ Uniform[0,(2^n \gamma CW_{min})-1] & r>1, n\geq 1 \end{cases}$$

where CW_{min} is the minimum size of the contention window. r is the rank in the scheduling table of the node's highest priority packet; n is the current number of transmission attempts made by the node; $nmax$ is the maximum number of retransmissions permitted; α is a constant; and γ is a constant that is used to control the congestion in the second attempt for the highest ranked nodes.

Multi-Hop Coordination

By means of the multi-hop coordination mechanism, the excess delay incurred by a packet at the upstream nodes is compensated for at the downstream nodes. When a node receives a packet, it would have already received the priority index of the packet piggy-backed on the previous RTS packet. In case the node is an intermediate node which has to further forward the packet, the node calculates the new priority index of the DATA packet in a recursive fashion, based on the received value of the priority index. If ${d_{i,j}}^k$ is the priority index assigned to the kth packet of flow i with size ${l_i}^k$ at its jth hop, and if ${t_i}^k$ is the time at which the kth packet of flow i arrives at its first hop (the next hop node to the source node on the path to the destination), then the new priority index assigned to the received packet at intermediate node j is given as

$$ {d_{i,j}}^k = \begin{cases} {t_i}^k + {\delta_{i,1}}^k, & j = 1 \\ {d_{i,j-1}}^k + {\delta_{i,j}}^k, & j > 1 \end{cases} $$

where the increment of the priority index ${\delta_{i,j}}^k$ is a non-negative function of $i, j, {l_i}^k$, and ${t_i}^k$. Because of this mechanism, if a packet suffers due to excess delay at the upstream nodes, the downstream nodes increase the priority of the packet so that the packet is able to meet its end-to-end delay target. Similarly, if a packet arrives very early due to lack of contention at the upstream nodes, then the priority of that packet would be reduced at the downstream nodes. Any suitable scheme can be used for obtaining the values for ${\delta_{i,j}}^k$. One simple scheme, called uniform delay budget allocation, works as follows. For a flow i with an end-to-end delay target of D, the increment in priority index value for a packet belonging to that flow, at hop j, is given as, ${\delta_{i,j}}^k = \frac{D}{K}$, where K is the length of the flow's path.

The distributed priority scheduling and multi-hop coordination schemes described above are fully distributed schemes. They can be utilized for carrying time-sensitive traffic on ad hoc wireless networks.

6.7.2 Distributed Wireless Ordering Protocol

The distributed wireless ordering protocol (DWOP) [19] consists of a media access scheme along with a scheduling mechanism. It is based on the distributed priority scheduling scheme proposed in [18]. DWOP ensures that packets access the medium according to the order specified by an ideal reference scheduler such as first-in-first-out (FIFO), virtual clock, or earliest deadline first. In this discussion, FIFO is chosen as the reference scheduler. In FIFO, packet priority indices are set to the arrival times of packets. Similar to DPS [18], control packets are used in DWOP to piggy-back priority information regarding head-of-line packets of nodes. As the targeted FIFO schedule would transmit packets in order of the arrival times, each node builds up a scheduling table (ST) ordered according to the overheard arrival times.

The key concept in DWOP is that a node is made eligible to contend for the channel only if its locally queued packet has a smaller arrival time compared to all other arrival times in its ST (all other packets queued at its neighbor nodes), that is, only if the node finds that it holds the next region-wise packet in the hypothetical FIFO schedule. Two additional table management techniques, receiver participation and stale entry elimination, are used in order to keep the actual schedule close to the reference FIFO schedule. DWOP may not suffer due to information asymmetry. Since in most networks all nodes are not within the radio range of each other, a transmitting node might not be aware of the arrival times of packets queued at another node which is not within its direct transmission range. This information asymmetry might affect the fair sharing of bandwidth. For example, in Figure 6.26 (a), the sender of flow B would be aware of the packets to be transmitted by the sender of flow A, and so it defers its transmission whenever a higher priority packet is queued at the sender of flow A. But the sender of flow A is not aware of the arrival times of packets queued at the sender of flow B and hence it concludes that it has the highest priority packet in its neighborhood. Therefore, node 1 unsuccessfully tries to gain access to the channel continuously. This would result in flow B receiving an unfair higher share of the available bandwidth. In order to overcome this information asymmetry problem, the *receiver participation* mechanism is used.

In the receiver participation mechanism, a receiver node, when using its ST information, finds that the sender is transmitting out of order, that is, the reference FIFO schedule is being violated, an *out-of-order notification* is piggy-backed by the receiver on the control packets (CTS/ACK) it sends to the sender. In essence, information regarding the transmissions taking place in the two-hop neighborhood of the sender is propagated by the receiver node whenever it detects a FIFO schedule violation. Since the notification is sent only when a FIFO violation is detected, the actual transmission may not strictly follow the FIFO schedule; rather, it approximates the FIFO schedule. On receiving an out-of-order packet from a sender node, the receiver node transmits a notification to the sender node carrying the actual rank R of the sender with respect to the receiver's local ST. On receiving this out-of-order notification, the sender node goes into a back-off state after completing the transmission of its current packet. The back-off period $T_{back-off}$ is given by

$$T_{back-off} = R \times (EIFS + DIFS + T_{success} + CW_{min})$$

where $T_{success}$ is the longest possible time required to transmit a packet successfully, including the RTS-CTS-DATA-ACK handshake. Thus the node backs off, allowing higher priority packets in the neighborhood of the receiver to get transmitted first. In order to obtain a perfect FIFO schedule, the receiver can very well be made not to reply to the out-of-order requests (RTS) of the sender. This would cause the sender

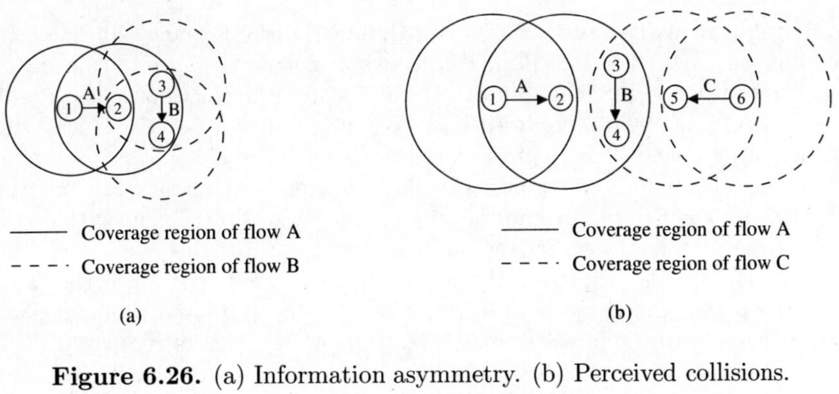

Figure 6.26. (a) Information asymmetry. (b) Perceived collisions.

to time out and back off, thereby avoiding any out-of-order transmission. But since the sender has already expended its resources in transmitting the RTS successfully, it is allowed to complete the transmission of its current packet. This is a trade-off between achieving perfect FIFO scheduling and high system utilization. Since in DWOP a node's access to the medium is dependent on its rank in the receiver node's ST (the rank of a node denotes the position of the node's entry in the receiver node's ST as per the priority of the corresponding packet), information maintained in the ST must be consistent with the actual network scenario. The stale entry elimination mechanism makes sure that the STs are free of stale entries. An entry is deleted from the ST only after an ACK packet for the corresponding entry is heard by the node. In case the ACK packet collides at the node, the corresponding entry in the ST will never be removed. This may cause a large deviation from the ideal FIFO schedule. Figure 6.26 (b) shows an example-perceived collisions scenario. The sender and receiver of flow B might have stale entries because of collisions caused by packets belonging to flow A and flow C at the sender and receiver of flow B. It can be observed that, in case there is a stale entry in the ST of a node, the node's own head-of-line packet position remains fixed, while other entries below the head-of-line entry keep changing. The above observation is used as a stale entry detection method. Thus, when a node observes that its rank remains fixed while packets whose priorities are below the priority of its head-of-line packet are being transmitted, it concludes that it may have one or more stale entries in its ST. The node simply deletes the oldest entry from its ST, assuming it to be the stale entry. This mechanism thus eliminates stale entries from the STs of nodes.

In summary, DWOP tries to ensure that packets get access to the channel according to the order defined by a reference scheduler. The above discussion was with respect to the FIFO scheduler. Though the actual schedule deviates from the ideal FIFO schedule due to information asymmetry and stale information in STs, the receiver participation and the stale entry elimination mechanisms try to keep the actual schedule as close as possible to the ideal schedule.

6.7.3 Distributed Laxity-Based Priority Scheduling Scheme

The distributed laxity-based priority scheduling (DLPS) scheme [20] is a packet scheduling scheme, where scheduling decisions are made taking into consideration the states of neighboring nodes and the feedback from destination nodes regarding packet losses. Packets are reordered based on their uniform laxity budgets (ULBs) and the packet delivery ratios of the flows to which they belong.

Each node maintains two tables: scheduling table (ST) and packet delivery ratio table (PDT). The ST contains information about packets to be transmitted by the node and packets overheard by the node, sorted according to their *priority index* values. Priority index expresses the priority of a packet. The lower the priority index, the higher the packet's priority. The PDT contains the count of packets transmitted and the count of acknowledgment (ACK) packets received for every flow passing through the node. This information is used for calculating current packet delivery ratio of flows (explained later in this section).

A node keeps track of packet delivery ratios (used for calculating priority index of packets) of all flows it is aware of by means of a feedback mechanism. Figure 6.27 depicts the overall functioning of the feedback mechanism. Incoming packets to a node are queued in the node's input queue according to their arrival times. The scheduler sorts them according to their priority values and inserts them into the transmission queue. The highest priority packet from this queue is selected for transmission. The node, after transmitting a packet, updates the count of packets transmitted so far in its PDT. The destination node of a flow, on receiving data packets, initiates a feedback by means of which the count of DATA packets received by it is conveyed to the source through ACK packets traversing the reverse path. These two pieces of information, together denoted by S_i in Figure 6.27, are received by the feedback information handler (FIH). The FIH, in parallel, also sends the previous state information S_{i-1} to the priority function module (PFM). The ULB of each packet in ST is available at the node (ULB calculation will be explained later in this section). This information is also sent to PFM, which uses the information fed to it to calculate the priority indices of packets in the ST.

Using the count of DATA packets transmitted (*pktsSent*) and count information carried by ACK packets (*acksRcvd*), available in PDT, packet delivery ratio (PDR) of the flow at any given time is computed as

$$PDR = \frac{acksRcvd}{pktsSent} \tag{6.7.1}$$

Priority index of a packet (PI) is defined as

$$PI = \frac{PDR}{M} \times ULB \tag{6.7.2}$$

Here, $ULB = \frac{deadline - currentTime}{remHops}$ is the uniform laxity budget of the packet, and M is a user-defined parameter representing the desired packet delivery ratio for the flow. *deadline* is the end-to-end deadline target of the packet and is equal to

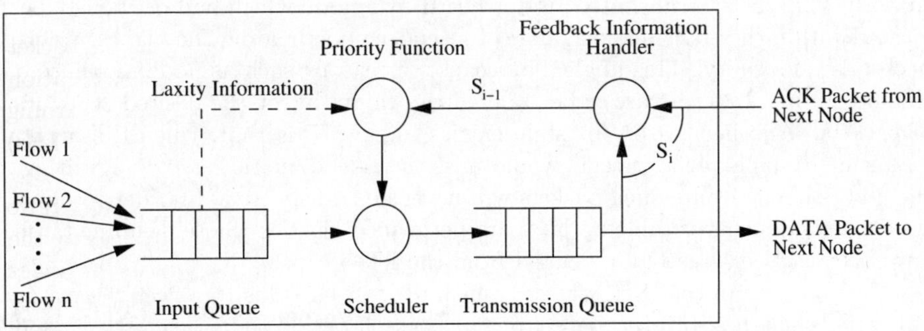

Figure 6.27. Feedback mechanism. *Reproduced with permission from [20], © Elsevier, 2004.*

(*packet creation time+end-to-end delay target*). *currentTime* denotes the current time according to the node's local clock.

When greater numbers of packets belonging to a flow meet their delay targets, the term $\frac{PDR}{M}$ would have a high value. Hence priority index would be high for packets of that flow, and therefore the actual priority of the packets would be low. When very few packets of a flow meet their delay targets, the value of $\frac{PDR}{M}$ would be much less, thereby lowering the priority index and increasing the priority of packets of that flow. ULB also plays an equally important role. Since *remHops*, the number of hops remaining to be traversed, is in the denominator of the expression for ULB, when a packet is near its source and needs to traverse several hops to reach its destination, its priority index value will be lowered, thereby increasing its priority. When it nears its destination, the fewer number of hops to be traversed tends to increase the priority index, thereby lowering its priority.

RTS and CTS packets transmitted by a node are modified to carry piggy-backed information regarding the *highest priority packet queued at the node.* Similarly, DATA and ACK packets transmitted by a node carry piggy-backed information corresponding to the *highest priority packet entry in the ST of the node.* A node hearing any packet retrieves the piggy-backed priority information, calculates the priority index of the corresponding packet, and adds a corresponding entry in its ST.

A DATA packet also carries information about itself. The end-to-end delay target, remaining number of hops, actual source ID, and the flow ID constitute this information. A node, on receiving a DATA packet, using the above information and the information maintained in the PDT, can obtain the priority index of the packet (PI), as given in Equation 6.7.2. Since the priority index of a packet keeps changing with time, it needs to be updated constantly. Each node, before calculating its back-off period and before inserting a new entry into its ST, recalculates and updates the priority index of each entry in its ST.

When a node hears a DATA packet, if an entry for the corresponding packet exists in its ST, then that entry is deleted. The sender node deletes its entry from the ST only when an ACK for the transmitted DATA packet is received. It may happen that a DATA packet transmitted is not heard by a node which had previously been located within the transmission range of a sender node holding the highest priority packet in its locality. This might be because of reasons such as node mobility and channel errors. In such cases, the stale entries might affect the desired scheduling of packets. Another reason for stale entries in the ST is that, when the network load is high, some of the packets would miss their deadline targets while waiting in the node's queue itself. Such packets will never be transmitted. In order to remove stale entries, whenever table updates are performed, entries whose deadline targets have been missed already are deleted from the ST.

The objective of the back-off mechanism used in DLPS is to reflect the priority of the node's highest priority packet on the back-off period to be taken by the node. If r is the rank (rank of an entry is the position of that entry in the scheduling table of the node), in ST of the node, of the current packet to be sent, n is the number of retransmission attempts made for the packet, and $nmax$ is the maximum number of retransmission attempts permitted, then the back-off interval is given by

$$\text{back} - \text{off} = \begin{cases} Uniform[0, (2^n \times CW_{min}) - 1] & (6.7.3) \\ \quad \text{if } r = 1 \text{ and } n \leq nmax & \\ \frac{PDR}{M} \times CW_{min} + \ Uniform[0, CW_{min} - 1] & (6.7.4) \\ \quad \text{if } r > 1 \text{ and } n = 0 & \\ ULB \times CW_{min} + \ Uniform[0, (2^n \times CW_{min}) - 1] & (6.7.5) \\ \quad \text{otherwise} & \end{cases}$$

where CW_{min} is the minimum size of the contention window, and M is the desired packet delivery ratio.

This means that if the packet has the highest rank in the broadcast region of the node, then it has the lowest back-off period according to Equation 6.7.3 and faces much less contention. Else, if it is the first time the packet is being transmitted, then the back-off distribution follows the second scheme as in Equation 6.7.4, where the back-off is more than that for the first case. Here the current PDR of the flow affects the back-off period. If PDR is considerably less, then the first term would be less, and if it is high, then the first term would be high and the node would have to wait for a longer time. Finally, if the packet does not fit into these two categories, then the back-off value is as per the third scheme in Equation 6.7.5, and is the longest of the three cases. The higher the value of ULB, the longer the back-off period.

DLPS delivers a higher percentage of packets within their delay targets and has lower average end-to-end delay in packet delivery when compared to the 802.11 DCF and the DPS [18] schemes.

6.8 MAC PROTOCOLS THAT USE DIRECTIONAL ANTENNAS

MAC protocols that use directional antennas for transmissions have several advantages over those that use omnidirectional transmissions. The advantages include reduced signal interference, increase in the system throughput, and improved channel reuse that leads to an increase in the overall capacity of the channel. In this section, some of the MAC layer protocols that make use of directional antennas are discussed.

6.8.1 MAC Protocol Using Directional Antennas

The MAC protocol for mobile ad hoc networks using directional antennas that was proposed in [21] makes use of directional antennas to improve the throughput in ad hoc wireless networks. The mobile nodes do not have any location information by means of which the direction of the receiver and sender nodes could be determined. The protocol makes use of an RTS/CTS exchange mechanism, which is similar to the one used in MACA [1]. The nodes use directional antennas for transmitting and receiving data packets, thereby reducing their interference to other neighbor nodes. This leads to an increase in the throughput of the system.

Each node is assumed to have only one radio transceiver, which can transmit and receive only one packet at any given time. The transceiver is assumed to be equipped with M directional antennas, each antenna having a conical radiation pattern, spanning an angle of $\frac{2\Pi}{M}$ radians (Figure 6.28). It is assumed that the transmissions by adjacent antennas never overlap, that is, the complete attenuation of the transmitted signal occurs outside the conical pattern of the directional antenna. The MAC protocol is assumed to be able to switch every antenna individually or all the antennas together to the *active* or *passive* modes. The radio transceiver uses only the antennas that are in the active mode. If a node transmits when all its antennas are active, then the transmission's radiation pattern is similar to that of an omnidirectional antenna. The receiver node uses receiver diversity while receiving on all antennas. This means that the receiver node uses the signal from the antenna which receives the incoming signal at maximum power. In the normal case, this selected antenna would be the one whose conical pattern is directed toward the source node whose signal it is receiving. It is assumed that the radio range is the same for all directional antennas of the nodes. In order to detect the presence of a signal, a threshold signal power value is used. A node concludes that the channel is active only if the received signal strength is higher than this threshold value.

The protocol works as follows. The packet-exchange mechanism followed for transmitting each data packet is depicted in Figure 6.29. In the example shown

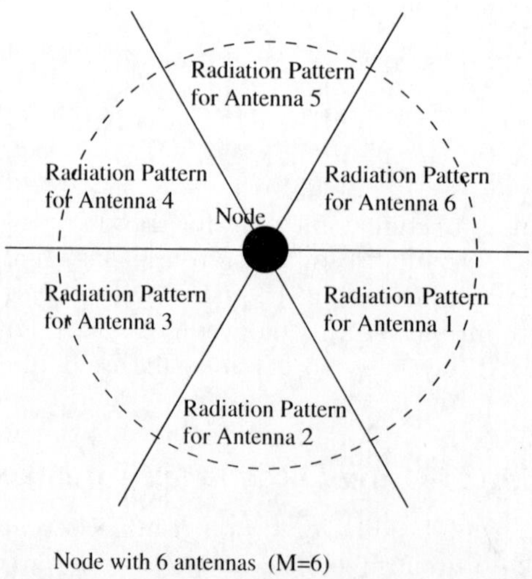

Figure 6.28. Radiation patterns of directional antennas.

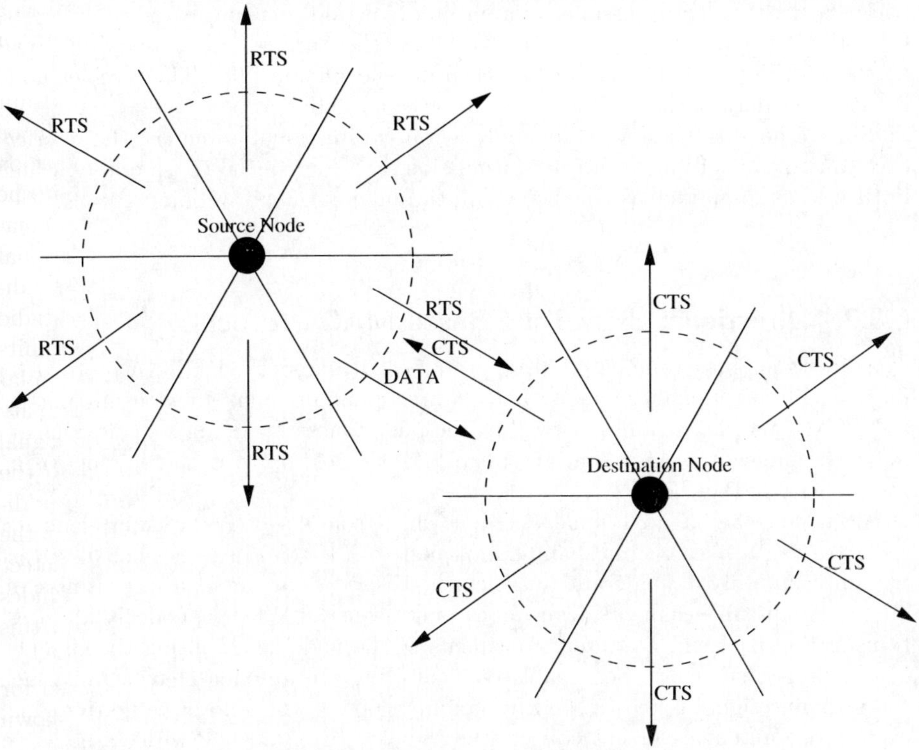

Figure 6.29. Packet transmission.

in the figure, each node is assumed to have six directional antennas. The main concept in this protocol is the mechanism used by the transmitting and receiving nodes to determine the directions of each other. The MAC layer at the source node must be able to find the direction of the intended next-hop receiver node so that the data packet could be transmitted through a directional antenna. It is the same case with the receiver. The receiver node must be able to determine the direction of the sender node before starting to receive data packets. This is performed in the following manner. An idle node is assumed to be listening to the on-going transmissions on all its antennas. The sender node first transmits an RTS packet addressed to the receiver. This RTS is transmitted through all the antennas of the node (omnidirectional transmission). The intended receiver node, on receiving this RTS packet, responds by transmitting a CTS packet, again on all its antennas (omnidirectional transmission). The receiver node also notes down the direction of the sender by identifying the antenna that received the RTS packet with maximum power. The source, on receiving the CTS packet, determines the direction of the receiver node in a similar manner. The neighbor nodes that receive the RTS or CTS packets defer their transmissions for appropriate periods of time. After receiving the CTS, the source node transmits the next data packet through the chosen directional antenna. All other antennas are switched off and remain idle. The receiver node receives this data packet only through its selected antenna.

Since a node transmits packets only through directional antennas, the interference caused to nodes in its direct transmission range is reduced considerably, which in turn leads to an increase in the overall throughput of the system.

6.8.2 Directional Busy Tone-Based MAC Protocol

The directional busy tone-based MAC protocol [22] adapts the DBTMA protocol [5] for use with directional antennas. It uses directional antennas for transmitting the RTS, CTS, and data frames, as well as the busy tones. By doing so, collisions are reduced significantly. Also, spatial reuse of the channel improves, thereby increasing the capacity of the channel.

Each node has a directional antenna which consists of N antenna elements, each covering a fixed sector spanning an angle of $(360/N)$ degrees. For a unicast transmission, only a single antenna element is used. For broadcast transmission, all the N antenna elements transmit simultaneously. When a node is idle (not transmitting packets), all antenna elements of the node keep sensing the channel. The node is assumed to be capable of identifying the antenna element on which the incoming signal is received with maximum power. Therefore, while receiving, exactly one antenna element collects the signals. In an ad hoc wireless network, nodes may be mobile most of the time. It is assumed that the orientation of sectors of each antenna element remains fixed. The protocol uses the same two busy tones BT_t and BT_r used in the DBTMA protocol. The purpose of the busy tones is the same. Before transmitting an RTS packet, the sender makes sure that the BT_r tone is not active in its neighborhood, so that its transmissions do not interfere with packets being received at a neighboring receiver node. Similarly, a receiver node, before transmitting a CTS, verifies that a BT_t is not active in its neighborhood. This is done to make sure that the data the node is expected to receive does not collide with any other on-going transmission. The modified directional DBTMA protocol operates as follows.

A node that receives a data packet for transmission first transmits an RTS destined to the intended receiver in all directions (omnidirectional transmission). On receiving this RTS, the receiver node determines the antenna element on which the RTS is received with maximum gain. The node then sends back a directional CTS to the source using the selected antenna element (which points toward the direction of the sender). It also turns on the busy tone BT_r in the direction toward the sender. On receiving the CTS packet, the sender node turns on the BT_t busy tone in the direction of the receiver node. It then starts transmitting the data packet through the antenna element on which the previous CTS packet was received with maximum gain. Once the packet transmission is over, it turns off the BT_t signal. The receiver node, after receiving the data packet, turns off the BT_r signal.

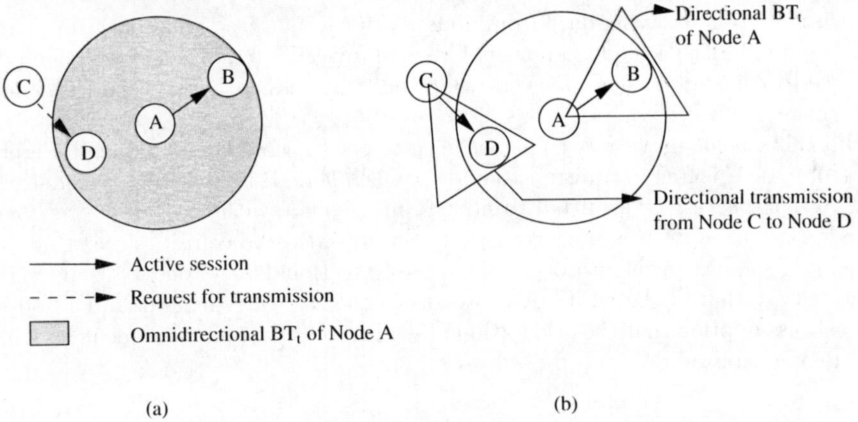

Figure 6.30. Directional DBTMA: Example 1.

The directional busy tones can permit simultaneous transmissions in the neighborhood of a transmitting or a receiving node. For example, in Figure 6.30 (a), where omnidirectional busy tones are being used, when a transmission is going on from node A to node B, node D is not permitted to receive any data as it hears the BT_t tone transmitted by node A. But when directional busy tone transmissions are used (Figure 6.30 (b)), it can be seen that node D can simultaneously receive data from node C. Another example is shown in Figure 6.31. When omnidirectional busy tones are used (Figure 6.31 (a)), node C is not permitted to transmit to node

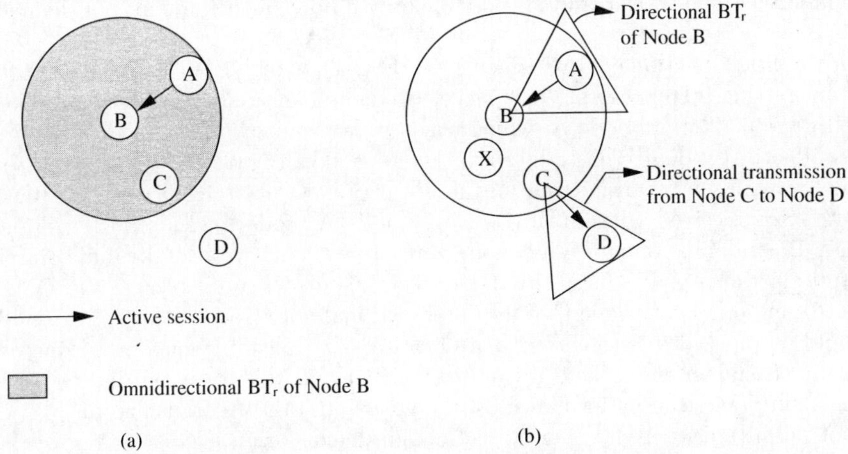

Figure 6.31. Directional DBTMA: Example 2.

D when node B is receiving data from node A. But when directional busy tones are used (Figure 6.31 (b)), node C does not receive any BT_r tone from node B and so it is free to transmit to node D even while the transmission between nodes A and B is going on.

But this scheme may cause collisions in certain scenarios. For example, in Figure 6.31 (b), node C is free to transmit to node X when node B is receiving packets from node A. The packets transmitted to node X may collide with those being received at node B. Therefore, this protocol is not guaranteed to be collision-free. But the overall performance of the protocol (which uses directional busy tone transmissions) is better than that of the DBTMA protocol. In the case of vehicle-mounted nodes, the basic assumption that the orientation of sectors of each antenna element remains fixed may not be valid.

6.8.3 Directional MAC Protocols for Ad Hoc Wireless Networks

Two MAC schemes using directional antennas are proposed in [23]. It is assumed that each node knows about the location of its neighbors as well as its own location. The physical location information can be obtained by a node using the global positioning system (GPS). In the IEEE 802.11 DCF scheme, a node that is aware of a nearby on-going transmission will not participate in a transmission itself. In the directional MAC (D-MAC) protocols proposed in [23], a similar logic is applied on a per-antenna basis. If a node has received an RTS or CTS packet related to an on-going transmission on a particular antenna, then that particular antenna is not used by the node till the other transmission is completed. This antenna stays blocked for the duration of that transmission. The key concept here is, though a particular antenna of a node may remain blocked, the remaining antennas of the node can be used for transmissions. This improves the throughput of the system. An omnidirectional transmission is possible only if none of the antennas of the node is blocked.

In the first directional MAC scheme (DMAC-1), a directional antenna is used for transmitting RTS packets. CTS packet transmissions are omnidirectional. Consider Figure 6.32. Here node A, which needs to transmit a packet to node B, first transmits a directional RTS (DRTS) packet to node B. Node B, on receiving this packet, responds by transmitting an omnidirectional CTS (OCTS) packet. Once the OCTS is received without any error by node A, node A sends a data packet using a directional antenna. When node B receives the data packet, it immediately transmits a directional ACK (DACK) packet. Node C would receive the OCTS packet from node B. At node C, only the directional antenna pointing toward node B would be blocked due to this. Node C can freely transmit to node D using another directional antenna. Thus it can be seen that in DMAC-1, usage of directional antennas improves the performance by allowing simultaneous transmissions, which are not permitted when only omnidirectional antennas are used.

In the second directional MAC scheme (DMAC-2) proposed in [23], both directional RTS (DRTS) as well as omnidirectional RTS (ORTS) transmissions are used. In DMAC-1, the usage of DRTS may increase the probability of control

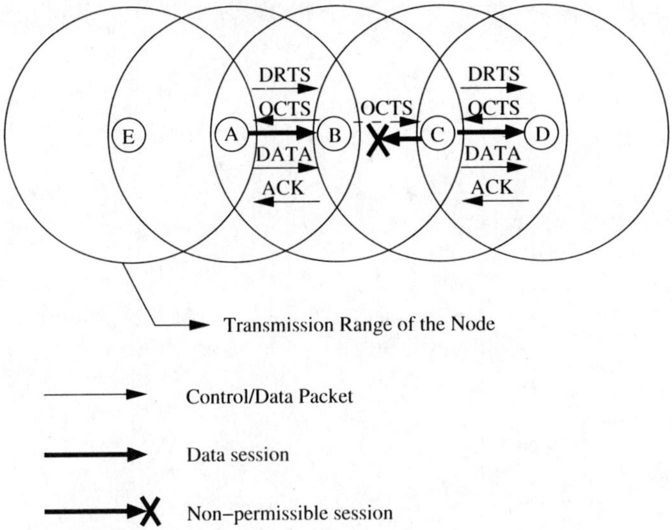

Figure 6.32. Operation of DMAC protocol.

packet collisions. For example, consider Figure 6.32. Here node A initiates a DRTS transmission to node B. The DRTS packet is not heard by node E, and so it is not aware of the transmission between nodes A and B. Suppose node E sends a packet to node A, then that packet may collide with the OCTS or DACK packets transmitted by node B. The probability of control packet collisions is reduced in DMAC-2. In DMAC-2, a node that wants to initiate a data transfer may send an ORTS or a DRTS as per the following two rules. (1) If none of the directional antennas at the node are blocked, then the node sends an ORTS packet. (2) Otherwise, the node sends a DRTS packet, provided the desired directional antenna is not blocked. Consider the same example in Figure 6.32. Here when node A initiates data transfer to node B, assuming all its antennas are not blocked, it sends an ORTS packet to node B. Node E would now receive this packet, and the antenna on which the ORTS packet was received would remain blocked for the duration of the transmission from node A to node B. If node E wants to send a packet to node A, it needs to wait for the duration of the transmission between nodes A and B, so that its directional antenna pointing toward node A becomes unblocked; only then can it start transmitting packets to node A. Thus, the combination of ORTS and DRTS packets in DMAC-2 reduces collisions between control packets.

Consider Figure 6.33. Node B sends an OCTS packet to node A after receiving DRTS from node A. Node C would be aware of the transmission and its antenna pointing toward node B would remain blocked for the duration of the transmission. Now suppose node D sends an ORTS to node C. Since one of node C's antennas is blocked currently, it would not respond to the ORTS packet. This results in

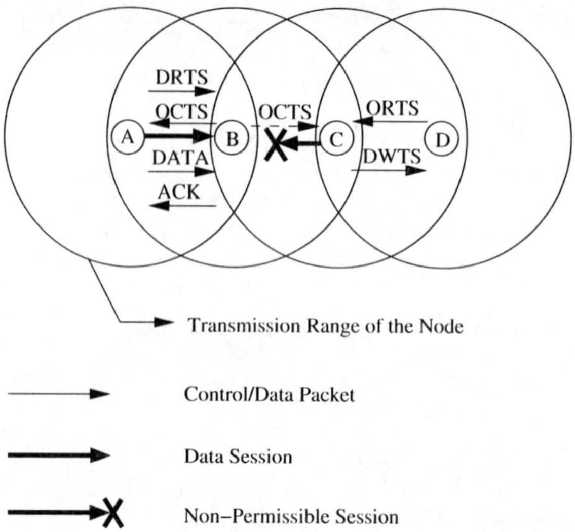

Figure 6.33. Operation of DMAC protocol.

node D timing out and unnecessary retransmissions of ORTS packets to node C. To avoid this situation, another packet called directional wait-to-send (DWTS) is used. On receiving the ORTS from node D, node C transmits the DWTS packet using a directional antenna toward node D. This DWTS packet carries the expected duration of the on-going transmission between nodes A and B. Node D, on receiving this packet, waits for the specified interval of time and then tries again.

By enabling simultaneous collision-free transmissions, the directional MAC schemes proposed in [23] improve channel reuse, thereby increasing the capacity of the channel. This leads to a significant increase in the throughput of the system.

6.9 OTHER MAC PROTOCOLS

There are several other MAC protocols that do not strictly fall under the categories discussed above. Some of these MAC protocols are described in this section.

6.9.1 Multichannel MAC Protocol

The multichannel MAC protocol (MMAC) [24] uses multiple channels for data transmission. There is no dedicated control channel. N channels that have enough spectral separation between each other are available for data transmission.

Each node maintains a data structure called *PreferableChannelList* (PCL). The usage of the channels within the transmission range of the node is maintained in the PCL. Based on their usage, channels can be classified into three types.

- High preference channel (HIGH): The channel has been selected by the current node and is being used by the node in the current beacon interval (beacon interval mechanism will be explained later). Since a node has only one transceiver, there can be only one HIGH channel at a time.

- Medium preference channel (MID): A channel which is free and is not being currently used in the transmission range of the node is said to be a medium preference channel. If there is no HIGH channel available, a MID channel would get the next preference.

- Low preference channel (LOW): Such a channel is already being used in the transmission range of the node by other neighboring nodes. A counter is associated with each LOW state channel. For each LOW state channel, the count of source-destination pairs which have chosen the channel for data transmission in the current beacon interval is maintained.

Time is divided into beacon intervals and every node is synchronized by periodic beacon transmissions. So, for every node, the beacon interval starts and ends almost at the same time. At the start of every beacon interval, there exists a time interval called the ad hoc traffic indication messages (ATIM) window. This window is used by the nodes to negotiate for channels for transmission during the current beacon interval. ATIM messages such as ATIM, ATIM-ACK (ATIM-acknowledgment), and ATIM-RES (ATIM-reservation) are used for this negotiation. The exchange of ATIM messages takes place on a particular channel called the *default channel*. The default channel is one of the multiple available channels. This channel is used for sending DATA packets outside the ATIM window, like any other channel. A node that wants to transmit in the current beacon interval sends an ATIM packet to the intended destination node. The ATIM message carries the PCL of the transmitting node. The destination node, upon receiving the packet, uses the PCL carried on the packet and its own PCL to select a channel. It includes this channel information in the ATIM-ACK packet it sends to the source node. The source node, on receiving the ATIM-ACK packet, determines whether it can transmit on the channel indicated in the ATIM-ACK message. If so, it responds by sending the destination node an ATIM-RES packet. The ATIM-ACK and ATIM-RES packets are also used to notify the neighbor nodes of the receiver and sender nodes, respectively, about the channel that is going to be used for transmission in the current beacon interval. The nodes that hear these packets update their PCLs accordingly. At the end of the ATIM window, the source and destination nodes switch to the agreed-upon channel and start communicating by exchanging RTS/CTS control packets. If the source node is not able to use the channel selected by the destination, it cannot transmit packets to that destination in the current beacon interval. It has to wait for the next beacon interval for again negotiating channels. The ATIM packets themselves may be lost due to collisions; in order to prevent this, each node waits for a randomly chosen back-off period (between 0 and CW_{min}) before transmitting the ATIM packet.

Operation of the MMAC protocol is illustrated in Figure 6.34. At the beginning of the beacon interval, source node S1 sends an ATIM message to receiver R1.

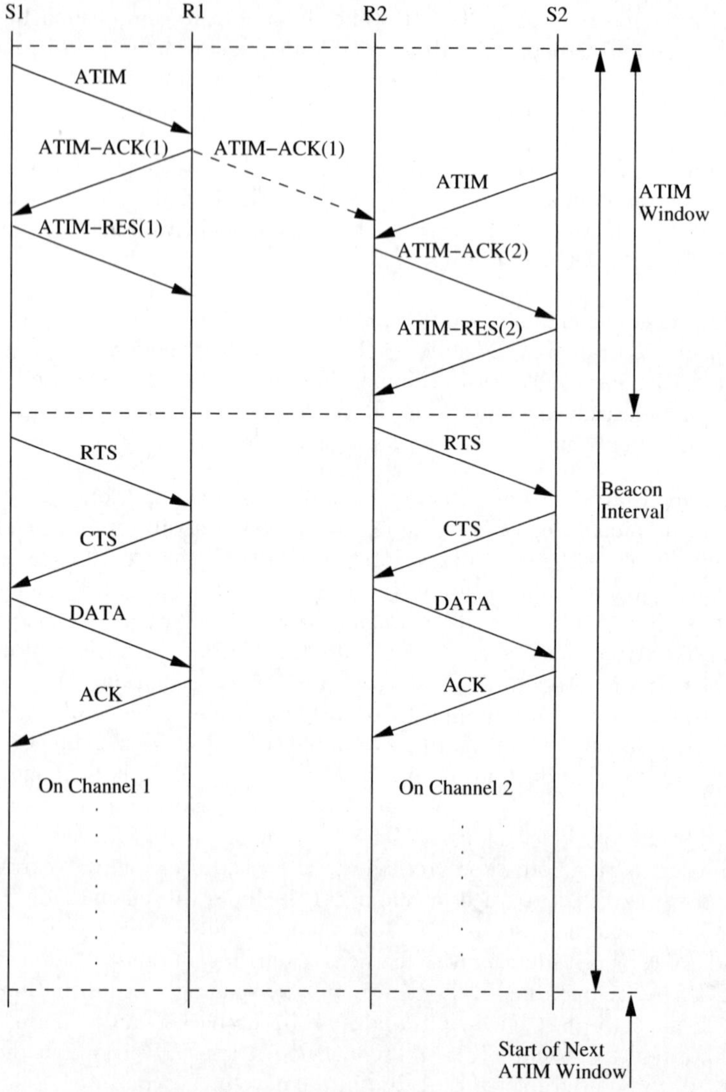

Figure 6.34. Operation of MMAC protocol.

Receiver R1 responds by sending an ATIM-ACK packet (ATIM-ACK(1)) carrying the ID 1 of the channel it prefers (in Figure 6.34 the number within parentheses indicates the ID of the preferred channel). Node S1, on receiving this packet, confirms the reservation by sending an ATIM-RES packet (ATIM-RES(1)) for channel 1. The ATIM-ACK(1) packet sent by receiver R1 is also overheard by node R2.

When node R2 receives an ATIM packet from source S2, it chooses a different channel with ID 2, and sends the channel information to source S2 through the ATIM-ACK packet (ATIM-ACK(2)). Since channel 2 is agreeable to node S2, it responds by sending the ATIM-RES(2) packet, and the reservation gets established. Once the ATIM window finishes, the data transmission (through RTS-CTS-DATA-ACK packet exchange) between node pairs S1-R1 and S2-R2 takes place on the corresponding reserved channels, channel 1 and channel 2, respectively.

In this protocol, it is the receiver node that plays a dominant role in channel selection. In case all channels are in use at the receiver, even then the receiver selects one of the channels. Since the actual data packet transmissions are protected by the RTS/CTS control packet exchange, the nodes transmitting packets on the same channel need to contend for the channel, as in IEEE 802.11 for transmitting packets. The protocol also employs a power-saving mode. In case a node realizes after the ATIM window that it is neither going to transmit packets nor going to receive packets, then the node goes into a power-saving *doze* mode.

Channel selection is done at the receiver in the following manner. The receiver node uses the PCL on the received ATIM packet and its own PCL for selecting the best possible channel for communication with the source node. The channel selection procedure tries to balance the network load on the channels. If a receiver node R receives an ATIM packet from a source node S, it selects a channel as below.

- If there exists a HIGH state channel in node R's PCL, then that channel is selected.

- Else if there exists a HIGH state channel in the PCL of node S, then this channel is selected.

- Else if there exists a common MID state channel in the PCLs of both node S and node R, then that channel is selected. If many such channels exists, one of them is selected randomly.

- Else if there exists a channel which is in the MID state at only one of the two nodes, then that channel is chosen. If many such channels exist, one of them is selected randomly.

- If all channels in both PCLs are in the LOW state, the counters of the corresponding channels at nodes S and R are added, and the channel with the least count is selected. Ties are broken arbitrarily.

MMAC uses simple hardware. It requires only a single transceiver. It does not have any dedicated control channel. The throughput of MMAC is higher than that of IEEE 802.11 when the network load is high. This higher throughput is in spite of the fact that in MMAC only a single transceiver is used at each node. Unlike other protocols, the packet size in MMAC need not be increased in order to take advantage of the presence of an increased number of channels.

6.9.2 Multichannel CSMA MAC Protocol

In the multichannel CSMA MAC protocol (MCSMA) [25], the available bandwidth is divided into several channels. A node with a packet to be transmitted selects an idle channel randomly. The protocol also employs the notion of *soft* channel reservation, where preference is given to the channel that was used for the previous successful transmission. Though the principle used in MCSMA is similar to the frequency division multiple access (FDMA) schemes used in cellular networks, the major difference here is that there is no centralized infrastructure available, and channel assignment is done in a distributed fashion using carrier-sensing. The operation of the protocol is discussed below.

The total available bandwidth is divided into N non-overlapping channels (N is independent of the number of hosts in the network), each having a bandwidth of (W/N), where W is the total bandwidth available for communication. The channels may be created in the frequency domain (FDMA) or in the code domain (CDMA). Since global synchronization between nodes is not available in ad hoc wireless networks, channel division in the time domain (TDMA) is not used.

An idle node (which is not transmitting packets) continuously monitors all the N channels. A channel whose total received signal strength[1] (TRSS) is below the sensing threshold (ST) of the node is marked IDLE by the node. The time at which TRSS drops below ST is also noted for each IDLE channel. Such channels are put in the *free-channels* list.

When an idle node receives a packet to be transmitted, it does the following. If the free-channels list is empty, it waits for any channel to become IDLE. It then waits for an additional long inter-frame space (LIFS) time, and for another random access back-off period. If the channel remains idle for this entire wait period, then the node starts transmitting its packets on this channel. In case the free-channels list is non-empty, the node first checks whether the channel it used for its most recent successful transmission is included in the list. If so, the node uses this channel for its new transmission. Otherwise, one among the IDLE channels available in the free-channels list is randomly chosen (using a uniform random number generator).

Before the actual packet transmission, the node checks the TRSS of the chosen channel. If it had remained below ST for at least LIFS period of time, then the node immediately initiates packet transmission. Otherwise, the node initiates back-off delay after the LIFS time period. During the back-off period, if the TRSS of the chosen channel goes above ST, then the back-off is immediately canceled. A new back-off delay is scheduled when the TRSS again goes below the ST. After successfully transmitting a packet (indicated by an acknowledgment from the receiver), the sender node notes the ID of the channel used. This channel would be given preference when a new channel is to be selected for its next transmission.

When the number of channels N is sufficiently large, each node tends to *reserve* a channel for itself. This is because a node prefers the channel used in its last successful transmission for its next transmission also. This reduces the probability

[1]The total received signal strength of a signal is calculated by the sum of contributions arising from the various individual multipath components of the signal.

of two contending nodes choosing the same channel for transmission. Nodes are expected to dynamically select channels for transmissions in a mutually exclusive manner, so as to enable parallel interference-free transmissions. Even at high traffic loads, due to the tendency of every node to choose a *reserved* channel for itself, the chances of collisions are greatly reduced.

The number of channels into which the available bandwidth is split is a very important factor affecting the performance of the protocol. If the number of channels is very large, then the protocol results in very high packet transmission times.

6.9.3 Power Control MAC Protocol for Ad Hoc Networks

The power control MAC protocol (PCM) [26] allows nodes to vary their transmission power levels on a per-packet basis. The PCM protocol is based on the power control protocol used in [27], which is referred to as the *BASIC* protocol in this section. In what follows, the working of the BASIC power control protocol is briefly described. This is followed by a discussion of the PCM protocol.

In the BASIC scheme, the RTS and CTS packets are transmitted with maximum power p_{max}. The RTS-CTS handshake is used for deciding upon the transmission power for the subsequent DATA and ACK packet transmissions. This can be done using two methods. In the first method, source node A transmits the RTS with maximum power p_{max}. This RTS is received at the receiver with signal level p_r. The receiver node B can calculate the minimum required transmission power level $p_{desired}$ for the DATA packet, based on the received power level p_r, the transmitted power level p_{max}, and the noise level at receiver B. Node B then specifies this $p_{desired}$ in the CTS packet it transmits to node A. Node A transmits the DATA packet using power level $p_{desired}$. In the second method, when the receiver node B receives an RTS packet, it responds with a CTS packet at the usual maximum power level p_{max}. When the source node receives this CTS packet, it calculates $p_{desired}$ based on the received power level p_r and transmitted power level p_{max} as

$$p_{desired} = \frac{p_{max}}{p_r} \times Rx_{thresh} \times c$$

where Rx_{thresh} is the minimum necessary received signal strength and c is a constant. The source node uses power level $p_{desired}$ to transmit the DATA packet. Similarly, the receiver uses the signal power of the received RTS packet to determine the power level to be used, $p_{desired}$, for the ACK packet. This method assumes the attenuation between the sender and receiver nodes to be the same in both directions. It also assumes the noise level at the nodes to be below a certain predefined threshold value.

Thus, the BASIC scheme uses maximum transmit power for RTS and CTS packets, and only necessary power levels for the DATA and ACK packets. But this scheme has a drawback. Consider Figure 6.35. Node A sends an RTS to node B, for which node B sends a CTS packet. Since these packets are sent at maximum power, nodes X and Y that are located in the carrier-sensing zones of nodes A and B, respectively (when a node N1 is in the carrier-sensing zone of node N2, node N1 can sense the signal from node N2, but the received signal strength is not high enough

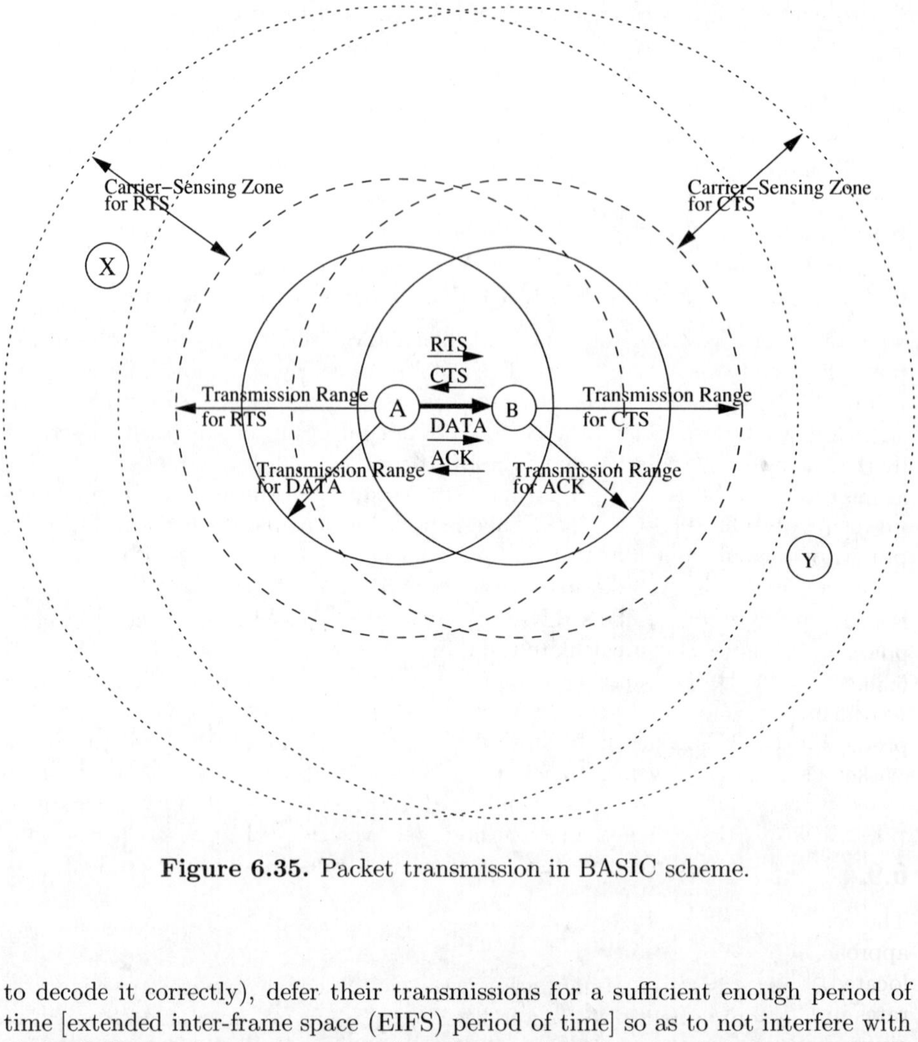

Figure 6.35. Packet transmission in BASIC scheme.

to decode it correctly), defer their transmissions for a sufficient enough period of time [extended inter-frame space (EIFS) period of time] so as to not interfere with the RTS-CTS exchange. But since the DATA and ACK transmissions use only the minimum necessary power, the DATA transmitted by node A cannot be sensed by node X, and the ACK packet transmitted by node B cannot be sensed by node Y. So if nodes X and Y transmit after the EIFS period (which is set in their NAVs on sensing the RTS or CTS packets), the packet transmitted by node X would collide at node A with the ACK packet from node B, and the packet transmitted by node Y would collide with the DATA packet at node B.

PCM modifies this scheme so as to minimize the probability of collisions. The source and receiver nodes transmit the RTS and CTS packets, as usual, with maximum power p_{max}. Nodes in the carrier-sensing zones of the source and receiver nodes set their NAVs for EIFS duration when they sense the signal but are not able to decode it. The source node generally transmits with minimum necessary power,

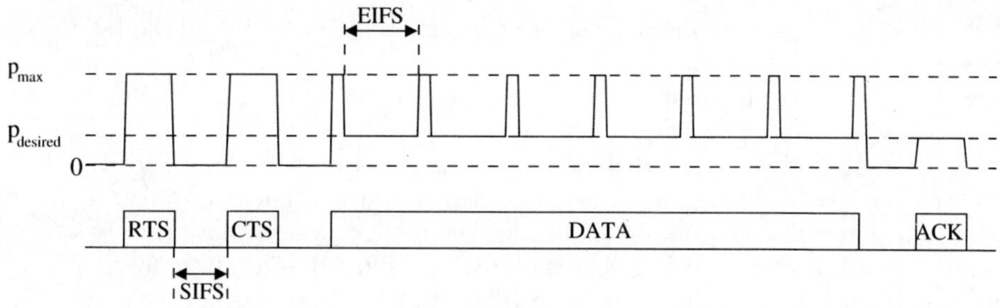

Figure 6.36. Transmission power pattern in PCM.

as in the BASIC scheme. But, in order to avoid collisions with packets transmitted by the nodes in its carrier-sensing zone, the source node transmits the DATA packet at maximum power level p_{max} periodically. The duration of each such transmission must be larger than the time required for physical carrier-sensing. Since the nodes in the carrier-sensing zone defer their transmissions for EIFS duration if they are not able to decode the received signal, the transmit power for the DATA packet is increased (and brought down back to original level) every EIFS duration. The power level changes for RTS-CTS-DATA-ACK transmissions is depicted in Figure 6.36. Thus this protocol prevents collisions of ACK packets at the sender node.

Hence with the above simple modification, the PCM protocol overcomes the problems faced in the BASIC scheme. PCM achieves throughput very close to that of the 802.11 protocol while using much less energy.

6.9.4 Receiver-Based Autorate Protocol

The receiver-based autorate protocol (RBAR) [28] uses a novel rate adaptation approach. The rate adaptation mechanism is at the receiver node instead of being located at the sender. Rate adaptation is the process of dynamically switching data rates in order to match the channel conditions so that optimum throughput for the given channel conditions is achieved. Rate adaptation consists of two processes, namely, channel quality estimation and rate selection. The accuracy of the channel quality estimates significantly influences the effectiveness of the rate adaptation process. Therefore, it is important that the best available channel quality estimates are used for rate selection. Since it is the channel quality at the receiver node which determines whether a packet can be received or not, it can be concluded that the best channel quality estimates are available at the receiver. The estimates must be used as early as possible before they get stale. If the sender is to implement the rate adaptation process, significant delay would be involved in communicating the channel quality estimates from the receiver to the sender, which may result in the estimates becoming stale before being used. Therefore, the RBAR protocol advocates for rate adaptation at the receiver node rather than at the sender.

Rate selection is done at the receiver on a per-packet basis during the RTS-CTS packet exchange. Since rate selection is done *during* the RTS-CTS exchange, the channel quality estimates are very close to the actual transmission times of the data packets. This improves the effectiveness of the rate selection process. The RTS and CTS packets carry the chosen modulation rate and the size of the data packet, instead of carrying the duration of the reservation. The packet transmission process is depicted in Figure 6.37. The sender node chooses a data rate based on some heuristic and inserts the chosen data rate and the size of the data packet into the RTS. When a neighbor node receives this RTS, it calculates the duration of the reservation D_{RTS} using the data rate and packet size carried on the RTS. The neighbor node then updates its NAV accordingly to reflect the reservation. While receiving the data packet, the receiver node generates an estimate of the channel conditions for the impending data transfer. Based on this estimate, it chooses an appropriate data rate. It stores the chosen data rate and the size of the packet on the CTS packet and transmits the CTS to the sender. Neighbor nodes receiving the CTS calculate the expected duration of the transmission and update their NAVs accordingly. The source node, on receiving the CTS packet, responds by transmitting the data packet at the rate chosen by the receiver node.

If the rates chosen by the sender and receiver are different, then the reservation duration D_{RTS} calculated by the neighbor nodes of the sender would not be valid. D_{RTS} time period, which is calculated based on the information carried initially by the RTS packet, is referred to as *tentative reservation.* In order to overcome this problem, the sender node sends the data packet with a special MAC header containing a *reservation subheader* (RSH). The RSH contains a subset of header fields already present in the IEEE 802.11 data frame, along with a check sequence for protecting the subheader. The fields in the RSH contain control information for determining the duration of the transmission. A neighbor node with tentative reservation entries in its NAV, on hearing the data packet, calculates D_{RSH}, the new reservation period, and updates its NAV to account for the difference between D_{RTS} and D_{RSH}.

For the channel quality estimation and prediction algorithm, the receiver node uses a sample of the instantaneous received signal strength at the end of RTS reception. For the rate selection algorithm, a simple threshold-based technique is

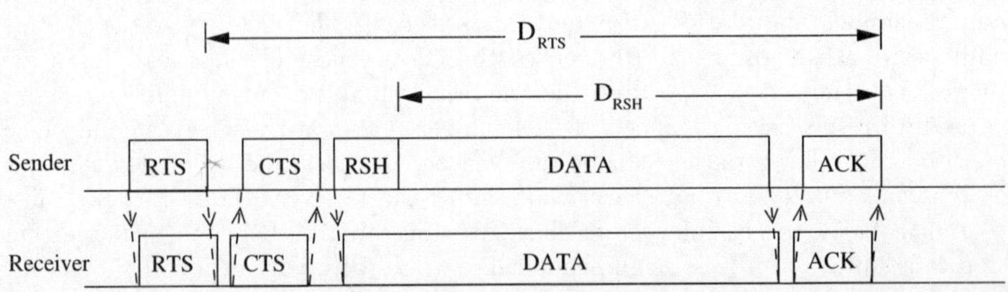

Figure 6.37. Packet transmission in RBAR.

used. Here the rate is chosen by comparing the channel quality estimate [*e.g.*, signal to noise ratio (SNR)] against a series of thresholds representing the desired performance bounds (*e.g.*, a series of SNR thresholds). The modulation scheme with the highest data rate, satisfying the performance objective for the channel quality estimate, is chosen.

RBAR employs an efficient quality estimation mechanism, which leads to a high overall system throughput. RBAR can be easily incorporated into many existing medium access control protocols.

6.9.5 Interleaved Carrier-Sense Multiple Access Protocol

The interleaved carrier-sense multiple access protocol (ICSMA) [29] efficiently overcomes the exposed terminal problem faced in ad hoc wireless networks. The inability of a source node to transmit, even though its transmission may not affect other ongoing transmissions, is referred to as the exposed terminal problem. For example, consider the topology shown in Figure 6.38. Here, when a transmission is going from node A to node B, nodes C and F would not be permitted to transmit to nodes D and E, respectively. Node C is called a sender-exposed node, and node E is called a receiver-exposed node. The exposed terminal problem reduces the bandwidth efficiency of the system.

In ICSMA, the total available bandwidth is split into two equal channels (say, channel 1 and channel 2). The handshaking process is interleaved between the two channels, hence the name interleaved carrier-sense multiple access. The working of ICSMA is very simple. It uses the basic RTS-CTS-DATA-ACK exchange mechanism used in IEEE 802.11 DCF. If the source node transmits the RTS packet on channel 1, the receiver node, if it is ready to accept packets from the sender, responds by transmitting the CTS packet on channel 2. Each node maintains a data

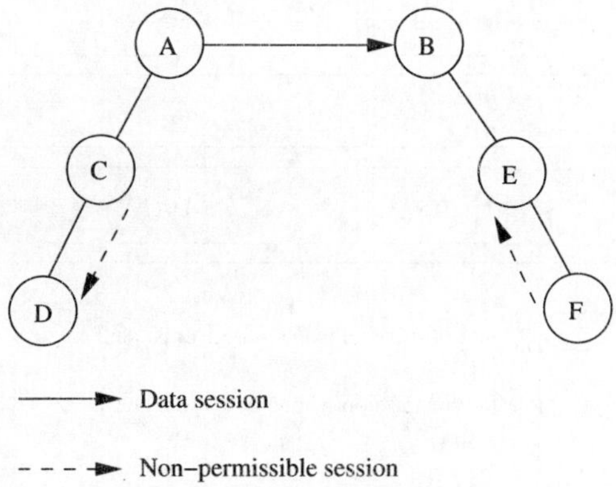

Figure 6.38. Exposed terminal problem. *Reproduced with permission from [29], © IEEE, 2003.*

structure called extended network allocation vector (E-NAV), which is analogous to the network allocation vector (NAV) used in IEEE 802.11 DCF. On receiving an RTS packet, the receiver node checks its E-NAV and finds out whether free time slots are available. It sends the CTS only if free slots are available. The source node, on receiving this CTS, transmits the DATA packet on channel 1. The receiver acknowledges the reception of the DATA packet by transmitting the ACK on channel 2. The ICSMA channel access mechanism is illustrated in Figure 6.39. Figure 6.39 (a) shows the RTS-CTS-DATA-ACK exchange of 802.11 DCF. Figure 6.39 (b) shows simultaneous data packet transmissions between two nodes in ICSMA. After transmitting an RTS or a DATA packet on a channel, the sender node waits on the other channel for the CTS or ACK packet. If it does not receive any packet on the other channel, it assumes that the RTS or DATA packet it transmitted was lost, and retries again. Similarly at the receiver, after transmitting a CTS frame on one of the channels, the receiver node waits on the other channel for the DATA packet.

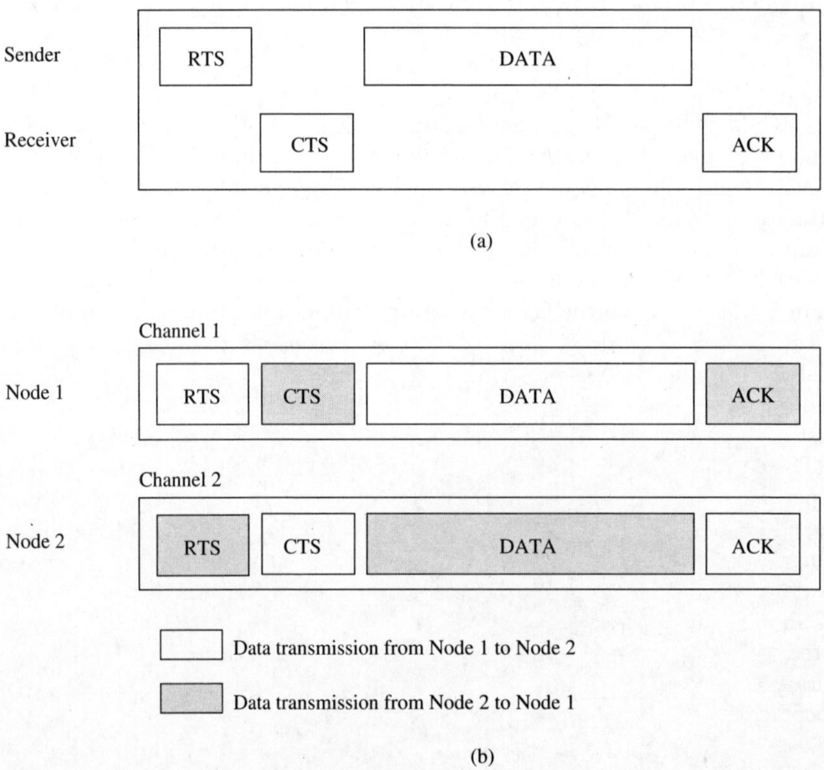

Figure 6.39. Packet transmissions in (a) 802.11 DCF and (b) ICSMA. *Reproduced with permission from [29], © IEEE, 2003.*

If the DATA packet is not received within the timeout period, it retransmits the CTS packet.

The performance improvement of ICSMA is attributed to the following facts:

- Nodes that hear RTS in a particular channel (say, channel 1) and do not hear the corresponding CTS on the other channel (channel 2) conclude that they are only sender-exposed in channel 1. Therefore, if they have packets to send, they can use channel 1 to transmit RTS to other nodes. This would not have been possible in 802.11 DCF, where transmissions by a sender-exposed node would have collided with the corresponding currently active sender node.

- Nodes that hear only the CTS in a particular channel (say, channel 1) and had not heard the corresponding RTS on the other complementary channel (channel 2) realize that they are only receiver-exposed on channel 1 to the on-going transmission. If they receive any RTS on channel 2, they would not refrain from sending a CTS on channel 1 for the received RTS. This would also not have been possible in 802.11 DCF, where there would have been collision at the receiver of the on-going session between the CTS packet transmitted by this node and the DATA packets belonging to the on-going session. Also, if this CTS transmission is successful, then there might have been collisions between DATA packets belonging to the two sessions at both the receiver nodes.

The E-NAV used in ICSMA is implemented as two linked lists of blocks, namely, the *SEL*ist and the *REL*ist. Each block in each linked list has a start time and an end time. A typical list looks like $s_1, f_1; s_2, f_2; ...; s_k, f_k$ where s_i denotes the start time of the ith block in the list and f_i denotes the finish time of the ith block in the list. The *SEL*ist is used to determine if the node would be sender-exposed at any given instant of time in the future. A node is predicted to be sender-exposed at any time t if there is a block s_j, f_j in the *SEL*ist such that $s_j < t < f_j$. Similarly, the *REL*ist tells if the node would be receiver-exposed at any time in the future. A node is predicted to be receiver-exposed at any time t if there exists a block s_j, f_j in the *REL*ist such that $s_j < t < f_j$. The *SEL*ist and the *REL*ist are updated whenever the RTS and CTS packets are received by the node. The *SEL*ist is modified when an RTS packet is received, and the *REL*ist is modified when a CTS packet is received by the node. The modification in the list might be adding a new block, modifying an existing block, or merging two or more existing blocks and modifying the resulting block.

ICSMA is a simple two-channel MAC protocol for ad hoc wireless networks that reduces the number of exposed terminals and tries to maximize the number of simultaneous sessions. ICSMA was found to perform better than the 802.11 DCF protocol is terms of metrics such as throughput and channel access delay.

6.10 SUMMARY

In this chapter, the major issues involved in the design of a MAC protocol for ad hoc wireless networks were identified. The different classifications of the existing MAC protocols were provided. The chapter also gave detailed descriptions of the major MAC protocols that exist for ad hoc wireless networks.

The main issues to be considered while designing the MAC protocol are bandwidth efficiency, QoS support, time synchronization, error-prone nature of the shared broadcast transmission channel, and node mobility. The MAC protocols are classified into several types based on criteria such as availability of bandwidth reservation mechanisms, data transfer initiation approach, usage of single or multiple transmission channels, and requirement of time synchronization. Protocols falling under the above categories were explained in detail.

6.11 PROBLEMS

1. Consider Figure 6.40. Transmission is going on between nodes A and B that are separated by a distance d. R is the transmission range of each node. Mark in the figure the regions that could contain exposed terminals and hidden terminals. Also calculate the area of each region.

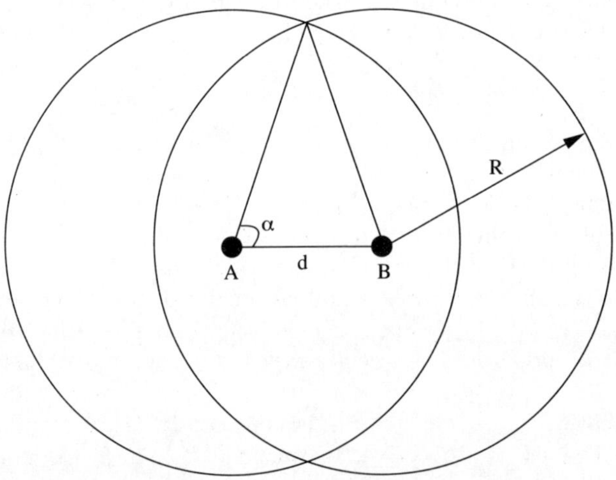

Figure 6.40. Hidden and exposed regions.

2. What are the advantages of reservation-based MAC protocols over contention-based MAC protocols?

3. Give application scenarios where contention-based, reservation-based, and packet scheduling-based MAC protocols can be used.

4. Compare the pros and cons of using scheduling-based MAC protocols over reservation-based MAC protocols.

5. What are the disadvantages of the binary exponential back-off mechanism used in MACA? How are they overcome in MACAW?

6. How does the packet queuing mechanism of MACA differ from that of MACAW? Which one of them is better? Why?

7. Calculate the probability of data packet collision in the MACA protocol. Assume that T_c is the control packet transmission and propagation delay, T_w is the optimal maximum back-off time, β is the percentage of ready nodes, and R is the transmission range of each node.

8. What are the disadvantages of the BTMA protocol? How are they overcome in the DBTMA protocol?

9. What is the main difference between the receiver-initiated MAC protocols MACA-BI and MARCH?

10. How is synchronization between nodes achieved in the HRMA protocol?

11. Which protocol is more bandwidth efficient, RTMAC or MACA/PR? Explain.

12. Explain the back-off calculation mechanism used in DWOP. Is it guaranteed to be accurate at all times? If not, explain why.

13. What is meant by the carrier-sensing zone of a transmission? Does it have any effect on the performance of a MAC protocol?

14. Channel quality estimation can be done both at the sender and the receiver. Which is more advantageous? Why?

15. What are the pros and cons of using multichannel MAC protocols over single-channel MAC protocols?

16. In FPRP, can a situation occur where a requesting node is not able to detect collisions that have occurred in the reservation request phase? If so, suggest simple modifications to solve the problem.

17. What are the advantages and disadvantages of MAC protocols using directional antennas?

BIBLIOGRAPHY

[1] P. Karn, "MACA – A New Channel Access Method for Packet Radio," *Proceedings of ARRL/CRRL Amateur Radio Computer Networking Conference 1990*, pp. 134-140, September 1990.

[2] V. Bharghavan, A. Demers, S. Shenker, and L. Zhang, "MACAW: A Media Access Protocol for Wireless LANs," *Proceedings of ACM SIGCOMM 1994*, pp. 212-225, August 1994.

[3] C. L. Fullmer and J. J. Garcia-Luna-Aceves, "Floor Acquisition Multiple Access (FAMA) for Packet-Radio Networks," *Proceedings of ACM SIGCOMM 1995*, pp. 262-273, September 1995.

[4] F. A. Tobagi and L. Kleinrock, "Packet Switching in Radio Channels: Part II–The Hidden Terminal Problem in Carrier Sense Multiple Access and the Busy Tone Solution," *IEEE Transactions on Communications*, vol. 23, no. 12, pp. 1417-1433, December 1975.

[5] J. Deng and Z. J. Haas, "Dual Busy Tone Multiple Access (DBTMA): A New Medium Access Control for Packet Radio Networks," *Proceedings of IEEE ICUPC 1998*, vol. 1, pp. 973-977, October 1998.

[6] C. S. Wu and V. O. K. Li, "Receiver-Initiated Busy Tone Multiple Access in Packet Radio Networks," *ACM Computer Communication Review*, vol. 17, no. 5, pp. 336- 42, August 1987.

[7] F. Talucci and M. Gerla, "MACA-BI (MACA by Invitation): A Wireless MAC Protocol for High Speed Ad Hoc Networking," *Proceedings of IEEE ICUPC 1997*, vol. 2, pp. 913-917, October 1997.

[8] C. K. Toh, V. Vassiliou, G. Guichal, and C. H. Shih, "MARCH: A Medium Access Control Protocol for Multi-Hop Wireless Ad Hoc Networks," *Proceedings of IEEE MILCOM 2000*, vol. 1, pp. 512-516, October 2000.

[9] S. Jiang, J. Rao, D. He, and C. C. Ko, "A Simple Distributed PRMA for MANETs," *IEEE Transactions on Vehicular Technology*, vol. 51, no. 2, pp. 293-305, March 2002.

[10] D. J. Goodman, R. A. Valenzuela, K. T. Gayliard, and B. Ramamurthi, "Packet Reservation Multiple Access for Local Wireless Communications," *IEEE Transactions on Communications*, vol. 37, no. 8, pp. 885-890, August 1989.

[11] Z. Tang and J. J. Garcia-Luna-Aceves, "A Protocol for Topology-Dependent Transmission Scheduling in Wireless Networks," *Proceedings of IEEE WCNC 1999*, vol. 3, no. 1, pp. 1333-1337, September 1999.

[12] Z. Tang and J. J. Garcia-Luna-Aceves, "Hop-Reservation Multiple Access (HRMA) for Ad Hoc Networks," *Proceedings of IEEE INFOCOM 1999*, vol. 1, pp. 194-201, March 1999.

[13] C. W. Ahn, C. G. Kang, and Y. Z. Cho, "Soft Reservation Multiple Access with Priority Assignment (SRMA/PA): A Novel MAC Protocol for QoS-Guaranteed Integrated Services in Mobile Ad Hoc Networks," *Proceedings of IEEE Fall VTC 2000*, vol. 2, pp. 942-947, September 2000.

[14] C. Zhu and M. S. Corson, "A Five-Phase Reservation Protocol (FPRP) for Mobile Ad Hoc Networks," *ACM/Baltzer Journal of Wireless Networks*, vol. 7, no. 4, pp. 371-384, July 2001.

[15] C. R. Lin and M. Gerla, "Real-Time Support in Multi-Hop Wireless Networks," *ACM/Baltzer Journal of Wireless Networks*, vol. 5, no. 2, pp. 125-135, March 1999.

[16] C. Perkins and P. Bhagwat, "Highly Dynamic Destination-Sequenced Distance-Vector Routing (DSDV) for Mobile Computers," *Proceedings of ACM SIGCOMM 1994*, pp. 234-244, August 1994.

[17] B. S. Manoj and C. Siva Ram Murthy, "Real-Time Traffic Support for Ad Hoc Wireless Networks," *Proceedings of IEEE ICON 2002*, pp. 335-340, August 2002.

[18] V. Kanodia, C. Li, A. Sabharwal, B. Sadeghi, and E. Knightly, "Distributed Priority Scheduling and Medium Access in Ad Hoc Networks," *ACM/Baltzer Journal of Wireless Networks*, vol. 8, no. 5, pp. 455-466, September 2002.

[19] V. Kanodia, C. Li, A. Sabharwal, B. Sadeghi, and E. Knightly, "Ordered Packet Scheduling in Wireless Ad Hoc Networks: Mechanisms and Performance Analysis," *Proceedings of ACM MOBIHOC 2002*, pp. 58-70, June 2002.

[20] I. Karthigeyan, B. S. Manoj, and C. Siva Ram Murthy, "A Distributed Laxity-Based Priority Scheduling Scheme for Time-Sensitive Traffic in Mobile Ad Hoc Networks," to appear in *Ad Hoc Networks Journal*.

[21] A. Nasipuri, S. Ye, J. You, and R. E. Hiromoto, "A MAC Protocol for Mobile Ad Hoc Networks Using Directional Antennas," *Proceedings of IEEE WCNC 2000*, vol. 1, pp. 1214-1219, September 2000.

[22] Z. Huang, C. C. Shen, C. Srisathapornphat, and C. Jaikaeo, "A Busy Tone-Based Directional MAC Protocol for Ad Hoc Networks," *Proceedings of IEEE MILCOM 2002*, October 2002.

[23] Y. B. Ko, V. Shankarkumar, and N. H. Vaidya, "Medium Access Control Protocols Using Directional Antennas in Ad Hoc Networks," *Proceedings of IEEE INFOCOM 2000*, vol. 1, pp. 13-21, March 2000.

[24] J. So and N. H. Vaidya, "A Multi-Channel MAC Protocol for Ad Hoc Wireless Networks," *Technical Report: http://www.crhc.uiuc.edu/~nhv/papers/jungmin-tech.ps*, January 2003.

[25] A. Nasipuri, J. Zhuang, and S. R. Das, "A Multi-Channel CSMA MAC Protocol for Multi-Hop Wireless Networks," *Proceedings of IEEE WCNC 1999*, vol. 1, pp. 1402-1406, September 1999.

[26] E. S. Jung and N. H. Vaidya, "A Power Control MAC Protocol for Ad Hoc Networks," *Proceedings of ACM MOBICOM 2002*, pp. 36-47, September 2002.

[27] J. Gomez, A. T. Campbell, M. Naghshineh, and C. Bisdikian, "Conserving Transmission Power in Wireless Ad Hoc Networks," *Proceedings of ICNP 2001*, pp. 11-14, November 2001.

[28] G. Holland, N. Vaidya, and P. Bahl, "A Rate-Adaptive MAC Protocol for Multi-Hop Wireless Networks," *Proceedings of ACM MOBICOM 2001*, pp. 236-251, July 2001.

[29] S. Jagadeesan, B. S. Manoj, and C. Siva Ram Murthy, "Interleaved Carrier Sense Multiple Access: An Efficient MAC Protocol for Ad Hoc Wireless Networks," *Proceedings of IEEE ICC 2003*, vol. 2, pp. 11-15, May 2003.

[30] I. Karthigeyan and C. Siva Ram Murthy, "Medium Access Control in Ad Hoc Wireless Networks: A Survey of Issues and Solutions," *Technical Report*, Department of Computer Science and Engineering, Indian Institute of Technology, Madras, India, March 2003.

Chapter 7

ROUTING PROTOCOLS FOR AD HOC WIRELESS NETWORKS

7.1 INTRODUCTION

An ad hoc wireless network consists of a set of mobile nodes (hosts) that are connected by wireless links. The network topology (the physical connectivity of the communication network) in such a network may keep changing randomly. Routing protocols that find a path to be followed by data packets from a source node to a destination node used in traditional wired networks cannot be directly applied in ad hoc wireless networks due to their highly dynamic topology, absence of established infrastructure for centralized administration (*e.g.*, base stations or access points), bandwidth-constrained wireless links, and resource (energy)-constrained nodes. A variety of routing protocols for ad hoc wireless networks has been proposed in the recent past. This chapter first presents the issues involved in designing a routing protocol and then the different classifications of routing protocols for ad hoc wireless networks. It then discusses the working of several existing routing protocols with illustrations.

7.2 ISSUES IN DESIGNING A ROUTING PROTOCOL FOR AD HOC WIRELESS NETWORKS

The major challenges that a routing protocol designed for ad hoc wireless networks faces are mobility of nodes, resource constraints, error-prone channel state, and hidden and exposed terminal problems. A detailed discussion on each of the following is given below.

7.2.1 Mobility

The network topology in an ad hoc wireless network is highly dynamic due to the movement of nodes, hence an on-going session suffers frequent path breaks. Disruption occurs either due to the movement of the intermediate nodes in the path or due to the movement of end nodes. Such situations do not arise because of reliable links in wired networks where all the nodes are stationary. Even though the wired network protocols find alternate routes during path breaks, their convergence is very slow. Therefore, wired network routing protocols cannot be used in ad hoc wireless networks where the mobility of nodes results in frequently changing network topologies. Routing protocols for ad hoc wireless networks must be able to perform efficient and effective mobility management.

7.2.2 Bandwidth Constraint

Abundant bandwidth is available in wired networks due to the advent of fiber optics and due to the exploitation of wavelength division multiplexing (WDM) technologies. But in a wireless network, the radio band is limited, and hence the data rates it can offer are much less than what a wired network can offer. This requires that the routing protocols use the bandwidth optimally by keeping the overhead as low as possible. The limited bandwidth availability also imposes a constraint on routing protocols in maintaining the topological information. Due to the frequent changes in topology, maintaining a consistent topological information at all the nodes involves more control overhead which, in turn, results in more bandwidth wastage. As efficient routing protocols in wired networks require the complete topology information, they may not be suitable for routing in the ad hoc wireless networking environment.

7.2.3 Error-Prone Shared Broadcast Radio Channel

The broadcast nature of the radio channel poses a unique challenge in ad hoc wireless networks. The wireless links have time-varying characteristics in terms of link capacity and link-error probability. This requires that the ad hoc wireless network routing protocol interacts with the MAC layer to find alternate routes through better-quality links. Also, transmissions in ad hoc wireless networks result in collisions of data and control packets. This is attributed to the hidden terminal problem [1]. Therefore, it is required that ad hoc wireless network routing protocols find paths with less congestion.

7.2.4 Hidden and Exposed Terminal Problems

The hidden terminal problem refers to the collision of packets at a receiving node due to the simultaneous transmission of those nodes that are not within the direct transmission range of the sender, but are within the transmission range of the receiver. Collision occurs when both nodes transmit packets at the same time without knowing about the transmission of each other. For example, consider Figure 7.1. Here, if both node A and node C transmit to node B at the same time, their packets collide at node B. This is due to the fact that both nodes A and C are hidden from each other, as they are not within the direct transmission range of each other and hence do not know about the presence of each other. Solutions for this problem include medium access collision avoidance (MACA) [2], medium ac-

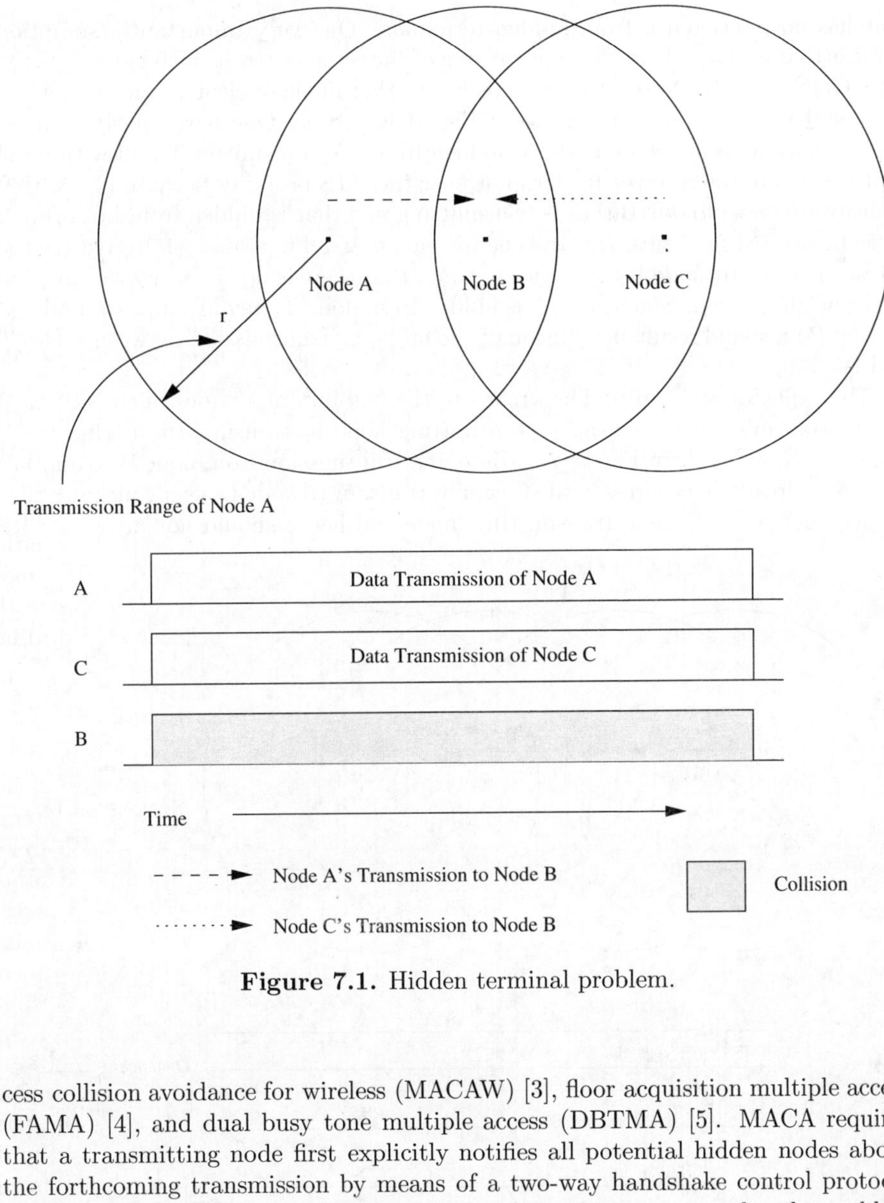

Figure 7.1. Hidden terminal problem.

cess collision avoidance for wireless (MACAW) [3], floor acquisition multiple access (FAMA) [4], and dual busy tone multiple access (DBTMA) [5]. MACA requires that a transmitting node first explicitly notifies all potential hidden nodes about the forthcoming transmission by means of a two-way handshake control protocol called the RTS-CTS protocol exchange. Note that this may not solve the problem completely, but it reduces the probability of collisions. To increase the efficiency, an improved version of the MACA protocol known as MACAW [3] has been proposed. This protocol requires that the receiver acknowledges each successful reception of a data packet. Hence, successful transmission is a four-way exchange mechanism, namely, RTS-CTS-Data-ACK. Even in the absence of bit errors and mobility, the RTS-CTS control packet exchange cannot ensure collision-free data transmission that has no interference from hidden terminals. One very important assumption made is that every node in the capture area of the receiver (transmitter) receives the CTS (RTS) cleanly. Nodes that do not hear either of these clearly can disrupt the successful transmission of the Data or the ACK packet. One particularly troublesome situation occurs when node A, hidden from the transmitter T and within the capture area of the receiver R, does not hear the CTS properly because it is within the capture area of node B that is transmitting and that is hidden from both R and T, as illustrated in Figure 7.2. In this case, node A did not successfully receive the CTS originated by node R and hence assumes that there is no on-going transmission in the neighborhood. Since node A is hidden from node T, any attempt to originate its own RTS would result in collision of the on-going transmission between nodes T and R.

The exposed terminal problem refers to the inability of a node which is blocked due to transmission by a nearby transmitting node to transmit to another node. Consider the example in Figure 7.3. Here, if a transmission from node B to another node A is already in progress, node C cannot transmit to node D, as it concludes that its neighbor, node B, is in transmitting mode and hence should not interfere with

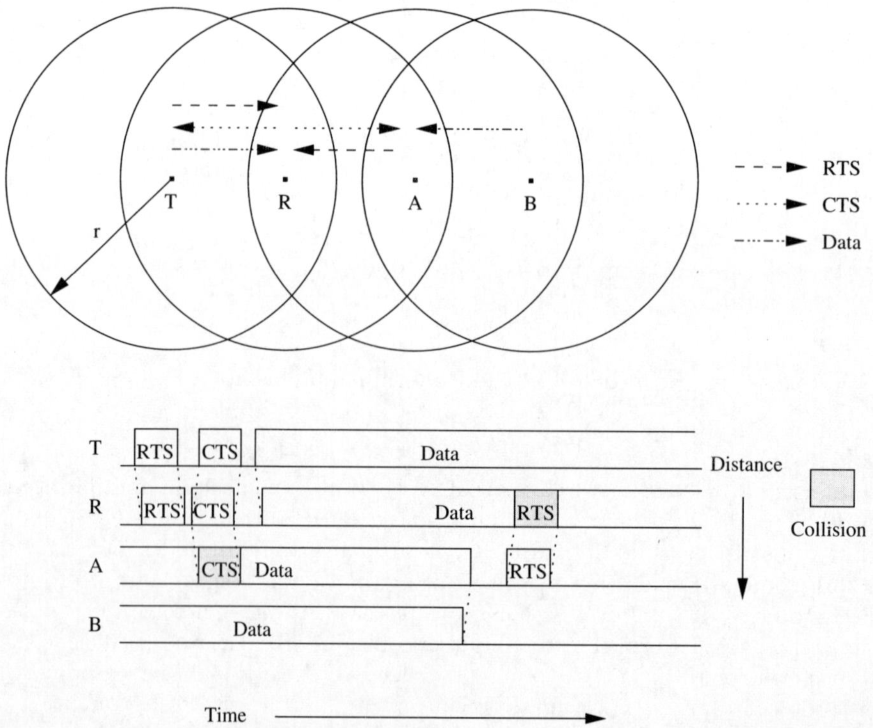

Figure 7.2. Hidden terminal problem with RTS-CTS-Data-ACK scheme.

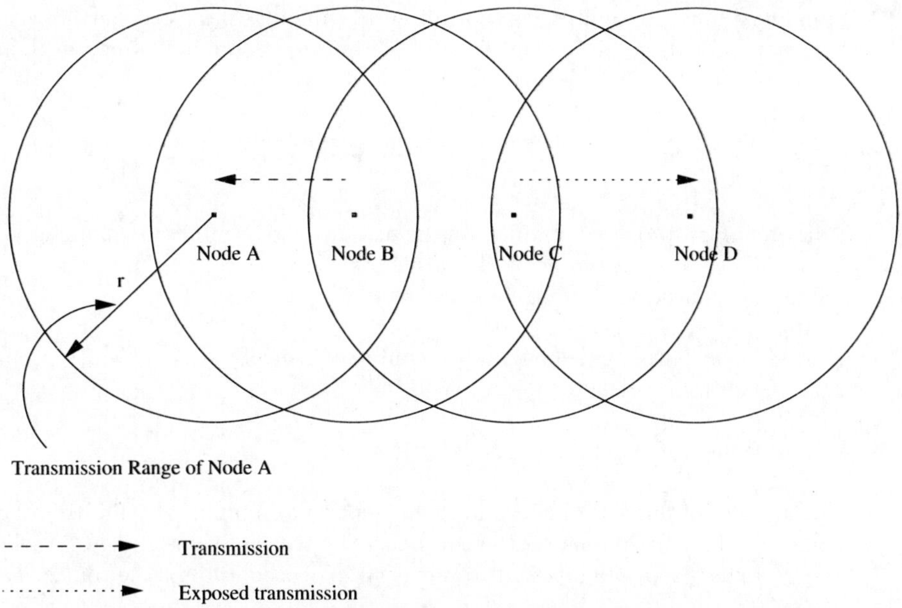

Figure 7.3. Exposed terminal problem.

the on-going transmission. Thus, reusability of the radio spectrum is affected. For node C to transmit simultaneously when node B is transmitting, the transmitting frequency of node C must be different from its receiving frequency.

7.2.5 Resource Constraints

Two essential and limited resources that form the major constraint for the nodes in an ad hoc wireless network are battery life and processing power. Devices used in ad hoc wireless networks in most cases require portability, and hence they also have size and weight constraints along with the restrictions on the power source. Increasing the battery power and processing ability makes the nodes bulky and less portable. Thus ad hoc wireless network routing protocols must optimally manage these resources.

7.2.6 Characteristics of an Ideal Routing Protocol for Ad Hoc Wireless Networks

Due to the issues in an ad hoc wireless network environment discussed so far, wired network routing protocols cannot be used in ad hoc wireless networks. Hence ad hoc wireless networks require specialized routing protocols that address the challenges described above. A routing protocol for ad hoc wireless networks should have the following characteristics:

1. It must be fully distributed, as centralized routing involves high control overhead and hence is not scalable. Distributed routing is more fault-tolerant than centralized routing, which involves the risk of single point of failure.

2. It must be adaptive to frequent topology changes caused by the mobility of nodes.

3. Route computation and maintenance must involve a minimum number of nodes. Each node in the network must have quick access to routes, that is, minimum connection setup time is desired.

4. It must be localized, as global state maintenance involves a huge state propagation control overhead.

5. It must be loop-free and free from stale routes.

6. The number of packet collisions must be kept to a minimum by limiting the number of broadcasts made by each node. The transmissions should be reliable to reduce message loss and to prevent the occurrence of stale routes.

7. It must converge to optimal routes once the network topology becomes stable. The convergence must be quick.

8. It must optimally use scarce resources such as bandwidth, computing power, memory, and battery power.

9. Every node in the network should try to store information regarding the stable local topology only. Frequent changes in local topology, and changes in the topology of parts of the network with which the node does not have any traffic correspondence, must not in any way affect the node, that is, changes in remote parts of the network must not cause updates in the topology information maintained by the node.

10. It should be able to provide a certain level of quality of service (QoS) as demanded by the applications, and should also offer support for time-sensitive traffic.

7.3 CLASSIFICATIONS OF ROUTING PROTOCOLS

Routing protocols for ad hoc wireless networks can be classified into several types based on different criteria. A classification tree is shown in Figure 7.4. Some of the classifications, their properties, and the basis of classifications are discussed below. The classification is not mutually exclusive and some protocols fall in more than one class. The deviation from the traditional routing metrics and path-finding processes that are employed in wired networks makes it worth further exploration

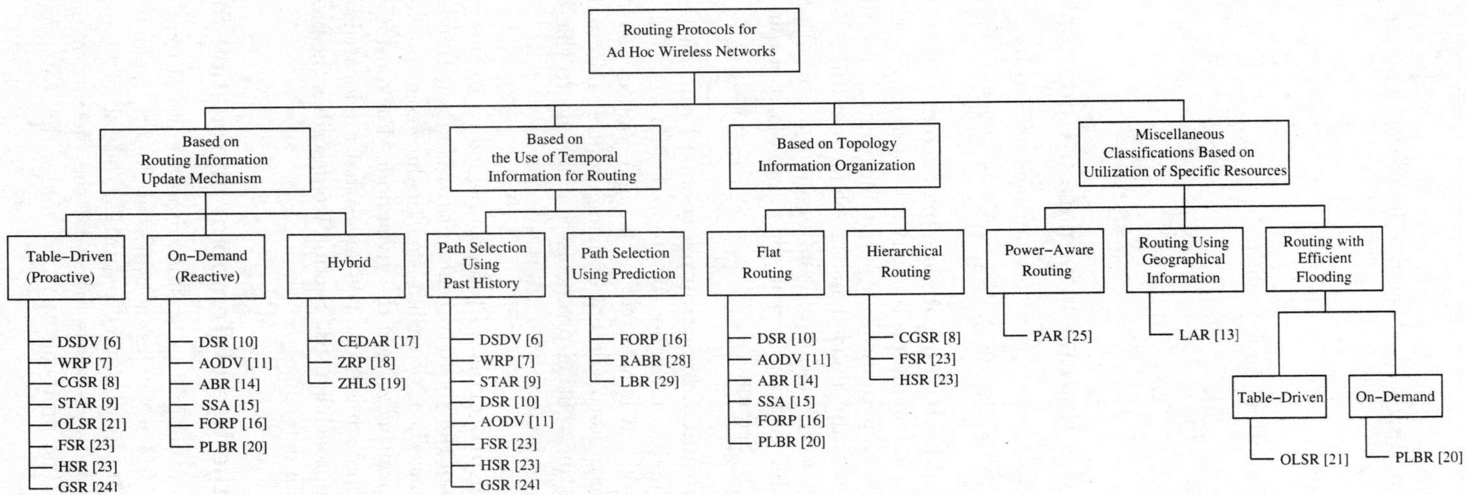

Figure 7.4. Classifications of routing protocols.

in this direction. The routing protocols for ad hoc wireless networks can be broadly classified into four categories based on

- Routing information update mechanism
- Use of temporal information for routing
- Routing topology
- Utilization of specific resources

7.3.1 Based on the Routing Information Update Mechanism

Ad hoc wireless network routing protocols can be classified into three major categories based on the routing information update mechanism. They are:

1. **Proactive or table-driven routing protocols:** In table-driven routing protocols, every node maintains the network topology information in the form of routing tables by periodically exchanging routing information. Routing information is generally flooded in the whole network. Whenever a node requires a path to a destination, it runs an appropriate path-finding algorithm on the topology information it maintains. Table-driven routing protocols are further explored in Section 7.4.

2. **Reactive or on-demand routing protocols:** Protocols that fall under this category do not maintain the network topology information. They obtain the necessary path when it is required, by using a connection establishment process. Hence these protocols do not exchange routing information periodically. Some of the existing routing protocols that belong to this category are discussed in Section 7.5.

3. **Hybrid routing protocols:** Protocols belonging to this category combine the best features of the above two categories. Nodes within a certain distance from the node concerned, or within a particular geographical region, are said to be within the routing zone of the given node. For routing within this zone, a table-driven approach is used. For nodes that are located beyond this zone, an on-demand approach is used. Section 7.6 describes the protocols belonging to this category in detail.

7.3.2 Based on the Use of Temporal Information for Routing

This classification of routing protocols is based on the use of temporal information used for routing. Since ad hoc wireless networks are highly dynamic and path breaks are much more frequent than in wired networks, the use of temporal information regarding the lifetime of the wireless links and the lifetime of the paths selected assumes significance. The protocols that fall under this category can be further classified into two types:

1. **Routing protocols using past temporal information:** These routing protocols use information about the past status of the links or the status of links at the time of routing to make routing decisions. For example, the routing metric based on the availability of wireless links (which is the current/present information here) along with a shortest path-finding algorithm, provides a path that may be efficient and stable at the time of path-finding. The topological changes may immediately break the path, making the path undergo a resource-wise expensive path reconfiguration process.

2. **Routing protocols that use future temporal information:** Protocols belonging to this category use information about the expected future status of the wireless links to make approximate routing decisions. Apart from the lifetime of wireless links, the future status information also includes information regarding the lifetime of the node (which is based on the remaining battery charge and discharge rate of the non-replenishable resources), prediction of location, and prediction of link availability.

7.3.3 Based on the Routing Topology

Routing topology being used in the Internet is hierarchical in order to reduce the state information maintained at the core routers. Ad hoc wireless networks, due to their relatively smaller number of nodes, can make use of either a flat topology or a hierarchical topology for routing.

1. **Flat topology routing protocols:** Protocols that fall under this category make use of a flat addressing scheme similar to the one used in IEEE 802.3 LANs. It assumes the presence of a globally unique (or at least unique to the connected part of the network) addressing mechanism for nodes in an ad hoc wireless network.

2. **Hierarchical topology routing protocols:** Protocols belonging to this category make use of a logical hierarchy in the network and an associated addressing scheme. The hierarchy could be based on geographical information or it could be based on hop distance.

7.3.4 Based on the Utilization of Specific Resources

1. **Power-aware routing:** This category of routing protocols aims at minimizing the consumption of a very important resource in the ad hoc wireless networks: the battery power. The routing decisions are based on minimizing the power consumption either locally or globally in the network.

2. **Geographical information assisted routing:** Protocols belonging to this category improve the performance of routing and reduce the control overhead by effectively utilizing the geographical information available.

The following section further explores the above classifications and discusses specific routing protocols belonging to each category in detail.

7.4 TABLE-DRIVEN ROUTING PROTOCOLS

These protocols are extensions of the wired network routing protocols. They maintain the global topology information in the form of tables at every node. These tables are updated frequently in order to maintain consistent and accurate network state information. The destination sequenced distance-vector routing protocol (DSDV), wireless routing protocol (WRP), source-tree adaptive routing protocol (STAR), and cluster-head gateway switch routing protocol (CGSR) are some examples for the protocols that belong to this category.

7.4.1 Destination Sequenced Distance-Vector Routing Protocol

The destination sequenced distance-vector routing protocol (DSDV) [6] is one of the first protocols proposed for ad hoc wireless networks. It is an enhanced version of the distributed Bellman-Ford algorithm where each node maintains a table that contains the shortest distance and the first node on the shortest path to every other node in the network. It incorporates table updates with increasing sequence number tags to prevent loops, to counter the count-to-infinity problem, and for faster convergence.

As it is a table-driven routing protocol, routes to all destinations are readily available at every node at all times. The tables are exchanged between neighbors at regular intervals to keep an up-to-date view of the network topology. The tables are also forwarded if a node observes a significant change in local topology. The table updates are of two types: incremental updates and full dumps. An incremental update takes a single network data packet unit (NDPU), while a full dump may take multiple NDPUs. Incremental updates are used when a node does not observe significant changes in the local topology. A full dump is done either when the local topology changes significantly or when an incremental update requires more than a single NDPU. Table updates are initiated by a destination with a new sequence number which is always greater than the previous one. Upon receiving an updated table, a node either updates its tables based on the received information or holds it for some time to select the best metric (which may be the lowest number of hops) received from multiple versions of the same update table from different neighboring nodes. Based on the sequence number of the table update, it may forward or reject the table. Consider the example as shown in Figure 7.5 (a). Here node 1 is the source node and node 15 is the destination. As all the nodes maintain global topology information, the route is already available as shown in Figure 7.5 (b). Here the routing table of node 1 indicates that the shortest route to the destination node (node 15) is available through node 5 and the distance to it is 4 hops, as depicted in Figure 7.5 (b).

The reconfiguration of a path used by an on-going data transfer session is handled by the protocol in the following way. The end node of the broken link initiates a table update message with the broken link's weight assigned to infinity (∞) and with a sequence number greater than the stored sequence number for that destination. Each node, upon receiving an update with weight ∞, quickly disseminates it to its neighbors in order to propagate the broken-link information to the whole net-

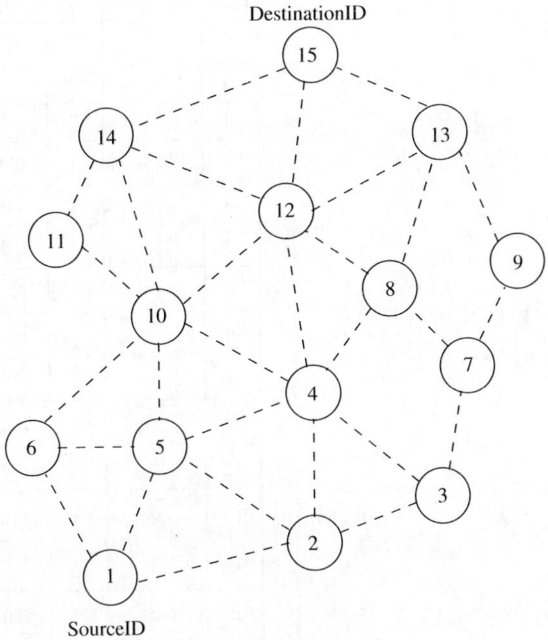

(a) Topology graph of the network

Dest	NextNode	Dist	SeqNo
2	2	1	22
3	2	2	26
4	5	2	32
5	5	1	134
6	6	1	144
7	2	3	162
8	5	3	170
9	2	4	186
10	6	2	142
11	6	3	176
12	5	3	190
13	5	4	198
14	6	3	214
15	5	4	256

(b) Routing table for Node 1

Figure 7.5. Route establishment in DSDV.

work. Thus a single link break leads to the propagation of table update information to the whole network. A node always assigns an odd sequence number to the link break update to differentiate it from the even sequence number generated by the destination. Consider the case when node 11 moves from its current position, as shown in Figure 7.6. When a neighbor node perceives the link break, it sets all the paths passing through the broken link with distance as ∞. For example, when node 10 knows about the link break, it sets the path to node 11 as ∞ and broadcasts its routing table to its neighbors. Those neighbors detecting significant changes in their routing tables rebroadcast it to their neighbors. In this way, the broken link information propagates throughout the network. Node 1 also sets the distance to node 11 as ∞. When node 14 receives a table update message from node 11, it informs the neighbors about the shortest distance to node 11. This information is also propagated throughout the network. All nodes receiving the new update message with the higher sequence number set the new distance to node 11 in their corresponding tables. The updated table at node 1 is shown in Figure 7.6, where the current distance from node 1 to node 11 has increased from three to four hops.

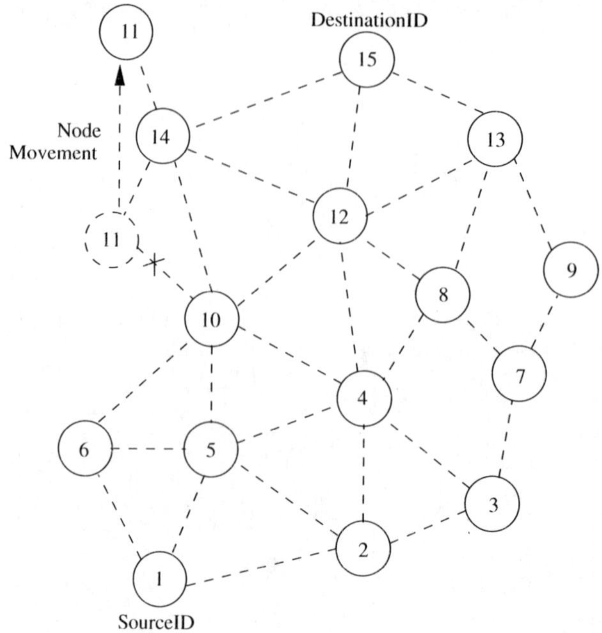

Routing Table for Node 1

Dest	NextNode	Dist	SeqNo
2	2	1	22
3	2	2	26
4	5	2	32
5	5	1	134
6	6	1	144
7	2	3	162
8	5	3	170
9	2	4	186
10	6	2	142
11	5	4	180
12	5	3	190
13	5	4	198
14	6	3	214
15	5	4	256

Figure 7.6. Route maintenance in DSDV.

Advantages and Disadvantages

The availability of routes to all destinations at all times implies that much less delay is involved in the route setup process. The mechanism of incremental updates with sequence number tags makes the existing wired network protocols adaptable to ad hoc wireless networks. Hence, an existing wired network protocol can be applied to ad hoc wireless networks with many fewer modifications. The updates are propagated throughout the network in order to maintain an up-to-date view of the network topology at all the nodes. The updates due to broken links lead to a heavy control overhead during high mobility. Even a small network with high mobility or a large network with low mobility can completely choke the available bandwidth. Hence, this protocol suffers from excessive control overhead that is proportional to the number of nodes in the network and therefore is not scalable in ad hoc wireless networks, which have limited bandwidth and whose topologies are highly dynamic. Another disadvantage of DSDV is that in order to obtain information about a particular destination node, a node has to wait for a table update message initiated by the same destination node. This delay could result in stale routing information at nodes.

7.4.2 Wireless Routing Protocol

The wireless routing protocol (WRP) [7], similar to DSDV, inherits the properties of the distributed Bellman-Ford algorithm. To counter the count-to-infinity problem and to enable faster convergence, it employs a unique method of maintaining information regarding the shortest distance to every destination node in the network and the penultimate hop node on the path to every destination node. Since WRP, like DSDV, maintains an up-to-date view of the network, every node has a readily available route to every destination node in the network. It differs from DSDV in table maintenance and in the update procedures. While DSDV maintains only one topology table, WRP uses a set of tables to maintain more accurate information. The tables that are maintained by a node are the following: distance table (DT), routing table (RT), link cost table (LCT), and a message retransmission list (MRL).

The DT contains the network view of the neighbors of a node. It contains a matrix where each element contains the distance and the penultimate node reported by a neighbor for a particular destination. The RT contains the up-to-date view of the network for all known destinations. It keeps the shortest distance, the *predecessor* node (penultimate node), the *successor* node (the next node to reach the destination), and a flag indicating the status of the path. The path status may be a simple path (correct), or a loop (error), or the destination node not marked (null). The LCT contains the cost (*e.g.*, the number of hops to reach the destination) of relaying messages through each link. The cost of a broken link is ∞. It also contains the number of update periods (intervals between two successive periodic updates) passed since the last successful update was received from that link. This is done to detect link breaks. The MRL contains an entry for every update message that is to be retransmitted and maintains a counter for each entry. This counter is decremented after every retransmission of an update message. Each update message contains a list of updates. A node also marks each node in the RT that has to acknowledge the update message it transmitted. Once the counter reaches zero, the entries in the update message for which no acknowledgments have been received are to be retransmitted and the update message is deleted. Thus, a node detects a link break by the number of update periods missed since the last successful transmission. After receiving an update message, a node not only updates the distance for transmitted neighbors but also checks the other neighbors' distance, hence convergence is much faster than DSDV. Consider the example shown in Figure 7.7, where the source of the route is node 1 and the destination is node 15. As WRP proactively maintains the route to all the destinations, the route to any destination node is readily available at the source node. From the routing table shown in Figure 7.7, the route from node 1 to node 15 has the next node as node 2. The predecessor node of 15 corresponding to this route is node 12. The predecessor information helps WRP to converge quickly during link breaks.

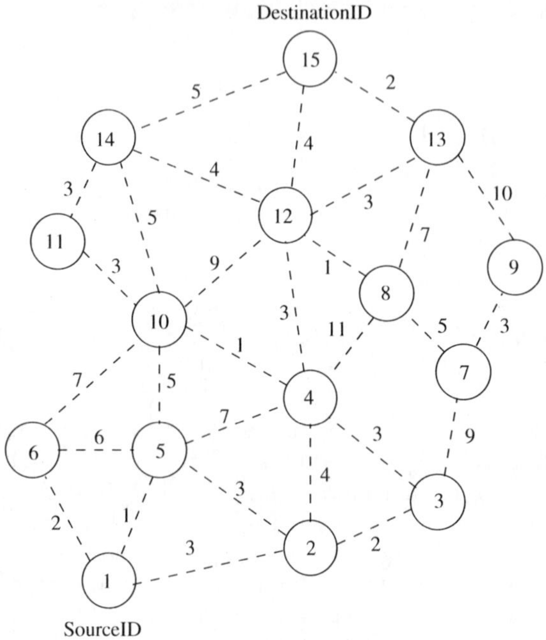

Routing Entry at Each Node for DestinationID 15

Node	NextNode	Pred	Cost
15	15	15	0
14	15	14	5
13	15	13	2
12	15	12	4
11	14	14	8
10	4	12	8
9	13	13	12
8	12	12	5
7	8	12	10
6	10	12	15
5	10	12	13
4	12	12	10
3	4	12	7
2	4	12	11
1	2	12	14

Figure 7.7. Route establishment in WRP.

When a node detects a link break, it sends an update message to its neighbors with the link cost of the broken link set to ∞. After receiving the update message, all affected nodes update their minimum distances to the corresponding nodes (including the distance to the destination). The node that initiated the update message then finds an alternative route, if available from its DT. Note that this new route computed will not contain the broken link. Consider the scenario shown in Figure 7.8. When the link between nodes 12 and 15 breaks, all nodes having a route to the destination with predecessor as node 12 delete their corresponding routing entries. Both node 12 and node 15 send update messages to their neighbors indicating that the cost of the link between nodes 12 and 15 is ∞. If the nodes have any other alternative route to the destination node 15, they update their routing tables and indicate the changed route to their neighbors by sending an update message. A neighbor node, after receiving an update message, updates its routing table only if the new path is better than the previously existing paths. For example, when node 12 finds an alternative route to the destination through node 13, it broadcasts an

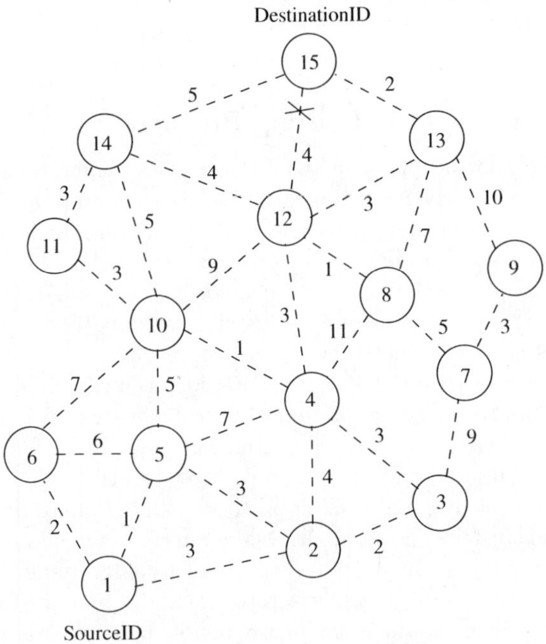

Routing Entry at Each Node for DestinationID 15

Node	NextNode	Pred	Cost
15	15	15	0
14	15	14	5
13	15	13	2
12	15	13	5
11	14	14	8
10	4	13	9
9	13	13	12
8	12	13	6
7	8	13	11
6	10	13	16
5	10	13	14
4	12	13	8
3	4	13	11
2	4	13	12
1	2	13	15

Figure 7.8. Route maintenance in WRP.

update message indicating the changed path. After receiving the update message from node 12, neighboring nodes 8, 14, 15, and 13 do not change their routing entry corresponding to destination 15 while node 4 and node 10 modify their entries to reflect the new updated path. Nodes 4 and 10 again send an update message to indicate the correct path to the destination for their respective neighbors. When node 10 receives node 4's update message, it again modifies its routing entry to optimize the path to the destination node (15) while node 4 discards the update entry it received from node 10.

Advantages and Disadvantages

WRP has the same advantages as that of DSDV. In addition, it has faster convergence and involves fewer table updates. But the complexity of maintenance of multiple tables demands a larger memory and greater processing power from nodes in the ad hoc wireless network. At high mobility, the control overhead involved in updating table entries is almost the same as that of DSDV and hence is not suitable for highly dynamic and also for very large ad hoc wireless networks.

7.4.3 Cluster-Head Gateway Switch Routing Protocol

The cluster-head gateway switch routing protocol (CGSR) [8] uses a hierarchical network topology, unlike other table-driven routing approaches that employ flat topologies. CGSR organizes nodes into clusters, with coordination among the members of each cluster entrusted to a special node named *cluster-head*. This cluster-head is elected dynamically by employing a *least cluster change (LCC)* algorithm [8]. According to this algorithm, a node ceases to be a cluster-head only if it comes under the range of another cluster-head, where the tie is broken either using the lowest ID or highest connectivity algorithm. Clustering provides a mechanism to allocate bandwidth, which is a limited resource, among different clusters, thereby improving reuse. For example, different cluster-heads could operate on different spreading codes on a CDMA system. Inside a cluster, the cluster-head can coordinate the channel access based on a token-based polling protocol. All member nodes of a cluster can be reached by the cluster-head within a single hop, thereby enabling the cluster-head to provide improved coordination among nodes that fall under its cluster. A token-based scheduling (assigning access token to the nodes in a cluster) is used within a cluster for sharing the bandwidth among the members of the cluster. CGSR assumes that all communication passes through the cluster-head. Communication between two clusters takes place through the common member nodes that are members of both the clusters. These nodes which are members of more than one cluster are called *gateways*. A gateway is expected to be able to listen to multiple spreading codes that are currently in operation in the clusters in which the node exists as a member. A gateway conflict is said to occur when a cluster-head issues a token to a gateway over a spreading code while the gateway is tuned to another code. Gateways that are capable of simultaneously communicating over two interfaces can avoid gateway conflicts.

The performance of routing is influenced by token scheduling and code scheduling (assigning appropriate spreading codes to two different clusters) that are handled at cluster-heads and gateways, respectively. The routing protocol used in CGSR is an extension of DSDV. Every member node maintains a routing table containing the destination cluster-head for every node in the network. In addition to the cluster member table, each node maintains a routing table which keeps the list of next-hop nodes for reaching every destination cluster. The *cluster (hierarchical) routing protocol* is used here. As per this protocol, when a node with packets to be transmitted to a destination gets the token from its cluster-head, it obtains the destination cluster-head and the next-hop node from the cluster member table and the routing table, respectively. CGSR improves the routing performance by routing packets through the cluster-heads and gateways. A path from any node a to any node b will be similar to $a - C_1 - G_1 - C_2 - G_2 - \ldots C_i - G_j \ldots G_n - b$, where G_i and C_j are the ith gateway and the jth cluster-head, respectively, in the path. Figure 7.9 shows the cluster-heads, *cluster gateways*, and normal cluster member

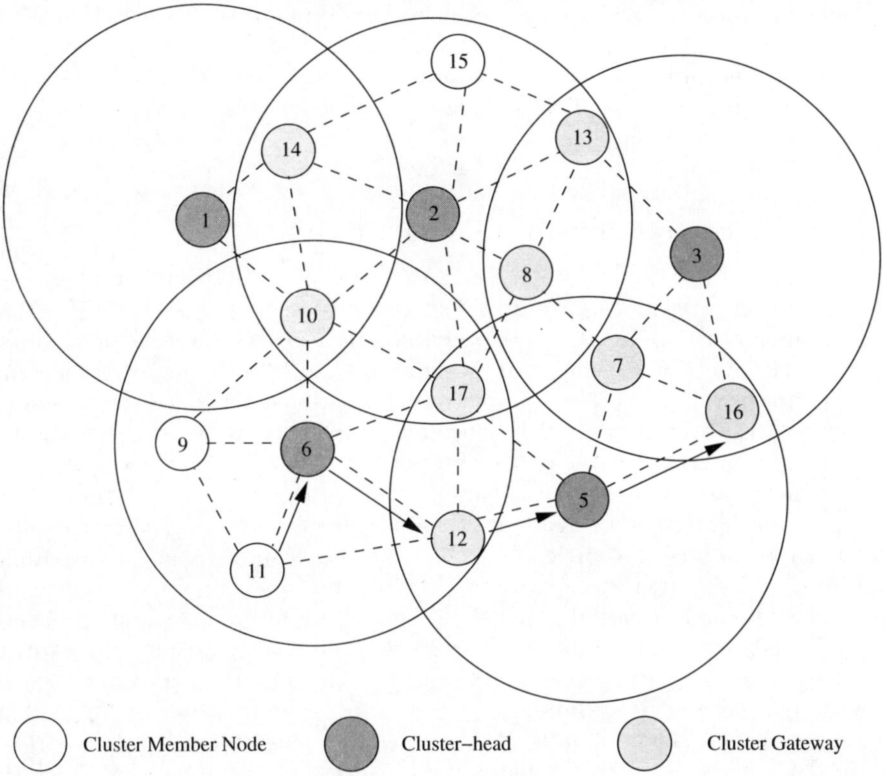

Figure 7.9. Route establishment in CGSR.

nodes in an ad hoc wireless network. A path between node 11 and node 16 would follow 11 – 6 – 12 – 5 – 16. Since the cluster-heads gain more opportunities for transmission, the cluster-heads, by means of a dynamic scheduling mechanism, can make CGSR obtain better delay performance for real-time flows. Route reconfiguration is necessitated by mainly two factors: firstly, the change in cluster-head and secondly, the stale entries in the cluster member table and routing table. CGSR depends on the table update mechanism to handle the latter problem, while the least cluster change algorithm [8] handles the former.

Advantages and Disadvantages

CGSR is a hierarchical routing scheme which enables partial coordination between nodes by electing cluster-heads. Hence, better bandwidth utilization is possible. It is easy to implement priority scheduling schemes with token scheduling and gateway code scheduling. The main disadvantages of CGSR are increase in path length and instability in the system at high mobility when the rate of change of cluster-heads is high. In order to avoid gateway conflicts, more resources (such as additional interfaces) are required. The power consumption at the cluster-head node is also a matter of concern because the battery-draining rate at the cluster-head is higher than that at a normal node. This could lead to frequent changes in the cluster-head, which may result in multiple path breaks.

7.4.4 Source-Tree Adaptive Routing Protocol

Source-tree adaptive routing protocol (STAR) [9] proposed by Garcia-Luna-Aceves and Spohn is a variation of table-driven routing protocols, with the least overhead routing approach (LORA) as the key concept rather than the optimum routing approach (ORA) that was employed by earlier table-driven routing protocols. The ORA protocols attempt to update routing information quickly enough to provide optimum paths with respect to the defined metric (which may be the lowest number of hops), but with LORA, the routing protocol attempts to provide feasible paths that are not guaranteed to be optimal, but involve much less control overhead. In STAR protocol, every node broadcasts its *source-tree* information. The source-tree of a node consists of the wireless links used by the node in its preferred path to destinations. Every node, using its adjacent links and the source-tree broadcast by its neighbors, builds a partial graph of the topology. During initialization, a node sends an update message to its neighbors. Also, every node is required to originate update messages about new destinations, the chances of routing loops, and the cost of paths exceeding a given threshold. Hence, each node will have a path to every destination node. The path, in most cases, would be sub-optimal.

In the absence of a reliable link layer broadcast mechanism, STAR uses the following path-finding approach. When a node s has data packets to send to a particular destination d, for which no path exists in its source-tree, it originates an update message to all its neighbors indicating the absence of a path to d. This update message triggers another update message from a neighbor which has a path to d. Node s retransmits the update message as long as it does not have a path to d with increasing intervals between successive retransmissions. After getting the source-tree update from a neighbor, the node s updates its source-tree and, using this, it finds a path to all nodes in the network. The data packet contains information about the path to be traversed in order to prevent the possibility of routing loop formation.

In the presence of a reliable broadcast mechanism, STAR assumes implicit route maintenance. The link update message about the unavailability of a next-hop node triggers an update message from a neighbor which has an alternate source tree indicating an alternate next-hop node to the destination. In addition to path breaks, the intermediate nodes are responsible for handling the routing loops. When an intermediate node k receives a data packet to destination d, and one of the nodes in the packet's traversed path is present in node k's path to the destination d, then it discards the packet and a *RouteRepair* update message is reliably sent to the node in the head of the route repair path. The route repair path corresponds to the path k to x, where x is the last router in the data packet's traversed path that is first found in the path k to d, that belongs to the source tree of k. The *RouteRepair* packet contains the complete source tree of node k and the traversed path of the packet.

When an intermediate node receives a *RouteRepair* update message, it removes itself from the top of the route repair path and reliably sends it to the head of the route repair path.

Advantages and Disadvantages

STAR has very low communication overhead among all the table-driven routing protocols. The use of the LORA approach in this table-driven routing protocol reduces the average control overhead compared to several other on-demand routing protocols.

7.5 ON-DEMAND ROUTING PROTOCOLS

Unlike the table-driven routing protocols, on-demand routing protocols execute the path-finding process and exchange routing information only when a path is required by a node to communicate with a destination. This section explores some of the existing on-demand routing protocols in detail.

7.5.1 Dynamic Source Routing Protocol

Dynamic source routing protocol (DSR) [10] is an on-demand protocol designed to restrict the bandwidth consumed by control packets in ad hoc wireless networks by eliminating the periodic table-update messages required in the table-driven approach. The major difference between this and the other on-demand routing protocols is that it is *beacon-less* and hence does not require periodic *hello* packet (*beacon*) transmissions, which are used by a node to inform its neighbors of its presence. The basic approach of this protocol (and all other on-demand routing protocols) during the route construction phase is to establish a route by flooding *RouteRequest* packets in the network. The destination node, on receiving a *RouteRequest* packet, responds by sending a *RouteReply* packet back to the source, which carries the route traversed by the *RouteRequest* packet received.

Consider a source node that does not have a route to the destination. When it has data packets to be sent to that destination, it initiates a *RouteRequest* packet. This *RouteRequest* is flooded throughout the network. Each node, upon receiving a *RouteRequest* packet, rebroadcasts the packet to its neighbors if it has not forwarded already or if the node is not the destination node, provided the packet's time to live (TTL) counter has not exceeded. Each *RouteRequest* carries a sequence number generated by the source node and the path it has traversed. A node, upon receiving a *RouteRequest* packet, checks the sequence number on the packet before forwarding it. The packet is forwarded only if it is not a duplicate *RouteRequest*. The sequence number on the packet is used to prevent loop formations and to avoid multiple transmissions of the same *RouteRequest* by an intermediate node that receives it through multiple paths. Thus, all nodes except the destination forward

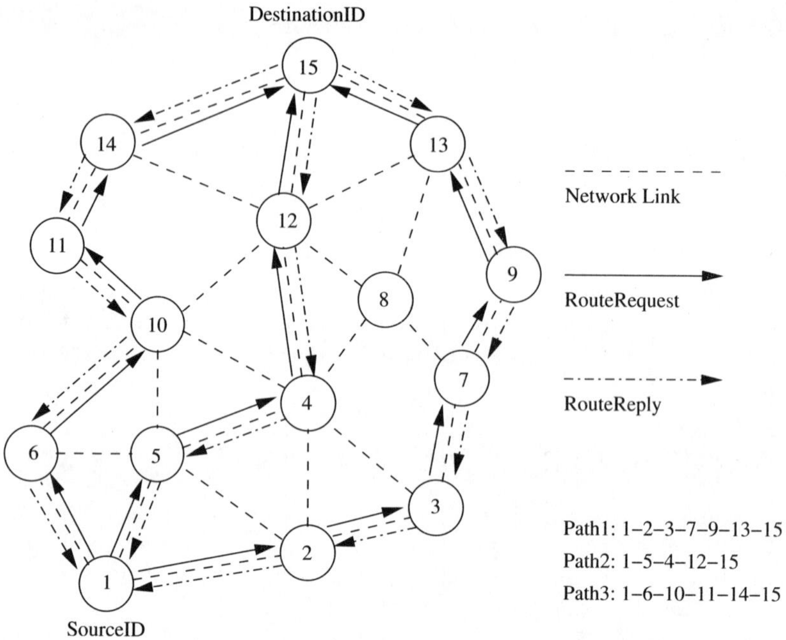

Figure 7.10. Route establishment in DSR.

a *RouteRequest* packet during the route construction phase. A destination node, after receiving the first *RouteRequest* packet, replies to the source node through the reverse path the *RouteRequest* packet had traversed. In Figure 7.10, source node 1 initiates a *RouteRequest* packet to obtain a path for destination node 15. This protocol uses a route cache that stores all possible information extracted from the source route contained in a data packet. Nodes can also learn about the neighboring routes traversed by data packets if operated in the promiscuous mode (the mode of operation in which a node can receive the packets that are neither broadcast nor addressed to itself). This route cache is also used during the route construction phase. If an intermediate node receiving a *RouteRequest* has a route to the destination node in its route cache, then it replies to the source node by sending a *RouteReply* with the entire route information from the source node to the destination node.

Optimizations

Several optimization techniques have been incorporated into the basic DSR protocol to improve the performance of the protocol. DSR uses the route cache at intermediate nodes. The route cache is populated with routes that can be extracted from the information contained in data packets that get forwarded. This cache information is used by the intermediate nodes to reply to the source when they receive a *RouteRequest* packet and if they have a route to the corresponding destination.

By operating in the promiscuous mode, an intermediate node learns about route breaks. Information thus gained is used to update the route cache so that the active routes maintained in the route cache do not use such broken links. During network partitions, the affected nodes initiate *RouteRequest* packets. An exponential backoff algorithm is used to avoid frequent *RouteRequest* flooding in the network when the destination is in another disjoint set. DSR also allows piggy-backing of a data packet on the *RouteRequest* so that a data packet can be sent along with the *RouteRequest.*

If optimization is not allowed in the DSR protocol, the route construction phase is very simple. All the intermediate nodes flood the *RouteRequest* packet if it is not redundant. For example, after receiving the *RouteRequest* packet from node 1 (refer to Figure 7.10), all its neighboring nodes, that is, nodes 2, 5, and 6, forward it. Node 4 receives the *RouteRequest* from both nodes 2 and 5. Node 4 forwards the first *RouteRequest* it receives from any one of the nodes 2 and 5 and discards the other redundant/duplicate *RouteRequest* packets. The *RouteRequest* is propagated till it reaches the destination which initiates the *RouteReply.* As part of optimizations, if the intermediate nodes are also allowed to originate *RouteReply* packets, then a source node may receive multiple replies from intermediate nodes. For example, in Figure 7.11, if the intermediate node 10 has a route to the destination via node 14, it also sends the *RouteReply* to the source node. The source node selects the latest

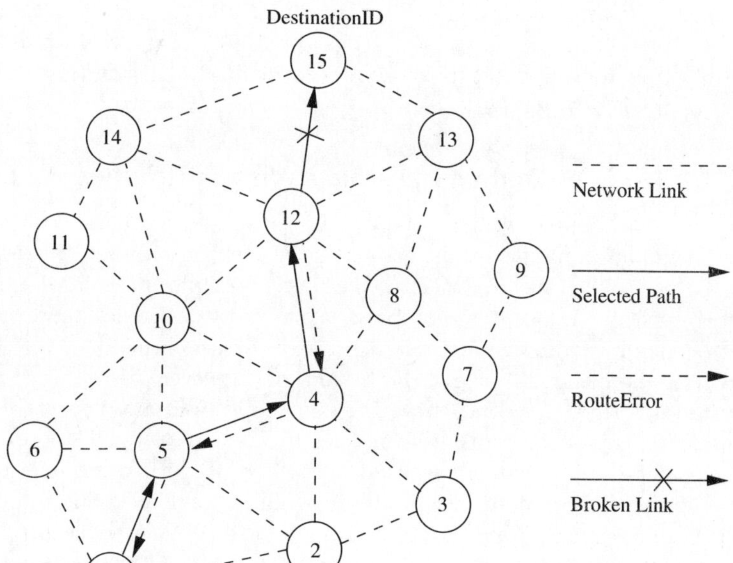

Figure 7.11. Route maintenance in DSR.

and best route, and uses that for sending data packets. Each data packet carries the complete path to its destination.

When an intermediate node in the path moves away, causing a wireless link to break, for example, the link between nodes 12 and 15 in Figure 7.11, a *RouteError* message is generated from the node adjacent to the broken link to inform the source node. The source node reinitiates the route establishment procedure. The cached entries at the intermediate nodes and the source node are removed when a *RouteError* packet is received. If a link breaks due to the movement of edge nodes (nodes 1 and 15), the source node again initiates the route discovery process.

Advantages and Disadvantages

This protocol uses a reactive approach which eliminates the need to periodically flood the network with table update messages which are required in a table-driven approach. In a reactive (on-demand) approach such as this, a route is established only when it is required and hence the need to find routes to all other nodes in the network as required by the table-driven approach is eliminated. The intermediate nodes also utilize the route cache information efficiently to reduce the control overhead. The disadvantage of this protocol is that the route maintenance mechanism does not locally repair a broken link. Stale route cache information could also result in inconsistencies during the route reconstruction phase. The connection setup delay is higher than in table-driven protocols. Even though the protocol performs well in static and low-mobility environments, the performance degrades rapidly with increasing mobility. Also, considerable routing overhead is involved due to the source-routing mechanism employed in DSR. This routing overhead is directly proportional to the path length.

7.5.2 Ad Hoc On-Demand Distance-Vector Routing Protocol

Ad hoc on-demand distance vector (AODV) [11] routing protocol uses an on-demand approach for finding routes, that is, a route is established only when it is required by a source node for transmitting data packets. It employs destination sequence numbers to identify the most recent path. The major difference between AODV and DSR stems out from the fact that DSR uses source routing in which a data packet carries the complete path to be traversed. However, in AODV, the source node and the intermediate nodes store the next-hop information corresponding to each flow for data packet transmission. In an on-demand routing protocol, the source node floods the *RouteRequest* packet in the network when a route is not available for the desired destination. It may obtain multiple routes to different destinations from a single *RouteRequest*. The major difference between AODV and other on-demand routing protocols is that it uses a destination sequence number (DestSeqNum) to determine an up-to-date path to the destination. A node updates its path information only if the DestSeqNum of the current packet received is greater than the last DestSeqNum stored at the node.

A *RouteRequest* carries the source identifier (SrcID), the destination identifier (DestID), the source sequence number (SrcSeqNum), the destination sequence number (DestSeqNum), the broadcast identifier (BcastID), and the time to live (TTL) field. DestSeqNum indicates the freshness of the route that is accepted by the source. When an intermediate node receives a *RouteRequest*, it either forwards it or prepares a *RouteReply* if it has a valid route to the destination. The validity of a route at the intermediate node is determined by comparing the sequence number at the intermediate node with the destination sequence number in the *RouteRequest* packet. If a *RouteRequest* is received multiple times, which is indicated by the BcastID-SrcID pair, the duplicate copies are discarded. All intermediate nodes having valid routes to the destination, or the destination node itself, are allowed to send *RouteReply* packets to the source. Every intermediate node, while forwarding a *RouteRequest*, enters the previous node address and its BcastID. A timer is used to delete this entry in case a *RouteReply* is not received before the timer expires. This helps in storing an active path at the intermediate node as AODV does not employ source routing of data packets. When a node receives a *RouteReply* packet, information about the previous node from which the packet was received is also stored in order to forward the data packet to this next node as the next hop toward the destination.

Consider the example depicted in Figure 7.12. In this figure, source node 1 initiates a path-finding process by originating a *RouteRequest* to be flooded in the network for destination node 15, assuming that the *RouteRequest* contains the des-

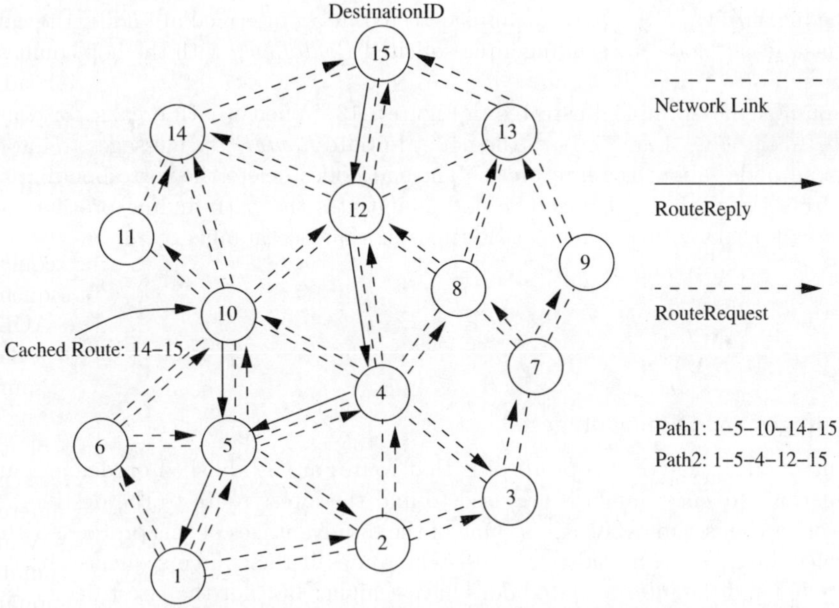

Figure 7.12. Route establishment in AODV.

tination sequence number as 3 and the source sequence number as 1. When nodes 2, 5, and 6 receive the *RouteRequest* packet, they check their routes to the destination. In case a route to the destination is not available, they further forward it to their neighbors. Here nodes 3, 4, and 10 are the neighbors of nodes 2, 5, and 6. This is with the assumption that intermediate nodes 3 and 10 already have routes to the destination node, that is, node 15 through paths 10-14-15 and 3-7-9-13-15, respectively. If the destination sequence number at intermediate node 10 is 4 and is 1 at intermediate node 3, then only node 10 is allowed to reply along the cached route to the source. This is because node 3 has an older route to node 15 compared to the route available at the source node (the destination sequence number at node 3 is 1, but the destination sequence number is 3 at the source node), while node 10 has a more recent route (the destination sequence number is 4) to the destination. If the *RouteRequest* reaches the destination (node 15) through path 4-12-15 or any other alternative route, the destination also sends a *RouteReply* to the source. In this case, multiple *RouteReply* packets reach the source. All the intermediate nodes receiving a *RouteReply* update their route tables with the latest destination sequence number. They also update the routing information if it leads to a shorter path between source and destination.

AODV does not repair a broken path locally. When a link breaks, which is determined by observing the periodical *beacons* or through link-level acknowledgments, the end nodes (*i.e.,* source and destination nodes) are notified. When a source node learns about the path break, it reestablishes the route to the destination if required by the higher layers. If a path break is detected at an intermediate node, the node informs the end nodes by sending an unsolicited *RouteReply* with the hop count set as ∞.

Consider the example illustrated in Figure 7.13. When a path breaks, for example, between nodes 4 and 5, both the nodes initiate *RouteError* messages to inform their end nodes about the link break. The end nodes delete the corresponding entries from their tables. The source node reinitiates the path-finding process with the new BcastID and the previous destination sequence number.

Advantages and Disadvantages

The main advantage of this protocol is that routes are established on demand and destination sequence numbers are used to find the latest route to the destination. The connection setup delay is less. One of the disadvantages of this protocol is that intermediate nodes can lead to inconsistent routes if the source sequence number is very old and the intermediate nodes have a higher but not the latest destination sequence number, thereby having stale entries. Also multiple *RouteReply* packets in response to a single *RouteRequest* packet can lead to heavy control overhead. Another disadvantage of AODV is that the periodic *beacon*ing leads to unnecessary bandwidth consumption.

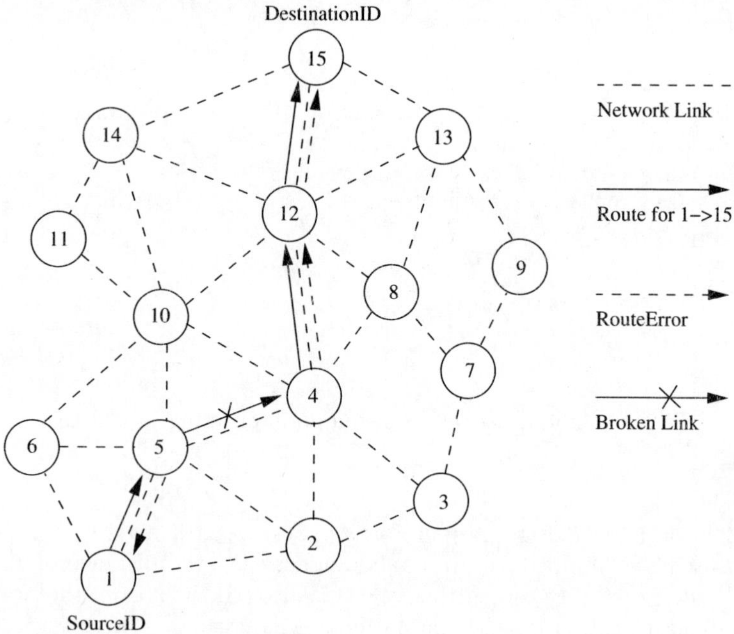

Figure 7.13. Route maintenance in AODV.

7.5.3 Temporally Ordered Routing Algorithm

Temporally ordered routing algorithm (TORA) [12] is a source-initiated on-demand routing protocol which uses a *link reversal algorithm* and provides loop-free multi-path routes to a destination node. In TORA, each node maintains its one-hop local topology information and also has the capability to detect partitions. TORA has the unique property of limiting the control packets to a small region during the reconfiguration process initiated by a path break. Figure 7.14 shows the distance metric used in TORA which is nothing but the length of the path, or the height from the destination. H(N) denotes the height of node N from the destination. TORA has three main functions: establishing, maintaining, and erasing routes.

The route establishment function is performed only when a node requires a path to a destination but does not have any directed link. This process establishes a destination-oriented directed acyclic graph (DAG) using a *Query/Update* mechanism. Consider the network topology shown in Figure 7.14. When node 1 has data packets to be sent to the destination node 7, a *Query* packet is originated by node 1 with the destination address included in it. This *Query* packet is forwarded by intermediate nodes 2, 3, 4, 5, and 6, and reaches the destination node 7, or any other node which has a route to the destination. The node that terminates (in this case, node 7) the *Query* packet replies with an *Update* packet containing its distance from the destination (it is zero at the destination node). In the example,

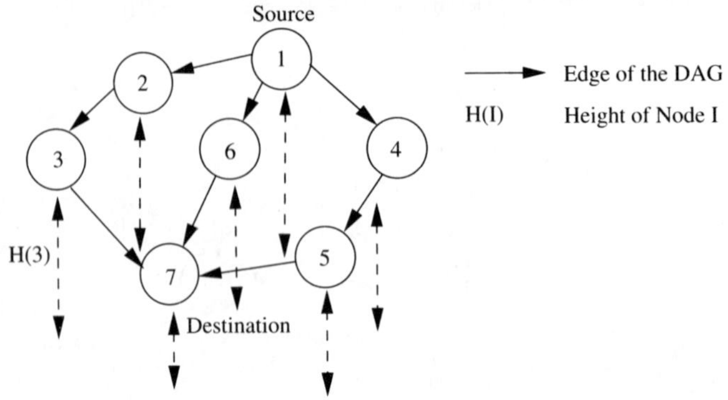

Figure 7.14. Illustration of temporal ordering in TORA.

the destination node 7 originates an *Update* packet. Each node that receives the *Update* packet sets its distance to a value higher than the distance of the sender of the *Update* packet. By doing this, a set of directed links from the node which originated the *Query* to the destination node 7 is created. This forms the DAG depicted in Figure 7.14. Once a path to the destination is obtained, it is considered to exist as long as the path is available, irrespective of the path length changes due to the reconfigurations that may take place during the course of the data transfer session.

When an intermediate node (say, node 5) discovers that the route to the destination node is invalid, as illustrated in Figure 7.15, it changes its distance value

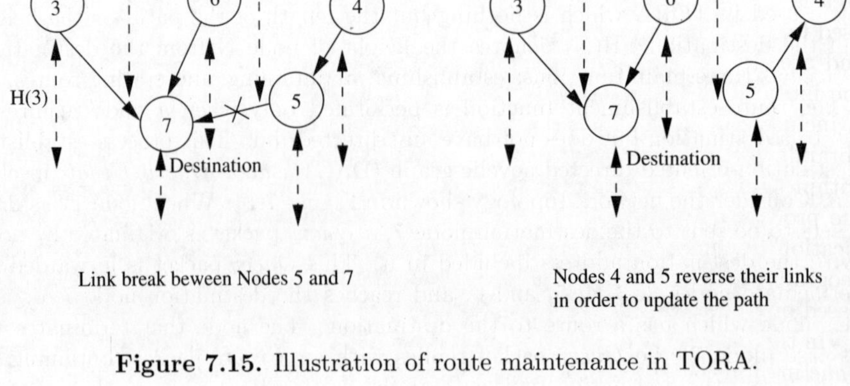

Figure 7.15. Illustration of route maintenance in TORA.

to a higher value than its neighbors and originates an *Update* packet. The neighboring node 4 that receives the *Update* packet reverses the link between 1 and 4 and forwards the *Update* packet. This is done to update the DAG corresponding to destination node 7. This results in a change in the DAG. If the source node has no other neighbor that has a path to the destination, it initiates a fresh *Query/Update* procedure. Assume that the link between nodes 1 and 4 breaks. Node 4 reverses the path between itself and node 5, and sends an update message to node 5. Since this conflicts with the earlier reversal, a partition in the network can be inferred. If the node detects a partition, it originates a *Clear* message, which erases the existing path information in that partition related to the destination.

Advantages and Disadvantages

By limiting the control packets for route reconfigurations to a small region, TORA incurs less control overhead. Concurrent detection of partitions and subsequent deletion of routes could result in temporary oscillations and transient loops. The local reconfiguration of paths results in non-optimal routes.

7.5.4 Location-Aided Routing

Location-aided routing protocol (LAR) [13] utilizes the location information for improving the efficiency of routing by reducing the control overhead. LAR assumes the availability of the global positioning system (GPS) for obtaining the geographical position information necessary for routing. LAR designates two geographical regions for selective forwarding of control packets, namely, *ExpectedZone* and *RequestZone*. The *ExpectedZone* is the region in which the destination node is expected to be present, given information regarding its location in the past and its mobility information (refer to Figure 7.16). In the event of non-availability of past information about the destination, the entire network area is considered to be the *ExpectedZone* of the destination. Similarly, with the availability of more information about its mobility, the *ExpectedZone* of the destination can be determined with more accuracy and improved efficiency. The *RequestZone* is a geographical region within which the path-finding control packets are permitted to be propagated. This area is determined by the sender of a data transfer session. The control packets used for path-finding are forwarded by nodes which are present in the *RequestZone* and are discarded by nodes outside the zone. In situations where the sender or the intermediate relay nodes are not present in the *RequestZone*, additional area is included for forwarding the packets. This is done when the first attempt for obtaining a path to a destination using the initial *RequestZone* fails to yield a path within a sufficiently long waiting time. In this case, the second attempt repeats the process with increased *RequestZone* size to account for mobility and error in location estimation. LAR uses flooding, but here flooding is restricted to a small geographical region. The nodes decide to forward or discard the control packets based on two algorithms, namely, LAR1 and LAR2.

In the LAR1 algorithm, the source node (say, S) explicitly specifies the *RequestZone* in the *RouteRequest* packet. As per LAR1, as illustrated in Figure 7.16, the

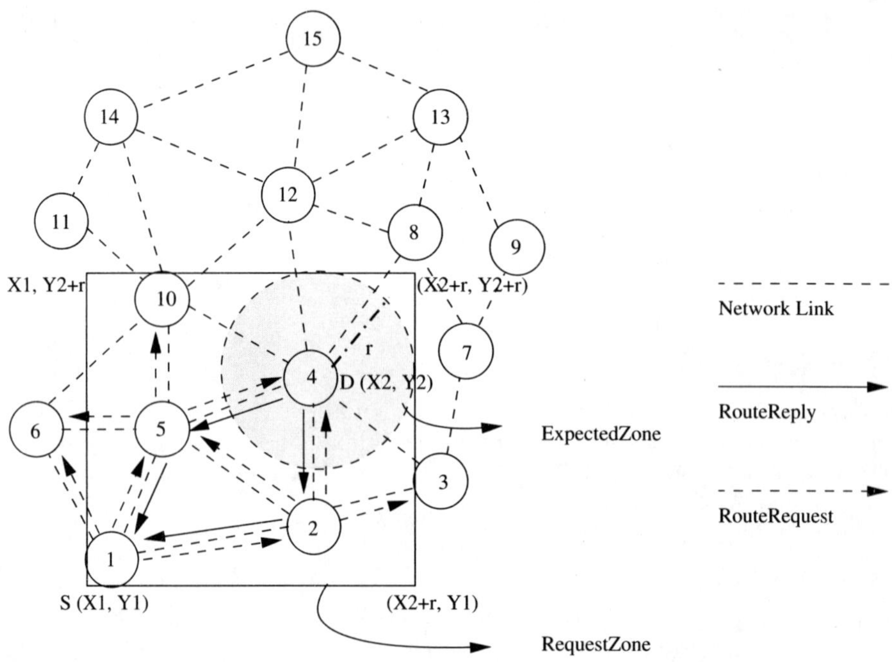

Figure 7.16. *RequestZone* and *ExpectedZone* in LAR1.

RequestZone is the smallest rectangle that includes the source node (S) and the *ExpectedZone*, the sides of which are parallel to the X and Y axes, when the node S is outside the *ExpectedZone*. When node S is within the *ExpectedZone*, then the *RequestZone* is reduced to the *ExpectedZone* itself. Every intermediate node that receives the *RouteRequest* packet verifies the *RequestZone* information contained in the packet and forwards it further if the node is within the *RequestZone*; otherwise, the packet is discarded. In Figure 7.16, the source node (node 1) originates a *RouteRequest*, which is broadcast to its neighbors (2, 5, and 6). These nodes verify their own geographical locations to check whether they belong to the *ExpectedZone*. Nodes 2 and 5 find that they are inside the *ExpectedZone* and hence they forward the *RouteRequest*. But node 6 discards the packet. Finally, when the *RouteRequest* reaches the destination node (node 4), it originates a *RouteReply* that contains the current location and current time of the node. Also, as an option, the current speed of movement can be included in the *RouteReply* if that information is available with the node. Such information included in the *RouteReply* packet is used by the source node for future route establishment procedures.

In LAR2 algorithm (Figure 7.17), the source node S (node 5) includes the distance between itself and the destination node D (node 8) along with the (X, Y) coordinates of the destination node D in the *RouteRequest* packet instead of the explicit information about the *Expected Region*. When an intermediate node re-

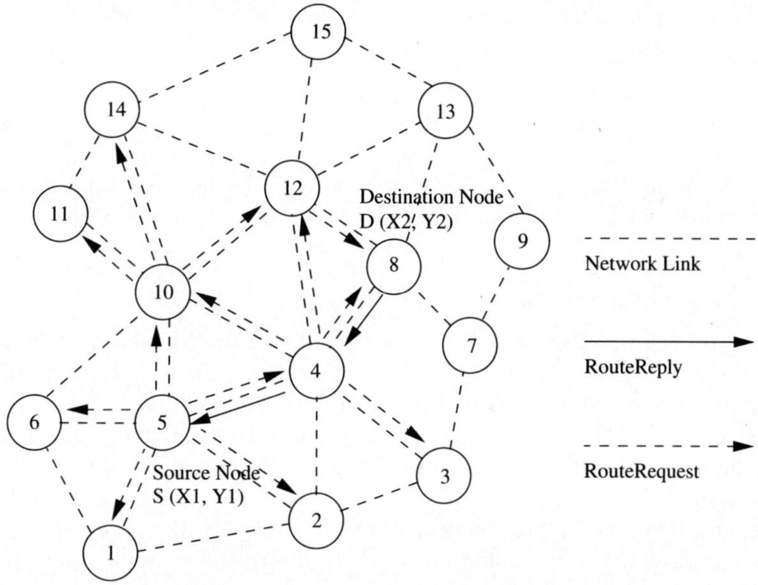

Figure 7.17. Route establishment in LAR2.

ceives this *RouteRequest* packet, it computes the distance to the node D. If this distance is less than the distance from S to node $D + \delta$, where δ is a parameter of the algorithm decided based on the error in location estimation and mobility, then the *RouteRequest* packet is forwarded. Otherwise, the *RouteRequest* is discarded. Consider the example illustrated in Figure 7.17. Assume that the value of δ is 0 here. The *RouteRequest* packet originated by node 5 is received by nodes 1, 2, 4, 6, and 10. Only nodes 4 and 10 find that the distance between them and the destination is less than the distance between the node 5 and the destination node; other nodes discard the *RouteRequest.* A *RouteRequest* packet is forwarded only once and the distance between the forwarding node and D is updated in the *RouteRequest* packet for further relaying. When node 4 forwards the *RouteRequest* packet, it updates the packet with the distance between itself and the destination node D. This packet, after being received at neighbor node 3, is discarded due to the fact that the distance between node 3 and the node 8 is greater than the distance between nodes 4 and 8. Once the *RouteRequest* reaches node 8, it originates a *RouteReply* packet back to the source node 5, containing the path through which future data packets are to be propagated. In order to compensate for the location error (due to the inaccuracy of GPS information or due to changes in the mobility of the nodes), a larger *RequestZone* that can accommodate the amount of error that occurred is considered.

Advantages and Disadvantages

LAR reduces the control overhead by limiting the search area for finding a path. The efficient use of geographical position information, reduced control overhead, and increased utilization of bandwidth are the major advantages of this protocol. The applicability of this protocol depends heavily on the availability of GPS infrastructure or similar sources of location information. Hence, this protocol cannot be used in situations where there is no access to such information.

7.5.5 Associativity-Based Routing

Associativity-based routing (ABR) [14] protocol is a distributed routing protocol that selects routes based on the stability of the wireless links. It is a *beacon*-based, on-demand routing protocol. A link is classified as stable or unstable based on its temporal stability. The temporal stability is determined by counting the periodic *beacons* that a node receives from its neighbors. Each node maintains the count of its neighbors' *beacons* and classifies each link as stable or unstable based on the *beacon* count corresponding to the neighbor node concerned. The link corresponding to a stable neighbor is termed as a stable link, while a link to an unstable neighbor is called an unstable link.

A source node floods *RouteRequest* packets throughout the network if a route is not available in its route cache. All intermediate nodes forward the *RouteRequest* packet. A *RouteRequest* packet carries the path it has traversed and the *beacon* count for the corresponding node in the path. When the first *RouteRequest* reaches the destination, the destination waits for a time period $T_{RouteSelectTime}$ to receive multiple *RouteRequests* through different paths. After this time duration, the destination selects the path that has the maximum proportion of stable links. If two paths have the same proportion of stable links, the shorter of them is selected. If more than one shortest path is available, then a random path among them is selected as the path between source and destination.

Consider Figure 7.18, in which the source node (node 1) initiates the *RouteRequest* to be flooded for finding a route to the destination node (node 15). The solid lines represent the *stable* links that are classified based on the *beacon* count, while dotted lines represent *unstable* links. ABR does not restrict any intermediate node from forwarding a *RouteRequest* packet based on the stable or unstable link criterion. It uses stability information only during the route selection process at the destination node. As depicted in Figure 7.18, the *RouteRequest* reaches the destination through three different routes. Route 1 is 1-5-10-14-15, route 2 is 1-5-4-12-15, and route 3 is 1-2-4-8-13-15. ABR selects route 3 as it contains the highest percentage of stable links compared to route 1 and route 2. ABR gives more priority to stable routes than to shorter routes. Hence, route 3 is selected even though the length of the selected route is more than that of the other two routes.

If a link break occurs at an intermediate node, the node closer to the source, which detects the break, initiates a local route repair process. In this process, the node locally broadcasts a route repair packet, termed the local query (LQ) broadcast, with a limited time to live (TTL), as illustrated in Figure 7.19 where a

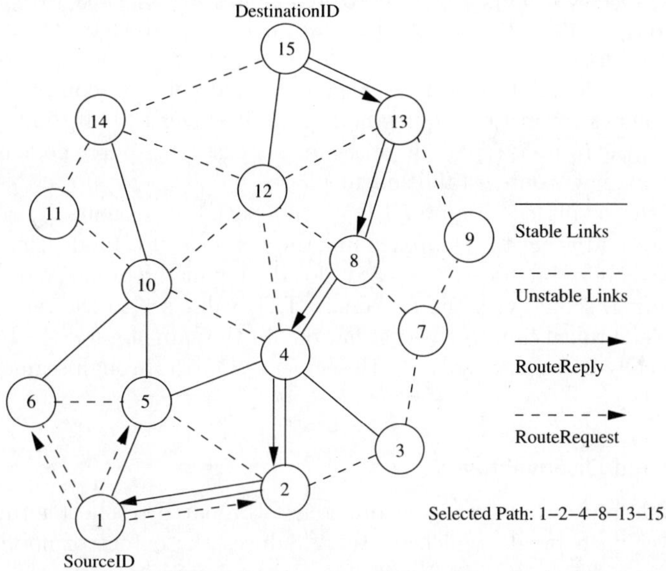

Figure 7.18. Route establishment in ABR.

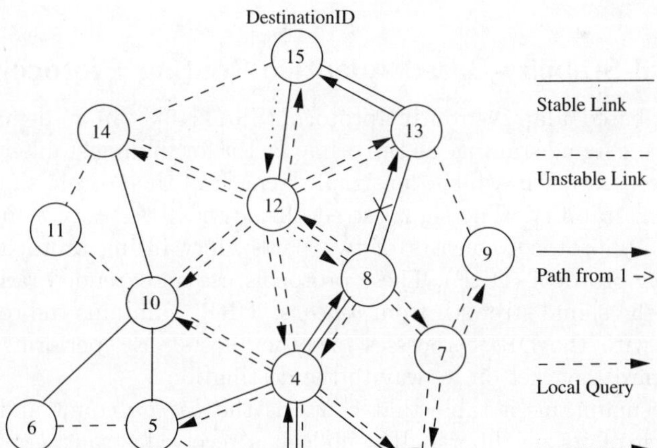

Figure 7.19. Route maintenance in ABR.

TTL value of 2 is used. This way a broken link is bypassed locally without flooding a new *RouteRequest* packet in the whole network. If a node fails to repair the broken link, then its uplink node (the previous node in the path which is closer to the source node) reinitiates the LQ broadcast. This route repair process continues along the intermediate nodes toward the source node until it traverses half the length of the broken path or the route is repaired. In the former case, the source node is informed, which initiates a new route establishment phase.

Consider the example in Figure 7.19. When a path break occurs between nodes 8 and 13, the node adjacent to the broken link toward the source node, that is, node 8, initiates a local query broadcast in order to locally repair the broken path. The local query has limited scope with the maximum TTL value set to the remaining path length from the broken link to the destination. In the same figure, the broken path is repaired locally by bypassing the path segment 8-13-15 through segment 8-12-15.

Advantages and Disadvantages

In ABR, stable routes have a higher preference compared to shorter routes. They result in fewer path breaks which, in turn, reduces the extent of flooding due to reconfiguration of paths in the network. One of the disadvantages of this protocol is that the chosen path may be longer than the shortest path between the source and destination because of the preference given to stable paths. Another disadvantage is that repetitive LQ broadcasts may result in high delays during route repairs.

7.5.6 Signal Stability-Based Adaptive Routing Protocol

Signal stability-based adaptive routing protocol (SSA) [15] is an on-demand routing protocol that uses signal stability as the prime factor for finding stable routes. This protocol is *beacon*-based, in which the signal strength of the *beacon* is measured for determining link stability. The signal strength is used to classify a link as *stable* or *unstable*. This protocol consists of two parts: forwarding protocol (FP) and dynamic routing protocol (DRP). These protocols use an extended radio interface that measures the signal strength from *beacons*. DRP maintains the routing table by interacting with the DRP processes on other hosts. FP performs the actual routing to forward a packet on its way to the destination.

Every node maintains a table that contains the *beacon* count and the signal strength of each of its neighbors. If a node has received strong *beacons* for the past few *beacons*, the node classifies the link as a *strong/stable* link. The link is otherwise classified as a *weak/unstable* link. Each node maintains a table called the signal stability table (SST), which is based on the signal strengths of its neighbors' *beacons*. This table is used by the nodes in the path to the destination to forward the incoming *RouteRequest* over strong links for finding the most stable end-to-end path. If the attempt of forwarding a *RouteRequest* over the stable links fails to obtain any path to a destination, the protocol floods the *RouteRequest* throughout the network without considering the stability of links as the forwarding criterion.

A source node which does not have a route to the destination floods the network with *RouteRequest* packets. But unlike other routing protocols, nodes that employ the SSA protocol process a *RouteRequest* only if it is received over a strong link. A *RouteRequest* received through a weak link is dropped without being processed. The destination selects the first *RouteRequest* packet received over strong links. The destination initiates a *RouteReply* packet to notify the selected route to the source.

Consider Figure 7.20, where the source node (node 1) broadcasts a *RouteRequest* for finding the route to the destination node (node 15). In Figure 7.20, solid lines represent the *stable* links, while the dotted lines represent *weak* links. Unlike ABR, SSA restricts intermediate nodes from forwarding a *RouteRequest* packet if the packet had been received over a weak link. It forwards only *RouteRequest* packets received over stable links. In Figure 7.20, when the *RouteRequest* is initiated by the source, it is to be processed by all its neighbors. But before processing, each neighbor node checks whether the *RouteRequest* packet was received through a stable link. If the *RouteRequest* had been received through a stable link and had not been sent already (*i.e.,* it is not a duplicate *RouteRequest*), it is forwarded by the node; otherwise, it is dropped. For example, when the *RouteRequest* from node 1 reaches nodes 2, 5, and 6, it is forwarded only by nodes 2 and 5 as the link between nodes 1 and 6 is weak. Similarly, the *RouteRequest* forwarded by node 2 is rejected by nodes 3 and 5, while node 4 forwards it to its neighbors, provided it

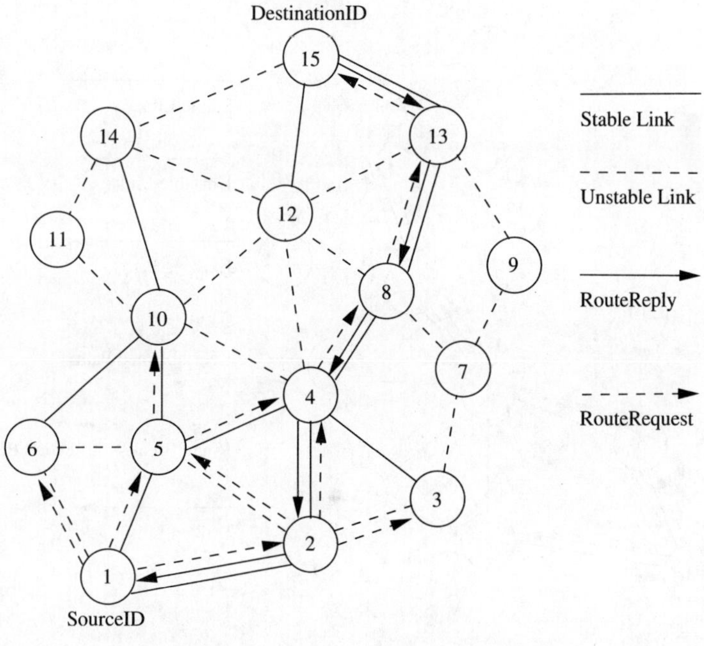

Figure 7.20. Route establishment in SSA.

has not already been forwarded. In this way, nodes forward the *RouteRequest* until it reaches the destination. In this example, only one route is established based on the strong link criterion. A *RouteRequest* propagated through path 1-5-10-14-15 is rejected by the destination as it receives a *RouteRequest* from node 14 on a weak link. The stable path consisting of strong links is 1- 2 (or 5)-4-8-13-15. The first *RouteRequest* packet that reaches the destination over a stable path is selected by the destination.

As shown in Figure 7.21, when a link breaks, the end nodes of the broken link (*i.e.*, nodes 2 and 4) notify the corresponding end nodes of the path (*i.e.*, nodes 1 and 15). A source node, after receiving a route break notification packet, rebroadcasts the *RouteRequest* to find another stable path to the destination. Stale entries are removed only if data packets that use the stale route information fail to reach the next node. If the link between nodes 2 and 4 breaks, a new strong path is established through 1-5-4-8-13-15. If no strong path is available when a link gets broken (*e.g.*, link 8-13), then the new route is established by considering weak links also. This is done when multiple *RouteRequest* attempts fail to obtain a path to the destination using only the stable links.

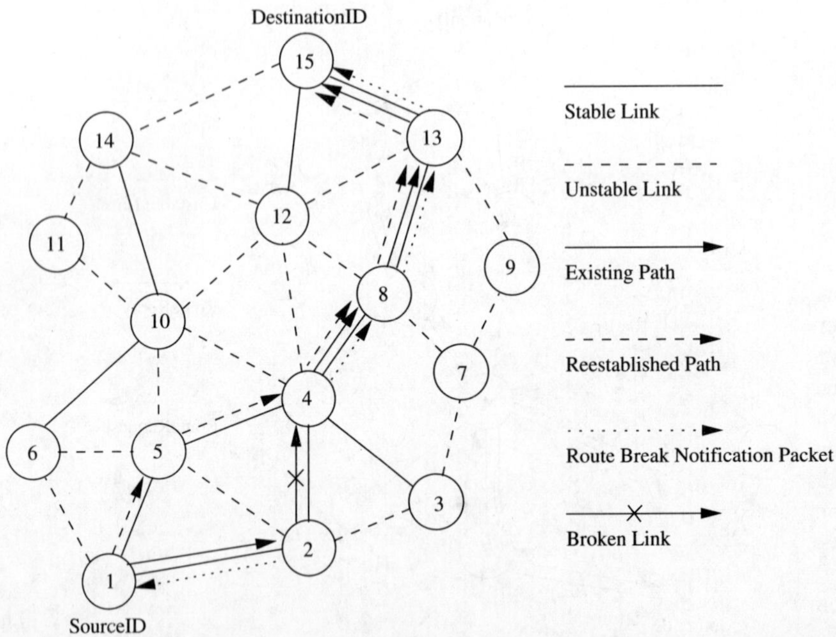

Figure 7.21. Route maintenance in SSA.

Advantages and Disadvantages

The main advantage of this protocol is that it finds more stable routes when compared to the shortest path route selection protocols such as DSR and AODV. This protocol accommodates temporal stability by using *beacon* counts to classify a link as stable or weak. The main disadvantage of this protocol is that it puts a strong *RouteRequest* forwarding condition which results in *RouteRequest* failures. A failed *RouteRequest* attempt initiates a similar path-finding process for a new path without considering the stability criterion. Such multiple flooding of *RouteRequest* packets consumes a significant amount of bandwidth in the already bandwidth-constrained network, and also increases the path setup time. Another disadvantage is that the strong links criterion increases the path length, as shorter paths may be ignored for more stable paths.

7.5.7 Flow-Oriented Routing Protocol

Flow-oriented routing protocol (FORP) [16] is an on-demand routing protocol that employs a prediction-based *multi-hop-handoff* mechanism for supporting time-sensitive traffic in ad hoc wireless networks. This protocol has been proposed for IPv6-based ad hoc wireless networks where quality of service (QoS) needs to be provided. The *multi-hop-handoff* is aimed at alleviating the effects of path breaks on the real-time packet flows. A sender or an intermediate node initiates the route maintenance process only after detecting a link break. This reactive route maintenance procedure may result in high packet loss, leading to a low quality of service provided to the user. FORP uses a unique prediction-based mechanism that utilizes the mobility and location information of nodes to estimate the link expiration time (LET). LET is the approximate lifetime of a given wireless link. The minimum of the LET values of all wireless links on a path is termed as the route expiry time (RET). Every node is assumed to be able to predict the LET of each of its links with its neighbors. The LET between two nodes can be estimated using information such as current position of the nodes, their direction of movement, and their transmission ranges. FORP requires the availability of GPS information in order to identify the location of nodes.

When a sender node needs to set up a real-time flow to a particular destination, it checks its routing table for the availability of a route to that destination. If a route is available, then that is used to send packets to the destination. Otherwise, the sender broadcasts a *Flow-REQ* packet carrying information regarding the source and the destination nodes. The *Flow-REQ* packet also carries a flow identification number/sequence number which is unique for every session. A neighbor node, on receiving this packet, first checks if the sequence number of the received *Flow-REQ* is higher than the sequence number corresponding to a packet belonging to the same session that had been previously forwarded by the node. If so, then it updates its address on the packet and extracts the necessary state information out of the packet. If the sequence number on the packet is less than that of the previously forwarded packet, then the packet is discarded. This is done to avoid looping of *Flow-REQ* packets. A *Flow-REQ* with the same sequence number as that of a *Flow-*

REQ belonging to the same session which had been forwarded already by the node, would be broadcast further only if it has arrived through a shorter (and therefore better) path. Before forwarding a *Flow-REQ*, the intermediate node appends its node address and the LET of the last link the packet had traversed onto the packet. The *Flow-REQ* packet, when received at the destination node, contains the list of nodes on the path it had traversed, along with the LET values of every wireless link on that path. FORP assumes all the nodes in the network to be synchronized to a common time by means of GPS information. If the calculated value of RET, corresponding to the new *Flow-REQ* packet arrived at the destination, is better than the RET value of the path currently being used, then the destination originates a *Flow-SETUP* packet. The LET of a link can be estimated given the information about the location, velocity, and transmission range of the nodes concerned. The LET of the wireless link between two nodes a and b with transmission range T_x, which are moving at velocity V_a and V_b at angles T_a and T_b, respectively (refer to Figure 7.22 (a)), can be estimated as described below:

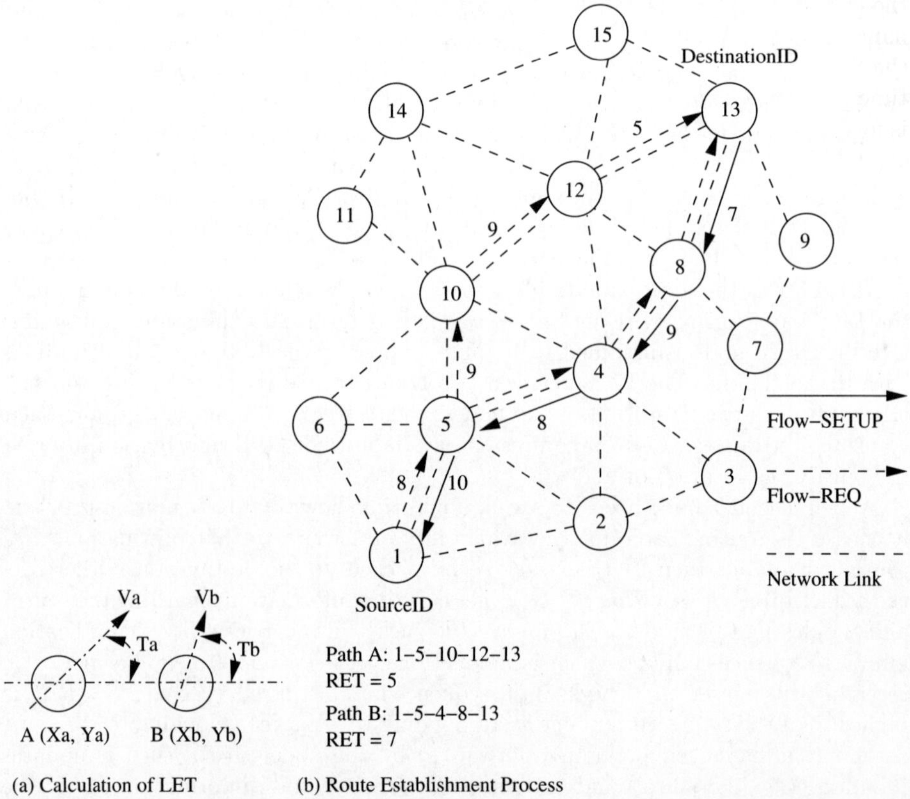

Figure 7.22. Route establishment in FORP.

$$LET_{ab} = \frac{-(pq + rs) + (p^2 + r^2)T_x{}^2 - (ps - qr)^2}{p^2 + q^2} \tag{7.5.1}$$

where

$$p = V_a \cos T_a - V_b \cos T_b \tag{7.5.2}$$

$$q = X_a - X_b \tag{7.5.3}$$

$$r = V_a \sin T_a - V_b \sin T_b \tag{7.5.4}$$

$$s = Y_a - Y_b \tag{7.5.5}$$

The route establishment procedure is shown in Figure 7.22 (b). In this case, the path 1-5-4-8-13 (path 1) has a RET value of 7, whereas the path 1-5-10-12-13 (path 2) has a RET value of 5. This indicates that path 1 may last longer than path 2. Hence the sender node originates a *Flow-SETUP* through the reverse path 13-8-4-5-1.

FORP employs a proactive route maintenance mechanism which makes use of the expected RET of the current path available at the destination. Route maintenance is illustrated in Figure 7.23. When the destination node determines (using the RET of the current path) that a route break is about to occur within a critical time period (t_c), it originates a *Flow-HANDOFF* packet to the source node, which is forwarded by the intermediate nodes. The mechanism by which *Flow-HANDOFF*

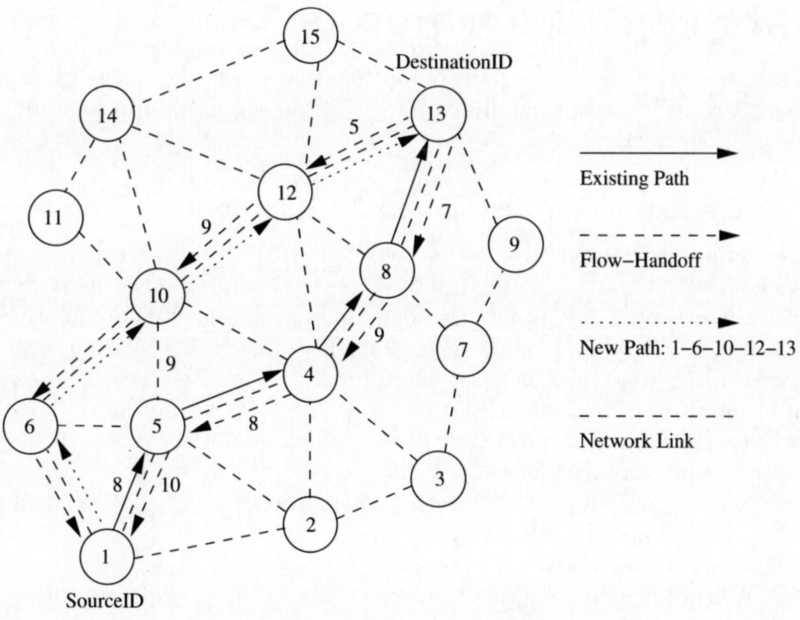

Figure 7.23. Route maintenance in FORP.

packets are forwarded is similar to the *Flow-REQ* forwarding mechanism. When many *Flow-HANDOFF* packets arrive at the source node, the source node calculates the RET values of paths taken by each of them, selects the best path, and uses this new path for sending packets to the destination. In the example shown in Figure 7.23, the *Flow-HANDOFF* packets are forwarded by every intermediate node after appending the LET information of the previous link traversed onto the packet. The existing path 1-5-4-8-13 is erased and a new path is selected by the source node based on the RETs corresponding to different paths traversed by the *Flow-HANDOFF* packets. In this case, the path 1-6-10-12-13 is chosen. The critical time (t_c) is taken as the difference between the RET and delay encountered by the latest packet which has traversed through the existing path from the source to the destination.

Advantages and Disadvantages

The use of LET and RET estimates reduces path breaks and their associated ill effects such as reduction in packet delivery, increase in the number of out-of-order packets, and non-optimal paths resulting from local reconfiguration attempts. The proactive route reconfiguration mechanism adopted here works well when the topology is highly dynamic. The requirements of time synchronization increases the control overhead. Dependency on the GPS infrastructure affects the operability of this protocol in environments where such infrastructure may not be available.

7.6 HYBRID ROUTING PROTOCOLS

In this section, we discuss the working of routing protocols termed as hybrid routing protocols. Here, each node maintains the network topology information up to m hops. The different existing hybrid protocols are presented below.

7.6.1 Core Extraction Distributed Ad Hoc Routing Protocol

Core extraction distributed ad hoc routing (CEDAR) [17] integrates routing and support for QoS. It is based on extracting core nodes (also called as dominator nodes) in the network, which together approximate the minimum dominating set. A dominating set (DS) of a graph is defined as a set of nodes in the graph such that every node in the graph is either present in the DS or is a neighbor of some node present in the DS. There exists at least one core node within three hops. The DS of the least cardinality in a graph is called the minimum dominating set. Nodes that choose a core node as their dominating node are called core member nodes of the core node concerned. The path between two core nodes is termed a virtual link. CEDAR employs a distributed algorithm to select core nodes. The selection of core nodes represents the core extraction phase.

CEDAR uses the core broadcast mechanism to transmit any packet throughout the network in the unicast mode, involving as minimum number of nodes as possible. These nodes that take part in the core broadcast process are called core nodes. In order to carry out a core broadcast efficiently, each core node must know about its neighboring core nodes. The transfer of information about neighboring core nodes results in significant control overhead at high mobility. When a core node to which many nodes are attached moves away from them, each node has to reselect a new core node. The selection of core nodes, which is similar to the distributed leader election process, involves substantial control overhead.

Each core node maintains its neighborhood local topology information. CEDAR employs an efficient link state propagation mechanism in which information regarding the presence of high bandwidth and stable links is propagated to several more nodes, compared to the propagation of information regarding low bandwidth and unstable links, which is suppressed and kept local. To propagate link information, slow-moving increase-waves and fast-moving decrease-waves are used. An increase-wave carrying update information is originated when the bandwidth on the link concerned increases above a certain threshold. The fast-moving decrease-waves are propagated in order to quickly notify the nearby nodes (core nodes which are at most separated by three hops) about the reduction in available bandwidth. As bandwidth increase information moves slowly, only stable high-bandwidth link state information traverses long distances. If the high-bandwidth link is unstable, then the corresponding increase-wave is overtaken by fast-moving decrease-waves which represent the decrease in available bandwidth on the corresponding link. These waves are very adaptive to the dynamic topology of ad hoc wireless networks. Increase- and decrease-waves are initiated only when changes in link capacity cross certain thresholds, that is, only when there is a significant change in link capacity. Fast-moving decrease-waves are prevented from moving across the entire network, thereby suppressing low bandwidth unstable information to the local nodes only. Route establishment in CEDAR is carried out in two phases. The first phase finds a core path from the source to the destination. The core path is defined as a path from the dominator of the source node (source core) to the dominator of the destination node (destination core). In the second phase, a QoS feasible path is found over the core path. A node initiates a *RouteRequest* if the destination is not in the local topology table of its core node; otherwise, the path is immediately established. For establishing a route, the source core initiates a core broadcast in which the *RouteRequest* is sent to all neighboring core nodes as unicast data. Each of these recipient core nodes in turn forwards the *RouteRequest* to its neighboring core nodes if the destination is not its core member. A core node which has the destination node as its core member replies to the source core. Once the core path is established, a path with the required QoS support is then chosen.

To find a path that can provide the required QoS, the source core first finds a path to the domain of the farthest possible core node in the core path, which can provide the bandwidth required. Among the available paths to this domain, the source core chooses the shortest-widest path (shortest path with highest bandwidth). Assume *MidCore* is the farthest possible core node found by the source core. In the next iteration, *MidCore* becomes the new source core and finds another *MidCore* node that satisfies the QoS support requirements. This iterative process repeats until a path to the destination with the required bandwidth is found. If no feasible path is available, the source node is informed about the non-availability of a QoS path.

Consider Figure 7.24 where the source is node 1 and the destination is node 15. The core nodes in the network are nodes 3, 5, 11, 12, and 13. In this figure, node 5 is the dominator of nodes 1 and 6. Similarly, node 12 is the dominator of node 15. When node 1 initiates a *RouteRequest* to be flooded throughout the network, it intimates its core node the <source id, destination id> pair information. If the core node 5 does not have any information about the dominator of node 15, which is the destination node, it initiates a core broadcast. Due to this, all nearby core nodes receive the request in the unicast transmission mode. This unicast transmission is done on the virtual links. For core node 5, the virtual link with core node 3 comprises of the links 5-2 and 2-3, while the virtual link between core nodes 5 and 13 is represented by path 5-4-8-13. When a core node receives the core broadcast message, it checks whether the destination is its core member. A core node having the destination as one of its core members replies to the source core node. In our case, core node 12 replies to core node 5. The path between core nodes 12 and 5 constitutes a core path. Once a core path is established, the feasibility of the path in terms of the availability of the required bandwidth is checked. If the required

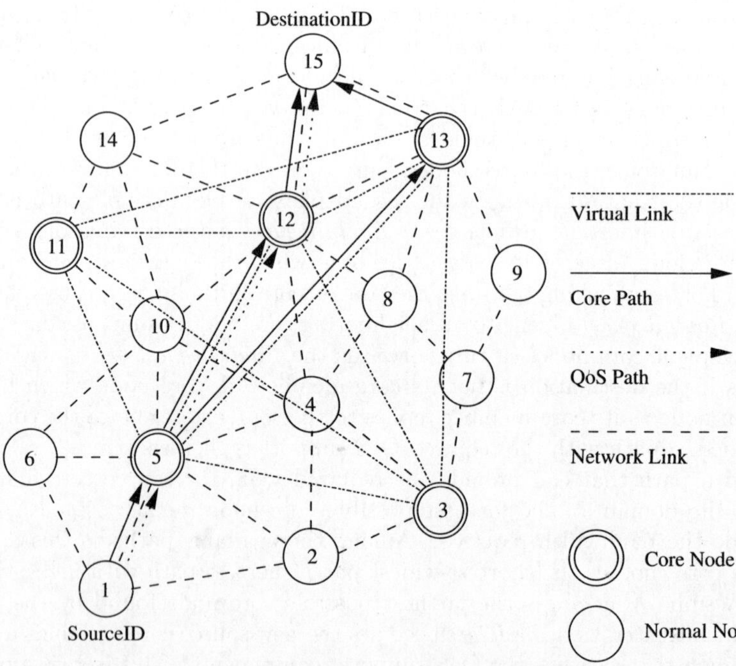

Figure 7.24. Route establishment in CEDAR.

bandwidth is available on the path, the connection is established; otherwise, the core path is rejected.

CEDAR attempts to repair a broken route locally when a path break occurs. When a link u-v on the path from source to the destination fails, node u sends back a notification to the source and starts recomputation of a route from itself to the destination node. Until the route recomputation gets completed, node u drops every subsequent packet it receives. Once the source node receives the notification sent by node u, it immediately stops transmitting packets belonging to the corresponding flow, and starts computing a new route to the destination. If the link break occurs near the source, route recomputation at node u may take a long time, but the notification sent by node u reaches the source node very rapidly and prevents large packet losses. If the broken link is very close to the destination, the notification sent by node u might take a longer time to reach the source, but the route recomputation time at node u is small and hence large packet losses are again prevented. If the link break occurs somewhere near the middle of the path, then both the local route recomputation mechanism and the route break notification mechanism are not fast enough, and hence a considerable amount of packet loss occurs in this case.

Consider the network topology shown in Figure 7.25. When the link between nodes 12 and 15 breaks, node 12 tries to reconnect to the destination using an alternate path that satisfies the bandwidth requirement. It also notifies the source

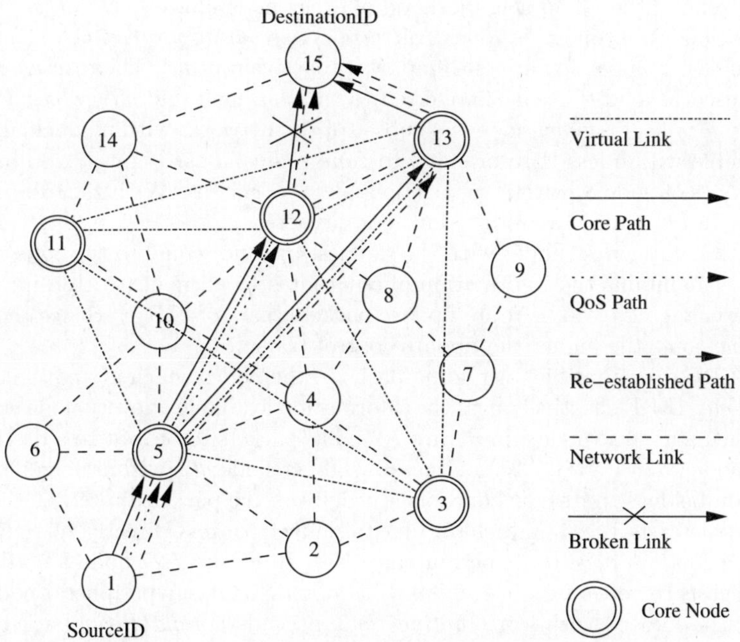

Figure 7.25. Route maintenance in CEDAR.

node about the link break. The source node tries to reconnect to the destination by reinitiating the route establishment process. In case node 12 does not have any other feasible path, then the alternate path 1-5-4-8-13-15 found by the source node is used for the further routing of packets.

Advantages and Disadvantages

The main advantage of CEDAR is that it performs both routing and QoS path computation very efficiently with the help of core nodes. The increase- and decrease-waves help in appropriate propagation of the stable high-bandwidth link information and the unstable low-bandwidth link information, respectively. Core broadcasts provide a reliable mechanism for establishing paths with QoS support. A disadvantage of this protocol is that since route computation is carried out at the core nodes only, the movement of the core nodes adversely affects the performance of the protocol. Also, the core node update information could cause a significant amount of control overhead.

7.6.2 Zone Routing Protocol

Zone routing protocol (ZRP) [18] is a hybrid routing protocol which effectively combines the best features of both proactive and reactive routing protocols. The key concept employed in this protocol is to use a proactive routing scheme within a limited zone in the r-hop neighborhood of every node, and use a reactive routing scheme for nodes beyond this zone. An *intra-zone routing protocol* (IARP) is used in the zone where a particular node employs proactive routing. The reactive routing protocol used beyond this zone is referred to as *inter-zone routing protocol* (IERP). The *routing zone* of a given node is a subset of the network, within which all nodes are reachable within less than or equal to *zone radius* hops. Figure 7.26 illustrates *routing zones* of node 8, with $r = 1$ hop and $r = 2$ hops. With *zone radius* $= 2$, the nodes 7, 4, 12, and 13 are interior nodes, whereas nodes 2, 3, 5, 9, 10, 13, and 15 are peripheral nodes (nodes with the shortest distance equal to the *zone radius*). Each node maintains the information about routes to all nodes within its *routing zone* by exchanging periodic route update packets (part of IARP). Hence the larger the *routing zone*, the higher the update control traffic.

The IERP is responsible for finding paths to the nodes which are not within the *routing zone*. IERP effectively uses the information available at every node's *routing zone*. When a node s (node 8 in Figure 7.27) has packets to be sent to a destination node d (node 15 in Figure 7.27), it checks whether node d is within its zone. If the destination belongs to its own zone, then it delivers the packet directly. Otherwise, node s bordercasts (uses unicast routing to deliver packets directly to the border nodes) the *RouteRequest* to its peripheral nodes. In Figure 7.27 node 8 bordercasts *RouteRequests* to nodes 2, 3, 5, 7, 9, 10, 13, 14, and 15. If any peripheral node finds node d to be located within its *routing zone*, it sends a *RouteReply* back to node s indicating the path; otherwise, the node rebordercasts the *RouteRequest* packet to the peripheral nodes. This process continues until node d is located. Nodes 10 and 14 find the information about node 16 to be available in their intra-zone

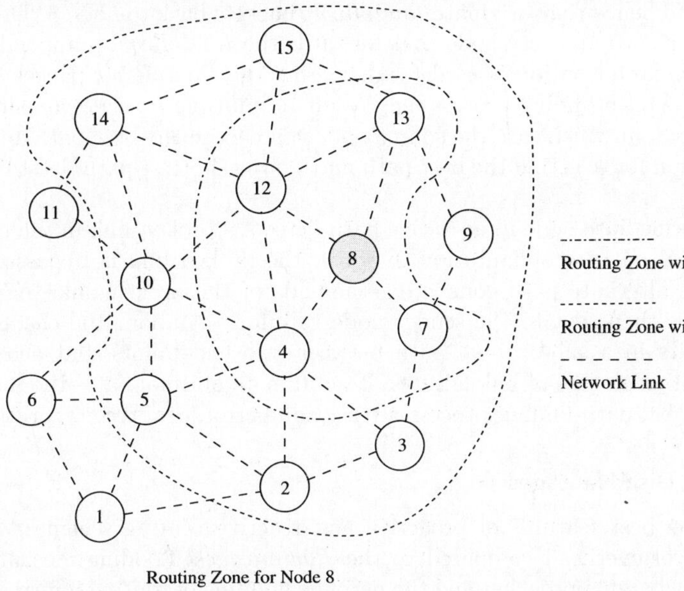

Figure 7.26. Routing zone for node 8 in ZRP.

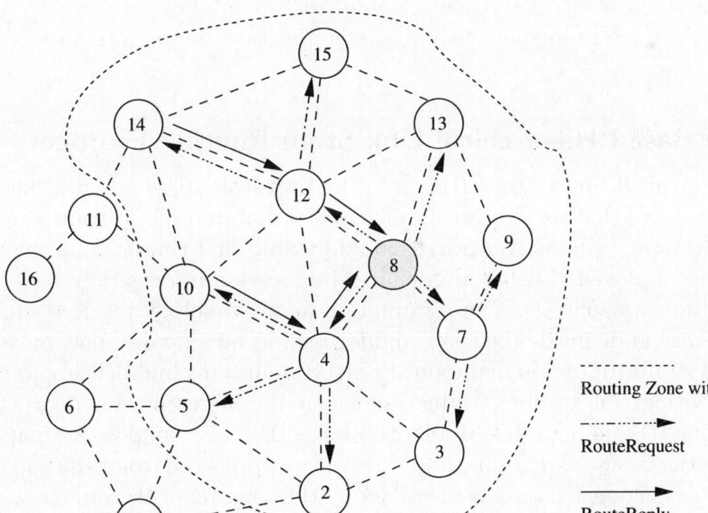

Figure 7.27. Path finding between node 8 and node 16.

routing tables, and hence they originate *RouteReply* packets back to node 8. During *RouteRequest* propagation, every node that forwards the *RouteRequest* appends its address to it. This information is used for delivering the *RouteReply* packet back to the source. The path-finding process may result in multiple *RouteReply* packets reaching the source, in which case the source node can choose the best path among them. The criterion for selecting the best path may be the shortest path, least delay path, etc.

When an intermediate node in an active path detects a broken link in the path, it performs a local path reconfiguration in which the broken link is bypassed by means of a short alternate path connecting the ends of the broken link. A path update message is then sent to the sender node to inform it about the change in path. This results in a sub-optimal path between two end points, but achieves quick reconfiguration in case of link failures. To obtain an optimal path, the sender reinitiates the global path-finding process after a number of local reconfigurations.

Advantages and Disadvantages

By combining the best features of proactive and reactive routing schemes, ZRP reduces the control overhead compared to the *RouteRequest* flooding mechanism employed in on-demand approaches and the periodic flooding of routing information packets in table-driven approaches. But in the absence of a query control, ZRP tends to produce higher control overhead than the aforementioned schemes. This can happen due to the large overlapping of nodes' *routing zones*. The query control must ensure that redundant or duplicate *RouteRequests* are not forwarded. Also, the decision on the zone radius has a significant impact on the performance of the protocol.

7.6.3 Zone-Based Hierarchical Link State Routing Protocol

Zone-based hierarchical link state (ZHLS) routing protocol [19] is a hybrid hierarchical routing protocol that uses the geographical location information of the nodes to form non-overlapping zones. A hierarchical addressing that consists of a zone ID and a node ID is employed. Each node requires its location information, based on which it can obtain its zone ID. The information about topology inside a zone is maintained at every node inside the zone, and for regions outside the zone, only the zone connectivity information is maintained. ZHLS maintains high-level hierarchy for inter-zone routing. Packet forwarding is aided by the hierarchical address comprising of the zone ID and node ID. Similar to ZRP, ZHLS also employs a proactive approach inside the geographical zone and a reactive approach beyond the zone. A destination node's current location is identified by the zone ID of the zone in which it is present and is obtained by a route search mechanism.

In ZHLS, every node requires GPS or similar infrastructure support for obtaining its own geographical location that is used to map itself onto the corresponding zone. The assignment of zone addresses to geographical areas is important and is done during a phase called the network design phase or network deployment phase. The area of the zone is determined by several factors such as the coverage of a single node, application scenario, mobility of the nodes, and size of the network. For example, the ad hoc network formed by a set of hand-held devices with a limited mobility may require a zone radius of a few hundred meters, whereas the zone area required in the network formed by a set of ships, airplanes, or military tanks may be much larger.

The intra-zone routing table is updated by executing the shortest path algorithm on the node-level topology of the zone. The node-level topology is obtained by using the *intra-zone clustering* mechanism, which is similar to the link state updates limited to the nodes present in the zone. Each node builds a one-hop topology by means of a link request and link response mechanism. Once the one-hop topology is available, each node prepares link state packets and propagates them to all nodes in the zone. These update packets contain the node IDs of all nodes that belong to the zone, and node IDs and zone IDs of all nodes that belong to other zones. The nodes that receive link responses from nodes belonging to other zones are called *gateway* nodes. Data traffic between two zones will be relayed through these gateway nodes. For example, nodes 5, 8, 10, and 12 are the gateway nodes for zone A in Figure 7.28 (a). Every node in a zone is aware of the neighboring zones connected to its zone and the gateway nodes to be used for reaching them. Once the node-level link state packets are exchanged and the node-level topology is updated, every node in a zone generates a zone link state packet. The zone link state packet contains information about the zone-level connectivity. These zone link state packets are propagated

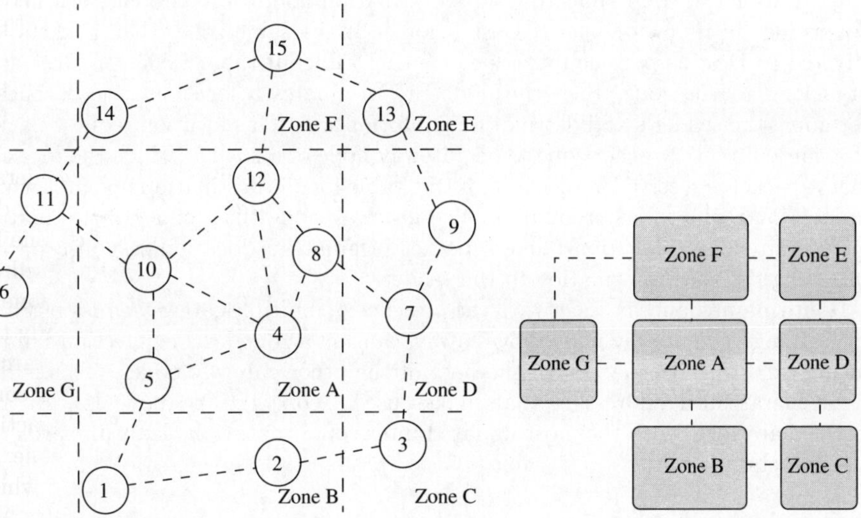

(a) Node–level topology in ZHLS

(b) Zone topology for the node–level topology in (a)

Figure 7.28. Zone-based hierarchical link state routing protocol.

Table 7.1. Zone link state packets

Source Zone	Zone Link State Packet
A	B, D, F, and G
B	C and A
C	B and D
D	A, C, and E
E	A, D, and F
F	A, E, and G
G	A and F

throughout the network by the gateway nodes. The zone-level topology is shown in Figure 7.28 (b). The zone link state packets originated by every zone are shown in Table 7.1. Using the information obtained from zone link state packets, a node can build the zone topology. The zone routing table can be formed for any destination zone by executing the shortest path algorithm on the zone-level topology. The zone link state packets are source sequence numbered and a time-stamp field is included to avoid stale link state packets. The association of the nodes to the respective zones helps in reducing routing overhead as in ZRP, but it includes the additional requirement of determining a given destination node's present location. If a source node Src wants to communicate with a destination node Dest, Src checks whether Dest resides in its own zone. If Dest belongs to the same zone, then packets are delivered to Dest as per the intra-zone routing table. If the destination Dest does not belong to the zone, then the node Src originates a location request packet containing the sender's and destination's information. This location request packet is forwarded to every other zone. The gateway node of a zone at which the location request packet is received verifies its routing table for the destination node for which the location request was originated. The gateway node that finds the destination node required by a location request packet originates a location response packet containing the zone information to the sender.

Route maintenance is easier with the presence of multiple gateway nodes between zones. If a given gateway node moves away, causing a zone-level connection failure, routing can still take place with the help of the other gateway nodes. This is due to the hierarchical addressing that makes use of zone ID and node ID. At any intermediate zone, with the most updated inter-zonal routing table, it forwards the data packets.

Advantages and Disadvantages

The hierarchical approach used in this protocol significantly reduces the storage requirements and the communication overhead created because of mobility. The zone-level topology is robust and resilient to path breaks due to mobility of nodes. Intra-zonal topology changes do not generate network-wide control packet transmissions. A main disadvantage of this protocol is the additional overhead incurred in the creation of the zone-level topology. Also the path to the destination is suboptimal. The geographical information required for the creation of the zone level topology may not be available in all environments.

7.7 ROUTING PROTOCOLS WITH EFFICIENT FLOODING MECHANISMS

Many of the existing on-demand routing protocols employ flooding of *RouteRequest* packets in order to obtain a feasible path with the required packet-forwarding constraints. Flooding of control packets results in a significant amount of redundancy, wastage of bandwidth, increase in number of collisions, and broadcast storms[1] at times of frequent topological changes. Existing routing protocols that employ efficient flooding mechanisms to counter the requirement of flooding include preferred link-based routing (PLBR) protocols [20] and optimized link state routing (OLSR) protocol [21]. The former belongs to the on-demand routing protocols category and the latter belongs to the table-driven routing protocols category. These protocols utilize algorithms that require a minimum number of transmissions in order to flood the entire network.

7.7.1 Preferred Link-Based Routing Protocols

SSA [15] uses the preferred link approach in an implicit manner by processing a *RouteRequest* packet only if it is received through a strong link. Wired networks also employ preferred links mechanisms [22], but restrict themselves by selecting a single preferred link, based on heuristics that satisfy multiple constraints, for example, minimum cost and least delay required by the route. In ad hoc networks, the single preferred link model is not suitable due to reasons such as lack of topology information, continuously changing link characteristics, broadcast nature of the radio channel, and mobility of nodes.

Sisodia *et al.* proposed two algorithms in [20] known as preferred link-based routing (PLBR) protocols that employ a different approach in which a node selects a subset of nodes from its neighbors list (NL). This subset is referred to as the *preferred list* (PL). Selection of this subset may be based on link or node characteristics. Every *RouteRequest* packet carries the list of a selected subset of neighbors. All neighbors receive *RouteRequest* packets because of the broadcast radio channel, but only neighbors present in the PL forward them further. The packet is forwarded by K neighbors, where K is the maximum number of neighbors allowed in the PL. PLBR aims to minimize control overhead in the ad hoc wireless network. All nodes operate in the promiscuous mode. Each node maintains information about its neighbors and their neighbors in a table called neighbor's neighbor table (NNT). It periodically transmits a *beacon* containing the changed neighbor's

[1] Broadcast storm refers to the presence/origination of a large number of broadcast control packets for routing due to the high topological instability occurring in the network as a result of mobility.

information. PLBR has three main phases: route establishment, route selection, and route maintenance.

The route establishment phase starts when a source node (Src) receives packets from the application layer, meant for a destination node (Dest) for which no route is currently available. If Dest is in Src's NNT, the route is established directly. Otherwise, Src transmits a *RouteRequest* packet containing the source node's address (SrcID), destination node's address (DestID), a unique sequence number (SeqNum) (which prevents formation of loops and forwarding of multiple copies of the same *RouteRequest* packet received from different neighbors), a traversed path (*TP*) (containing the list of nodes through which the packet has traversed so far and the weight assigned to the associated links), and a PL. It also contains a time to live (TTL) field that is used to avoid packets being present in the network forever, and a $NoDelay$ flag, the use of which will be described later in this section. Before forwarding a *RouteRequest* packet, each eligible node recomputes the preferred list table (PLT) that contains the list of neighbors in the order of preference. The node inserts the first K entries of the PLT onto the PL field of the *RouteRequest* packet (K is a global parameter that indicates the maximum size of PL). The old PL of a received packet is replaced every time with a new PL by the forwarding node. A node is eligible for forwarding a *RouteRequest* only if it satisfies all the following criteria: the node ID must be present in the received *RouteRequest* packet's PL, the *RouteRequest* packet must not have been already forwarded by the node, and the TTL on the packet must be greater than zero. If Dest is in the eligible node's NNT, the *RouteRequest* is forwarded as a unicast packet to the neighbor, which might either be Dest or whose NL contains Dest. If there are multiple neighbors whose NL have Dest, the *RouteRequest* is forwarded to only one randomly selected neighbor. Otherwise, the packet is broadcast with a new PL computed from the node's NNT. PLT is computed by means of one of the two algorithms discussed later in this section. If the computed PLT is empty, that is, there are no eligible neighbors, the *RouteRequest* packet is discarded and marked as *sent*. If the *RouteRequest* packet reaches the destination, the route is selected by the route selection procedure given below.

When multiple *RouteRequest* packets reach Dest, the route selection procedure selects the best route among them. The criterion for selecting the best route can be the shortest path, or the least delay path, or the most stable path. Dest starts a timer after receiving the first *RouteRequest* packet. The timer expires after a certain *RouteSelectWait* period, after which no more *RouteRequest* packets would be accepted. From the received *RouteRequest* packets, a route is selected as follows.

For every *RouteRequest* i that reached Dest during the *RouteSelectWait* period, Max(W^i_{min}) is selected, where W^i_{min} is the minimum weight of a link in the path followed by i. If two or more paths have the same value for $Max(W^i_{min})$, the shortest path is selected. After selecting a route, all subsequent *RouteRequest* packets from the same Src with a SeqNum less than or equal to the SeqNum of the selected *RouteRequest* are discarded. If the $NoDelay$ flag is set, the route selection procedure is omitted and TP of the first *RouteRequest* reaching the Dest is selected as the route. The $NoDelay$ flag can be set if a fast connection setup is needed.

Mobility of nodes causes frequent path breaks that should be locally repaired to reduce broadcast of the *RouteRequest*. The local route repair broadcast mechanism used in ABR [14] has a high failure rate due to the use of restricted TTL, which increases the average delay in route reestablishment. PLBR uses a quick route repair mechanism which bypasses the down link (Dest side) node from the broken path, using information about the next two hops from NNT.

Algorithms for Preferred Links Computation

Two different algorithms have been proposed by Sisodia *et al.* in [20], for finding preferred links. The first algorithm selects the route based on degree of neighbor nodes (degree of a node is the number of neighbors). Preference is given to nodes whose neighbor degree is higher. As higher degree neighbors cover more nodes, only a few of them are required to cover all the nodes of the NNT. This reduces the number of broadcasts. The second algorithm gives preference to stable links. Links are not explicitly classified as stable or unstable. The notion of stability is based on the weight given to links.

Neighbor Degree-Based Preferred Link Algorithm (NDPL)

Let d be the node that calculates the preferred list table PLT. TP is the traversed path and OLD_{PL} is the preferred list of the received *RouteRequest* packet. The NNT of node d is denoted by NNT_d. $N(i)$ denotes the neighbors of node i and itself. Include list (INL) is a set containing all neighbors reachable by transmitting the *RouteRequest* packet after execution of the algorithm, and the exclude list (EXL) is a set containing all neighbors that are unreachable by transmitting the *RouteRequest* packet after execution of the algorithm.

1. In this step, node d marks the nodes that are not eligible for further forwarding the *RouteRequest* packet.

    (a) If a node i of TP is a neighbor of node d, mark all neighbors of i as reachable, that is, add $N(i)$ to INL.

    (b) If a node i of OLD_{PL} is a neighbor of node d, and $i < d$, mark all neighbors of node i as reachable, that is, include $N(i)$ in INL.

    (c) If neighbor i of node d has a neighbor n present in TP, mark all neighbors of i as reachable by adding $N(i)$ to INL.

    (d) If neighbor i of node d has a neighbor n present in OLD_{PL}, and $n < d$, here again add $N(i)$ to INL, thereby marking all neighbors of node i as reachable.

2. If neighbor i of node d is not in INL, put i in preferred list table PLT and mark all neighbors of i as reachable. If i is present in INL, mark the neighbors of i as unreachable by adding them to EXL, as $N(i)$ may not be included in this step. Here neighbors i of d are processed in decreasing order of their degrees. After execution of this step, the *RouteRequest* is guaranteed to reach

all neighbors of d. If EXL is not empty, some neighbor's neighbors n of node d are currently unreachable, and they are included in the next step.

3. If neighbor i of d has a neighbor n present in EXL, put i in PLT and mark all neighbors of i as reachable. Delete all neighbors of i from EXL. Neighbors are processed in decreasing order of their degrees. After execution of this step, all the nodes in NNT_d are reachable. Apply reduction steps to remove overlapping neighbors from PLT without compromising on reachability.

4. Reduction steps are applied here in order to remove overlapping neighbors from PLT without compromising on reachability.

   (a) Remove each neighbor i from PLT if $N(i)$ is covered by remaining neighbors of PLT. Here the minimum degree neighbor is selected every time.

   (b) Remove neighbor i from PLT whose $N(i)$ is covered by node d itself.

Weight-Based Preferred Link Algorithm (WBPL)

In this algorithm, a node finds the preferred links based on stability, which is indicated by a weight, which in turn is based on its neighbors' temporal and spatial stability.

1. Let $BCnt_i$ be the count of *beacons* received from a neighbor i and TH_{bcon} is the number of beacons generated during a time period equal to that required to cover twice the transmission range ($TH_{bcon} = \frac{2 \times transmission\ range}{maximum\ velocity \times period\ of\ beacon}$). Weight given to i based on time stability (WT^i_{time}) is

$$WT_{time} = \begin{cases} 1 & \text{if } BCnt_i > TH_{bcon} \\ BCnt_i/TH_{bcon} & \text{otherwise.} \end{cases}$$

2. Estimate the distance (D^i_{Est}) to i from the received power of the last few packets using appropriate propagation models. The weight based on spatial stability is $WT^i_{spatial} = \frac{R - D_{Est}}{R}$.

3. The weight assigned to the link i is the combined weight given to time stability and spatial stability. $W_i = WT^i_{time} + WT^i_{spatial}$.

4. Arrange the neighbors in a non-increasing order of their weights. The nodes are put into the PLT in this order.

5. If a link is overloaded, delete the associated neighbor from PLT. Execute *Step* 1 of NDPL and delete $\forall i,\ i \in PLT \cap i \in INL$. Also, delete those neighbors from PLT that satisfy *Step* 4 of NDPL.

Consider, for example, Figure 7.29, where the node 3 is the source and node 8 is the destination. S and U denote stable and unstable links. In WBPL and NDPL, the source that initiates that *RouteRequest* as Dest is not present in NNT and computes the preferred link table (PLT). Let $K = 2$ be the preferred list's

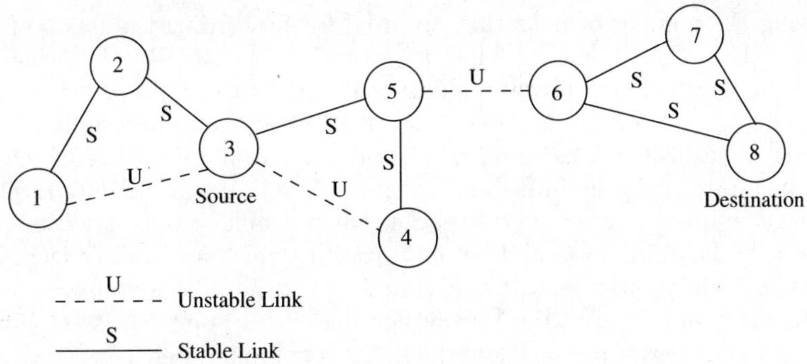

Figure 7.29. Example network. *Reproduced with permission from [20], © Korean Institute of Communication Sciences, 2002.*

size. In NDPL, after *Step* 2 the *PLT* becomes {5,1}, and after *Step* 3 also the *PLT* remains {5,1}. In reduction *Step* 4*b*, neighbor 1 is deleted from *PLT* and hence node 3 sends the *RouteRequest* only to neighbor 5. In WBPL, the weights are assigned to all neighbors according to *Steps* 1, 2, and 3, and all neighbors are in *PLT*. In *Step* 5, neighbors 1, 4, and 2 are deleted from *PLT* due to *Step* 4*a* and 4*b* of NDPL and hence only node 5 remains. Now the *RouteRequest* can be sent as a unicast packet to avoid broadcast. If it is broadcast, all the nodes receive the packet, but only node 5 can further forward it. As Dest 8 is present in node 5's *NNT*, it directly sends it to node 6 which forwards it to Dest. Here only three packets are transmitted for finding the route and the path length is 3. Now consider SSA. After broadcasts by nodes 3 and 5, the *RouteRequest* packet reaches node 6, where it is rejected and hence the *RouteRequest* fails to find a route. After timeout, it sets a flag indicating processed by *all* and hence finds the same route as WBPL and NDPL.

Advantages and Disadvantages

The efficient flooding mechanism employed in this protocol minimizes the broadcast storm problem prevalent in on-demand routing protocols. Hence this protocol has higher scalability compared to other on-demand routing protocols. The reduction in control overhead results in a decrease in the number of collisions and improvement in the efficiency of the protocol. PLBR achieves bandwidth efficiency at the cost of increased computation. Both NDPL and WBPL are computationally more complex than other *RouteRequest* forwarding schemes.

7.7.2 Optimized Link State Routing

The optimized link state routing (OLSR) protocol [21] is a proactive routing protocol that employs an efficient link state packet forwarding mechanism called *multipoint relaying.* This protocol optimizes the pure link state routing protocol. Optimizations are done in two ways: by reducing the size of the control packets and by reducing the number of links that are used for forwarding the link state packets. The reduction in the size of link state packets is made by declaring only a subset of the links in the link state updates. This subset of links or neighbors that are designated for link state updates and are assigned the responsibility of packet forwarding are called *multipoint relays.* The optimization by the use of *multipoint relaying* facilitates periodic link state updates. The link state update mechanism does not generate any other control packet when a link breaks or when a link is newly added. The link state update optimization achieves higher efficiency when operating in highly dense networks. Figure 7.30 (a) shows the number of message transmissions required when the typical flooding-based approach is employed. In this case, the number of message transmissions is approximately equal to the number of nodes that constitute the network. The set consisting of nodes that are *multipoint relays* is referred to as *MPRset.* Each node (say, *P*) in the network selects an *MPRset* that processes and forwards every link state packet that node *P* originates. The neighbor nodes that do not belong to the *MPRset* process the link state packets originated by node *P* but do not forward them. Similarly, each node

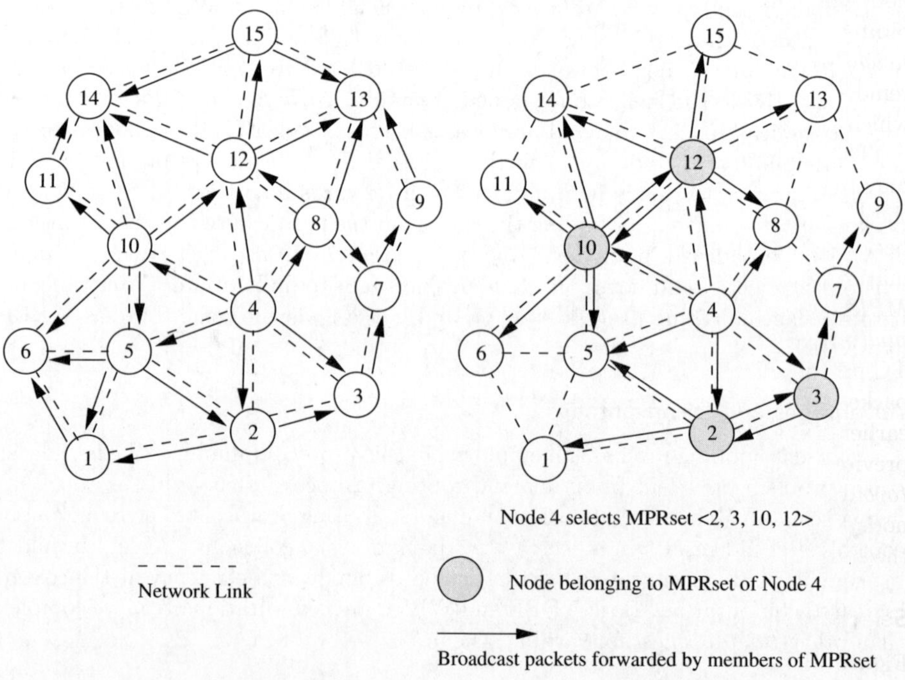

(a) Flooding the network takes as many transmissions as the number of nodes

(b) Flooding the entire network with six transmissions using MPR scheme

Figure 7.30. An example selection of MPRset in OLSR.

maintains a subset of neighbors called *MPR selectors*, which is nothing but the set of neighbors that have selected the node as a *multipoint relay*. A node forwards packets that are received from nodes belonging to its *MPRSelector* set. The members of both *MPRset* and *MPRSelectors* keep changing over time. The members of the *MPRset* of a node are selected in such a manner that every node in the node's two-hop neighborhood has a bidirectional link with the node. The selection of nodes that constitute the *MPRset* significantly affects the performance of OLSR because a node calculates routes to all destinations only through the members of its *MPRset*. Every node periodically broadcasts its *MPRSelector* set to nodes in its immediate neighborhood. In order to decide on the membership of the nodes in the *MPRset*, a node periodically sends *Hello* messages that contain the list of neighbors with which the node has bidirectional links and the list of neighbors whose transmissions were received in the recent past but with whom bidirectional links have not yet been confirmed. The nodes that receive this *Hello* packet update their own two-hop topology tables. The selection of *multipoint relays* is also indicated in the *Hello* packet. A data structure called *neighbor table* is used to store the list of neighbors, the two-hop neighbors, and the status of neighbor nodes. The neighbor nodes can be in one of the three possible link status states, that is, uni-directional, bidirectional, and *multipoint relay*. In order to remove the stale entries from the *neighbor table*, every entry has an associated timeout value, which, when expired, removes the table entry. Similarly a sequence number is attached with the *MPRset* which gets incremented with every new *MPRset*.

The *MPRset* need not be optimal, and during initialization of the network it may be same as the neighbor set. The smaller the number of nodes in the *MPRset*, the higher the efficiency of protocol compared to link state routing. Every node periodically originates *topology control* (TC) packets that contain topology information with which the routing table is updated. These TC packets contain the *MPRSelector* set of every node and are flooded throughout the network using the *multipoint relaying* mechanism. Every node in the network receives several such TC packets from different nodes, and by using the information contained in the TC packets, the *topology table* is built. A TC message may be originated by a node earlier than its regular period if there is a change in the *MPRSelector* set after the previous transmission and a minimal time has elapsed after that. An entry in the *topology table* contains a destination node which is the *MPRSelector* and a last-hop node to that destination, which is the node that originates the TC packet. Hence, the routing table maintains routes for all other nodes in the network.

Selection of Multipoint Relay Nodes

Figure 7.30 (b) shows the forwarding of TC packets using the *MPRset* of node 4. In this example, node 4 selects the nodes 2, 3, 10, and 12 as members of its *MPRset*. Forwarding by these nodes makes the TC packets reach all nodes within the transmitting node's two-hop local topology. The selection of the optimal *MPRset* is NP-complete [21]. In [21], a heuristic has been proposed for selecting the *MPRset*. The notations used in this heuristic are as follows: $N_i(x)$ refers to the ith hop neighbor set of node x and $MPR(x)$ refers to the *MPRset* of node x.

1. $MPR(x) \leftarrow \phi$ /* Initializing empty *MPRset* */

2. $MPR(x) \leftarrow$ { Those nodes that belong to $N_1(x)$ and which are the only neighbors of nodes in $N_2(x)$ }

3. While there exists some node in $N_2(x)$ which is not covered by $MPR(x)$

   (a) For each node in $N_1(x)$, which is not in $MPR(x)$, compute the maximum number of nodes that it covers among the uncovered nodes in the set $N_2(x)$.

   (b) Add to $MPR(x)$ the node belonging to $N_1(x)$, for which this number is maximum.

A node updates its *MPRset* whenever it detects a new bidirectional link in its neighborhood or in its two-hop topology, or a bidirectional link gets broken in its neighborhood.

Advantages and Disadvantages

OLSR has several advantages that make it a better choice over other table-driven protocols. It reduces the routing overhead associated with table-driven routing, in addition to reducing the number of broadcasts done. Hence OLSR has the advantages of low connection setup time and reduced control overhead.

7.8 HIERARCHICAL ROUTING PROTOCOLS

The use of routing hierarchy has several advantages, the most important ones being reduction in the size of routing tables and better scalability. This section discusses the existing hierarchical routing protocols for ad hoc wireless networks.

7.8.1 Hierarchical State Routing Protocol

The hierarchical state routing (HSR) protocol [23] is a distributed multi-level hierarchical routing protocol that employs clustering at different levels with efficient membership management at every level of clustering. The use of clustering enhances resource allocation and management. For example, the allocation of different frequencies or spreading codes to different clusters can improve the overall spectrum reuse. HSR operates by classifying different levels of clusters. Elected leaders at every level form the members at the immediate higher level. Different clustering algorithms, such as the one proposed in [8], are employed for electing leaders at every level. The first level of physical clustering is done among the nodes that are reachable in a single wireless hop. The next higher level of physical clustering is done among the nodes that are elected as leaders of each of these first-level clusters. In addition to the *physical clustering*, a *logical clustering* scheme has been proposed in HSR, which is based on certain relations among the nodes rather than on their geographical positions, as in the case of *physical clustering*.

Figure 7.31 illustrates the multilayer clustering defined by the HSR protocol. At the lowest level ($L = 0$), there are six cluster leaders (nodes 1, 2, 3, 4, 5, and 6). Nodes are classified as cluster leaders, or gateway nodes, or normal member nodes. A cluster leader is entrusted with responsibilities such as slot/frequency/code allocation, call admission control, scheduling of packet transmissions, exchange of routing information, and handling route breaks. In Figure 7.31, node 5 is a cluster-head marked as $L0-5$, which refers to the level of clustering ($L = 0$) and node ID (5). Similarly, each of the higher-level cluster leaders is also marked (*e.g.*, $L1-6$, $L2-6$, and $L3-6$ refer to the same node 6, but acting as leader with the given leader IDs at levels 1, 2, and 3, respectively). The spectrum reuse schemes, including spreading code assignment, can be used among the cluster leaders of the $L = 0$ clusters.

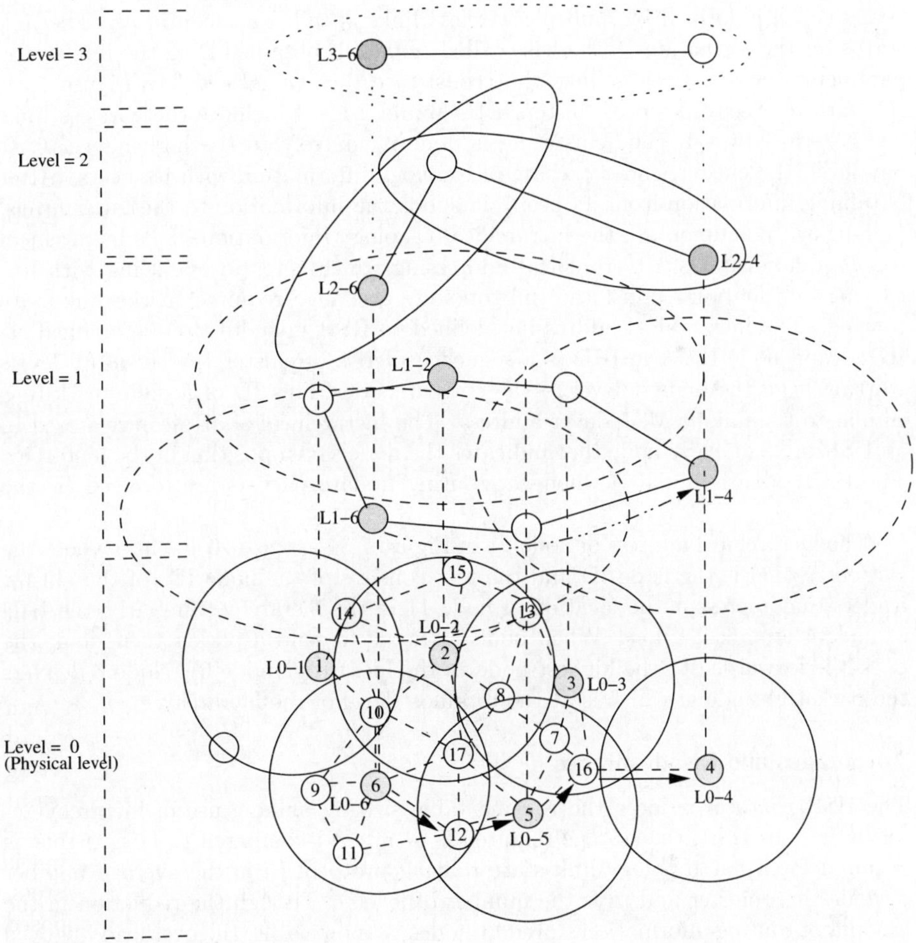

Figure 7.31. Example of HSR multi-level clustering.

For the nodes under the leadership of node 6 at level 0, the cluster members are nodes 9, 10, 11, 12, and 17. Those nodes that belong to multiple clusters are referred to as cluster gateway nodes. For the level 0 cluster whose leader is node 6, the cluster gateways are nodes 10, 12, and 17. The second level of clustering is done among the leaders of the first level, that is, the leaders of 0th level clusters, $L0-1, L0-2, L0-3, L0-4, L0-5$, and $L0-6$, form the members of the first-level cluster.

Every node maintains information about all its neighbors and the status of links to each of them. This information is broadcast within the cluster at regular intervals. The cluster leader exchanges the topology and link state routing information among its peers in the neighborhood clusters, using which the next-higher-level clustering is performed. This exchange of link state information is done over multiple hops that consist of gateway nodes and cluster-heads. The path between two cluster-heads which is formed by multiple wireless links is called a *virtual link*. The link status for the *virtual link* (otherwise called *tunnel*) is obtained from the link status parameters of the wireless links that constitute the *virtual link*. In Figure 7.31, the path between first-level clusters $L1-6$ and $L1-4$ includes the wireless links $6-12-5-16-4$. The clustering is done recursively to the higher levels. At any level, the cluster leader exchanges topology information with its peers. After obtaining information from its peers, it floods the information to the lower levels, making every node obtain the hierarchical topology information. This hierarchical topology necessitates a hierarchical addressing which helps in operating with less routing information against the full topology exchange required in the link state routing. The hierarchical addressing defined in HSR includes the hierarchical ID (HID) and node ID. The HID is a sequence of IDs of cluster leaders of all levels starting from the highest level to the current node. This ID of a node in HSR is similar to the unique MAC layer address. The hierarchical addresses are stored in an HSR table at every node that indicates the node's own position in the hierarchy. The HSR table is updated whenever routing update packets are received by the node.

The hierarchical address of node 11 in Figure 7.31 is $< 6,6,6,6,11 >$, where the last entry (11) is the node ID and the rest consists of the node IDs of the cluster leaders that represent the location of node 11 in the hierarchy. Similarly, the HID of node 4 is $< 6,4,4,4 >$. When node 11 needs to send a packet to node 4, the packet is forwarded to the highest node in the hierarchy (node 6). Node 6 delivers the packet to node 4, which is at the top-most level of the hierarchy.

Advantages and Disadvantages

The HSR protocol reduces the routing table size by making use of hierarchy information. In HSR, the storage required is $O(n \times m)$ compared to $O(n^m)$ that is required for a flat topology link state routing protocol (n is the average number of nodes in a cluster and m is the number of levels). Though the reduction in the amount of routing information stored at nodes is appreciable, the overhead involved in exchanging packets containing information about the multiple levels of hierarchy and the leader election process make the protocol unaffordable in the ad hoc wireless networks context. Besides, the number of nodes that participate in an ad hoc wireless network does not grow to the dimensions of the number of nodes in the Internet where the hierarchical routing is better suited. In the military applications of ad hoc wireless networks, the hierarchy of routing assumes significance where devices with higher capabilities of communication can act as the cluster leaders.

7.8.2 Fisheye State Routing Protocol

The table-driven routing protocols generate routing overhead that is dependent on the size of the network and mobility of the nodes, whereas the routing overhead generated by on-demand routing protocols are dependent on the number of connections present in the system in addition to the above two factors. Hence, as the number of senders in the network increases, the routing overhead also increases. ZRP uses an intra-zone proactive approach and an inter-zone reactive approach to reduce control overhead. The fisheye state routing (FSR) protocol [23] is a generalization of the GSR [24] protocol. FSR uses the *fisheye* technique to reduce information required to represent graphical data, to reduce routing overhead. The basic principle behind this technique is the property of a fish's eye that can capture pixel information with greater accuracy near its eye's focal point. This accuracy decreases with an increase in the distance from the center of the focal point. This property is translated to routing in ad hoc wireless networks by a node, keeping accurate information about nodes in its local topology, and not-so-accurate information about far-away nodes, the accuracy of the network information decreasing with increasing distance. FSR maintains the topology of the network at every node, but does not flood the entire network with the information, as is done in link state routing protocols. Instead of flooding, a node exchanges topology information only with its neighbors. A sequence numbering scheme is used to identify the recent topology changes. This constitutes a hybrid approach comprising of the link-level information exchange of distance vector protocols and the complete topology information exchange of link state protocols. The complete topology information of the network is maintained at every node and the desired shortest paths are computed as required. The topology information exchange takes place periodically rather than being driven by an event. This is because instability of the wireless links may cause excessive control overhead when event-driven updates are employed. FSR defines routing scope, which is the set of nodes that are reachable in a specific number of hops. The scope of a node at two hops is the set of nodes that can be reached in two hops. Figure 7.32 shows the scope of node 5 with one hop and two hops. The routing overhead is significantly reduced by adopting different frequencies of updates for nodes belonging to different scopes.

The link state information for the nodes belonging to the smallest scope is exchanged at the highest frequency. The frequency of exchanges decreases with an increase in scope. This keeps the immediate neighborhood topology information maintained at a node more precise compared to the information about nodes farther away from it. Thus the message size for a typical topology information update packet is significantly reduced due to the removal of topology information regarding

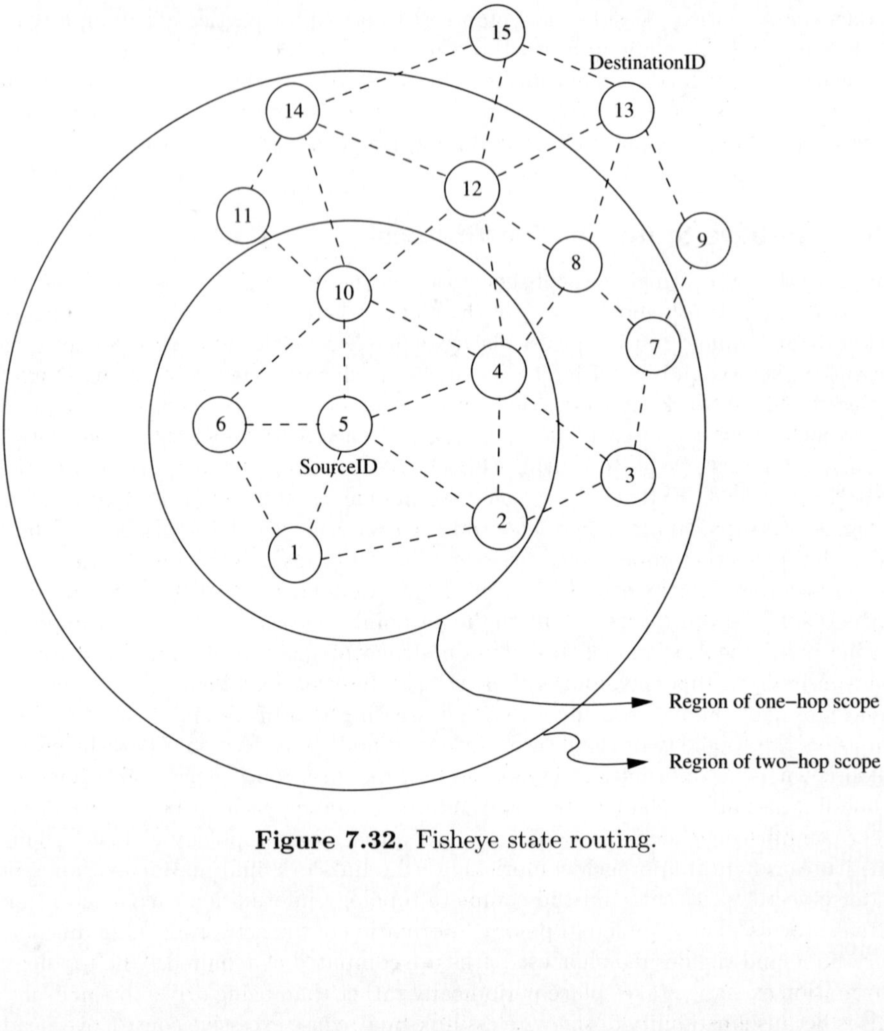

Figure 7.32. Fisheye state routing.

the far-away nodes. The path information for a distant node may be inaccurate as there can be staleness in the information. But this is compensated by the fact that the route gets more and more accurate as the packet nears its destination. FSR scales well for large ad hoc wireless networks because of the reduction in routing overhead due to the use of the above-described mechanism, where varying frequencies of updates are used.

Figure 7.33 illustrates an example depicting the network topology information maintained at nodes in a network. The routing information for the nodes that are one hop away from a node are exchanged more frequently than the routing information about nodes that are more than one hop away. Information regarding nodes that are more than one hop away from the current node are listed below the dotted line in the topology table.

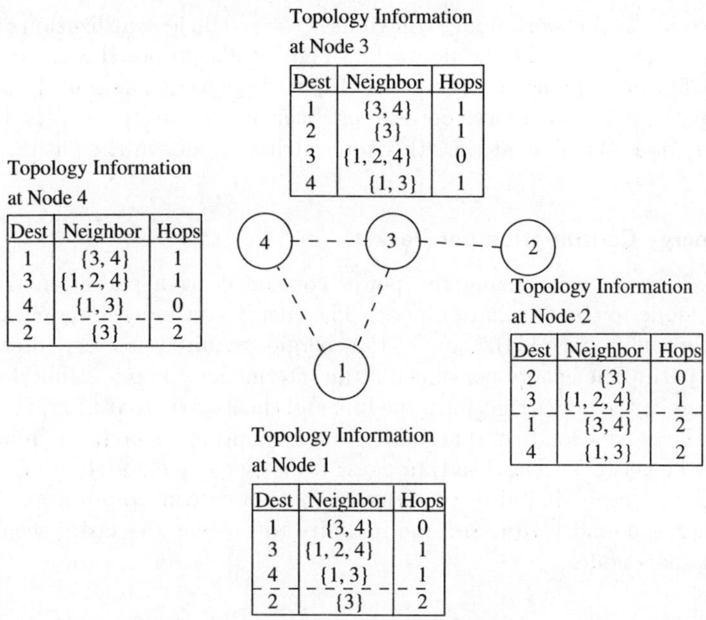

Figure 7.33. An illustration of routing tables in FSR.

Advantages and Disadvantages

The notion of multi-level scopes employed by FSR significantly reduces the bandwidth consumed by link state update packets. Hence, FSR is suitable for large and highly mobile ad hoc wireless networks. The choice of the number of hops associated with each scope level has a significant influence on the performance of the protocol at different mobility values, and hence must be carefully chosen.

7.9 POWER-AWARE ROUTING PROTOCOLS

In a deviation from the traditional wired network routing and cellular wireless network routing, power consumption by the nodes is a serious factor to be taken into consideration by routing protocols for ad hoc wireless networks. This is because, in ad hoc wireless networks, the routers are also equally power-constrained just as the nodes are. This section discusses some of the important routing metrics that take into consideration this energy factor.

7.9.1 Power-Aware Routing Metrics

The limitation on the availability of power for operation is a significant bottleneck, given the requirements of portability, weight, and size of commercial hand-held devices. Hence, the use of routing metrics that consider the capabilities of the power sources of the network nodes contributes to the efficient utilization of energy and increases the lifetime of the network. Singh *et al.* proposed a set of routing metrics in [25] that supports conservation of battery power. The routing protocols that select paths so as to conserve power must be aware of the states of the batteries at the given node as well as at the other intermediate nodes in the path.

Minimal Energy Consumption per Packet

This metric aims at minimizing the power consumed by a packet in traversing from source node to the destination node. The energy consumed by a packet when traversing through a path is the sum of the energies required at every intermediate hop in that path. The energy consumed at an intermediate hop is a function of the distance between the nodes that form the link and the load on that link. This metric does not balance the load so that uniform consumption of power is maintained throughout the network. The disadvantages of this metric include selection of paths with large hop length, inability to measure the power consumption at a link in advance when the load varies, and the inability to prevent the fast discharging of batteries at some nodes.

Maximize Network Connectivity

This metric attempts to balance the routing load among the *cut-set* (the subset of the nodes in the network, the removal of which results in network partitions). This assumes significance in environments where network connectivity is to be ensured by uniformly distributing the routing load among the *cut-set*. With a variable traffic origination rate and unbounded contention in the network, it is difficult to achieve a uniform battery draining rate for the *cut-set*.

Minimum Variance in Node Power Levels

This metric proposes to distribute the load among all nodes in the network so that the power consumption pattern remains uniform across them. This problem is very complex when the rate and size of data packets vary. A nearly optimal performance can be achieved by routing packets to the least-loaded next-hop node.

Minimum Cost per Packet

In order to maximize the life of every node in the network, this routing metric is made as a function of the state of the node's battery. A node's cost decreases with an increase in its battery charge and vice versa. Translation of the remaining battery charge to a cost factor is used for routing. With the availability of a battery discharge pattern, the cost of a node can be computed. This metric has the advantage of ease in the calculation of the cost of a node and at the same time congestion handling is done.

Minimize Maximum Node Cost

This metric minimizes the maximum cost per node for a packet after routing a number of packets or after a specific period. This delays the failure of a node, occurring due to higher discharge because of packet forwarding.

7.10 SUMMARY

In this chapter, the major issues involved in the design of a routing protocol and the different classifications of routing protocols for ad hoc wireless networks were described. The major challenges that an ad hoc wireless routing protocol must address are the mobility of nodes, rapid changes in topology, limited bandwidth, hidden and exposed terminal problems, limited battery power, time-varying channel properties, and location-dependent contention. The different approaches upon which the protocols can be classified include the classification based on the type of topology maintenance approach, the routing topology used, the use of temporal information, and the type of specific resource utilization considered for making routing decisions. The protocols belonging to each of these categories were explained in detail with illustrations.

7.11 PROBLEMS

1. Discuss the differences in the maintenance of topology information in various protocols such as CGSR, HSR, SSA, ABR, PLBR, OLSR, and CEDAR.

2. Discuss the differences in topology reorganization in DSDV and CGSR routing protocols.

3. How is the cluster-head selected in the CGSR protocol? In the CGSR protocol, the resources of the node chosen as the cluster-head get drained very quickly, more rapidly than the other nodes in the cluster. How can this problem be overcome?

4. Is a table-driven routing protocol suitable for high-mobility environments?

5. Both ABR and SSA use stability information for routing. How do they differ in using the stability information?

6. What are the advantages of hierarchical topology-based protocols over protocols that use flat topologies?

7. Consider the topology given in Figure 7.34. Simulate DSR, SSA, and ABR protocols for path establishment from node 1 to node 10, find the paths found and the ratio of the number of *RouteRequest* packets sent in the network. (Links labeled "U" refer to unstable ones.)

8. For the sample topology given in Figure 7.35, find the Zone Link State packets for the various zones marked.

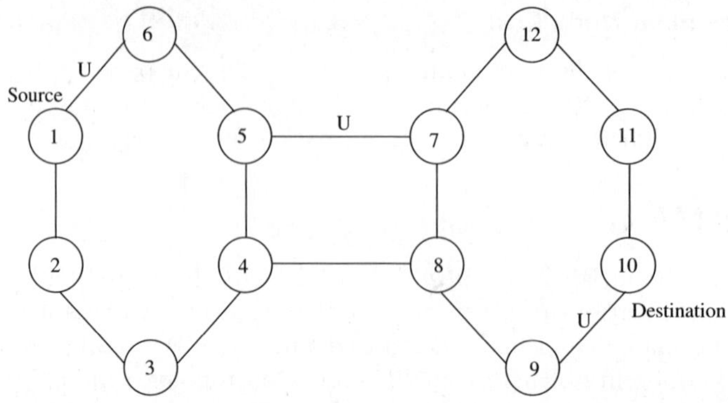

Figure 7.34. Topology for Problem 7.

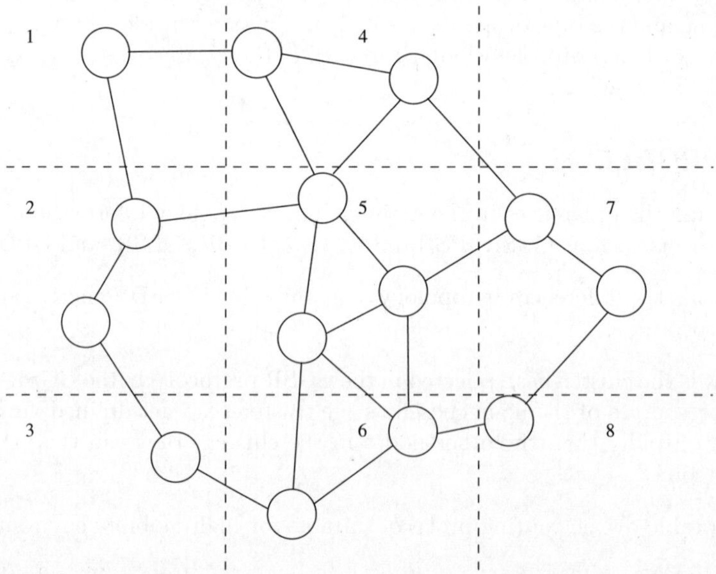

Figure 7.35. Topology for Problem 8.

9. Does the LCC algorithm (when run consistently with node degrees or node IDs) give a deterministic result? If so, prove the above fact. Otherwise, give a counter-example.
10. What are the key differences between the LAR1 and the LAR2 algorithms?
11. For the network shown in Figure 7.36, determine the fisheye routing table for nodes 7 and 5.

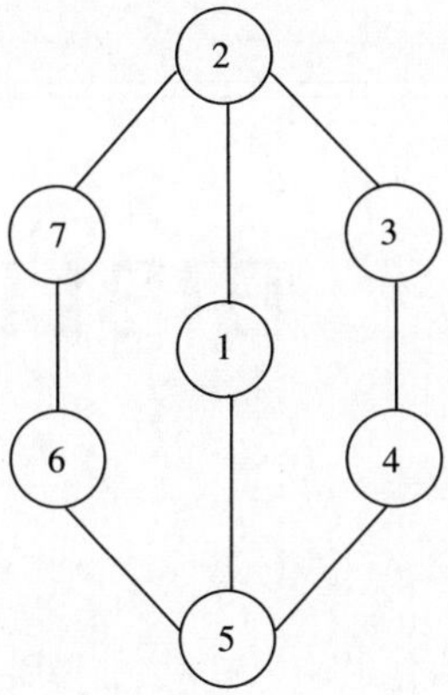

Figure 7.36. Topology for Problem 11.

12. For the topology shown in Figure 7.37, create a DAG for node 1 as the source and node 7 as the destination in TORA. If the link between nodes 4 and 6 breaks as shown in the figure, find the change in the DAG. (Also mark the distance from the destination.)

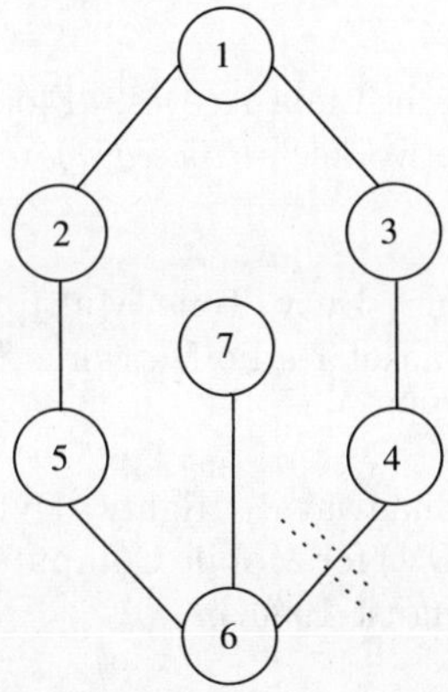

Figure 7.37. Topology for Problem 12.

13. Identify some of the key issues involved in QoS routing in ad hoc networks.

BIBLIOGRAPHY

[1] C. L. Fullmer and J. J. Garcia-Luna-Aceves, "Solutions to Hidden Terminal Problems in Wireless Networks," *Proceedings of ACM SIGCOMM 1997*, pp. 39-49, September 1997.

[2] P. Karn, "MACA – A New Channel Access Method for Packet Radio," *Proceedings of ARRL/CRRL Amateur Radio Computer Networking Conference 1990*, pp. 134-140, September 1990.

[3] V. Bharghavan, A. Demers, S. Shenker, and L. Zhang, "MACAW: A Media Access Protocol for Wireless LANs," *Proceedings of ACM SIGCOMM 1994*, pp. 212-225, August 1994.

[4] C. L. Fullmer and J. J. Garcia-Luna-Aceves, "Floor Acquisition Multiple Access (FAMA) for Packet-Radio Networks," *Proceedings of ACM SIGCOMM 1995*, pp. 262-273, August 1995.

[5] J. Deng and Z. Haas, "Dual Busy Tone Multiple Access (DBTMA): A New Medium Access Control for Packet Radio Networks," *Proceedings of ICUPC 1998*, vol. 1, pp. 973-977, October 1998.

[6] C. E. Perkins and P. Bhagwat, "Highly Dynamic Destination-Sequenced Distance-Vector Routing (DSDV) for Mobile Computers," *Proceedings of ACM SIGCOMM 1994*, pp. 234-244, August 1994.

[7] S. Murthy and J. J. Garcia-Luna-Aceves, "An Efficient Routing Protocol for Wireless Networks," *ACM Mobile Networks and Applications Journal, Special Issue on Routing in Mobile Communication Networks*, vol. 1, no. 2, pp. 183-197, October 1996.

[8] C. C. Chiang, H. K. Wu, W. Liu, and M. Gerla, "Routing in Clustered Multi-Hop Mobile Wireless Networks with Fading Channel," *Proceedings of IEEE SICON 1997*, pp. 197-211, April 1997.

[9] J. J. Garcia-Luna-Aceves and M. Spohn, "Source-Tree Routing in Wireless Networks," *Proceedings of IEEE ICNP 1999*, pp. 273-282, October 1999.

[10] D. B. Johnson and D. A. Maltz, "Dynamic Source Routing in Ad Hoc Wireless Networks," *Mobile Computing, Kluwer Academic Publishers*, vol. 353, pp. 153-181, 1996.

[11] C. E. Perkins and E. M. Royer, "Ad Hoc On-Demand Distance Vector Routing," *Proceedings of IEEE Workshop on Mobile Computing Systems and Applications 1999*, pp. 90-100, February 1999.

[12] V. D. Park and M. S. Corson, "A Highly Adaptive Distributed Routing Algorithm for Mobile Wireless Networks," *Proceedings of IEEE INFOCOM 1997*, pp. 1405-1413, April 1997.

[13] Y. Ko and N. H. Vaidya, "Location-Aided Routing (LAR) in Mobile Ad Hoc Networks," *Proceedings of ACM MOBICOM 1998*, pp. 66-75, October 1998.

[14] C. K. Toh, "Associativity-Based Routing for Ad Hoc Mobile Networks," *Wireless Personal Communications*, vol. 4, no. 2, pp. 1-36, March 1997.

[15] R. Dube, C. D. Rais, K. Y. Wang, and S. K. Tripathi, "Signal Stability-Based Adaptive Routing for Ad Hoc Mobile Networks," *IEEE Personal Communications Magazine*, pp. 36-45, February 1997.

[16] W. Su and M. Gerla, "IPv6 Flow Handoff in Ad Hoc Wireless Networks Using Mobility Prediction," *Proceedings of IEEE GLOBECOM 1999*, pp. 271-275, December 1999.

[17] P. Sinha, R. Sivakumar, and V. Bharghavan, "CEDAR: A Core Extraction Distributed Ad Hoc Routing Algorithm," *IEEE Journal on Selected Areas in Communications*, vol. 17, no. 8, pp. 1454-1466, August 1999.

[18] Z. J. Haas, "The Routing Algorithm for the Reconfigurable Wireless Networks," *Proceedings of ICUPC 1997*, vol. 2, pp. 562-566, October 1997.

[19] M. Joa-Ng and I. T. Lu, "A Peer-to-Peer Zone-Based Two-Level Link State Routing for Mobile Ad Hoc Networks," *IEEE Journal on Selected Areas in Communications,* vol. 17, no. 8, pp. 1415- 1425, August 1999.

[20] R. S. Sisodia, B. S. Manoj, and C. Siva Ram Murthy, "A Preferred Link-Based Routing Protocol for Ad Hoc Wireless Networks," *Journal of Communications and Networks*, vol. 4, no. 1, pp. 14-21, March 2002.

[21] T. H. Clausen, G. Hansen, L. Christensen, and G. Behrmann, "The Optimized Link State Routing Protocol, Evaluation Through Experiments and Simulation," *Proceedings of IEEE Symposium on Wireless Personal Mobile Communications 2001*, September 2001.

[22] R. Sriram, G. Manimaran, and C. Siva Ram Murthy, "Preferred Link-Based Delay-Constrained Least-Cost Routing in Wide Area Networks," *Computer Communications*, vol. 21, pp. 1655-1669, November 1998.

[23] A. Iwata, C. C. Chiang, G. Pei, M. Gerla, and T. W. Chen, "Scalable Routing Strategies for Ad Hoc Wireless Networks," *IEEE Journal on Selected Areas in Communications*, vol. 17, no. 8, pp. 1369-1379, August 1999.

[24] T. W. Chen and M. Gerla, "Global State Routing: A New Routing Scheme for Ad Hoc Wireless Networks," *Proceedings of IEEE ICC 1998,* pp. 171-175, June 1998.

[25] S. Singh, M. Woo, and C. S. Raghavendra, "Power-Aware Routing in Mobile Ad Hoc Networks," *Proceedings of ACM MOBICOM 1998*, pp. 181-190, October 1998.

[26] D. L. Gu, G. Pei, M. Gerla, and X. Hong, "UAV-Aided Intelligent Routing for Ad Hoc Wireless Networks," *Proceedings of IEEE WCNC 2000*, pp. 1220-1225, September 2000.

[27] M. Gerla and J. T. C. Tsai, "Multicluster, Mobile, Multimedia Radio Network," *ACM/Baltzer Wireless Networks Journal,* vol. 1, no. 3, pp. 255-265, 1995.

[28] S. Agarwal, A. Ahuja, J. P. Singh, and R. Shorey, "Route-Lifetime Assessment-Based Routing (RABR) Protocol for Mobile Ad Hoc Networks," *Proceedings of IEEE ICC 2000,* vol. 3, pp. 1697-1701, June 2000.

[29] B. S. Manoj, R. Ananthapadmanabha, and C. Siva Ram Murthy, "Link Life-Based Routing Protocol for Ad Hoc Wireless Networks," *Proceedings of IEEE IC-CCN 2001*, pp. 573-576, October 2001.

Chapter 8

MULTICAST ROUTING IN AD HOC WIRELESS NETWORKS

8.1 INTRODUCTION

Ad hoc wireless networks find applications in civilian operations (collaborative and distributed computing), emergency search-and-rescue, law enforcement, and warfare situations, where setting up and maintaining a communication infrastructure may be difficult or costly. In all these applications, communication and coordination among a given set of nodes are necessary. Multicast routing protocols play an important role in ad hoc wireless networks to provide this communication. It is always advantageous to use multicast rather than multiple unicast, especially in the ad hoc environment, where bandwidth comes at a premium.

Conventional wired network Internet protocol (IP) multicast routing protocols, such as DVMRP [1], MOSPF [2], CBT [3], and PIM [4], do not perform well in ad hoc networks because of the dynamic nature of the network topology. The dynamically changing topology, coupled with relatively low bandwidth and less reliable wireless links, causes long convergence times and may give rise to formation of transient routing loops which rapidly consume the already limited bandwidth.

In a wired network, the basic approach adopted for multicasting consists of establishing a routing tree for a group of routing nodes that constitute the multicast session. Once the routing tree (or the spanning tree, which is an acyclic connected subgraph containing all the nodes in the tree) is established, a packet sent to all nodes in the tree traverses each node and each link in the tree only once. Such a multicast structure is not appropriate for ad hoc networks because the tree could easily break due to the highly dynamic topology.

Multicast tree structures are not stable and need to be reconstructed continuously as connectivity changes. Maintaining a routing tree for the purpose of multicasting packets, when the underlying topology keeps changing frequently, can incur substantial control traffic. The multicast trees used in the conventional wired network multicast protocols require a global routing sub-structure such as a link state or a distance vector sub-structure. The frequent exchange of routing vectors or link state tables, triggered by continuous topology changes, yields excessive control and processing overhead. Further, periods of routing table instability lead to instability of the multicast tree, which in turn results in increased buffering time for packets, higher packet losses, and an increase in the number of retransmissions. Therefore, multicast protocols used in static wired networks are not suitable for ad hoc wireless networks. This chapter addresses the multicast routing problem (the problem of determining which nodes in the network should participate for targeting of multicast data packets, transmitted from a source, to a select set of receivers) and presents several multicast routing protocols for ad hoc wireless networks.

8.2 ISSUES IN DESIGNING A MULTICAST ROUTING PROTOCOL

Limited bandwidth availability, an error-prone shared broadcast channel, the mobility of nodes with limited energy resources, the hidden terminal problem [5], and limited security make the design of a multicast routing protocol for ad hoc networks a challenging one. There are several issues involved here which are discussed below.

- **Robustness:** Due to the mobility of the nodes, link failures are quite common in ad hoc wireless networks. Thus, data packets sent by the source may be dropped, which results in a low packet delivery ratio. Hence, a multicast routing protocol should be robust enough to sustain the mobility of the nodes and achieve a high packet delivery ratio.

- **Efficiency:** In an ad hoc network environment, where the bandwidth is scarce, the efficiency of the multicast protocol is very important. Multicast efficiency is defined as the ratio of the total number of data packets received by the receivers to the total number of (data and control) packets transmitted in the network.

- **Control overhead:** In order to keep track of the members in a multicast group, the exchange of control packets is required. This consumes a considerable amount of bandwidth. Since bandwidth is limited in ad hoc networks, the design of a multicast protocol should ensure that the total number of control packets transmitted for maintaining the multicast group is kept to a minimum.

- **Quality of service:** One of the important applications of ad hoc networks is in military/strategic applications. Hence, provisioning quality of service (QoS) is an issue in ad hoc multicast routing protocols. The main parameters which are taken into consideration for providing the required QoS are throughput, delay, delay jitter, and reliability.

- **Dependency on the unicast routing protocol:** If a multicast routing protocol needs the support of a particular routing protocol, then it is difficult for the multicast protocol to work in heterogeneous networks. Hence, it is desirable if the multicast routing protocol is independent of any specific unicast routing protocol.

- **Resource management:** Ad hoc networks consist of a group of mobile nodes, with each node having limited battery power and memory. An ad hoc multicast routing protocol should use minimum power by reducing the number of packet transmissions. To reduce memory usage, it should use minimum state information.

8.3 OPERATION OF MULTICAST ROUTING PROTOCOLS

Based on the type of operation, multicast protocols for ad hoc wireless networks are broadly classified into two types: *source-initiated* protocols and *receiver-initiated* protocols. There exist certain other multicast protocols (such as MCEDAR [6] and AMRoute [7]) which may not strictly fall under the above two types. In this section, a general framework for understanding the operation of multicast routing protocols is discussed.

8.3.1 Source-Initiated Protocols

Figure 8.1 (a) shows the events as they occur in a source-initiated protocol that uses a soft state maintenance approach. In the soft state maintenance approach, the multicast tree or mesh is periodically updated by means of control packets. In such protocols, the source(s) of the multicast group periodically floods a *JoinRequest* (*JoinReq*) packet throughout the network. This *JoinReq* packet is propagated by other nodes in the network, and it eventually reaches all the receivers of the group. Receivers of the multicast group express their wish to receive packets for the group by responding with a *JoinReply* (*JoinRep*) packet, which is propagated along the reverse path of that followed by the *JoinReq* packet. This *JoinRep* packet establishes forwarding states (forwarding state refers to the information regarding the multicast group maintained at nodes in the multicast tree or mesh, which aids the nodes in forwarding multicast packets to the appropriate next-hop neighbor nodes) in the intermediate nodes (either in the tree or mesh), and finally reaches the source. Thus, this is a two-pass protocol for establishing the tree (or mesh). There is no explicit procedure for route repair. In soft state protocols, the source periodically (once every *refresh period*) initiates the above procedure.

In Figure 8.1 (b), the operation of a hard state source-initiated protocol is shown. It is similar to that of a soft state source-initiated protocol, except that there is an explicit route repair procedure that is initiated when a link break (in the tree or mesh) is detected. The route repair procedure shown in Figure 8.1 (b) is a simple solution: The upstream node which detects that one of its downstream nodes has moved away, initiates a tree construction procedure (similar to the one initiated by the source). Different protocols adopt different strategies for route repair. Some protocols choose to have the downstream node search for its former parent (in the

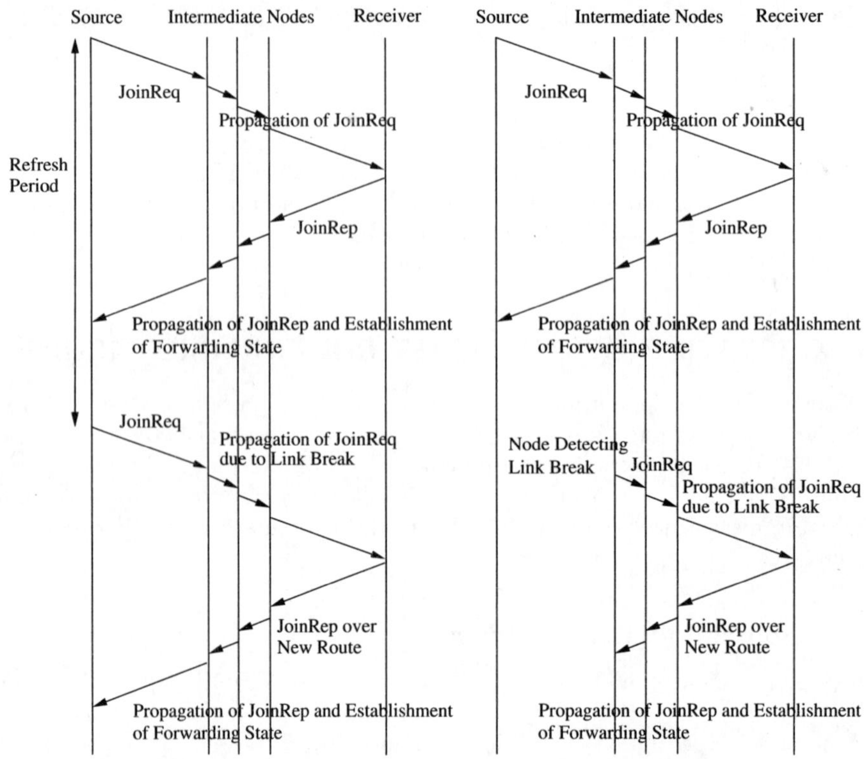

(a) Source–Initiated, Soft State Multicast Protocols (b) Source–Initiated, Hard State Multicast Protocols

Figure 8.1. Source-initiated multicast protocols.

tree or mesh) by means of limited flooding, while others impose this responsibility on the upstream node.

8.3.2 Receiver-Initiated Protocols

In the receiver-initiated multicasting protocols, the receiver uses flooding to search for paths to the sources of the multicast groups to which it belongs. The soft state variant is illustrated in Figure 8.2 (a). The tree (or mesh) construction is a three-phase process. First, the receiver floods a *JoinReq* packet, which is propagated by the other nodes. Usually, the sources of the multicast group and/or nodes which are already part of the multicast tree (or mesh), are allowed to respond to the *JoinReq* packet with a *JoinRep* packet, indicating that they would be able to send data packets for that multicast group. The receiver chooses the *JoinRep* with the smallest hop count (or some other criterion) and sends a *JoinAcknowledgment*

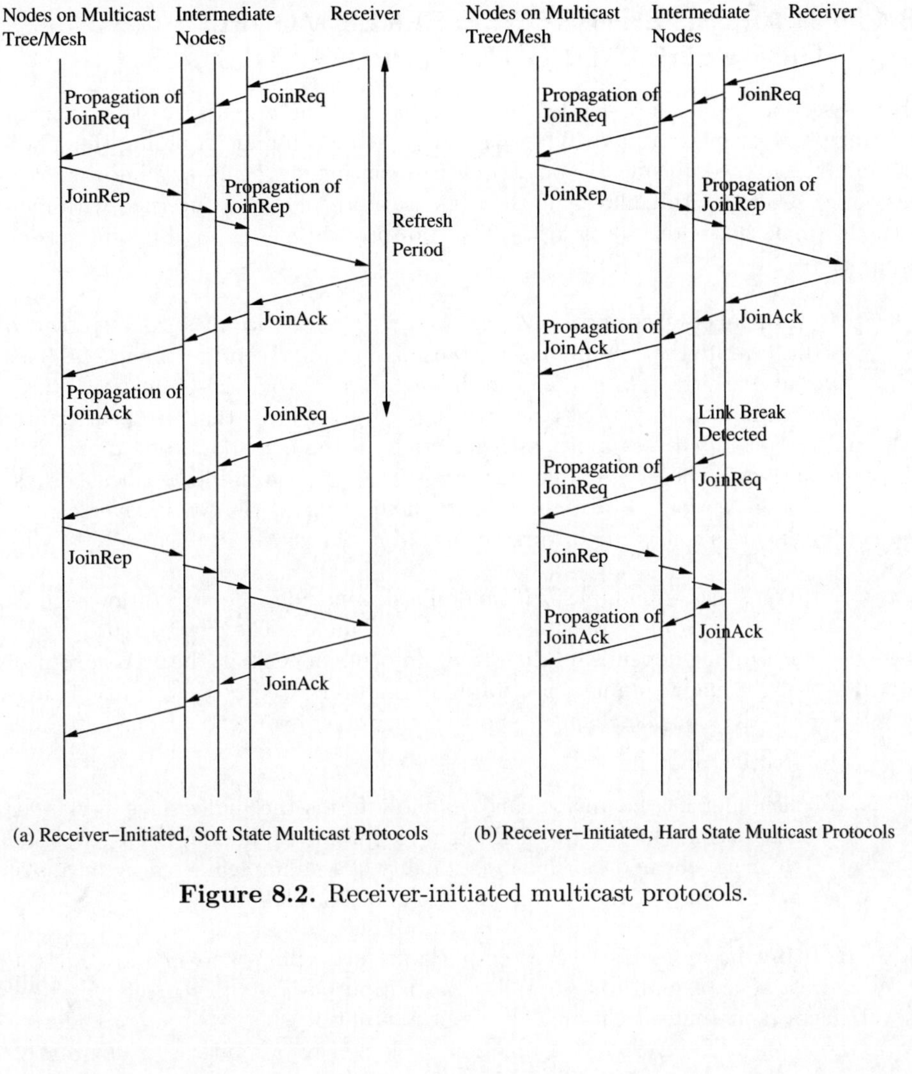

Figure 8.2. Receiver-initiated multicast protocols.

(*JoinAck*) packet along the reverse path (taken by the *JoinRep*). Route maintenance is accomplished by the periodic initiation of this procedure by the receiver.

In Figure 8.2 (b), the hard state variant is illustrated. The initial tree- (or mesh-) joining phase is the same as that in the corresponding soft state protocol. The route repair mechanism comes into play when a link break is detected: The responsibility to restore the multicast topology can be ascribed to either the downstream or to the upstream node. In Figure 8.2 (b), the downstream node searches for a route to the multicast tree (or mesh) through a procedure similar to the initial topology construction procedure.

8.4 AN ARCHITECTURE REFERENCE MODEL FOR MULTICAST ROUTING PROTOCOLS

In this section, a reference model for understanding the architecture of multicast routing protocols is presented. This will aid the reader in understanding the different modules in the implementation of multicast routing protocols for ad hoc wireless networks. There are three layers in the network protocol stack concerned with multicasting in ad hoc wireless networks (The transport layer is ignored for the sake of simplicity):

1. **Medium access control (MAC) layer:** The important services provided by this layer to the ones above are transmission and reception of packets. This layer also arbitrates access to the channel. Apart from these functions, three other important functions are performed by this layer that are particularly important in wireless multicast: detecting all the neighbors (nodes at a hop distance of 1), observing link characteristics, and performing broadcast transmission/reception. Corresponding to these services, the MAC layer can be thought of as consisting of three principal modules:

   (a) Transmission module: This module also includes the arbitration module which schedules transmissions on the channel. The exact nature of this scheduling depends on the MAC protocol. In general, the MAC protocol might maintain multicast state information based on past transmissions observed on the channel, and the scheduling is dependent on that state.

   (b) Receiver module.

   (c) Neighbor list handler: This module informs the higher layers whether a particular node is a neighbor node or not. It maintains a list of all the neighbor nodes. This functionality can be implemented by means of beacons or by overhearing all packets on the channel.

2. **Routing layer:** This layer is responsible for forming and maintaining the unicast session/multicast group. For this purpose, it uses a set of tables, timers, and route caches. The important multicast services it provides to the application layer are the functions to join/leave a multicast group and to transmit/receive multicast packets. Most of the multicast routing protocols operate in the routing layer. Other layers have been touched upon here in order to clarify the interactions in which the routing layer is involved. The routing layer uses the following components/modules:

   (a) Unicast routing information handler: This serves to discover unicast routes (by an on-demand or a table-driven mechanism).

   (b) Multicast information handler: This maintains all the pertinent information related to the state of the current node with respect to the multicast groups of which it is a part, in the form of a table. This state might include a list of its downstream nodes, the address of its upstream node(s), sequence number information, etc. This table might be maintained per group or per source per group.

(c) Forwarding module: This uses the information provided by the multicast information handler to decide whether a received multicast packet should be broadcast, or be forwarded to a neighbor node, or be sent to the application layer.

(d) Tree/mesh construction module: This module is used to construct the multicast topology. It can use information provided by the unicast routing information handler for this purpose; for example, this module might initiate flooding on being requested to join a group by the application layer. Also, when the application layer process (through module 10) sends session termination messages to this module, this module transmits the appropriate messages to the network for terminating the participation of the current node in the multicast session.

(e) Session maintenance module: This module initiates route repair on being informed of a link break by the lower layer. It might use information from the multicast and unicast routing tables to perform a search (possibly localized) for the node (upstream or downstream) in order to restore the multicast topology.

(f) Route cache maintenance module: The purpose of this module is to glean information from routing packets overheard on the channel for possible use later. Such information might be the addresses of nodes which have requested for a route to a multicast group source, etc. The route cache is updated as newer information is obtained from the more recent packets heard on the channel. This module is usually optional in most multicast protocols. It increases efficiency by reducing the control overhead.

3. **Application layer:** This layer utilizes the services of the routing layer to satisfy the multicast requirements of applications. It primarily consists of two modules:

   (a) Data packet transmit/receive controller

   (b) Multicast session initiator/terminator

All the above modules and the interactions between them are illustrated in Figure 8.3. The interactions between these modules can be understood by considering some actions that take place during the lifetime of a multicast session:

1. *Joining a group*: Module 10, which exists in the application layer, makes a request to join a group to module 5 present in the routing layer, which can use cached information from module 4 and the unicast route information from module 9. It then initiates flooding of *JoinReq* packets (or other mechanisms) by using module 2 of the MAC layer. These *JoinReq* packets are passed by module 3 of other nodes to their forwarding module, which updates the multicast table and propagates this message. During the reply phase, the forwarding states in the multicast tables of intermediate nodes are established.

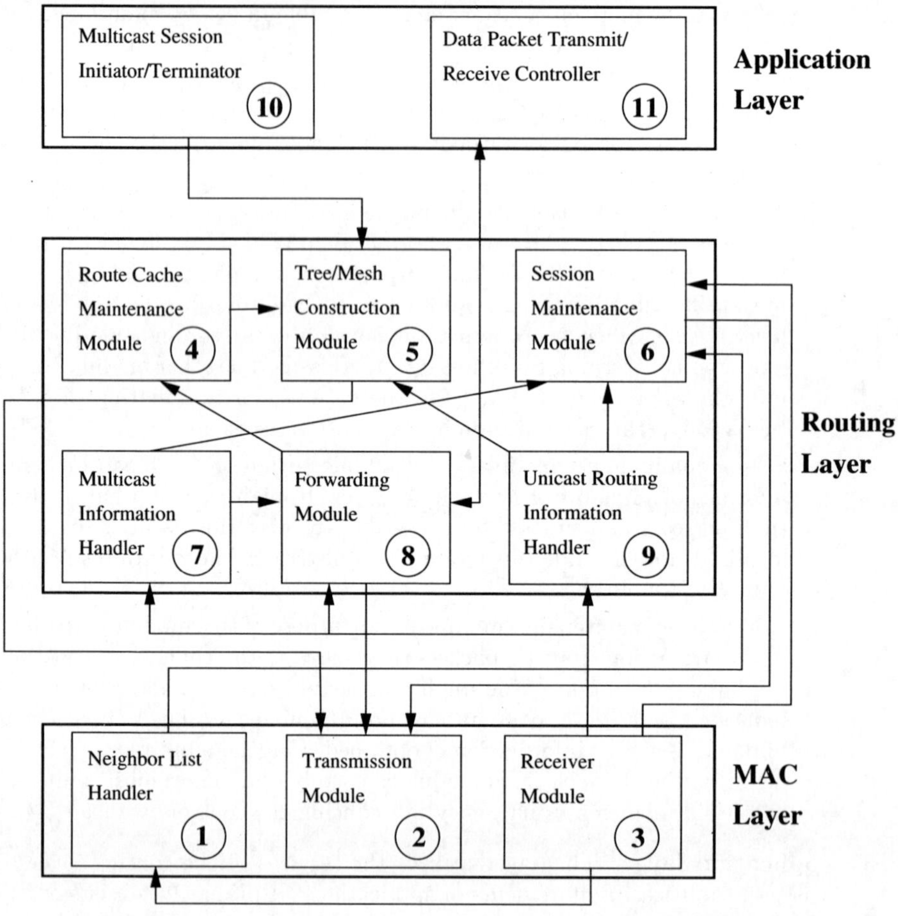

Figure 8.3. Architectural framework of an ad hoc multicast protocol.

2. *Data packet propagation*: Data packets are handled by module 11 in the application layer, which passes them on to module 8 (forwarding module), which makes the decision on whether to broadcast the packets after consulting module 7 (multicast information handler). A similar process occurs in all nodes belonging to the multicast topology until eventually the data packets are sent by the forwarding module of the receivers to the application layer.

3. *Route repair*: Route repair is handled by module 6 on being informed by module 1 of link breaks. It uses the unicast and multicast routing tables to graft the node back into the multicast topology.

Table 8.1. Active modules in different nodes

Type of Node	Active Modules
Source/Receiver	All modules
Intermediate Nodes	All modules of the MAC and routing layers
Other Nodes	Modules 2, 3, 8, optionally module 4

All modules do not operate in all the nodes at any given time. Table 8.1 indicates the different modules which are in operation at different nodes.

8.5 CLASSIFICATIONS OF MULTICAST ROUTING PROTOCOLS

Multicast routing protocols for ad hoc wireless networks can be broadly classified into two types: application-independent/generic multicast protocols and application-dependent multicast protocols (refer to Figure 8.4). While application-independent multicast protocols are used for conventional multicasting, application-dependent multicast protocols are meant only for specific applications for which they are designed. Application-independent multicast protocols can be classified along three different dimensions.

1. **Based on topology:** Current approaches used for ad hoc multicast routing protocols can be classified into two types based on the multicast topology: *tree-based* and *mesh-based*. In tree-based multicast routing protocols, there exists only a single path between a source-receiver pair, whereas in mesh-based multicast routing protocols, there may be more than one path between a source-receiver pair. Tree-based multicast protocols are more efficient compared to mesh-based multicast protocols, but mesh-based multicast protocols are robust due to the availability of multiple paths between the source and receiver. Tree-based multicast protocols can be further divided into two types: *source-tree-based* and *shared-tree-based*. In source-tree-based multicast protocols, the tree is rooted at the source, whereas in shared-tree-based multicast protocols, a single tree is shared by all the sources within the multicast group and is rooted at a node referred to as the *core node*. The source-tree-based multicast protocols perform better than the shared-tree-based protocols at heavy loads because of efficient traffic distribution. But the latter type of protocols are more scalable. The main problem in a shared-tree-based multicast protocol is that it heavily depends on the core node, and hence, a single point failure at the core node affects the performance of the multicast protocol.

2. **Based on initialization of the multicast session:** The multicast group formation can be initiated by the source as well as by the receivers. In a multicast protocol, if the group formation is initiated only by the source node, then

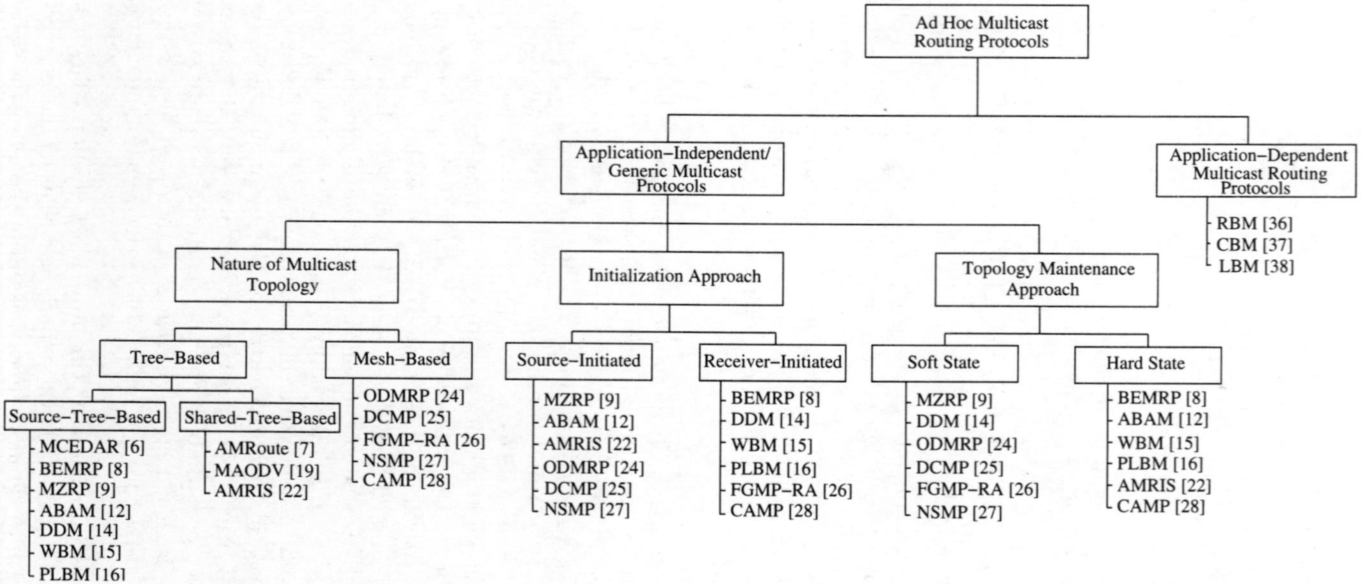

Figure 8.4. Classifications of multicast routing protocols.

it is called a *source-initiated* multicast routing protocol, and if it is initiated by the receivers of the multicast group, then it is called a *receiver-initiated* multicast routing protocol. Some multicast protocols do not distinguish between source and receiver for initialization of the multicast group. We call these *source-or-receiver-initiated* multicast routing protocols.

3. **Based on the topology maintenance mechanism:** Maintenance of the multicast topology can be done either by the *soft state approach* or by the *hard state approach*. In the soft state approach, control packets are flooded periodically to refresh the route, which leads to a high packet delivery ratio at the cost of more control overhead, whereas in the hard state approach, the control packets are transmitted (to maintain the routes) only when a link breaks, resulting in lower control overhead, but at the cost of a low packet delivery ratio.

8.6 TREE-BASED MULTICAST ROUTING PROTOCOLS

Tree-based multicasting is a well-established concept used in several wired multicast protocols to achieve high multicast efficiency. In tree-based multicast protocols, there is only one path between a source-receiver pair. The main drawback of these protocols is that they are not robust enough to operate in highly mobile environments. Tree-based multicast protocols can be classified into two types: source-tree-based multicast routing protocols and shared-tree-based multicast routing protocols (refer to Figure 8.4). In a source-tree-based protocol, a single multicast tree is maintained per source, whereas in a shared-tree-based protocol, a single tree is shared by all the sources in the multicast group. Shared-tree-based multicast protocols are more scalable compared to source-tree-based multicast protocols. By scalability, we mean the ability of the protocol to work well without any degradation in performance when the number of sources in a multicast session or the number of multicast sessions is increased. In source-tree-based multicast routing protocols, an increase in the number of sources gives rise to a proportional increase in the number of source-trees. This results in a significant increase in bandwidth consumption in the already-bandwidth-constrained network. But in a shared-tree-based multicast protocol, this increase in bandwidth usage is not as high as in source-tree-based protocols because, even when the number of sources for multicast sessions increases, the number of trees remains the same. Another factor that affects the scalability of source-tree-based protocols is the memory requirement. When the multicast group size is large with a large number of multicast sources, in a source-tree-based multicast protocol, the state information that is maintained per source per group consumes a large amount of memory at the nodes. But in a shared-tree-based multicast protocol, since the state information is maintained per group, the additional memory required when the number of sources increases is not very high. Hence shared-tree-based multicast protocols are more scalable compared to source-tree-based multicast protocols. The rest of this section describes some of the existing tree-based multicast routing protocols for ad hoc wireless networks.

8.6.1 Bandwidth-Efficient Multicast Routing Protocol

Ad hoc networks operate in a highly bandwidth-scarce environment, and hence bandwidth efficiency is one of the key design criteria for multicast protocols. Bandwidth efficient multicast routing protocol (BEMRP) [8] tries to find the nearest forwarding node, rather than the shortest path between source and receiver. Hence, it reduces the number of data packet transmissions. To maintain the multicast tree, it uses the hard state approach, that is, to rejoin the multicast group, a node transmits the required control packets only after the link breaks. Thus, it avoids periodic transmission of control packets and hence bandwidth is saved. To remove unwanted forwarding nodes, route optimization is done, which helps in further reducing the number of data packet transmissions. The multicast tree initialization phase and the tree maintenance phase are discussed in the following sections.

Tree Initialization Phase

In BEMRP, the multicast tree construction is initiated by the receivers. When a receiver wants to join the group, it initiates flooding of *Join* control packets. The existing members of the multicast tree, on receiving these packets, respond with *Reply* packets. When many such *Reply* packets reach the requesting node, it chooses one of them and sends a *Reserve* packet on the path taken by the chosen *Reply* packet. When a new receiver R3 (Figure 8.5) wants to join the multicast group, it floods the *Join* control packet. The nodes S, I1, and R2 of the multicast tree may receive more than one *Join* control packet. After waiting for a specific time, each of these tree nodes chooses one *Join* packet with the smallest hop count traversed. It sends back a *Reply* packet along the reverse path which the selected *Join* packet had traversed. When tree node I1 receives *Join* packets from the previous nodes I9 and I2, it sends a *Reply* packet to receiver R3 through node I2. The receiver may receive more than one *Reply* packet. In this case, it selects the *Reply* packet which has the lowest hop count, and sends a *Reserve* packet along the reverse path that the selected *Reply* packet had traversed. Here, in Figure 8.5, receiver R3 receives *Reply* packets from source S, receiver R2, and intermediate node I1. Since the *Reply* packet sent by intermediate node I1 has the lowest hop count (which is 3), it sends a *Reserve* packet to node I3, and thus joins the multicast group.

Tree Maintenance Phase

To reduce the control overhead, in BEMRP, tree reconfiguration is done only when a link break is detected. There are two schemes to recover from link failures.

1. Broadcast-multicast scheme: In this scheme, the upstream node is responsible for finding a new route to the previous downstream node. This is shown in Figure 8.6. When receiver R3 moves from A to B, it gets isolated from the remaining part of the tree. The upstream node I3 now floods broadcast-multicast packets (with limited TTL). After receiving this packet, receiver R3 sends a *Reserve* packet and joins the group again.

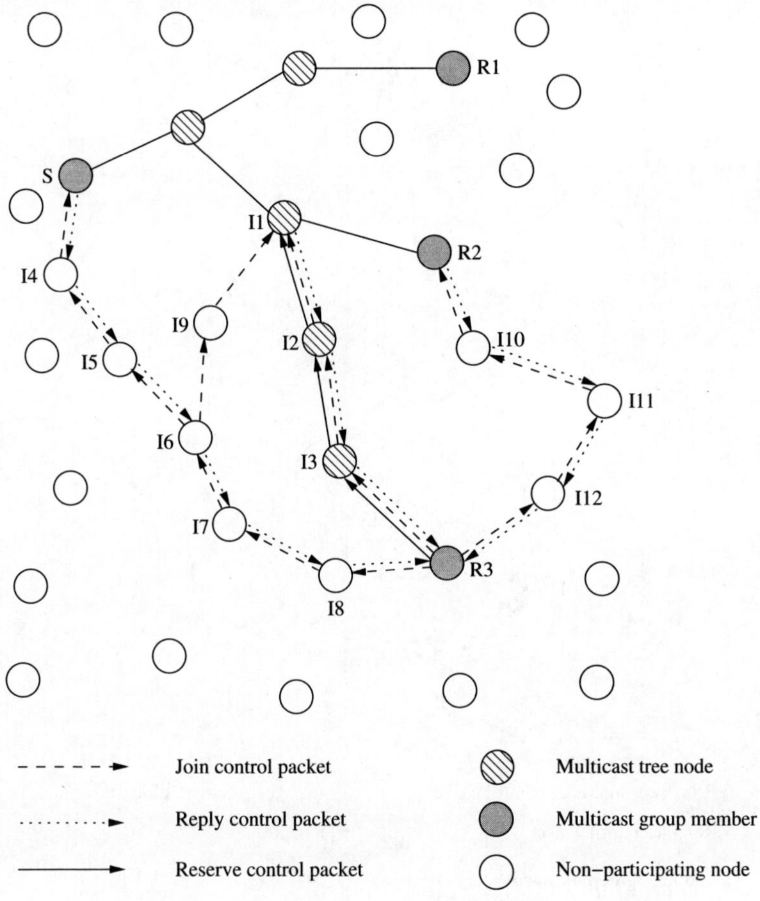

Figure 8.5. Multicast tree initialization in BEMRP.

2. Local rejoin scheme: In this scheme, the downstream node of the broken link tries to rejoin the multicast group by means of limited flooding of the *Join* packets. In Figure 8.7, when the link between receiver R3 and its upstream node I3 fails (due to movement of node R3), then R3 floods the *Join* control packet with a certain TTL value (depending on the topology, this value can be tuned). When tree nodes receive the *Join* control packet, they send back the *Reply* packet. After receiving the *Reply* packet, the downstream node R3 rejoins the group by sending a *Reserve* packet to the new upstream node I4.

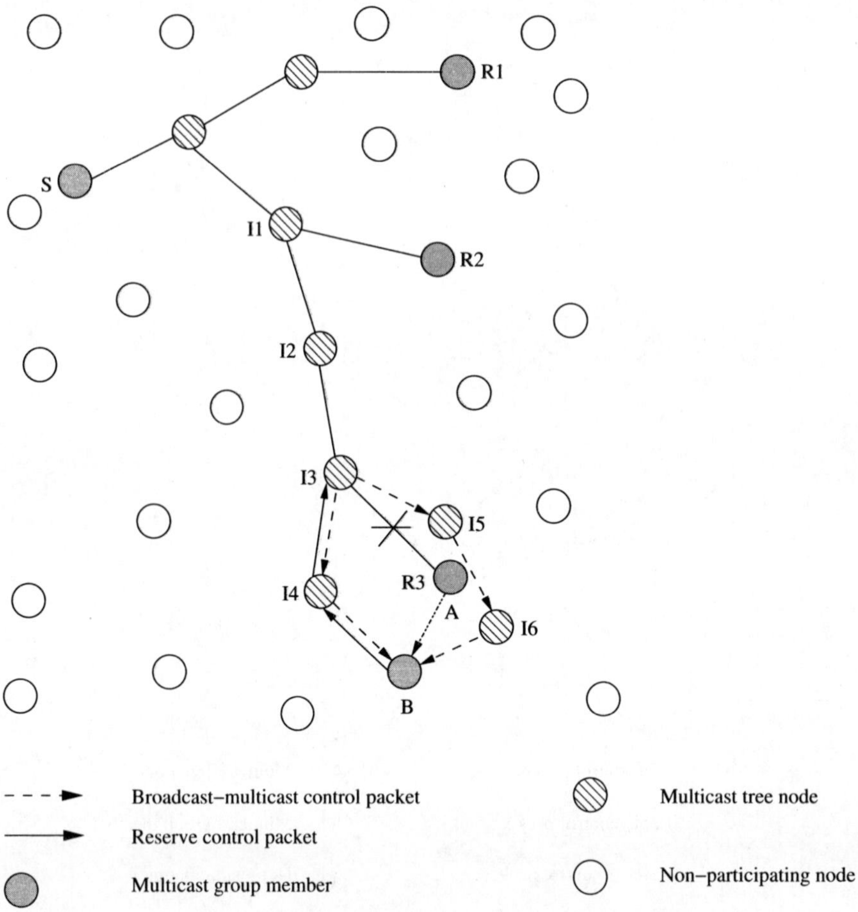

Figure 8.6. Multicast tree maintenance in broadcast-multicast scheme.

Route Optimization Phase

When a tree node or a receiver node comes within the transmission range of other tree nodes, then unwanted tree nodes are pruned by sending the *Quit* message. In Figure 8.8, when receiver R3 comes within the transmission range of the intermediate node I2, it will receive a multicast packet from node I2 earlier than from node I5. When node R3 receives a multicast packet from node I2, it sends a *Reserve* packet to node I2 to set up a new route directly to node I2, and sends a *Quit* packet to node I5. Since node R3 is no more its downstream node, node I5 sends a *Quit* packet to node I4, node I4 sends a *Quit* packet to node I3, and node I3 in turn sends a *Quit* packet to node I2. Thus unnecessary forwarding nodes are pruned. This mechanism helps to reduce the number of data packet transmissions.

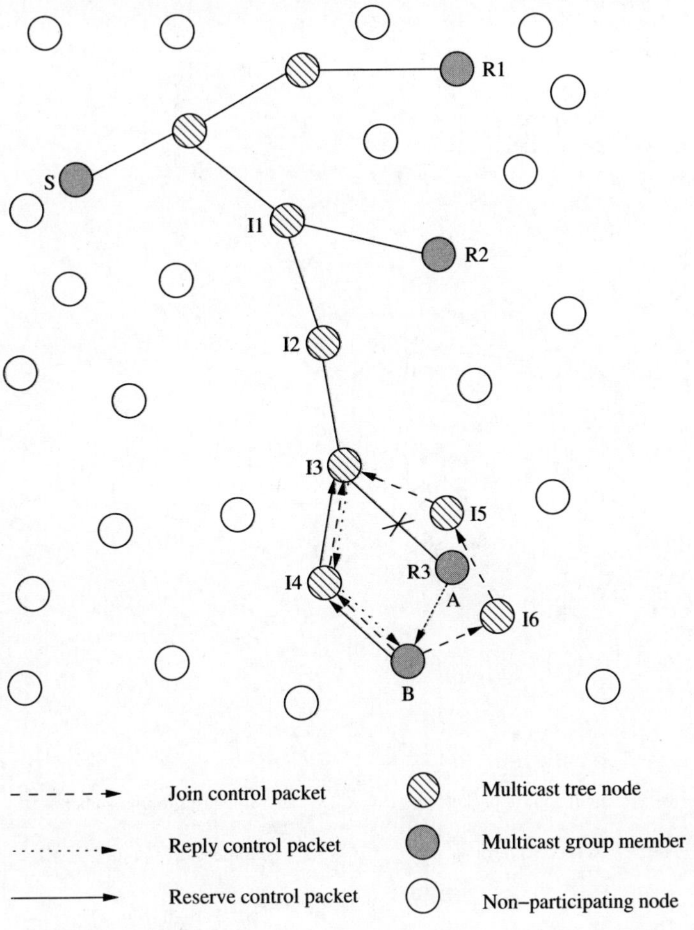

Figure 8.7. Multicast tree maintenance in local rejoin scheme.

Advantages and Disadvantages

The main advantage of this multicast protocol is that it saves bandwidth due to the reduction in the number of data packet transmissions and the hard state approach being adopted for tree maintenance. Since a node joins the multicast group through its nearest forwarding node, the distance between source and receiver increases. This increase in distance increases the probability of path breaks, which in turn gives rise to an increase in delay and reduction in the packet delivery ratio. Also, since the protocol uses the hard state approach for route repair, a considerable amount of time is spent by the node in reconnecting to the multicast session, which adds to the delay in packet delivery.

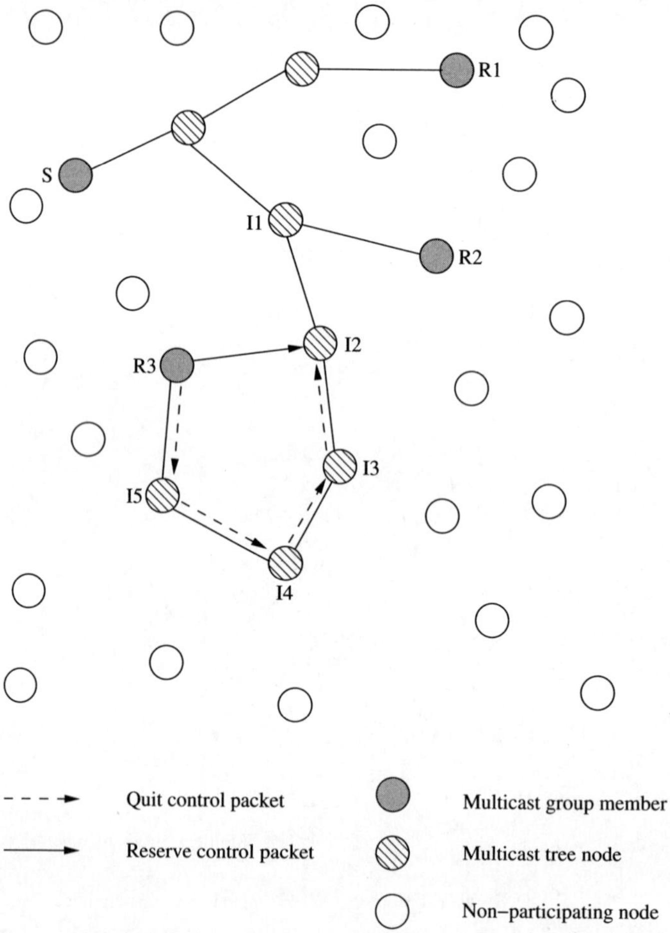

Figure 8.8. Multicast tree optimization in BEMRP.

8.6.2 Multicast Routing Protocol Based on Zone Routing

In multicast zone routing protocol (MZRP) [9], the flooding of control packets by each node which searches for members of the multicast group is controlled by using the *zone routing mechanism* [10]. In zone routing, each node is associated with a routing zone. For routing, a pro-active approach is used inside the zone (the node maintains the topology inside the zone, using a *table-driven* routing protocol), whereas a reactive approach is used across zones. In a nutshell, it attempts to combine the best of both on-demand and table-driven routing approaches.

Tree Initialization Phase

To create a multicast delivery tree over the network, the source initiates a two-stage process. In the first stage, the source tries to form the tree inside the zone, and then in the second stage it extends the tree to the entire network. In Figure 8.9, to create the tree, initially source S sends a TREE-CREATE control packet to nodes within its zone through unicast routing as it is aware of the topology within its zone, then node R1, which is interested in joining the group, replies with a TREE-CREATE-ACK packet and forms the route (for the sake of clarity, routing zones of nodes have been represented as circular regions in the figure; however, they need not always take the shape of circles). To extend the tree outside the zone, source S sends a TREE-PROPAGATE packet to all the border nodes of the zone. When node B

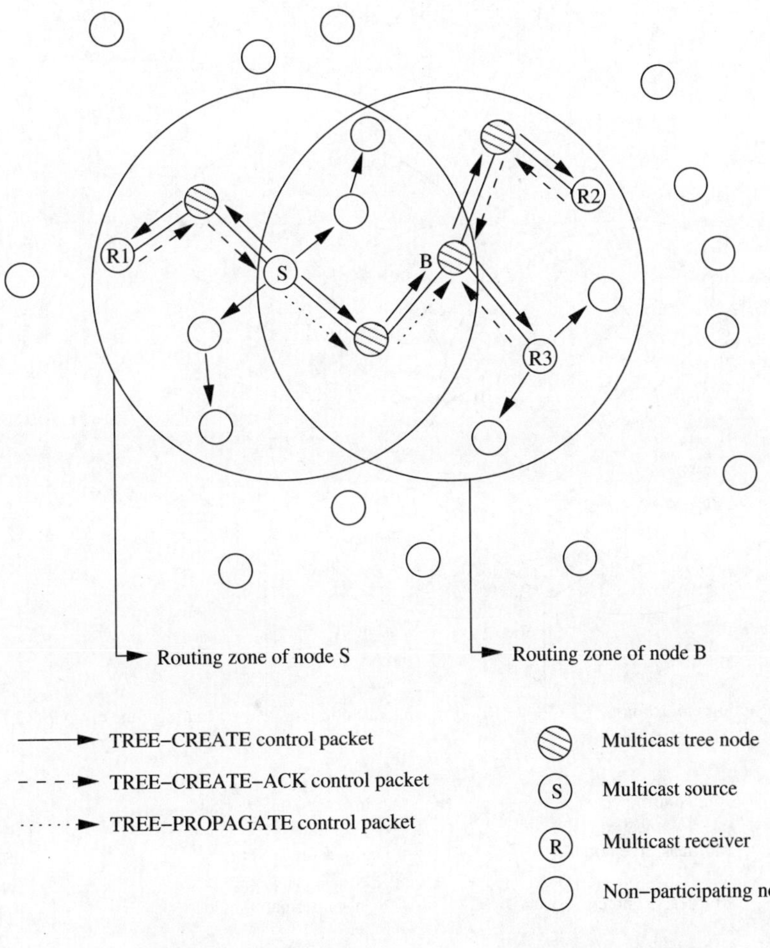

Figure 8.9. Multicast tree initialization in MZRP.

(which is at the border of the zone of node S) receives the TREE-PROPAGATE packet, it sends a TREE-CREATE packet to each of its zone nodes. Thus receivers R2 and R3 receive the TREE-CREATE packets and join the multicast session by sending TREE-CREATE-ACK packets.

Tree Maintenance Phase

Once the multicast tree is created, the source node periodically transmits TREE-REFRESH packets down the tree to refresh the multicast tree. If any tree node does not receive a TREE-REFRESH packet within a specific time period, it removes the corresponding stale multicast route entry. When a link in the multicast tree breaks, downstream nodes are responsible for detecting link breaks and rejoining the multicast group. Due to movement of the intermediate node I (Figure 8.10), receiver R2 gets isolated from the rest of the multicast tree. Node R2 first unicasts a

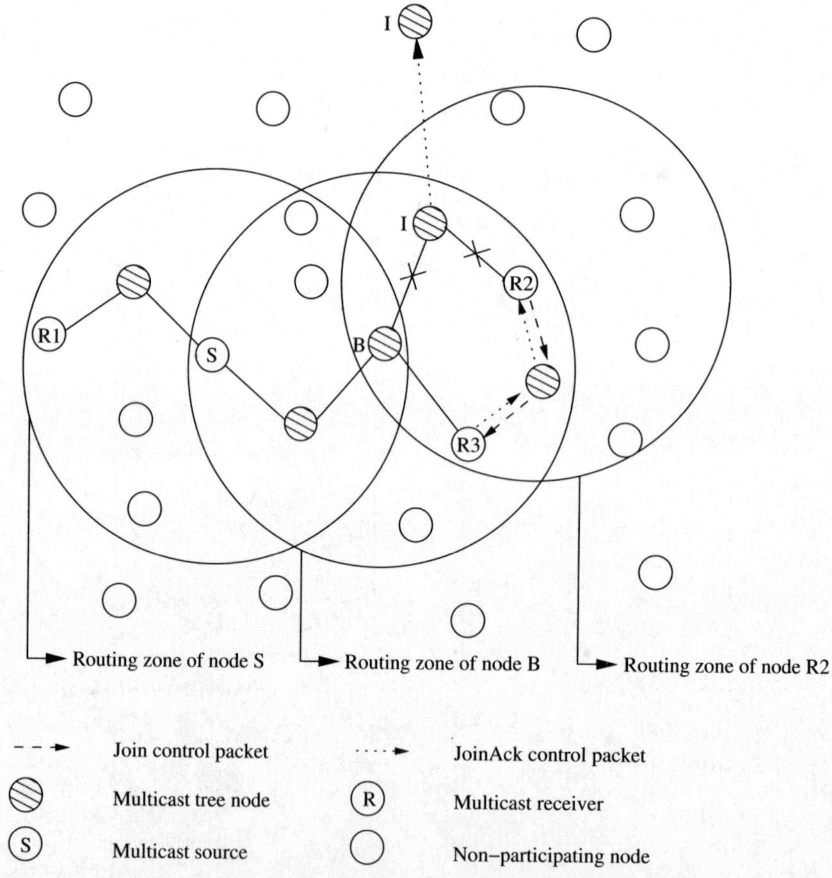

Figure 8.10. Multicast tree maintenance in MZRP.

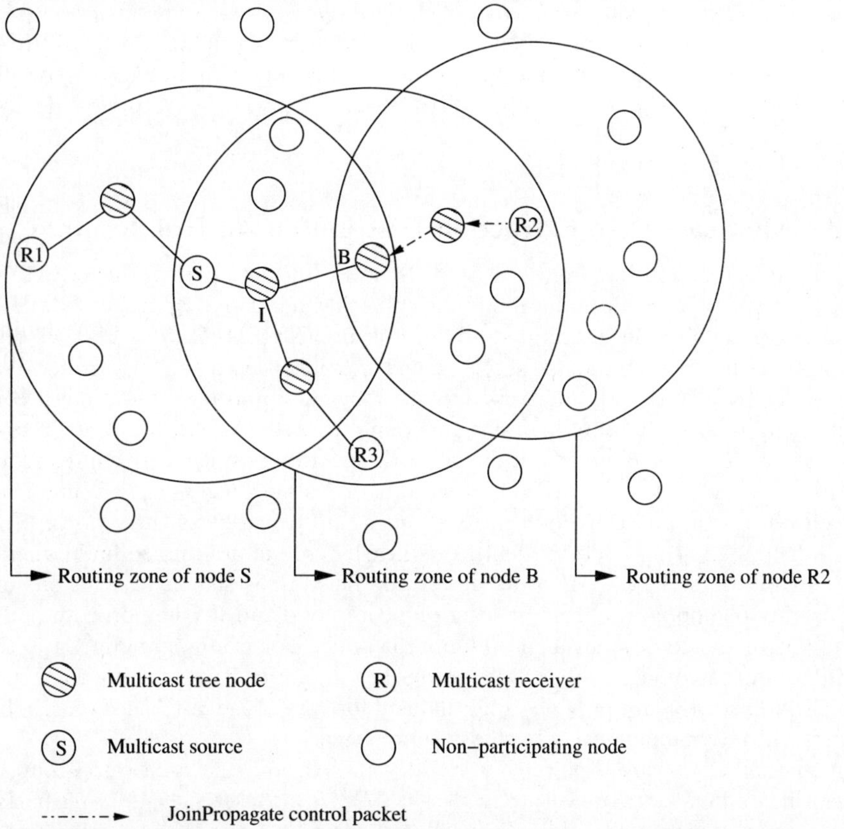

Figure 8.11. Multicast tree maintenance in MZRP.

Join packet to all zone nodes. Since a tree node R3 is already in the zone, it replies back by sending a *JoinAck* to node R2. Thus receiver R2 again joins the multicast group. It may be that there are no tree nodes in the zone of receiver R2. In this case, receiver R2 does not get any reply from zone nodes, it sends *JoinPropagate* control packets to border nodes (node B), and it joins the tree through intermediate node I (see Figure 8.11).

Advantages and Disadvantages

MZRP has reduced control overhead as it runs over ZRP. The fact that the unicast and multicast routing protocols can exchange information with each other is another advantage of MZRP. MZRP is important as it shows the efficacy of the zone-based approach to multicast routing. The size of the zone is very important in MZRP. The size should be neither too large nor too small. The optimum value for the zone radius is likely to vary with multicast group size, network load conditions, etc. A disadvantage of this protocol is that a receiver node which is located far off from the source needs to wait for a long time before it can join the multicast session, because the propagation of the TREE-PROPAGATE message takes considerable time.

8.6.3 Multicast Core-Extraction Distributed Ad Hoc Routing

It is known that tree-based multicast protocols are efficient but less robust. To increase the robustness while maintaining the efficiency, a different approach is used in multicast core-extraction distributed ad hoc routing (MCEDAR) [6]. A source-tree over an underlying mesh infrastructure called *mgraph* is used for forwarding data packets. The CEDAR [11] architecture is used by this protocol for the mesh construction. In this architecture, a minimum dominating set (MDS), which consists of certain nodes (called core nodes) in the network, is formed using a core computation algorithm. After joining the MDS, each core node issues a piggy-backed broadcast through its beacon packet to inform its presence up to the next three hops. This process helps each core node to identify its nearby core nodes and to build virtual links. In Figure 8.12, the MDS of the ad hoc network is shown. Each non-core node is only one hop away from at least one core node and it (the non-core node) selects one of the core nodes as its dominator node. For example, node C4 is the dominator of the non-core node R3 (Figure 8.12). In addition to creating the MDS, the CEDAR architecture provides a mechanism for core broadcast based on reliable unicast, which dynamically establishes a source-tree.

In Figure 8.12, when a new receiver R3 wants to join the multicast group, it requests its dominator (node C4) to transmit a *JoinReq* packet. The *JoinReq* packet consists of an option called *JoinID*, which is used to prevent any loop formation in the *mgraph*. Initially, the value of *JoinID* is set to infinity. When a tree node (node C1) of the multicast group receives this *JoinReq* packet, it replies with a *JoinAck* packet if its *JoinID* is less than the *JoinID* of the requesting node. Before sending the *JoinAck*, the node sets its own *JoinID* in the *JoinAck* packet. When an intermediate node on the reverse path receives the *JoinAck* packet, it decides on whether to accept or reject it based on one parameter, the *robustness factor*. If the number of *JoinAck* packets received by the intermediate node C3 is less than the robustness factor, then it accepts the *JoinAck* packet and adds the upstream *mgraph* member to its parent set; otherwise, it rejects it. Afterwards, intermediate node C3 forwards the packet to its downstream core node C4 and adds the downstream node to its child set.

Now, the dominator node C4 may receive more than one *JoinAck* packet from the members of the multicast group. For example, if the robustness factor is 2, and if core node C4 receives three *JoinAck* packets, it accepts only two *JoinAck*s and rejects the third. In this way, the receiver joins the multicast group. Although the *mgraph* for the multicast group is a mesh structure, the forwarding of data packets is done only on a source-tree due to the core broadcast mechanism (Figure 8.13). If the core node C4 loses connectivity with all its parents due to movement, then

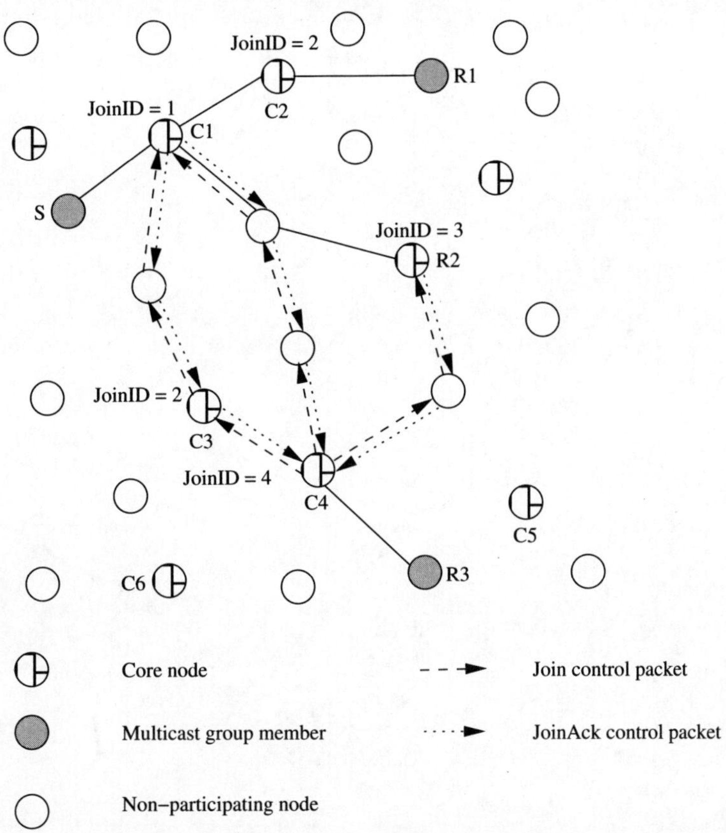

Figure 8.12. *JoinReq* sent by dominating core node.

it issues a *JoinReq* again (with its current *JoinID*) and joins the multicast group. To reduce the overhead on *mgraphs*, optimization is performed by decoupling the control infrastructure from the data-forwarding infrastructure.

Advantages and Disadvantages

Due to the underlying mesh structure, this multicast routing protocol is robust, and using source-tree over mesh for forwarding the data packets makes it as efficient as other tree-based multicast routing protocols. Depending on the *robustness factor* parameter, the dominator node of a receiver has multiple paths to the multicast session. So even if the current path breaks, the dominator node always has an alternate path to the multicast session. But the disadvantage of MCEDAR is that it is more complex compared to other multicast routing protocols. The increase in complexity may not result in a proportional increase in the performance of the

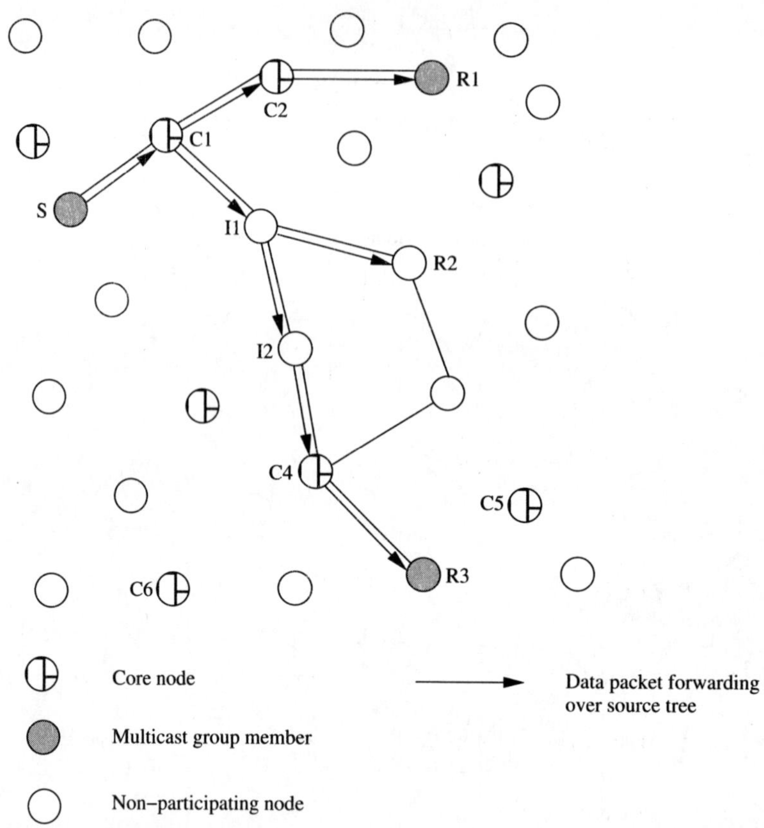

Figure 8.13. Data forwarded over source-tree.

protocol. When the mobility is very high, the nodes need to frequently change their core nodes, which increases the control overhead.

8.6.4 Associativity-Based Ad Hoc Multicast Routing

Associativity-based ad hoc multicast routing (ABAM) [12] is an on-demand source-tree-based multicast protocol in which a path (from source to receiver) is constructed based on link stability rather than hop distance. Hence, this multicast protocol is adaptive to the network mobility.

Tree Initialization Phase

The source node initiates the multicast tree construction phase. Joining a group is a three-step process: flooding by the source, replies along the stable path by the receivers, and path setup by the source. The tree initialization phase is illustrated in

Figure 8.14. For creating the multicast tree, initially the source broadcasts a multicast broadcast query (*MBQ*) packet in the network to inform all potential receivers. When an intermediate node receives the *MBQ* packet, it appends its ID, associativity ticks [13] (the number of beacons continuously received from neighboring nodes which reflects the stability of the link), along with QoS information, to the *MBQ* packet, and rebroadcasts it. After receiving *MBQ* packets through different paths, each of the receivers R1, R2, and R3 responds by sending an *MBQ-Reply* packet through the most stable path. After receiving a number of *MBQ-Reply* packets, the source sends *MC-Setup* packets to all receivers in order to establish the multicast tree.

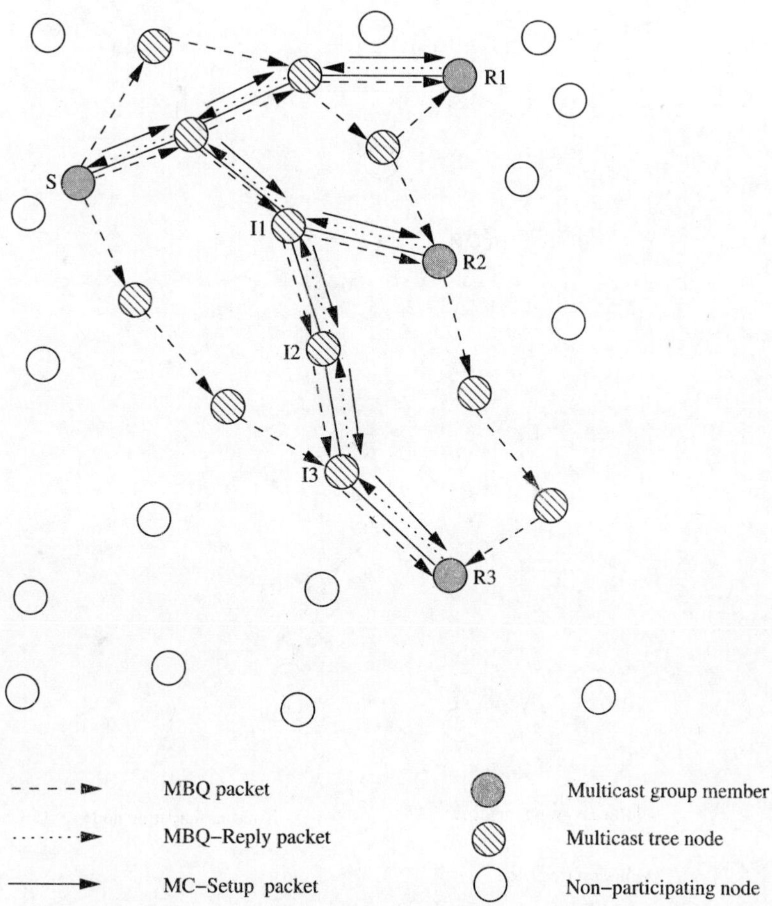

Figure 8.14. Multicast tree initialization in ABAM.

Tree Maintenance Phase

Due to the movement of the nodes, the links between nodes break frequently. ABAM uses different procedures to repair the multicast tree, depending on the type of the moved node. If a leaf receiver R3 moves and the link breaks (Figure 8.15), then the upstream node I3 of the broken link tries to find a new route to the receiver by transmitting a *LocalQuery* packet with the TTL field set as 1. When receiver R3 receives a *LocalQuery* packet from the upstream node, it replies by sending a *LocalQuery-Reply* message. Receiver R3 now rejoins the multicast group after receiving an *MC-Setup* packet from the upstream node. If the receiver fails to join the group again, then the immediate upstream node I2 of node I3 takes the responsibility of finding a route to R3 by transmitting a *LocalQuery* message with the TTL field value set to 2. This backtracking process terminates at branch node

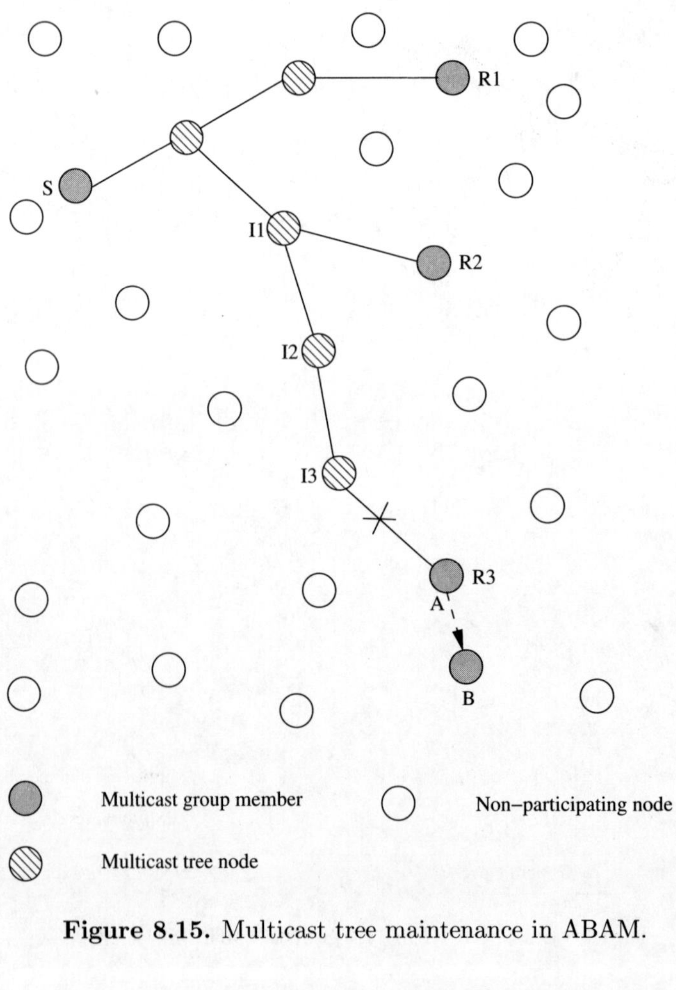

Figure 8.15. Multicast tree maintenance in ABAM.

I1, if all previous attempts fail, or the timer (this timer is set immediately after a link failure is detected) expires at node R3. After the timer expiry, receiver R3 sends a *JoinQuery* packet to join the multicast group.

When a branch node moves away, then many receivers connected through that branch node get affected. Here in Figure 8.15, when branch node I1 moves, then receivers R2 and R3 are cut off from the multicast group. In this case, the branch node I1 broadcasts a *LocalQuery* packet with TTL field value of 3 (the hop distance between branch node I1 and the farthest affected receiver, R3). ABAM allows a new receiver to join the multicast group (if it has not joined the multicast group at the time of tree establishment) by sending a *JoinQuery* packet. To leave the group, a receiver sends the *Leave* message to its upstream node, and that branch gets pruned if there are no other receivers on that branch.

Advantages and Disadvantages

In ABAM, the path between a source and receiver is more stable compared to other multicast protocols, and hence it achieves a higher packet delivery ratio. Also, the control overhead is less due to a fewer number of link failures. But increased hop distance between the source-receiver pair makes the protocol less efficient. When there are a lot of receivers belonging to the same multicast session close by, it results in congestion of the most stable path, which in turn may result in increased delay and reduction in the packet delivery ratio. As such, the protocol is not scalable. For scalability, it needs to employ load-balancing techniques.

8.6.5 Differential Destination Multicast Routing Protocol

The multicast protocols discussed above follow *distributed group membership management* and *distributed multicast routing state maintenance* policies. A totally different approach to ad hoc multicast routing is proposed in differential destination multicast (DDM) [14] routing protocol, which is a stateless multicast routing protocol that avoids maintaining multicast states in the nodes. It is particularly applicable where the group size is small.

Tree Initialization Phase

To join a particular multicast session, each interested destination node unicasts a *Join* control packet to the source. When the source receives a *Join* packet from a destination, it checks its validity (for security reasons) and sends an *ACK* control packet to the destination after storing the destination address in its member list (ML). Each destination periodically sends *Join* control packets to the source to show its interest in the multicast session. These *Join* control packets refresh the ML table at the source. The source removes stale member information if it does not receive any *Join* message from that particular destination for a certain time period.

Tree Maintenance Phase

DDM can operate in *stateless mode* as well as in *soft state* mode to maintain the tree. Stateless mode is straightforward, where source S inserts the destination address in a field called DDM block of the data packet and unicasts it to the next node I1, using the underlying unicast routing protocol (see Figure 8.16). When node I1 receives the data packet, it gets the next hop address from the DDM block of the received data packet and unicasts the packet to nodes I2 and I3. DDM blocks in the data packets sent to nodes I2 and I3 contain the address of destinations D1 and D2, respectively. In the soft state mode, each node along the forwarding path remembers the destination address by storing it in the forwarding set (FS). By caching this information, there is no need to list all the destination addresses in every data packet. When changes occur, the upstream node informs the downstream

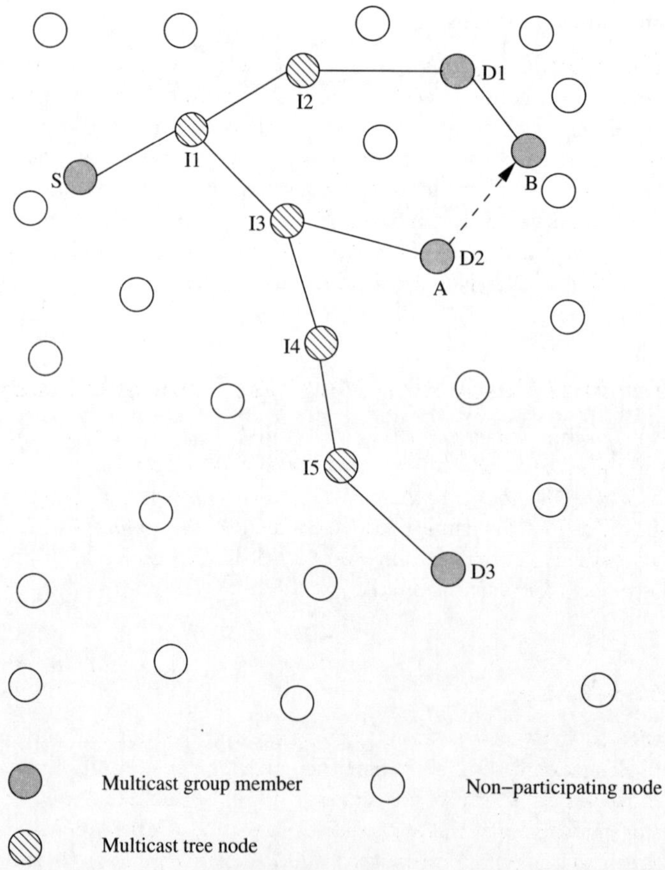

Figure 8.16. Multicast tree maintenance in DDM.

node about the difference through a *D-type* DDM block. In Figure 8.16, due to the movement of destination D2 from location A to B, D2 loses contact with I3. Since there exists an alternate route to D2 through intermediate node I2, node I1 now sends a D-type DDM block to nodes I2 and I3 informing them about the changes. To detect the loss of data packets which contain the D-type DDM block, both the upstream node and the downstream node maintain the DDM block sequence number. When there is a loss of these data packets, the downstream node sends an RSYNC message to its upstream node to get the complete list of destination addresses.

Advantages and Disadvantages

Since DDM does not maintain the multicast state, it uses minimum memory resources. Due to the centralized admission control policy, security is assured. Any shift in the location of a receiver results in the automatic rerouting of packets to its new location. But the main drawback is that it is not scalable when the multicast group size increases. DDM involves periodic control packet transmissions from the receiver nodes to the source node. This results in a significant consumption of bandwidth when the number of receivers is high. As the number of receivers increases, the size of the DDM block carried in the data packets becomes large, leading to more bandwidth consumption. Hence, this multicast protocol is suitable only when the multicast group size is small.

8.6.6 Weight-Based Multicast Protocol

The weight-based multicast (WBM) [15] protocol uses the concept of weight when deciding upon the entry point in the multicast tree where a new multicast member node is to join. Before joining a multicast group, a node takes into consideration not only the number of newly added forwarding nodes (here, forwarding nodes refers to the nodes which are currently not part of the multicast tree, but which need to be added to the tree in order to connect the new multicast receiver to the multicast session), but also the distance between the source node and itself in the multicast tree. The weight concept provides flexibility for a receiver to join either the nearest node in the multicast tree or the node nearest to the multicast source.

Tree Initialization Phase

The main aim here is to find the best point of entry for a new node joining the multicast group. A receiver-initiated approach is adopted here. When a new receiver R5 (Figure 8.17) intends to join the group, it broadcasts a *JoinReq* packet with a certain time-to-live (TTL) value set. These *JoinReq* packets are forwarded until they are received by a tree node. Consider the example illustrated in Figure 8.17. Upon receiving a *JoinReq* packet, a tree node for example I1, sends a *JoinReply* (*Reply*) packet. There can be several such replier nodes which send *Reply* packets. The *Reply* packet initially contains the distance of the node I1 from the source S. As *Reply* packets are forwarded, the count of hops taken from I1 is maintained in the *Reply* packet. Thus, the *Reply* packet upon its receipt at node R5, will have the

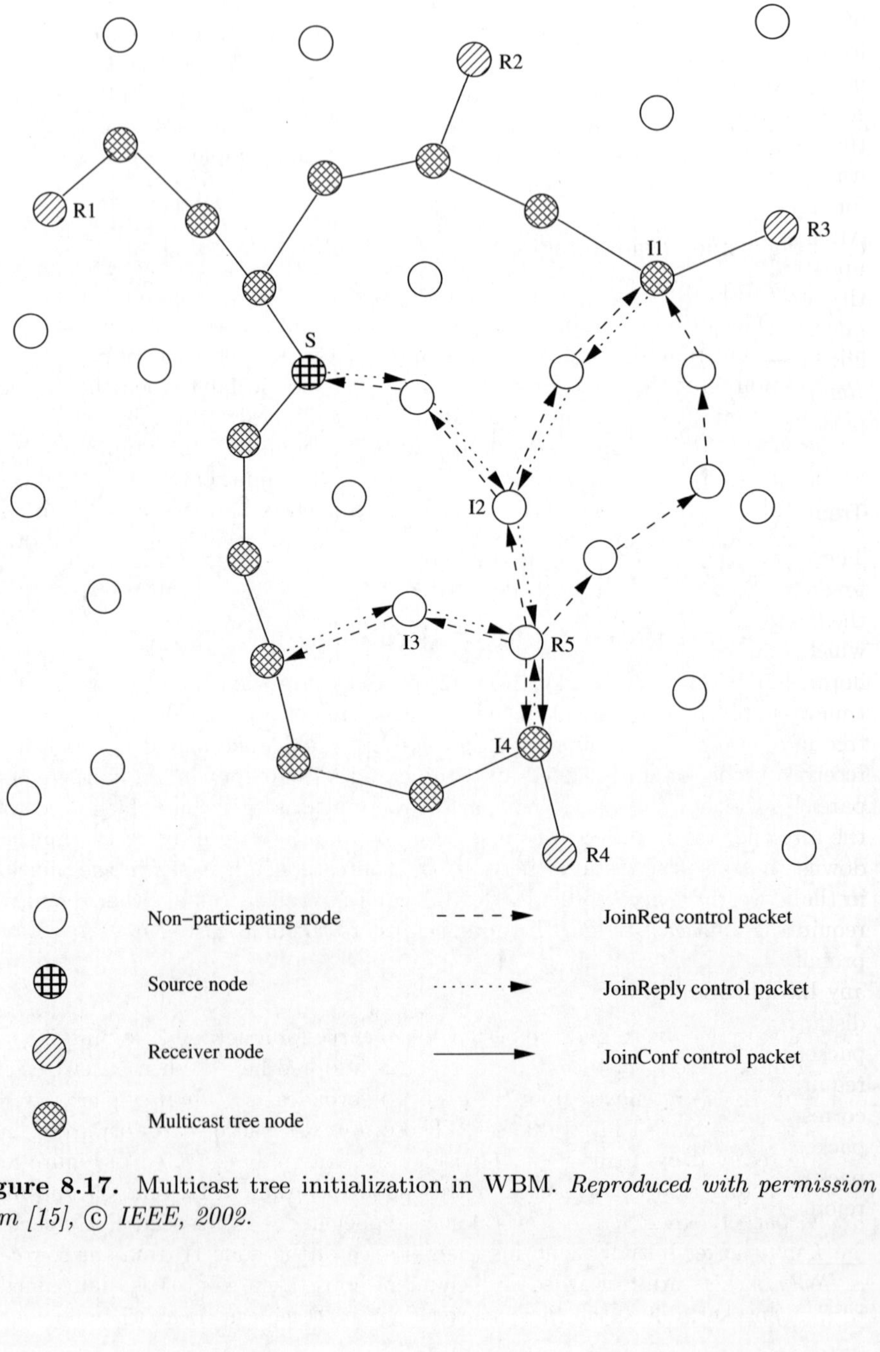

Figure 8.17. Multicast tree initialization in WBM. *Reproduced with permission from [15], © IEEE, 2002.*

firm packet from that node. Otherwise, it forwards the *JoinQuery* using the same procedure described above. The *JoinReply* packet follows the route traversed by the *JoinQuery* packet, in the reverse order. When an intermediate node receives a *JoinReply* packet, a check is made to verify that the packet had not been processed already. Any processed *JoinReply*, identified through the *multicast group ID*, *multicast source ID*, and *sequence number* fields on the packet is discarded, while unprocessed *JoinReply* packets are forwarded. Similar to the *JoinQuery* transmission, the *JoinReply* message is transmitted using a unicast approach. All neighboring nodes process the *JoinReply* in their promiscuous mode and mark the *JoinReply* as processed. This prevents multiple *JoinReply* packets from reaching the *JoinQuery* source node (new receiver node) and also reduces the transmission of redundant *JoinReply* packets. When the first *JoinReply* reaches the *JoinQuery* source node, the *JoinQuery* source node confirms its connectivity to the sender of this *JoinReply* by sending back a *JoinConfirm* packet. Each intermediate node receiving the *JoinConfirm* packet marks itself as connected and forwards the *JoinConfirm* packet. It also stores, in its CT, the path information regarding the next two hops on both sides, obtained using the path information carried by the *JoinConfirm* packet. The path information toward the multicast source node and the path information toward the member node are termed uplink and downlink information, respectively.

In Figure 8.18, when node 1 wants to connect to the multicast source (node 18), it first computes the PL using PLBA. Nodes 3 and 4 are the nodes in the preferred list, as determined using PLBA. When node 3 and node 4 receive the *JoinQuery* packet from node 1, they in turn compute their preferred neighbors using PLBA. Node 3 sends *JoinQuery* to nodes 10 and 11, while node 4 forwards *JoinQuery* to nodes 5, 8, and 9. The *JoinQuery* is dropped at nodes 5 and 11 as no preferred nodes are available at these nodes. Nodes 8, 9, and 10 further forward *JoinQuery* to nodes 16, 16, and 9, respectively. Finally, a single *JoinQuery* reaches the multicast source, which then sends a *JoinReply* to node 1 through path 18-16-9-4-1 (path 1) [or 18-16-8-4-1 (path 2)]. Node 1 finally confirms its connectivity to the multicast source node, that is, to node 18. When member node 6 wants to connect to the multicast source, it first checks whether any tree node (forwarding nodes or connected member nodes) is in its NNT. As node 4 is a forwarding node for the multicast group, node 6 connects directly to the multicast group by sending *JoinConfirm* to node 4 without flooding *JoinQuery*. For node 12, which wants to join the multicast group, no tree nodes are present in its NNT. It initiates a *JoinQuery* to be flooded in a limited manner in the network using the PLBA algorithm. It forwards *JoinQuery* to node 13 and node 11. Nodes 13 and 11 have tree nodes in their NNTs and hence they forward *JoinQuery* directly to those nodes. Node 9 receives *JoinQuery* through node 10, forwarded by node 13. Similarly, node 4 receives *JoinQuery* forwarded by node 11 through node 3. Node 12 receives two *JoinReply* packets, one from node 4 and one from node 9. It connects to the forwarding node whose *JoinReply* comes first, by sending a *JoinConfirm* message.

Instead of transmitting the data packet in the broadcast mode, which is more prone to collisions, the PLBM protocol, taking advantage of the promiscuous mode support at the nodes, transmits each data packet as a unicast packet, which provides

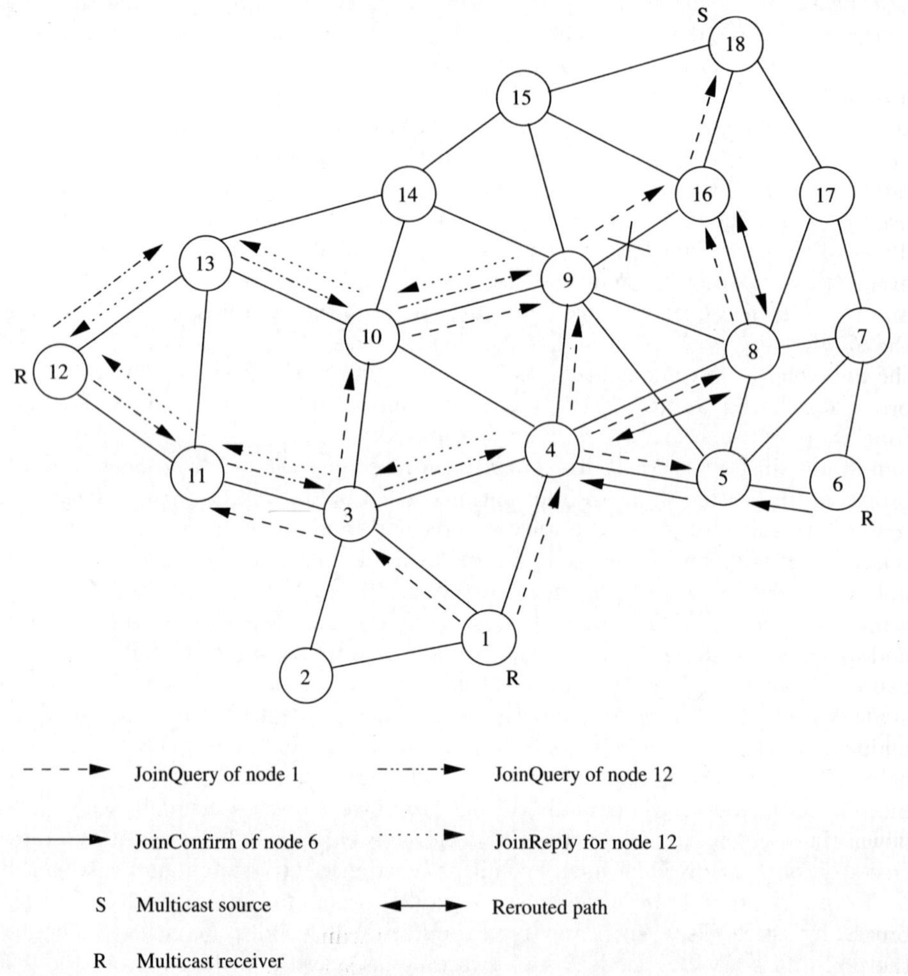

Figure 8.18. Multicast tree initialization in PLBM. *Reproduced with permission from [16], © IEEE, 2003.*

reliable data packet delivery at the cost of slightly higher processing overhead. A node piggy-backs the list of downlink nodes on each multicast data packet it transmits. This serves the following purposes. It explicitly specifies the downlink nodes that are eligible to further forward the data packet, thereby eliminating the possibility of the formation of loops. It also helps in informing nodes hearing the packet about tree nodes that are located within their neighborhood. Thirdly, it helps in removing inconsistencies in the CTs maintained at the nodes.

Tree Maintenance Phase

In the PLBM multicast protocol, link breaks can be detected in two ways. Since the PLBM protocol is beacon-based, a node can detect a link break if it has not received a beacon for a certain duration of time, called the T_{CkBcon} period. The selection of the T_{CkBcon} period is very important. If it is too long, then link break detection is delayed, resulting in loss of data packets. If it is too short, it results in detection of false link breaks as beacons are also prone to collisions. The other mechanism that is used to detect a link break is based on unicast packet transmission characteristics. Transmission of each unicast packet is preceded by the RTS-CTS control packet exchange. A link is assumed to be broken if the sender node does not receive any CTS packet in response to multiple retransmissions of RTS packets. Since in the PLBM protocol each data packet is transmitted as a unicast packet to one of the preferred neighbor nodes (other preferred nodes receive the data packet in the promiscuous mode), a link break can be detected quickly. Using the two-hop local topology information maintained in the NNT, and the two-hop tree information maintained in the CT, the broken link is bypassed quickly. The end node of the broken link toward the multicast source is termed the uplink node, while the one toward the member node is termed the downlink node. The following steps are taken by a node when it detects a link break. It deletes the neighbor (moved uplink node) and its neighbors' information from its NNT and CT. If the node is a multicast group member or has multiple downlink nodes or member nodes, the node tries to repair the tree by reconnecting to any of the tree nodes in its local two-hop topology. If no tree nodes are available, it initiates a *JoinQuery* and sends a *StickToMe* message to its downlink nodes. The purpose of this message is to maintain intact the subtree for which the current downlink node is the root. This helps in reducing the flooding of *JoinQuery* packets by multiple nodes when a single link breaks. If only a single downlink node is present, the node informs the single downlink member node about the link break, which then reconnects to the multicast tree as described in the tree initialization phase.

In Figure 8.18, consider the case when node 9 moves away. This causes a link break between node 16 and node 9. On detecting this link break, node 16 reroutes the data packets using the two-hop topology information from its NNT and the two-hop path information from its CT. Node 9 is bypassed, and the new path between multicast source node 18 and receiver node 1 is 18-16-8-4-1. When node 10 is connected to the tree through node 9 and does not receive data packets for a specified waiting period, it assumes a link break. Each forwarding node tries to locally repair the tree using its NNT and CT information. In case no tree node is present in its NNT and more than one downlink node is present in its CT, it broadcasts the *JoinQuery* packet. Otherwise, the link break information is propagated to the corresponding receiver node (node 12). Node 12 refloods the *JoinQuery* using PLBA algorithm in order to reconnect to the multicast tree.

Advantages and Disadvantages

The PLBM protocol makes use of two-hop local topology information for efficient multicast routing. PLBM provides better flexibility and adaptability through the preferred link concept. The criterion for selecting the preferred list need not be restricted to the neighbor degree alone (PLBA algorithm); any other node or link characteristic (such as link stability, link load, residual bandwidth, and link delay) can be used for computing the preferred links. One of the shortcomings of PLBM is the usage of periodic beacons. Transmission of periodic beacons by each and every node in the network incurs significant control overhead in the already bandwidth-constrained ad hoc wireless networks.

Preferred Link-Based Unified Routing Protocol

The preferred link-based unified routing protocol (PLBU) [18], which is an extension of the PLBM protocol described above, is the only known protocol of its kind to date. PLBU routes both unicast as well as multicast traffic simultaneously in a transparent manner. Using separate unicast and multicast routing protocols has several disadvantages. The actual traffic in an ad hoc wireless network consists of simultaneous unicast and multicast sessions. Separate unicast and multicast protocols generate separate overheads/control packets; the overheads which are redundant in most cases lead to wastage of bandwidth and decrease in the overall efficiency of the system. A unicast session (*e.g.,* a voice session) may need to be converted into a multicast session at any time. This would be very complex if separate unicast and multicast protocols are used. PLBU overcomes the above disadvantages. PLBU handles unicast as well as multicast traffic seamlessly. The advantages of a unified approach are as follows. The need for a separate protocol for each type of traffic is eliminated, which reduces complexity and the memory requirements in resource-constrained ad hoc wireless networks. The path established by multicast sessions can be used by unicast sessions, and vice versa. This reduces the control overhead in the network.

PLBU has two phases: *connect phase* and *reconfiguration phase.* For a unicast session, the source node initiates the connect phase for getting connected to the destination node, whereas for a multicast session, receiver nodes invoke the connect phase in order to connect to the multicast source. A node that initiates a connect phase is termed *ConnectSource* node. A receiver-initiated approach is used for multicast tree construction. It is performed in the same manner as is done in PLBM. In the connect phase, the *ConnectSource* node finds a route to the appropriate destination node, which is the multicast source for a multicast session and the final destination node for a unicast session. During the connect phase, the *ConnectSource* node initiates the transmission of *RouteProbe* packets. For a unicast session, only the destination node is allowed to respond to this *RouteProbe* packet. In case of a multicast session, any multicast tree node receiving the *RouteProbe* can send back a *RouteReply* packet. This *RouteReply* packet is received by the *ConnectSource* node. For a unicast session, the *ConnectSource* node immediately starts transmission of data packets. In case of a multicast session, the *ConnectSource* node, on receiving the *RouteReply* packet, transmits a *ConnectConfirm*, and gets connected to the multicast group. Once route establishment gets completed, the actual data transfer can begin.

PLBU does not differentiate between unicast and multicast traffic, and hence it finds routes for both types of sessions in almost the same way. The only difference is that the multicast connect session has to confirm the connectivity to one of the tree nodes as it may receive *RouteReply* packets from many tree nodes. This difference is also eliminated if only the multicast source is allowed to send a reply to the *RouteProbe* packet, but this is at the cost of high control overhead. The transmission of data packets is also done in the same manner for both types of traffic. Source routing is not used; the data packet is transmitted in a hop-by-hop manner. A data packet carries the path it has traversed so far. This traversed path information is used by the intermediate nodes to refresh their existing route cache entries. This refreshing helps in maintaining an up-to-date view of the path in the route cache tables of the intermediate nodes. The cache table is used by multicast sessions to obtain paths to the multicast source nodes, thereby helping in reducing the control overhead. The routing of a data packet is done as follows. The source node sends the data packet to the next node, the address of which is obtained from the route cache table. The reason for not using source routing is to maintain homogeneity while routing data packets for multicast sessions, where source routing is not possible. As mentioned above, the data packet carries the path traversed by it so far. After receiving a data packet, a forwarding node appends its address to the traversed path field of the packet, and forwards the packet to the next hop node, the identifier (ID) of which is available in the route cache table. Data packets belonging to multicast sessions are also routed in the same way. When a data packet transmitted by the source arrives at an intermediate node, it is forwarded as a unicast packet. The data packet carries the list of nodes to which it is intended. All downlink nodes process the data packet in the promiscuous mode. This list carried by the data packet gives the intermediate nodes information about the current multicast session and thereby helps in removing inconsistencies in the multicast tree information maintained at the node.

PLBU makes use of the route cache. The main motivation behind maintaining route cache information is to reduce the control overhead. Lack of route cache information at intermediate nodes does not affect the functioning of PLBU. It is just an enhancement made to the protocol for reducing control overhead. The key information that PLBU maintains about any session at the intermediate nodes is the two-hop path information toward both the sender and the receiver, which is used to quickly reconfigure a broken path.

Dynamic movement of nodes in an ad hoc wireless network causes frequent link breaks. Such broken links are dealt with in the reconfiguration phase. These link breaks, if not repaired locally, result in the flooding of *RouteProbe* packets by the *ConnectSource* nodes for reestablishing the path. Local route repair can be performed if the local topology information is available at each node. The usual approach is to inform the source node(s) about the link break, which floods the network and finds alternate paths. The other approaches are to bypass the broken link by using the local one-hop topology information regarding nearby alternate nodes, or to use a local broadcast route maintenance scheme. Since in PLBU the two-hop local topology information is maintained, the broken link is bypassed efficiently.

As mentioned above, the PLBU protocol offers several advantages over separate unicast and multicast routing. The unified approach makes minimum differentiation between unicast and multicast traffic and is very suitable for practical networks where both unicast and multicast traffic are generated simultaneously.

8.6.8 Multicast Ad Hoc On-Demand Distance Vector Routing Protocol

Multicast ad hoc on-demand distance vector (MAODV) routing protocol [19], [20] is an extension of the AODV [21] protocol. MAODV adds multicast capability to the AODV protocol; multicast, unicast, and broadcast features have been streamlined into MAODV. MAODV uses sequence numbers to ensure that the most recent route to the multicast group is used.

Tree Initialization Phase

MAODV uses the notion of a *group leader* which updates the sequence number of the group periodically and broadcasts it using *group hellos* (GRPHs). The group leader is typically the first node to join the group. Nodes which wish to join the group, if they have the address of the group leader (recorded by the node when the group leader joined the group), unicast a *route request* (RREQ) to the group leader. If they do not have any record of the address of the leader for the group, they broadcast a *RREQ* packet. This *RREQ* is rebroadcast by nodes which are not members of the multicast tree (which also establish the reverse path and keep the state information consisting of the group address, requesting node id, and next-hop information). The *RREQ* is answered with a *route reply (RREP)* by a member of the multicast group. This *RREP*, containing the distance of the replying node from the group leader and the current sequence number of the multicast group, is unicast to the requesting node (which establishes the forward path). Note that only the nodes which have recorded a sequence number greater than that in the *RREQ* packet can reply. The receiver node selects the most recent and the shortest path from all the *RREPs* it receives and sends a *multicast activation (MACT)* message. *MACT* confirms to the intermediate relaying nodes that they are part of the tree. Only after an *MACT* is received is the forward path, established during the *RREP* propagation, activated. Nodes that wish to send data to the source use a similar procedure, which differs from the previously outlined procedure in only one respect: Any node with a *recent* (in terms of sequence numbers) route to the multicast group can reply to a non-join *RREQ*. In Figure 8.19, R3 attempts to join the multicast tree rooted at S by flooding *RREQ* packets, which invites replies from S, I1, and R2. R3 chooses the reply sent by I1 and sends a *MACT* message along the reverse path taken by the *RREP*.

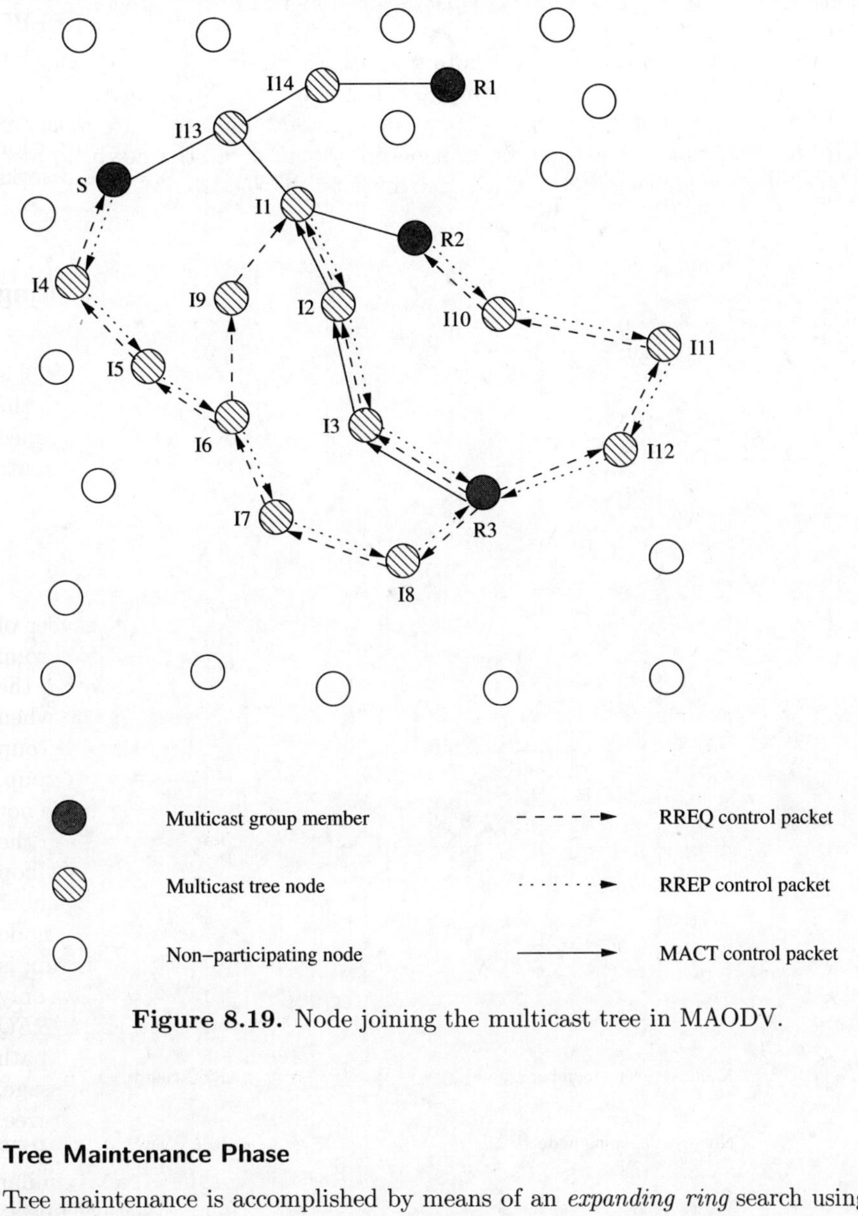

Figure 8.19. Node joining the multicast tree in MAODV.

Tree Maintenance Phase

Tree maintenance is accomplished by means of an *expanding ring* search using the *RREQ, RREP, MACT* cycle. The downstream node is responsible for issuing a fresh *RREQ* for the group. This *RREQ* contains the hop count of the requesting node from the group leader and the last known sequence number for that group. This *RREQ* can be answered only by member-nodes whose recorded sequence number is greater than that indicated in the *RREQ* and whose hop distance is lesser (to ensure that nodes on the same side of the break do not reply to this *RREQ*). When a leaf-node wishes to leave the group, it sends a *prune* message upstream which might be propagated further up the tree. A non-leaf node continues to be a member of the multicast tree and forwards packets for other multicast receivers, even after it has left the multicast group. Partition repair is facilitated by the *GRPH* message: Any node which overhears *GRPH* messages from different nodes initiates a group leader election protocol, which finally elects a single group leader for the newly formed component. Figure 8.20 shows the tree repair initiated by the R3–I3 link break.

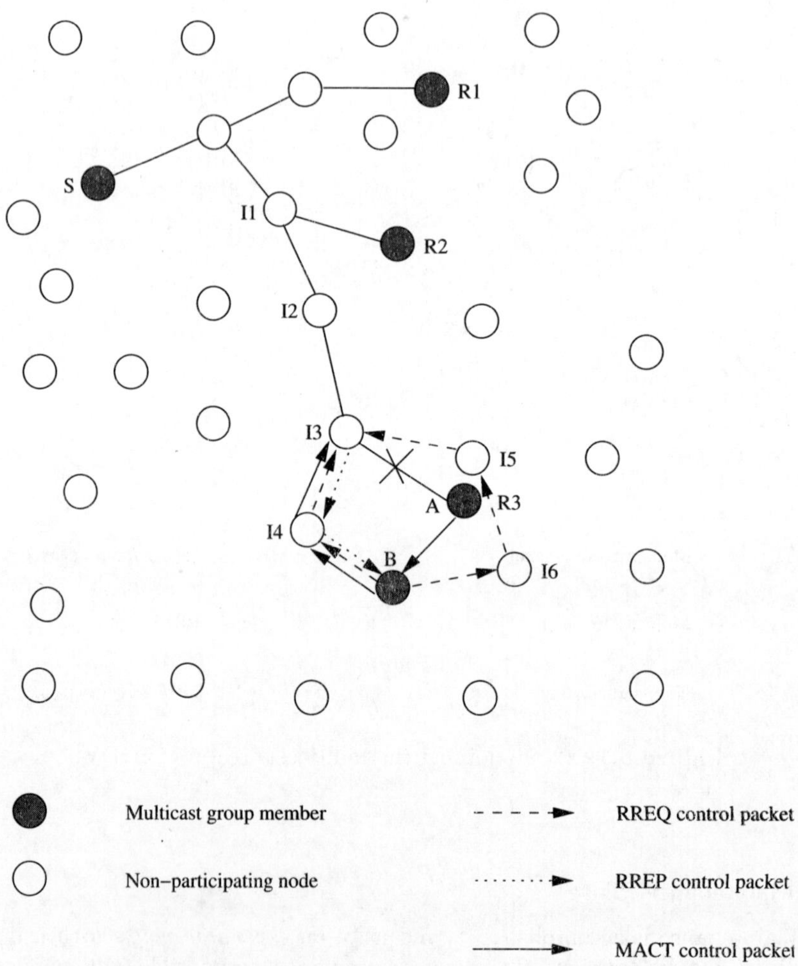

Figure 8.20. Tree repair in MAODV.

Advantages and Disadvantages

One of the advantages of MAODV is the integration of unicast and multicast into a unified framework. Thus, information gleaned during unicast route discovery can be used in the multicast route discovery and vice versa. This sharing of information helps in reducing the control overhead, which is one of the aims of an ad hoc multicast protocol. The protocol is also free from loops. The disadvantage stems from its tree-based multicast topology: poor packet delivery under mobility, congestion along links in the tree, etc. This protocol uses a shared-tree, which is not efficient when the number of multicast sessions is high. Usage of a shared-tree always carries along with it the risk of single-point failure of the group leader, which might severely affect all multicast sessions in progress.

8.6.9 Ad Hoc Multicast Routing Protocol Utilizing Increasing ID-Numbers

Ad hoc multicast routing protocol utilizing increasing id-numbers (AMRIS) [22] is a source-initiated multicast routing protocol in which a shared-tree is constructed to support multiple sources and receivers. The main idea in this protocol is that each tree node has a session-specific multicast session member identifier (MSM-ID) which indicates its logical height in the shared tree. The purpose of MSM-ID is to avoid any loop formation and repair the broken links locally.

Tree Initialization Phase

To initialize the multicast tree, one of the sources in the multicast group (which is called Sid) broadcasts the *NewSession* message. The *NewSession* message contains Sid's MSM-ID, multicast session ID, and routing metrics. When all the nodes receive the *NewSession* message, they store the information derived from the *NewSession* message in their neighbor-status tables up to some fixed amount of time and generate their own MSM-ID. The new MSM-ID is larger than the received MSM-ID and it should not be consecutive in order to make use of the successive sequence numbers for quick local repair. In AMRIS, each node maintains a neighbor-status table which stores the list of existing neighbors and their MSM-IDs. The nodes in the ad hoc network periodically transmit beacons (which contain the MSM-ID of the node) to indicate their presence. In Figure 8.21, the number inside the circle represents the MSM-ID of the corresponding node. Note that some nodes may not have a valid MSM-ID due to their absence during *NewSession* propagation, or because the MSM-ID stored in the neighbor-status table might have been erased due to timeout.

When a node wants to join the multicast group, it gets the list of *potential parent nodes* from its neighbor-status table and sends the *JoinReq* control packet to one of its neighbor nodes. The potential parents are those neighbor nodes which have the MSM-ID less than the MSM-ID of the node which has to send the *JoinReq* control packets. Here in Figure 8.21, node S1 sends the *JoinReq* to its potential parent node I2. Since node I2 is not in the multicast tree, it repeats the same process by sending the *JoinReq* to node I1. After a successful search of the path, node I2 sends

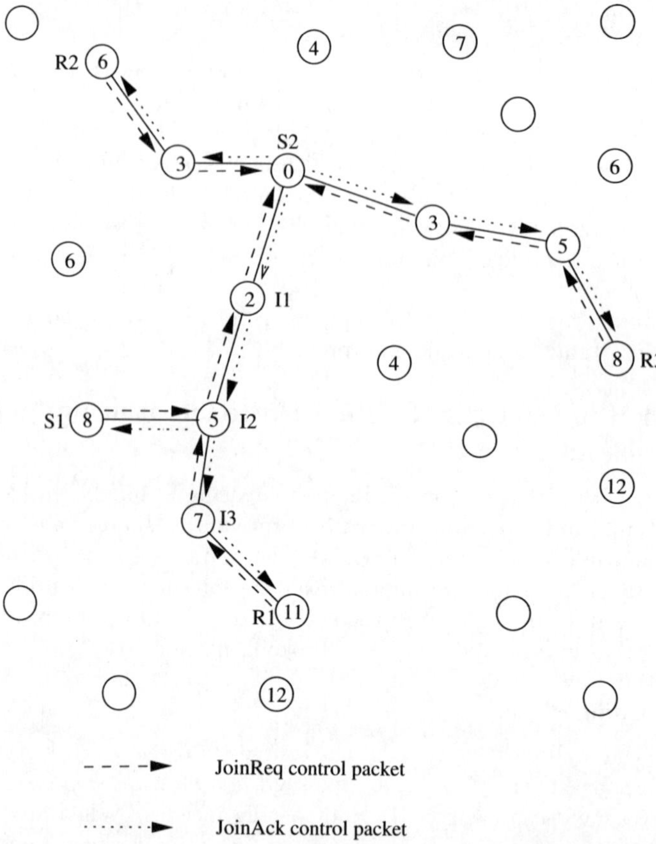

Figure 8.21. Node joining the multicast tree in AMRIS.

JoinAck to its downstream nodes to establish the route. Other members also join the multicast group in the same way.

Tree Maintenance Phase

Once the tree is formed, then the process of tree maintenance comes into the picture. Tree maintenance is required to repair the broken links. In AMRIS, when a node is isolated from the tree due to link failure, it rejoins the tree by executing branch reconstruction (BR), which has two subroutines: BR1 and BR2. The subroutine BR1 is executed if the node has neighboring potential parent nodes. If the node does not have any neighboring potential parent node, then the BR2 subroutine is executed. In Figure 8.22, when node R1 moves from A to B, the link I3-R1 fails. Since node R1 has neighboring potential node I4, it sends the *JoinReq* packet to node I4. The node I4 is not in the tree and hence it repeats the same process to join

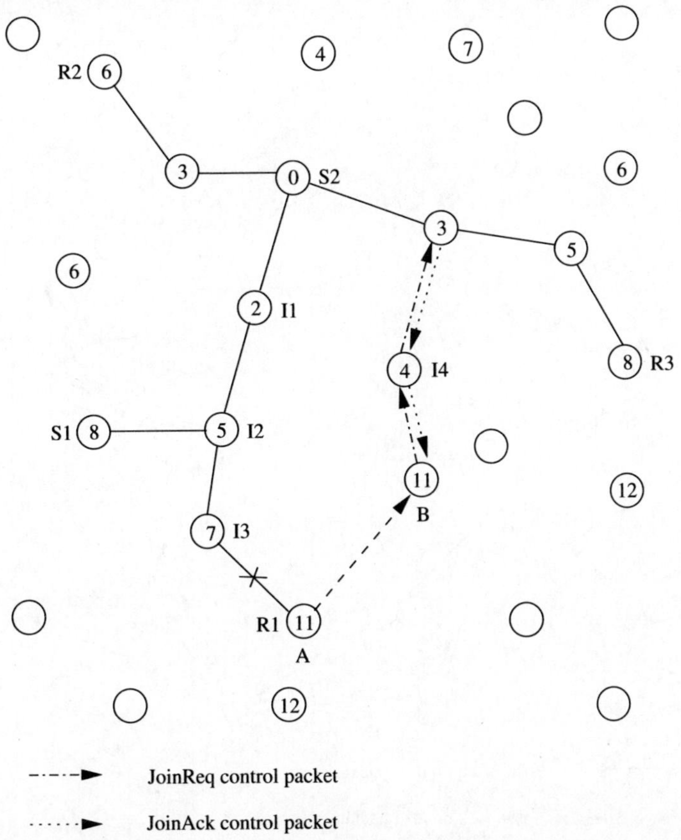

Figure 8.22. Tree maintenance by BR1 subroutine.

the tree. If node I4 succeeds, then it sends *JoinAck* to node R1; otherwise, it sends *JoinNack*. If node R1 receives *JoinNack* or times out on the reply, then it selects another neighboring potential parent node and repeats the same process. If none is available, node R1 executes the BR2 subroutine. The other case is that node R1 moves from A to C (Figure 8.23). Since node R1 does not have any neighboring potential parent node, it executes BR2 subroutine in which it floods the *JoinReq* packet with some TTL value. Hence it receives a number of *JoinAck* packets. It chooses one of the routes by sending a *JoinConf* packet.

Advantages and Disadvantages

Since AMRIS uses MSM-ID, it avoids loop formations, and link breaks are repaired locally. Hence, the control overhead is less. Another advantage of AMRIS is its simplicity as compared to other multicast protocols. The main disadvantages of

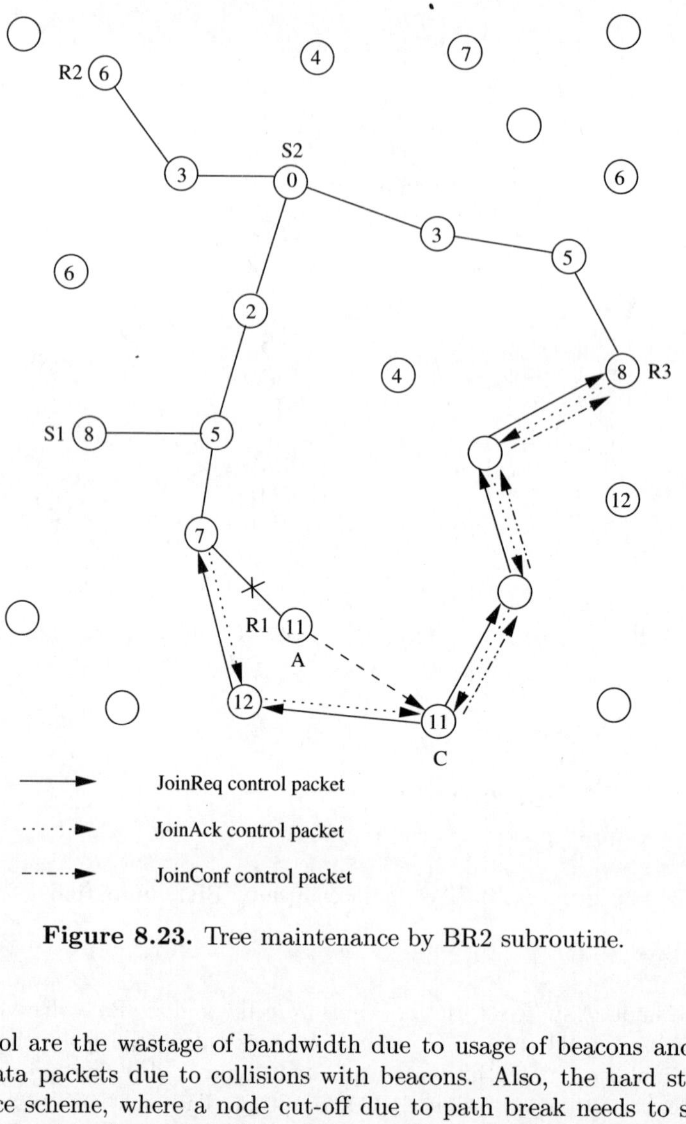

Figure 8.23. Tree maintenance by BR2 subroutine.

this protocol are the wastage of bandwidth due to usage of beacons and the loss of many data packets due to collisions with beacons. Also, the hard state route maintenance scheme, where a node cut-off due to path break needs to search for a suitable parent node, wastes a lot of time and contributes toward an increased end-to-end delay in packet delivery and loss of packets. The selection of potential parent nodes based on MSM-ID tends to increase the average hop-length between the receivers and the source, leading to increased delay and increased probability of packet losses.

8.6.10 Ad Hoc Multicast Routing Protocol

A different approach is used in ad hoc multicast routing protocol (AMRoute) [7] in order to enhance the robustness of the tree-based approach. This multicast protocol emphasizes robustness even with high mobility by creating a multicast tree over the underlying mesh. It assumes the existence of an underlying unicast routing protocol (but does not depend on any particular unicast routing protocol) in the network environment, which is responsible for keeping track of the network dynamics. It has two main phases: mesh creation phase and virtual user-multicast tree creation phase. The *logical core* node initiates mesh creation and tree creation. Here the logical core is not the central point for all data as in core-based trees [3] and the logical core changes dynamically.

Tree Initialization Phase

When members (either source or receiver) want to join the multicast group, they declare themselves as logical cores of a one-node mesh and flood *JoinReq* control packets with increasing TTL to discover other members. When node A detects node B (Figure 8.24), a bidirectional tunnel is established between node A and node B. Thus, they form the mesh segment (of two member nodes) indicated by I. Since each segment should have only one logical core, by using a *core resolution algorithm* only one node remains as the logical-core node in the segment. In Figure 8.24, node

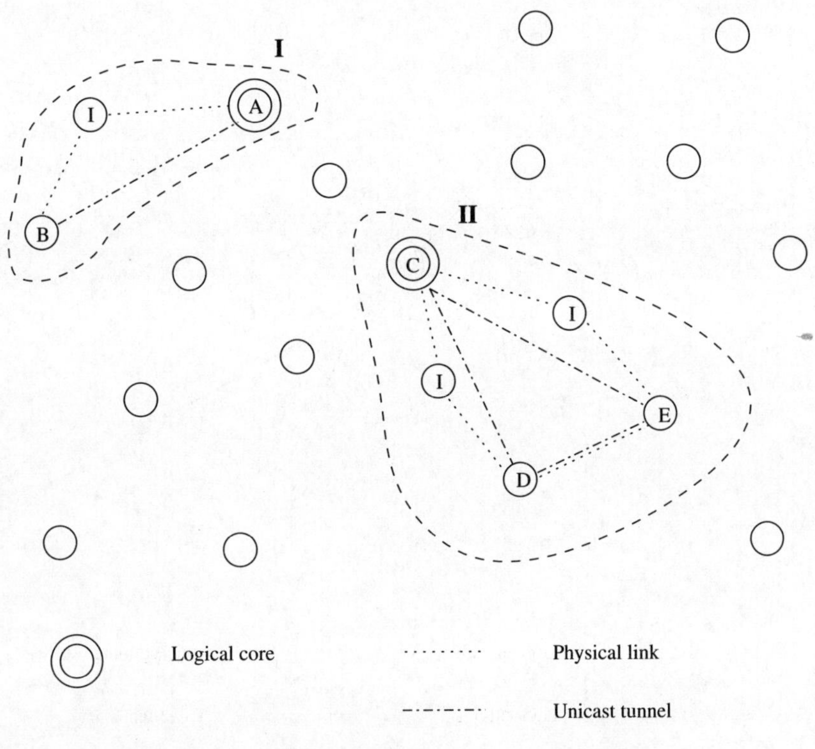

Figure 8.24. Formation of mesh segments.

A is the logical-core node in segment I. Now, node A again floods a *JoinReq* control packet to discover other group members and disjoint mesh segments. Keeping a single logical core in each segment to discover members and disjoint mesh segments makes the protocol scalable, as the control overhead is reduced. When member C in mesh segment II receives *JoinReq*, it responds with a *JoinAck* and a new bidirectional tunnel is established between nodes A and C. Note that any member, either core or non-core in the mesh segment, can respond to the *JoinReq* discovery message to avoid adding many links to a core. Since mesh segments I and II merge, the resultant mesh contains more than one logical core. Finally, only one of the core nodes (say, node C here) will be the logical core as selected by the core resolution algorithm. The logical core C periodically transmits a *TreeCreate* message to create a virtual user-multicast tree (Figure 8.25) over the mesh which is shared by all members of the group. The underlying mesh structure and periodic transmissions of the *TreeCreate* message for creating the multicast tree make this protocol robust. As long as there exists a path among all the nodes connected by mesh links, the multicast tree need not change because of movement of the nodes. The members of the multicast group forward the data packets through the

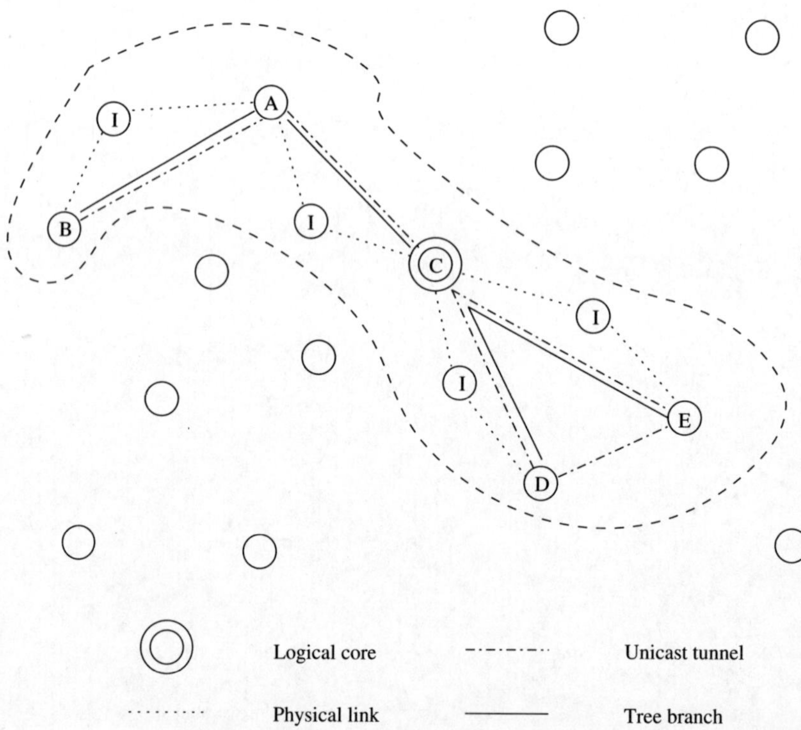

Figure 8.25. A virtual user-multicast tree.

unicast tunnel along the branches of the virtual multicast tree. Non-members are not required to replicate the data packets and support any IP multicast protocol, but this reduces the efficiency of the protocol.

Tree Maintenance Phase

Due to movement of the nodes or due to node failures, the mesh may split into parts, as shown in Figure 8.26. Nodes D and E in the fragmented part II expect a *TreeCreate* message from the logical core. After waiting for a random time period, one of the members (*e.g.*, node D) becomes the core node and initiates the process of discovering the other disjoint mesh (part I) as well as tree creation. Again when part I and part II join, then multiple cores are resolved by a core resolution algorithm.

Advantages and Disadvantages

AMRoute protocol is robust due to its tree structure over an underlying mesh. But it is less efficient due to loop formations in the multicast tree. Also, under mobility

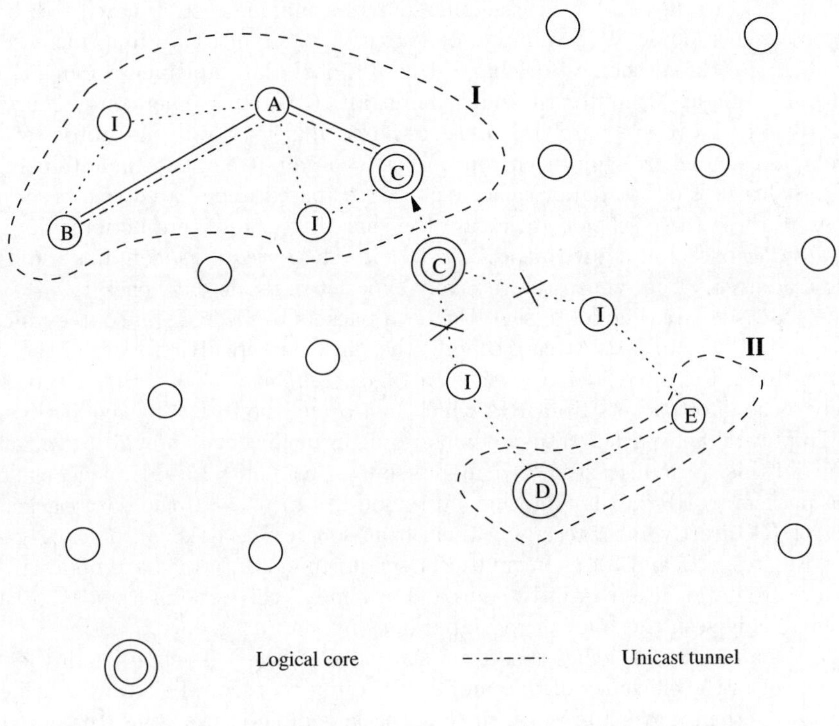

Figure 8.26. Tree maintenance in AMRoute.

conditions the hop count of the unicast tunnels increases, thereby decreasing the throughput. Similar to other shared-tree-based protocols, AMRoute also has the risk of single point failure of the core node. Failure of the core node causes loss of packets and increases the delay, as considerable time is wasted before one of the existing nodes is made as the core node. When the size of the network is large and the node mobility is high, the core nodes keep changing frequently. This leads to a continuous transfer of state information from the existing core nodes to the new core nodes, resulting in wastage of already scarce bandwidth.

8.6.11 Adaptive Shared-Tree Multicast Routing Protocol

The above described multicast routing protocols are either source-tree-based or shared-tree-based. A source-tree is rooted at the source node, whereas a shared-tree is rooted at the rendezvous point (RP) and is shared by multiple sources. Source-tree-based multicast protocols are better compared to shared-tree-based protocols because of the less delay involved (due to shortest path between source and receiver) and even traffic distribution. The main problem with source-tree-based protocols is that they are not stable in case the source is moving fast. Shared-tree-based protocols are scalable as the number of sources in the multicast group increases.

There are three schemes for shared-tree protocol [23] implementation: (1) unicast sender mode, (2) multicast sender mode, and (3) adaptive per source multicasting. The first two are purely shared-tree-based multicast routing protocols. But in adaptive shared-tree multicast routing protocol, an attempt is made to combine the best features of both approaches (source-tree-based and shared-tree-based).

In all these three schemes to create the shared-tree, receivers periodically send the *JoinReq* packet along with the *forward list*. The *forward list* contains the list of source addresses from which the receivers expect to get the data packets.

In unicast shared mode, to send the data packets to receivers, a source unicasts the data packets to RP. After receiving the data packets, RP forwards the data packets to all the interested receivers. In Figure 8.27, it is shown that receiver R1 receives the data packets from RP, which had been sent to RP by senders S1 and S2. Duplicate data packet transmissions result in inefficient bandwidth utilization. To avoid this, in multicast shared mode, sources S1 and S2 send unencapsulated data packets to RP so that intermediate node I2 can forward the data packets to receiver R1 directly, after receiving them from source S1. This can be clearly seen from Figures 8.27 and 8.28. In multicast sender mode, intermediate node I1 does not forward data packets (sent by source S1) when it receives them from RP because it has already seen those unencapsulated data packets.

In the above two schemes (based on shared-tree), there is scope for further improvement in the efficiency of the multicast routing protocol. This can be achieved by allowing the receiver to receive the data packets directly from the source, in case the data packets traverse (through RP) a path which is much longer than the shortest path between source and receiver. In Figure 8.29 receiver R1 receives the data packets from source S2 (through RP) with hop count eight, which is much longer than the actual distance between R1 and S2, which is two. In addition to that,

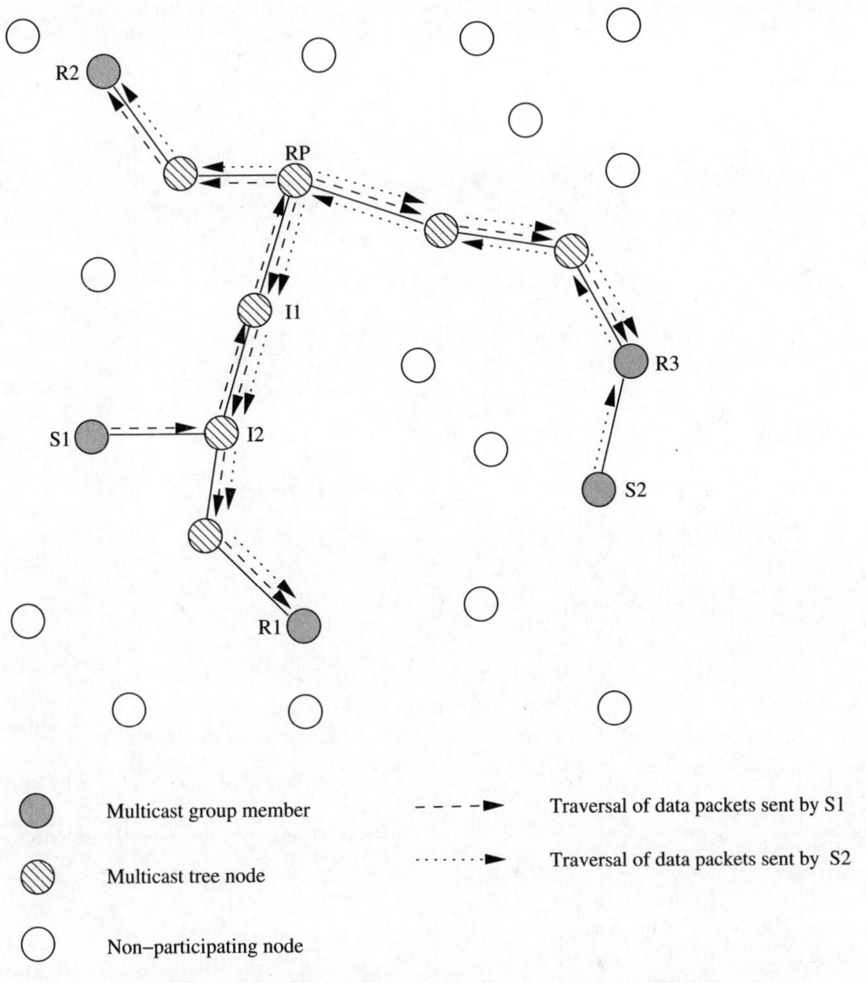

Figure 8.27. Data packets traversal in unicast sender mode.

the actual distance between R1 and S2 is less than the distance between receiver R1 and RP. Hence, receiver R1 connects to source S2 through the shortest path by sending *JoinReq* to S2. Now RP does not forward the data packets (which are sent by source S2) to receiver R1. Further due to movement of the nodes, if the distance between R1 and S2 increases, then it switches over to RP-rooted shared-tree from per source-tree.

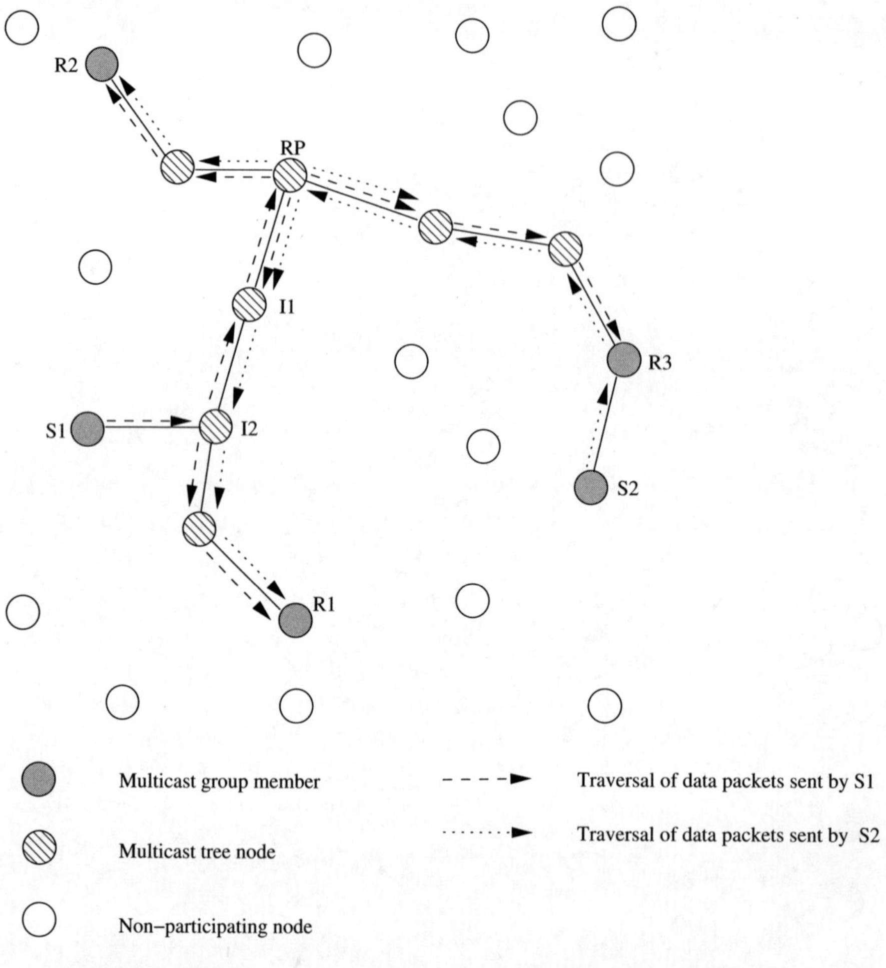

Figure 8.28. Data packets traversal in multicast sender mode.

Advantages and Disadvantages

Due to the adaptive nature of the tree configuration, it is scalable (due to shared-tree) and achieves high packet delivery ratio (due to per source-tree). The control overhead is reduced because of the shared-tree, and at the same time, it allows sources to send packets to receivers by the shortest path, resulting in better throughput. Similar to other shared-tree-based protocols, this scheme also carries with it the risk of single point failure of the RP. Failure of the RP affects multiple sessions and results in significant packet losses.

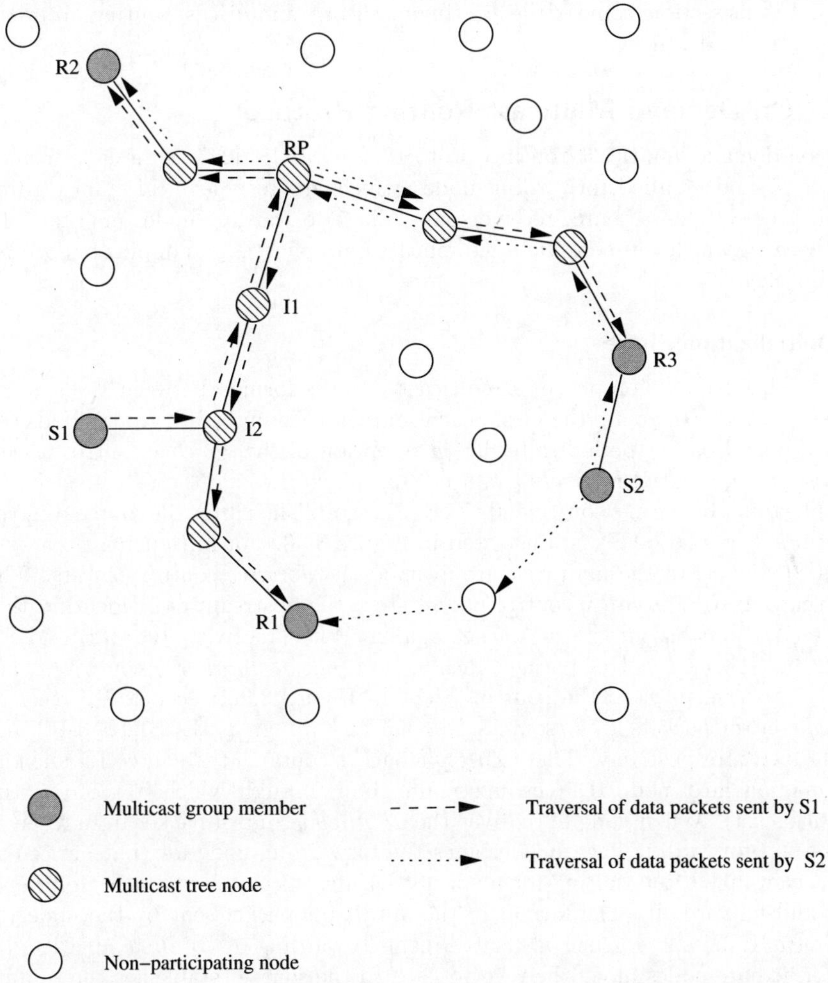

Figure 8.29. Data packets traversal in adaptive per source mode.

8.7 MESH-BASED MULTICAST ROUTING PROTOCOLS

In ad hoc wireless networks, wireless links break due to the mobility of the nodes. In the case of multicast routing protocols, the path between a source and receiver, which consists of multiple wireless hops, suffers very much due to link breaks. Multicast routing protocols which provide multiple paths between a source-receiver pair are classified as mesh-based multicast routing protocols. The presence of multiple paths adds to the robustness of the mesh-based protocols at the cost of multicast efficiency. In this section, some of the existing mesh-based multicast routing protocols are described in detail.

8.7.1 On-Demand Multicast Routing Protocol

In the on-demand multicast routing protocol (ODMRP) [24], a mesh is formed by a set of nodes called forwarding nodes which are responsible for forwarding data packets between a source-receiver pair. These forwarding nodes maintain the *message-cache* which is used to detect duplicate data packets and duplicate *JoinReq* control packets.

Mesh Initialization Phase

In the mesh initialization phase, a multicast mesh is formed between the sources and the receivers. To create the mesh, each source in the multicast group floods the *JoinReq* control packet periodically. Upon reception of the *JoinReq* control packet from a source, potential receivers can send *JoinReply* through the reverse shortest path. The route between a source and receiver is established after the source receives the *JoinReply* packet. This is illustrated in Figure 8.30. For initializing the mesh, sources S1 and S2 in the multicast group flood the *JoinReq* control packets. The nodes that receive a *JoinReq* control packet store the upstream node identification number (ID) and broadcast the packet again. When receivers R1, R2, and R3 receive the *JoinReq* control packet, each node sends a *JoinReply* control packet along the reverse path to the source. Here in Figure 8.30, receiver R2 receives *JoinReq* control packets from sources S1 and S2 through paths S1-I2-I3-R2 and S2-I6-I4-I5-R2, respectively. The *JoinReply* packet contains the source ID and the corresponding next node ID (the upstream node through which it received the *JoinReq* packet). When node I2 receives the *JoinReply* control packet from receiver R1, it sets a forwarding flag and becomes the forwarding node for that particular multicast group. After waiting for a specified time, it composes a new *JoinReply* packet and forwards it. The format of the *JoinReply* packet sent by the node R2 is shown in Table 8.2. In this way, subsequent forwarding of *JoinReply* packets by the intermediate nodes along the reverse path to the source establishes the route.

Table 8.2. Format of *JoinReply* packet sent by receiver R2

Source ID	Next Node ID
$S1$	$I3$
$S2$	$I5$

Mesh Maintenance Phase

In this phase, attempts are made to maintain the multicast mesh topology formed with sources, forwarding nodes, and receivers. To some extent, the multicast mesh

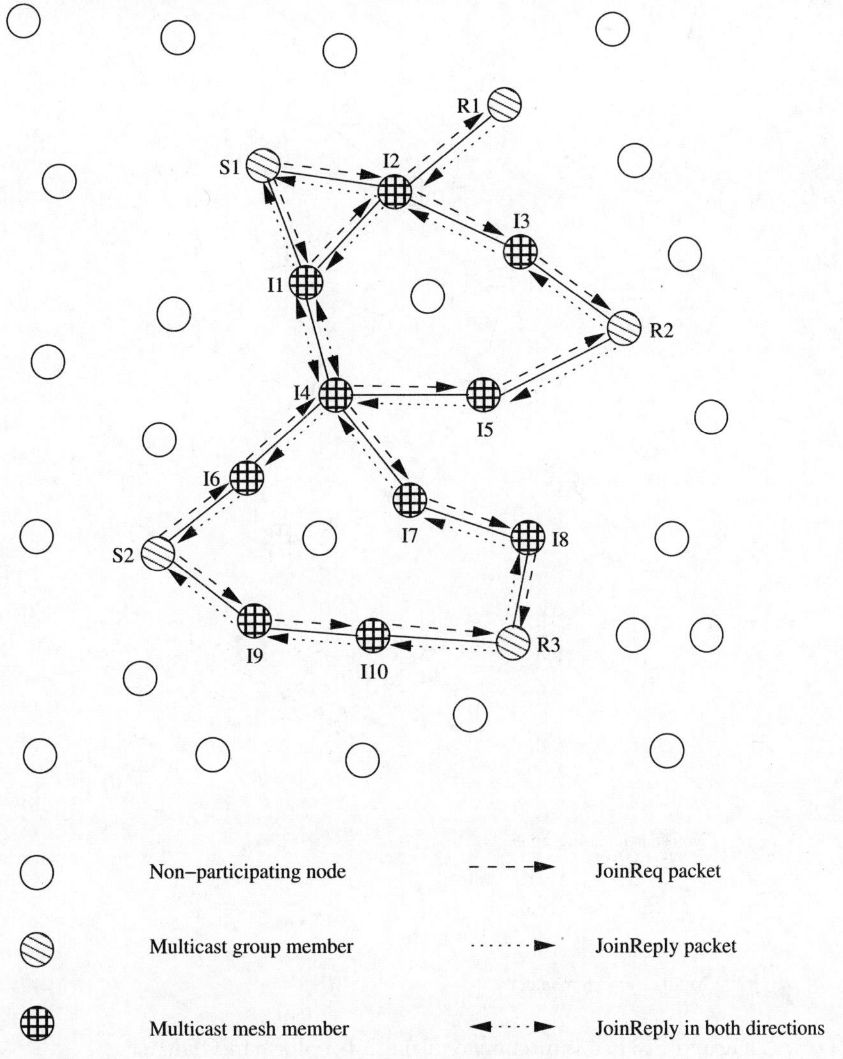

Figure 8.30. Mesh topology in ODMRP.

protects the session from being affected by mobility of nodes. For example, due to movement of the receiver R3 (from A to B), when the route S2-I9-I10-R3 breaks (Figure 8.31), R3 can still receive data packets through route S2-I6-I4-I7-I8-R3 and this contributes to the high packet delivery ratio. ODMRP uses a soft state approach to maintain the mesh, that is, to refresh the routes between the source and the receiver, the source periodically floods the *JoinReq* control packet. In Figure 8.31, when receiver R3 receives a new *JoinReq* control packet from node I11 (sent

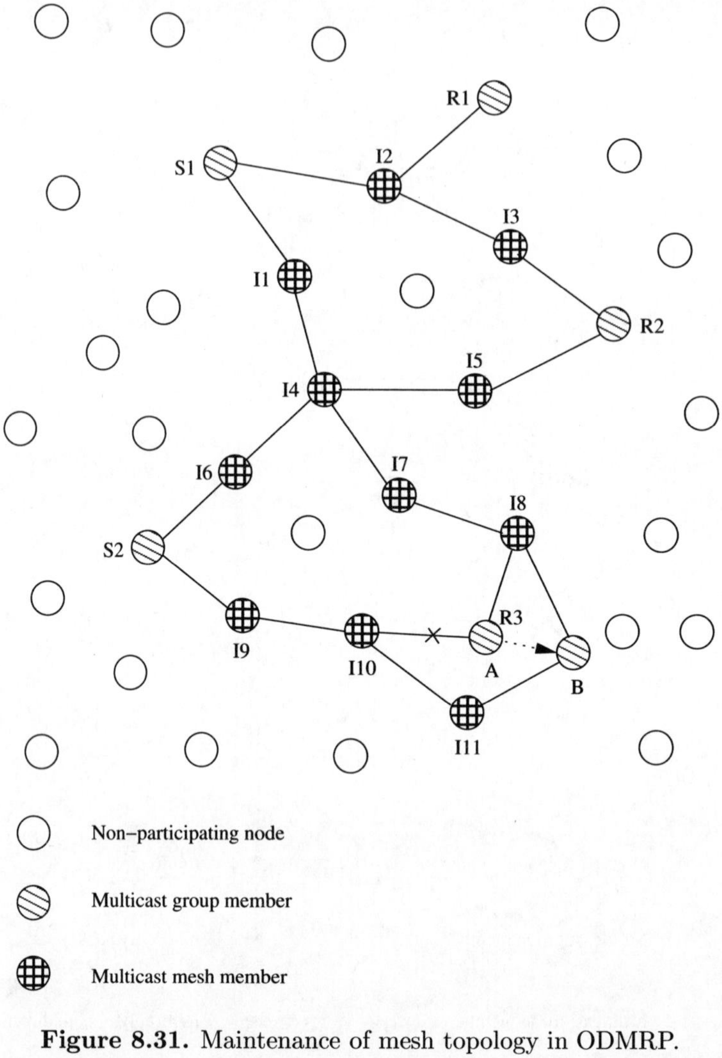

Figure 8.31. Maintenance of mesh topology in ODMRP.

by the source S2), it sends a *JoinReply* on this new shortest path R3-I11-I10-I9-S2, thereby maintaining the mesh structure.

Advantages and Disadvantages

Since ODMRP uses the soft state approach for maintaining the mesh, it exhibits robustness. But this robustness is at the expense of high control overhead. Another disadvantage is that the same data packet (from source S2 to receiver R3) propagates through more than one path to a destination node, resulting in an increased number of data packet transmissions, thereby reducing the multicast efficiency.

8.7.2 Dynamic Core-Based Multicast Routing Protocol

The dynamic core-based multicast routing protocol (DCMP) [25] attempts to improve the efficiency of the ODMRP multicast protocol by reducing control overhead and providing better packet delivery ratio. Mesh-based protocols, such as ODMRP, suffer from two disadvantages:

1. Excessive data forwarding: Too many nodes become forwarding nodes, resulting in an excessive number of retransmissions of data packets. In ODMRP, all nodes on the shortest path between each source and each receiver become forwarding nodes, resulting in too many forwarding nodes. (The advantage of such a mesh containing many forwarding nodes is, of course, the superior packet delivery ratio and robustness under mobility.)

2. High control overhead: In ODMRP, each source periodically floods its *JoinReq* packets and the mesh is reconstructed periodically. This leads to a heavy control overhead.

DCMP attempts to increase the packet delivery ratio and at the same time reduce the control overhead by reducing the number of sources which flood *JoinReq* packets, thereby reducing the number of forwarding nodes.

Mesh Initialization Phase

In the mesh initialization phase, DCMP attempts to reduce the number of sources flooding their *JoinReq* packets. In DCMP, there are three kinds of sources: passive sources, active sources, and core active sources. Each passive source is associated with a core active source, which plays the role of a proxy for the passive source, by forwarding data packets from the *passive* source, over the mesh created by its *JoinReq* packets. Passive sources do not flood *JoinReq* packets, unlike active and core active sources. Sources which flood *JoinReq* packets on their behalf as well as on the behalf of some passive source are called core active sources. The mesh establishment protocol is similar to that in ODMRP. Data packets of the active sources and core active sources are sent over the mesh created by themselves, while a passive source forwards the packet to its proxy core active node, which in turn sends it over its mesh. The control overhead is reduced, as compared to ODMRP, because there are a fewer number of sources which flood their *JoinReq* packets, and thus the number of forwarding nodes is also fewer. To allow the mesh to have enough forwarding nodes for robust operation, the number of passive sources has to be limited: This is done in DCMP by limiting the maximum number of passive sources a single core active source can serve (this maximum number is called *MaxPassSize*). To ensure that the packet delivery ratio is not reduced because of the fact that the average passive source-to-receiver distance is likely to be higher (because it is reusing the mesh created by its proxy core active node), the maximum hop distance between a passive source and its proxy core active node is also limited (this maximum hop distance is called *MaxHop*).

The example in Figure 8.32 shows mesh construction in DCMP, with the *MaxHop* and *MaxPassSize* parameters set to two. There are four sources in the multicast group, S1, S2, S3, and S4 and two receivers labeled R. Since S3 and S4 are at a hop distance of two from each other (which is equal to *MaxHop*), S3 goes passive and uses a proxy in the core active node S4. No other set of sources is within a hop distance of two from each other, so eventually S1 and S2 are the active sources, S3 is a passive source, and S4 is a core active source. The mesh consists of all the shortest paths between the sources S1, S2, and S3 and the two receivers R. Thus,

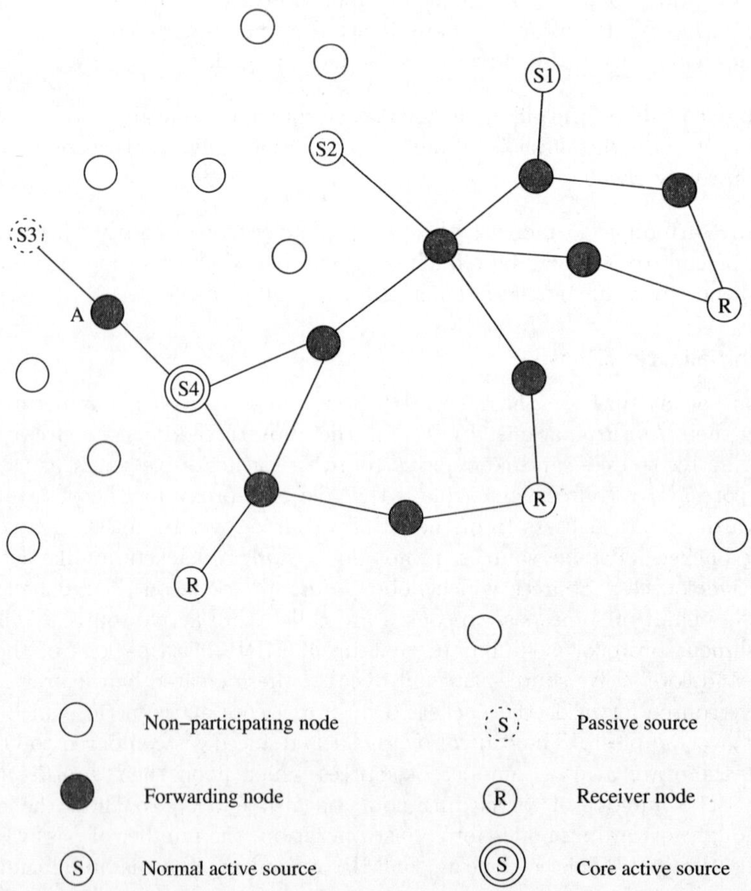

Figure 8.32. Construction of mesh in DCMP. *Reproduced with permission from [25], © ACM, 2002.*

the number of forwarding nodes is reduced, as compared to ODMRP, without much reduction in robustness and packet delivery ratio.

Mesh Maintenance Phase

DCMP's mesh maintenance is soft state, similar to that of ODMRP. Thus, the mesh is reconstructed periodically and forwarding nodes that are no longer part of the mesh, cancel their forwarding status after a timeout period. In the example shown in Figure 8.33, source S3 has moved away from its core active node S4, with the hop distance between them increasing to three (which is greater than the *MaxHop* parameter=2). Thus, S3 goes active, and begins flooding periodically with its own

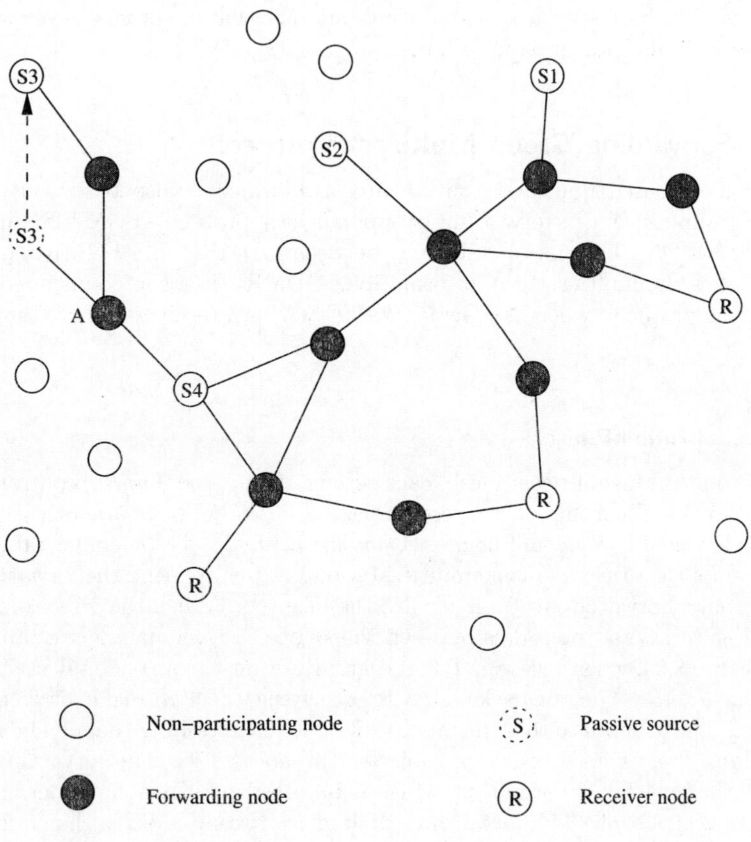

Figure 8.33. Maintenance of mesh topology in DCMP. *Reproduced with permission from [25], © ACM, 2002.*

JoinReq packets. Therefore, more nodes attain forwarding status and are grafted onto the mesh.

Advantages and Disadvantages

The primary advantage of DCMP is its scalability due to decreased control overhead and its superior packet delivery ratio. The performance improvement of DCMP over ODMRP increases with the number of sources in the multicast group (though they start performing on par beyond a certain number of sources, when almost all nodes are part of the mesh). One of the drawbacks of DCMP is that the parameters associated with it, *MaxPassSize* and *MaxHop*, are likely to depend on the network load conditions, group size, and number of sources, and optimal values of these parameters may even vary from one node to another. Failure of an active core node might result in multiple session failures.

8.7.3 Forwarding Group Multicast Protocol

Another mesh-based multicast routing protocol, which is also based on the forwarding group concept, is forwarding group multicast protocol (FGMP-RA [receiver advertising]) [26]. The major difference between ODMRP and FGMP-RA lies in who initiates the multicast group formation. ODMRP is a source-initiated multicast routing protocol, whereas FGMP (FGMP-RA) is a receiver-initiated multicast routing protocol.

Mesh Initialization Phase

In order to form the multicast mesh, each receiver floods the *JoinReq* control packet in the network. When the sources receive these *JoinReq* control packets, each source updates its *member table* and creates a *forwarding table*. The member table keeps the IDs of all the receivers in the multicast group. After creating the forwarding table, the source forwards this table toward the receivers. When the forwarding table reaches the receivers, the routes between the source-receiver pairs get established.

In Figure 8.34, when receivers R1, R2, and R3 want to join the multicast group, they send *JoinReq* control packets. After receiving the *JoinReq* control packets, sources S1 and S2 update their member tables. After refreshing the member tables, each source creates its forwarding table and broadcasts it. The forwarding table contains the next-hop information, which is obtained by using the underlying unicast routing protocol. The forwarding table at source node S2 is shown in Table 8.3. After receiving the forwarding table, neighbor nodes I6 and I9, whose node ID matches with the entry in the forwarding table, set their forwarding flags and become the forwarding nodes for that particular multicast group. Now nodes I6 and I9 build their own forwarding tables and forward them again. In this way, the forwarding tables reach receivers R1, R2, and R3 along the reverse shortest path and the route is established.

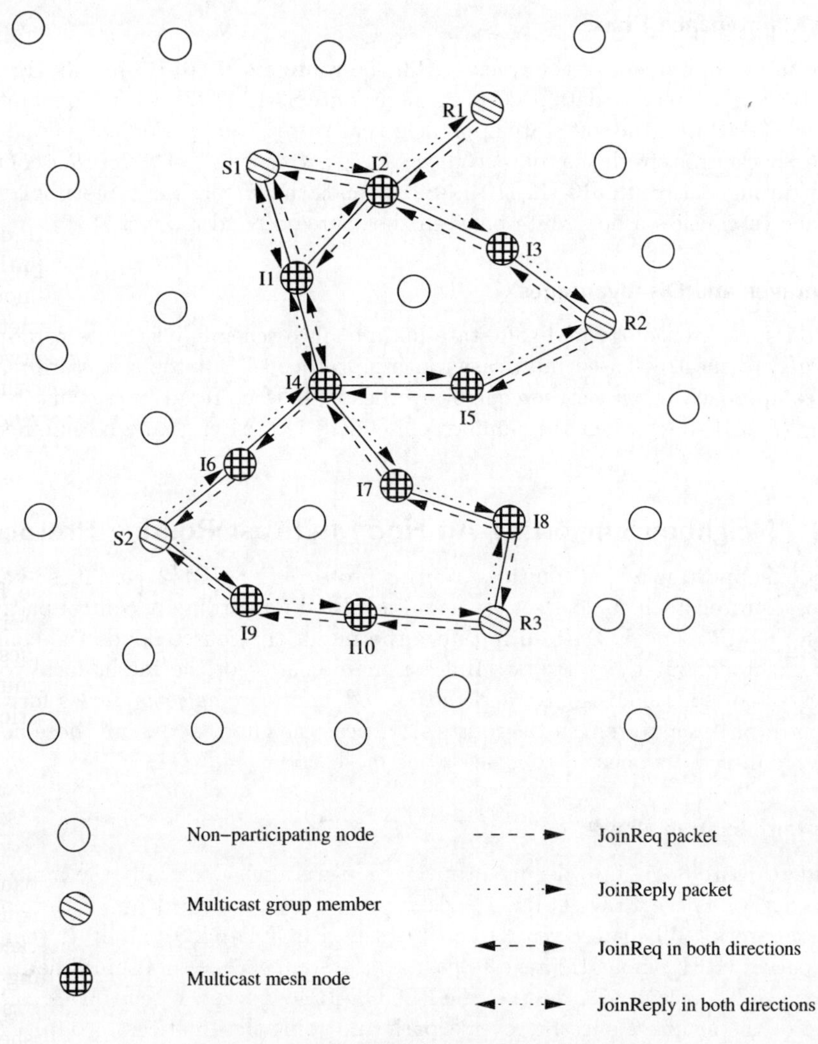

Figure 8.34. Mesh topology in FGMP.

Table 8.3. Format of forwarding table sent by source S2

Receiver ID	Next Node ID
$R1$	$I6$
$R2$	$I6$
$R3$	$I9$

Mesh Maintenance Phase

If, due to the movement of the receiver R3, the route S2-I9-I10-R3 breaks (Figure 8.31), R3 can still receive data packets through route S2-I6-I4-I7-I8-R3. To maintain a route, FGMP uses the soft state approach, that is, receivers periodically flood the *JoinReq* packet to refresh the route. When source S2 receives a new *JoinReq* control packet through the route R3-I11-I10-19-S2, it sends the forwarding table along that path and establishes a new route between the source S2 and receiver R3.

Advantages and Disadvantages

Due to its mesh topology and soft state maintenance scheme, it is more robust as compared to the tree-based multicast routing protocols. But the soft state maintenance approach increases the control overhead. It is more advantageous to use FGMP (FGMP-RA) when the number of senders is greater than the number of receivers.

8.7.4 Neighbor Supporting Ad Hoc Multicast Routing Protocol

Neighbor supporting ad hoc multicast routing protocol (NSMP) [27] is a mesh-based multicast protocol which does selective and localized forwarding of control packets. Like in ODMRP and FGMP, to initialize the mesh, the source floods the control message throughout the network. But for maintenance of the mesh, local route discovery is used, that is, only *mesh nodes* and *multicast neighbor nodes* forward the control message to refresh the routes. Multicast neighbor nodes are those nodes which are directly connected to at least one mesh node.

Mesh Initialization Phase

Consider Figure 8.35. In order to form the multicast mesh, initially the multicast source S1 floods the FLOOD-REQ packets which are forwarded by all the nodes in the network. When receiver R1 receives this FLOOD-REQ packet, it sends a reply packet (REP) along the reverse path to the source S1, establishing the route. Similarly, when receiver R2 receives the FLOOD-REQ packet, it sends back a REP packet to the source along the reverse path and joins the multicast group. Each source node periodically transmits a LOCAL-REQ packet, which is relayed by all mesh nodes and multicast neighbor nodes. If new receiver R3 wants to join the multicast group, it waits for a specified time for a LOCAL-REQ packet from any mesh node or multicast neighbor node. When receiver R3 receives a LOCAL-REQ packet from the multicast neighbor node I2, it joins the group by sending a REP control packet. This REP packet is relayed back to the multicast source through nodes I2 and I1. Some receivers may not be within the range of any multicast mesh node or multicast neighbor node. In Figure 8.35, receiver R4, which is more than two hops away from the mesh, does not get any LOCAL-REQ packet. Hence it floods MEM-REQ control packets. Any node that is part of the multicast mesh which receives the MEM-REQ packet sends back a reply route discovery packet. Node R4 receives route discovery packets sent by nodes I5 and R1, through paths

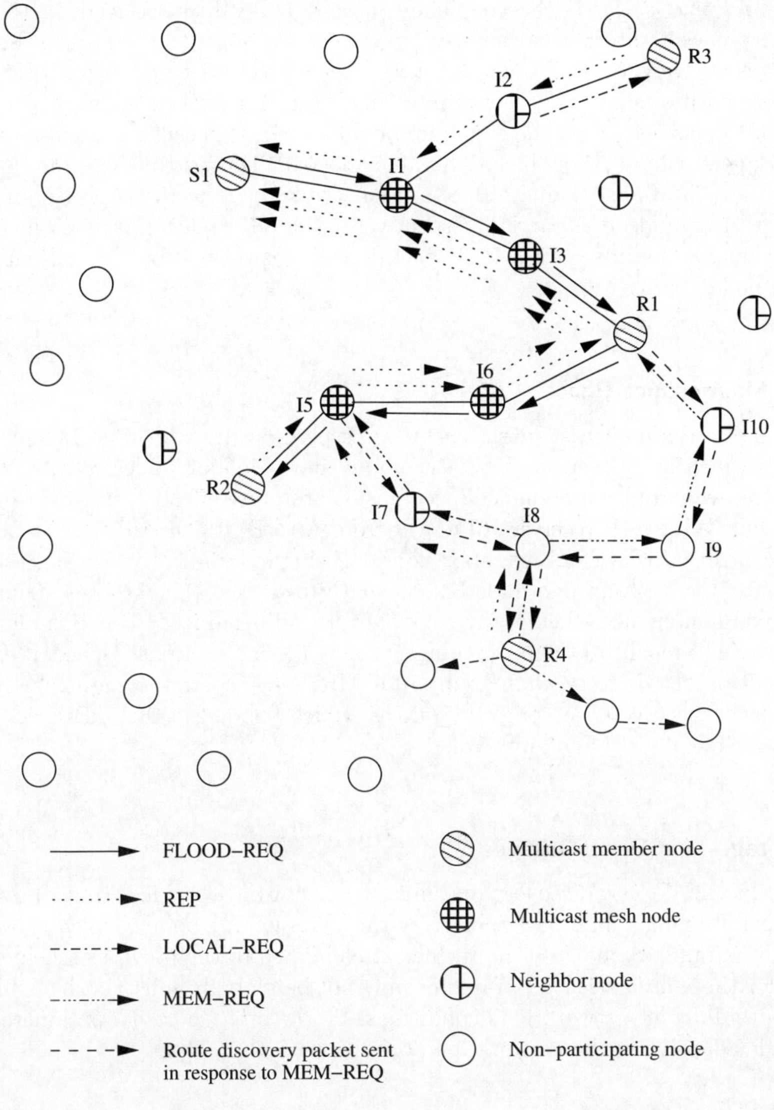

Figure 8.35. Mesh initialization in NSMP.

I5-I7-I8-R4 (path I) and R1-I10-I9-I8-R4 (path II), respectively. When any new receiver node receives multiple route discovery packets, NSMP uses a *relative weight* metric for selecting one of the multiple routes. The *relative weight* metric is given as $((1-\alpha) \times FC + \alpha \times NC)$, where FC is the count of forwarding nodes on the path from the source node till the current node, and NC is the count of non-forwarding nodes from the source node till the current node. A path with the lowest value for the *relative weight* is chosen. The NC and FC values received on the route discovery packet through path I are five and two, respectively, and are three and three, respectively, on the route discovery packet received through path II. NSMP chooses a path which has a larger number of existing forwarding nodes, which makes the protocol efficient since the total number of multicast packet transmissions in the network gets reduced. For this, an α value above 0.5 is used. In the example shown in Figure 8.35, if an α value of 0.8 is used, the *relative weight* metric on path I and path II would be 2.6 and 3, respectively. Node R4 selects path I which has a lower value, and sends back a REP packet, which is relayed back to the multicast source node.

Mesh Maintenance Phase

To maintain the multicast mesh, each source periodically transmits LOCAL-REQ packets which are forwarded by mesh nodes and neighbor nodes only. The local route discovery and maintenance mechanisms are quite effective in repairing most of the link failures. If receiver R3 moves from A to B, it can still receive LOCAL-REQ control packets from neighbor node I12 (Figure 8.36). By sending the REP packet to source S2 through node I12, the path to source S2 is repaired. Due to the localized maintenance scheme, the control overhead is reduced. But this mesh may not be able to repair all the link failures causing mesh partitions. Hence, it exhibits less data packet delivery compared to ODMRP. To repair the mesh partition, a group leader is selected among the sources, which floods the FLOOD-REQ packet at every FLOOD-PERIOD interval.

Advantages and Disadvantages

Due to localized route discovery and maintenance, NSMP reduces control overhead while maintaining a high packet delivery ratio. Its joining policy, which is based on a weight parameter, makes it more efficient compared to the previous schemes. The value for the *relative weight* parameter may not be globally constant. It is likely to vary with different network load conditions. So the *relative weight* parameter must be made adaptive to the network load conditions.

8.7.5 Core-Assisted Mesh Protocol

It is known that mesh-based multicast protocols deliver more data packets compared to tree-based multicast protocols, due to the existence of multiple paths between the source and receiver. But some of the existing mesh-based multicast protocols such as ODMRP and FGMP use the flooding approach to create and maintain the mesh topology, which results in reduced efficiency in bandwidth utilization. To eliminate flooding of control packets, core-assisted mesh protocol (CAMP) [28] uses core nodes in the mesh.

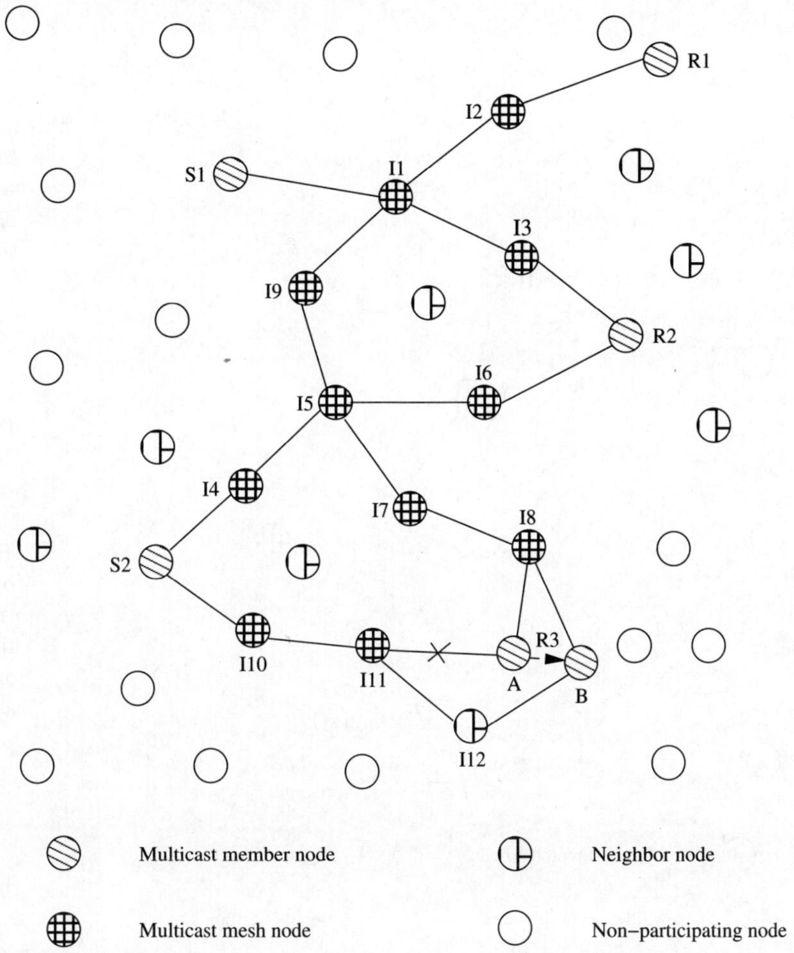

Figure 8.36. Mesh maintenance in NSMP.

Mesh Initialization Phase

CAMP is a receiver-initiated multicast routing protocol. Initially, to join the shared mesh, a receiver extracts the core node ID from its core-to-group address mapping (CAM) table (the CAM table contains the core node IDs of the multicast group) and unicasts a *JoinReq* packet toward this core node. In Figure 8.37, when a new receiver R3 wants to join the multicast group, it gets the core node ID I2 from its CAM table and sends a *JoinReq* toward it. To forward the *JoinReq* toward the core node, the next node I6 is obtained by using the underlying unicast routing protocol. CAMP assumes the existence of an underlying unicast routing protocol in the network environment which provides the next node ID. When the mesh node I1 receives this

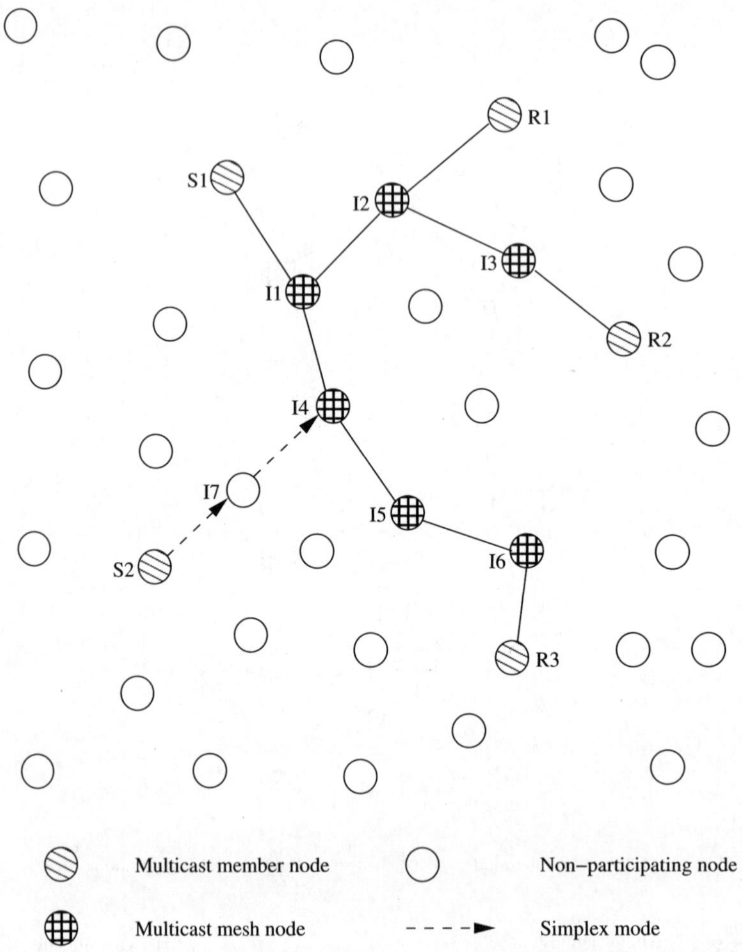

Figure 8.37. Joining of the source S2 to the multicast group.

JoinReq, it sends an ACK to receiver R3, and thus it becomes a part of the multicast group. Unlike in the core-based tree [3], if receiver R1 (which has a neighbor node I2 already present in the mesh) wants to join the group, then there is no need to send a *JoinReq* packet explicitly. In this case, receiver R2 joins the multicast group by announcing its membership using a *MulticastRoutingUpdate* message and modifies the multicast routing table (MRT). MRT contains the multicast group address of which the node is the member.

In CAMP, there is a provision for a sender-only node to join in the *simplex mode*, that is, there will be a flow of data in only one direction (from sender-only node to mesh). In Figure 8.37, source S2 has joined the group in the simplex mode. When node I7 receives a data packet (sent by the source S1) from node I4, it does not forward the data packet toward node S2, due to its simplex mode behavior. After source S2 joins the multicast group, receiver R2 receives data packets (sent by the source S2) through path S2-I7-I4-I1-I2-I3-R2, which is not the shortest path between source S2 and receiver R2 (Figure 8.38). Hence, receiver R2 sends a *HeartBeat* or *PushJoin* control packet periodically to node I8 (This next node ID I8 is obtained through the underlying unicast routing protocol). When node I8 receives this *HeartBeat* message, it forwards it to node I4. Thus, a route through the shortest path between source S2 and receiver R2 gets established. In this way, CAMP makes sure that all the shortest paths are part of the mesh.

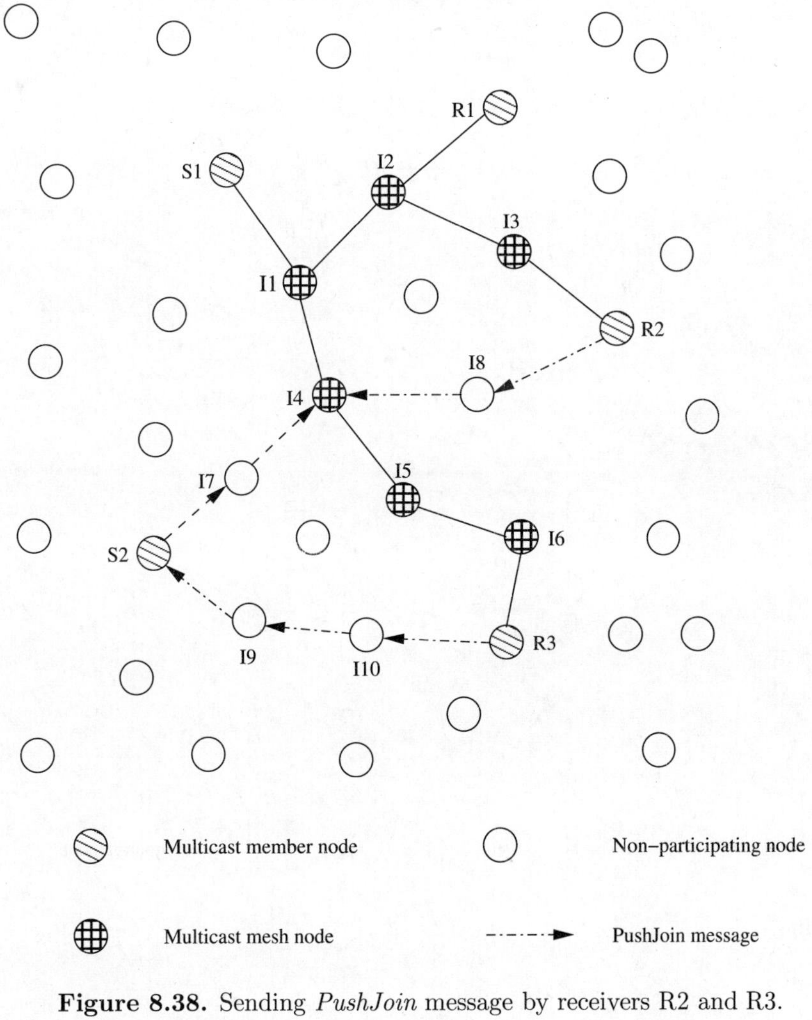

Figure 8.38. Sending *PushJoin* message by receivers R2 and R3.

Mesh Maintenance Phase

Link failures are not very critical in CAMP. For example, in Figure 8.39, when the link I10-R3 breaks (caused due to movement of node R3), receiver R3 can still receive data packets through the next shortest route S2-I7-I4-I5-I6-R3. Now receiver R3 sends a *PushJoin* message to node I15, thereby establishing a new shortest path S2-I9-I10-I15-R3. Due to the mobility of nodes, the multicast mesh may become partitioned. CAMP ensures partition repair by means of the *CoreExplicitJoin* message which is sent by each active core in the mesh component to the cores in the

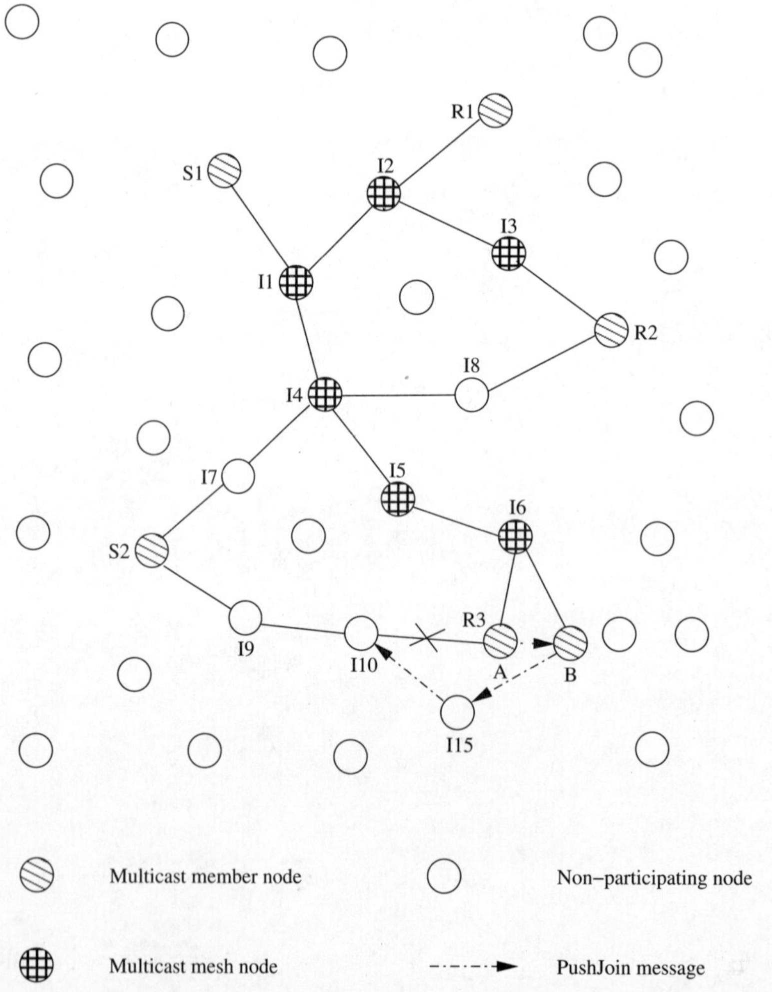

Figure 8.39. Mesh maintenance in CAMP.

other mesh components. The partition is repaired when a core receives the *Core-ExplicitJoin* message and replies with an *Ack* message.

Advantages and Disadvantages

The main improvement in CAMP is that it exhibits a high packet delivery ratio while keeping its total control overhead less, compared to other multicast routing protocols, by avoiding the flooding of control packets. Core node failures cause significant packet losses. Another drawback of CAMP is that it needs the support of a routing protocol that can work correctly in the presence of router failures and network partitions. As such, not all routing protocols are compatible with CAMP.

8.8 SUMMARY OF TREE- AND MESH-BASED PROTOCOLS

Table 8.4 summarizes the characteristics of the various multicast routing protocols for ad hoc wireless networks described so far. It helps in characterizing and identifying the qualitative behavior of multicasting protocols. However, a quantitative comparison study in terms of their performance under a wide range of parameters such as connectivity and size of the network, mobility of nodes, number of multicast sources, and multicast group size requires extensive analytical and/or simulation studies.

8.9 ENERGY-EFFICIENT MULTICASTING

Since the nodes in an ad hoc wireless network are battery-operated, it is important to find energy-conserving mechanisms and protocols that optimize the use of battery power in order to increase the lifetime of the network. This section describes some energy-efficient multicasting protocols for ad hoc wireless networks.

8.9.1 Energy-Efficient Reliable Broadcast and Multicast Protocols

In [29], the authors propose a set of power-efficient schemes which are based upon the existing power-aware broadcast algorithms for wireless networks such as broadcast incremental power (BIP) algorithm, broadcast least unicast (BLU) algorithm, and broadcast link-based minimum spanning tree (BLiMST) algorithm. These schemes also take into consideration the packet-error probability on a link to determine the expected energy required for reliably transmitting a packet on that link. The expected energy required for the reliable transmission of a packet from node i to node j is given by $E_{ij(reliable)} = \frac{E_{ij}}{(1-p_{ij})}$, where p_{ij} is the packet-error probability and $\frac{1}{(1-p_{ij})}$ is the expected number of retransmissions required from node i to node j. E_{ij} refers to the energy required to transmit a packet once from node i to node j. The modified BIP, BLU, and the BLiMST algorithms take this $E_{ij(reliable)}$ value into consideration in their respective multicast tree formation algorithms in order to construct energy-efficient trees.

Table 8.4. Summary of multicast routing protocols

Multicast Protocols	Multicast Topology	Initialization	Independent of Routing Protocol	Dependency on Specific Routing Protocol	Maintenance Approach	Loop Free	Flooding of Control Packets	Periodic Control Messaging
ABAM	Source-tree	Source	Yes	No	Hard state	Yes	Yes	No
BEMRP	Source-tree	Receiver	Yes	No	Hard state	Yes	Yes	No
DDM	Source-tree	Receiver	No	No	Soft state	Yes	Yes	Yes
MCEDAR	Source-tree over Mesh	Source or Receiver	No	Yes (CEDAR)	Hard state	Yes	Yes	No
MZRP	Source-tree	Source	Yes	No	Hard state	Yes	Yes	Yes
WBM	Source-tree	Receiver	Yes	No	Hard state	Yes	Yes	No
PLBM	Source-tree	Receiver	Yes	No	Hard state	Yes	No	Yes
MAODV	Shared-tree	Receiver	Yes	No	Hard state	Yes	Yes	Yes
Adaptive Shared	Combination of Shared- and Source-trees	Receiver	Yes	No	Soft state	Yes	Yes	Yes
AMRIS	Shared-tree	Source	Yes	No	Hard state	Yes	Yes	Yes
AMRoute	Shared-tree over Mesh	Source or Receiver	No	No	Hard state	No	Yes	Yes
ODMRP	Mesh	Source	Yes	No	Soft state	Yes	Yes	Yes
DCMP	Mesh	Source	Yes	No	Soft state	Yes	Yes	Yes
FGMP	Mesh	Receiver	Yes	No	Soft state	Yes	Yes	Yes
CAMP	Mesh	Source or Receiver	No	No	Hard state	Yes	No	No
NSMP	Mesh	Source	Yes	No	Soft state	Yes	Yes	Yes

8.9.2 A Distributed Power-Aware Multicast Routing Protocol

In [30], an underlying unicast routing protocol is expected to be present which can give the least-cost path from the current node to all other nodes in the network. The information gained from the unicast routing protocol is used to construct the multicast tree. The authors of [30] propose schemes which take the power factor into consideration to find these minimum cost paths. To find the minimum cost path (consisting of nodes $1, 2, 3, ..., j-1, j$, where j is the destination), the cost C is given by $C = \frac{(P_{1,2}+P_{2,3}+...+P_{j-1,j})}{min(K_1,K_2,...,K_j)}$, where K_i is the number of free transceivers available at node i, and $P_{i,j}$ is the power required for transmitting a packet from node i to node j. The denominator in this expression takes care of the congestion factor, and the numerator covers the power requirements. A strictly additive metric to determine the cost is also given as follows. The distance $D_{i,j}$ between nodes i and j is given as $D_{i,j} = \frac{P_{i,j}}{min(K_i,K_j)}$. The end-to-end path distance D is given by $D = \sum_{i=0}^{n-1} D_{i,i+1}$, where 0 is the ID of the source node, and n is the ID of the destination node. The lowest-cost path corresponds to the path with the smallest value for D.

8.9.3 Energy-Efficient Multicast Routing Protocol

Energy-efficient multicast routing protocol (E^2MRP) [31] is a mesh-based protocol. The protocol operates in two phases. The first phase uses a heuristic called minimum energy consumed per packet (MECP). The received power at a node is directly proportional to the quantity $d^{-\theta}$, where d is the distance of the receiving node from the sender and θ is a constant. If $P_{i,j}$ is the power required to transmit a packet from node i to node j, and $d_{i,j}$ is the distance between nodes i and j, then R_k, the cost of power for a route from a source with ID 1, to a destination with ID n, is given as $C_{R_k} = \sum_{i=1}^{n-1} P_{i,i+1} = \sum_{i=1}^{n-1} d^2(n_i, n_{i+1})$. θ is taken as 2 here. The route selected is the one with the lowest C_{R_k} value. The main idea here is to minimize the total end-to-end power consumed in sending a packet from the source to its final destination.

The second phase uses a heuristic called the minimum maximum node cost (MMNC). Here the cost function used is $C_i(V_i(t)) = \frac{Y}{V_i(t)}$, where Y is a constant, and $V_i(t)$ is the energy consumed by node i until time t. C_{R_k}, the cost of a route, consisting of nodes $1, 2, ..., n$, is given as $C_{R_k} = max(C_i(V_i(t))), i = (1..n)$, and the route with minimum C_{R_k} is chosen. This heuristic gives more importance to the power available at nodes on the route.

The protocol switches between two phases: the MECP phase and the MMNC phase. First, in the MECP phase, a set of routes is selected based on the MECP heuristic and a mesh is formed. Once the mesh establishment is completed, a timer, called cost function switcher (CFS), is set. Just before the CFS times out, the system shifts to the MMNC phase. Here again, a mesh based on the MMNC heuristic is established and the CFS timer is set. As before, just before the timer expires, the system goes back to the MECP phase, and the whole process repeats continuously, alternately switching between the MECP and the MMNC phases.

8.9.4 Energy-Efficient Cluster Adaptation of Multicast Protocol

In [32], a cluster-based scheme for power optimization is proposed. There exist several clusters of nodes in the network. Each cluster has a cluster-head, and all cluster-heads are connected through a super-node network. In a cluster-based scheme, when the number of cluster-heads is less, then more number of nodes can be reached by a cluster-head in a single hop, but the cluster-heads need to expend more energy and energy gets wasted when the cluster-head tries to reach a lone farthest node within its cluster. On the other hand, when the number of clusters is greater, energy again gets wasted on trying to reach a greater number of nodes in the super-node network. Hence, a balance between these two cases needs to be achieved.

In this scheme, each node initially starts off as a cluster-head. The node periodically sends beacons with its highest power level. In the cluster-building phase, each node increases its power level by one step at a time until the incremental change in the number of nodes seen reduces, or the power level reaches a specific maximum level. It then sends a recruiting message to nodes within its reach, with the list of nodes in its range, and its power level. A node joins the first cluster it hears, whose cluster-head ID is greater than its own id. The nodes that still remain as cluster-heads send cluster-forming messages to nodes within their cluster. The cluster member nodes respond and the cluster-heads finally respond with finalizing messages. The objective of the cluster-forming and finalizing messages is to make sure that each node within the cluster is able to reach all other nodes within the same cluster. Nodes that are not part of any cluster finally act as singleton clusters. The ODMRP protocol [24] is applied on the super-node topology to establish a mesh network among the cluster-heads. The mesh comes into play only when the destination node is not within the same cluster as the source node. In order to balance energy depletion, nodes within a cluster take turns to become cluster-heads.

8.10 MULTICASTING WITH QUALITY OF SERVICE GUARANTEES

Provisioning quality of service (QoS) implies providing guarantees such as deterministic end-to-end delay, availability of a fixed amount of bandwidth, buffers, and computational resources to the multicast session. Two multicast protocols, the wireless ad hoc real-time multicasting protocol and the multicast priority scheduling protocol, that attempt to provide QoS guarantees are described below.

8.10.1 Wireless Ad Hoc Real-Time Multicasting Protocol

Wireless ad hoc real-time multicasting (WARM) protocol [33] enables spatial bandwidth reuse along a multicast mesh. Bandwidth is guaranteed for real-time [constant bit rate (CBR)] traffic. The protocol uses periodic message exchanges, but the messaging is localized to the neighborhood of the receiving multicast member and hence the control overhead is low. It mainly deals with the transmission scheduling problem for a multicast session. A receiver node reserves time division multiple access

(TDMA) time-slots and attaches itself to the multicast mesh through a neighbor node which is a multicast member.

The protocol deals with two types of traffic: CBR traffic and variable bit rate (VBR) traffic. Time is slotted and is grouped into superframes. Each superframe consists of two frames, one for transmitting data and the other for receiving data. Each frame consists of two parts: a reserved part and a random-access part, as shown in Figure 8.40. Both the frames are slotted. Packets transmitted by a node are numbered sequentially for that frame. Each node marks its receive (transmit) slots with the frame-sequence numbers of the packets to be received (transmitted) in those slots. A node may transmit the same packet in more than one slot, for supporting different children that may not be able to receive the packet in the same slot (In Figure 8.40, it can be seen that the same packet P1 is transmitted in slots 1 and 2). Packets in a frame that are in excess of the reserved number of slots are transmitted or received in the random-access portion of the frame.

Each node maintains the following information for a particular multicast session: ID (the node's unique identifier), HC (the node's hop count, *i.e.*, the number of hops from the multicast source), F (a bit which indicates which of the two frames in the superframe is the transmit frame), RS (the number of receive slots currently reserved by the node), RS_{des} (the desired number of receive slots as determined by the node, based on the current traffic load), RS_{min} (the minimum acceptable number of reserved slots), PID (a vector representing the parents of the node), $RxSlot$ (a vector which lists the receive slots of the node), $RxSeq$ (a vector indicating the frame-sequence in which packets are received in the receive slots), SIR (a vector containing the signal-to-interference ratios of the node in its receive slots), G_p (a vector representing the path losses between the node and each of its parents), $TxSlot$ (a vector which lists the transmit slots of the node), $TxSeq$ (a vector indicating the frame-sequence in which packets are transmitted in the transmit slots), and US (a vector listing the slot numbers of the transmit slots unusable by the node). A node considers itself to be connected to the multicast mesh as long as $RS \geq RS_{min}$. If the node receives fewer than RS_{min} packets in its reserved receive slots, it becomes

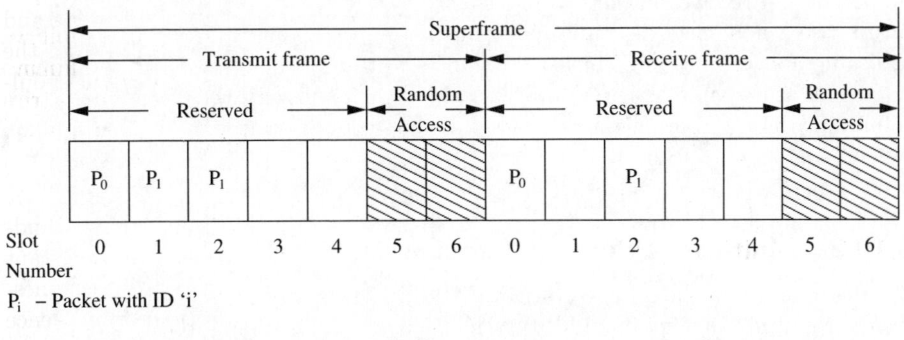

Figure 8.40. TDMA frame structure in WARM.

disconnected and sets its RS to zero. It then attempts to reconnect to the multicast session. Each node periodically transmits its status information consisting of ID, HC, F, RS, $RxSlot$, $RxSeq$, PID, $TxSlot$, $TxSeq$, and US. Transmission of this information is done on a separate signaling channel in a round-robin fashion.

In case RS is less than RS_{des}, the node attempts to receive more receive slots through the signaling message. Here the node appends the required extra receive slots, the potential parents from which it can receive in these slots, and also the frame-sequence numbers of packets it is missing, to the signaling message. Each node maintains a neighborhood database containing information regarding all nodes from which it receives signaling information.

Connection establishment when $RS_{min} \leq RS < RS_{des}$ is done as below. First, the node determines the frame-sequence numbers of the packets it is missing. It then looks into the neighborhood database to find nodes that are already transmitting those packets and whose hop-count is less than its own by one. For each such neighbor, it determines the SIR for the concerned slot. If the SIR is above a certain threshold value, then the node can receive data from the new parent node in that slot, and so it adds the new parent node's ID, slot number, and the frame-sequence to the appropriate vectors PID, $RxSlot$, and $RxSeq$, respectively. If, even after this process, the node misses packets, it identifies neighboring nodes with hop-length one less than its own and that can add transmit slots to their $TxSlots$ in order to relay the missing packets. If the SIR it computes on these slots is acceptable, it makes these nodes its parents. Consider Figure 8.41, where node S is the source. Packets transmitted by node A are received without any collisions at node B and node C. Node B transmits packets P1, P2, and P3 in slots 0, 1, and 2, respectively. Node C transmits the same packets P1, P2, and P3 received from node A in slots 2, 4, and 5, respectively. Node B is the parent of node D. Packets P1 and P2 transmitted by node B are received by node D in slots 0 and 1 without any error. But in slot 2, both nodes B and C transmit packets simultaneously. Hence, the packets transmitted by them collide during slot 2 at node D. Node D therefore is unable to receive packet P3 from node B. It searches for other possible parent nodes transmitting packet P3. It finds that node C transmits packet P3 in slot 5. If slot 5 at node B is free and if the SIR on slot 5 is acceptable, it makes node C its parent and receives packet P3 in slot 5.

In case a node gets disconnected from the multicast mesh ($RS = 0$), it follows the same above procedure, but it tries to use neighbor nodes with the minimum hop-count as its parents. If RS_{min} slots cannot be found with these parents, it tries to obtain packets from neighbor nodes with hop-counts greater than the minimum by one, and so on.

8.10.2 Multicast Priority Scheduling Protocol

Multicast priority scheduling protocol (MPSP) [34] is a packet scheduling mechanism for multicast traffic in ad hoc wireless networks. The main objective of MPSP is to provide improved multicast packet delivery with bounded end-to-end delays. It is based on the DLPS protocol [35] (described in detail in Section 6.7.3) which

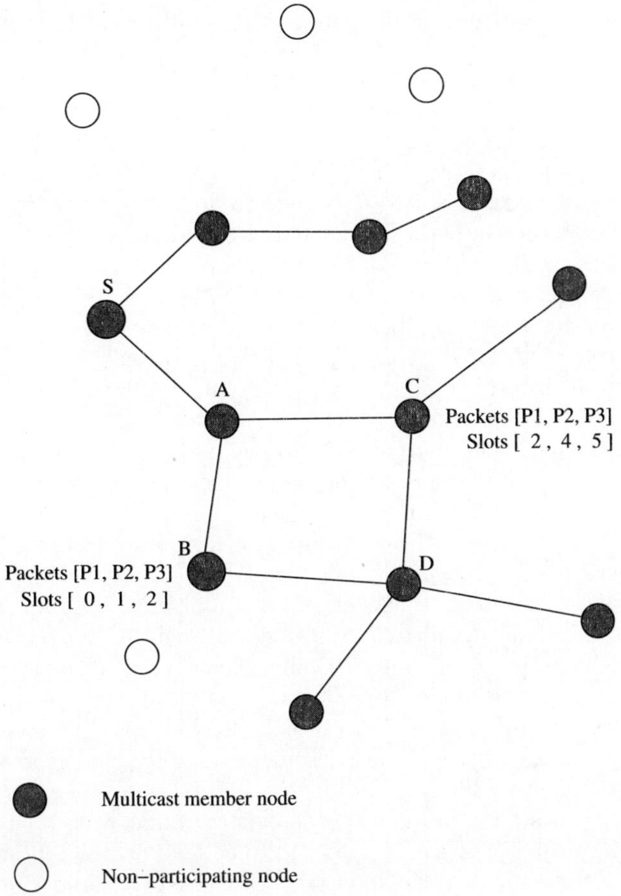

Figure 8.41. Example of WARM.

uses laxity-based priority scheduling and guarantees higher packet delivery for unicast traffic under bounded end-to-end delay conditions. Packet transmissions in a multicast session are broadcast in nature and are not protected by the RTS-CTS exchange. Therefore, the RTS, CTS, and ACK packets are not available for carrying piggy-backed priority information as in DLPS. MPSP modifies the DLPS protocol to suit the characteristics of multicast transmissions.

MPSP can work with both tree-based as well as mesh-based multicast protocols. But since tree-based protocols are more efficient when compared to mesh-based protocol, the following discussion uses a tree-based multicast protocol for describing the operation of MPSP. Each node maintains a scheduling table (ST) which contains information about packets to be transmitted by the node and information about packets in the neighbor nodes (which is obtained by overhearing data packets carrying such information transmitted by neighbor nodes), sorted according to their *priority index* values. Priority index expresses the priority of a packet. The lower the priority index, the higher the packet's priority. The performance of MPSP was studied in [34] using the WBM protocol [15], which was explained in Section 8.6.6. MPSP consists of four main components, namely, feedback mechanism, priority index calculation, scheduling table updates, and a back-off mechanism.

The feedback mecha·.ism is used to convey, to the multicast source, information regarding the percentage of packets delivered at the multicast receiver nodes. This task would be relatively simple in unicast routing as there exists only one receiver node, receiving packets from the source node through a single path. MPSP uses *soft acknowledgments* (described below) for this purpose. A leaf node l, on receiving a data packet from its parent node p, increments the count of packets received $countPkts$ for the corresponding session, maintained by it. It sends back an acknowledgment (ACK) packet to its parent node p carrying the value of $countPkts$. If multiple leaf nodes are connected to a parent node, then that parent node receives an ACK from each of the leaf nodes. Now, parent p computes $avgCount$, the average of the received $countPkts$ values. The value of $avgCount$ is piggy-backed on each data packet transmitted by node p. This data packet is heard by node pp, which is the parent of node p. Node pp hears data packets transmitted by each of its child nodes. It computes the new $avgCount$ from the $avgCount$ values on the data packets heard. This new $avgCount$ is piggy-backed by node pp on each DATA packet it sends. Thus, in this manner the value of $avgCount$ moves up the multicast tree and reaches the source node after a small initial delay. This $avgCount$ value is used in the calculation of priority index of packets, which is explained below.

After transmitting each multicast packet, a node increments the count of multicast packets transmitted for that session $txCount$. The average count of packets received at the multicast receiver nodes $avgCount$ is available at each multicast tree node. Each transmitted packet carries its end-to-end delay target *deadline*, that is, the time by which it should reach the destination. Using the above quantities, the priority index PI of a packet is calculated. It is given by

$$PI = \frac{PDR}{M} \times ULB \qquad (8.10.1)$$

Here, $PDR = \frac{avgCount}{txCount}$ is the packet delivery ratio of the flow to which the packet belongs, $ULB = \frac{deadline-currentTime}{remHops}$ is the uniform laxity budget of the packet ($currentTime$ denotes the current time according to the node's local clock), and M is a user-defined parameter representing the desired packet delivery ratio for the multicast session. $remHops$ is the maximum number of hops remaining to be traversed by the multicast packet. It is obtained in the following manner.

In WBM, when a new receiver node wants to join the multicast group, it initiates a *JoinReq*. A multicast tree node, on receiving this *JoinReq*, responds by sending back a *JoinReply* packet. The receiver node may receive multiple *JoinReply* packets. It chooses one of them and confirms its selection by transmitting a *JoinConf* to the corresponding multicast tree node. This *JoinConf* packet is utilized by MPSP. Before transmitting the *JoinConf* packet, the new receiver node initializes a *hopCount* field on the *JoinConf* packet to zero. An intermediate node forwarding the *JoinConf* increments the *hopCount* by one. Hence, all nodes on the path to the node at which the new receiver node joins the multicast group (the node which initiated the *JoinReply* packet) know the remaining number of hops to be traversed by a multicast packet transmitted by it. The remaining number of hops *remHops* at node n is given by the maximum of *hopCount* to each receiver node through node n. A node piggy-backs this *remHops* value on each data packet it transmits. The node's parent, on hearing the transmitted data packet, extracts the piggy-backed value and computes its own *remHops* value, which it piggy-backs on the data packets it sends. Thus, the value of *remHops* moves up the multicast tree and finally reaches the multicast source node. Since each node in the multicast tree has a *remHops* value for the multicast session, the *ULB* of a packet queued at a node can now be computed. The above procedure is illustrated in Figure 8.42. Here, the new receiver node NR (node 1) initiates route discovery by transmitting

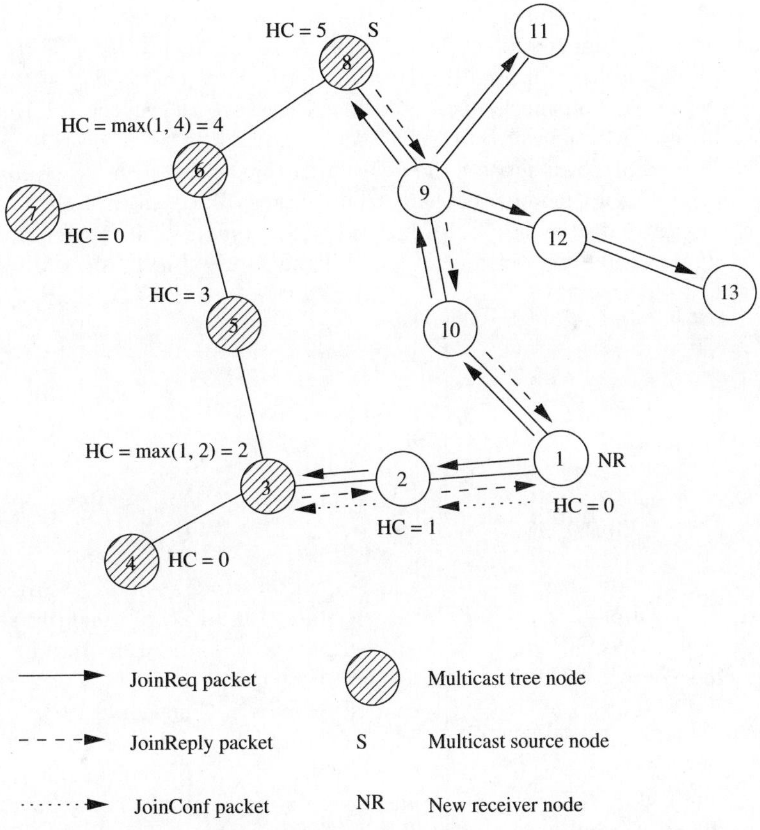

Figure 8.42. Propagation of *hopCount* in MPSP.

a *JoinReq* packet. It reaches multicast tree nodes 3 and 8. Each of these two nodes responds by sending back a *JoinReply* packet. Node 1 chooses the *JoinReply* sent by node 3. It sends a *JoinConf* destined to node 3, with the *hopCount* value (HC) set to zero. When the packet reaches node 2, node 2 increments the HC by one and piggy-backs it on the next data packet it transmits. Node 3, on hearing this packet, retrieves the HC value one and computes the new HC value for node 1 as two. It is already connected to node 4, and so the HC from node 3 to node 4 is one. The new HC value for the multicast session at node 3 is computed as max(1, 2) = 2. This value is piggy-backed by node 3 on the next multicast packet it transmits, which would be heard by node 5. Nodes 5 and 6 compute the HC values in a similar fashion. When node 6 transmits a packet with the HC set to 4, node 8, the multicast source, would hear the packet and calculate the maximum number of hops remaining though node 6 as five.

Priority information regarding the highest priority ready packet to be transmitted next is piggy-backed on each data packet. A neighbor node, on hearing this packet, updates its scheduling table with the piggy-backed priority information. Hence, each node would know the priority of its own packet compared to the priorities of packets queued at its neighbor nodes.

The back-off mechanism used in MPSP is the same as that used in DLPS. The objective of the back-off mechanism used in DLPS is to reflect the priority of the node's highest priority packet on the back-off period to be taken by the node. If r is the rank (the rank of an entry is the position of that entry in the scheduling table of the node), in ST of the node, of the current packet to be sent, n is the number of retransmission attempts made for the packet, and $nmax$ is the maximum number of retransmission attempts permitted, then the back-off interval is given by

$$\text{back}-\text{off} = \begin{cases} Uniform[0, (2^n \times CW_{min}) - 1] & (8.10.2) \\ \quad \text{if } r = 1 \text{ and } n \leq nmax & \\ \frac{PDR}{M} \times CW_{min} + \; Uniform[0, CW_{min} - 1] & (8.10.3) \\ \quad \text{if } r > 1 \text{ and } n = 0 & \\ ULB \times CW_{min} + \; Uniform[0, (2^n \times CW_{min}) - 1] & (8.10.4) \\ \quad \text{otherwise} & \end{cases}$$

where CW_{min} is the minimum size of the contention window, and M is the desired packet delivery ratio. If the packet has the highest rank in the broadcast region of the node, then it has the lowest back-off period according to Equation 8.10.2 and faces very less contention. Else, if it is the first time the packet is being transmitted, the back-off distribution follows the second scheme as in Equation 8.10.3, where the back-off is more than that for the first case. Here the current PDR of the flow affects the back-off period. If PDR is very low, then the first term would be low, and if it is high, then the first term would be high and the node would have to wait for a longer time. Finally, if the packet does not fit into these two categories, then the back-off value is as per the third scheme in Equation 8.10.4 and is the longest of the three cases. The higher the value of ULB, the longer the back-off period.

The above-described mechanisms aid MPSP in providing better multicast packet delivery compared to other regular multicast protocols, under bounded end-to-end delay conditions.

8.11 APPLICATION-DEPENDENT MULTICAST ROUTING

There are some multicast routing protocols that cater to different needs of a user depending on the scenarios in which they are used. Some of them are described in this section.

8.11.1 Role-Based Multicast

Role-based multicast [36] is a multicast scheme designed to meet the special needs of inter-vehicle communication. The multicast group changes dynamically based on the location, speed, driving direction, and time. An example application, described and discussed in detail in [36], is the accident situation on highways where information about the accident needs to be disseminated to the relevant vehicles (receivers). A modified flooding technique floods information about the accident. Taking into consideration the speed, direction of movement, and distance from the source, a multicast group is calculated. This group includes all vehicles for which the information could be useful, so that the drivers can brake their vehicles before the accident zone. The information regarding the accident is flooded so that all such receivers receive it on time. Note that in this model the source is stationary and the receivers move at high speeds, mostly toward the source. This model cannot be used in situations where the sources are also in constant motion. The content-based multicast [37] model described below handles such situations.

8.11.2 Content-Based Multicast

Content-based multicasting [37] is used in areas where the source set and the receiver set for the information keep changing dynamically based on the content of the information and the mobility of the receivers themselves. An example of this kind of application is in a battlefield where the moving soldiers need to be continuously updated on the impending threats that may occur within a certain duration (say, in the next ten minutes) or that may be present at a certain distance (say, 5 Km) from the soldier. Information on the presence of their allies may also be of use to them. Autonomous sensors dropped in the forward area may be used to collect the information required. Thus, in the content-based multicast model, it can be assumed that nodes are interested in obtaining information about *threats* and *resources* that are (i) t time away from their current location and/or (ii) distance d away.

However, the main problem in such applications is that since the nodes, threats, and resources keep moving all the time, it is difficult for the information sources to determine their receiver sets, and it is also difficult for the receiver nodes to determine the identity of the senders. In [37], this problem is solved by means of a *sensor-push* and *receiver-pull* approach. The entire area is divided into geographical regions called blocks. Each block has a block leader.

In the *sensor-push* part of the scheme, sensor nodes generate information and pass them on to the leader of the block they are located in by means of *ThreatWarning* messages. Based on the location and velocity of the threat, this information is further pushed into other blocks by the block leader. In the *receiver-pull* part, if the time specification is t, then the receiver node sends a *PullRequest* to the leader of the block in which it is expected to be present after time period t. In case that leader has incomplete threat information, it further generates *PullRequest* messages to leaders of blocks in the direction of the threat, retrieves threat information if any, and forwards it to the requesting receiver.

Consider Figure 8.43. Here node I is moving at a speed s_i and is expected to reach block PQRS in time t_i. It sends a *PullRequest* message to node B which is the leader of block PQRS. This message contains node I's velocity along with a specification of threats about which the node needs to be warned. Node B sends back all information regarding the threats of which it is aware. In case node B has incomplete knowledge about threats that are time t_i away, which is very likely, it initiates additional *PullRequest* messages. If the maximum speed of the threats is s_{max}, then node B sends the *PullRequest* messages to the leaders of all blocks located at distance $s_{max}t_i$ away. These block leaders determine whether the threats they are aware of would affect nodes in block PQRS after time period t_i from the current point of time. If so, then they send the threat information to the node B, which then forwards the information received to node I.

8.11.3 Location-Based Multicast

Location-based multicasting or geocasting, which is a variant of the conventional multicasting, makes use of geographical information for multicasting. Here, a group of nodes present in a particular geographic region constitutes the multicast receiver group. The global positioning system (GPS) plays an integral part in all location-based schemes. A node uses GPS to obtain its latitude, longitude, and altitude coordinates. Geocasting has several applications, such as sending emergency messages to people within a small area such as buildings, paging a person who is known to be present within a small area such as a suburb in a city, and sending advertisements meant only for a particular region. A few of the location-based multicast schemes are described below.

In [38], two schemes that use a modified flooding approach for geocasting are proposed. These schemes use the concept of forwarding regions. In the first scheme, a rectangle encompassing the source and the multicast receiver region, and whose sides are parallel to the X and Y axes, constitutes the forwarding region. A node, receiving a multicast message, forwards it only if it lies within the forwarding region. Otherwise, the message is discarded. Figure 8.44 illustrates this scheme. The forwarding region is shown by the dotted rectangle. Packets transmitted by source S reach nodes A and C. Node C, which is within the forwarding region, forwards the packets to nodes D and E. But since node A is not located within the forwarding

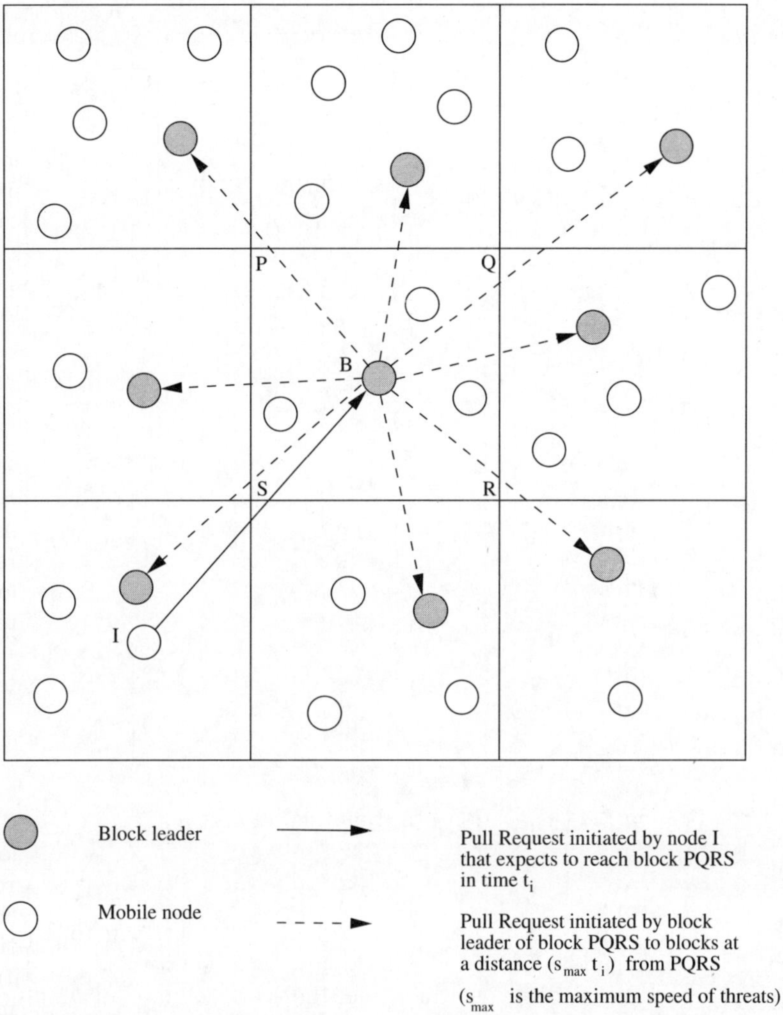

Figure 8.43. Content-based multicast.

region, it cannot further forward the packets, and so it just discards all incoming data packets for the multicast region. Similarly, packets transmitted by node E reach nodes F and G. Here again, node F, which is within the forwarding zone, forwards its incoming packets for the multicast region, while node G located outside the forwarding zone discards them.

In the second scheme, the forwarding region does not form any definite shape. When any node J receives a multicast data packet, which had been originated by a node S, from a node I, it forwards the packet only if it is, at most, δ distance

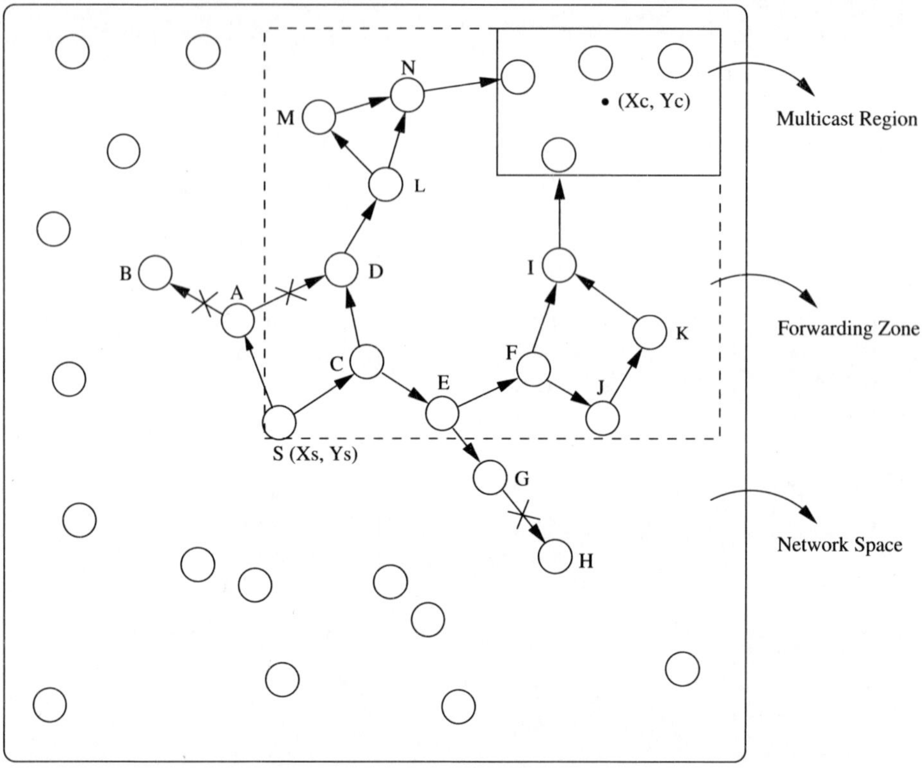

Figure 8.44. Location-based multicast: Scheme 1.

farther from the center of the multicast region, than the node I. The distance is calculated as below. Before transmitting a data packet, the source node inserts its own coordinates (Xs, Ys) and the coordinates of the center point of the multicast region (Xc, Yc) on the packet. A receiving node, since it knows its own coordinates through GPS, calculates its distance from (Xc, Yc), and also the distance between its previous node and (Xc, Yc). If its distance is, at most, δ more than the distance of its previous node (source node) from the multicast region center, it forwards the packet; otherwise, it discards the packet. Before forwarding, it replaces the coordinates of its previous node on the packet with its own coordinates, so that the immediate receiver node can compute the two distances by means of which the decision on whether or not to forward the packet further is taken. This scheme can be better understood by means of Figure 8.45. Here δ is assumed to be zero. The notation $DIST_i$ denotes the distance of node I from (Xc, Yc). Packets transmitted by source node S are received by both nodes A and C. Since both $DIST_a$ and $DIST_c$ are less than $DIST_s$, both node A and node C forward the packets. Packets transmitted by node C reach nodes D and E. But now, since $DIST_e$ is greater than

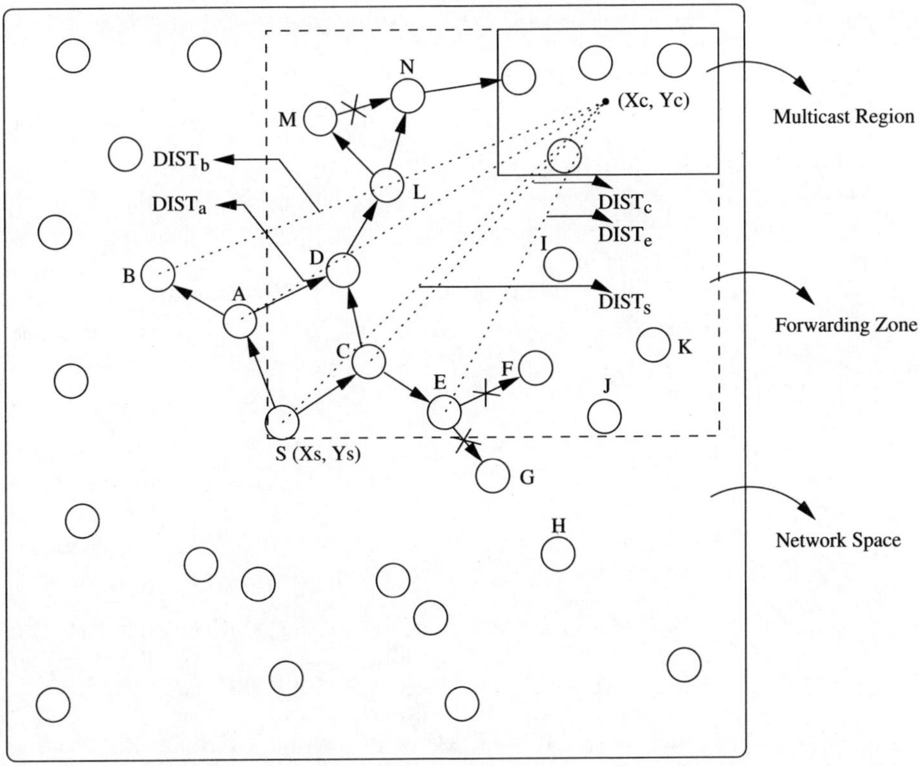

Figure 8.45. Location-based multicast: Scheme 2.

$DIST_c$, node E does not forward the packets further, but instead discards them. In case of node A, since $DIST_a$ is less than $DIST_s$, node A forwards its incoming multicast packets to nodes B and D. Similarly, packets transmitted by node L reach nodes M and N. But since node M is located farther from (Xc,Yc) compared to node L, it does not forward the packets further.

8.12 SUMMARY

The challenges faced by multicast routing protocols for ad hoc wireless networks are much more complex than those faced by their wired network counterparts. In this chapter, the problem of multicast routing in ad hoc wireless networks was studied. After identifying the main issues involved in the design of a multicast routing protocol, a classification of the existing multicasting protocols was given. Several of these multicast routing protocols were described in detail with suitable examples. The advantages and drawbacks involved in each protocol were also listed.

Some energy-conserving multicasting routing protocols were presented, as most of the nodes in ad hoc wireless networks are battery-operated.

Reliable multicasting has become indispensable for the successful deployment of ad hoc wireless networks as these networks support important applications such as military battlefields and emergency operations. Real-time multicasting that supports bounded delay delivery for streaming data is also a potential avenue for research. Security is another necessary requirement, which is still lacking in ad hoc multicast routing protocols, as multicast traffic of important and high security (*e.g.*, military) applications may pass through unprotected network components (routers/links). Thus, multicasting in ad hoc networks is a significant problem that merits further exploration.

8.13 PROBLEMS

1. Comment on the scaling properties of source-initiated and receiver-initiated multicast protocols with respect to the number of sources and receivers in the group. Which of them would be suitable for (a) a teacher multicasting his lectures to a set of students (assume the students do not interact with one another) and (b) a distributed file sharing system?

2. Is hop length always the best metric for choosing paths? In an ad hoc network with a number of nodes, each differing in mobility, load generation characteristics, interference level, etc., what other metrics are possible?

3. Link-level broadcast capability is assumed in many of the multicast routing protocols. Are such broadcasts reliable? Give some techniques that could be used to improve the reliability of broadcasts.

4. What are the two basic approaches for maintenance of the multicast tree in bandwidth efficient multicast protocol (BEMRP)? Which of the two performs better? Why?

5. Though both MCEDAR and AMRoute form trees over their underlying meshes, MCEDAR is more efficient compared to AMRoute. Why?

6. Why is ABM not efficient? How can its efficiency be increased?

7. Why is DDM not scalable with respect to the group size?

8. Sometimes, AMRIS may not exhibit high packet delivery ratio even when all nodes exhibit mobility confined to a small region. Why?

9. Calculate the approximate control overhead for the ODMRP protocol over a 200-second time period (refer to Figure 8.46). Assume that all nodes are stationary. Number of nodes: 50. Time period for sending a *JoinReq*: 2 secs.

10. Calculate the approximate control overhead in the dynamic core-based multicast routing protocol (DCMP) (refer to Figure 8.47), over a 200-second time

period. Assume that all nodes are stationary. S1 is a core active source and S2 is a passive source. Number of nodes: 50. Time period for sending a *JoinReq*: 2 secs.

11. Calculate the efficiency of the ODMRP and DCMP protocols from Figures 8.46 and 8.47, respectively.
12. The CAMP protocol has been inspired from CBT. But it is more efficient compared to CBT. Give the reasons behind this.
13. What makes CAMP an efficient protocol?
14. What are the two different topology maintenance approaches? Which of the two approaches is better when the topology is highly dynamic? Give reasons for your choice.
15. Provide a timing diagram similar to that in Figure 8.1 or Figure 8.2 for a core-based multicast protocol.

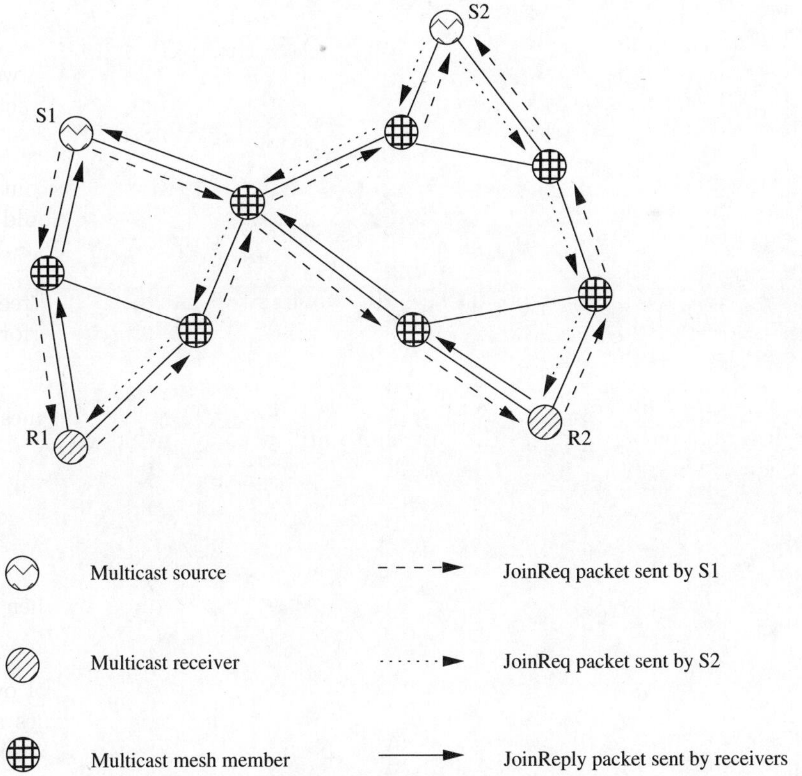

Figure 8.46. ODMRP example.

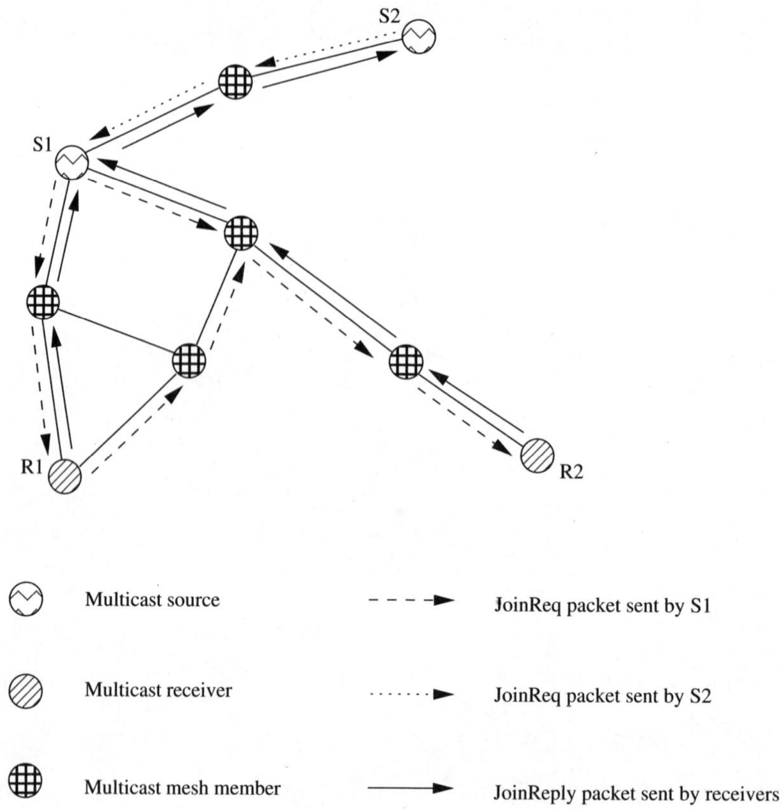

Figure 8.47. DCMP example.

16. Comment on how content-based multicasting (CBM) could be advantageous or disadvantageous as far as the bandwidth utilization of the network is concerned.

BIBLIOGRAPHY

[1] D. Waitzman, C. Partridge, and S. Deering, "Distance Vector Multicast Routing Protocol," *Request For Comments 1075*, November 1988.

[2] J. Moy, "Multicast Routing Extensions for OSPF," *Communications of the ACM*, vol. 37, no. 8, pp. 61-66, August 1994.

[3] A. Ballardie, P. Francis, and J. Crowcroft, "Core-Based Trees (CBT): An Architecture for Scalable Multicast Routing," *Proceedings of ACM SIGCOMM 1993*, pp. 85-95, September 1993.

[4] S. Deering, D. L. Estrin, D. Farinacci, V. Jacobson, C. G. Liu, and L. Mei, "The PIM Architecture for Wide-Area Multicast Routing," *IEEE/ACM Transactions on Networking*, vol. 4, no. 2, pp. 153-162, April 1996.

[5] C. L. Fullmer and J. J. Garcia-Luna-Aceves, "Solutions to Hidden Terminal Problems in Wireless Networks," *Proceedings of ACM SIGCOMM 1997*, pp. 39-49, September 1997.

[6] P. Sinha, R. Sivakumar, and V. Bharghavan, "MCEDAR: Multicast Core Extraction Distributed Ad Hoc Routing," *Proceedings of IEEE WCNC 1999*, pp. 1313-1317, September 1999.

[7] E. Bommaiah, M. Liu, A. McAuley, and R. Talpade, "AMRoute: Ad Hoc Multicast Routing Protocol," *Internet draft (work in progress)*, draft-talpade-manet-amroute-00.txt, August 1998.

[8] T. Ozaki, J. B. Kim, and T. Suda, "Bandwidth Efficient Multicast Routing Protocol for Ad Hoc Networks," *Proceedings of IEEE ICCCN 1999*, pp. 10-17, October 1999.

[9] V. Devarapalli, A. A. Selcuk, D. Sidhu, "MZR: A Multicast Protocol for Mobile Ad Hoc Networks," *Internet draft (work in progress)*, draft-vijay-manet-mzr-01.txt, July 2001.

[10] Z. J. Haas, M. R. Pearlman, and P. Samar, "Zone Routing Protocol (ZRP)," *Internet draft (work in progress)*, draft-ietf-manet-zrp-04.txt, January 2001.

[11] P. Sinha, R. Sivakumar, and V. Bharghavan, "CEDAR: A Core-Extraction Distributed Ad Hoc Routing Algorithm," *Proceedings of IEEE INFOCOM 1999*, pp. 202-209, March 1999.

[12] C. K. Toh, G. Guichala, and S. Bunchua. "ABAM: On-Demand Associativity-Based Multicast Routing for Ad Hoc Mobile Networks," *Proceedings of IEEE VTC 2000*, pp. 987-993, September 2000.

[13] C. K. Toh, "Associativity-Based Routing for Ad Hoc Mobile Networks," *Wireless Personal Communications Journal*, vol. 4, no. 2, pp. 1-36, March 1997.

[14] L. Ji and M. S. Corson, "Differential Destination Multicast (DDM) Specification," *Internet draft (work in progress)*, draft-ietf-manet-ddm-00.txt, July 2000.

[15] S. K. Das, B. S. Manoj, and C. Siva Ram Murthy, "Weight-Based Multicast Routing Protocol for Ad Hoc Wireless Networks," *Proceedings of IEEE GLOBECOM 2002*, vol. 1, pp. 17-21, November 2002.

[16] R. S. Sisodia, I. Karthigeyan, B. S. Manoj, and C. Siva Ram Murthy, "A Preferred Link-Based Multicast Protocol for Wireless Mobile Ad Hoc Networks," *Proceedings of IEEE ICC 2003*, vol. 3, pp. 2213-2217, May 2003.

[17] R. S. Sisodia, B. S. Manoj, and C. Siva Ram Murthy, "A Preferred Link-Based Routing Protocol for Ad Hoc Wireless Networks," *Journal of Communications and Networks*, vol. 4, no. 1, pp. 14-21, March 2002.

[18] R. S. Sisodia, I. Karthigeyan, B. S. Manoj, and C. Siva Ram Murthy, "A Novel Scheme for Supporting Integrated Unicast and Multicast Traffic in Ad Hoc Wireless Networks," revised version submitted to *Journal of Parallel and Distributed Computing*, 2004.

[19] E. M. Royer and C. E. Perkins, "Multicast Operation of the Ad Hoc On-Demand Distance Vector Routing Protocol," *Proceedings of ACM MOBICOM 1999*, pp. 207-218, August 1999.

[20] E. M. Royer and C. E. Perkins, "Multicast Ad Hoc On-Demand Distance Vector (MAODV) Routing," *Internet draft (work in progress)*, draft-ietf-manet-maodv-00.txt, July 2000.

[21] C. E. Perkins and E. M. Royer, "Ad Hoc On-Demand Distance Vector Routing," *Proceedings of IEEE WMCSA 1999*, pp. 90-100, February 1999.

[22] C. W. Wu, Y. C. Tay, and C. K. Toh, "Ad Hoc Multicast Routing Protocol Utilizing Increasing id-numberS (AMRIS) Functional Specification," *Internet draft (work in progress)*, draft-ietf-manet-amris-spec-00.txt, November 1998.

[23] C. C. Chiang, M. Gerla, and L. Zhang, "Adaptive Shared Tree Multicast in Mobile Wireless Networks," *Proceedings of GLOBECOM 1998*, pp. 1817-1822, November 1998.

[24] S. J. Lee, M. Gerla, and C. C. Chiang, "On-Demand Multicast Routing Protocol," *Proceedings of IEEE WCNC 1999*, pp. 1298-1302, September 1999.

[25] S. K. Das, B. S. Manoj, and C. Siva Ram Murthy, "A Dynamic Core-Based Multicast Routing Protocol for Ad Hoc Wireless Networks," *Proceedings of ACM MOBIHOC 2002*, pp. 24-35, June 2002.

[26] C. C. Chiang, M. Gerla, and L. Zhang, "Forwarding Group Multicasting Protocol for Multi-Hop, Mobile Wireless Networks," *ACM/Baltzer Journal of Cluster Computing: Special Issue on Mobile Computing*, vol. 1, no. 2, pp. 187-196, 1998.

[27] S. Lee and C. Kim, "Neighbor Supporting Ad Hoc Multicast Routing Protocol," *Proceedings of ACM MOBIHOC 2000*, pp. 37-50, August 2000.

[28] J. J. Garcia-Luna-Aceves and E. L. Madruga, "The Core-Assisted Mesh Protocol," *IEEE Journal on Selected Areas in Communications*, vol. 17, no. 8, pp. 1380-1994, August 1999.

[29] S. Banerjee, A. Misra, J. Yeo, and A. Agrawala, "Energy-Efficient Broadcast and Multicast Trees for Reliable Wireless Communication," *http://www.umiacs.umd.edu/mind/documents/wirelesscom.pdf*

[30] J. E. Wieselthier, G. D. Nguyen, and A. Ephremides, "Multicasting in Energy-Limited Ad Hoc Wireless Networks," *Proceedings of IEEE MILCOM 1998*, pp. 18-21, October 1998.

[31] H. Jiang, S. Cheng, Y. He, and B. Sun, "Multicasting along Energy-Efficient Meshes in Mobile Ad Hoc Networks," *Proceedings of IEEE WCNC 2002*, vol. 2, pp. 807-811, March 2002.

[32] C. Tang, C. S. Raghavendra, and V. Prasanna, "Energy-Efficient Adaptation of Multicast Protocols in Power-Controlled Wireless Ad Hoc Networks," *Proceedings of IEEE International Symposium on Parallel Architectures, Algorithms, and Networks 2002*, pp. 80-85, May 2002.

[33] G. D. Kondylis, S. V. Krishnamurthy, S. K. Dao, and Gregory J. Pottie, "Multicasting Sustained CBR and VBR Traffic in Wireless Ad Hoc Networks," *Proceedings of IEEE ICC 2000*, pp. 543-549, June 2000.

[34] I. Karthigeyan, B. S. Manoj, and C. Siva Ram Murthy, "Multicast Priority Scheduling Protocol for Ad Hoc Wireless Networks," *Technical Report*, Department of Computer Science and Engineering, Indian Institute of Technology, Madras, India, January 2004.

[35] I. Karthigeyan, B. S. Manoj, and C. Siva Ram Murthy, "A Distributed Laxity-Based Priority Scheduling Scheme for Time-Sensitive Traffic in Mobile Ad Hoc Networks," to appear in *Ad Hoc Networks Journal*.

[36] L. Briesemeister and G. Hommel, "Role-Based Multicast in Highly Mobile But Sparsely Connected Ad Hoc Networks," *Proceedings of ACM MOBIHOC 2000*, pp. 45-50, August 2000.

[37] H. Zhou and S. Singh, "Content-Based Multicast (CBM) in Ad Hoc Networks," *Proceedings of ACM MOBIHOC 2000*, pp. 51-60, August 2000.

[38] Y. B. Ko and N. H. Vaidya, "Geocasting in Mobile Ad Hoc Networks: Location-Based Multicast Algorithms," *Proceedings of IEEE WMCSA 1999*, pp. 101-110, February 1999.

[39] C. C. Chiang, H. K. Wu, W. Liu, and M. Gerla, "Routing in Clustered Multi-hop, Mobile Wireless Networks with Fading Channel," *Proceedings of IEEE SICON 1997*, pp. 197-211, April 1997.

[40] R. Dube, C. D. Rais, K. Y. Wang, and S. K. Tripathi, "Signal Stability-Based Adaptive Routing for Ad Hoc Mobile Networks," *IEEE Personal Communications Magazine*, vol. 4, no. 1, pp. 36-45, February 1997.

[41] D. B. Johnson and D. A. Maltz, "Dynamic Source Routing in Ad Hoc Wireless Networks," *Mobile Computing, Kluwer Academic Publishers*, vol. 353, pp. 153-181, 1996.

[42] S. Murthy and J. J. Garcia-Luna-Aceves, "An Efficient Routing Protocol for Wireless Networks," *ACM Mobile Networks and Applications Journal: Special Issue on Routing in Mobile Communication Networks*, vol. 1, no. 2, pp. 183-197, October 1996.

[43] V. D. Park and M. S. Corson, "A Highly Adaptive Distributed Routing Algorithm for Mobile Wireless Networks," *Proceedings of IEEE INFOCOM 1997*, pp. 1405-1413, April 1997.

[44] C. E. Perkins and P. Bhagwat, "Highly Dynamic Destination-Sequenced Distance-Vector Routing (DSDV) for Mobile Computers," *Proceedings of ACM SIGCOMM 1994*, pp. 234-244, August 1994.

Chapter 9

TRANSPORT LAYER AND SECURITY PROTOCOLS FOR AD HOC WIRELESS NETWORKS

9.1 INTRODUCTION

The objectives of a transport layer protocol include the setting up of an end-to-end connection, end-to-end delivery of data packets, flow control, and congestion control. There exist simple, unreliable, and connection-less transport layer protocols such as UDP, and reliable, byte-stream-based, and connection-oriented transport layer protocols such as TCP for wired networks. These traditional wired transport layer protocols are not suitable for ad hoc wireless networks due to the inherent problems associated with the latter. The first half of this chapter discusses the issues and challenges in designing a transport layer protocol for ad hoc wireless networks, the reasons for performance degradation when TCP is employed in ad hoc wireless networks, and it also discusses some of the existing TCP extensions and other transport layer protocols for ad hoc wireless networks.

The previous chapters discussed various networking protocols for ad hoc wireless networks. However, almost all of them did not take into consideration one very important aspect of communication: security. Due to the unique characteristics of ad hoc wireless networks, which have been mentioned in the previous chapters, such networks are highly vulnerable to security attacks compared to wired networks or infrastructure-based wireless networks (such as cellular networks). Therefore, security protocols being used in the other networks (wired networks and infrastructure-based wireless networks) cannot be directly applied to ad hoc wireless networks. The second half of this chapter focuses on the security aspect of communication in ad hoc wireless networks. Some of the recently proposed protocols for achieving secure communication are discussed.

9.2 ISSUES IN DESIGNING A TRANSPORT LAYER PROTOCOL FOR AD HOC WIRELESS NETWORKS

In this section, some of the issues to be considered while designing a transport layer protocol for ad hoc wireless networks are discussed.

- **Induced traffic:** Unlike wired networks, ad hoc wireless networks utilize multi-hop radio relaying. A link-level transmission affects the neighbor nodes of both the sender and receiver of the link. In a path having multiple links, transmission at a particular link affects one upstream link and one downstream link. This traffic at any given link (or path) due to the traffic through neighboring links (or paths) is referred to as induced traffic. This is due to the broadcast nature of the channel and the location-dependent contention on the channel. This induced traffic affects the throughput achieved by the transport layer protocol.

- **Induced throughput unfairness:** This refers to the throughput unfairness at the transport layer due to the throughput/delay unfairness existing at the lower layers such as the network and MAC layers. For example, an ad hoc wireless network that uses IEEE 802.11 DCF as the MAC protocol may experience throughput unfairness at the transport layer as well. A transport layer protocol should consider these in order to provide a fair share of throughput across contending flows.

- **Separation of congestion control, reliability, and flow control:** A transport layer protocol can provide better performance if end-to-end reliability, flow control, and congestion control are handled separately. Reliability and flow control are end-to-end activities, whereas congestion can at times be a local activity. The transport layer flow can experience congestion with just one intermediate link under congestion. Hence, in networks such as ad hoc wireless networks, the performance of the transport layer may be improved if these are separately handled. While separating these, the most important objective to be considered is the minimization of the additional control overhead generated by them.

- **Power and bandwidth constraints:** Nodes in ad hoc wireless networks face resource constraints including the two most important resources: (i) power source and (ii) bandwidth. The performance of a transport layer protocol is significantly affected by these constraints.

- **Misinterpretation of congestion:** Traditional mechanisms of detecting congestion in networks, such as packet loss and retransmission timeout, are not suitable for detecting the network congestion in ad hoc wireless networks. This is because the high error rates of wireless channel, location-dependent contention, hidden terminal problem, packet collisions in the network, path breaks due to the mobility of nodes, and node failure due to a drained battery can also lead to packet loss in ad hoc wireless networks. Hence, interpretation of network congestion as used in traditional networks is not appropriate in ad hoc wireless networks.

- **Completely decoupled transport layer:** Another challenge faced by a transport layer protocol is the interaction with the lower layers. Wired network transport layer protocols are almost completely decoupled from the lower layers. In ad hoc wireless networks, the cross-layer interaction between the transport layer and lower layers such as the network layer and the MAC layer is important for the transport layer to adapt to the changing network environment.

- **Dynamic topology:** Some of the deployment scenarios of ad hoc wireless networks experience rapidly changing network topology due to the mobility of nodes. This can lead to frequent path breaks, partitioning and remerging of networks, and high delay in reestablishment of paths. Hence, the performance of a transport layer protocol is significantly affected by the rapid changes in the network topology.

9.3 DESIGN GOALS OF A TRANSPORT LAYER PROTOCOL FOR AD HOC WIRELESS NETWORKS

The following are the important goals to be met while designing a transport layer protocol for ad hoc wireless networks:

- The protocol should maximize the throughput per connection.

- It should provide throughput fairness across contending flows.

- The protocol should incur minimum connection setup and connection maintenance overheads. It should minimize the resource requirements for setting up and maintaining the connection in order to make the protocol scalable in large networks.

- The transport layer protocol should have mechanisms for congestion control and flow control in the network.

- It should be able to provide both reliable and unreliable connections as per the requirements of the application layer.

- The protocol should be able to adapt to the dynamics of the network such as the rapid change in topology and changes in the nature of wireless links from uni-directional to bidirectional or vice versa.

- One of the important resources, the available bandwidth, must be used efficiently.

- The protocol should be aware of resource constraints such as battery power and buffer sizes and make efficient use of them.

- The transport layer protocol should make use of information from the lower layers in the protocol stack for improving the network throughput.
- It should have a well-defined cross-layer interaction framework for effective, scalable, and protocol-independent interaction with lower layers.
- The protocol should maintain end-to-end semantics.

9.4 CLASSIFICATION OF TRANSPORT LAYER SOLUTIONS

Figure 9.1 shows a classification tree for some of the transport layer protocols discussed in this chapter. The top-level classification divides the protocols as extensions of TCP for ad hoc wireless networks and other transport layer protocols which are not based on TCP. The solutions for TCP over ad hoc wireless networks can further be classified into split approaches and end-to-end approaches.

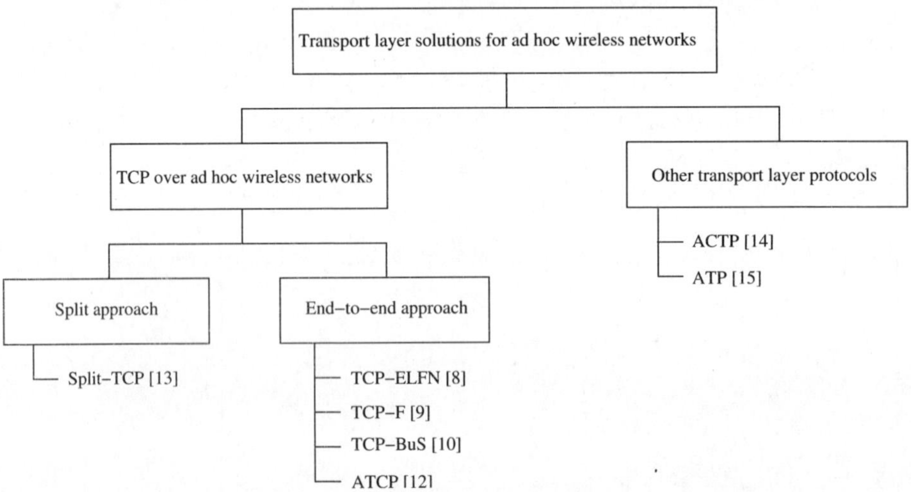

Figure 9.1. Classification of transport layer solutions.

9.5 TCP OVER AD HOC WIRELESS NETWORKS

The transmission control protocol (TCP) is the most predominant transport layer protocol in the Internet today. It transports more than 90% percent of the traffic on the Internet. Its reliability, end-to-end congestion control mechanism, byte-stream transport mechanism, and, above all, its elegant and simple design have not only contributed to the success of the Internet, but also have made TCP an influencing protocol in the design of many of the other protocols and applications. Its adaptability to the congestion in the network has been an important feature leading to graceful degradation of the services offered by the network at times of extreme congestion. TCP in its traditional form was designed and optimized only for wired networks. Extensions of TCP that provide improved performance across wired and single-hop wireless networks were discussed in Chapter 4. Since TCP is widely used today and the efficient integration of an ad hoc wireless network with the Internet is paramount wherever possible, it is essential to have mechanisms that can improve TCP's performance in ad hoc wireless networks. This would enable the seamless operation of application-level protocols such as FTP, SMTP, and HTTP across the integrated ad hoc wireless networks and the Internet.

This section discusses the issues and challenges that TCP experiences when used in ad hoc wireless networks as well as some of the existing solutions for overcoming them.

9.5.1 A Brief Revisit to Traditional TCP

TCP [1] is a reliable, end-to-end, connection-oriented transport layer protocol that provides a byte-stream-based service [the stream of bytes from the application layer is split into TCP segments,[1] the length of each segment limited by a maximum segment size (MSS)]. The major responsibilities of TCP include congestion control, flow control, in-order delivery of packets, and reliable transportation of packets. Congestion control deals with excess traffic in the network which may lead to degradation in the performance of the network, whereas flow control controls the per-flow traffic such that the receiver capacity is not exceeded. TCP regulates the number of packets sent to the network by expanding and shrinking the congestion window. The TCP sender starts the session with a congestion window value of one MSS. It sends out one MSS and waits for the ACK. Once the ACK is received within the retransmission timeout (RTO) period, the congestion window is doubled and two MSSs are originated. This doubling of the congestion window with every successful acknowledgment of all the segments in the current congestion window, is called *slow-start* (a more appropriate name would be *exponential start*, as it actually grows exponentially) and it continues until the congestion window reaches the *slow-start threshold* (the slow-start threshold has an initial value of 64 KB). Figure 9.2 shows the variation of the congestion window in TCP; the slow start phase is between points A-B. Once it reaches the slow-start threshold (in Figure 9.2, the slow-start threshold is initially taken as 16 for illustration), it grows linearly, adding one MSS to the congestion window on every ACK received. This linear growth, which continues until the congestion window reaches the receiver window (which is advertised by the TCP receiver and carries the information about the receiver's buffer size), is called *congestion avoidance*, as it tries to avoid increasing the congestion window exponentially, which will surely worsen the congestion in the network. TCP updates the RTO period with the current round-trip delay calculated on the arrival of every

[1] TCP does not maintain packet boundaries, and hence multiple application layer packets belonging to the same TCP connection, containing stream of bytes, may be combined into a single packet, or a single packet may be split into multiple packets, but delivered as a stream of bytes. Hence, a TCP packet is considered as a segment containing several bytes rather than a packet. However, segment and packet are used interchangeably in this chapter.

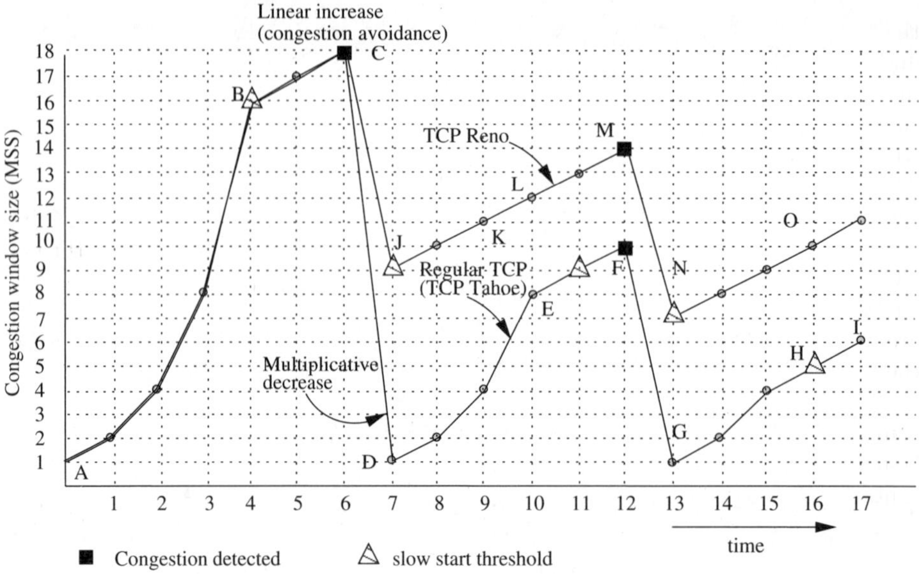

Figure 9.2. Illustration of TCP congestion window.

ACK packet. If the ACK packet does not arrive within the RTO period, then it assumes that the packet is lost. TCP assumes that the packet loss is due to the congestion in the network and it invokes the congestion control mechanism. The TCP sender does the following during congestion control: (i) reduces the slow-start threshold to half the current congestion window or two MSSs whichever is larger, (ii) resets the congestion window size to one MSS, (iii) activates the slow-start algorithm, and (iv) resets the RTO with an exponential back-off value which doubles with every subsequent retransmission. The slow-start process further doubles the congestion window with every successfully acknowledged window and, upon reaching the slow-start threshold, it enters into the congestion avoidance phase.

The TCP sender also assumes a packet loss if it receives three consecutive duplicate ACKs (DUPACKs) [repeated acknowledgments for the same TCP segment that was successfully received in-order at the receiver]. Upon reception of three DUPACKs, the TCP sender retransmits the oldest unacknowledged segment. This is called the *fast retransmit* scheme. When the TCP receiver receives out-of-order packets, it generates DUPACKs to indicate to the TCP sender about the sequence number of the last in-order segment received successfully.

Among the several extensions of TCP, some of the important schemes are discussed below. The regular TCP which was discussed above is also called as TCP Tahoe [2] (in most of the existing literature). TCP Reno [3] is similar to TCP Tahoe with *fast recovery*. On timeout or arrival of three DUPACKs, the TCP Reno sender enters the fast recovery during which (refer to points C-J-K in Figure 9.2) the TCP Reno sender retransmits the lost packet, reduces the slow-start threshold and congestion window size to half the size of the current congestion window, and increments the congestion window linearly (one MSS per DUPACK) with every subsequent DUPACK. On reception of a new ACK (not a DUPACK, *i.e.*, an ACK with a sequence number higher than the highest seen sequence number so far), the TCP Reno resets the congestion window with the slow-start threshold and enters the congestion avoidance phase similar to TCP Tahoe (points K-L-M in Figure 9.2).

J. C. Hoe proposed TCP-New Reno [4] extending the TCP Reno in which the TCP sender does not exit the fast-recovery state, when a new ACK is received. Instead it continues to remain in the fast-recovery state until all the packets originated are acknowledged. For every intermediate ACK packet, TCP-New Reno assumes the next packet after the last acknowledged one is lost and is retransmitted.

TCP with selective ACK (SACK) [5], [6] improves the performance of TCP by using the selective ACKs provided by the receiver. The receiver sends a SACK instead of an ACK, which contains a set of SACK blocks. These SACK blocks contain information about the recently received packets which is used by the TCP sender while retransmitting the lost packets.

9.5.2 Why Does TCP Not Perform Well in Ad Hoc Wireless Networks?

The major reasons behind throughput degradation that TCP faces when used in ad hoc wireless networks are the following:

- **Misinterpretation of packet loss:** Traditional TCP was designed for wired networks where the packet loss is mainly attributed to network congestion. Network congestion is detected by the sender's packet RTO period. Once a packet loss is detected, the sender node assumes congestion in the network and invokes a congestion control algorithm. Ad hoc wireless networks experience a much higher packet loss due to factors such as high bit error rate (BER) in the wireless channel, increased collisions due to the presence of hidden terminals, presence of interference, location-dependent contention, uni-directional links, frequent path breaks due to mobility of nodes, and the inherent fading properties of the wireless channel.

- **Frequent path breaks:** Ad hoc wireless networks experience dynamic changes in network topology because of the unrestricted mobility of the nodes in the network. The topology changes lead to frequent changes in the connectivity of wireless links and hence the route to a particular destination may need to be recomputed very often. The responsibility of finding a route and reestablishing it once it gets broken is attached to the network layer (Chapter 7 discusses network layer routing protocols in detail). Once a path is broken, the routing protocol initiates a route reestablishment process. This route reestablishment process takes a significant amount of time to obtain a new route to the destination. The route reestablishment time is a function of the number of nodes in the network, transmission ranges of nodes, current topology of the network, bandwidth of the channel, traffic load in the network, and the nature of the routing protocol. If the route reestablishment time is greater than the RTO period of the TCP sender, then the TCP sender assumes congestion in the network, retransmits the lost packets, and initiates the congestion control algorithm. These retransmissions can lead to wastage of bandwidth and battery power. Eventually, when a new route is found, the TCP throughput continues to be low for some time, as it has to build up the congestion window since the traditional TCP undergoes a slow start.

- **Effect of path length:** It is found that the TCP throughput degrades rapidly with an increase in path length in string (linear chain) topology ad hoc wireless networks [7], [8]. This is shown in Figure 9.3. The possibility of a path break increases with path length. Given that the probability of a link break is p_l, the probability of a path break (p_b) for a path of length k can be obtained as $p_b = 1 - (1 - p_l)^k$. Figure 9.4 shows the variation of p_b with path length for $p_l = 0.1$. Hence as the path length increases, the probability of a path break increases, resulting in the degradation of the throughput in the network.

- **Misinterpretation of congestion window:** TCP considers the congestion window as a measure of the rate of transmission that is acceptable to the network and the receiver. In ad hoc wireless networks, the congestion control mechanism is invoked when the network gets partitioned or when a path break occurs. This reduces the congestion window and increases the RTO period. When the route is reconfigured, the congestion window may not reflect the transmission rate acceptable to the new route, as the new route may actually accept a much higher transmission rate. Hence, when there are frequent path

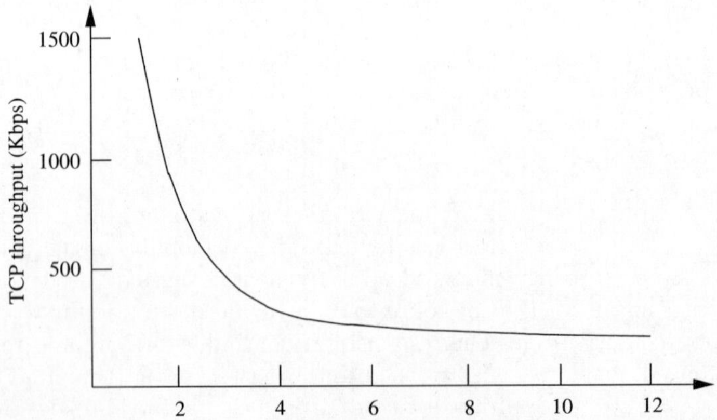

Figure 9.3. Variation of TCP throughput with path length.

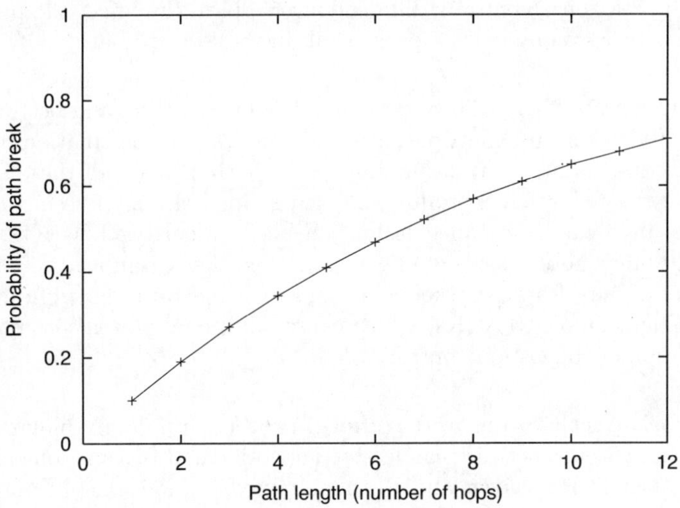

Figure 9.4. Variation of p_b with path length ($p_l = 0.1$).

breaks, the congestion window may not reflect the maximum transmission rate acceptable to the network and the receiver.

- **Asymmetric link behavior:** The radio channel used in ad hoc wireless networks has different properties such as location-dependent contention, environmental effects on propagation, and directional properties leading to asymmetric links. The directional links can result in delivery of a packet to a node, but failure in the delivery of the acknowledgment back to the sender. It is possible for a bidirectional link to become uni-directional for a while. This can also lead to TCP invoking the congestion control algorithm and several retransmissions.

- **Uni-directional path:** Traditional TCP relies on end-to-end ACK for ensuring reliability. Since the ACK packet is very short compared to a data segment, ACKs consume much less bandwidth in wired networks. In ad hoc wireless networks, every TCP ACK packet requires RTS-CTS-Data-ACK exchange in case IEEE 802.11 is used as the underlying MAC protocol. This can lead to an additional overhead of more than 70 bytes if there are no retransmissions. This can lead to significant bandwidth consumption on the reverse path, which may or may not contend with the forward path. If the reverse path contends with the forward path, it can lead to the reduction in the throughput of the forward path. Some routing protocols select the forward path to be also used as the reverse path, whereas certain other routing protocols may use an entirely different or partially different path for the ACKs.

A path break on an entirely different reverse path can affect the performance of the network as much as a path break in the forward path.

- **Multipath routing:** There exists a set of QoS routing and best-effort routing protocols that use multiple paths between a source-destination pair. There are several advantages in using multipath routing. Some of these advantages include the reduction in route computing time, the high resilience to path breaks, high call acceptance ratio, and better security. For TCP, these advantages may add to throughput degradation. These can lead to a significant amount of out-of-order packets, which in turn generates a set of duplicate acknowledgments (DUPACKs) which cause additional power consumption and invocation of congestion control.

- **Network partitioning and remerging:** The randomly moving nodes in an ad hoc wireless network can lead to network partitions. As long as the TCP sender, the TCP receiver, and all the intermediate nodes in the path between the TCP sender and the TCP receiver remain in the same partition, the TCP connection will remain intact. It is likely that the sender and receiver of the TCP session will remain in different partitions and, in certain cases, that only the intermediate nodes are affected by the network partitioning. Figure 9.5 illustrates the effect of network partitions in ad hoc wireless networks. A network with two TCP sessions A and B is shown in Figure 9.5 (a) at time instant t1. Due to dynamic topological changes, the network gets partitioned into two as in Figure 9.5 (b) at time t2. Now the TCP session A's sender and receiver belong to two different partitions and the TCP session B experiences a path break. These partitions could merge back into a single network at time t3 (refer to Figure 9.5 (c)).

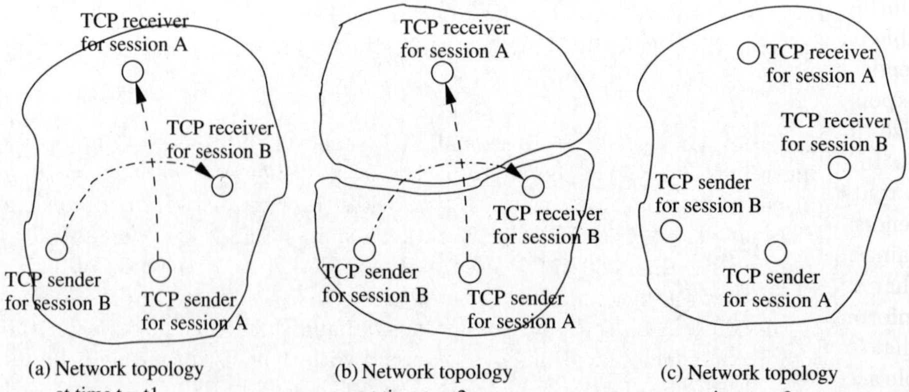

Figure 9.5. Effect of partitioning and merging of network.

- **The use of sliding-window-based transmission**: TCP uses a sliding window for flow control. The transmission of packets is decided by the size of the window, and when the ACKs arrive from a destination, further packets are transmitted. This avoids the use of individual fine-grained timers for transmission of each TCP flow. Such a design is preferred in order to improve scalability of the protocol in high-bandwidth networks such as the Internet where millions of TCP connections may be established with some heavily loaded servers. The use of a sliding window can also contribute to degraded performance in bandwidth-constrained ad hoc wireless networks where the MAC layer protocol may not exhibit short-term and long-term fairness. For example, the popular MAC protocols such as CSMA/CA protocol show short-term unfairness, where a node that has captured the channel has a higher probability of capturing the channel again. This unfairness can lead to a number of TCP ACK packets being delivered to the TCP sender in succession, leading to a burstiness in traffic due to the subsequent transmission of TCP segments.

The enhancements to TCP that improve the performance of TCP in ad hoc wireless networks are discussed in the following sections.

9.5.3 Feedback-Based TCP

Feedback-based TCP [also referred to as TCP feedback (TCP-F)] [9] proposes modifications to the traditional TCP for improving performance in ad hoc wireless networks. It uses a feedback-based approach. TCP-F requires the support of a reliable link layer and a routing protocol that can provide feedback to the TCP sender about the path breaks. The routing protocol is expected to repair the broken path within a reasonable time period. TCP-F aims to minimize the throughput degradation resulting from the frequent path breaks that occur in ad hoc wireless networks. During a TCP session, there could be several path breaks resulting in considerable packet loss and path reestablishment delay. Upon detection of packet loss, the sender in a TCP session invokes the congestion control algorithm leading to the exponential back-off of retransmission timers and a decrease in congestion window size. This was discussed earlier in this chapter.

In TCP-F, an intermediate node, upon detection of a path break, originates a route failure notification (RFN) packet. This RFN packet is routed toward the sender of the TCP session. The TCP sender's information is expected to be obtained from the TCP packets being forwarded by the node. The intermediate node that originates the RFN packet is called the failure point (FP). The FP maintains information about all the RFNs it has originated so far. Every intermediate node that forwards the RFN packet understands the route failure, updates its routing table accordingly, and avoids forwarding any more packets on that route. If any of the intermediate nodes that receive RFN has an alternate route to the same destination, then it discards the RFN packet and uses the alternate path for forwarding further data packets, thus reducing the control overhead involved in the route reconfiguration process. Otherwise, it forwards the RFN toward the source node. When a TCP sender receives an RFN packet, it goes into a state called *snooze*. In the snooze state, a sender stops sending any more packets to the destination, cancels all the timers, freezes its congestion window, freezes the retransmission timer, and sets up a route failure timer. This route failure timer is dependent on the routing protocol, network size, and the network dynamics and is to be taken as the worst-case route reconfiguration time. When the route failure timer expires, the TCP sender changes from the snooze state to the *connected* state. Figure 9.6 shows the operation of the TCP-F protocol. In the figure, a TCP session is set up between node A and node D over the path A-B-C-D [refer to Figure 9.6 (a)]. When the intermediate link between node C and node D fails, node C originates an RFN packet and forwards it on the reverse path to the source node [see Figure 9.6 (b)]. The sender's TCP state is changed to the snooze state upon receipt of an RFN packet. If the link CD rejoins, or if any of the intermediate nodes obtains a path to destination node D, a route reestablishment notification (RRN) packet is sent to node A and the TCP state is updated back to the connected state [Figure 9.6 (c)].

As soon as a node receives an RRN packet, it transmits all the packets in its buffer, assuming that the network is back to its original state. This can also take care of all the packets that were not acknowledged or lost during transit due to the path break. In fact, such a step avoids going through the slow-start process that would otherwise have occurred immediately after a period of congestion. The route failure timer set after receiving the RFN packet ensures that the sender does not remain in the snooze state indefinitely. Once the route failure timer expires, the sender goes back to the connected state in which it reactivates the frozen timers and starts sending the buffered and unacknowledged packets. This can also take care of the loss of the RRN packet due to any possible subsequent congestion. TCP-F permits the TCP congestion control algorithm to be in effect when the sender is not in the snooze state, thus making it sensitive to congestion in the network.

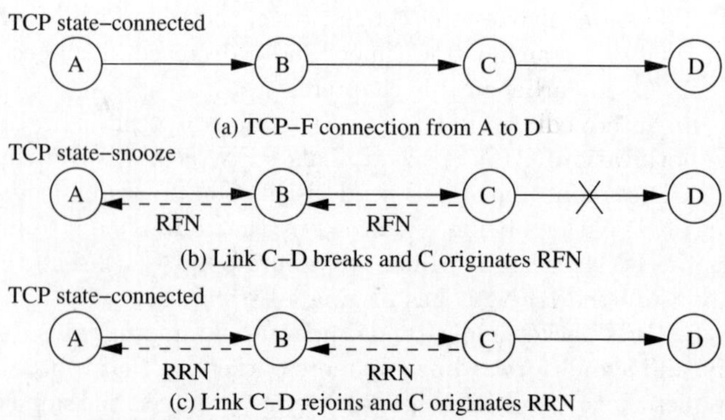

Figure 9.6. Operation of TCP-F.

Advantages and Disadvantages

TCP-F provides a simple feedback-based solution to minimize the problems arising out of frequent path breaks in ad hoc wireless networks. At the same time, it also permits the TCP congestion control mechanism to respond to congestion in the network. TCP-F depends on the intermediate nodes' ability to detect route failures and the routing protocols' capability to reestablish a broken path within a reasonably short duration. Also, the FP should be able to obtain the correct path (the path which the packet traversed) to the TCP-F sender for sending the RFN packet. This is simple with a routing protocol that uses source routing [*i.e.*, dynamic source routing (DSR)]. If a route to the sender is not available at the FP, then additional control packets may need to be generated for routing the RFN packet. TCP-F has an additional state compared to the traditional TCP state machine, and hence its implementation requires modifications to the existing TCP libraries. Another disadvantage of TCP-F is that the congestion window used after a new route is obtained may not reflect the achievable transmission rate acceptable to the network and the TCP-F receiver.

9.5.4 TCP with Explicit Link Failure Notification

Holland and Vaidya proposed the use of TCP with explicit link failure notification (TCP-ELFN) [8] for improving TCP performance in ad hoc wireless networks. This is similar to TCP-F, except for the handling of explicit link failure notification (ELFN) and the use of TCP probe packets for detecting the route reestablishment. The ELFN is originated by the node detecting a path break upon detection of a link failure to the TCP sender. This can be implemented in two ways: (i) by sending an ICMP[2] destination unreachable (DUR) message to the sender, or (ii) by piggy-backing this information on the *RouteError*[3] message that is sent to the sender.

Once the TCP sender receives the ELFN packet, it disables its retransmission timers and enters a *standby* state. In this state, it periodically originates probe packets to see if a new route is reestablished. Upon reception of an ACK by the TCP receiver for the probe packets, it leaves the standby state, restores the retransmission timers, and continues to function as normal.

Advantages and Disadvantages

TCP-ELFN improves the TCP performance by decoupling the path break information from the congestion information by the use of ELFN. It is less dependent on the routing protocol and requires only link failure notification about the path break. The disadvantages of TCP-ELFN include the following: (i) when the network is temporarily partitioned, the path failure may last longer and this can lead

[2]Internet control message protocol (IETF RFC 792) is used for defining control messages for aiding routing in the Internet.

[3]Certain routing protocols for ad hoc wireless networks have explicit *RouteError* messages to inform the sender about path breaks so that the sender can recompute a fresh route to the destination. This is especially used in on-demand routing protocols such as DSR.

to the origination of periodic probe packets consuming bandwidth and power and (ii) the congestion window used after a new route is obtained may not reflect the achievable transmission rate acceptable to the network and the TCP receiver.

9.5.5 TCP-BuS

TCP with buffering capability and sequence information (TCP-BuS) [10] is similar to the TCP-F and TCP-ELFN in its use of feedback information from an intermediate node on detection of a path break. But TCP-BuS is more dependent on the routing protocol compared to TCP-F and TCP-ELFN. TCP-BuS was proposed, with associativity-based routing (ABR) [11] protocol as the routing scheme. Hence, it makes use of some of the special messages such as localized query (LQ) and REPLY, defined as part of ABR for finding a partial path. These messages are modified to carry TCP connection and segment information. Upon detection of a path break, an upstream intermediate node [called pivot node (PN)] originates an explicit route disconnection notification (ERDN) message. This ERDN packet is propagated to the TCP-BuS sender and, upon reception of it, the TCP-BuS sender stops transmission and freezes all timers and windows as in TCP-F. The packets in transit at the intermediate nodes from the TCP-BuS sender to the PN are buffered until a new partial path from the PN to the TCP-BuS receiver is obtained by the PN. In order to avoid unnecessary retransmissions, the timers for the buffered packets at the TCP-BuS sender and at the intermediate nodes up to PN use timeout values proportional to the round-trip time (RTT). The intermediate nodes between the TCP-BuS sender and the PN can request the TCP-BuS sender to selectively retransmit any of the lost packets. Upon detection of a path break, the downstream node originates a route notification (RN) packet to the TCP-BuS receiver, which is forwarded by all the downstream nodes in the path. An intermediate node that receives an RN packet discards all packets belonging to that flow. The ERDN packet is propagated to the TCP-BuS sender in a reliable way by using an implicit acknowledgment and retransmission mechanism. The PN includes the sequence number of the TCP segment belonging to the flow that is currently at the head of its queue in the ERDN packet. The PN also attempts to find a new partial route to the TCP-BuS receiver, and the availability of such a partial path to destination is intimated to the TCP-BuS sender through an explicit route successful notification (ERSN) packet. TCP-BuS utilizes the route reconfiguration mechanism of ABR to obtain the partial route to the destination. Due to this, other routing protocols may require changes to support TCP-BuS. The LQ and REPLY messages are modified to carry TCP segment information, including the last successfully received segment at the destination. The LQ packet carries the sequence number of the segment at the head of the queue buffered at the PN and the REPLY carries the sequence number of the last successful segment the TCP-BuS receiver received. This enables the TCP-BuS receiver to understand the packets lost in transition and those buffered at the intermediate nodes. This is used to avoid fast retransmission requests usually generated by the TCP-BuS receiver when it notices an out-of-order packet delivery. Upon a successful LQ-REPLY process to obtain a new route to the

TCP-BuS receiver, PN informs the TCP-BuS sender of the new partial path using the ERSN packet. When the TCP-BuS sender receives an ERSN packet, it resumes the data transmission.

Since there is a chance for ERSN packet loss due to congestion in the network, it needs to be sent reliably. The TCP-BuS sender also periodically originates probe packets to check the availability of a path to the destination. Figure 9.7 shows an illustration of the propagation of ERDN and RN messages when a link between nodes 4 and 12 fails.

When a TCP-BuS sender receives the ERSN message, it understands, from the sequence number of the last successfully received packet at the destination and the sequence number of the packet at the head of the queue at PN, the packets lost in transition. The TCP-BuS receiver understands that the lost packets will be delayed further and hence uses a selective acknowledgment strategy instead of fast retransmission. These lost packets are retransmitted by the TCP-BuS sender. During the retransmission of these lost packets, the network congestion between the TCP-BuS sender and PN is handled in a way similar to that in traditional TCP.

Advantages and Disadvantages

The advantages of TCP-BuS include performance improvement and avoidance of fast retransmission due to the use of buffering, sequence numbering, and selective

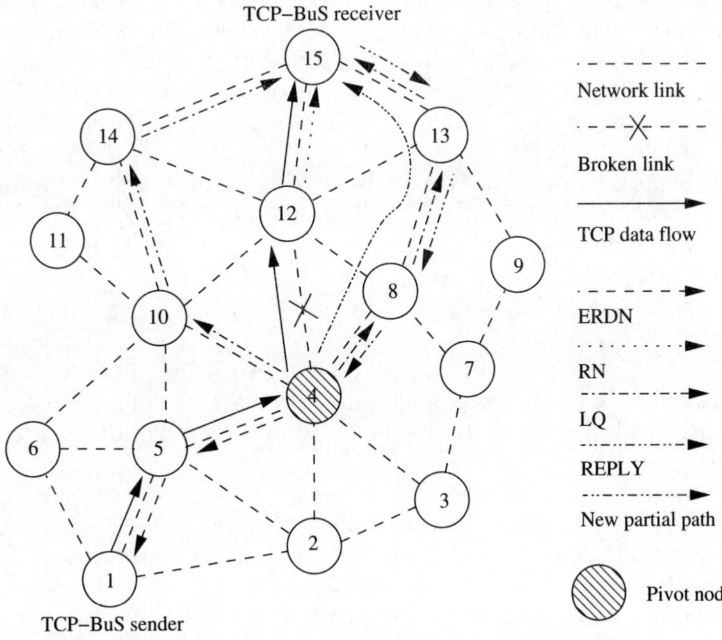

Figure 9.7. Operation of TCP-BuS.

acknowledgment. TCP-BuS also takes advantage of the underlying routing protocols, especially the on-demand routing protocols such as ABR. The disadvantages of TCP-BuS include the increased dependency on the routing protocol and the buffering at the intermediate nodes. The failure of intermediate nodes that buffer the packets may lead to loss of packets and performance degradation. The dependency of TCP-BuS on the routing protocol may degrade its performance with other routing protocols that do not have similar control messages as in ABR.

9.5.6 Ad Hoc TCP

Similar to TCP-F and TCP-ELFN, ad hoc TCP (ATCP) [12] also uses a network layer feedback mechanism to make the TCP sender aware of the status of the network path over which the TCP packets are propagated. Based on the feedback information received from the intermediate nodes, the TCP sender changes its state to the *persist* state, *congestion control* state, or the *retransmit* state. When an intermediate node finds that the network is partitioned, then the TCP sender state is changed to the persist state where it avoids unnecessary retransmissions. When ATCP puts TCP in the persist state, it sets TCP's congestion window size to one in order to ensure that TCP does not continue using the old congestion window value. This forces TCP to probe the correct value of the congestion window to be used for the new route. If an intermediate node loses a packet due to error, then the ATCP at the TCP sender immediately retransmits it without invoking the congestion control algorithm. In order to be compatible with widely deployed TCP-based networks, ATCP provides this feature without modifying the traditional TCP. ATCP is implemented as a thin layer residing between the IP and TCP protocols. The ATCP layer essentially makes use of the explicit congestion notification (ECN) for maintenance of the states.

Figure 9.8 (a) shows the thin layer implementation of ATCP between the traditional TCP layer and the IP layer. This does not require changes in the existing TCP protocol. This layer is active only at the TCP sender. The major function of the ATCP layer is to monitor the packets sent and received by the TCP sender, the state of the TCP sender, and the state of the network. Figure 9.8 (b) shows the state transition diagram for the ATCP at the TCP sender. The four states in the ATCP are (i) NORMAL, (ii) CONGESTED, (iii) LOSS, and (iv) DISCONN. When a TCP connection is established, the ATCP sender state is in NORMAL. In this state, ATCP does not interfere with the operation of TCP and it remains invisible.

When packets are lost or arrive out-of-order at the destination, it generates duplicate ACKs. In traditional TCP, upon reception of duplicate ACKs, the TCP sender retransmits the segment under consideration and shrinks the contention window. But the ATCP sender counts the number of duplicate ACKs received and if it reaches three, instead of forwarding the duplicate ACKs to TCP, it puts TCP in the persist state and ATCP in the LOSS state. Hence, the TCP sender avoids invoking congestion control. In the LOSS state, ATCP retransmits the unacknowledged segments from the TCP buffer. When a new ACK comes from the TCP receiver,

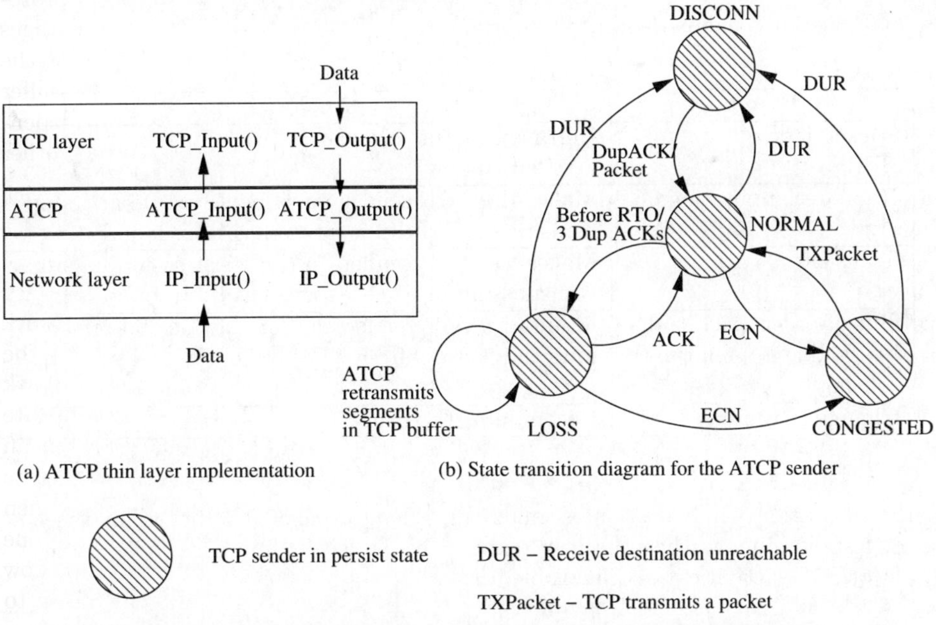

(a) ATCP thin layer implementation

(b) State transition diagram for the ATCP sender

Figure 9.8. An illustration of ATCP thin layer and ATCP state diagram.

it is forwarded to TCP and the TCP sender is removed from the persist state and then the ATCP sender changes to the NORMAL state.

When the ATCP sender is in the LOSS state, the receipt of an ECN message or an ICMP *source quench* message changes it to the CONGESTED state. Along with this state transition, the ATCP sender removes the TCP sender from the persist state. When the network gets congested, the ECN[4] flag is set in the data and the ACK packets. When the ATCP sender receives this ECN message in the normal state, it changes to the CONGESTED state and just remains invisible, permitting TCP to invoke normal congestion control mechanisms. When a route failure or a transient network partition occurs in the network, ATCP expects the network layer to detect these and inform the ATCP sender through an ICMP destination unreachable (DUR) message. Upon reception of the DUR message, ATCP puts the TCP sender into the persist state and enters into the DISCONN state. It remains in the DISCONN state until it is connected and receives any data or duplicate ACKs. On the occurrence of any of these events, ATCP changes to the NORMAL state. The connected status of the path can be detected by the acknowledgments for the periodic probe packets generated by the TCP sender. The receipt of an ICMP DUR message in the LOSS state or the CONGESTED state causes a transition to the DISCONN state. When ATCP puts TCP into the persist state, it sets

[4]ECN is currently under consideration by IETF and is now a standard (IETF RFC 3168).

Table 9.1. The actions taken by ATCP

Event	Action
Packet loss due to high BER	Retransmits the lost packets without reducing congestion window
Route recomputation delay	Makes the TCP sender go to persist state and stop transmission until new route has been found
Transient partitions	Makes the TCP sender go to persist state and stop transmission until new route has been found
Out-of-order packet delivery due to multipath routing	Maintains TCP sender unaware of this and retransmits the packets from TCP buffer
Change in route	Recomputes the congestion window

the congestion window to one segment in order to make TCP probe for the new congestion window when the new route is available. In summary, ATCP tries to perform the activities listed in Table 9.1.

Advantages and Disadvantages

Two major advantages of ATCP are (i) it maintains the end-to-end semantics of TCP and (ii) it is compatible with traditional TCP. These advantages permit ATCP to work seamlessly with the Internet. In addition, ATCP provides a feasible and efficient solution to improve throughput of TCP in ad hoc wireless networks. The disadvantages of ATCP include (i) the dependency on the network layer protocol to detect the route changes and partitions, which not all routing protocols may implement and (ii) the addition of a thin ATCP layer to the TCP/IP protocol stack that requires changes in the interface functions currently being used.

9.5.7 Split TCP

One of the major issues that affects the performance of TCP over ad hoc wireless networks is the degradation of throughput with increasing path length, as discussed early in this chapter. The short (*i.e.,* in terms of path length) connections generally obtain much higher throughput than long connections. This can also lead to unfairness among TCP sessions, where one session may obtain much higher throughput than other sessions. This unfairness problem is further worsened by the use of MAC protocols such as IEEE 802.11, which are found to give a higher throughput for certain link-level sessions, leading to an effect known as *channel capture* effect. This effect leads to certain flows capturing the channel for longer time durations, thereby reducing throughput for other flows. The channel capture effect can also lead to low overall system throughput. The reader can refer to Chapter 6 for more details on MAC protocols and throughput fairness.

Split-TCP [13] provides a unique solution to this problem by splitting the transport layer objectives into congestion control and end-to-end reliability. The congestion control is mostly a local phenomenon due to the result of high contention and high traffic load in a local region. In the ad hoc wireless network environment, this demands local solutions. At the same time, reliability is an end-to-end requirement and needs end-to-end acknowledgments.

In addition to splitting the congestion control and reliability objectives, split-TCP splits a long TCP connection into a set of short concatenated TCP connections (called *segments* or *zones*) with a number of selected intermediate nodes (known as *proxy* nodes) as terminating points of these short connections. Figure 9.9 illustrates the operation of split-TCP where a three segment split-TCP connection exists between source node 1 and destination node 15. A proxy node receives the TCP packets, reads its contents, stores it in its local buffer, and sends an acknowledgment to the source (or the previous proxy). This acknowledgment called local acknowledgment (LACK) does not guarantee end-to-end delivery. The responsibility of further delivery of packets is assigned to the proxy node. A proxy node clears a buffered packet once it receives LACK from the immediate successor proxy node for that packet. Split-TCP maintains the end-to-end acknowledgment mechanism intact, irrespective of the addition of zone-wise LACKs. The source node clears the buffered packets only after receiving the end-to-end acknowledgment for those packets.

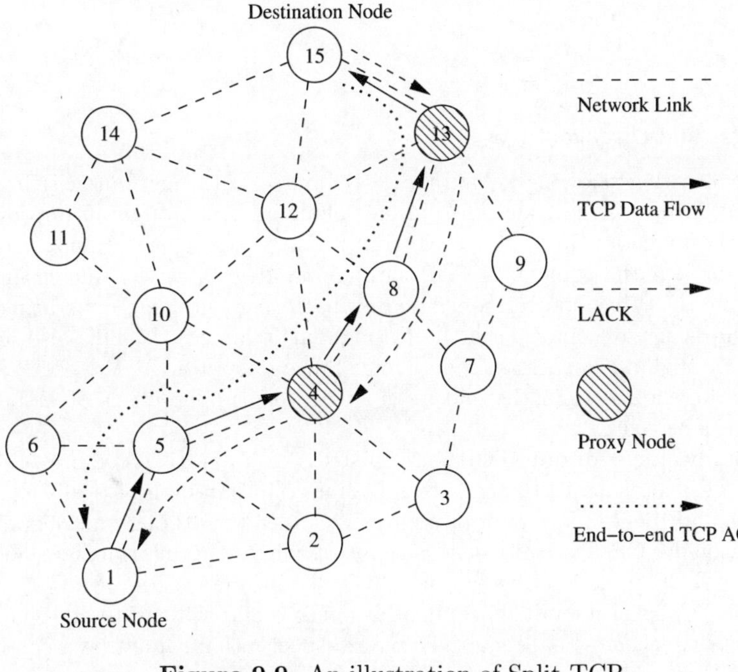

Figure 9.9. An illustration of Split-TCP.

In Figure 9.9, node 1 initiates a TCP session to node 15. Node 4 and node 13 are chosen as proxy nodes. The number of proxy nodes in a TCP session is determined by the length of the path between source and destination nodes. Based on a distributed algorithm, the intermediate nodes that receive TCP packets determine whether to act as a proxy node or just as a simple forwarding node. The most simple algorithm makes the decision for acting as proxy node if the packet has already traversed more than a predetermined number of hops from the last proxy node or the sender of the TCP session. In Figure 9.9, the path between node 1 and node 4 is the first zone (segment), the path between nodes 4 and 13 is the second zone (segment), and the last zone is between node 13 and 15.

The proxy node 4, upon receipt of each TCP packet from source node 1, acknowledges it with a LACK packet, and buffers the received packets. This buffered packet is forwarded to the next proxy node (in this case, node 13) at a transmission rate proportional to the arrival of LACKs from the next proxy node or destination. The transmission control window at the TCP sender is also split into two windows, that is, the congestion window and the end-to-end window. The congestion window changes according to the rate of arrival of LACKs from the next proxy node and the end-to-end window is updated based on the arrival of end-to-end ACKs. Both these windows are updated as per traditional TCP except that the congestion window should stay within the end-to-end window. In addition to these transmission windows at the TCP sender, every proxy node maintains a congestion window that governs the segment level transmission rate.

Advantages and Disadvantages

Split-TCP has the following advantages: (i) improved throughput, (ii) improved throughput fairness, and (iii) lessened impact of mobility. Throughput improvement is due to the reduction in the effective transmission path length (number of hops in a zone or a path segment). TCP throughput degrades with increasing path length. Split-TCP has shorter concatenated path segments, each operating at its own transmission rate, and hence the throughput is increased. This also leads to improved throughput fairness in the system. Since in split-TCP, the path segment length can be shorter than the end-to-end path length, the effect of mobility on throughput is lessened.

The disadvantages of split-TCP can be listed as follows: (i) It requires modifications to TCP protocol, (ii) the end-to-end connection handling of traditional TCP is violated, and (iii) the failure of proxy nodes can lead to throughput degradation. The traditional TCP has end-to-end semantics, where the intermediate nodes do not process TCP packets, whereas in split-TCP, the intermediate nodes need to process the TCP packets and hence, in addition to the loss of end-to-end semantics, certain security schemes that require IP payload encryption cannot be used. During frequent path breaks or during frequent node failures, the performance of split-TCP may be affected.

9.5.8 A Comparison of TCP Solutions for Ad Hoc Wireless Networks

Table 9.2 compares how various issues are handled in the TCP extensions discussed so far in this chapter.

9.6 OTHER TRANSPORT LAYER PROTOCOLS FOR AD HOC WIRELESS NETWORKS

The performance of a transport layer protocol can be enhanced if it takes into account the nature of the network environment in which it is applied. Especially in wireless environments, it is important to consider the properties of the physical layer and the interaction of the transport layer with the lower layers. This section discusses some of the transport layer protocols that were designed specifically for ad hoc wireless networks. Even though interworking with TCP is very important, there exist several application scenarios such as military communication where a radically new transport layer protocol can be used.

9.6.1 Application Controlled Transport Protocol

Unlike the TCP solutions discussed earlier in this chapter, application controlled transport protocol (ACTP[5]) [14] is a light-weight transport layer protocol. It is not an extension to TCP. ACTP assigns the responsibility of ensuring reliability to the application layer. It is more like UDP with feedback of delivery and state maintenance. ACTP stands in between TCP and UDP where TCP experiences low performance with high reliability and UDP provides better performance with high packet loss in ad hoc wireless networks.

The key design philosophy of ACTP is to leave the provisioning of reliability to the application layer and provide a simple feedback information about the delivery status of packets to the application layer. ACTP supports the priority of packets to be delivered, but it is the responsibility of the lower layers to actually provide a differentiated service based on this priority.

Figure 9.10 shows the ACTP layer and the API functions used by the application layer to interact with the ACTP layer. Each API function call to send a packet [*SendTo()*] contains the additional information required for ACTP such as the maximum delay the packet can tolerate (delay), the message number of the packet, and the priority of the packet. The message number is assigned by the application layer, and it need not to be in sequence. The priority level is assigned for every packet by the application. It can be varied across packets in the same flow with increasing numbers referring to higher priority packets. The non-zero value in the message number field implicitly conveys that the application layer expects a delivery status information about the packet to be sent. This delivery status is maintained at the ACTP layer, and is available to the application layer for verification through another API function *IsACKed*<*message number*>. The delivery status returned by

[5] Originally called ATP, for differentiating with ad hoc transport protocol it is referred to as ACTP in this chapter.

Table 9.2. A comparison of TCP solutions for ad hoc wireless networks

Issue	TCP-F	TCP-ELFN	TCP-BuS	ATCP	Split-TCP
Packet loss due to BER or collision	Same as TCP	Same as TCP	Same as TCP	Retransmits the lost packets without invoking congestion control	Same as TCP
Path breaks	RFN is sent to the TCP sender and state changes to snooze	ELFN is sent to the TCP sender and state changes to standby	ERDN is sent to the TCP sender, state changes to snooze, ICMP DUR is sent to the TCP sender, and ATCP puts TCP into persist state	Same as TCP	Same as TCP
Out-of-order packets	Same as TCP	Same as TCP	Out-of-order packets reached after a path recovery are handled	ATCP reorders packets and hence TCP avoids sending duplicates	Same as TCP
Congestion	Same as TCP	Same as TCP	Explicit messages such as ICMP source quench are used	ECN is used to notify TCP sender. Congestion control is same as TCP	Since connection is split, the congestion control is handled within a zone by proxy nodes
Congestion window after path reestablishment	Same as before the path break	Same as before the path break	Same as before the path break	Recomputed for new route	Proxy nodes maintain congestion window and handle congestion
Explicit path break notification	Yes	Yes	Yes	Yes	No
Explicit path reestablishment notification	Yes	No	Yes	No	No
Dependency on routing protocol	Yes	Yes	Yes	Yes	No
End-to-end semantics	Yes	Yes	Yes	Yes	No
Packets buffered at intermediate nodes	No	No	Yes	No	Yes

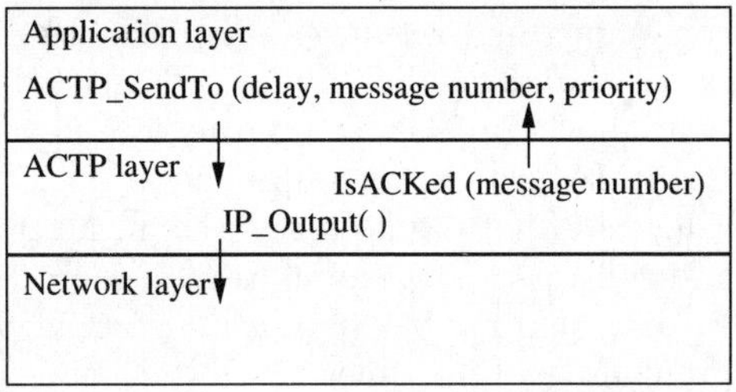

Figure 9.10. An illustration of the interface functions used in ACTP.

IsACKed<message number> function call can reflect (i) a successful delivery of the packet (ACK received), (b) a possible loss of the packet (no ACK received and the deadline has expired), (iii) remaining time for the packet (no ACK received but the deadline has not expired), and (iv) no state information exists at the ACTP layer regarding the message under consideration. A zero in the delay field refers to the highest priority packet, which requires immediate transmission with minimum possible delay. Any other value in the delay field refers to the delay that the message can experience. On getting the information about the delivery status, the application layer can decide on retransmission of a packet with the same old priority or with an updated priority. Well after the packet's lifetime expires, ACTP clears the packet's state information and delivery status. The packet's lifetime is calculated as 4×retransmit timeout (RTO) and is set as the lifetime when the packet is sent to the network layer. A node estimates the RTO interval by using the round-trip time between the transmission time of a message and the time of reception of the corresponding ACK. Hence, the RTO value may not be available if there are no existing reliable connections to a destination. A packet without any message number (*i.e.*, no delivery status required) is handled exactly the same way as in UDP without maintaining any state information.

Advantages and Disadvantages

One of the most important advantages of ACTP is that it provides the freedom of choosing the required reliability level to the application layer. Since ACTP is a light-weight transport layer protocol, it is scalable for large networks. Throughput is not affected by path breaks as much as in TCP as there is no congestion window for manipulation as part of the path break recovery. One disadvantage of ACTP is that it is not compatible with TCP. Use of ACTP in a very large ad hoc wireless network can lead to heavy congestion in the network as it does not have any congestion control mechanism.

9.6.2 Ad Hoc Transport Protocol

Ad hoc transport protocol (ATP) [15] is specifically designed for ad hoc wireless networks and is not a variant of TCP. The major aspects by which ATP defers from

TCP are (i) coordination among multiple layers, (ii) rate based transmissions, (iii) decoupling congestion control and reliability, and (iv) assisted congestion control. Similar to other TCP variants proposed for ad hoc wireless networks, ATP uses services from network and MAC layers for improving its performance. ATP uses information from lower layers for (i) estimation of the initial transmission rate, (ii) detection, avoidance, and control of congestion, and (iii) detection of path breaks.

Unlike TCP, ATP utilizes a timer-based transmission, where the transmission rate is decided by the granularity of the timer which is dependent on the congestion in the network. The congestion control mechanism is decoupled from the reliability and flow control mechanisms. The network congestion information is obtained from the intermediate nodes, whereas the flow control and reliability information are obtained from the ATP receiver. The intermediate nodes attach the congestion information to every ATP packet and the ATP receiver collates it before including it in the next ACK packet. The congestion information is expressed in terms of the weighted averaged[6] queuing delay (D_Q) and contention delay (D_C) experienced by the packets at every intermediate node. The field in which this delay information is included is referred to as the *rate feedback field* and the transmission rate is the inverse of the delay information contained in the rate feedback field. Intermediate nodes attach the current delay information to every ATP data packet if the already existing value is smaller than the current delay. The ATP receiver collects this delay information and the weighted average value is attached in the periodic ACK (ATP uses SACK mechanism, hence ACK refers to SACK) packet sent back to the ATP sender. During a connection startup process or when ATP recovers from a path break, the transmission rate to be used is determined by a process called *quick start*. During the quick start process, the ATP sender propagates a probe packet to which the intermediate nodes attach the transmission rate (in the form of current delay), which is received by the ATP receiver, and an ACK is sent back to the ATP sender. The ATP sender starts using the newly obtained transmission rate by setting the data transmission timers. During a connection startup, the connection request and the ACK packets are used as probe packets in order to reduce control overhead. When there is no traffic around an intermediate node, the transmission delay is approximated as $\beta \times (D_Q + D_C)$, where β is the factor that considers the induced traffic load. This is to consider the induced load (load on a particular link due to potential contention introduced by the upstream and downstream nodes in the path) when the actual transmission begins. A default value of 3 is used for β. ATP uses SACK packets periodically to ensure the selective retransmission of lost packets, which ensures the reliability of packet delivery. The SACK period is chosen such that it is more than the round-trip time and can track the network dynamics. The receiver performs a weighted average of the delay/transmission rate information for every incoming packet to obtain the transmission rate for an ATP flow and this value is included in the subsequent SACK packet it sends. In addition to the rate feedback, the ATP receiver includes flow control information in the SACK packets.

[6]Originally called "exponentially averaged," renamed here with a more appropriate term, "weighted average." An example for this is $\bar{Q}_{delay} = \alpha \times Q_{delay_{new}} + (1 - \alpha) \times Q_{delay_{old}}$, where α is an appropriate weight factor and the other terms are self-explanatory.

Unlike TCP, which employs either a decrease of the congestion window or an increase of the congestion window after a congestion, ATP has three phases, namely, increase, decrease, and maintain. If the new transmission rate (R) fed back from the network is beyond a threshold (γ) greater than the current transmission rate (S) [*i.e.*, $R > S(1+\gamma)$], then the current transmission rate is increased by a fraction (k) of the difference between the two transmission rates (*i.e.*, $S = S + \frac{R-S}{k}$). The fraction and threshold are taken to avoid rapid fluctuations in the transmission rate and induced load. The current transmission rate is updated to the new transmission rate if the new transmission rate is lower than the current transmission rate. In the maintain phase, if the new transmission rate is higher than the current transmission rate, but less than the above mentioned threshold, then the current transmission rate is maintained without any change.

If an ATP sender has not received any ACK packets for two consecutive feedback periods, it undergoes a multiplicative decrease of the transmission rate. After a third such period without any ACK, the connection is assumed to be lost and the ATP sender goes to the connection initiation phase during which it periodically generates probe packets. When a path break occurs, the network layer detects it and originates an ELFN packet toward the ATP sender. The ATP sender freezes the sender state and goes to the connection initiation phase. In this phase also, the ATP sender periodically originates probe packets to know the status of the path. With a successful probe, the sender begins data transmission again.

Advantages and Disadvantages

The major advantages of ATP include improved performance, decoupling of the congestion control and reliability mechanisms, and avoidance of congestion window fluctuations. ATP does not maintain any per flow state at the intermediate nodes. The congestion information is gathered directly from the nodes that experience it.

The major disadvantage of ATP is the lack of interoperability with TCP. As TCP is a widely used transport layer protocol, interoperability with TCP servers and clients in the Internet is important in many applications. For large ad hoc wireless networks, the fine-grained per-flow timer used at the ATP sender may become a scalability bottleneck in resource-constrained mobile nodes.

9.7 SECURITY IN AD HOC WIRELESS NETWORKS

As mentioned earlier, due to the unique characteristics of ad hoc wireless networks, such networks are highly vulnerable to security attacks compared to wired networks or infrastructure-based wireless networks. The following sections discuss the various security requirements in ad hoc wireless networks, the different types of attacks possible in such networks, and some of the solutions proposed for ensuring network security.

9.8 NETWORK SECURITY REQUIREMENTS

A security protocol for ad hoc wireless networks should satisfy the following requirements. The requirements listed below should in fact be met by security protocols for other types of networks also.

- **Confidentiality:** The data sent by the sender (source node) must be comprehensible only to the intended receiver (destination node). Though an intruder might get hold of the data being sent, he/she must not be able to derive any useful information out of the data. One of the popular techniques used for ensuring confidentiality is data encryption.
- **Integrity:** The data sent by the source node should reach the destination node as it was sent: unaltered. In other words, it should not be possible for any malicious node in the network to tamper with the data during transmission.
- **Availability:** The network should remain operational all the time. It must be robust enough to tolerate link failures and also be capable of surviving various attacks mounted on it. It should be able to provide the guaranteed services whenever an authorized user requires them.
- **Non-repudiation:** Non-repudiation is a mechanism to guarantee that the sender of a message cannot later deny having sent the message and that the recipient cannot deny having received the message. Digital signatures, which function as unique identifiers for each user, much like a written signature, are used commonly for this purpose.

9.9 ISSUES AND CHALLENGES IN SECURITY PROVISIONING

Designing a foolproof security protocol for ad hoc wireless is a very challenging task. This is mainly because of certain unique characteristics of ad hoc wireless networks, namely, shared broadcast radio channel, insecure operating environment, lack of central authority, lack of association among nodes, limited availability of resources, and physical vulnerability. A detailed discussion on how each of the above mentioned characteristics causes difficulty in providing security in ad hoc wireless networks is given below.

- **Shared broadcast radio channel:** Unlike in wired networks where a separate dedicated transmission line can be provided between a pair of end users, the radio channel used for communication in ad hoc wireless networks is broadcast in nature and is shared by all nodes in the network. Data transmitted by a node is received by all nodes within its direct transmission range. So a malicious node could easily obtain data being transmitted in the network. This problem can be minimized to a certain extent by using directional antennas.
- **Insecure operational environment:** The operating environments where ad hoc wireless networks are used may not always be secure. One important

application of such networks is in battlefields. In such applications, nodes may move in and out of hostile and insecure enemy territory, where they would be highly vulnerable to security attacks.

- **Lack of central authority:** In wired networks and infrastructure-based wireless networks, it would be possible to monitor the traffic on the network through certain important central points (such as routers, base stations, and access points) and implement security mechanisms at such points. Since ad hoc wireless networks do not have any such central points, these mechanisms cannot be applied in ad hoc wireless networks.
- **Lack of association:** Since these networks are dynamic in nature, a node can join or leave the network at any point of the time. If no proper authentication mechanism is used for associating nodes with a network, an intruder would be able to join into the network quite easily and carry out his/her attacks.
- **Limited resource availability:** Resources such as bandwidth, battery power, and computational power (to a certain extent) are scarce in ad hoc wireless networks. Hence, it is difficult to implement complex cryptography-based security mechanisms in such networks.
- **Physical vulnerability:** Nodes in these networks are usually compact and hand-held in nature. They could get damaged easily and are also vulnerable to theft.

9.10 NETWORK SECURITY ATTACKS

Attacks on ad hoc wireless networks can be classified into two broad categories, namely, *passive* and *active* attacks. A passive attack does not disrupt the operation of the network; the adversary snoops the data exchanged in the network without altering it. Here, the requirement of confidentiality can be violated if an adversary is also able to interpret the data gathered through snooping. Detection of passive attacks is very difficult since the operation of the network itself does not get affected. One way of overcoming such problems is to use powerful encryption mechanisms to encrypt the data being transmitted, thereby making it impossible for eavesdroppers to obtain any useful information from the data overheard.

An active attack attempts to alter or destroy the data being exchanged in the network, thereby disrupting the normal functioning of the network. Active attacks can be classified further into two categories, namely, *external* and *internal* attacks. External attacks are carried out by nodes that do not belong to the network. These attacks can be prevented by using standard security mechanisms such as encryption techniques and firewalls.[7] Internal attacks are from compromised nodes that are

[7]A firewall is used to separate a local network from the outside world. It is a software which works closely with a router program and filters all packets entering the network to determine whether or not to forward those packets toward their intended destinations. A firewall protects the resources of a private network from malicious intruders on foreign networks such as the Internet. In an ad hoc wireless network, the firewall software could be installed on each node on the network.

actually part of the network. Since the adversaries are already part of the network as authorized nodes, internal attacks are more severe and difficult to detect when compared to external attacks.

Figure 9.11 shows a classification of the different types of attacks possible in ad hoc wireless networks. The following sections describe the various attacks listed in the figure.

9.10.1 Network Layer Attacks

This section lists and gives brief descriptions of the attacks pertaining to the network layer in the network protocol stack.

- **Wormhole attack:** In this attack, an attacker receives packets at one location in the network and tunnels them (possibly selectively) to another location in the network, where the packets are resent into the network [16]. This tunnel between two colluding attackers is referred to as a wormhole. It could be established through a single long-range wireless link or even through a wired link between the two colluding attackers. Due to the broadcast nature of the radio channel, the attacker can create a wormhole even for packets not addressed to itself. Though no harm is done if the wormhole is used properly for efficient relaying of packets, it puts the attacker in a powerful position compared to other nodes in the network, which the attacker could use in a manner that could compromise the security of the network. If proper mechanisms are not employed to defend the network against wormhole attacks, most of the existing routing protocols for ad hoc wireless networks may fail to find valid routes.

- **Blackhole attack:** In this attack, a malicious node falsely advertises good paths (*e.g.*, shortest path or most stable path) to the destination node during the path-finding process (in on-demand routing protocols) or in the route update messages (in table-driven routing protocols). The intention of the malicious node could be to hinder the path-finding process or to intercept all data packets being sent to the destination node concerned.

- **Byzantine attack:** Here, a compromised intermediate node or a set of compromised intermediate nodes works in collusion and carries out attacks such as creating routing loops, routing packets on non-optimal paths, and selectively dropping packets [17]. Byzantine failures are hard to detect. The network would seem to be operating normally in the viewpoint of the nodes, though it may actually be exhibiting Byzantine behavior.

- **Information disclosure:** A compromised node may leak confidential or important information to unauthorized nodes in the network. Such information may include information regarding the network topology, geographic location of nodes, or optimal routes to authorized nodes in the network.

- **Resource consumption attack:** In this attack, a malicious node tries to consume/waste away resources of other nodes present in the network. The

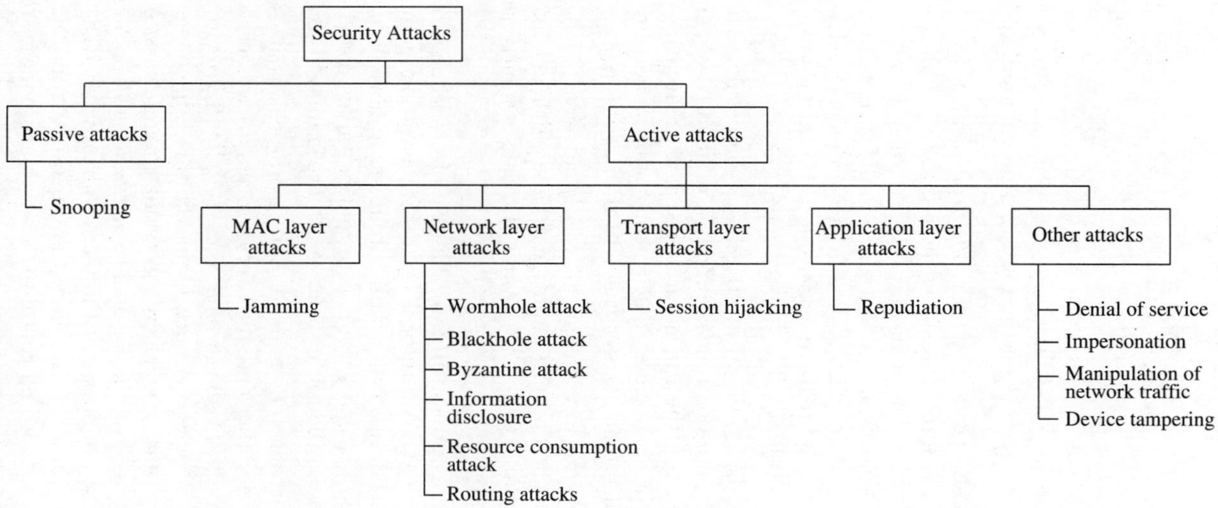

Figure 9.11. Classifications of attacks.

resources that are targeted are battery power, bandwidth, and computational power, which are only limitedly available in ad hoc wireless networks. The attacks could be in the form of unnecessary requests for routes, very frequent generation of beacon packets, or forwarding of stale packets to nodes. Using up the battery power of another node by keeping that node always busy by continuously pumping packets to that node is known as a sleep deprivation attack.

- **Routing attacks:** There are several types attacks mounted on the routing protocol which are aimed at disrupting the operation of the network. In what follows, the various attacks on the routing protocol are described briefly.
  - **Routing table overflow:** In this type of attack, an adversary node advertises routes to non-existent nodes, to the authorized nodes present in the network. The main objective of such an attack is to cause an overflow of the routing tables, which would in turn prevent the creation of entries corresponding to new routes to authorized nodes. Proactive routing protocols are more vulnerable to this attack compared to reactive routing protocols.
  - **Routing table poisoning:** Here, the compromised nodes in the networks send fictitious routing updates or modify genuine route update packets sent to other uncompromised nodes. Routing table poisoning may result in sub-optimal routing, congestion in portions of the network, or even make some parts of the network inaccessible.
  - **Packet replication:** In this attack, an adversary node replicates stale packets. This consumes additional bandwidth and battery power resources available to the nodes and also causes unnecessary confusion in the routing process.
  - **Route cache poisoning:** In the case of on-demand routing protocols (such as the AODV protocol [18]), each node maintains a route cache which holds information regarding routes that have become known to the node in the recent past. Similar to routing table poisoning, an adversary can also poison the route cache to achieve similar objectives.
  - **Rushing attack:** On-demand routing protocols that use duplicate suppression during the route discovery process are vulnerable to this attack [19]. An adversary node which receives a *RouteRequest* packet from the source node floods the packet quickly throughout the network before other nodes which also receive the same *RouteRequest* packet can react. Nodes that receive the legitimate *RouteRequest* packets assume those packets to be duplicates of the packet already received through the adversary node and hence discard those packets. Any route discovered by the source node would contain the adversary node as one of the intermediate nodes. Hence, the source node would not be able to find secure routes, that is, routes that do not include the adversary node. It is extremely difficult to detect such attacks in ad hoc wireless networks.

9.10.2 Transport Layer Attacks

This section discusses an attack which is specific to the transport layer in the network protocol stack.

- **Session hijacking:** Here, an adversary takes control over a session between two nodes. Since most authentication processes are carried out only at the start of a session, once the session between two nodes gets established, the adversary node masquerades as one of the end nodes of the session and hijacks the session.

9.10.3 Application Layer Attacks

This section briefly describes a security flaw associated with the application layer in the network protocol stack.

- **Repudiation:** In simple terms, repudiation refers to the denial or attempted denial by a node involved in a communication of having participated in all or part of the communication. As mentioned in Section 9.8, non-repudiation is one of the important requirements for a security protocol in any communication network.

9.10.4 Other Attacks

This section discusses security attacks that cannot strictly be associated with any specific layer in the network protocol stack.

Multi-layer Attacks

Multi-layer attacks are those that could occur in any layer of the network protocol stack. Denial of service and impersonation are some of the common multi-layer attacks. This section discusses some of the multi-layer attacks in ad hoc wireless networks.

- **Denial of Service:** In this type of attack, an adversary attempts to prevent legitimate and authorized users of services offered by the network from accessing those services. A denial of service (DoS) attack can be carried out in many ways. The classic way is to flood packets to any centralized resource (*e.g.*, an access point) used in the network so that the resource is no longer available to nodes in the network, resulting in the network no longer operating in the manner it was designed to operate. This may lead to a failure in the delivery of guaranteed services to the end users. Due to the unique characteristics of ad hoc wireless networks, there exist many more ways to launch a DoS attack in such a network, which would not be possible in wired networks. DoS attacks can be launched against any layer in the network protocol stack [20]. On the physical and MAC layers, an adversary could employ jamming signals which disrupt the on-going transmissions on the wireless channel. On the network layer, an adversary could take part in the routing process and exploit the routing protocol to disrupt the normal functioning of the network. For example, an adversary node could participate in a session but simply drop a certain number of packets, which may lead to degradation in the QoS being offered by the network. On the higher layers, an adversary could bring down critical services such as the key management service (key management will be described in detail in the next section). Some of the DoS attacks are described below.

- **Jamming:** In this form of attack, the adversary initially keeps monitoring the wireless medium in order to determine the frequency at which the receiver node is receiving signals from the sender. It then transmits signals on that frequency so that error-free reception at the receiver is hindered. Frequency hopping spread spectrum (FHSS) and direct sequence spread spectrum (DSSS) (described in detail in the first chapter of this book) are two commonly used techniques that overcome jamming attacks.
- **SYN flooding:** Here, an adversary sends a large number of SYN packets[8] to a victim node, spoofing the return addresses of the SYN packets. On receiving the SYN packets, the victim node sends back acknowledgment (SYN-ACK) packets to nodes whose addresses have been specified in the received SYN packets. However, the victim node would not receive any ACK packet in return. In effect, a half-open connection gets created. The victim node builds up a table/data structure for holding information regarding all pending connections. Since the maximum possible size of the table is limited, the increasing number of half-open connections results in an overflow in the table. Hence, even if a connection request comes from a legitimate node at a later point of time, because of the table overflow, the victim node would be forced to reject the call request.
- **Distributed DoS attack:** A more severe form of the DoS attack is the distributed DoS (DDoS) attack. In this attack, several adversaries that are distributed throughout the network collude and prevent legitimate users from accessing the services offered by the network.

• **Impersonation:** In impersonation attacks, an adversary assumes the identity and privileges of an authorized node, either to make use of network resources that may not be available to it under normal circumstances, or to disrupt the normal functioning of the network by injecting false routing information into the network. An adversary node could masquerade as an authorized node using several methods. It could by chance guess the identity and authentication details of the authorized node (target node), or it could snoop for information regarding the identity and authentication of the target node from a previous communication, or it could circumvent or disable the authentication mechanism at the target node. A *man-in-the-middle* attack is another

[8]SYN packets are used to establish an end-to-end session between two nodes at the transport layer.

type of impersonation attack. Here, the adversary reads and possibly modifies, messages between two end nodes without letting either of them know that they have been attacked. Suppose two nodes X and Y are communicating with each other; the adversary impersonates node Y with respect to node X and impersonates node X with respect to node Y, exploiting the lack of third-party authentication of the communication between nodes X and Y.

Device Tampering

Unlike nodes in a wired network, nodes in ad hoc wireless networks are usually compact, soft, and hand-held in nature. They could get damaged or stolen easily.

9.11 KEY MANAGEMENT

Having seen the various kinds of attacks possible on ad hoc wireless networks, we now look at various techniques employed to overcome the attacks. Cryptography is one of the most common and reliable means to ensure security. Cryptography is not specific to ad hoc wireless networks. It can be applied to any communication network. It is the study of the principles, techniques, and algorithms by which information is transformed into a disguised version which no unauthorized person can read, but which can be recovered in its original form by an intended recipient. In the parlance of cryptography, the original information to be sent from one person to another is called *plaintext*. This plaintext is converted into *ciphertext* by the process of encryption, that is, the application of certain algorithms or functions. An authentic receiver can decrypt/decode the ciphertext back into plaintext by the process of decryption. The processes of encryption and decryption are governed by *keys*, which are small amounts of information used by the cryptographic algorithms. When the key is to be kept secret to ensure the security of the system, it is called a secret key. The secure administration of cryptographic keys is called key management.

The four main goals of cryptography are confidentiality, integrity, authentication (the receiver should be able to identify the sender and verify that the message actually came from that sender), and non-repudiation. A detailed study of cryptography is presented in [21].

There are two major kinds of cryptographic algorithms: symmetric key algorithms, which use the same key for encryption and decryption, and asymmetric key algorithms, which use two different keys for encryption and decryption. Symmetric key algorithms are usually faster to execute electronically, but require a secret key to be shared between the sender and receiver. When communication needs to be established among a group of nodes, each sender-receiver pair should share a key, which makes the system non-scalable. If the same key is used among more than two parties, a breach of security at any one point makes the whole system vulnerable. The asymmetric key algorithms are based on some mathematical principles which make it infeasible or impossible to obtain one key from another; therefore, one of the keys can be made public while the other is kept secret (private). This is called public key cryptography. Such systems are used extensively in practice, but are not provably secure. They rely upon the difficulty of solving certain mathematical problems, and the network would be open to attacks once the underlying mathematical problem is solved.

9.11.1 Symmetric Key Algorithms

Symmetric key algorithms rely on the presence of the shared key at both the sender and receiver, which has been exchanged by some previous arrangement. There are two kinds of symmetric key algorithms, one involving block ciphers and the other stream ciphers. A block cipher is an encryption scheme in which the plaintext is broken into fixed-length segments called blocks, and the blocks are encrypted one at a time. The simplest examples include substitution and transposition. In substitution, each alphabet of the plaintext is substituted by another in the ciphertext, and this table mapping the original and the substituted alphabet is available at both the sender and receiver. A transposition cipher permutes the alphabet in the plaintext to produce the ciphertext. Figure 9.12 (a) illustrates the encryption using

Original Alphabet	A B C D E F G H I J K L M N O P Q R S T U V W X Y Z
Substitution	E F G H I J K L M N O P Q R S T U V W X Y Z A B C D
Plaintext	EVERYDAY CREATES A HISTORY EVERY DAYCR EATES AHIST ORY
Ciphertext	IZIVC HECGV IEXIW ELMWX SVC

(a)

Transposition	1 2 3 4 5 ↓ 3 5 1 4 2
Plaintext	EVERYDAY CREATES A HISTORY EVERY DAYCR EATES AHIST ORY
Ciphertext	EYERV YRDCA TSEEA ITASH YOR

(b)

Figure 9.12. Substitution and transposition.

substitution, and Figure 9.12 (b) shows a transposition cipher. The block length used is five.

A stream cipher is, in effect, a block cipher of block length one. One of the simplest stream ciphers is the Vernam cipher, which uses a key of the same length as the plaintext for encryption. For example, if the plaintext is the binary string 10010100, and the key is 01011001, then the encrypted string is given by the XOR of the plaintext and key, to be 11001101. The plaintext is again recovered by XOR-ing the ciphertext with the same key. If the key is randomly chosen, transported securely to the receiver, and used for only one communication, this forms the one-time pad which has proven to be the most secure of all cryptographic systems. The only bottleneck here is to be able to securely send the key to the receiver.

9.11.2 Asymmetric Key Algorithms

Asymmetric key (or public key) algorithms use different keys at the sender and receiver ends for encryption and decryption, respectively. Let the encryption process be represented by a function E, and decryption by D. Then the plaintext m is transformed into the ciphertext c as $c = E(m)$. The receiver then decodes c by applying D. Hence, D is such that $m = D(c) = D(E(m))$. When this asymmetric key concept is used in public key algorithms, the key E is made public, while D is private, known only to the intended receiver. Anyone who wishes to send a message to this receiver encrypts it using E. Though c can be overheard by adversaries, the function E is based on a computationally difficult mathematical problem, such as the factorization of large prime numbers. Hence, it is not possible for adversaries to derive D given E. Only the receiver can decrypt c using the private key D.

A very popular example of public key cryptography is the RSA system [21] developed by Rivest, Shamir, and Adleman, which is based on the integer factorization problem.

Digital signatures schemes are also based on public key encryption. In these schemes, the functions E and D are chosen such that $D(E(m)) = E(D(m)) = m$ for any message m. These are called reversible public key systems. In this case, the person who wishes to sign a document encrypts it using his/her private key D, which is known only to him/her. Anybody who has his/her public key E can decrypt it and obtain the original document, if it has been signed by the corresponding sender. In practice, a trusted third party (TTP) is agreed upon in advance, who is responsible for issuing these digital signatures (D and E pairs) and for resolving any disputes regarding the signatures. This is usually a governmental or business organization.

9.11.3 Key Management Approaches

The primary goal of key management is to share a secret (some information) among a specified set of participants. There are several methods that can be employed to perform this operation, all of them requiring varying amounts of initial configuration, communication, and computation. The main approaches to key management are key predistribution, key transport, key arbitration, and key agreement [22].

Key Predistribution

Key predistribution, as the name suggests, involves distributing keys to all interested parties before the start of communication. This method involves much less communication and computation, but all participants must be known *a priori*, during the initial configuration. Once deployed, there is no mechanism to include new members in the group or to change the key. As an improvement over the basic predistribution scheme, sub-groups may be formed within the group, and some communication can be restricted to a subgroup. However, the formation of sub-groups is also an *a priori* decision with no flexibility during the operation.

Key Transport

In key transport systems, one of the communicating entities generates keys and transports them to the other members. The simplest scheme assumes that a shared key already exists among the participating members. This prior shared key is used to encrypt a new key and is transmitted to all corresponding nodes. Only those nodes which have the prior shared key can decrypt it. This is called the key encrypting key (KEK) method. However, the existence of a prior key cannot always be assumed. If the public key infrastructure (PKI) is present, the key can be encrypted with each participant's public key and transported to it. This assumes the existence of a TTP, which may not be available for ad hoc wireless networks.

An interesting method for key transport without prior shared keys is the Shamir's three-pass protocol [22]. The scheme is based on a special type of encryption called commutative encryption schemes [which are reversible and composable (composition of two functions f and g is defined as $f(g(x))$)]. Consider two nodes X and Y which wish to communicate. Node X selects a key K which it wants to use in its communication with node Y. It then generates another random key k_x, using which it encrypts K with f, and sends to node Y. Node Y encrypts this with a random key k_y using g, and sends it back to node X. Now, node X decrypts this message with its key k_x, and after applying the inverse function f^{-1}, sends it to node Y. Finally, node Y decrypts the message using k_y and g^{-1} to obtain the key K. The message exchanges of the protocol are illustrated in Figure 9.13.

Key Arbitration

Key arbitration schemes use a central arbitrator to create and distribute keys among all participants. Hence, they are a class of key transport schemes. Networks which have a fixed infrastructure use the AP as an arbitrator, since it does not have stringent power or computation constraints. In ad hoc wireless networks, the problem with implementation of arbitrated protocols is that the arbitrator has to be powered on at all times to be accessible to all nodes. This leads to a power drain on that particular node. An alternative would be to make the keying service distributed, but simple replication of the arbitration at different nodes would be expensive for resource-constrained devices and would offer many points of vulnerability to attacks. If any one of the replicated arbitrators is attacked, the security of the whole system breaks down.

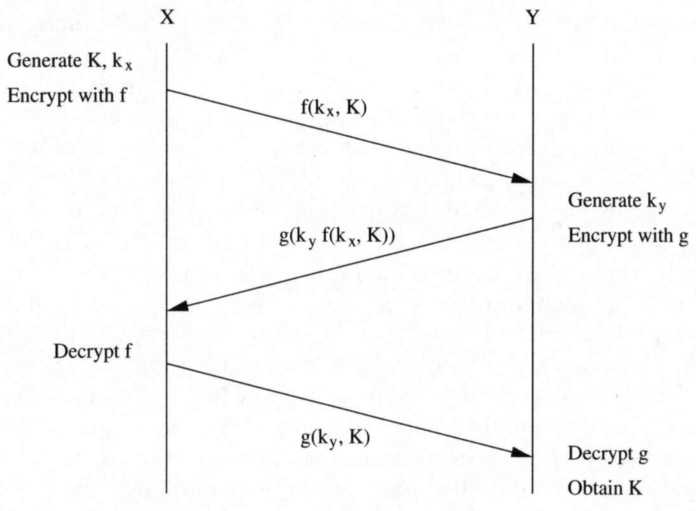

Figure 9.13. Shamir's three-pass protocol.

Key Agreement

Most key agreement schemes are based on asymmetric key algorithms. They are used when two or more people want to agree upon a secret key, which will then be used for further communication. Key agreement protocols are used to establish a secure context over which a session can be run, starting with many parties who wish to communicate and an insecure channel. In group key agreement schemes, each participant contributes a part to the secret key. These need the least amount of preconfiguration, but such schemes have high computational complexity. The most popular key agreement schemes use the Diffie-Hellman exchange [21], an asymmetric key algorithm based on discrete logarithms.

9.11.4 Key Management in Ad Hoc Wireless Networks

Ad hoc wireless networks pose certain specific challenges in key management due to the lack of infrastructure in such networks. Three types of infrastructure have been identified in [23], which are absent in ad hoc wireless networks. The first is the network infrastructure, such as dedicated routers and stable links, which ensure communication with all nodes. The second missing infrastructure is services such as name resolution, directory, and TTPs. The third missing infrastructure in ad hoc wireless networks is the administrative support of certifying authorities.

Password-Based Group Systems

Several solutions for group keying in ad hoc wireless networks have been suggested in [23]. The example scenario for implementation is a meeting room, where different mobile devices want to start a secure session. Here, the parties involved in the session are to be identified based on their location, that is, all devices in the room can be part of the session. Hence, relative location is used as the criterion for access control. If a TTP which knows the location of the participants exists, then it can implement location-based access control. A prior shared secret can be obtained by a physically more secure medium such as a wired network. This secret can be obtained by plugging onto a wired network first, before switching to the wireless mode.

A password-based system has been explored where, in the simplest case, a long string is given as the password for users for one session. However, human beings tend to favor natural language phrases as passwords, over randomly generated strings. Such passwords, if used as keys directly during a session, are very weak and open to attack because of high redundancy, and the possibility of reuse over different sessions. Hence, protocols have been proposed to derive a strong key (not vulnerable to attacks) from the weak passwords given by the participants. This password-based system could be two-party, with a separate exchange between any two participants, or it could be for the whole group, with a leader being elected to preside over the session. Leader election is a special case of establishing an order among all participants. The protocol used is as follows. Each participant generates a random number, and sends it to all others. When every node has received the random number of every other node, a common predecided function is applied on all the numbers to calculate a *reference value*. The nodes are ordered based on the difference between their random number and the reference value.

Threshold Cryptography

Public key infrastructure (PKI) enables the easy distribution of keys and is a scalable method. Each node has a public/private key pair, and a certifying authority (CA) can bind the keys to the particular node. But the CA has to be present at all times, which may not be feasible in ad hoc wireless networks. It is also not advisable to simply replicate the CA at different nodes. In [20], a scheme based on threshold cryptography has been proposed by which n servers exist in the ad hoc wireless network, out of which any $(t+1)$ servers can jointly perform any arbitration or authorization successfully, but t servers cannot perform the same. Hence, up to t compromised servers can be tolerated. This is called an $(n, t+1)$ configuration, where $n \geq 3t+1$.

To sign a certificate, each server generates a partial signature using its private key and submits it to a combiner. The combiner can be any one of the servers. In order to ensure that the key is combined correctly, $t+1$ combiners can be used to account for at most t malicious servers. Using $t+1$ partial signatures (obtained from itself and t other servers), the combiner computes a signature and verifies its validity using a public key. If the verification fails, it means that at least one of the $t+1$ keys is not valid, so another subset of $t+1$ partial signatures is tried. If the combiner itself is malicious, it cannot get a valid key, because the partial signature of itself is always invalid.

The scheme can be applied to asynchronous networks, with no bound on message delivery or processing times. This is one of the strengths of the scheme, as the requirement of synchronization makes the system vulnerable to DoS attacks. An adversary can delay a node long enough to violate the synchrony assumption, thereby disrupting the system.

Sharing a secret in a secure manner alone does not completely fortify a system. Mobile adversaries can move from one server to another, attack them, and get hold of their private keys. Over a period of time, an adversary can have more than t private keys. To counter this, *share refreshing* has been proposed, by which servers create a new independent set of shares (the partial signatures which are used by the servers) periodically. Hence, to break the system, an adversary has to attack and capture more than t servers within the period between two successive refreshes; otherwise, the earlier share information will no longer be valid. This improves protection against mobile adversaries.

Self-Organized Public Key Management for Mobile Ad Hoc Networks

The authors of [24] have proposed a completely self-organized public key system for ad hoc wireless networks. This makes use of absolutely no infrastructure – TTP, CA, or server – even during initial configuration. The users in the ad hoc wireless network issue certificates to each other based on personal acquaintance. A certificate is a binding between a node and its public key. These certificates are also stored and distributed by the users themselves. Certificates are issued only for a specified period of time and contain their time of expiry along with them. Before it expires, the certificate is updated by the user who had issued the certificate.

Initially, each user has a local repository consisting of the certificates issued by him and the certificates issued by other users to him. Hence, each certificate is initially stored twice, by the issuer and by the person for whom it is issued. Periodically, certificates from neighbors are requested and the repository is updated by adding any new certificates. If any of the certificates are conflicting (*e.g.,* the same public key to different users, or the same user having different public keys), it is possible that a malicious node has issued a false certificate. A node then labels such certificates as *conflicting* and tries to resolve the conflict. Various methods exist to compare the confidence in one certificate over another. For instance, another set of certificates obtained from another neighbor can be used to take a majority decision. This can be used to evaluate the trust in other users and detect malicious nodes. If the certificates issued by some node are found to be wrong, then that node may be assumed to be malicious.

The authors of [24] define a certificate graph as a graph whose vertices are public keys of some nodes and whose edges are public-key certificates issued by users. When a user X wants to obtain the public key of another user Y, he/she finds a chain of valid public key certificates leading to Y. The chain is such that the first hop uses an edge from X, that is, a certificate issued by X, the last hop leads into Y (this is a certificate issued to Y), and all intermediate nodes are trusted through the previous certificate in the path. The protocol assumes that trust is transitive, which may not always be valid.

Having seen the various key management techniques employed in ad hoc wireless networks, we now move on to discuss some of the security-aware routing schemes for ad hoc wireless networks.

9.12 SECURE ROUTING IN AD HOC WIRELESS NETWORKS

Unlike the traditional wired Internet, where dedicated routers controlled by the Internet service providers (ISPs) exist, in ad hoc wireless networks, nodes act both as regular terminals (source or destination) and also as routers for other nodes. In the absence of dedicated routers, providing security becomes a challenging task in these networks. Various other factors which make the task of ensuring secure communication in ad hoc wireless networks difficult include the mobility of nodes, a promiscuous mode of operation, limited processing power, and limited availability of resources such as battery power, bandwidth, and memory. Section 9.10.1 has pointed out some of the possible security attacks at the network layer. In the following sections, we show how some of the well-known traditional routing protocols for ad hoc networks fail to provide security. Some of the mechanisms proposed for secure routing are also discussed.

9.12.1 Requirements of a Secure Routing Protocol for Ad Hoc Wireless Networks

The fundamental requisites of a secure routing protocol for ad hoc wireless networks are listed as follows:

- **Detection of malicious nodes:** A secure routing protocol should be able to detect the presence of malicious nodes in the network and should avoid the participation of such nodes in the routing process. Even if such malicious nodes participate in the route discovery process, the routing protocol should choose paths that do not include such nodes.

- **Guarantee of correct route discovery:** If a route between the source and the destination nodes exists, the routing protocol should be able to find the route, and should also ensure the correctness of the selected route.

- **Confidentiality of network topology:** As explained in Section 9.10.1, an information disclosure attack may lead to the discovery of the network topology by the malicious nodes. Once the network topology is known, the attacker may try to study the traffic pattern in the network. If some of the nodes are found to be more active compared to others, the attacker may try to mount (*e.g.,* DoS) attacks on such bottleneck nodes. This may ultimately affect the on-going routing process. Hence, the confidentiality of the network topology is an important requirement to be met by the secure routing protocols.

- **Stability against attacks:** The routing protocol must be self-stable in the sense that it must be able to revert to its normal operating state within a finite amount of time after a passive or an active attack. The routing protocol should take care that these attacks do not permanently disrupt the routing process. The protocol must also ensure Byzantine robustness, that is, the protocol should work properly even if some of the nodes, which were earlier participating in the routing process, turn out to become malicious at a later point of time or are intentionally damaged.

In the following sections, some of the security-aware routing protocols proposed for ad hoc wireless networks are discussed.

9.12.2 Security-Aware Ad Hoc Routing Protocol

The security-aware ad hoc routing (SAR) protocol [25] uses security as one of the key metrics in path finding. A framework for enforcing and measuring the attributes of the security metric has been provided in [25]. This framework also enables the use of different levels of security for different applications that use SAR for routing. In ad hoc wireless networks, communication between end nodes through possibly multiple intermediate nodes is based on the fact that the two end nodes trust the intermediate nodes. SAR defines *level of trust* as a metric for routing and as one of the attributes for security to be taken into consideration while routing. The routing protocol based on the level of trust is explained using Figure 9.14. As shown in Figure 9.14, two paths exist between the two officers $O1$ and $O2$ who want to communicate with each other. One of these paths is a shorter path which runs through private nodes whose trust levels are very low. Hence, the protocol chooses a longer but secure path which passes through other secure (officer) nodes.

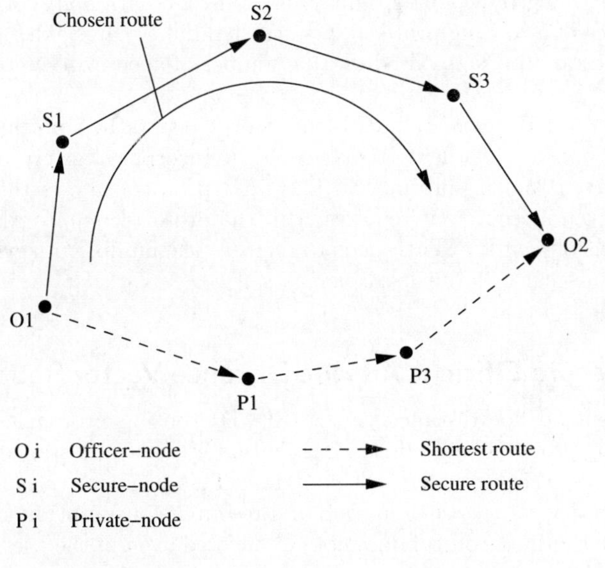

Figure 9.14. Illustration of the level of trust metric.

The SAR protocol can be explained using any one of the traditional routing protocols. This section explains SAR using the AODV protocol [18] discussed in detail in Chapter 7. In the AODV protocol, the source node broadcasts a *RouteRequest* packet to its neighbors. An intermediate node, on receiving a *RouteRequest* packet, forwards it further if it does not have a route to the destination. Otherwise, it initiates a *RouteReply* packet back to the source node using the reverse path traversed by the *RouteRequest* packet. In SAR, a certain level of security is incorporated into the packet-forwarding mechanism. Here, each packet is associated with a security level which is determined by a number calculation method (explained later in this section). Each intermediate node is also associated with a certain level of security. On receiving a packet, the intermediate node compares its level of security with that defined for the packet. If the node's security level is less than that of the packet, the *RouteRequest* is simply discarded. If it is greater, the node is considered to be a secure node and is permitted to forward the packet in addition to being able to view the packet. If the security levels of the intermediate node and the received packet are found to be equal, then the intermediate node will not be able to view the packet (which can be ensured using a proper authentication mechanism); it just forwards the packet further.

Nodes of equal levels of trust distribute a common key among themselves and with those nodes having higher levels of trust. Hence, a hierarchical level of security could be maintained. This ensures that an encrypted packet can be decrypted (using the common key) only by nodes of the same or higher levels of security compared to the level of security of the packet. Different levels of trust can be defined using a number calculated based on the level of security required. It can be calculated using many methods. Since timeliness, in-order delivery of packets, authenticity, authorization, integrity, confidentiality, and non-repudiation are some of the desired characteristics of a routing protocol, a suitable number can be defined for the trust level for nodes and packets based on the number of such characteristics taken into account.

The SAR mechanism can be easily incorporated into the traditional routing protocols for ad hoc wireless networks. It could be incorporated into both on-demand and table-driven routing protocols. The SAR protocol allows the application to choose the level of security it requires. But the protocol requires different keys for different levels of security. This tends to increase the number of keys required when the number of security levels used increases.

9.12.3 Secure Efficient Ad Hoc Distance Vector Routing Protocol

Secure efficient ad hoc distance vector (SEAD) routing protocol [26], is a secure ad hoc routing protocol based on the destination-sequenced distance vector (DSDV) routing protocol [27] discussed in Chapter 7. This protocol is mainly designed to overcome security attacks such as DoS and resource consumption attacks. The operation of the routing protocol does not get affected even in the presence of multiple uncoordinated attackers corrupting the routing tables. The protocol uses a one-way hash function and does not involve any asymmetric cryptographic operation.

Distance Vector Routing

Distance vector routing protocols belong to the category of table-driven routing protocols. Each node maintains a routing table containing the list of all known routes to various destination nodes in the network. The metric used for routing is the distance measured in terms of hop-count. The routing table is updated periodically by exchanging routing information. An alternative to this approach is *triggered updates*, in which each node broadcasts routing updates only if its routing table gets altered. The DSDV protocol for ad hoc wireless networks uses *sequence number* tags to prevent the formation of loops, to counter the count-to-infinity problem, and for faster convergence. When a new route update packet is received for a destination, the node updates the corresponding entry in its routing table only if the sequence number on the received update is greater than that recorded with the corresponding entry in the routing table. If the received sequence number and the previously recorded sequence number are both equal, but if the routing update has a new value for the routing metric (distance in number of hops), then in this case also the update is effected. Otherwise, the received update packet is discarded. DSDV uses triggered updates (for important routing changes) in addition to the regular periodic updates. A slight variation of DSDV protocol known as DSDV-SQ (DSDV for sequence numbers) initiates triggered updates on receiving a new sequence number update.

One-Way Hash Function

SEAD uses authentication to differentiate between updates that are received from non-malicious nodes and malicious nodes. This minimizes resource consumption attacks caused by malicious nodes. SEAD uses a one-way hash function for authenticating the updates. A one-way hash function (H) generates a one-way hash chain ($h_1, h_2, \ldots$). The function H maps an input bit-string of any length to a fixed length bit-string, that is, $H : (0,1)^* \rightarrow (0,1)^\rho$, where ρ is the length in bits of the output bit-string. To create a one-way hash chain, a node generates a random number with initial value $x \in (0,1)^\rho$. h_0, the first number in the hash chain is initialized to x. The remaining values in the chain are computed using the general formula, $h_i = H(h_{i-1})$ for $0 \leq i \leq n$, for some n. Now we shall see how the one-way hash function incorporates security into the existing DSDV-SQ routing protocol. The SEAD protocol assumes an upper bound on the metric used. For example, if the metric used is distance, then the upper bound value $m - 1$ defines the maximum diameter (maximum of lengths of all the routes between a pair of nodes) of the ad hoc wireless network. Hence, the routing protocol ensures that no route of length greater than m hops exists between any two nodes.

If the sequence of values calculated by a node using the hash function H is given by $(h_1, h_2, \ldots, h_n)$, where n is divisible by m, then for a routing table entry with sequence number i, let $k = \frac{k}{m} - i$. If the metric j (distance) used for that routing table entry is $0 \leq j \leq m - 1$, then the value h_{km+j} is used to authenticate the routing update entry for that sequence number i and that metric j. Whenever a route update message is sent, the node appends the value used for authentication along with it. If the authentication value used is h_{km+j}, then the attacker who tries to modify this value can do so only if he/she knows h_{km+j-1}. Since it is a one-way hash chain, calculating h_{km+j-1} becomes impossible. An intermediate node, on receiving this authenticated update, calculates the new hash value based on the earlier updates (h_{km+j-1}), the value of the metric, and the sequence number. If the calculated value matches with the one present in the route update message, then the update is effected; otherwise, the received update is just discarded.

SEAD avoids routing loops unless the loop contains more than one attacker. This protocol could be implemented easily with slight modifications to the existing distance vector routing protocols. The protocol is robust against multiple uncoordinated attacks. The SEAD protocol, however, would not be able to overcome attacks where the attacker uses the same metric and sequence number which were used by the recent update message, and sends a new routing update.

9.12.4 Authenticated Routing for Ad Hoc Networks

Authenticated routing for ad hoc networks (ARAN) routing protocol [28], based on cryptographic certificates, is a secure routing protocol which successfully defeats all identified attacks in the network layer. It takes care of authentication, message integrity, and non-repudiation, but expects a small amount of prior security coordination among nodes. In [28], vulnerabilities and attacks specific to AODV and DSR protocols are discussed and the two protocols are compared with the ARAN protocol.

During the route discovery process of ARAN, the source node broadcasts *RouteRequest* packets. The destination node, on receiving the *RouteRequest* packets, responds by unicasting back a reply packet on the selected path. The ARAN protocol uses a preliminary cryptographic certification process, followed by an end-to-end route authentication process, which ensures secure route establishment.

Issue of Certificates

This section discusses the certification process in which the certificates are issued to the nodes in the ad hoc wireless network. There exists an authenticated trusted server whose public key is known to all legal nodes in the network. The ARAN protocol assumes that keys are generated *a priori* by the server and distributed to all nodes in the network. The protocol does not specify any specific key distribution algorithm. On joining the network, each node receives a certificate from the trusted server. The certificate received by a node A from the trusted server T looks like the following:

$$T \rightarrow A: \; cert_A = [IP_A, K_{A+}, t, e]K_{T-} \tag{9.12.1}$$

Here, IP_A, K_{A+}, t, e, and K_{T-} represent the IP address of node A, the public key of node A, the time of creation of the certificate, the time of expiry of the certificate, and the private key of the server, respectively.

End-to-End Route Authentication

The main goal of this end-to-end route authentication process is to ensure that the correct intended destination is reached by the packets sent from the source node. The source node S broadcasts a *RouteRequest/RouteDiscovery* packet destined to the destination node D. The *RouteRequest* packet contains the packet identifier [route discovery process (RDP)], the IP address of the destination (IP_D), the certificate of the source node S ($Cert_S$), the current time (t), and nonce N_S. The process can be denoted as below. Here, K_{S-} is the private key of the source node S.

$$S \rightarrow broadcasts := \ [RDP, IP_D, Cert_S, N_S, t]K_{S-} \tag{9.12.2}$$

Whenever the source sends a route discovery message, it increments the value of nonce. Nonce is a counter used in conjunction with the time-stamp in order to make the nonce recycling easier. When a node receives an RDP packet from the source with a higher value of the source's nonce than that in the previously received RDP packets from the same source node, it makes a record of the neighbor from which it received the packet, encrypts the packet further with its own certificate, and broadcasts it further. The process can be denoted as follows:

$$A \rightarrow broadcasts := \ [[RDP, IP_D, Cert_S, N_S, t]K_{S-}]K_{A-}, Cert_A \tag{9.12.3}$$

An intermediate node B, on receiving an RDP packet from a node A, removes its neighbor's certificate, inserts its own certificate, and broadcasts the packet further. The destination node, on receiving an RDP packet, verifies node S's certificate and the tuple (N_S, t) and then replies with the *RouteReply* packet (REP). The destination unicasts the REP packet to the source node along the reverse path as follows:

$$D \rightarrow X := \ [REP, IP_S, Cert_D, N_S, t]K_{D-} \tag{9.12.4}$$

where node X is the neighbor of the destination node D, which had originally forwarded the RDP packet to node D. The REP packet follows the same procedure on the reverse path as that followed by the route discovery packet. An error message is generated if the time-stamp or nonce do not match the requirements or if the certificate fails. The error message looks similar to the other packets except that the packet identifier is replaced by the ERR message.

Table 9.3 shows a comparison between the AODV, DSR, and ARAN protocols with respect to their security-related features. ARAN remains robust in the presence of attacks such as unauthorized participation, spoofed route signaling, fabricated routing messages, alteration of routing messages, securing shortest paths, and replay attacks.

9.12.5 Security-Aware AODV Protocol

This section discusses security solutions that address a particular security flaw in the AODV routing protocol [18]. AODV is an on-demand routing protocol where

Table 9.3. Comparison of vulnerabilities of ARAN with DSR and AODV protocols

Attacks	Protocols		
	AODV	**DSR**	**ARAN**
Modifications required during remote redirection	Sequence number and hop-counts	Source routes	None
Tunneling during remote redirection	Yes	Yes	Yes
Spoofing	Yes	Yes	No
Cache poisoning	No	Yes	No

the route discovery process is initiated by sending *RouteRequest* packets only when data packets arrive at a node for transmission. A malicious intermediate node could advertise that it has the shortest path to the destination, thereby redirecting all the packets through itself. This is known as a blackhole attack, as explained in Section 9.10.1. The blackhole attack is illustrated in Figure 9.15. Let node M be the malicious node that enters the network. It advertises that it has the shortest path to the destination node D when it receives the *RouteRequest* packet sent by node S. The attacker may not be able to succeed if node A, which also receives the *RouteRequest* packet from node S, replies earlier than node M. But a major advantage for the malicious node is that it does not have to search its routing table

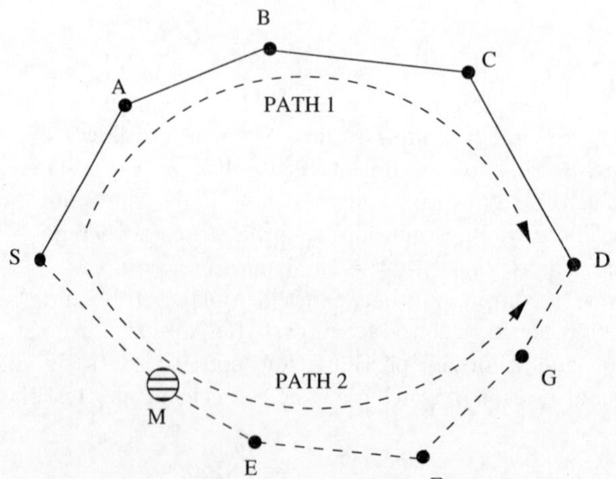

Figure 9.15. Illustration of blackhole problem.

for a route to the destination. Also, the *RouteReply* packets originate directly from the malicious node and not from the destination node. Hence, the malicious node would be able to reply faster than node A, which would have to search its routing table for a route to the destination node. Thus, node S may tend to establish a route to destination D through the malicious node M, allowing node M to listen to all packets meant for the destination node.

Solutions for the Blackhole Problem

One of the solutions for the blackhole problem is to restrict the intermediate nodes from originating *RouteReply* packets. Only the destination node would be permitted to initiate *RouteReply* packets. Security is still not completely assured, since the malicious node may lie in the path chosen by the destination node. Also, the delay involved in the route discovery process increases as the size of the network increases. In another solution to this problem, suggested in [29], as soon as the *RouteReply* packet is received from one of the intermediate nodes, another *RouteRequest* packet is sent from the source node to the neighbor node of the intermediate node in the path. This is to ensure that such a path exists from the intermediate node to the destination node. For example, let the source node send *RouteRequest* packets and receive *RouteReply* through the intermediate malicious node M. The *RouteReply* packet of node M contains information regarding its next-hop neighbor nodes. Let it contain information about the neighbor node E. Then, as shown in Figure 9.16, the source node S sends *FurtherRouteRequest* packets to this neighbor node E. Node E responds by sending a *FurtherRouteReply* packet to source node S. Since

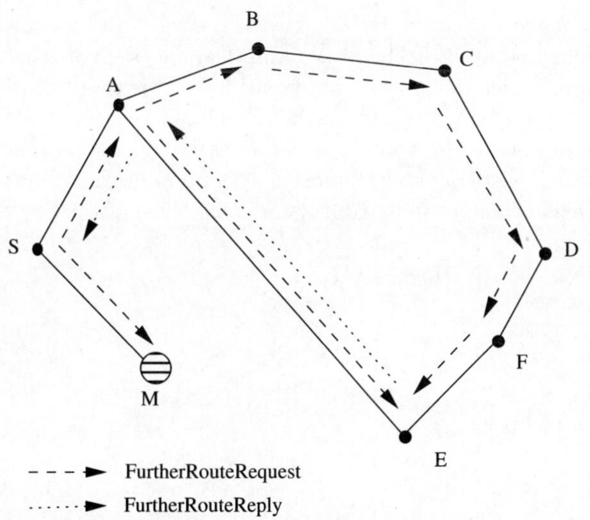

Figure 9.16. Propagation of *FurtherRouteRequest* and *FurtherRouteReply*.

node M is a malicious node which is not present in the routing list of node E, the *FurtherRouteReply* packet sent by node E will not contain a route to the malicious node M. But if it contains a route to the destination node D, then the new route to the destination through node E is selected, and the earlier selected route through node M is rejected. This protocol completely eliminates the blackhole attack caused by a single attacker. The major disadvantage of this scheme is that the control overhead of the routing protocol increases considerably. Also, if the malicious nodes work in a group, this protocol fails miserably.

9.13 SUMMARY

This chapter discussed the major challenges that a transport layer protocol faces in ad hoc wireless networks. The major design goals of a transport layer protocol were listed and a classification of existing transport layer solutions was provided. TCP is the most widely used transport layer protocol and is considered to be the backbone of today's Internet. It provides end-to-end, reliable, byte-streamed, in-order delivery of packets to nodes. Since TCP was designed to handle problems present in traditional wired networks, many of the issues that are present in dynamic topology networks such as ad hoc wireless networks are not addressed. This causes reduction of throughput when TCP is used in ad hoc wireless networks. It is very important to employ TCP in ad hoc wireless networks as it is important to seamlessly communicate with the Internet whenever and wherever it is available. This chapter provided a discussion on the major reasons for the degraded performance of traditional TCP in ad hoc wireless networks and explained a number of recently proposed solutions to improve TCP's performance. Other non-TCP solutions were also discussed in detail.

The second half of this chapter dealt with the security aspect of communication in ad hoc wireless networks. The issues and challenges involved in provisioning security in ad hoc wireless networks were identified. This was followed by a layer-wise classification of the various types of attacks. Detailed discussions on key management techniques and secure routing techniques for ad hoc wireless networks were provided. Table 9.4 lists out the various attacks possible in ad hoc wireless networks along with the solutions proposed for countering those attacks.

Table 9.4. Defense against attacks

Attack	Targeted Layer in the Protocol Stack	Proposed Solutions
Jamming	Physical and MAC layers	FHSS, DSSS
Wormhole attack	Network layer	Packet Leashes [16]
Blackhole attack	Network layer	[25], [29]
Byzantine attack	Network layer	[17]
Resource consumption attack	Network layer	SEAD [26]
Information disclosure	Network layer	SMT [30]
Location disclosure	Network layer	SRP [30], NDM [31]
Routing attacks	Network layer	[19], SEAD [26], ARAN [28], ARIADNE [32]
Repudiation	Application layer	ARAN [28]
Denial of Service	Multi-layer	SEAD [26], ARIADNE [32]
Impersonation	Multi-layer	ARAN [28]

9.14 PROBLEMS

1. Assume that when the current size of the congestion window is 48 KB, the TCP sender experiences a timeout. What will be the congestion window size if the next three transmission bursts are successful? Assume that MSS is 1 KB. Consider (a) TCP Tahoe and (b) TCP Reno.

2. Find out the probability of a path break for an eight-hop path, given that the probability of a link break is 0.2.

3. Discuss the effects of multiple breaks on a single path at the TCP-F sender.

4. What additional state information is to be maintained at the FP in TCP-F?

5. Mention one advantage and one disadvantage of using probe packets for detection of a new path.

6. Mention one advantage and one disadvantage of using LQ and REPLY for finding partial paths in TCP-BuS.

7. What is the impact of the failure of proxy nodes in split-TCP?

8. During a research discussion, one of your colleagues suggested an extension of split-TCP where every intermediate node acts as a proxy node. What do you think would be the implications of such a protocol?

9. What are the pros and cons of assigning the responsibility of end-to-end reliability to the application layer?

10. What is the default value of β used for handling induced traffic in ATP and why is such a value chosen?

11. Explain how network security requirements vary in the following application scenarios of ad hoc wireless networks:

    (a) Home networks

    (b) Classroom networks

    (c) Emergency search-and-rescue networks

    (d) Military networks

12. Explain how security provisioning in ad hoc wireless networks differs from that in infrastructure-based networks?

13. Explain the key encrypting key (KEK) method.

14. Nodes A and B want to establish a secure communication, and node A generates a random key 11001001. Suppose the function used by both nodes A and B for encryption is XOR, and let node A generate a random transport key 10010101, and let node B generate 00101011. Explain the three-pass Shamir protocol exchanges.

15. Why is it not advisable to use natural-language passwords directly for cryptographic algorithms?

16. Consider the certificate graph shown in Figure 9.17, with the local certificate repositories of nodes A and B as indicated. Find the possible paths of trust from node A to node B which can be obtained using a chain of keys.

17. List a few inherent security flaws present in the following types of routing protocols: (a) table-driven and (b) on-demand routing.

18. List and explain how some of the inherent properties of the wireless ad hoc networks introduce difficulties while implementing security in routing protocols.

19. Mark the paths chosen by the following secure-routing protocols for the network topology shown in Figure 9.18: (a) Shortest path routing and (b) SAR protocol. Assume that node 2 is a secure node. (c) If node 2 (which lies in the path chosen by SAR protocol) is suddenly attacked and becomes a malicious node, then mark an alternative path chosen by SAODV protocol.

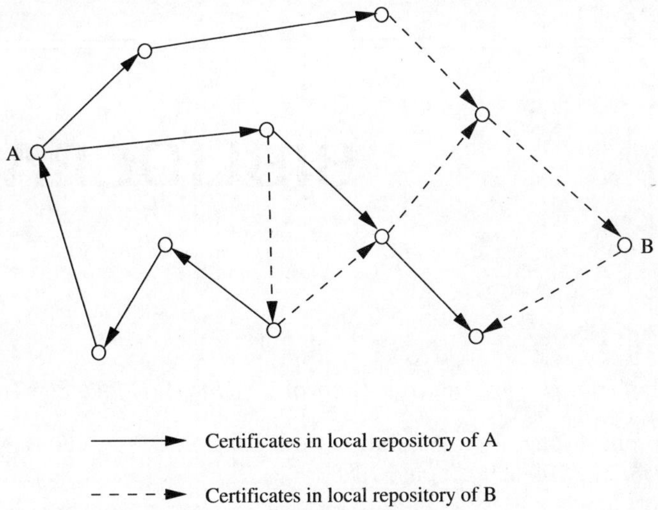

Figure 9.17. Certificate graph.

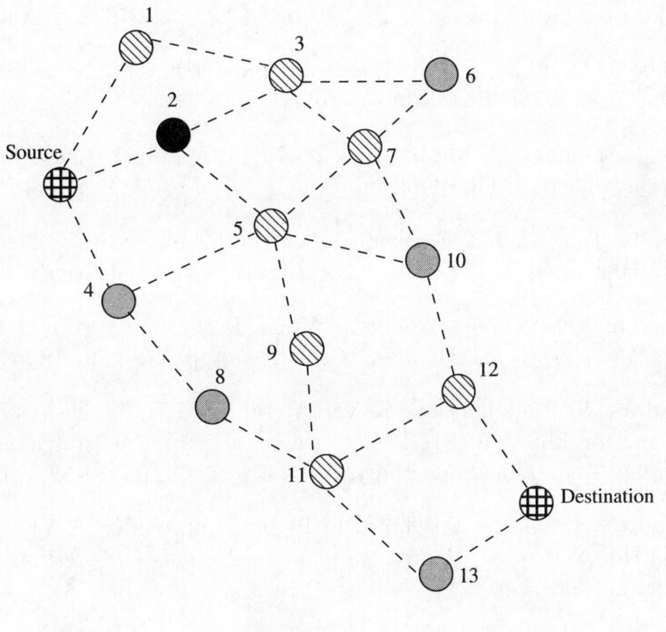

Figure 9.18. Example network topology.

BIBLIOGRAPHY

[1] J. Postel, "Transmission Control Protocol," *IETF RFC 793*, September 1981.

[2] V. Jacobson, "Congestion Avoidance and Control," *Proceedings of ACM SIGCOMM 1988*, pp. 314–329, August 1988.

[3] W. R. Stevens, "TCP Slow Start, Congestion Avoidance, Fast Retransmission, and Fast Recovery Algorithms," *IETF RFC 2001*, January 1997.

[4] J. C. Hoe, "Improving the Start-Up Behavior of a Congestion Control Scheme for TCP," *Proceedings of the ACM SIGCOMM 1996*, pp. 270–280, August 1996.

[5] M. Mathis, J. Mahdavi, S. Floyd, and A. Romanow, "TCP Selective Acknowledgment Options," *IETF RFC 2018*, October 1996.

[6] S. Floyd, J. Mahdavi, M. Mathis, and M. Podolsky, "An Extension to the Selective Acknowledgment (SACK) Option for TCP," *IETF RFC 2883*, July 2000.

[7] M. Gerla, K. Tang, and R. Bagrodia, "TCP Performance in Wireless Multi-Hop Networks," *Proceedings of IEEE WMCSA 1999*, pp. 41-50, February 1999.

[8] G. Holland and N. Vaidya, "Analysis of TCP Performance over Mobile Ad Hoc Networks," *Proceedings of ACM MOBICOM 1999*, pp. 219-230, August 1999.

[9] K. Chandran, S. Raghunathan, S. Venkatesan, and R. Prakash, "A Feedback-Based Scheme for Improving TCP Performance in Ad Hoc Wireless Networks," *IEEE Personal Communications Magazine*, vol. 8, no. 1 , pp. 34-39, February 2001.

[10] D. Kim, C. K. Toh, and Y. Choi, "TCP-BuS: Improving TCP Performance in Wireless Ad Hoc Networks," *Journal of Communications and Networks*, vol. 3, no. 2, pp. 1-12, June 2001.

[11] C. K. Toh, "Associativity-Based Routing for Ad Hoc Mobile Networks," *Wireless Personal Communications*, vol. 4, no. 2, pp. 1-36, March 1997.

[12] J. Liu and S. Singh,"ATCP: TCP for Mobile Ad Hoc Networks," *IEEE Journal on Selected Areas in Communications*, vol. 19, no. 7, pp. 1300-1315, July 2001.

[13] S. Kopparty, S. V. Krishnamurthy, M. Faloutsos, and S. K. Tripathi, "Split TCP for Mobile Ad Hoc Networks," *Proceedings of IEEE GLOBECOM 2002*, vol. 1, pp. 138-142, November 2002.

[14] J. Liu and S. Singh, "ATP: Application Controlled Transport Protocol for Mobile Ad Hoc Networks," *Proceedings of IEEE WCMC 1999*, vol. 3, pp. 1318-1322, September 1999.

[15] K. Sundaresan, V. Anantharaman, H. Y. Hsieh, and R. Sivakumar, "ATP: A Reliable Transport Protocol for Ad Hoc Networks," *Proceedings of ACM MOBIHOC 2003*, pp. 64-75, June 2003.

[16] Y. Hu, A. Perrig, and D. B. Johnson, "Packet Leashes: A Defense Against Wormhole Attacks in Wireless Ad Hoc Networks," *Proceedings of IEEE INFOCOM 2003*, vol. 3, pp. 1976-1986, April 2003.

[17] B. Awerbuch, D. Holmer, C. Nita-Rotaru, and H. Rubens, "An On-Demand Secure Routing Protocol Resilient to Byzantine Failures," *Proceedings of the ACM Workshop on Wireless Security 2002*, pp. 21-30, September 2002.

[18] C. E. Perkins and E. M. Royer, "Ad Hoc On-Demand Distance Vector Routing," *Proceedings of IEEE Workshop on Mobile Computing Systems and Applications*, pp. 90-100, February 1999.

[19] Y. Hu, A. Perrig, and D. B. Johnson, "Rushing Attacks and Defense in Wireless Ad Hoc Network Routing Protocols," *Proceedings of the ACM Workshop on Wireless Security 2003*, pp. 30-40, September 2003.

[20] L. Zhou and Z. J. Haas, "Securing Ad Hoc Networks," *IEEE Network Magazine*, vol. 13, no. 6, pp. 24-30, December 1999.

[21] A. J. Menezes, P. C. V. Oorschot, and S. A. Vanstone, *Handbook of Applied Cryptography*, CRC Press, 1996.

[22] A. Khalili, W. A. Arbaugh, "Security of Wireless Ad Hoc Networks," http://www.cs.umd.edu/~aram/wireless/survey.pdf.

[23] N. Asokan and P. Ginzboorg, "Key-Agreement in Ad Hoc Networks," *Computer Communications*, vol. 23, no. 17, pp. 1627-1637, 2000.

[24] S. Capkun, L. Buttyan, and J. P. Hubaux, "Self-Organized Public-Key Management for Mobile Ad Hoc Networks," *IEEE Transactions on Mobile Computing*, vol. 2, no. 1, pp. 52-64, January-March 2003.

[25] S. Yi, P. Naldurg, and R. Kravets, "Security-Aware Ad Hoc Routing for Wireless Networks," *Proceedings of ACM MOBIHOC 2001*, pp. 299-302, October 2001.

[26] Y. Hu, D. B. Johnson, and A. Perrig, "SEAD: Secure Efficient Distance Vector Routing for Mobile Wireless Ad Hoc Networks," *Proceedings of IEEE WMCSA 2002*, pp. 3-13, June 2002.

[27] C. E. Perkins and P. Bhagwat, "Highly Dynamic Destination-Sequenced Distance-Vector Routing (DSDV) for Mobile Computers," *Proceedings of ACM SIGCOMM 1994*, pp. 234-244, August 1994.

[28] K. Sanzgiri, B. Dahill, B. N. Levine, C. Shields, and E. M. B. Royer, "A Secure Routing Protocol for Ad Hoc Networks," *Proceedings of IEEE ICNP 2002*, pp. 78-87, November 2002.

[29] H. Deng, W. Li, and D. P. Agrawal, "Routing Security in Wireless Ad Hoc Networks," *IEEE Communications Magazine*, vol. 40, no. 10, pp. 70-75, October 2002.

[30] P. Papadimitratos and Z. J. Haas, "Secure Routing: Secure Data Transmission in Mobile Ad Hoc Networks," *Proceedings of ACM Workshop on Wireless Security 2003*, pp. 41-50, September 2003.

[31] A. Fasbender, D. Kesdogan, and O. Kubitz, "Variable and Scalable Security: Protection of Location Information in Mobile IP," *Proceedings of IEEE VTC 1996*, vol. 2, pp. 963-967, May 1996.

[32] Y. Hu, A. Perrig, and D. B. Johnson, "Ariadne: A Secure On-Demand Routing for Ad Hoc Networks," *Proceedings of ACM MOBICOM 2002*, pp. 12-23, September 2002.

Chapter 10

QUALITY OF SERVICE IN AD HOC WIRELESS NETWORKS

10.1 INTRODUCTION

Quality of service (QoS) is the performance level of a service offered by the network to the user. The goal of QoS provisioning is to achieve a more deterministic network behavior, so that information carried by the network can be better delivered and network resources can be better utilized. A network or a service provider can offer different kinds of services to the users. Here, a service can be characterized by a set of measurable prespecified service requirements such as minimum bandwidth, maximum delay, maximum delay variance (jitter), and maximum packet loss rate. After accepting a service request from the user, the network has to ensure that the service requirements of the user's flow are met, as per the agreement, throughout the duration of the flow (a packet stream from the source to the destination). In other words, the network has to provide a set of service guarantees while transporting a flow.

After receiving a service request from the user, the first task is to find a suitable loop-free path from the source to the destination that will have the necessary resources available to meet the QoS requirements of the desired service. This process is known as QoS routing. After finding a suitable path, a resource reservation protocol is employed to reserve necessary resources along that path. QoS guarantees can be provided only with appropriate resource reservation techniques. For example, consider the network shown in Figure 10.1. The attributes of each link are shown in a tuple $< BW, D >$, where BW and D represent available bandwidth in Mbps and delay[1] in milliseconds. Suppose a packet-flow from node B to node G requires a bandwidth guarantee of 4 Mbps. Throughout the chapter, the terms "node" and "station" are used interchangeably. QoS routing searches for a path that has sufficient bandwidth to meet the bandwidth requirement of the flow. Here, six paths are available between nodes B and G as shown in Table 10.1. QoS routing selects

[1]Delay includes transmission delay, propagation delay, and queuing delay.

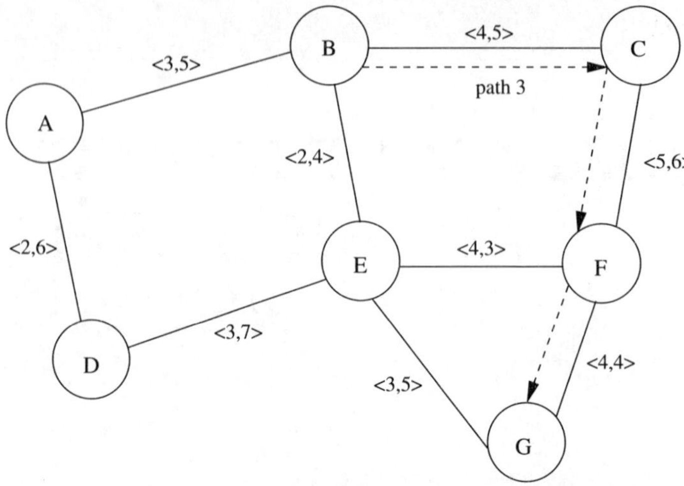

Figure 10.1. An example of QoS routing in ad hoc wireless network.

path 3 (*i.e.*, $B \rightarrow C \rightarrow F \rightarrow G$) because, out of the available paths, path 3 alone meets the bandwidth constraint of 4 Mbps for the flow. The end-to-end bandwidth of a path is equal to the bandwidth of the bottleneck link (*i.e.*, the link having minimum bandwidth among all the links of a path). The end-to-end delay of a path is equal to the sum of delays of all the links of a path. Clearly, path 3 is not optimal in terms of hop count and/or end-to-end delay parameters, while path 1 is optimal in terms of both hop count and end-to-end delay parameters. Hence, QoS routing has to select a suitable path that meets the QoS constraints specified in the service request made by the user. QoS routing has been described in detail in Section 10.5.1.

Table 10.1. Available paths from node B to node G

No.	Path	Hop Count	End-to-end Bandwidth (Mbps)	End-to-end Delay (milliseconds)
1	$B \rightarrow E \rightarrow G$	2	2	9
2	$B \rightarrow E \rightarrow F \rightarrow G$	3	2	11
3	$B \rightarrow C \rightarrow F \rightarrow G$	3	4	15
4	$B \rightarrow C \rightarrow F \rightarrow E \rightarrow G$	4	3	19
5	$B \rightarrow A \rightarrow D \rightarrow E \rightarrow G$	4	2	23
6	$B \rightarrow A \rightarrow D \rightarrow E \rightarrow F \rightarrow G$	5	2	25

QoS provisioning often requires negotiation between host and network, call admission control, resource reservation, and priority scheduling of packets. QoS can be rendered in ad hoc wireless networks through several ways, namely, per flow, per link, or per node. In ad hoc wireless networks, the boundary between the service provider (network) and the user (host) is not defined clearly, thus making it essential to have better coordination among the hosts to achieve QoS. Characteristics of ad hoc wireless networks such as lack of central coordination, mobility of hosts, and limited availability of resources make QoS provisioning very challenging.

10.1.1 Real-Time Traffic Support in Ad Hoc Wireless Networks

Real-time applications require mechanisms that guarantee bounded delay and delay jitter. The end-to-end delay in packet delivery includes the queuing delay at the source and intermediate nodes, the processing time at the intermediate nodes, and the propagation duration over multiple hops from the source node to the destination node. Real-time applications can be classified as hard real-time applications and soft real-time applications. A hard real-time application requires strict QoS guarantees. Some of the hard real-time applications include nuclear reactor control systems, air traffic control systems, and missile control systems. In these applications, failure to meet the required delay constraints may lead to disastrous results. On the other hand, soft real-time applications can tolerate degradation in the guaranteed QoS to a certain extent. Some of the soft real-time applications are voice telephony, video-on-demand, and video conferencing. In these applications, the loss of data and variation in delay and delay jitter may degrade the service but do not produce hazardous results. Providing hard real-time guarantees in ad hoc wireless networks is extremely difficult due to reasons such as the unrestricted mobility of nodes, dynamically varying network topology, time-varying channel capacity, and the presence of hidden terminals. The research community is currently focusing on providing QoS support for applications that require soft real-time guarantees.

10.1.2 QoS Parameters in Ad Hoc Wireless Networks

As different applications have different requirements, the services required by them and the associated QoS parameters differ from application to application. For example, in case of multimedia applications, bandwidth, delay jitter, and delay are the key QoS parameters, whereas military applications have stringent security requirements. For applications such as emergency search-and-rescue operations, availability of the network is the key QoS parameter. Applications such as group communication in a conference hall require that the transmissions among nodes consume as little energy as possible. Hence, battery life is the key QoS parameter here.

Unlike traditional wired networks, where the QoS parameters are mainly characterized by the requirements of multimedia traffic, in ad hoc wireless networks the QoS requirements are more influenced by the resource constraints of the nodes. Some of the resource constraints are battery charge, processing power, and buffer space.

10.2 ISSUES AND CHALLENGES IN PROVIDING QOS IN AD HOC WIRELESS NETWORKS

Providing QoS support in ad hoc wireless networks is an active research area. Ad hoc wireless networks have certain unique characteristics that pose several difficulties in provisioning QoS. Some of the characteristics are dynamically varying network topology, lack of precise state information, lack of a central controller, error-prone shared radio channel, limited resource availability, hidden terminal problem, and insecure medium. A detailed discussion on how each of the above-mentioned characteristics affects QoS provisioning in ad hoc wireless networks is given below.

- **Dynamically varying network topology:** Since the nodes in an ad hoc wireless network do not have any restriction on mobility, the network topology changes dynamically. Hence, the admitted QoS sessions may suffer due to frequent path breaks, thereby requiring such sessions to be reestablished over new paths. The delay incurred in reestablishing a QoS session may cause some of the packets belonging to that session to miss their delay targets/deadlines, which is not acceptable for applications that have stringent QoS requirements.

- **Imprecise state information:** In most cases, the nodes in an ad hoc wireless network maintain both the link-specific state information and flow-specific state information. The link-specific state information includes bandwidth, delay, delay jitter, loss rate, error rate, stability, cost, and distance values for each link. The flow-specific information includes session ID, source address, destination address, and QoS requirements of the flow (such as maximum bandwidth requirement, minimum bandwidth requirement, maximum delay, and maximum delay jitter). The state information is inherently imprecise due to dynamic changes in network topology and channel characteristics. Hence, routing decisions may not be accurate, resulting in some of the real-time packets missing their deadlines.

- **Lack of central coordination:** Unlike wireless LANs and cellular networks, ad hoc wireless networks do not have central controllers to coordinate the activity of nodes. This further complicates QoS provisioning in ad hoc wireless networks.

- **Error-prone shared radio channel:** The radio channel is a broadcast medium by nature. During propagation through the wireless medium, the radio waves suffer from several impairments such as attenuation, multipath propagation, and interference (from other wireless devices operating in the vicinity) as discussed in Chapter 1.

- **Hidden terminal problem:** The hidden terminal problem is inherent in ad hoc wireless networks. This problem occurs when packets originating from two or more sender nodes, which are not within the direct transmission range of each other, collide at a common receiver node. It necessitates the retransmission of the packets, which may not be acceptable for flows that have stringent QoS requirements. The RTS/CTS control packet exchange mechanism, proposed in [1] and adopted later in the IEEE 802.11 standard [2], reduces the hidden terminal problem only to a certain extent. BTMA and DBTMA provide two important solutions for this problem, which are described in Chapter 6.

- **Limited resource availability:** Resources such as bandwidth, battery life, storage space, and processing capability are limited in ad hoc wireless networks. Out of these, bandwidth and battery life are critical resources, the availability of which significantly affects the performance of the QoS provisioning mechanism. Hence, efficient resource management mechanisms are required for optimal utilization of these scarce resources.

- **Insecure medium:** Due to the broadcast nature of the wireless medium, communication through a wireless channel is highly insecure. Therefore, security is an important issue in ad hoc wireless networks, especially for military and tactical applications. Ad hoc wireless networks are susceptible to attacks such as eavesdropping, spoofing, denial of service, message distortion, and impersonation. Without sophisticated security mechanisms, it is very difficult to provide secure communication guarantees.

Some of the design choices for providing QoS support are described below.

- **Hard state versus soft state resource reservation:** QoS resource reservation is one of the very important components of any QoS framework (a QoS framework is a complete system that provides required/promised services to each user or application). It is responsible for reserving resources at all intermediate nodes along the path from the source to the destination, as requested by the QoS session. QoS resource reservation mechanisms can be broadly classified into two categories: *hard state* and *soft state* reservation mechanisms. In hard state resource reservation schemes, resources are reserved at all intermediate nodes along the path from the source to the destination throughout the duration of the QoS session. If such a path is broken due to network dynamics, these reserved resources have to be explicitly released by a deallocation mechanism. Such a mechanism not only introduces additional control overhead, but may also fail to release resources completely in case a node previously belonging to the session becomes unreachable. Due to these problems, soft state resource reservation mechanisms, which maintain reservations only for small time intervals, are used. These reservations get refreshed if packets belonging to the same flow are received before the timeout period. The soft state reservation timeout period can be equal to packet inter-arrival time or a multiple of the packet inter-arrival time. If no data packets are received for the specified time interval, the resources are deallocated in a decentralized manner without incurring any additional control overhead. Thus no explicit teardown is required for a flow. The hard state schemes reserve resources explicitly and hence, at high network loads, the call blocking ratio will be high, whereas soft state schemes provide high call acceptance at a gracefully degraded fashion.

- **Stateful versus stateless approach:** In the stateful approach, each node maintains either *global state* information or only *local state* information, while in the case of a stateless approach, no such information is maintained at the nodes. State information includes both the topology information and the flow-specific information. If global state information is available, the source node can use a centralized routing algorithm to route packets to the destination. The performance of the routing protocol depends on the accuracy of the global state information maintained at the nodes. Significant control overhead is incurred in gathering and maintaining global state information. On the other hand, if mobile nodes maintain only local state information (which is more accurate), distributed routing algorithms can be used. Even though control overhead incurred in maintaining local state information is low, care must be taken to obtain loop-free routes. In the case of the stateless approach, neither flow-specific nor link-specific state information is maintained at the nodes. Though the stateless approach solves the scalability problem permanently and reduces the burden (storage and computation) on nodes, providing QoS guarantees becomes extremely difficult.

- **Hard QoS versus soft QoS approach:** The QoS provisioning approaches can be broadly classified into two categories: *hard QoS* and *soft QoS* approaches. If QoS requirements of a connection are guaranteed to be met for the whole duration of the session, the QoS approach is termed a hard QoS approach. If the QoS requirements are not guaranteed for the entire session, the QoS approach is termed a soft QoS approach. Keeping network dynamics of ad hoc wireless networks in mind, it is very difficult to provide hard QoS guarantees to user applications. Thus, QoS guarantees can be given only within certain statistical bounds. Almost all QoS approaches available in the literature provide only soft QoS guarantees.

10.3 CLASSIFICATIONS OF QOS SOLUTIONS

The QoS solutions can be classified in two ways. One classification is based on the QoS approach employed, while the other one classifies QoS solutions based on the layer at which they operate in the network protocol stack.

10.3.1 Classifications of QoS Approaches

As shown in Figure 10.2, several criteria are used for classifying QoS approaches. The QoS approaches can be classified based on the interaction between the routing protocol and the QoS provisioning mechanism, based on the interaction between the network and the MAC layers, or based on the routing information update mechanism. Based on the interaction between the routing protocol and the QoS provisioning mechanism, QoS approaches can be classified into two categories: *coupled* and

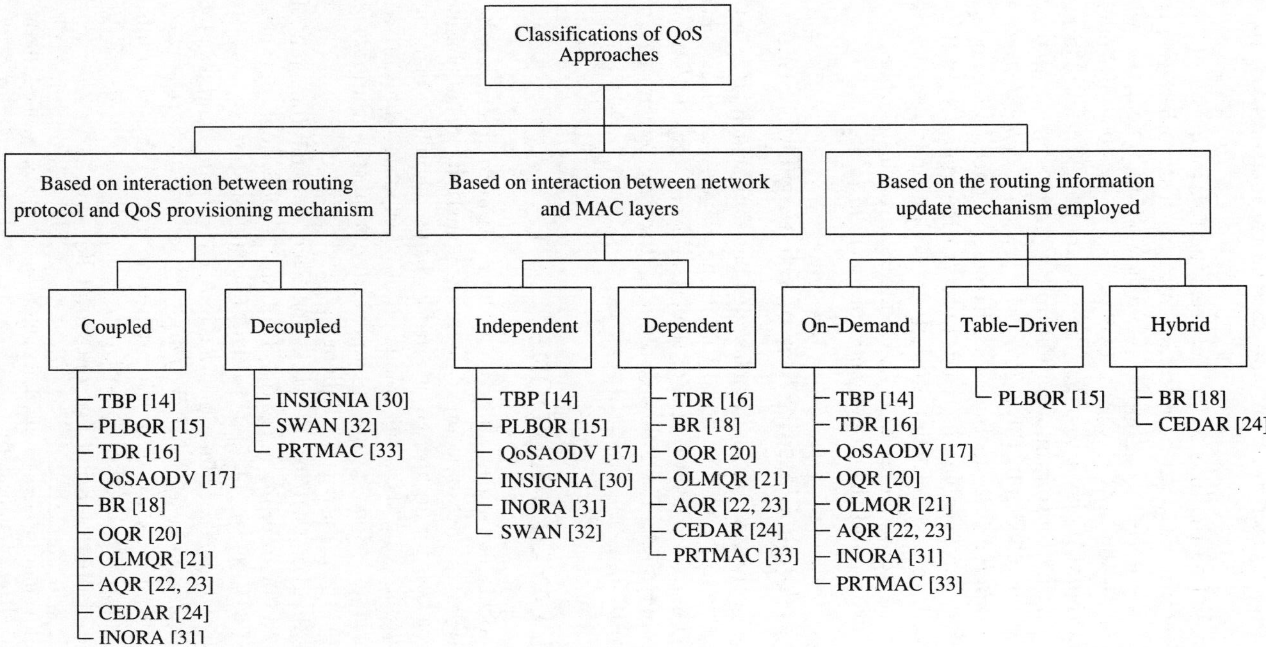

Figure 10.2. Classifications of QoS approaches.

decoupled QoS approaches. In the case of the coupled QoS approach, the routing protocol and the QoS provisioning mechanism closely interact with each other for delivering QoS guarantees. If the routing protocol changes, it may fail to ensure QoS guarantees. But in the case of the decoupled approach, the QoS provisioning mechanism does not depend on any specific routing protocol to ensure QoS guarantees.

Similarly, based on the interaction between the routing protocol and the MAC protocol, QoS approaches can be classified into two categories: *independent* and *dependent* QoS approaches. In the independent QoS approach, the network layer is not dependent on the MAC layer for QoS provisioning. The dependent QoS approach requires the MAC layer to assist the routing protocol for QoS provisioning. Finally, based on the routing information update mechanism employed, QoS approaches can be classified into three categories, namely, *table-driven*, *on-demand*, and *hybrid* QoS approaches. In the table-driven approach, each node in the network maintains a routing table which aids in forwarding packets. In the on-demand approach, no such tables are maintained at the nodes, and hence the source node has to discover the route on the fly. The hybrid approach incorporates features of both the table-driven and the on-demand approaches.

10.3.2 Layer-Wise Classification of Existing QoS Solutions

The existing QoS solutions can also be classified based on which layer in the network protocol stack they operate in. Figure 10.3 gives a layer-wise classification of QoS solutions. The figure also shows some of the cross-layer QoS solutions proposed for ad hoc wireless networks. The following sections describe the various QoS solutions listed in Figure 10.3.

10.4 MAC LAYER SOLUTIONS

The MAC protocol determines which node should transmit next on the broadcast channel when several nodes are competing for transmission on that channel. The existing MAC protocols for ad hoc wireless networks use channel sensing and random back-off schemes, making them suitable for best-effort data traffic. Real-time traffic (such as voice and video) requires bandwidth guarantees. Supporting real-time traffic in these networks is a very challenging task.

In most cases, ad hoc wireless networks share a common radio channel operating in the ISM band[2] or in military bands. The most widely deployed medium access technology is the IEEE 802.11 standard [2]. The 802.11 standard has two modes of operation: a distributed coordination function (DCF) mode and a point coordination function (PCF) mode. The DCF mode provides best-effort service, while the PCF mode has been designed to provide real-time traffic support in infrastructure-based wireless network configurations. Due to lack of fixed infrastructure support, the PCF mode of operation is ruled out in ad hoc wireless networks. Currently,

[2]ISM refers to the industrial, scientific, and medical band. The frequencies in this band (from 2.4 GHz to 2.4835 GHz) are unlicensed.

the IEEE 802.11 Task Group e (TGe) is enhancing the legacy 802.11 standard to support real-time traffic. The upcoming 802.11e standard has two other modes of operation, namely, enhanced DCF (EDCF) and hybrid coordination function (HCF) to support QoS in both infrastructure-based and infrastructure-less network configurations. These two modes of operation are discussed later in this section. In addition to these standardized MAC protocols, several other MAC protocols that provide QoS support for applications in ad hoc wireless networks have been proposed. Some of these protocols are described below.

10.4.1 Cluster TDMA

Gerla and Tsai proposed Cluster TDMA [3] for supporting real-time traffic in ad hoc wireless networks. In bandwidth-constrained ad hoc wireless networks, the limited resources available need to be managed efficiently. To achieve this goal, a dynamic clustering scheme is used in Cluster TDMA. In this clustering approach, nodes are split into different groups. Each group has a cluster-head (elected by members of that group), which acts as a regional broadcast node and as a local coordinator to enhance the channel throughput. Every node within a cluster is one hop away from the cluster-head. The formation of clusters and selection of cluster-heads are done in a distributed manner. Clustering algorithms split the nodes into clusters so that they are interconnected and cover all the nodes. Three such algorithms used are lowest-ID algorithm, highest-degree (degree refers to the number of neighbors which are within transmission range of a node) algorithm, and least cluster change (LCC) algorithm. In the lowest-ID algorithm, a node becomes a cluster-head if it has the lowest ID among all its neighbors. In the highest-degree algorithm, a node with a

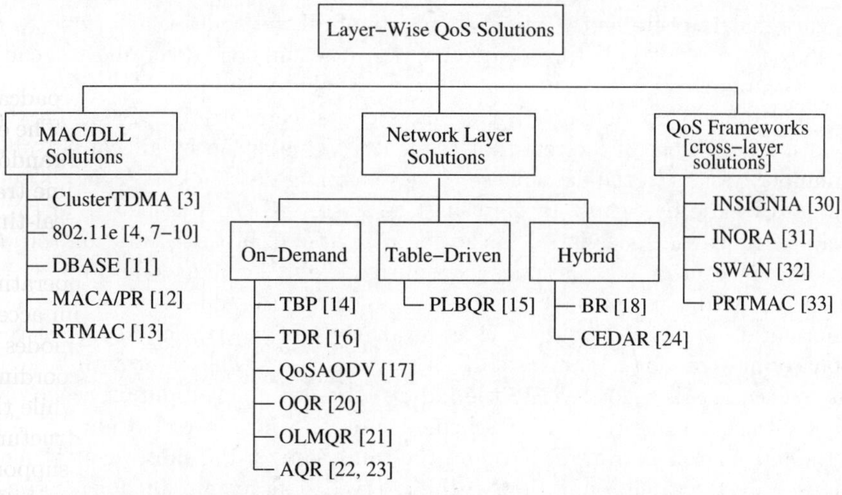

Figure 10.3. Layer-wise classification of QoS solutions.

degree greater than the degrees of all its neighbors becomes the cluster-head. In the LCC algorithm, cluster-head change occurs only if a change in network causes two cluster-heads to come into one cluster or one of the nodes moves out of the range of all the cluster-heads.

The time division multiple access (TDMA) scheme is used within a cluster for controlling access to the channel. Further, it is possible for multiple sessions to share a given TDMA slot via code division multiple access (CDMA). Across clusters, either spatial reuse of the time-slots or different spreading codes can be used to reduce the effect of inter-cluster interference. A synchronous time division frame is defined to support TDMA access within a cluster and to exchange control information. Each synchronous time division frame is divided into slots. Slots and frames are synchronized throughout the network. A frame is split into a control phase and a data phase. In the control phase, control functions such as frame and slot synchronization, routing, clustering, power management, code assignment, and virtual circuit (VC) setup are done.

The cluster-head does the reservation for the VC by assigning the slot(s) and code(s) to be used for that connection. The number of slots per frame to be assigned to a VC is determined by the bandwidth requirement of the VC. Each station broadcasts the routing information it has, the ID of its cluster-head, the power gain[3] list (the power gain list consists of the power gain values corresponding to each of the single-hop neighbors of the node concerned) it maintains, reservation status of the slots present in its data phase, and ACKs for frames that are received in the last data phase. Upon receiving this information, a node updates its routing table, calculates power gains for its neighbors, updates the power gain matrix, selects its cluster-head, records the slot reservation status of its neighbors, obtains ACKs for frames that are transmitted in the last data phase, and reserves slot(s). In each cluster, the corresponding cluster-head maintains a power gain matrix. The power gain matrix contains the power gain lists of all the nodes that belong to a particular cluster. It is useful for controlling the transmission power and the code division within a cluster.

The data phase supports both real-time and best-effort traffic. Based on the bandwidth requirement of the real-time session, a VC is set up by allocating sufficient number of slots in the data phase. The remaining data slots (*i.e.*, free slots) can be used by the best-effort traffic using the slotted-ALOHA scheme. For each node, a predefined slot is assigned in the control phase to broadcast its control information. The control information is transmitted over a common code throughout the network. At the end of the control phase, each node would have learned from the information broadcast by the cluster-head, the slot reservation status of the data phase and the power gain lists of all its neighbors. This information helps a node to schedule free slots, verify the failure of reserved slots, and drop expired real-time packets. A fast reservation scheme is used in which a reservation is made when the first packet is transmitted, and the same slots in the subsequent frames can be used for the same connection. If the reserved slots remain idle for a certain timeout period, then they are released.

[3]Power gain is the power propagation loss from the transmitter to the receiver.

10.4.2 IEEE 802.11e

In this section, the IEEE 802.11 MAC protocol is first described. Then, the recently proposed mechanisms for QoS support, namely, enhanced distributed coordination function (EDCF) and hybrid coordination function (HCF), defined in the IEEE 802.11e draft, are discussed.

IEEE 802.11 MAC Protocol

The 802.11 MAC protocol [2], which is discussed in Chapter 2, describes how a station present in a WLAN should access the broadcast channel for transmitting data to other stations. It supports two modes of operation, namely, distributed coordination function (DCF) and point coordination function (PCF). The DCF mode does not use any kind of centralized control, while the PCF mode requires an access point (AP, *i.e.*, central controller) to coordinate the activity of all nodes in its coverage area. All implementations of the 802.11 standard for WLANs must provide the DCF mode of operation, while the PCF mode of operation is optional.

The time interval between the transmission of two consecutive frames is called the inter-frame space (IFS). There are four IFSs defined in the IEEE 802.11 standard, namely, short IFS (SIFS), PCF IFS (PIFS), DCF IFS (DIFS), and extended IFS (EIFS). The relationship among them is as follows:

$$SIFS < PIFS < DIFS < EIFS$$

Distributed Coordination Function

In the DCF mode, all stations are allowed to contend for the shared medium simultaneously. CSMA/CA mechanism and random back-off scheme are used to reduce frame collisions. Each unicast frame is acknowledged immediately after being received. If the acknowledgment is not received within the timeout period, the data frame is retransmitted. Broadcast frames do not require acknowledgments from the receiving stations.

If a station A wants to transmit data to station B, station A listens to the channel. If the channel is busy, it waits until the channel becomes idle. After detecting the idle channel, station A further waits for a DIFS period and invokes a back-off procedure. The back-off time is given by

$$\text{Back} - \text{off Time} = rand(0, CW) \times slottime$$

where *slottime* includes the time needed for a station to detect a frame, the propagation delay, the time needed to switch from the receiving state to the transmitting state, and the time to signal to the MAC layer the state of the channel. The function $rand(0, CW)$ returns a pseudo-random integer from a uniform distribution over an interval $[0,\ CW]$. The current value of the contention window (CW) plays an important role in determining the back-off period of the station. The initial value of CW is CW_{min}. If a collision occurs, the value of CW is doubled. As the number of collisions increases, the value of CW is increased exponentially in order to reduce the chance of collision occurrence. The maximum value of CW is CW_{max}. The values of CW_{min} and CW_{max} specified by the IEEE 802.11 standard are presented in Chapter 2.

After detecting the channel as being idle for a DIFS period, station A starts decrementing the back-off counter. If it senses the channel as busy during this count-down process, it suspends the back-off counter till it again detects the channel as being idle for a DIFS period. Station A then continues the count-down process, where it suspended the back-off counter. Once the back-off counter reaches zero, station A transmits a request-to-send (RTS) frame and waits for a clear-to-send (CTS) frame from the receiver B. If other stations do not cause any interference, station B acknowledges the RTS frame by sending a CTS frame. Upon receiving the CTS frame, station A transmits its data frame, the reception of which is acknowledged by receiver B by sending an ACK frame. In the above scenario, if another station C apart from station A also senses the channel as being idle (*i.e.*, stations A and C sense the channel as being idle and the back-off counters set by them expire at the same time) and transmits an RTS frame, a collision occurs and both the stations initiate back-off procedures.

If the size of the MAC frame, that is, MAC service data unit (MSDU),[4] is greater than the fragmentation threshold, it is fragmented into smaller frames, that is, MAC protocol data units (MPDUs),[5] before transmission, and each MPDU has to be acknowledged separately. Once an MSDU is transmitted successfully, CW is reset to CW_{min}. The RTS/CTS control frame exchange helps in reducing the hidden terminal problem inherent in CSMA-based ad hoc wireless networks.

Point Coordination Function

The IEEE 802.11 standard incorporates an optional access method known as PCF to let stations have priority access to the wireless medium. This access method uses a point coordinator (PC), which operates at an AP. Hence PCF is usable only in infrastructure-based network configurations. A station which requires the PCF mode of operation sends an association message to the PC to register in its polling list and gets an association identifier (AID). The PC polls the stations registered in its polling list in ascending order of AIDs to allow them contention-free access to the medium. The role of the PC is to determine which station should gain access to the channel. The stations requesting the PCF mode of operation get associated with the PC during the contention period (CP). With PCF, the channel access alternates between the contention-free period (CFP) and the contention period (CP) for the PCF and DCF modes of operation, respectively.

A CFP and the following CP form a superframe. The PC generates a beacon frame at regular beacon frame intervals called target beacon transmission time (TBTT). The value of TBTT is announced in the beacon frame. Each superframe starts with a beacon frame, which is used to maintain synchronization among local timers in the stations and to deliver protocol-related parameters. The PC uses contention-free poll (*CF-Poll*) packets to ask stations to transmit their frames. A

[4]MSDU is the information that is delivered as a unit between MAC service access points.

[5]MPDU is the unit of data exchanged between two peer MAC entities using the services of the physical layer.

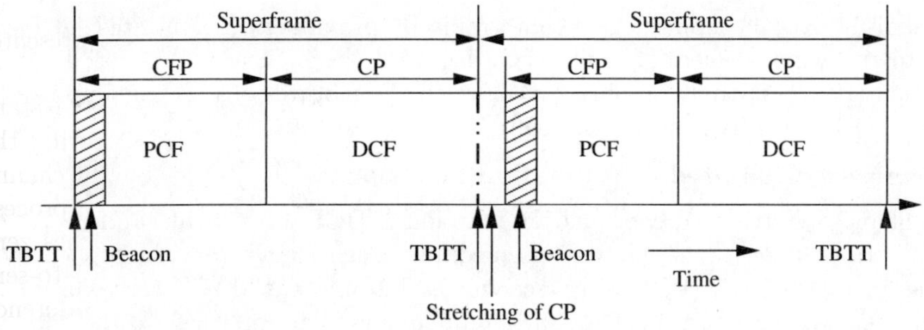

Figure 10.4. PCF and DCF frame sharing.

station that is able to respond to *CF-Poll* frames is said to be *CF-Pollable*. It is optional for a *CF-Pollable* station to respond to a *CF-Poll* frame received from the PC. If the PC receives no response from the polled station for a PIFS period, it polls the next station in the polling list (in case the remaining duration of CFP is long enough for at least one CFP transmission) or ends the CFP by transmitting *CF-End* control frame. The PC and the *CF-Pollable* stations do not use the RTS/CTS control frame exchange in the CFP. Figure 10.4 shows the operation of the network in the combined PCF and DCF modes. The channel access switches alternately between the PCF mode and the DCF mode, but the CFP may shrink due to stretching when DCF takes more time than expected. This happens when an MSDU is fragmented into several MPDUs, hence giving priority to these fragments over the PCF mode of operation.

PCF has certain shortcomings which make it unsuitable for supporting real-time traffic [4]. At TBTT, the PC has to sense the medium idle for at least PIFS before transmitting the beacon frame. If the medium is busy around TBTT, the beacon is delayed, thereby delaying the transmission of real-time traffic that has to be delivered in the following CFP. Further, polled stations' transmission durations are unknown to the PC. The MAC frame (*i.e.*, MSDU) of the polled station may have to be fragmented and may be of arbitrary length. Further, the transmission time of an MSDU is not under the control of the PC because of different modulation and coding schemes specified in the IEEE 802.11 standard. PCF is not scalable to support real-time traffic for a large number of users, as discussed in [5] and [6]. Due to these reasons, several mechanisms have been proposed to enhance the IEEE 802.11 standard to provide QoS support. The QoS mechanisms that are proposed as part of the IEEE 802.11e standard are described below.

QoS Support Mechanisms of IEEE 802.11e

The IEEE 802.11 WLAN standard supports only best-effort service. The IEEE 802.11 Task Group e (TGe) has been set up to enhance the current 802.11 MAC protocol so that it is able to support multimedia applications. The TGe has chosen the virtual DCF (VDCF) [7] proposal as the enhanced DCF (EDCF) access mechanism. EDCF supports real-time traffic by providing differentiated DCF access to the wireless medium. The TGe has also specified a hybrid coordination function (HCF) [8] that combines EDCF with the features of PCF to simplify the QoS provisioning. HCF operates during both the CFP and the CP.

Enhanced Distributed Coordination Function

Enhanced distributed coordination function (EDCF) [7] provides differentiated and distributed access to the wireless medium. Each frame from the higher layer carries its user priority (UP). After receiving each frame, the MAC layer maps it into an access category (AC). Each AC has a different priority of access to the wireless medium. One or more UPs can be assigned to each AC. EDCF channel access has up to eight ACs [9], to support UPs. EDCF supports eight UPs. Similar to the DCF, each AC has a set of access parameters, such as CW_{min}, CW_{max}, $AIFS$, and transmission opportunity (TXOP) limit, which would be described later in this section. Hence, each AC is an enhanced variant of the DCF. Flows that fall under the same AC are effectively given identical priority to access the channel. A station accesses the channel based on the AC of the frame to be transmitted. An access point that provides QoS is called QoS access point (QAP). Each QAP will provide at least four ACs. Each station contends for transmission opportunities (TXOPs) using a set of EDCF channel access parameters that are unique to the AC of the packet to be transmitted. The TXOP is defined as an interval of time during which a station has the right to initiate transmissions. It is characterized by a starting time and a maximum duration called TXOPLimit. Depending on the duration of TXOP, a station may transmit one or more MSDUs. Priority of an AC refers to the lowest UP assigned to that AC.

During CP, each AC (of priority i) of the station contends for a TXOP and independently starts a back-off counter after detecting the channel being idle for an arbitration inter-frame space ($AIFS[i]$) as specified in [10]. $AIFS[i]$ is set as given below.

$$AIFS[i] = SIFS + AIFSN[i] \times \textit{slottime}$$

where $AIFSN[i]$ is the *AIFS slot count* (*i.e.*, the number of time-slots a station has to sense the channel as idle before initiating the back-off process) for priority class i and takes values greater than zero. For high-priority classes, low $AIFSN$ values are assigned to give higher priorities for them. After waiting for $AIFS[i]$, each back-off counter is set to a random integer drawn from the range:

$$[1,\ CW[i] + 1]\ \textit{for each class } i \textit{ with } AIFSN[i] = 1;$$

$$[0,\ CW[i]]\ \textit{for other classes } i \textit{ with } AIFSN[i] > 1.$$

The reason for having a different range for classes with $AIFSN[i] = 1$ is to avoid transmissions initiated by stations that are operating in the EDCF mode from colliding with the hybrid coordinator's (HC, which is explained later in this section) poll packets. The HC operates at QAP and controls QoS basic service set (QBSS)

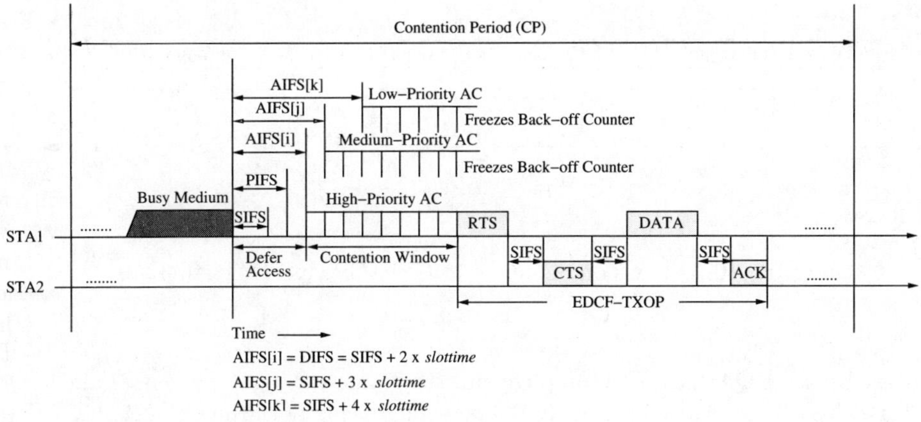

Figure 10.5. An example of EDCF access mechanism.

operation under the HCF. Figure 10.5 illustrates the relationship between SIFS, PIFS, DIFS, and various AIFS values. As in legacy DCF, if a station detects the channel to be busy before the back-off counter reaches zero, the back-off counter is suspended. The station has to wait for the channel to become idle again for an AIFS period, before continuing to decrement the counter. In this figure, it is assumed that station $STA1$ has traffic that belongs to three different ACs. The back-off counter of the highest priority AC expires first, which causes the corresponding AC to seize an EDCF-TXOP for initiating data transmission. The other ACs suspend their back-off counters and wait for the channel to become idle again. When the back-off counter of a particular AC reaches zero, the corresponding station initiates a TXOP and transmits frame(s) that have the highest priority. TXOPs are allocated via contention (EDCF-TXOP) or granted through HCF (polled-TXOP) [4]. The duration of EDCF-TXOP is limited by a QBSS-wide TXOPLimit transmitted in beacons by the HC, while during the CFP the starting time and maximum duration of each polled-TXOP is specified in the corresponding *CF-Poll* frame by the HC. If the back-off counters of two or more ACs in a single station reach zero at the same time, a scheduler inside the station avoids the *virtual collision* by granting the TXOP to the highest priority AC, while low-priority ACs behave as if there was an external collision on the wireless medium.

Hybrid Coordination Function

The hybrid coordination function (HCF) [8] combines features of EDCF and PCF to provide the capability of selectively handling MAC service data units (MSDUs), in a manner that has upward compatibility with both DCF and PCF. It uses a common set of frame exchange sequences during both the CP and the CFP. The HCF is usable only in infrastructure-based BSSs that provide QoS, that is, QBSSs. The HCF uses a QoS-aware point coordinator, called HC, which is typically col-

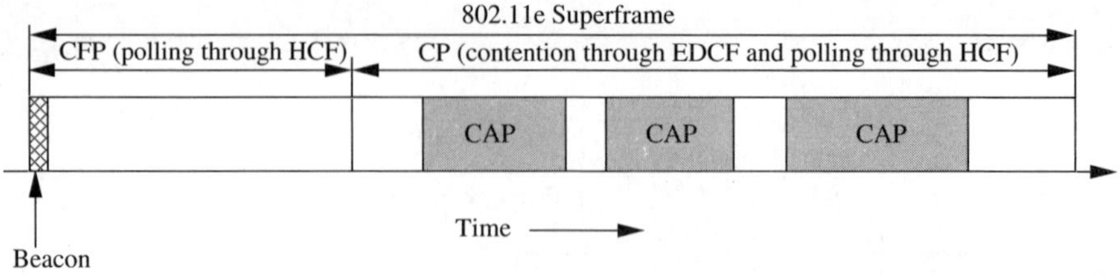

Figure 10.6. Division of time into CFP, CP, and CAP intervals.

located with a QAP. The HC implements the frame exchange sequences and the MSDU handling rules defined in HCF, operating during both the CP and the CFP. It allocates TXOPs to stations and initiates controlled contention periods for the stations to send reservation requests. When the HC needs access to the wireless medium, it senses the medium. If the medium remains idle for a PIFS period, it initiates MSDU deliveries. The HC can start contention-free controlled access periods (CAPs) at any time during a CP, after the medium is determined to be idle for at least one PIFS period.

A CAP may include one or more TXOPs. During the CAP, the HC may transmit frames and issue polls to stations which grant them TXOPs. At the end of the TXOP or when the station has no more frames to transmit, it explicitly hands over control of the medium back to the HC. Figure 10.6 shows an example of a superframe divided into CFP, CP, and three CAP intervals. During CP, each TXOP begins either when the medium is determined to be available under the EDCF rules (EDCF-TXOP) or when the station receives a QoS *CF-Poll* frame from the HC (polled-TXOP).

Figure 10.7 illustrates CFP in the HCF mode of operation. During CFP, the HC grants TXOPs to stations by sending QoS *CF-Poll* frames. The polled station can transmit one or more MSDUs in the allocated TXOP. If the size of an MSDU is too large, it can be divided into two or more fragments and transmitted sequentially with SIFS waiting periods in between them. These fragments have to be acknowledged individually. The CFP ends after the time announced in the beacon frame or by a *CF-End* frame from the HC.

10.4.3 DBASE

The distributed bandwidth allocation/sharing/extension (DBASE) protocol [11] supports multimedia traffic [both variable bit rate (VBR) and constant bit rate (CBR)] over ad hoc WLANs. In an ad hoc WLAN, there is no fixed infrastructure (*i.e.*, AP) to coordinate the activity of individual stations. The stations are part of a single-hop wireless network and contend for the broadcast channel in a distributed manner. For real-time traffic (*rt-traffic*), a contention-based process is used in order to gain access to the channel. Once a station gains channel access, a reservation-based process is used to transmit the subsequent frames. The non-real-

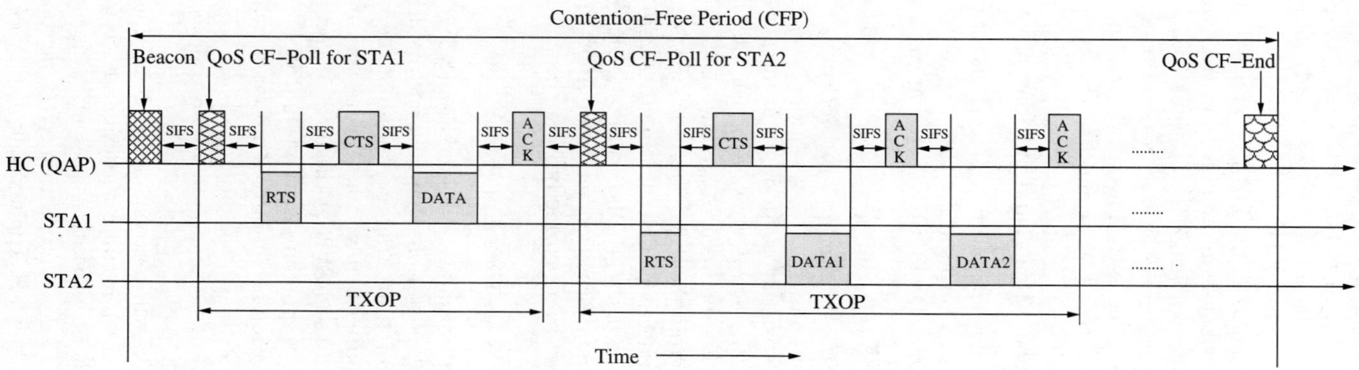

Figure 10.7. An example of HCF access mechanism.

time stations (*nrt*-stations) regulate their accesses to the channel according to the standard CSMA/CA protocol used in 802.11 DCF. The DBASE protocol permits real-time stations (*rt*-stations) to acquire excess bandwidth on demand. It is still compliant with the IEEE 802.11 standard [2].

Like the IEEE 802.11 standard, the DBASE protocol divides the frames into three priority classes. Frames belonging to different priority classes have to wait for different IFSs before they are transmitted. Stations have to wait for a minimum of PIFS before transmitting *rt*-frames such as reservation frame (RF) and request-to-send (RTS). The *nrt*-frames have the lowest priority, and hence stations have to wait for DIFS before transmitting such frames.

The Access Procedure for Non Real-Time Stations

The channel access method for *nrt*-stations is based on conventional DCF. An *nrt*-station with data traffic has to keep sensing the channel for an additional random time called data back-off time (DBT) after detecting the channel as being idle for a DIFS period. The DBT is given by

$$DBT = rand(a,\ b) \times slottime$$

The function $rand(a,\ b)$ returns a pseudo-random integer from a uniform distribution over an interval $[a,\ b]$, where b grows exponentially for each retransmission attempt, and the range of b is between b_{min} and b_{max}. DBASE adopts the contention window parameters from the IEEE 802.11 DSSS specification. If the channel is idle, the DBT counter is decremented till it reaches zero, but it is frozen while the channel becomes busy. Once the DBT counter reaches zero, the *nrt*-station transmits its *nrt*-frame. The destination sends an ACK to the source after SIFS period after receiving the *nrt*-frame correctly from the source.

The Access Procedure for Real-Time Stations

Each *rt*-station maintains a virtual reservation table (RSVT). In this virtual table, the information regarding all *rt*-stations that have successfully reserved the required bandwidth is recorded. Before initiating an *rt*-session, the *rt*-station sends an RTS in order to reserve the required bandwidth. Before transmitting the RTS, a corresponding entry is made in the RSVT of the node. Every station that hears this RTS packet also makes a corresponding entry in its RSVT. After recording into the RSVT successfully, an *rt*-station need not contend for the channel any more during its whole session.

Bandwidth Reservation

One of the *rt*-stations takes the responsibility of initiating the contention-free period (CFP) periodically. Such an *rt*-station is designated as CFP generator (CFPG). The CFP is utilized by the active *rt*-stations present in the network to transmit their *rt*-frames. The CFPG issues a reservation frame (RF) periodically and has the right to send its *rt*-frame first in the CFP. The maximum delay between any two consecutive RFs is D_{max}, where D_{max} is the minimum of maximum delay bounds among all active *rt*-connections. The RF is a broadcast frame that announces the beginning of the CFP. The RF contains the information about the number of active *rt*-stations and the information about all *rt*-stations recorded in the RSVT of the CFPG.

Assume that at time t an *rt*-station wants to transmit data. Then it monitors the channel for detecting the RF during the interval $(t,\ t + D_{max})$. If the *rt*-station detects the RF, it waits until the CFP finishes. After the CFP finishes, the *rt*-station keeps sensing the channel for a period of real-time back-off time (RBT) after detecting the channel as being idle for a PIFS period. The RBT of an *rt*-station is given by

$$RBT = rand(c,\ d) \times slottime$$

where $rand(c,\ d)$ returns a pseudo-random integer from a uniform distribution over an interval $[c,\ d]$. The values of c and d are set to 0 and 3, respectively. If the channel is idle, the RBT counter is decremented till it reaches zero, but it is frozen while the medium is sensed busy. Once the RBT counter reaches zero, the *rt*-station contends for its reservation by sending an RTS packet. If no collision occurs, it updates its tables and transmits its first *rt*-frame. If a collision occurs, the P-persistent scheme is used to resolve the contention. The *rt*-station involved in collision retransmits the RTS in the next time-slot (*i.e., slottime*) with a probability P. With probability $(1 - P)$, it defers for at least one time-slot and recalculates the RBT using the following equation:

$$RBTP = rand(c + 1,\ d) \times slottime$$

where RBTP is the recalculated RBT for the P-persistent scheme.

If an RF is not received during the interval $(t,\ t + D_{max})$, it means that there are no active *rt*-stations. If the channel is still idle in the interval $(t + D_{max} + \delta,\ t + D_{max} + \delta + PIFS)$ and no RF is detected, the *rt*-station that wants to transmit data at time instant t will execute the back-off scheme. Here δ represents the remaining transmitting time of the current frame at the time instant $t + D_{max}$. During the back-off process, the *rt*-station should keep monitoring the channel to check whether any *rt*-station has started acting as the CFP generator. If RBT reaches zero, the *rt*-station sends an RTS frame to the receiver. If no collision occurs, it gets CTS from the receiver and plays the role of CFPG in the network. If a collision occurs, the P-persistent scheme as mentioned above is used to decide when the stations are to transmit again.

The bandwidth reservation scheme is illustrated in Figure 10.8. Figure 10.8 (a) depicts a case in which no collision occurs, while Figure 10.8 (b) shows a scenario in which a collision occurs. In Figure 10.8 (a), stations A and C have *rt*-frames for transmission to stations B and D, respectively. Besides these, station E has *nrt*-frames to be transmitted to station D. After listening to the channel for D_{max} time period in order to detect the presence of an RF, stations A and C conclude that no CFPG exists in the network. Then, if they find the channel as being idle for a PIFS period, they initiate their back-off timers. In this case, assume that RBT_A is one slot and RBT_C is three slots. During the back-off process, once the channel

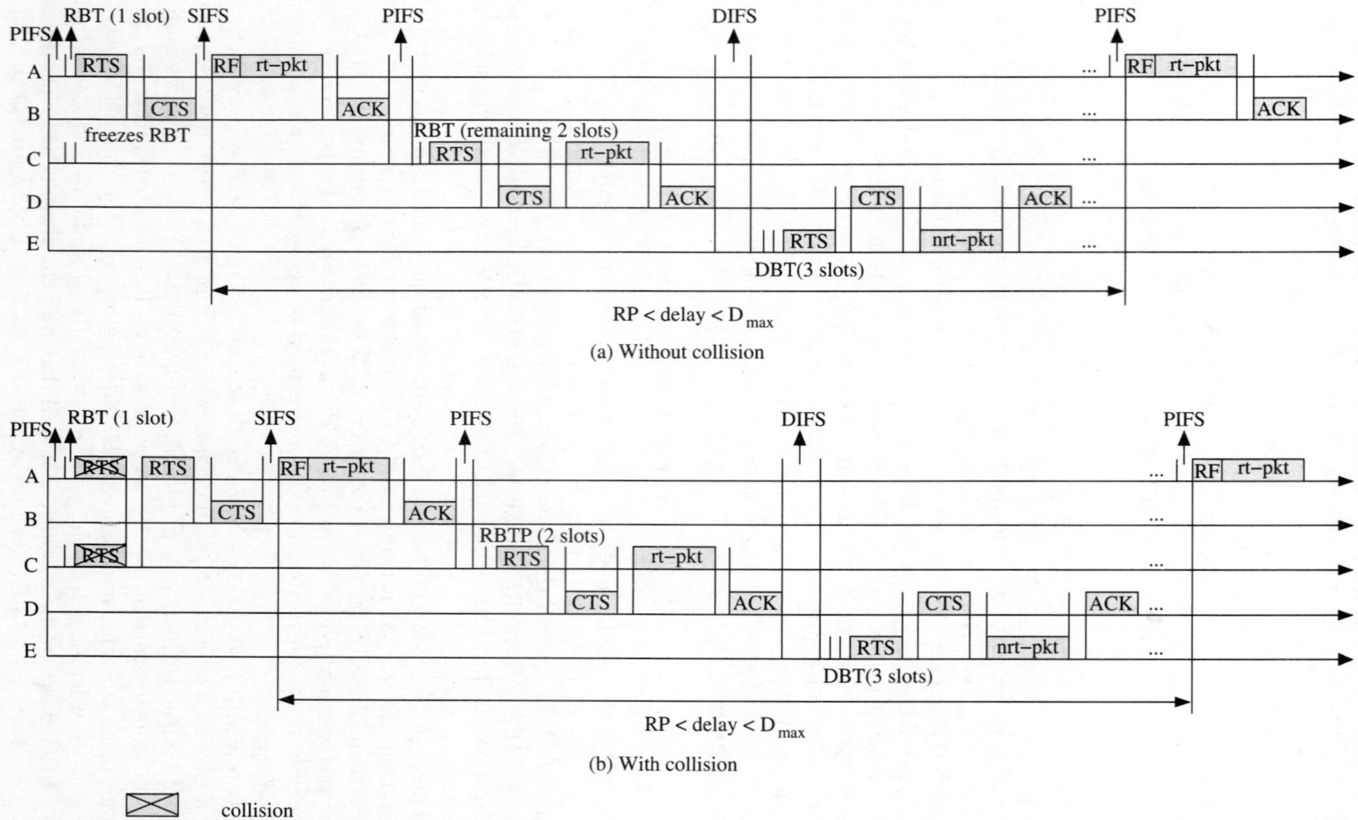

Figure 10.8. An example of new *rt*-stations joining the network.

becomes busy, the back-off timer of station C is paused as shown in Figure 10.8 (a). It is restarted from the same value once the channel becomes idle again. After RBT_A counts down to zero, station A seizes the channel and sends an RTS. When station A starts transmitting, station C pauses its back-off counter. If no collision occurs, station A receives a CTS within SIFS time duration. Then station A records its reservation information into the RSVT and becomes the CFPG. Since station A is currently playing the role of CFPG, it transmits an RF before transmitting its first *rt*-frame. Once station A completes its transmission, station C continues its back-off process. When RBT_C counts down to zero, station C reserves bandwidth by adding a corresponding entry into the RSVT and transmits its first *rt*-frame. When station E detects the channel as being idle for DIFS, it implies that no other *rt*-station wants to transmit currently, and hence station E sends its RTS as soon as DBT_E counts down to zero. By the end of a contention period whose length is limited by a parameter RP_{max} (maximum repetition period), bandwidth would be reserved for the *rt*-stations, and thereafter they need not exchange RTS/CTS control frames before transmitting their *rt*-frames. The delay between two RFs varies from real-time period (RP) to D_{max}, where RP is the sum of the CFP (*rt*-stations reserved period) and the CP for new *rt*-stations.

In Figure 10.8 (b), assume that both station A and station C generate RBT as one slot. After waiting for one time-slot, both transmit their RTS frames, which results in a collision. Then the P-persistent scheme is applied. Assume that station A gets access to the channel during the next slot itself, but station C does not. Then, station A will retransmit its RTS in the following slot, while station C initiates a new back-off time $RBTP_C$. If no collision occurs, station A gets a CTS within SIFS, and sends out an RF and its *rt*-frame. When $RBTP_C$ counts down to zero, station C seizes the channel to send an RTS. If any collision occurs, the *rt*-station uses the P-persistent scheme to resolve the collision. The collision resolution process is restricted from crossing the RP_{max} boundary.

The MAC layer solutions such as MACA/PR [12] and RTMAC [13] provide real-time traffic support in asynchronous ad hoc wireless networks. These solutions are discussed in Chapter 6.

10.5 NETWORK LAYER SOLUTIONS

The bandwidth reservation and real-time traffic support capability of MAC protocols can ensure reservation at the link level only, hence the network layer support for ensuring end-to-end resource negotiation, reservation, and reconfiguration is very essential. This section describes the existing network layer solutions that support QoS provisioning.

10.5.1 QoS Routing Protocols

QoS routing protocols search for routes with sufficient resources in order to satisfy the QoS requirements of a flow. The information regarding the availability of resources is managed by a resource management module which assists the QoS routing protocol in its search for QoS feasible paths. The QoS routing protocol should find paths that consume minimum resources. The QoS metrics can be classified as additive metrics, concave metrics, and multiplicative metrics.

An additive metric A_m is defined as $\sum_{i=1}^{h} L_i(m)$, where $L_i(m)$ is the value of metric m over link L_i and $L_i \epsilon$ P. The hop length of path P is h. A concave metric represents the minimum value over a path P and is formally defined as $C_m = min(L_i(m))$, $L_i(m) \epsilon$ P. A multiplicative metric represents the product of QoS metric values and is defined as $M_m = \prod_{i=1}^{h}(L_i(m))$, $L_i(m) \epsilon$ P. To find a QoS feasible path for a concave metric, the available resource on each link should be at least equal to the required value of the metric. Bandwidth is a concave metric, while cost, delay, and delay jitter are additive metrics. The reliability or availability of a link, based on some criteria such as link-break-probability, is a multiplicative metric. Finding an optimal path with multiple constraints may be an NP-complete problem if it involves two or more additive metrics. For example, finding a delay-constrained least-cost path is an NP-complete problem.

To assist QoS routing, the topology information can be maintained at the nodes of ad hoc wireless networks. The topology information needs to be refreshed frequently by sending link state update messages, which consume precious network resources such as bandwidth and battery power. Otherwise, the dynamically varying network topology may cause the topology information to become imprecise. This trade-off affects the performance of the QoS routing protocol. As path breaks occur frequently in ad hoc wireless networks, compared to wired networks where a link goes down very rarely, the path satisfying the QoS requirements needs to be recomputed every time the current path gets broken. The QoS routing protocol should respond quickly in case of path breaks and recompute the broken path or bypass the broken link without degrading the level of QoS. In the literature, numerous routing protocols have been proposed for finding QoS paths. In the following sections, some of these QoS routing protocols are described.

10.5.2 Ticket-Based QoS Routing Protocol

Ticket-based QoS routing [14] is a distributed QoS routing protocol for ad hoc wireless networks. This protocol has the following features:

- It can tolerate imprecise state information during QoS route computation and exhibits good performance even when the degree of imprecision is high.
- It probes multiple paths in parallel for finding a QoS feasible path. This increases the chance of finding such a path. The number of multiple paths searched is limited by the number of tickets issued in the probe packet by the source node. State information maintained at intermediate nodes is used for more accurate route probing. An intelligent hop-by-hop selection mechanism is used for finding feasible paths efficiently.
- The optimality of a path among several feasible paths is explored. A low-cost path that uses minimum resources is preferred when multiple feasible paths are available.

- A primary-backup-based fault-tolerant technique is used to reduce service disruption during path breaks that occur quite frequently in ad hoc wireless networks.

Protocol Overview

The basic idea of the ticket-based probing protocol is that the source node issues a certain number of tickets and sends these tickets in probe packets for finding a QoS feasible path. Each probe packet carries one or more tickets. Each ticket corresponds to one instance of the probe. For example, when the source node issues three tickets, it means that a maximum of three paths can be probed in parallel. The number of tickets generated is based on the precision of state information available at the source node and the QoS requirements of the connection request. If the available state information is not precise or if the QoS requirements are very stringent, more tickets are issued in order to improve the chances of finding a feasible path. If the QoS requirements are not stringent and can be met easily, fewer tickets are issued in order to reduce the level of search, which in turn reduces the control overhead. There exists a trade-off here between the performance of the QoS routing protocol and the control overhead.

The state information, at the source node, about intermediate nodes is useful in finding a much better QoS path, even if such information is not precise. The state information maintained at each node is comprised of estimations of end-to-end delay and available path bandwidth for every other node present in the network. When an intermediate node receives a probe packet, it is either split to explore more than one path or is forwarded to just one neighbor node based on the state information available at that intermediate node.

Based on the idea of ticket-based probing, two heuristic algorithms are proposed, one for delay-constrained QoS routing, and the other for bandwidth-constrained QoS routing. In delay-constrained QoS routing, each probe accumulates the delay of the path it has traversed so far. In other words, if an intermediate node A receives a probe packet (PKT) from a neighbor node B, node A updates the delay field in PKT by adding delay value of the link between nodes B and A. Then node A determines the list of candidate neighbors to which it has to send probe packets. It distributes tickets present in PKT among these new probe packets and then forwards these probe packets to the respective candidate neighbors. If multiple probe packets arrive at the destination node (with each carrying the list of intermediate nodes along its path), node A selects the path with least cost as the primary path and the other paths as the backup paths which will be used when the primary path is broken due to the mobility of intermediate nodes.

Optimizing Cost of a Feasible Path

This protocol searches for the lowest cost path among the feasible paths. This is done during the QoS path probing. The source node issues two types of tickets, yellow tickets and green tickets, and sends them along with probe packets. Yellow tickets prefer paths that satisfy the requirement of a probe in terms of QoS metrics.

For example, in delay-constrained QoS routing, yellow tickets are used to search for paths that have least delay, such that the end-to-end delay requirement is met. If the delay requirement is very large and can be met easily, only one yellow ticket is issued. If the delay requirement is too small to be met, then the source node does not issue any yellow ticket and rejects the connection request. Otherwise, more than one yellow ticket is issued to search multiple paths for finding a feasible QoS path. Green tickets are used to search for QoS paths with low costs. Similar to the manner in which the source node determines the number of yellow tickets, it also determines the number of green tickets to be issued on the basis of the delay requirement of the connection request. The distribution of yellow and green tickets (by an intermediate node to its candidate neighbors) is based on the delay and cost requirements of the connection request, respectively. The concept behind two types of tickets is to use the more aggressive green tickets to find a least cost feasible path, and use yellow tickets as a backup to maximize the probability of finding a feasible path.

Path Rerouting and Path Maintenance

This protocol suggests a primary-backup-based, fault-tolerant technique to cope up with the network dynamics. To tolerate faults, a multi-level redundancy scheme is proposed. For the highest level of redundancy, multiple paths (preferably disjoint) are probed and data is routed independently on all paths. The destination node selects the first data copy and discards other copies which arrive later. Another level of redundancy which requires less resources has also been proposed. Here, one path is selected as the primary path and other paths (having resources reserved) act as backup paths. The third type of redundancy incurs even less control overhead and consumes very few resources. Here, backup paths are available along with the primary path, but resources are not reserved in these backup paths. During path maintenance, in order to eliminate the broken link, the call is rerouted over a backup path which has enough resources to satisfy the QoS requirements of the call. In the case of the third type of redundancy, since no resource reservation has been done along backup paths, during path breaks it is extremely difficult to find a backup that has enough resources to satisfy the QoS requirements of the call.

Advantages and Disadvantages

The objective of ticket-based probing is to improve the average call acceptance ratio (ACAR) of ad hoc wireless networks. ACAR is the ratio of the number of calls accepted to the number of calls received by the network. The protocol adapts dynamically to the requirements of the application and the degree of imprecision of state information maintained. It offers a trade-off between control overhead incurred in finding a feasible path and the cost of a feasible path. As the maximum number of probes in the network is equal to the number of tickets issued, the control overhead is bound by the number of tickets. The performance of the protocol depends on the ticket-issuing mechanism at the source node and the ticket-splitting procedure at the intermediate nodes.

The protocol assumes that each node has global state information, but maintaining such information incurs huge control overhead in the already bandwidth-constrained ad hoc wireless networks. The proposed heuristic algorithms, which are based on an imprecise state information model, may fail in finding a feasible path in the extreme cases where the topology changes very rapidly. In delay-constrained QoS routing, the queuing delay and the processing delay at the intermediate nodes are not taken into consideration while measuring the delay experienced so far by the probe packet. This may cause some data packets to miss their deadlines. The routing algorithm works well only when the average lifetime of an established path is much longer than the average rerouting time. During the rerouting process, if QoS requirements are not met, data packets are transmitted as best-effort packets. This may not be acceptable for applications that have stringent QoS requirements.

10.5.3 Predictive Location-Based QoS Routing Protocol

The predictive location-based QoS routing protocol (PLBQR) [15] is based on the prediction of the location of nodes in ad hoc wireless networks. The prediction scheme overcomes to some extent the problem arising due to the presence of stale routing information. No resources are reserved along the path from the source to the destination, but QoS-aware admission control is performed. The network does its best to support the QoS requirements of the connection as specified by the application. The QoS routing protocol takes the help of an update protocol and location and delay prediction schemes. The update protocol aids each node in broadcasting its geographic location and resource information to its neighbors. Using the update messages received from the neighbors, each node updates its own view of the network topology. The update protocol has two types of update messages, namely, *Type 1 update* and *Type 2 update*. Each node generates a Type 1 update message periodically. A Type 2 update message is generated when there is a considerable change in the node's velocity or direction of motion. From its recent update messages, each node can calculate an expected geographical location where it should be located at a particular instant and then periodically checks if it has deviated by a distance greater than δ from this expected location. If it has deviated, a Type 2 update message is generated.

Location and Delay Predictions

In establishing a connection to the destination D, the source S first has to predict the geographic location of node D and the intermediate nodes, at the instant when the first packet reaches the respective nodes. Hence, this step involves location prediction as well as propagation delay prediction. The location prediction is used to predict the geographic location of the node at a particular instant t_f in the future when the packet reaches that node. The propagation delay prediction is used to estimate the value of t_f used in the above location prediction. These predictions are performed based on the previous update messages received from the respective nodes.

Location Prediction

Let (x_1, y_1) at t_1 and (x_2, y_2) at t_2 $(t_2 > t_1)$ be the latest two updates from the destination D to the source node S. Assume that the second update message also indicates v, which is the velocity of D at (x_2, y_2). Assume that node S wants to predict the location (x_f, y_f) of node D at some instant t_f in the future. This situation is depicted in Figure 10.9. The value of t_f has to be estimated first using the delay prediction scheme, which will be explained later in this section. From Figure 10.9, using similarity of triangles, the following equation is obtained:

$$\frac{y_2 - y_1}{y_f - y_1} = \frac{x_2 - x_1}{x_f - x_1} \tag{10.5.1}$$

By solving the above equation for y_f,

$$y_f = y_1 + \frac{(x_f - x_1)(y_2 - y_1)}{x_2 - x_1} \tag{10.5.2}$$

Using the above Equation 10.5.2, source S can calculate y_f if it knows x_f, which in turn can be calculated as follows. Using similarity of triangles again, the following equation is obtained:

$$y_f - y_2 = \frac{(y_2 - y_1)(x_f - x_2)}{x_2 - x_1} \tag{10.5.3}$$

By using the Pythagorean theorem,

$$(x_f - x_2)^2 + (y_f - y_2)^2 = v^2(t_f - t_2)^2 \tag{10.5.4}$$

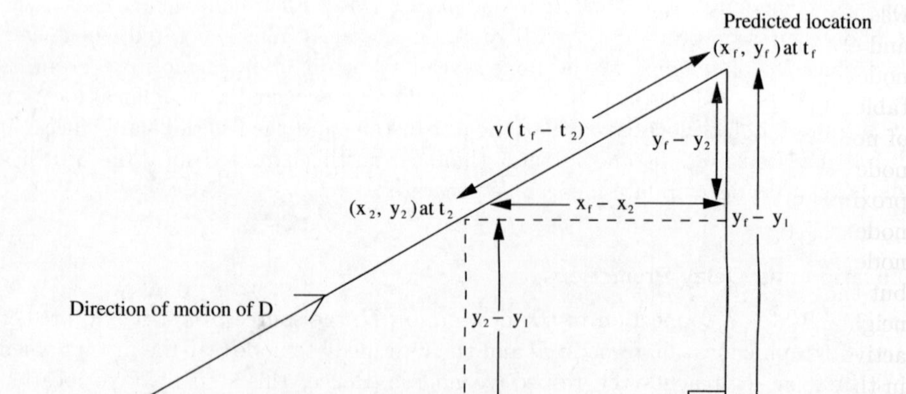

Figure 10.9. Prediction of location at a future time by node S using the last two updates.

Substituting for $y_f - y_2$ from Equation 10.5.3 in the above Equation 10.5.4 and solving for x_f, the following equation is obtained:

$$x_f = x_2 + \frac{v(t_f - t_1)(x_2 - x_1)}{\sqrt{[(x_2 - x_1)^2 + (y_2 - y_1)^2}} \quad (10.5.5)$$

If updates include the direction information of nodes, only one previous update is required to predict future location (x_f, y_f). The calculation of (x_f, y_f) is then exactly the same as that of the periodic calculation of expected location (x_e, y_e) by the update protocol [15].

Delay Prediction

The source node S has to predict the time instant t_f at which a packet reaches the given destination node or intermediate node D. This can be known only if the end-to-end delay between nodes S and D is known. It is assumed that the end-to-end delay for a data packet from node S to node D is equal to the delay experienced by the latest update message received by node S from node D.

QoS Routing

Each node in the network has information about the complete network topology, which is refreshed by means of update messages. Using this information, the source node performs source-routing. The network state information is maintained in two tables, namely, the *update table* and the *routing table.* When node A receives an update message from node B, node A updates the corresponding entry for node B in the update table. In that entry, node A stores the ID of node B, the time instant at which the update packet was sent, the time at which the update packet was received, the geographic coordinates, speed, resource parameters of node B, and optionally the direction of motion of node B. For each node N in the network, node A stores the last two update packets received from that node in its update table. For some nodes, node A also maintains proximity lists. The proximity list of node K is a list of all nodes lying within a distance 1.5×transmission range of node K. The proximity lists are used during route computation. By maintaining a proximity list rather than a neighbor list for node K (*i.e.*, list of nodes lying within node K's transmission range), node A also considers the nodes that were outside node K's transmission range at the time their respective last updates were sent, but that have since moved into node K's transmission range, while computing the neighbors of node K. The routing table at node A contains information about all active connections with node A as source. When an update message from any node in the network reaches node A, it checks if any of the routes in its routing table is broken or is about to be broken. In either case, route recomputation is initiated. Using the location prediction based on the updates, it is possible to predict whether any link on the path is about to break. Thus, route recomputation can be initiated even before the route actually breaks.

The routing algorithm given in [15] works as follows. The source node S first runs location and delay predictions on each node in its proximity list in order to obtain a list of its neighbors at present. It determines which of these neighbors have the resources to satisfy the QoS requirements of the connection (the neighbors that satisfy the QoS requirements are called candidates). Then it performs a depth-first search for the destination, starting with each of these candidate neighbors to find all candidate routes satisfying the QoS requirements of the connection request. From the resulting candidate routes, the geographically shortest route is chosen and the connection is established. Data packets are forwarded along this chosen route until the end of the connection or until the route is recomputed in anticipation of breakage. Note that only node S uses its view of the network for the entire computation.

Advantages and Disadvantages

PLBQR protocol uses location and delay prediction schemes which reduce to some extent the problem arising due to the presence of stale routing information. Using the prediction schemes, it estimates when a QoS session will experience path breaks and proactively finds an alternate path to reroute the QoS session quickly. But, as no resources are reserved along the route from the source to the destination, it is not possible to provide hard QoS guarantees using this protocol. Even soft QoS guarantees may be broken in cases when the network load is high. Since the location prediction mechanism inherently depends on the delay prediction mechanism, the inaccuracy in delay prediction adds to the inaccuracy of the location prediction. The end-to-end delay for a packet depends on several factors such as the size of the packet, current traffic load in the network, scheduling policy and processing capability of intermediate nodes, and capacity of links. As the delay prediction mechanism does not take into consideration some of the above factors, the predictions made by the location prediction mechanism may not be accurate, resulting in QoS violations for the real-time traffic.

10.5.4 Trigger-Based Distributed QoS Routing Protocol

The trigger-based (on-demand) distributed QoS routing (TDR) protocol [16] was proposed by De *et al.* for supporting real-time applications in ad hoc wireless networks. It operates in a distributed fashion. Every node maintains only the local neighborhood information in order to reduce computation overhead and storage overhead. To reduce control overhead, nodes maintain only the active routes. When a link failure is imminent, TDR utilizes the global positioning system-based (GPS) location information of the destination to localize the reroute queries only to certain neighbors of the nodes along the source-to-destination active route. For a quick rerouting with reduced control overhead, rerouting is attempted from the location of an imminent link failure, called intermediate node-initiated rerouting (INIR). If INIR fails, then in order to keep the flow state disruption to a minimum, rerouting is attempted from the source, which is termed source-initiated rerouting (SIRR).

Database Management

All nodes in the network maintain the local neighborhood information. For each neighbor, every node maintains *received power level*, current geographic coordinates, velocity, and direction of motion in the database.

Activity-Based Database

In addition to the local neighborhood information, node N maintains a source table ST_N, a destination table DT_N, or an intermediate table IT_N based on whether it actively participates in a session as the source (S), the destination (D), or as an intermediate node (I), respectively. These tables are referred to as the activity-based database. For a session, the source table contains the following fields: session ID, source ID, destination ID, maximum bandwidth demand (MaxBW), maximum acceptable delay (MaxDelay measured in terms of hop count), destination location (DLoc), next-node ID (NID) toward the destination, and activity flag (NodActv). An intermediate table contains the following fields: session ID, source ID, destination ID, source location (SLoc), MaxBW, MaxDelay, DLoc, NID, previous-node ID toward the source (PID), distance from the source (measured in terms of hop count), and NodActv. The destination table contains the following fields: session ID, source ID, destination ID, SLoc, MaxBW, MaxDelay, PID, distance from the source (hop count), and NodActv. At any time instant, a node may have to maintain one or more tables simultaneously for different on-going sessions. Each node N also maintains an updated residual bandwidth ($ResiBW_N$) which indicates its ability to participate in a session. A soft state approach is used to maintain the activity-based database. Hence, the database needs to be refreshed periodically. It is refreshed when data packets belonging to the on-going sessions are received by a node.

Routing Protocol

The messages that are exchanged for initiating, maintaining, and terminating a real-time session are described below.

Initial Route Discovery

If the source S has enough $ResiBW_S$ to satisfy the MaxBW for the session, the required bandwidth is temporarily reserved for a certain duration within which it expects an acknowledgment from the destination D. If the source knows the location of the destination, it performs route discovery through selective forwarding. In this approach, the source node takes advantage of location information of its neighbors and forwards route requests to only selective neighbors that are lying closely toward the destination node and satisfying QoS requirements of the connection request. Otherwise, the source initiates a flooding-based initial route discovery process. Before transmitting the route discovery packet, an entry is made in the source table ST_S for this session with NodActv flag set to zero (*i.e.*, idle). To ensure the stability of routes and in order to reduce the control overhead, only selected neighbors, from which packets were received with power level more than a threshold level (P_{th1}), are considered during route establishment. After receiving a route discovery packet, an intermediate node (IN) checks in its IT_{IN} whether any such packet was already received for the same session. If so, the current route discovery packet is rejected to ensure loop-free routing. Otherwise, it is the first discovery packet for a session. Then the intermediate node (IN) increments the hop-count field of the received packet by one and checks for $ResiBW_{IN}$. If it can meet the MaxBW requirement for the session and if the updated hop-count field is less than MaxDelay, the required bandwidth is temporarily reserved, and an entry is made into the activity table IT_{IN} for the session with NodActv flag set to zero. Then the packet is forwarded to its downstream neighbors with the updated NID field. If either or both of $ResiBW$ and MaxDelay criteria cannot be satisfied, the discovery packet is simply dropped. Upon receiving the first discovery packet, if the destination D is also able to satisfy both the $ResiBW$ and the MaxDelay criteria, the discovery packet and the corresponding route are accepted.

Route/Reroute Acknowledgment

After accepting the route, the destination node D builds DT_D table with the NodActv flag set to 1 (*i.e.*, active) and sends an ACK to the source S along the selected route. On receiving the ACK packet, all intermediate nodes and the source S set the NodActv flags in their respective tables to 1 and refresh their $ResiBW$ status. The packet transmission for the session follows immediately.

Alternate Route Discovery

In SIRR, when the received power level at an intermediate node falls below a threshold P_{th2}, the intermediate node sends a rerouting indication to the source S. Then the source S initiates the rerouting process through selective forwarding. But in INIR, when the power level of a packet received from the next node toward the destination falls below a threshold P_{th1} ($P_{th1} > P_{th2}$), it initiates a status query packet toward the source with appropriate identification fields and with a flag field called route repair status (RR_Stat) set to zero. If any upstream node is in the rerouting process, upon reception of the status query packet it sets the RR_Stat flag to 1 and sends the status reply packet to the querying node. On arriving at the source, the status query packet is discarded. If the querying node receives no status reply packet before its received power level from the downstream node goes below P_{th2}, it triggers the alternate route discovery process (*i.e.*, SIRR). Otherwise, it relinquishes control of rerouting. This query/reply process eliminates the chances of duplicate reroute discovery for a session. In both SIRR and INIR, the alternate route discovery process is similar to the initial route discovery except that the rerouting process takes advantage of the location information of the local neighbors and the approximate location of the destination, and forwards the rerouting requests to only selected neighbors that are close to the destination and that satisfy the delay and bandwidth constraints. The threshold parameters P_{th1} and P_{th2} have to be selected judiciously in order to avoid unnecessary rerouting.

Route Deactivation

In case of session completion or termination, the source node purges its corresponding ST table and sends a route deactivation packet toward the destination.

Upon receiving a deactivation request, each node which was part of that session updates its *ResiBW* and purges the activity table for that session. No explicit deactivation packet is sent in case of rerouting, as the new route could still consist of some nodes that were part of the old route.

Advantages and Disadvantages

In TDR protocol, if the source node knows the location of the destination node, it performs route discovery through selective forwarding to reduce the control overhead. For a quick rerouting with reduced control overhead and to reduce the packet loss during path breaks, it uses INRR and SIRR schemes. However, in this protocol a QoS session is rerouted if the received power level from a downstream node falls below a certain value (*i.e.*, threshold). Due to small-scale fading, the received power level may vary rapidly over short periods of time or distance traveled. Some of the factors that influence fading are multipath propagation, velocity of the nodes, and bandwidth of the channel. Even though the downstream node may be within the transmission range of the upstream node, due to fading the received power level at the upstream node may fall below the threshold value. This increases the control overhead because of initiation of the alternate route discovery process and false rerouting of some of the sessions.

10.5.5 QoS-Enabled Ad Hoc On-Demand Distance Vector Routing Protocol

Perkins *et al.* have extended the basic ad hoc on-demand distance vector (AODV) routing protocol to provide QoS support in ad hoc wireless networks [17]. To provide QoS, packet formats have been modified in order to specify the service requirements which must be met by the nodes forwarding a *RouteRequest* or a *RouteReply*.

QoS Extensions to AODV Protocol

Several modifications have been carried out for the routing table structure and *RouteRequest* and *RouteReply* messages in order to support QoS routing. Each routing table entry corresponds to a different destination node. The following fields are appended to each routing table entry:

- Maximum delay
- Minimum available bandwidth
- List of sources requesting delay guarantees
- List of sources requesting bandwidth guarantees

Maximum Delay Extension Field

The maximum delay extension field is interpreted differently for *RouteRequest* and *RouteReply* messages. In a *RouteRequest* message, it indicates the maximum time (in seconds) allowed for a transmission from the current node to the destination node. In a *RouteReply* message, it indicates the current estimate of cumulative delay from the current intermediate node forwarding the *RouteReply*, to the destination. Using this field the source node finds a path (if it exists) to the destination node satisfying the maximum delay constraint. Before forwarding the *RouteRequest*, an intermediate node compares its *node traversal time* (*i.e.*, the time it takes for a node to process a packet) with the (remaining) delay indicated in the maximum delay extension field. If the delay is less than the node traversal time, the node discards the *RouteRequest* packet. Otherwise, the node subtracts node traversal time from the delay value in the extension and processes the *RouteRequest* as specified in the AODV protocol.

The destination node returns a *RouteReply* with the maximum delay extension field set to zero. Each intermediate node forwarding the *RouteReply* adds its own node traversal time to the delay field and forwards the *RouteReply* toward the source. Before forwarding the *RouteReply* packet, the intermediate node records this delay value in the routing table entry for the corresponding destination node.

Minimum Bandwidth Extension Field

In a *RouteRequest* message, this field indicates the minimum bandwidth (in Kbps) that must be available along an acceptable path from the source to the destination. In a *RouteReply* message, it indicates the minimum bandwidth available on the route between the node forwarding the *RouteReply* and the destination node. Using this field, the source node finds a path (if it exists) to the destination node satisfying the minimum bandwidth constraint. Before forwarding the *RouteRequest*, an intermediate node compares its available bandwidth with the bandwidth field in the extension. If the requested amount of bandwidth is not available, the node discards the *RouteRequest* message. Otherwise, the node processes the *RouteRequest* as specified in the AODV protocol.

The destination node returns a *RouteReply* in response to a *RouteRequest* with the bandwidth field set to infinity (a very large number). Each node forwarding the *RouteReply* compares the bandwidth field in the *RouteReply* with its own link capacity and updates the bandwidth field of the *RouteReply* with the minimum of the two, before forwarding the *RouteReply*. This value is also stored in the routing table entry for the corresponding destination and indicates the minimum available bandwidth to the destination.

List of Sources Requesting QoS Guarantees

A *QoSLost* message is generated when an intermediate node experiences an increase in node traversal time or a decrease in the link capacity. The *QoSLost* message is forwarded to all sources potentially affected by the change in the QoS parameter. These are the sources to which *RouteReply*s with QoS extension have been forwarded by the node earlier.

Advantages and Disadvantages

The advantage of QoS AODV protocol is the simplicity of extension of the AODV protocol that can potentially enable QoS provisioning. However, as no resources are reserved along the path from the source to the destination, this protocol is not suitable for applications that require hard QoS guarantees. Further, node traversal time is only the processing time for the packet, so the major part of the delay at a node is contributed by packet queuing and contention at the MAC layer. Hence, a packet may experience much more delay than this when the traffic load is high in the network.

10.5.6 Bandwidth Routing Protocol

The bandwidth routing (BR) protocol [18] consists of an end-to-end path bandwidth calculation algorithm to inform the source node of the available bandwidth to any destination in the ad hoc network, a bandwidth reservation algorithm to reserve a sufficient number of free slots for the QoS flow, and a standby routing algorithm to reestablish the QoS flow in case of path breaks.

Here, only bandwidth is considered to be the QoS parameter. In TDMA-based networks, bandwidth is measured in terms of the number of free slots available at a node. The goal of the bandwidth routing algorithm is to find a shortest path satisfying the bandwidth requirement. The transmission time scale is organized into frames, each containing a fixed number of time-slots. The entire network is synchronized on a frame and slot basis. Each frame is divided into two phases, namely, the control phase and the data phase. The control phase is used to perform the control functions such as slot and frame synchronization, virtual circuit (VC) setup, and routing. The data phase is used for transmission/reception of data packets. For each node, a slot is assigned in the control phase for it to broadcast its routing information and slot requirements. At the end of the control phase, each node knows about the channel reservations made by its neighbors. This information helps nodes to schedule free slots, verify the failure of reserved slots, and drop expired real-time packets. The BR protocol assumes assumes a half-duplex CDMA-over-TDMA system in which only one packet can be transmitted in a given slot.

Bandwidth Calculation

Since the network is multi-hop in nature, the free slots recorded at each node may be different. The set of common free slots between two adjacent nodes denotes the link bandwidth between them. The path bandwidth between two nodes is the maximum bandwidth available in the path between them. If the two nodes are adjacent, the path bandwidth between them equals their link bandwidth. For example, consider two adjacent nodes, node A and node B, having free slots {2,5,6,8} and {1,2,4,5}, respectively. The link bandwidth $linkBW(A, B) = freeslot(A) \cap freeslot(B) = \{2, 5\}$. It means that only slots 2 and 5 can be used by nodes A and B for transmitting data packets to each other. The $freeslot(X)$ is defined as the set of slots which are not used by any adjacent node of node X (to receive or to send) from the point of view of node X.

To compute the end-to-end bandwidth for a path in a TDMA-based network, one has to know not only the available bandwidth on the individual links on the path, but also determine the scheduling of the free slots. The BR protocol also provides a heuristic-based hop-by-hop path bandwidth calculation algorithm to assign free slots at every hop along the path. The call admission control mechanism of the BR protocol uses the information regarding the availability of end-to-end bandwidth while making a decision on whether to admit or reject a new QoS session. The path bandwidth calculation algorithm is explained with the help of the example shown in Figure 10.10, where a path from source node S to destination node D is illustrated. The process of computing $pathBW(S, D)$ is explained below.

- $pathBW(S, A)$: Since node S and node A are adjacent, the $pathBW(S, A) = linkBW(A, S)$, which is four slots. The four slots are $\{2, 5, 6, 7\}$.

- $pathBW(S, B)$: Since $pathBW(S, A) = linkBW(A, B) = \{2, 5, 6, 7\}$, if S uses slots 6 and 7 to send packets to A, then A can use only slots 2 and 5 for transmission of packets to B. This is because a node cannot be in transmission and reception modes simultaneously. Hence $pathBW(S, B)$ is two slots, by assigning slots $\{6, 7\}$ on link(S, A) and slots $\{2, 5\}$ on link(A, B).

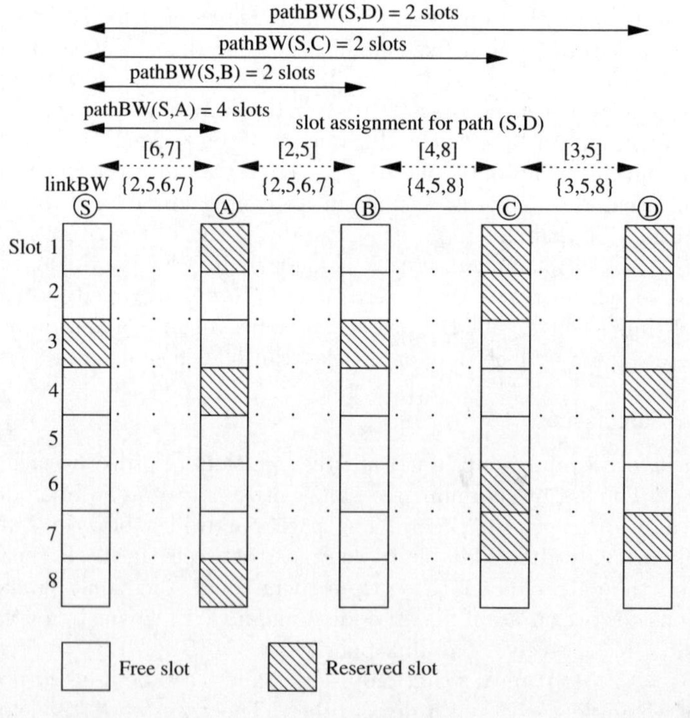

Figure 10.10. An example of path bandwidth calculation in BR protocol.

- *pathBW*(S, C): Here slots 4 and 8 are exclusively available for *linkBW*(B, C), slot 2 is exclusively available for *pathBW*(S, B), and slot 5 is common for both of them. So assign one of slots 4, 8 to link(B, C), for example, assign slot 4 to link(B, C), and slot 2 to path(S, B). For achieving maximum bandwidth, assign slot 8 to link (B, C) and slot 5 to path(S, B). Hence, *pathBW*(S, C) is 2 slots, by assigning slots $\{6, 7\}$ on link(S, A), slots $\{2, 5\}$ on link(A, B), and slots $\{4, 8\}$ on link(B, C).

- *pathBW*(S, D): This case is similar to the previous one. So slots 4 and 8 are assigned to path(S, C) and slots 3 and 5 are assigned to link(C, D) to get two slots for *pathBW*(S, D).

Slot Assignment

The path bandwidth calculation algorithm requires periodic exchange of bandwidth information. The slot assignment algorithm in each node assigns free slots during the call setup. When a node receives a call setup packet, it checks whether the slots that the immediate sender will use for transmission are free, and it also finds out if there are free slots that can be used for forwarding the incoming packets. If such free slots are available, the slot assignment algorithm reserves the required number of slots, updates the routing table, and then forwards the call setup packet to the next hop. If the required number of slots are not available at the node, all the reservations that have been made so far along the path from the source node to the current node have to be canceled in order to release the slots assigned for this connection. This is done by sending a *Reset* packet back to the source along the path that has been established so far. If reservations are made successfully along the path from the source to the destination, the destination sends a *Reply* packet back to the source to acknowledge having set up the connection. The reservations are soft state in nature in order to avoid resource lock-up at intermediate nodes due to path breaks.

Standby Routing Mechanism

The connections may get broken due to dynamic changes in the network topology. The standby routing mechanism has to reestablish such broken connections. Secondary paths are maintained in the routing table, which can be used when the primary path fails. The standby route is easily computed using the DSDV algorithm [19] without any extra overhead. Each node periodically exchanges routing information with its neighboring nodes. The neighbor with the shortest distance to the destination node becomes the next node on the primary path to the destination node. The neighbor node with the second shortest distance to the destination becomes the next node on the standby route to the destination. It is to be noted that this standby route is not guaranteed to be a link- or node-disjoint one. When a primary path fails, the upstream node that detects the link break will try to rebuild a new path immediately, using the standby route. If the standby route satisfies the

QoS requirements, the new path from the point of the path break is established by sending a call setup packet hop-by-hop to the destination through the standby path.

Since this scheme follows DSDV protocol, a table-driven routing protocol, and uses on-demand call admission control, similar to the on-demand routing protocols, it is classified into the category of hybrid solutions in the classifications in Figure 10.2.

Advantages and Disadvantages

The BR protocol provides an efficient bandwidth allocation scheme for CDMA-over-TDMA-based ad hoc wireless networks. The standby routing mechanism can reduce the packet loss during path breaks. But the CDMA-over-TDMA channel model that is used in this protocol requires assigning a unique control slot in the control phase of superframe for each node present in the network. This assignment has to be done statically before commissioning the network. Due to this, it is not possible for a new node to enter into the network at a later point of time. If a particular node leaves the network, the corresponding control slot remains unused and there is no way to reuse such a slot(s). Further, the network needs to be fully synchronized.

10.5.7 On-Demand QoS Routing Protocol

Lin proposed an admission control scheme over an on-demand QoS routing (OQR) protocol [20] to guarantee bandwidth for real-time applications. Since routing is on-demand in nature, there is no need to exchange control information periodically and maintain routing tables at each node. Similar to the bandwidth routing (BR) protocol, the network is time-slotted and bandwidth is the key QoS parameter. The path bandwidth calculation algorithm proposed in BR is used to measure the available end-to-end bandwidth. The on-demand QoS routing protocol is explained below.

Route Discovery

During the route discovery process, the source node that wants to find a QoS route to the destination floods a QoS route request (QRREQ) packet. A QRREQ packet contains the following fields: packet type, source ID, destination ID, sequence number, route list, slot array list, data, and TTL. The pair {source ID, sequence number} is used to uniquely identify a packet. For each QRREQ packet, the source node uses a new sequence number (which is monotonically increasing) in order to avoid multiple forwarding of the same packet by intermediate nodes. The route list records the nodes that have been visited by the QRREQ packet, whereas the slot array list records free slots available at each of these nodes. The TTL field limits the maximum length of the path to be found. A node N receiving a QRREQ packet performs the following operations:

1. If a QRREQ with the same {source ID, sequence number} had been received already, this QRREQ packet gets discarded.

2. Otherwise, the route list field is checked for the address of this node N. If it is present, node N discards this QRREQ packet.

3. Otherwise,

   - Node N decrements TTL by one. If TTL counts down to zero, it discards this QRREQ packet.
   - It calculates the path bandwidth from the source to this node. If it satisfies the QoS requirement, node N records the available free slots in the slot array list of the QRREQ packet. Otherwise, node N discards this QRREQ packet.
   - Node N appends the address of this node to the route list and re-broadcasts this QRREQ packet if it is not the destination.

For the example shown in Figure 10.10, assume that the source S floods a QR-REQ packet with bandwidth requirement of two time-slots. Here, the destination D receives a QRREQ packet with the following information in its fields. The route list field contains (S, A, B, C) and the slot array list contains ([A, {2, 5, 6, 7}], [B, {2, 5}], [C, {4, 5}], [D, {3, 8}]). The destination may receive more than one QRREQ packet, each giving a unique feasible QoS path from the source to the destination.

Bandwidth Reservation

The destination node may receive one or more QRREQ packets, each giving a feasible QoS path for the connection request. The destination node selects the least-cost path among them. Then it copies the fields {route list, slot array list} from the corresponding QRREQ packet to the QoS Route Reply (QRREP) packet and sends the QRREP packet to the source along the path recorded in the route list. As the QRREP traverses back to the source, each node recorded in the route list reserves the free slots that have been recorded in the slot array list field. Finally, when the source receives the QRREP, the end-to-end bandwidth reservation process is completed successfully. The reservations made are soft state in nature in order to avoid resource lock-up. The source can start sending data packets in the data phase. At the end of the session, all reserved slots are released.

Reservation Failure

The reservation of bandwidth may fail, either due to route breaks or because the free slots that are recorded in the slot array list get occupied by some other connection(s) before the QRREP packet sent by the destination reaches the corresponding intermediate nodes. In the second case, the node at which reservation fails, sends a *ReserveFail* packet to the destination node. The destination then restarts the reservation process along the next feasible path. All nodes on the path from the interrupted node to the destination free the reserved slots for this connection on receiving the *ReserveFail* packet. If no connection could be set up due to non-availability of feasible paths, the destination broadcasts a *NoRoute* packet to notify the source. Then the source either restarts the route discovery process, if it still needs a connection to the destination, or rejects the call.

Route Maintenance

When a route gets broken, the nodes detecting the link break send a *RouteBroken* packet to the source and the destination nodes. In other words, once the next hop becomes unreachable, the upstream node which is toward the source node sends a *RouteBroken* packet to the source, and the downstream node which is toward the destination sends another *RouteBroken* packet to the destination. The intermediate nodes relaying the *RouteBroken* packet release all reserved slots for this connection and drop all data packets of this connection which are still pending in their respective queues. After receiving the *RouteBroken* packet, the source restarts the route discovery process in order to reestablish the connection over a new path, while the destination releases resources reserved for that connection.

Advantages and Disadvantages

OQR protocol uses an on-demand resource reservation scheme and hence produces lower control overhead. Since it uses the CDMA-over-TDMA channel model, the network needs to be fully synchronized. Further, the on-demand nature of route discovery process leads to higher connection setup time.

10.5.8 On-Demand Link-State Multipath QoS Routing Protocol

Unlike the QoS routing protocols described above in this chapter which try to find a single path from the source to the destination satisfying the QoS requirements, the on-demand link-state multipath QoS routing (OLMQR) protocol [21] searches for multiple paths which collectively satisfy the required QoS. The original bandwidth requirement is split into sub-bandwidth requirements. Notably, the paths found by the multipath routing protocol are allowed to share the same sub-paths. OLMQR has better call acceptance rate in ad hoc wireless networks where finding a single path satisfying all the QoS requirements is very difficult.

In this protocol, the MAC layer is assumed to be using the CDMA-over-TDMA channel model similar to BR and OQR protocols. A mobile node in the network knows the bandwidth available to each of its neighbors. When the source node requires a QoS session with bandwidth BW to the destination, it floods a QoS route request (QRREQ) packet. Each packet carries the path history and link-state information from the source to the destination. The destination node collects all possible link-state information from different QRREQ packets received and constructs its own view of the current network topology. A multipath routing algorithm is applied at the destination to determine multiple paths which collectively fulfill the original bandwidth requirement BW of the QoS flow. Then the destination node sends reply packets along these paths, which reserve the corresponding resources (sub-bandwidth requirements) on the corresponding paths on their way back to the source. The operation of this protocol consists of three phases: Phase 1 is on-demand link-state discovery, phase 2 is unipath discovery, and phase 3 is multipath discovery and reply.

On-Demand Link-State Discovery

For each call request, the source node floods a QRREQ packet toward the destination. Each packet records the path history and all link-state information along its route. A QRREQ packet contains the following fields: source ID, destination ID, node history, free time-slot list, bandwidth requirement, and time to live (TTL). The node history field records the path from source to the current traversed node, the free time-slot list field contains a list of free time-slots of links, where each entry in the list records free time-slots between the current traversed node and the last node recorded in the node history, and TTL field limits the hop length of the search path.

The source S floods a QRREQ(S, D, node history $= \{S\}$, free time-slot list $= \phi$, BW, TTL) packet into the network toward the destination D, if the given requirement is BW. An intermediate node N receiving a QRREQ packet performs the following operations:

1. Node N checks the node history field of the QRREQ packet for its address. If it is present, the node discards this QRREQ packet.

2. Otherwise,

   - Node N decrements TTL by one. If TTL counts down to zero, it discards this QRREQ packet.
   - Node N adds itself into the node history field, appends the free time-slots of the link between itself and the last node recorded in the node history field into the free time-slot list field, and rebroadcasts this QRREQ packet.

The destination may receive many different QRREQ packets from the source. It constructs its own view of the current network topology. It also calculates the available bandwidths of the links present in that network topology. For example, consider the network shown in Figure 10.11. The source S floods the network with a QRREQ packet by setting BW and TTL fields to 3 and 4, respectively. The destination D receives six QRREQ packets, which have traversed along the paths: $S \rightarrow A \rightarrow B \rightarrow D$, $S \rightarrow E \rightarrow F \rightarrow D$, $S \rightarrow A \rightarrow C \rightarrow B \rightarrow D$, $S \rightarrow A \rightarrow C \rightarrow F \rightarrow D$, $S \rightarrow E \rightarrow C \rightarrow F \rightarrow D$, and $S \rightarrow E \rightarrow C \rightarrow B \rightarrow D$. Using this information, a partial view of the network is constructed at the destination D.

Unipath Discovery

Unlike the BR [18] and the OQR [20] protocols discussed earlier in this section, here the unipath discovery operation (*i.e.*, path bandwidth calculation algorithm) does not follow the traditional hop-by-hop approach to determine the end-to-end path

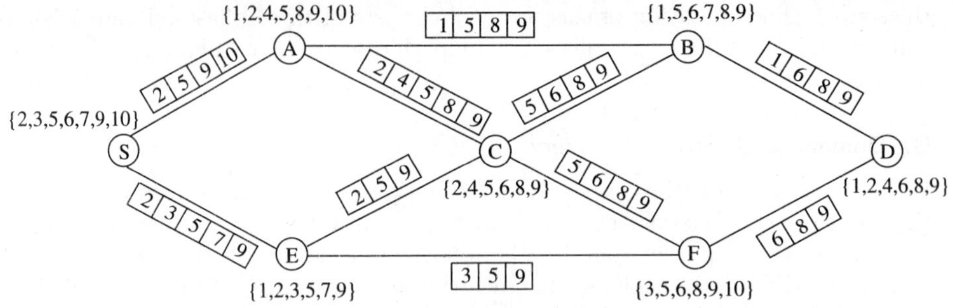

Figure 10.11. An example network.

bandwidth. The unipath discovery approach acquires higher end-to-end path bandwidth than that acquired through the hop-by-hop approach. For a given path (*i.e.*, unipath), the unipath discovery operation determines its maximum path bandwidth by constructing a least-cost-first time-slot reservation tree T_{LCF}. Before constructing T_{LCF}, a time-slot reservation tree T is constructed. The T_{LCF} and T trees are used to reserve time-slots efficiently for a given unipath.

A time-slot reservation tree T is constructed by the breadth-first-search approach as follows. Given a path $S \rightarrow A \rightarrow B \cdots K \rightarrow D$, let the root of T be represented as $abcd \cdots xy$, where a represents the bandwidth (*i.e.*, the set of free time-slots) of link(S, A) and b represents the bandwidth of link(A, B). Let $\underline{ab}cd \cdots xy$ denote the time-slots that are reserved on links a and b. Child nodes of the root are $\underline{ab}cd \cdots xy$, $a\underline{bc}d \cdots xy$, $ab\underline{cd} \cdots xy, \cdots$, and $abcd \cdots \underline{xy}$, which form the first level of tree T. The tree T recursively expands all child nodes of each node on each level of tree T, and follows the same rules as that of the first level of tree T until the leaf nodes are reached. Each path from the root to the leaf nodes gives a time-slot reservation pattern. This pattern is used to reserve time-slots from the source to the destination. To reduce the time needed to search a path satisfying a given bandwidth requirement BW, a least-cost-first time-slot reservation tree T_{LCF} is constructed from the time-slot reservation tree T as follows. To obtain the T_{LCF}, the child nodes on each level of tree T are sorted in ascending order from left to right by using the number of reserved time-slots in them. The unipath time-slot reservation algorithm performs depth-first-search on the T_{LCF} tree to determine a time-slot reservation pattern having maximum path bandwidth. The search is completed if either the tree traversal is completed or a reservation pattern is identified with a bandwidth $\bar{B}$, where $\bar{B} \geq BW$.

For example, consider the path $S \rightarrow A \rightarrow B \rightarrow D$ from the source S to the destination D in the network shown in Figure 10.11. Let a, b, c denote free time-slots of links (S, A), (A, B), and (B, D), respectively, as shown in Figure 10.12 (a). For this path, a time-slot reservation tree T can be constructed as shown in Figure 10.12 (b). It shows two reservation patterns: The first pattern is $\underline{ab}$, $\underline{c}$ and the second pattern is $\underline{bc}$, $\underline{a}$. In the first pattern, $\underline{ab}$ has three time-slots bandwidth

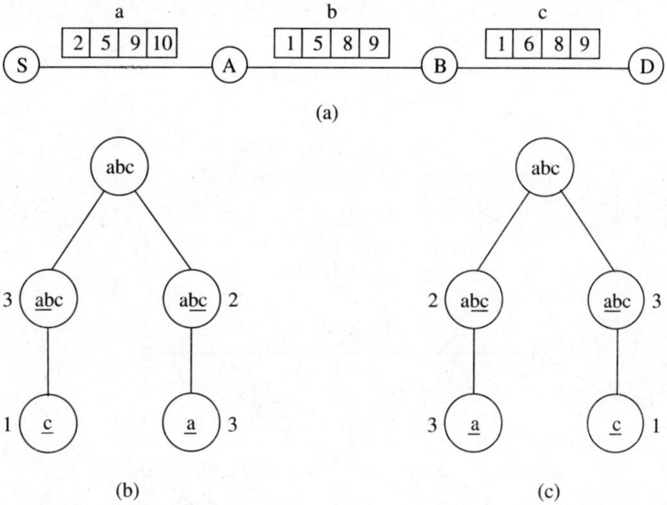

Figure 10.12. Example of T and T_{LCF} trees for a path.

(by assigning slots 2, 5, and 10 for the link a and slots 1, 8, and 9 for the link b) and $\underline{c}$ has one time-slot bandwidth (by assigning the remaining slot 6 for the link c). Hence, the first pattern $\underline{ab}$, $\underline{c}$ has one time-slot path bandwidth (which is the minimum of bandwidths of $\underline{ab}$ and $\underline{c}$). Similarly, in the second pattern, $\underline{bc}$ has two time-slots bandwidth (by assigning slots 1 and 5 for the link b and slots 6 and 8 for the link c), and $\underline{a}$ has three time-slots bandwidth (by assigning the remaining slots 2, 9, and 10 for the link a). Hence, the second pattern $\underline{bc}$, $\underline{a}$ has two time-slots path bandwidth. From T, a least-cost-first time-slot reservation tree T_{LCF} can be constructed as shown in Figure 10.12 (c). Comparing the T-tree traversal with the T_{LCF}-tree traversal scheme, the T_{LCF}-tree traversal scheme is more efficient than the T-tree traversal scheme as it reduces the time required to find a feasible QoS path.

Multipath Discovery and Reply

The destination initiates the multipath discovery operation by sequentially exploiting multiple unipaths such that the sum of path bandwidths fulfills the original bandwidth requirement BW. The destination applies the unipath discovery operation to each path in order to determine the maximum achievable path bandwidth of each path. After accepting a path, the destination updates the network state information it maintains in order to reflect the current bandwidth availability on the links. Finally, the destination sends reply packets along these paths, which reserve the corresponding resources (sub-bandwidth requirements) on the corresponding paths on their way back to the source. In the above example, the destination D

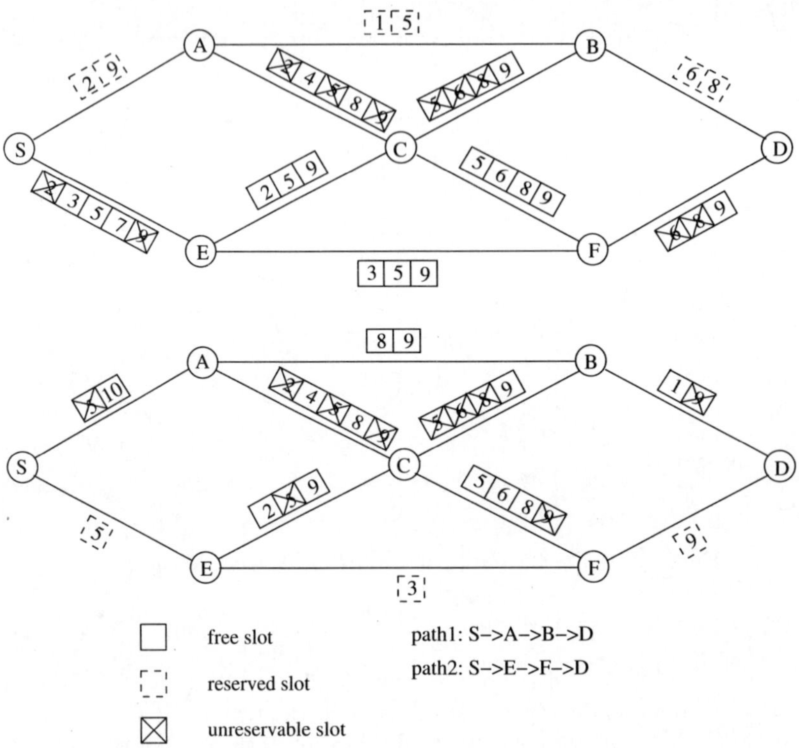

Figure 10.13. The unipaths found by multipath discovery algorithm.

finds two unipaths: $S \rightarrow A \rightarrow B \rightarrow D$ with two time-slots path bandwidth and $S \rightarrow E \rightarrow F \rightarrow D$ with one time-slot path bandwidth, as shown in Figure 10.13.

Advantages and Disadvantages

If the QoS requirements of a flow cannot be met by a single path from the source to the destination, multiple paths are checked which collectively satisfy the required QoS. Hence, OLMQR protocol has better ACAR. But the overhead of maintaining and repairing paths is very high compared to traditional unipath routing protocols because multiple paths are used to satisfy each flow's QoS requirements.

10.5.9 Asynchronous Slot Allocation Strategies

The QoS solutions discussed so far such as BR, OQR, and OLMQR assume a TDMA-based network or a CDMA-over-TDMA model for the network. This requires time synchronization across all nodes in the network. Time synchronization demands periodic exchange of control packets, which results in high bandwidth consumption. Ad hoc wireless networks experience rapid changes in topology leading

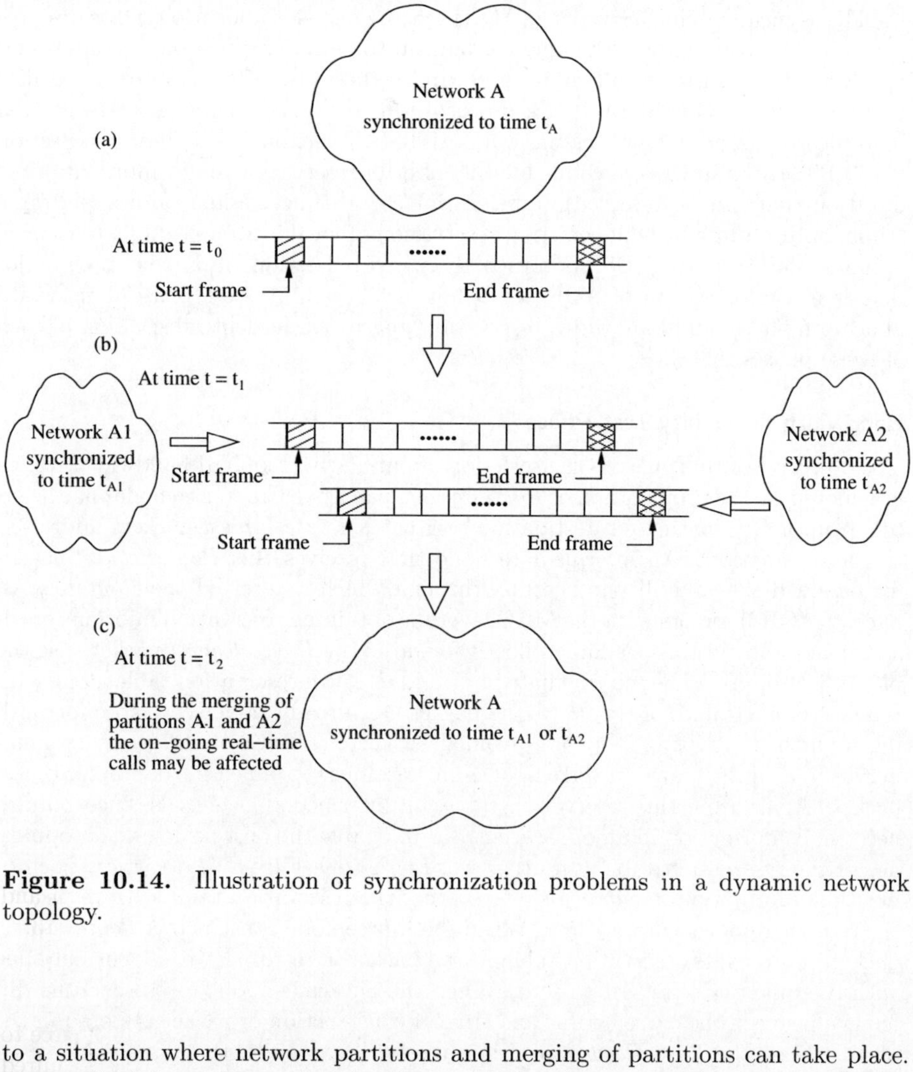

Figure 10.14. Illustration of synchronization problems in a dynamic network topology.

to a situation where network partitions and merging of partitions can take place. Figure 10.14 shows the synchronization problems arising out of dynamic topological changes in an ad hoc wireless network.

A completely connected and synchronized network A at time $t = t_0$ (shown in Figure 10.14 (a)) may be partitioned into two disjoint networks A1 and A2 at time $t = t_1$ (shown in Figure 10.14 (b)). These two networks may be synchronized to two different clock times as illustrated. Due to the dynamic topology experienced in an ad hoc wireless network, it is possible to have two separately synchronized networks A1 (synchronized to t_{A1}) and A2 (synchronized to t_{A2}) merge to form a combined network A (Figure 10.14 (c)). During the merging process, the real-time calls existing in the network may be affected while accommodating the changes in synchronization.

The asynchronous QoS routing (AQR) scheme and slot allocation strategies proposed in [22], [23] provide a unique mechanism to reserve asynchronous end-to-end bandwidth for real-time calls in ad hoc wireless networks. These strategies utilize the real-time MAC (RTMAC) [13] protocol that can effect bandwidth reservation in asynchronous ad hoc wireless networks. RTMAC is explained in detail in Section 6.6.7. RTMAC can reserve conn-slots [number of reservation slots (minimum time duration that can be reserved) sufficient for a real-time session] on a superframe (time duration in which the existing reservations repeat). AQR is an extension of dynamic source routing (DSR) protocol discussed in Section 7.5.1. The three major phases in the operation of AQR are bandwidth feasibility test phase, bandwidth allocation phase, and bandwidth reservation phase. An in-depth discussion of each of these phases follows.

Bandwidth Feasibility Test Phase

The objective of this phase is the selection of paths with required bandwidth, which is achieved by the propagation of *RouteRequest* packets. The source node that needs to set up a QoS path to a destination originates *RouteRequest* packets addressed to the destination. An intermediate node that receives this *RouteRequest* checks for bandwidth availability in the link through which it received the *RouteRequest* packet. AQR interacts with the MAC layer for obtaining reservation information. If sufficient bandwidth is available, then it forwards the *RouteRequest* packet, else the packet is dropped. The intermediate node adds its own reservation table along with the reservation tables of the nodes the packet has already traversed before forwarding it further. Routing loops are avoided by keeping track of the sequence number, source address, and traversed path informations contained in the *RouteRequest* packet. Apart from this reservation table, an intermediate node also incorporates necessary information in an *offset time* field to enable the destination node to make use of the reservation table. In other words, the offset time field carries synchronization information required for interpreting the reservation table with respect to the receiving node's current time. When the source node constructs a *RouteRequest* packet, it stores its reservation table in the packet with respect to its current time with the quantity offset set to zero. When the packet is about to be sent, the difference between the current time and time of construction of packet is stored in the offset. When the *RouteRequest* packet is received at a node, the offset is increased by the estimated propagation delay of transmission. Hence by using this offset time, the relative difference between the local clock and the time information contained in the reservation table carried in the *RouteRequest* can be incorporated and then used for synchronizing the reservation information. When the *RouteRequest* packet reaches its destination, it runs the slot allocation algorithm on a selected path, after constructing a data structure called *QoS Frame* for every link in that path. The *QoS Frame* is used to calculate, for every link, the free bandwidth slots in the superframe and unreservable slots due to reservations carried out by the neighborhood nodes (also referred to as unreservable slots due to hidden terminals). The destination node waits for a specific time interval, gathers a set of *RouteRequest* packets, and chooses a shortest path with necessary bandwidth.

Bandwidth Allocation Phase

In this phase, the destination node performs a bandwidth allocation strategy that assigns free slots to every intermediate link in the chosen path. The information about asynchronous slots assigned at every intermediate link is included in the *RouteReply* packet and propagated through the selected path back to the source. Slot allocation strategies such as early fit reservation (EFR), minimum bandwidth-based reservation (MBR), position-based hybrid reservation (PHR), and k-hopcount hybrid reservation (k-HHR) discussed later in this section can be used for allocation of bandwidth and positioning of slots in this phase.

Slot Allocation Strategies

The slot allocation strategies are used in the bandwidth allocation phase in order to decide upon the order of links in a chosen path and particular slot positions to be assigned. The order of links chosen for allocation and the position of assigned bandwidth-slots on each link influence the end-to-end delay of the path and the call acceptance rate.

- Early fit reservation (EFR): During the bandwidth allocation phase, the destination node runs the following steps for the EFR scheme:

  - Step 1: Order the links in the path from source to destination.
  - Step 2: Allocate the first available free slot for the first link in the path.
  - Step 3: For every subsequent link, allocate the first immediate free slot after the assigned slot in the previous link.
  - Step 4: Continue Step 3 until the last link in the chosen path is reached.

  EFR attempts to provide the least end-to-end delay. The average end-to-end delay can be obtained as $(n-1) \times \frac{t_{sf}}{2}$ where n is the number of hops in the path and t_{sf} is the duration of the superframe. Figure 10.15 (a) illustrates a simple string topology and Figure 10.15 (b) shows the slot allocation carried out for three real-time flows. In the example, the average delay experienced can be calculated as $\frac{8}{3}$ slots. The flow $E \rightarrow C$ experiences a delay of two slots, and flows $A \rightarrow D$ and $E \rightarrow B$ experience a delay of three slots each, making the average delay of $\frac{8}{3}$ slots.

- Minimum bandwidth-based reservation (MBR): The following steps are executed by the destination node for the MBR scheme:

  - Step 1: Order the links in the non-decreasing order of free bandwidth.
  - Step 2: Allocate the first free slot in the link with lowest free bandwidth.
  - Step 3: Reorder the links in the non-decreasing order of free bandwidth and assign the first free slot on the link with lowest bandwidth.
  - Step 4: Continue Step 3 until bandwidth is allotted for all the links.

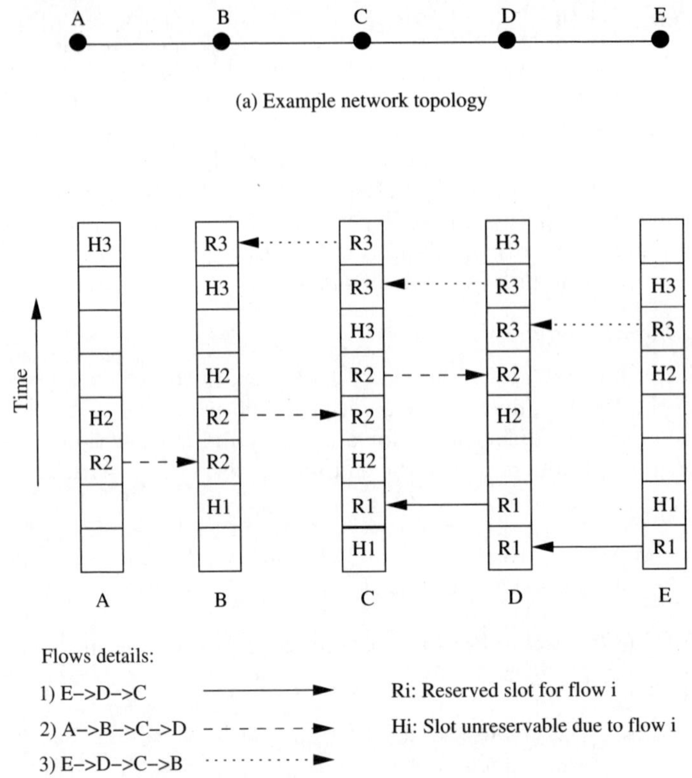

(a) Example network topology

(b) EFR scheme

Figure 10.15. Illustration of EFR scheme. *Reproduced with permission from [23], © John Wiley & Sons Limited, 2004.*

MBR allots bandwidth for the links in the increasing order of free bandwidth. In case a tie occurs, where two links exist with the same amount of free bandwidth, it is broken by choosing the link with lowest bandwidth in the neighboring links. Further ties are broken by choosing the link with lowest ID of the link-level sender. Figure 10.16 (b) shows the slot allocation carried out in the MBR scheme over a simple string topology network. The worst case end-to-end delay provided by MBR can be $(n-1) \times t_{sf}$. In the example in Figure 10.16 (b), the average delay experienced can be calculated as $\frac{33}{3}$ slots.

- Position-based hybrid reservation (PHR): Similar to EFR and MBR schemes, PHR also is executed at the destination node. The following are the steps in the PHR algorithm:

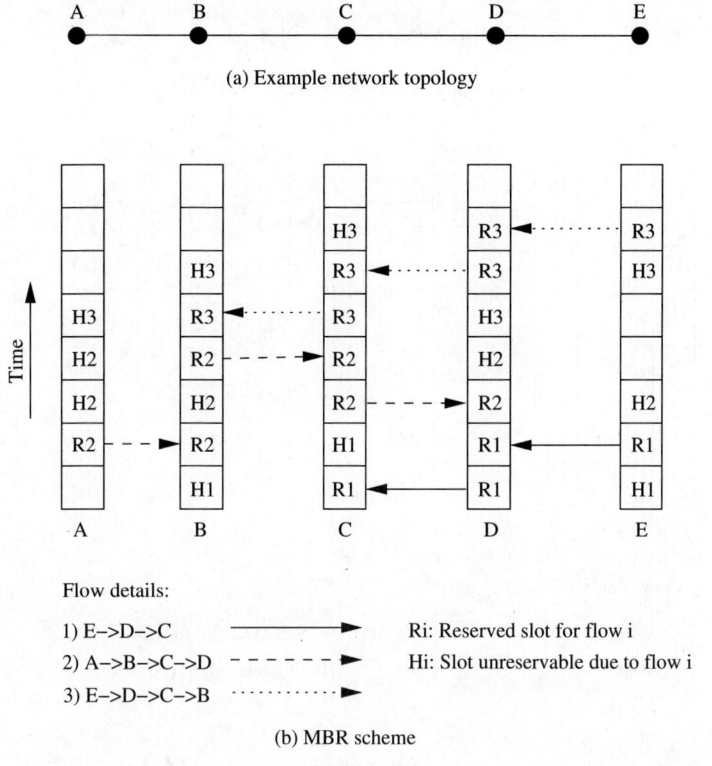

Figure 10.16. Illustration of MBR scheme. *Reproduced with permission from [23], © John Wiley & Sons Limited, 2004.*

- Step 1: List the links in the order of increasing bandwidth.
- Step 2: Assign a free slot for the link with least amount of bandwidth, such that the position of assignment of bandwidth is proportional to $\frac{i}{L_{path}}$ where i is the position of the link and L_{path} is the path length.
- Step 3: Repeat Step 2 until all the links are assigned with free slots.

Figure 10.17 shows the slot allocation done on a string topology for three flows. In the given example, the average delay experienced can be calculated as $\frac{18}{3}$ slots.

- k-hopcount hybrid routing (k-HHR): This is a hybrid slot allocation scheme in which either EFR or PHR is chosen dynamically by the destination node based on the hop length of the path. The k-HHR scheme is described below.

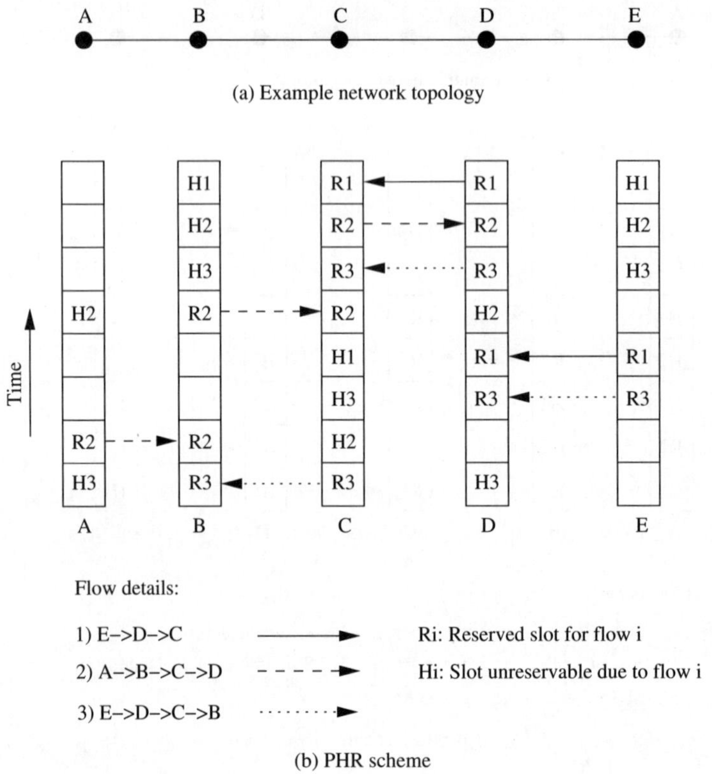

Figure 10.17. Illustration of PHR scheme. *Reproduced with permission from [23], © John Wiley & Sons Limited, 2004.*

if (*pathlength* $>$ *k*)

  Use EFR for slot allocation

else

  Use PHR for slot allocation

This takes the end-to-end delay advantage of the EFR scheme for long flows and the high call acceptance with medium end-to-end delay of the PHR scheme for flows with shorter length.

Bandwidth Reservation Phase

In this phase, a reservation of bandwidth at every link of a path is carried out. The reservation is effected by the intermediate nodes with the information carried in the *RouteReply* packet, in an asynchronous fashion using RTMAC protocol. Once the reservation at an intermediate link is successful in the designated time duration (the time duration for a free conn-slot, at which the reservation is to be carried out), the *RouteReply* packet is further forwarded. If the designated slot is not free at the time the intermediate node attempts the reservation (this can happen either due to the mobility of nodes or due to the staleness of the information), the intermediate node can try reserving any of the free slots available. If the intermediate node finds it impossible to reserve bandwidth, it drops the *RouteReply* and sends a control packet to the destination, which makes all the nodes in its way, those that have successfully reserved bandwidth, release the bandwidth and the destination node to find another path with the necessary bandwidth.

Advantages and Disadvantages

AQR has a unique advantage in that it can provide end-to-end bandwidth reservation in asynchronous networks. Also, the slot allocation strategies can be used to plan for the delay requirements and dynamically choose appropriate algorithms. AQR is an on-demand QoS routing scheme and hence the setup time and reconfiguration time of real-time calls are high. Also, the bandwidth efficiency of such an asynchronous system may not be as high as a fully synchronized TDMA system due to the formation of bandwidth holes (short free slots which cannot be used).

10.6 QOS FRAMEWORKS FOR AD HOC WIRELESS NETWORKS

A framework for QoS is a complete system that attempts to provide required/promised services to each user or application. All components within this system cooperate in providing the required services.

The key component of any QoS framework is the QoS service model which defines the way user requirements are met. The key design issue here is whether to serve users on a per session basis or on a per class basis. Each class represents an aggregation of users based on certain criteria. The other key components of the framework are QoS routing which is used to find all or some of the feasible paths in the network that can satisfy user requirements, QoS signaling for resource reservation, QoS medium access control, call admission control, and packet scheduling schemes. The QoS modules, namely, routing protocol, signaling protocol, and the resource management mechanism, should react promptly to changes in the network state (topology changes) and flow state (change in the end-to-end view of the service delivered). In what follows, each component's functionality and its role in providing QoS in ad hoc wireless networks will be described.

- *Routing protocol:* Similar to the QoS routing protocols, discussed earlier in this chapter, the routing protocol module in any QoS framework is used to find a path from the source to the destination and to forward the data packet to the next intermediate relay node. QoS routing describes the process of finding suitable path(s) that satisfy the QoS service requirements of an application. If multiple paths are available, the information regarding such paths helps to restore the service quickly when the service becomes disturbed due to a path break. The performance of the routing protocol, in terms of control overhead, affects the performance of the QoS framework. The routing protocol should be able to track changes in the network topology with minimum control overhead. The routing protocol needs to work efficiently with other components of the QoS framework such as signaling, admission control, and resource management mechanisms in order to provide end-to-end QoS guarantees. These mechanisms should consume minimal resources in operation and react rapidly to changes in the network state and the flow state.

- *QoS resource reservation signaling:* Once a path with the required QoS is found, the next step is to reserve the required resources along that path. This is done by the resource reservation signaling protocol. For example, for applications that require certain minimum bandwidth guarantees, signaling protocol communicates with the medium access control subsystem to find and reserve the required bandwidth. On completion/termination of a session, the previously reserved resources are released.
- *Admission control:* Even though a QoS feasible path may be available, the system needs to decide whether to actually serve the connection or not. If the call is to be served, the signaling protocol reserves the resources; otherwise, the application is notified of the rejection. When a new call is accepted, it should not jeopardize the QoS guarantees given to the already admitted calls. A QoS framework is evaluated based on the number of QoS sessions it serves and it is represented by the average call acceptance ratio (ACAR) metric. Admission control ensures that there is no perceivable degradation in the QoS being offered to the QoS sessions admitted already.
- *Packet scheduling:* When multiple QoS connections are active at the same time through a link, the decision on which QoS flow is to be served next is made by the scheduling scheme. For example, when multiple delay-constrained sessions are passing through a node, the scheduling mechanism decides on when to schedule the transmission of packets when packets belonging to more than one session are pending in the transmission queue of the node. The performance of a scheduling scheme is reflected by the percentage of packets that meet their deadlines.

10.6.1 QoS Models

A QoS model defines the nature of service differentiation. In wired network QoS frameworks, several service models have been proposed. Two of these models are the integrated services (IntServ) model [25] and the differentiated services (DiffServ) model [26]. The IntServ model provides QoS on a per flow basis, where a flow is an application session between a pair of end users. Each IntServ-enabled router maintains all the flow specific state information such as bandwidth requirements, delay bound, and cost. The different types of services offered in this model are guaranteed service, controlled load service, and best-effort service. The resource reservation protocol (RSVP) [27] is used for reserving the resources along the route.

The volume of information maintained at an IntServ-enabled router is proportional to the number of flows. Hence, the IntServ model is not scalable for the Internet, but it can be applied to small-sized ad hoc wireless networks. However, per flow information is difficult to maintain precisely at a node in an ad hoc wireless network due to reasons such as limited processing capability, limited battery energy, frequent changes in network topology, and continuously varying link capacity due to the time-varying characteristics of radio links. The DiffServ model was proposed in order to overcome the difficulty in implementing and deploying IntServ model and RSVP in the Internet. In this model, flows are aggregated into a limited number of service classes. Each flow belongs to one of the DiffServ classes of service. This solved the scalability problem faced by the IntServ model.

The above two service models cannot be directly applied to ad hoc wireless networks because of the inherent characteristics of ad hoc wireless networks such as continuously varying network topology, limited resource availability, and error-prone shared radio channel. Any service model proposed should first decide upon what types of services are feasible in such networks. A hybrid service model for ad hoc wireless networks called flexible QoS model for mobile ad hoc networks (FQMM) is described below. This model is based on the above two QoS service models.

Flexible QoS Model for Mobile Ad Hoc Networks

The flexible QoS model for mobile ad hoc networks (FQMM) [28] takes advantage of the per flow granularity of IntServ and aggregation of services into classes in DiffServ. In this model the nodes are classified into three different categories, namely, *ingress* node (source), *interior* node (intermediate relay node), and *egress* node (destination) on a per flow basis. A source node, which is the originator of the traffic, is responsible for traffic-shaping. Traffic-shaping is the process of delaying packets belonging to a flow so that packets conform to a certain defined traffic profile. The traffic profile contains a description of the temporal properties of a flow such as its mean rate (*i.e.,* the rate at which data can be sent per unit time on average) and burst size (which specifies in bits per burst how much traffic can be sent within a given unit of time without creating scheduling concerns). The FQMM model provides per flow QoS services for the high-priority flows while lower priority flows are aggregated into a set of service classes as illustrated in Figure 10.18. This hybrid service model is based on the assumption that the percentage of flows requiring per flow QoS service is much less than that of low-priority flows which can be aggregated into QoS classes. Based on the current traffic load in the network, the service level of a flow may change dynamically from per flow to per class and vice versa.

Advantages and Disadvantages

FQMM provides the ideal per flow QoS service and overcomes the scalability problem by classifying the low-priority traffic into service classes. This protocol addresses the basic problem faced by QoS frameworks and proposes a generic solution

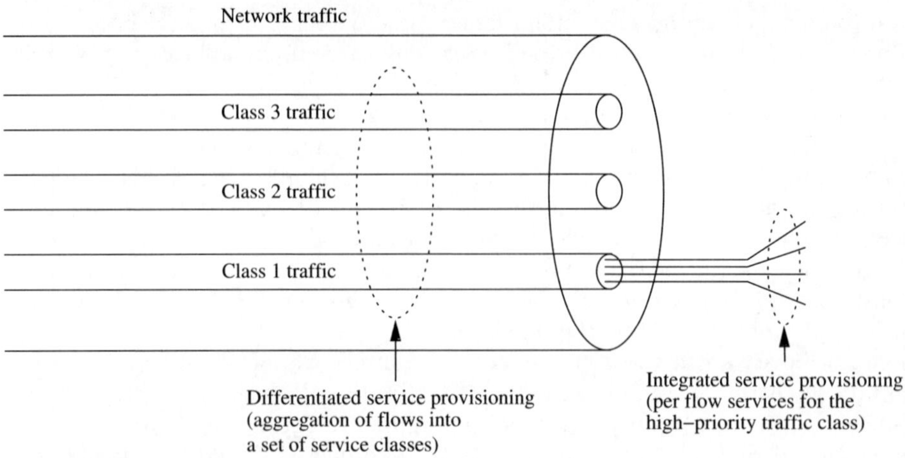

Figure 10.18. FQMM model.

for ad hoc wireless networks that can be a base for a better QoS model. However, several issues still remain unresolved, such as decision upon traffic classification, allotment of per flow or aggregated service for the given flow, amount of traffic belonging to per flow service, the mechanisms used by the intermediate nodes to get information regarding the flow, and scheduling or forwarding of the traffic by the intermediate nodes.

10.6.2 QoS Resource Reservation Signaling

The QoS resource reservation signaling scheme is responsible for reserving the required resources and informing the corresponding applications, which then initiate data transmission. Signaling protocol consists of three phases, namely, connection establishment, connection maintenance, and connection termination. On establishing a connection, it monitors the path and repairs/reconfigures it if the connection suffers from any violation in its QoS guarantees. On completion/termination of a session, it releases the resources that had been reserved for that session. In the wired networks, the RSVP protocol [27] is used for resource reservation, but it cannot be applied directly to ad hoc wireless networks due to the following reasons:

- The amount of control overhead generated during the connection maintenance phase of RSVP signaling is too heavy for bandwidth-constrained ad hoc wireless networks.

- It is not adaptive to network dynamics. In wired networks, once the resources are reserved, they are assumed to be available to applications throughout the session. But these assumptions are not true in ad hoc wireless networks due to the unrestricted mobility of nodes, which results in dynamic changes in the network topology.

MRSVP: A Resource Reservation Protocol for Cellular Networks

The MRSVP [29], as discussed in Chapter 4, is an extension of RSVP protocol for mobile hosts. It allows a mobile host to connect to the network through different points (base stations) in the course of time. It is basically proposed as an extension of RSVP for cellular networks to integrate them with the IP network. It is assumed that a mobile host predicts precisely the set of locations that the host is expected to visit during the lifetime of the flow. This information is provided in the form of mobility specifications to the network, so that reservations are made before that host uses the paths. The protocol proposes two types of reservations: *active* and *passive*. The reservation made over a path for a QoS flow is said to be active if data packets currently flow along that path. A reservation is said to be passive if the path on which resources have been reserved is to be used only in the future. Resources that are reserved passively for a flow can be utilized by other flows that require best-effort service.

MRSVP employs *proxy agents* (just as home agents and foreign agents in mobile IP protocol) to reserve resources along the path from the sender to the locations in the mobility specification of the mobile host. The *local proxy agent* (*i.e.*, the *proxy agent* present at the current location of the mobile host) makes an *active* reservation from the sender to the mobile host. The *remote proxy agents* (*i.e.*, *proxy agents* present at other locations in the mobility specification of the mobile host) make *passive* reservations on behalf of the mobile host.

Limitations of Adapting MRSVP for Ad Hoc Wireless Networks

MRSVP requires the future locations of mobile hosts in advance. Obtaining such location information is extremely difficult in ad hoc wireless networks because of the unrestricted mobility of the mobile hosts. Due to this reason, passive reservations fail in the case of ad hoc wireless networks. Secondly, even if future locations are known, finding a path and reserving resources on that path in advance may not be a viable and efficient solution. This inefficiency is because of the random and unpredictable movement of the intermediate nodes. It is also unknown which nodes should act as proxy agents due to the lack of infrastructure support in ad hoc wireless networks.

10.6.3 INSIGNIA

The INSIGNIA QoS framework [30] was developed to provide adaptive services in ad hoc wireless networks. Adaptive services support applications that require only a minimum quantitative QoS guarantee (such as minimum bandwidth) called *base QoS*. The service level can be extended later to *enhanced QoS* when sufficient resources become available. Here user sessions adapt to the available level of service without explicit signaling between the source-destination pairs. The key design issues in providing adaptive services are as follows:

- How fast can the application service level be switched from *base QoS* to *enhanced QoS* and vice versa in response to changes in the network topology and channel conditions?

- How and when is it possible to operate on the *base QoS* or *enhanced QoS* level for an adaptive application (*i.e.*, an application that can sustain variation in QoS levels)?

This framework can scale down, drop, or scale up user sessions adaptively based on network dynamics and user-supplied adaptation policies. A key component of this framework is the INSIGNIA in-band signaling system, which supports fast reservation, restoration, and adaptation schemes to deliver the adaptive services. The signaling system is light-weight and responds rapidly to changes in the network topology and end-to-end QoS conditions. As depicted in Figure 10.19, the INSIGNIA framework has the following key components for supporting adaptive real-time services:

- *Routing module:* The routing protocol finds a route from the source to the destination. It is also used to forward a data packet to the next intermediate

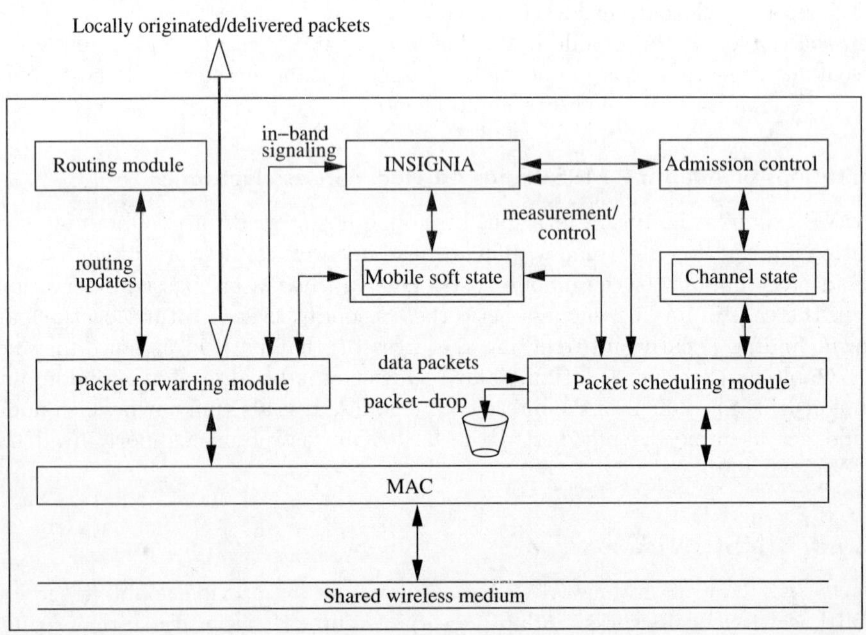

Figure 10.19. INSIGNIA QoS framework.

relay node. The routing module is independent of other components and hence any existing routing protocol can be used. INSIGNIA assumes that the routing protocol provides new routes in case of topology changes.

- *In-band signaling:* This module is used to establish, adapt, restore, and tear down adaptive services between source-destination pairs. It is not dependent on any specific link layer protocol. In in-band signaling systems, the control information is carried along with data packets and hence no explicit control channel is required. In the INSIGNIA framework, each data packet contains an optional QoS field (INSIGNIA option) to carry the control information. The signaling information is encoded into this optional QoS field. The in-band signaling system can operate at speeds close to that of packet transmissions and is therefore better suited for highly dynamic mobile network environments.

- *Admission control:* This module allocates bandwidth to flows based on the maximum/minimum bandwidth requested. Once the bandwidth is reserved, the reservation must be refreshed periodically by a soft state mechanism. Typically, the reception of data packets refreshes the reservations done.

- *Packet forwarding:* This module classifies the incoming packets and delivers them to the appropriate module. If the current node is the destination of the packet, then the packet is delivered to the local application. If the packet has an INSIGNIA option, it is delivered to the INSIGNIA signaling module. If the destination is some other node, it is relayed to the next intermediate node with the help of the routing and packet-scheduling modules.

- *Packet scheduling:* Packets that are to be routed to other nodes are handled by the packet-scheduling module. The packets to be transmitted by a node are scheduled by the scheduler based on the forwarding policy. INSIGNIA uses a weighted round-robin service discipline.

- *Medium access control (MAC):* The MAC protocol provides QoS-driven access to the shared wireless medium for adaptive real-time services. The INSIGNIA framework is transparent to any underlying MAC protocol.

The INSIGNIA framework uses a soft state resource management mechanism for efficient utilization of resources. When an intermediate node receives a data packet with a RES (reservation) flag set for a QoS flow and no reservation has been done until now, the admission control module allocates the resources based on availability. If the reservation has been done already, it is reconfirmed. If no data packets are received for a specified timeout period, the resources are deallocated in a distributed manner without incurring any control overhead. In setting the value for the timeout period, care should be taken to avoid *false restoration* (which occurs when the time interval is smaller than the inter-arrival time of packets) and resource lock-up (which occurs when the time interval is much greater than the inter-arrival time of packets).

Operation of INSIGNIA Framework

The INSIGNIA framework supports adaptive applications which can be applications requiring best-effort service or applications with *base QoS* requirements or those with *enhanced QoS* requirements. Due to the adaptation of the protocol to the dynamic behavior of ad hoc wireless networks, the service level of an application can be degraded in a distributed manner if enough resources are not available. For example, data packets belonging to the *enhanced QoS* service mode may have to be routed in the *base QoS* or best-effort service modes adaptively due to lack of enough resources along the path. If enough resources become available later during the lifetime of the connection, the application can be upgraded.

The INSIGNIA option field contains the following information: service mode, payload type, bandwidth indicator, and bandwidth request. These indicate the dynamic behavior of the path and the requirements of the application. The intermediate nodes take decisions regarding the flow state in a distributed manner based on the INSIGNIA option field. The service mode can be either best-effort (BE) or service requiring reservation (RES) of resources. The payload type indicates the QoS requirements of the application. It can be either *base QoS* for an application that requires minimum bandwidth, or *enhanced QoS* for an application which requires a certain maximum bandwidth but can operate with a certain minimum bandwidth below which they are useless. Examples of applications that require enhanced service mode are video applications that can tolerate packet loss and delay jitter to a certain extent. The bandwidth indicator flag has a value of MAX or MIN, which represents the bandwidth available for the flow. Table 10.2 shows how service mode, payload type, and bandwidth indicator flags reflect the current status of flows. It can be seen from the table that the best-effort (BE) packets are routed as normal data packets. If QoS is required by an application, it can opt for *base QoS* in which a certain minimum bandwidth is guaranteed. For that application, the bandwidth indicator flag is set to MIN. For *enhanced QoS*, the source sets the bandwidth indicator flag to MAX, but it can be downgraded at the intermediate nodes to MIN; the service mode flag is changed to BE from RES if sufficient bandwidth is not available. The downgraded service can be restored to RES if sufficient bandwidth becomes available. For *enhanced QoS*, the service can

Table 10.2. INSIGNIA flags reflecting the behavior of flows

Service Mode	Payload Type	Bandwidth Indicator	Downgrading	Upgrading
BE	-	-	-	-
RES	Base QoS	MIN	Base QoS → BE	BE → Base QoS
RES	Enhanced QoS (EQoS)	MAX	EQoS → BE EQoS → BQoS	BE → EQoS BQoS → EQoS

be downgraded either to BE service or RES service with *base QoS*. The downgraded *enhanced QoS* can be upgraded later, if all the intermediate nodes have the required (MAX) bandwidth.

The destination nodes actively monitor the on-going flows, inspecting the bandwidth indicator field of incoming data packets and measuring the delivered QoS (*e.g.*, packet loss, delay, and throughput). The destination nodes send QoS reports to the source nodes. The QoS reports contain information regarding the status of the on-going flows.

Releasing Resources in INSIGNIA

In order to release resources, the destination node sends a QoS report to the source so that the intermediate nodes release the extra resources. Assume that a source node transmits an *enhanced QoS* data packet with MAX requirements. If sufficient bandwidth is available, the intermediate nodes reserve the MAX bandwidth. Now assume that sufficient bandwidth is not available at an intermediate node (bottleneck node). Then the bottleneck node changes the bandwidth indicator flag of the incoming data packet from MAX to MIN. In this case, the intermediate nodes (from the source to the bottleneck node) would have allocated extra resources that remain unutilized, while downstream nodes from the bottleneck node allocate resources only for the downgraded service. Upon receiving the incoming data packet with bandwidth indicator flag set to MIN, the destination node sends a QoS report to inform the corresponding source node that the service level of the flow has been degraded. Further, the intermediate nodes receiving this QoS report release the extra unutilized resources.

Route Maintenance

Due to host mobility, an on-going session may have to be rerouted in case of a path break. The flow restoration process must reestablish the reservation as quickly and efficiently as possible. During restoration, INSIGNIA does not preempt resources from the existing flows for admitting the rerouted flows. INSIGNIA supports three types of flow restoration, namely, *immediate restoration*, which occurs when a rerouted flow immediately recovers to its original reservation; *degraded restoration*, which occurs when a rerouted flow is degraded for a period (T) before it recovers to its original reservation; and *permanent restoration*, which occurs when the rerouted flow never recovers to its original reservation.

Advantages and Disadvantages

INSIGNIA framework provides an integrated approach to QoS provisioning by combining in-band signaling, call admission control, and packet scheduling. The soft state reservation scheme used in this framework ensures that resources are quickly released at the time of path reconfiguration. However, this framework supports only adaptive applications, for example, multimedia applications. Since this framework is transparent to any MAC protocol, fairness and the reservation scheme of the MAC protocol have a significant influence in providing QoS guarantees. Also, because this framework assumes that the routing protocol provides new routes in the case of topology changes, the route maintenance mechanism of the routing protocol employed significantly affects the delivery of real-time traffic. If enough resources are not available because of the changing network topology, the *enhanced* QoS application may be downgraded to *base* QoS or even to best-effort service. As this framework uses in-band signaling, resources are not reserved before the actual data transmission begins. Hence, INSIGNIA is not suitable for real-time applications that have stringent QoS requirements.

10.6.4 INORA

INORA [31] is a QoS framework for ad hoc wireless networks that makes use of the INSIGNIA in-band signaling mechanism and the TORA routing protocol discussed in Section 7.5.3. The QoS resource reservation signaling mechanism interacts with the routing protocol to deliver QoS guarantees. The TORA routing protocol provides multiple routes between a given source-destination pair. The INSIGNIA signaling mechanism provides feedback to the TORA routing protocol regarding the route chosen and asks for alternate routes if the route provided does not satisfy the QoS requirements. For resource reservation, a soft state reservation mechanism is employed. INORA can be classified into two schemes: *coarse feedback scheme* and *class-based fine feedback scheme.*

Coarse Feedback Scheme

In this scheme, if a node fails to admit a QoS flow either due to lack of minimum required bandwidth (BW_{min}) or because of congestion at the node, it sends an out-of-band *admission control failure* (ACF) message to its upstream node. After receiving the ACF message, the upstream node reroutes the flow through another downstream node provided by the TORA routing protocol. If none of its neighbors is able to admit the flow, it in turn sends an ACF message to its upstream node. While INORA is trying to find a feasible path by searching the *directed acyclic graph* (DAG) following admission control failure at an intermediate node, the packets are transmitted as best-effort packets from the source to destination. In this scheme, different flows between the same source-destination pair can take different routes.

The operations of the coarse feedback scheme are explained through the following example. Here a QoS flow is being initiated by the source node S to the destination node D.

1. Let the DAG created by the TORA protocol be as shown in Figure 10.20. Let $S \rightarrow A \rightarrow B \rightarrow D$ be the path chosen by the TORA routing protocol.

2. INSIGNIA tries to establish soft state reservations for the QoS flow along the path. Assume that node A has admitted the flow successfully and node B fails to admit the flow due to lack of sufficient resources. Node B sends an ACF message to node A.

3. Node A tries to reroute the flow through neighbor node Y provided by TORA.

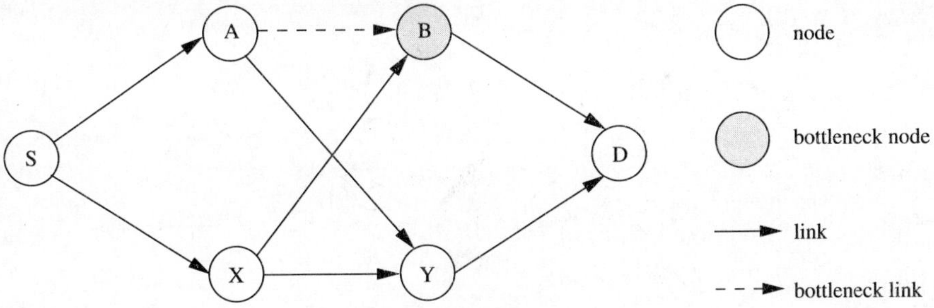

Figure 10.20. INORA coarse feedback scheme: admission control failure at node B.

4. If node Y admits the flow, the flow gets the required reservation all along the path. The new path is $S \rightarrow A \rightarrow Y \rightarrow D$.

5. If node Y fails to admit the flow, it sends an ACF message to node A, which in turn sends an ACF message to node S.

6. Node S tries with its other downstream neighbors to find a QoS path for the flow.

7. If no such neighbor is available, node S rejects the flow.

Class-Based Fine Feedback Scheme

In this scheme, the interval between BW_{min} and BW_{max} of a QoS flow is divided into N classes, where BW_{min} *and* BW_{max} are the minimum and maximum bandwidths required by the QoS flow. Consider a QoS flow being initiated by the source node S to destination node D. Let the flow be admitted with class m ($m < N$).

1. Let the DAG created by the TORA protocol be as shown in Figure 10.21. Let $S \rightarrow A \rightarrow B \rightarrow D$ be the path chosen by the TORA routing protocol.

2. INSIGNIA tries to establish soft state reservations for the QoS flow along the path. Assume that node A has admitted the flow with class m successfully and node B has admitted the flow with bandwidth of class l ($l < m$) only.

3. Node B sends an *Admission Report* message ($AR(l)$) to upstream node A, indicating its ability to give only class l bandwidth to the flow.

4. Node A splits the flow in the ratio of l to $m - l$ and forwards the flow to node B and node Y in that ratio.

5. If node Y is able to give class $(m - l)$ as requested by node A, then the flow of class m is split into two flows, one flow with bandwidth of class l along the path $S \rightarrow A \rightarrow B \rightarrow D$ and the other one with bandwidth of class $(m - l)$ along path $S \rightarrow A \rightarrow Y \rightarrow D$.

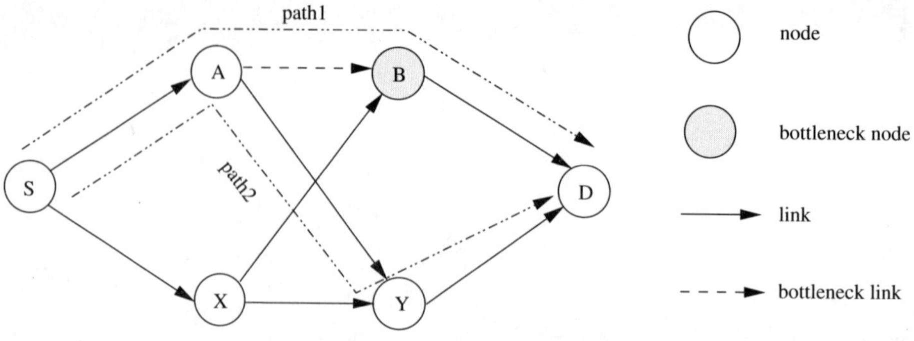

Figure 10.21. INORA fine feedback scheme: node A has admitted the flow with class m, but node B is able to give it class l ($l < m$).

6. If node Y gives only class n ($n < m - l$), it sends an $AR(n)$ message to the upstream node A.

7. Node A, realizing that its downstream neighbors are unable to give class m service, informs of its ability to provide service class of $(l + n)$ by sending an $AR(l + n)$ to node S.

8. Node S tries to find another downstream neighbor which might be able to accommodate the flow with class $(m - (l + n))$.

9. If no such neighbor is available, node S rejects the flow.

Advantages and Disadvantages

INORA is better than INSIGNIA in that it can search multiple paths with lesser QoS guarantees. It uses the INSIGNIA in-band signaling mechanism. Since no resources are reserved before the actual data transmission begins and since data packets have to be transmitted as best-effort packets in case of admission control failure at the intermediate nodes, this model may not be suitable for applications that require hard service guarantees.

10.6.5 SWAN

Ahn *et al.* proposed a distributed network model called stateless wireless ad hoc networks (SWAN) [32] that assumes a best-effort MAC protocol and uses feedback-based control mechanisms to support real-time services and service differentiation in ad hoc wireless networks. SWAN uses a local rate control mechanism for regulating injection of best-effort traffic into the network, a source-based admission control while accepting new real-time sessions, and an explicit congestion notification (ECN) mechanism for dynamically regulating admitted real-time sessions. In this model, intermediate nodes are relieved of the responsibility of maintaining per-flow or aggregate state information, unlike stateful QoS models such as INSIGNIA and INORA. Changes in topology and network conditions, even node and link failures, do not affect the operation of the SWAN control system. This makes the system simple, robust, and scalable.

SWAN Model

The SWAN model has several control modules which are depicted in Figure 10.22. Upon receiving a packet from the IP layer, the *packet classifier* module checks whether it is marked (*i.e.*, real-time packet) or not (*i.e.*, best-effort packet). If it is a best-effort packet, it is forwarded to the *traffic-shaper* for regulation. If it is a real-time packet, the module forwards it directly to the MAC layer, bypassing the *traffic shaper*. The *traffic shaper* represents a simple leaky bucket traffic policy. The traffic shaper delays best-effort packets in conformance with the rate calculated by the *traffic rate controller*. The *call admission controller* module is responsible for admitting or rejecting new real-time sessions. The decision on whether to admit or reject a real-time session is taken solely by the source node based on the result of an end-to-end request/response probe. The SWAN distributed control algorithms are described in the following sections.

Local Rate Control of Best-Effort Traffic

The SWAN model assumes that most of the traffic existing in the network is best-effort, which can serve as a "buffer zone" or absorber for real-time traffic bursts

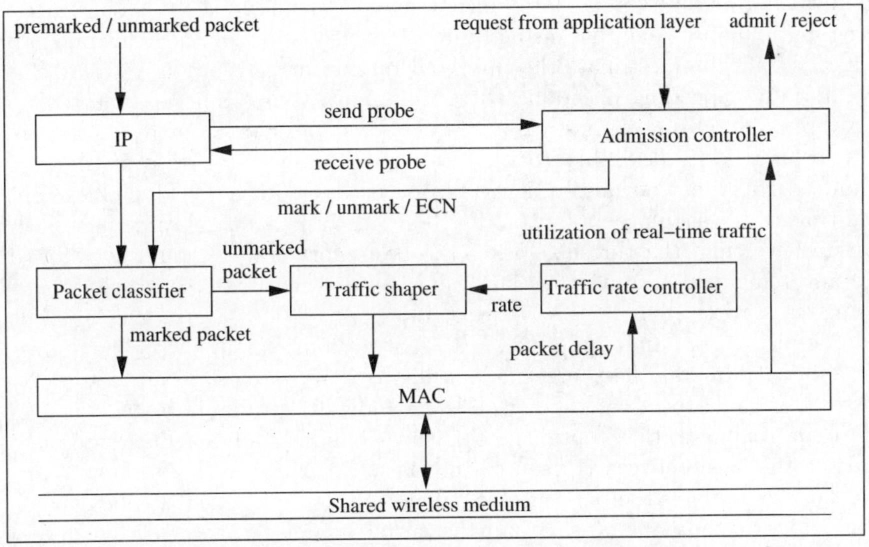

Figure 10.22. The SWAN model.

introduced by mobility (because of rerouting of the already admitted real-time sessions) or traffic variations (*e.g.*, bursty data). The best-effort traffic can be locally and rapidly rate-controlled in an independent manner at each node in order to yield the necessary low delays and stable throughput for real-time traffic. The best-effort traffic utilizes remaining bandwidth (if any) left out by real-time traffic. Hence, this model does not work in scenarios where most of the traffic is real-time in nature.

The *traffic rate controller* determines the departure rate of the traffic shaper using an additive increase multiplicative decrease (AIMD) rate-control algorithm which is based on packet delay feedback from the MAC layer. The SWAN AIMD rate-control algorithm works as follows. Every T seconds, each node increases its transmission rate gradually (additive increase with increment rate of c Kbps). If the packet delays exceed the threshold delay of d seconds, then the node decrements its transmission rate (multiplicative decrease by r percent). The shaping rate is adjusted every T seconds. The traffic rate controller monitors the actual transmission rate. When the difference between the shaping rate and the actual transmission rate is greater than g percent of the actual rate, then the traffic rate controller adjusts the shaping rate to be g percent above the actual rate. This gap allows the best-effort traffic to increase its actual rate gradually. The threshold delay d is based on the delay requirements of real-time applications in the network.

Source-Based Admission Control of Real-Time Traffic

At each node, the admission controller measures the rate of real-time traffic in bps. If the threshold rate that would trigger excessive delays is known, then the bandwidth availability in a shared media channel is simply the difference between the threshold rate and the current rate of transmission of the real-time traffic. But it is difficult to estimate the threshold rate accurately because it may change dynamically depending on traffic patterns. If real-time traffic is admitted up to the threshold rate, then best-effort traffic would be starved of resources. Also, there would be no flexibility to support any increase in the rate of the already admitted real-time sessions, which could occur due to channel dynamics. Hence real-time traffic should be admitted up to an admission control rate which is more conservative than the threshold rate; the best-effort traffic should be allowed to use any remaining bandwidth. The value for the admission control rate can be estimated statistically.

The process of admitting a new real-time session is as follows. The admission controller module at the source node sends a probing request packet toward the destination node to assess the end-to-end bandwidth availability. This is a best-effort control packet that contains a bottleneck bandwidth field. Each intermediate node on the path between the source-destination pair that receives the probing request packet updates the bottleneck bandwidth field in the packet if the bandwidth availability at the node is less than the current value of the field. On receiving the probing request packet, the destination node sends a probing response packet back to the source node with the bottleneck field copied from the received probing request packet. After receiving the response message, the source node admits the new real-time session only if sufficient end-to-end bandwidth is available. In this model, no bandwidth request is carried in the probing message, no admission control is done at intermediate nodes, and no resource allocation or reservation is done on behalf of the source node during the lifetime of an admitted session.

Impact of Mobility and False Admission

Host mobility and false admission pose a serious threat for fulfilling the service guarantees promised to the flows. Mobility necessitates dynamic rerouting of the admitted real-time flows. Since, due to mobility, nodes may be unaware of flow rerouting, resource conflicts can arise. The newly selected intermediate nodes may not have sufficient resources for supporting previously admitted real-time traffic. Take the case of multiple source nodes initiating admission control at the same instant and sharing common intermediate nodes on their paths to destination nodes. Since intermediate nodes do not maintain state information and since admission control is fully source-based, each source node may receive a response to its probe packet indicating that resources are available, even though the available resources may not be sufficient to satisfy all the requests. The source node, being unaware of this fact, falsely admits a new flow and starts transmitting real-time packets under the assumption that resources are available for meeting the flow's needs. If left unresolved, the rerouting of admitted real-time flows can cause excessive delays in delivery of real-time traffic since the admission control rate is violated by the falsely admitted calls. To resolve this problem, the SWAN AIMD rate control and source-based admission control algorithms were augmented with dynamic regulation of real-time traffic. The algorithms used for this dynamic regulation are described below.

Regulation Algorithms

The ECN-based regulation of real-time sessions operates as follows. Each node continuously estimates the locally available bandwidth. When a node detects congestion/overload conditions, it starts marking the ECN bits in the IP header of the real-time packets. If the destination receives a packet with ECN bits marked, it notifies the source using a regulate message. After receiving a regulate message, the source node initiates reestablishment of its real-time session based on its original bandwidth requirements by sending a probing request packet to the destination. A source node terminates the session if the available end-to-end bandwidth cannot meet its bandwidth requirements. If the node detecting violations marks (*i.e.*, sets) the ECN bits of all packets, then all sessions passing through this node are forced to reestablish their connections at the same instance. Since such an approach is inefficient, the SWAN model considered two approaches in which only a small number of sources are penalized.

Source-Based Regulation

In this scheme, the source node waits for a random amount of time after receiving a regulate message from a congested or overloaded intermediate node on the path to the destination node and then initiates the reestablishment process. This can avoid flash-crowd conditions. In this scheme, the rate of the real-time traffic will gradually decrease until it reaches below the admission control rate. Then the congested or overloaded nodes will stop marking packets. Even though this scheme is simple and source-based, it has the disadvantage that sources that regulate earlier than other sources are more likely to find the path overbooked and be forced to terminate their sessions.

Network-Based Regulation

Unlike the previous scheme, in this scheme, congested or overbooked nodes randomly select a *congestion set* of real-time sessions and mark only packets associated with that set. A congested node marks the congested set for a time period of T seconds and then calculates a new congested set. Hence, some intelligence is required at the intermediate nodes. As in the previous approach, nodes stop marking packets as *congested* when the measured rate of real-time traffic reaches below the admission control rate.

Advantages and Disadvantages

SWAN gives a framework for supporting real-time applications by assuming a best-effort MAC protocol and not making any resource reservation. It uses feedback-based control mechanisms to regulate real-time traffic at the time of congestion in the network. As best-effort traffic serves as a buffer zone for real-time traffic, this model does not work well in scenarios where most of the traffic is real-time in nature. Even though this model is scalable (because the intermediate nodes do not maintain any per-flow or aggregate state information), it cannot provide hard QoS guarantees due to lack of resource reservation at the intermediate nodes. An admitted real-time flow may encounter periodic violations in its bandwidth requirements. In the worst case, it may have to be dropped or be made to live with downgraded best-effort service. Hence, the local rate control of best-effort traffic mechanism alone may not be sufficient to fully support real-time traffic.

10.6.6 Proactive RTMAC

Proactive RTMAC (PRTMAC) [33] is a cross-layer framework, with an on-demand QoS extension of DSR routing protocol at the network layer and real-time MAC (RTMAC) [13] protocol at the MAC layer. PRTMAC is a tightly coupled solution which requires the bandwidth reservation and bandwidth availability estimation services from the underlying MAC protocol. It is designed to provide enhanced real-time traffic support and service differentiation to highly mobile ad hoc wireless networks such as that formed by military combat vehicles. The performance of real-time calls in ad hoc wireless networks are affected by the mobility of nodes in many different ways. The two major ways in which mobility affects real-time calls are *breakaways* and reservation *clashs*, which will be explained later in this section.

Reservation-based QoS solutions provide end-to-end bandwidth reservation for a real-time connection. This explicit reservation can be affected by the *breakaways* and *clashs*. Proactive RTMAC (PRTMAC) is a solution that operates in both network and MAC layers and uses an out-of-band signaling channel to gather additional information about the on-going real-time calls, so that proactive measures can be taken to protect these calls. A narrow-band control channel that operates over a transmission range with twice that of the data transmission range, is used as the out-of-band signaling channel.

Figures 10.23 (a) and 10.23 (b) show the *breakaway* problem where a path between node A to node D is established through nodes B and C. The intermediate link B-C can be broken due to the mobility of node C. Now, in a reservation-based real-time support scheme, the intermediate nodes that detect the broken link have to either repair the broken link or inform the sender and receiver about the path break. Since the *breakaways* are very frequent in ad hoc wireless networks, the control overhead generated as part of route reconfiguration can consume a significant amount of bandwidth, in addition to the annoying effect that each path reconfiguration can give to the end users. Figures 10.24 (a) and 10.24 (b) show the reservation *clash* problem in ad hoc wireless networks. Consider the bandwidth reservations done in a given slot (say, slot #1) between nodes A and B and between nodes C and D as illustrated in Figure 10.24 (a). This is a valid reservation because the node pairs are not overlapping in the reserved slot.

Now assume that the node D is mobile and moving toward node B, then at some point of time when the nodes B and D tend to get closer, *i.e.,* within each other's transmission range, the reservation overlapping occurs in slot #1. This is illustrated in Figure 10.24 (b). As a result of this, the packets scheduled to be transmitted by both nodes at this slot can get scrambled and lost. In such a case, traditional path reconfiguration processes identify the broken reservations and reconfigure both of them. This problem is referred to as *clash*.

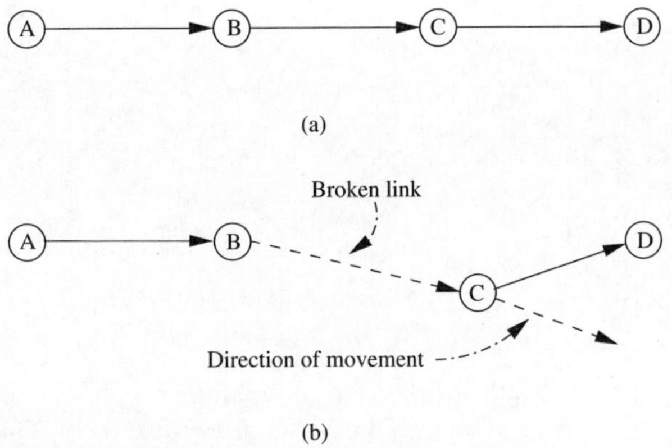

Figure 10.23. Illustration of *breakaway*.

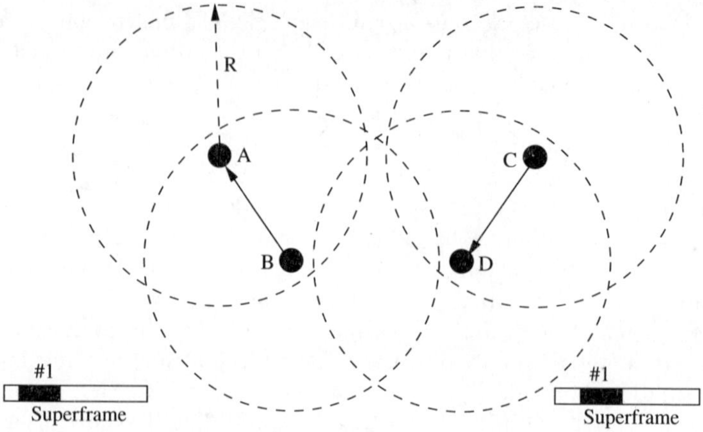

(a) Two overlapping reservations, but out of range: valid.

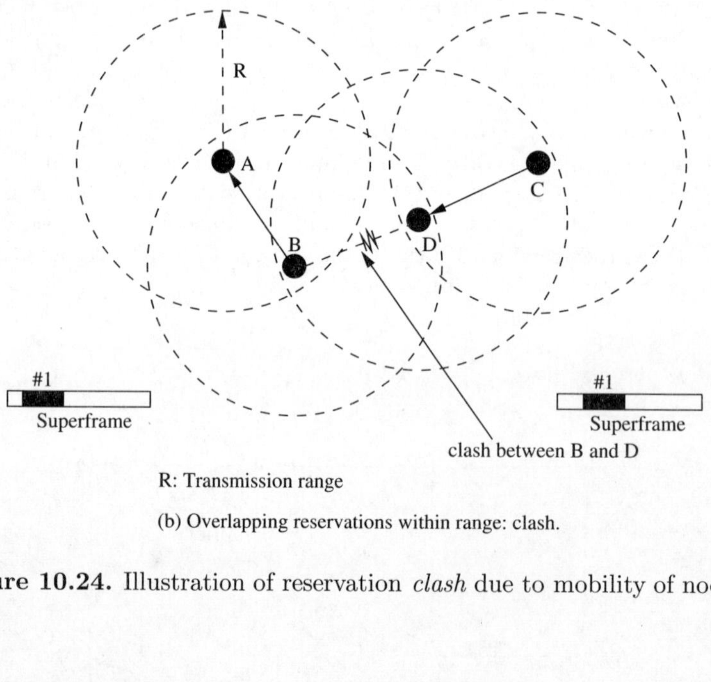

(b) Overlapping reservations within range: clash.

Figure 10.24. Illustration of reservation *clash* due to mobility of nodes.

Operation of PRTMAC

The PRTMAC framework is shown in Figure 10.25. This framework includes an out-of-band signaling module, a proactive call maintenance module, and a routing and call admission control module. The MAC protocol used is RTMAC, which is discussed in detail in Section 6.6.7. The operation of PRTMAC lies in collecting additional information about the real-time calls in the network to counter the effects of *breakaways* and *clashs*. This information is gathered over a narrow-band *control*

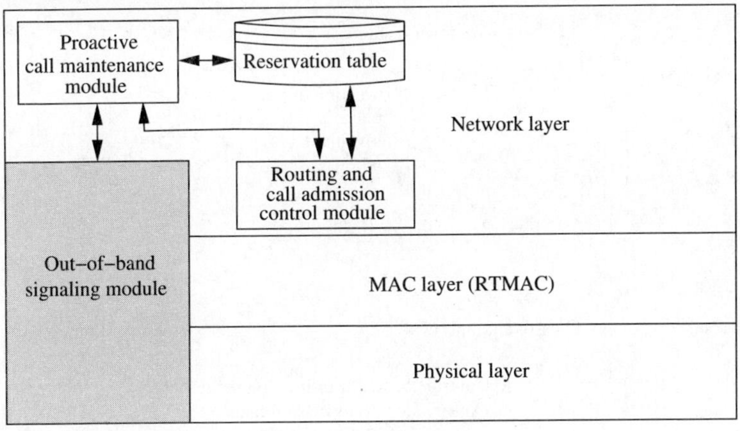

Figure 10.25. Modules in PRTMAC framework.

channel that has a greater transmission range than the data channel. Every node sends out control beacons (short fixed-sized packets) at regular intervals over the control channel. The information carried by the beacons, and the beacon itself, are used by the nodes to gather information about real-time calls. Firstly, the signal strength of the received beacon is used to gain an idea of the relative distance of the node which sent the beacon. Further, the information carried by the beacon is used in predicting *breakaways* and *clashs*. The beacons carry information about each of the calls that the originating node is carrying, and the slots in the superframe that have been reserved for them. Each node originates periodic beacons on the control channel. The beacon has information about all on-going real-time calls at the node. The information includes the start- and end-times of the reservation slot of each call, the sender and the receiver of the call, and the service class (service classes are used to provide differentiated services among the real-time calls existing in the system, for example, the command and control calls in a military communication system may require higher priority than the other calls) to which the call belongs. The range of the control channel must be sufficiently larger than that of the data channel so that all possible events that can cause a call to be interrupted can be discovered well in advance.

Crossover-Time Prediction

Crossover-time is defined as the time at which a node crosses another node's data transmission range r. This event is defined as *crossover*. As apparent from Figures 10.26 (a) and 10.26 (b), there are two different *crossover-times*, namely, *crossover-time-in* and *crossover-time-out*.

The *crossover-time-in* is the expected time at which node B in Figure 10.26 (a) reaches the crossover-point such that a bidirectional link forms between nodes A and B. Figure 10.26 (b) shows the *crossover-time-out*, which occurs at the instant node

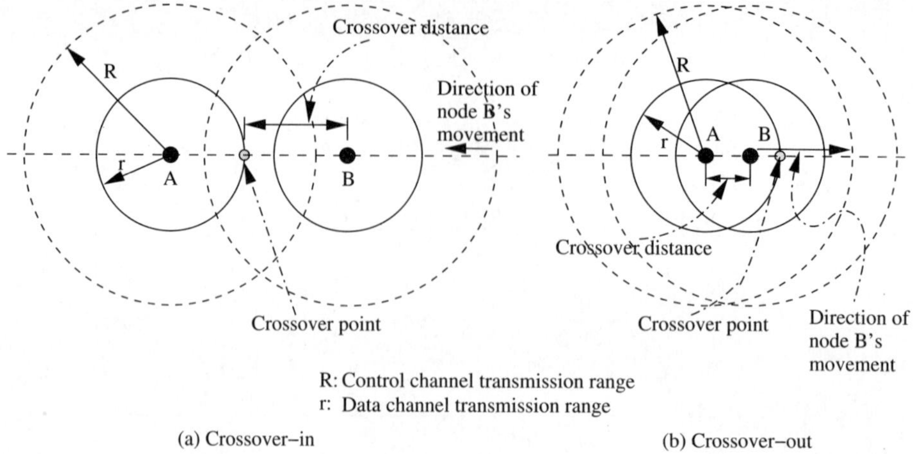

Figure 10.26. Illustration of *crossover-in* and *crossover-out.*

B moves away from node A so that the link between nodes A and B breaks. Each node (say, node A), upon reception of every new beacon from another node (say, node B), predicts the *crossover-time* based on the signal strength history obtained from past beacons, that is, if node B is inside the range of the data channel of node A, node A predicts the *crossover-time-out*, and if node B is outside the range of the data channel of node A, node A predicts the *crossover-time-in.*

The prediction of the *crossover-time-out* of node B with respect to node A is performed by keeping track of the signal strengths of the beacons previously sent by node B to node A (see Figure 10.27). A node stores a fixed number of $< time, signalstrength >$ tuples of the beacons received from any other node. Using this, it generates a polynomial on the variation of signal strength with time. The roots of the polynomial refer to the time at which the signal strength can cross a receiving threshold. When node A predicts that node B is going to cross the data channel range within the next beacon interval, it takes proactive actions described in the next section. If node B is already within the data channel range of node A, then the prediction will be for a *crossover-out* event, and all calls between nodes A and B will be interrupted. If node B is outside the range of node A, then it is a *crossover-in* event, and any packets belonging to existing real-time calls at node A and node B will collide if their reservation times overlap. Note that if the predicted time of entry is beyond the next beacon interval, no action needs to be taken as of now, since the event would be predicted again on receipt of the next beacon. When a node discovers that the *crossover* of another node is imminent, it takes proactive action to safeguard its privileged calls. If it is a *crossover-out*, it checks to see if it has any on-going traffic with the other node. If so, it has to resolve a case of *breakaway.* On the other hand, if it is a case of *crossover-in*, it examines the traffic

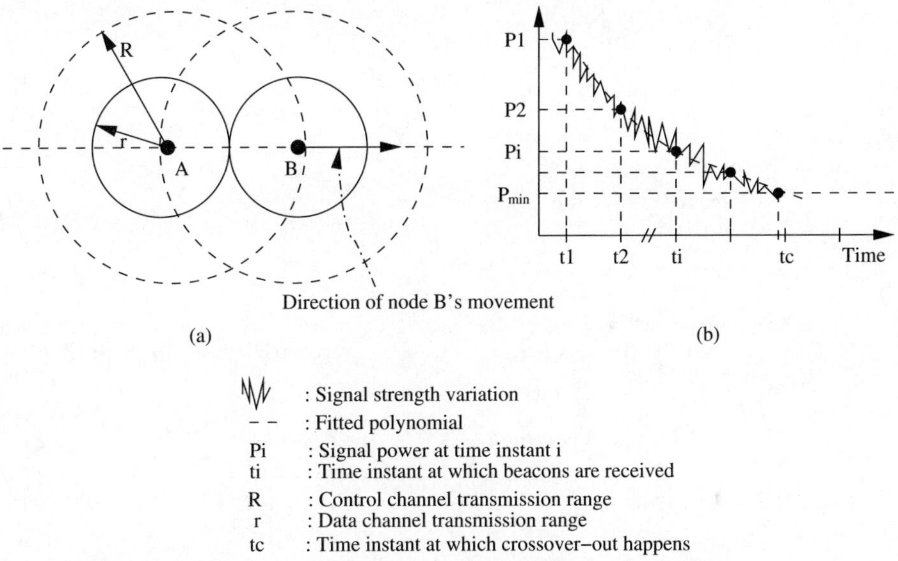

Figure 10.27. Illustration of prediction *crossover-time-out*: (a) Node B moves away from node A. (b) Signal strength variation of received beacons with time.

information of the other node and its own reservation tables to check for possible overlaps. If there is any such overlap, it has to resolve a case of *clash*.

Handling Breakaways

Figure 10.28 illustrates the handling of a broken path due to *breakaway*. The event of *breakaways* can be handled in two different ways: First is the local reconfiguration and second is the end-to-end reconfiguration.

The path reconfiguration is performed when a link breaks as in Figure 10.28 (a), where the path $A \rightarrow B \rightarrow C \rightarrow D \rightarrow E$ between nodes A and E is illustrated and the link formed between nodes C and D is broken. The local reconfiguration is also illustrated in Figure 10.28 (a), where the intermediate node, the link with whose downstream node is broken, holds the responsibility of finding a path to the destination node E. The intermediate node (node C) originates fresh route probe packets to obtain a path with reservation to the destination node. The end-to-end reconfiguration method is depicted in Figure 10.28 (b), where node C originates *RouteError* to the sender node in order to obtain a new path to the destination node. In this case, the reservation done in the old path may entirely be released. In PRTMAC a combination of the above two types is attempted, which is described as follows: Node C checks to see if its routing tables have another path toward the destination node (say, node F). If there exists such a node, then node C makes reservations on the link C-F for the on-going call. When the call is interrupted and

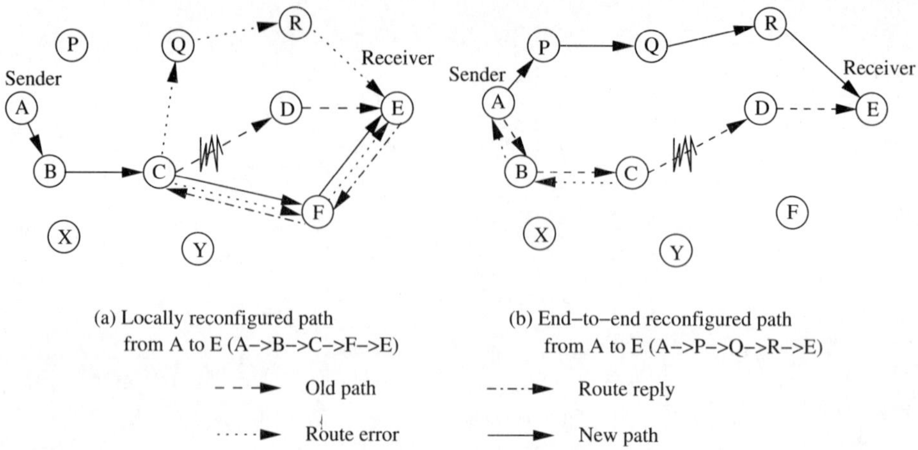

Figure 10.28. Illustration of route reconfiguration schemes for a path affected by breakaway.

reconfigured locally a number of times, as expected in an ad hoc wireless network, the end-to-end reconfiguration is attempted.

Handling Clashs

Figure 10.29 (a) illustrates how two nodes can reside safely within range of each other if the reserved slots do not overlap with each other. If the reservation slots *clash* for the two nodes, as indicated in Figure 10.29 (b), then PRTMAC handles it in such way that the flow between, say, node N and node C is assigned to a new slot (#5), as shown in Figure 10.30.

In the absence of any measures taken to resolve a *clash*, both the calls that experience a *clash* will be reconfigured from the source to the destination, resulting in degradation of performance. PRTMAC prevents such an occurrence to the extent possible by proactively shifting one of the calls to a new slot, so that the two calls do not *clash*. This benefit of *clash* resolution is more important when a higher priority call *clashs* with a lower priority call.

For the proactive slot shifting of one of the calls to happen, there is a need to decide unambiguously the node that will perform this reconfiguration. Since PRTMAC wants a high-priority call to continue undisturbed as long as possible, in case of a *clash* between a high-priority call and a low-priority call, it is the responsibility of the node having the low-priority call to reconfigure it to a new slot. If both the calls that *clash* are of high-priority, the one with the lower FlowID has to be reconfigured, and if both such calls have the same FlowID, the node with the lower node address (expected to be unique throughout the network) will have to reconfigure its call. The FlowID is a unique number associated with a call at a given node.

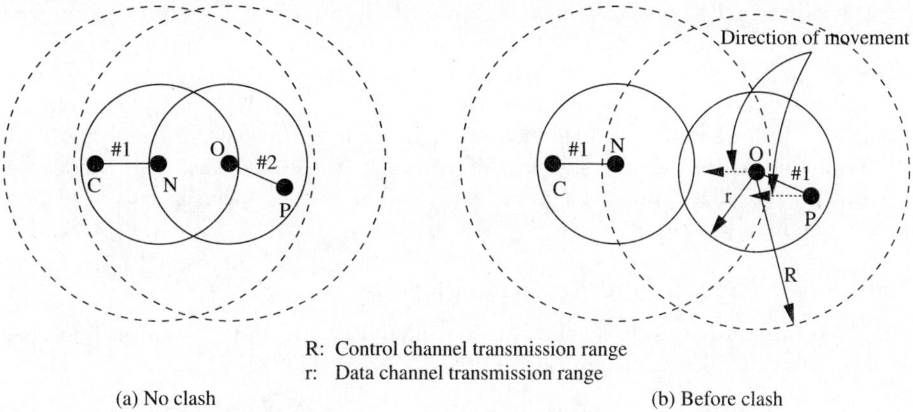

Figure 10.29. (a) Nodes N and O are within range, but reservation is safe. (b) Nodes N and O are out of range but are going to *clash* as nodes P and O are moving toward them.

As illustrated in Figure 10.30, the node whose responsibility it is to reconfigure the call is denoted by node N, the other node, whose call *clashs* with node N's call, is denoted by node O, and the counterpart of node N in its call by node C. Node N goes through its reservation tables and its neighbor reservation table corresponding to node C and tries to come up with a free reservation slot in both nodes N and C large enough to accommodate the call to be shifted. If it succeeds in finding such a free slot, the existing reservations for the call must be dropped and new reservations must be made for the call in the free slot. This is achieved when the originator of the call frees the earlier reservation and issues a request for the reservation of the slots belonging to the free slot.

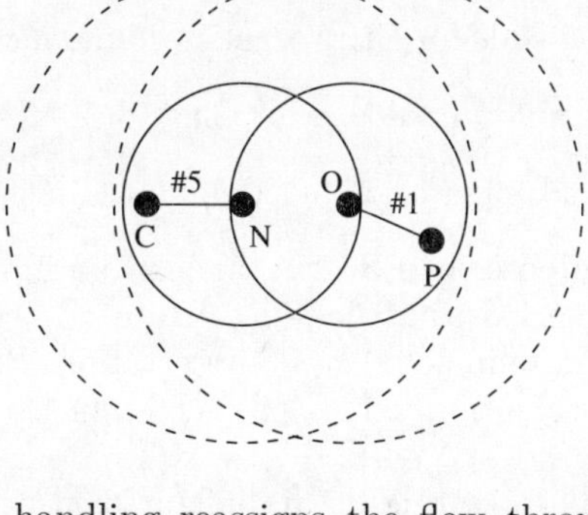

Figure 10.30. *Clash* handling reassigns the flow through the link N-C to a new slot.

If both the calls that *clash* have high priority and node N cannot come up with a slot free enough to accommodate the call, it informs node O about its failure in shifting the call. Now node O executes the above process with its counterpart and tries to shift the call. If one of the calls that *clash* is a high-priority call and the other a low-priority one, and the node that has a low-priority call (here it is node N) is unable to find a new slot to shift the call, the low-priority call undergoes end-to-end reconfiguration. This is to ensure that the low-priority call would not hinder the high-priority calls.

Differentiated Services Provisioning in PRTMAC

PRTMAC also provides class-based service differentiation among the priority classes. It supports the following three classes:

- *Class 1* is the class with the highest privilege. Calls belonging to this class are high-priority real-time calls. PRTMAC attempts to ensure that a Class 1 call is not interrupted as far as possible. If a situation occurs in which a Class 1 call can be sustained only at the expense of calls belonging to the other classes, PRTMAC utilizes the possibility and preempts the low-privilege call to preserve the Class 1 call.
- *Class 2* is the next privileged class. Calls belonging to this class are real-time calls with bandwidth reservation. A Class 2 call is sustained provided that there are no events in the network (such as *clash*s or *breakaway*s) that could cause the call to be terminated. Class 2 calls may be preempted in order to service Class 1 calls. A Class 2 call has an end-to-end bandwidth reservation at the time of accepting the call. If such a reservation is not possible, then the call is not accepted.
- The *Best-effort* class is the least privileged class. There are no guarantees made regarding any parameters corresponding to best-effort traffic.

The following are three major instances where priority is given to a Class 1 call over a Class 2 call:

- Preempting Class 2 calls for Class 1 calls while a new call is admitted.
- Handling of *clash*s and *breakaway*s for Class 1 calls.
- Prioritizing path-reconfiguration attempts for Class 1 calls.

During a new call admission, PRTMAC tries to admit a Class 1 call, even if it requires preempting an existing Class 2 call. In addition to privileged admission, upon detection of an imminent *clash* of reservation, PRTMAC provides higher priority to the Class 1 calls in order to protect them from being affected by the *clash*.

In addition to the above, during the reconfiguration process, the number of attempts made for reconfiguring a broken Class 1 call is higher than that of Class 2 calls. Among the equal priority calls, the differentiation is provided based on the node addresses and FlowIDs. Hence, by deliberately providing node addresses to the designated persons or nodes based on their rank or position in military establishments, and by choosing the FlowIDs for the calls originated by them, PRTMAC can provide very high chances of supporting guaranteed services. For example, the leader of the military group can be given the highest node address, and if the system is so designed that the FlowIDs for the calls he generates are higher than those for the rest of the nodes, then none of the other calls can terminate his calls if he decides to originate a Class 1 call.

Advantages and Disadvantages

PRTMAC is appropriate in providing better real-time traffic support and service differentiation in high mobility ad hoc wireless networks such as military networks formed by high-speed combat vehicles, fleets of ships, and fleets of aircrafts where the power resource is not a major concern. In ad hoc wireless networks, formed by low-power and resource-constrained handheld devices, having another channel may not be an economically viable solution.

10.7 SUMMARY

In this chapter, several solutions proposed in the literature for providing QoS support for applications in ad hoc wireless networks have been described. First, the issues and challenges in providing QoS in ad hoc wireless networks were discussed. Then the classifications of the existing QoS approaches under several criteria such as interaction between routing protocol and resource reservation signaling, interaction between network and MAC layer, and information update mechanism were discussed. The data link layer solutions such as cluster TDMA, IEEE 802.11e, and DBASE and the network layer solutions such as ticket-based probing, predictive location-based QoS routing, trigger-based QoS routing, QoS enabled AODV, bandwidth routing, on-demand routing, asynchronous QoS routing, and multipath QoS routing were described. Finally, QoS frameworks for ad hoc wireless networks such as INSIGNIA, INORA, SWAN, and PRTMAC were described.

10.8 PROBLEMS

1. What are the limitations of the IEEE 802.11 MAC protocol that prevent it from supporting QoS traffic?

2. Express various inter-frame spaces (IFSs) of the IEEE 802.11e MAC protocol in terms of $SIFS$ and *slottime*.

3. Compare and contrast the hybrid coordinator (HC) of the IEEE 802.11e MAC protocol with the point coordinator (PC) of the IEEE 802.11 MAC protocol.

4. What are the advantages of having transmission opportunities (TXOPs) in the IEEE 802.11e MAC protocol?

5. Compare and contrast the IEEE 802.11e MAC protocol with the DBASE protocol.

6. Discuss how a source node determines how many number of tickets (green and yellow tickets) are to be issued for a session in delay-constrained TBP protocol.

7. Discuss how a node estimates its expected location and under what circumstances the node generates a *Type2* update message in PLBQR protocol.

8. Consider the network topology shown in Figure 10.31 (a). Assume that free slots available at various nodes are as given in Figure 10.31 (b).

   (a) Using the hop-by-hop path bandwidth calculation algorithm proposed in the BR protocol, calculate the end-to-end path bandwidth for the paths given below.

      i. PATH1: $A \rightarrow B \rightarrow C \rightarrow D \rightarrow H \rightarrow L$

      ii. PATH2: $K \rightarrow G \rightarrow F \rightarrow E \rightarrow A$

      iii. PATH3: $J \rightarrow F \rightarrow B \rightarrow C \rightarrow D$

      iv. PATH4: $I \rightarrow J \rightarrow K \rightarrow L$

   (b) Further assume that four call setup packets are generated in the order given below. After admitting or rejecting a particular call, the next call is issued. Discuss which of these calls are admitted and rejected.

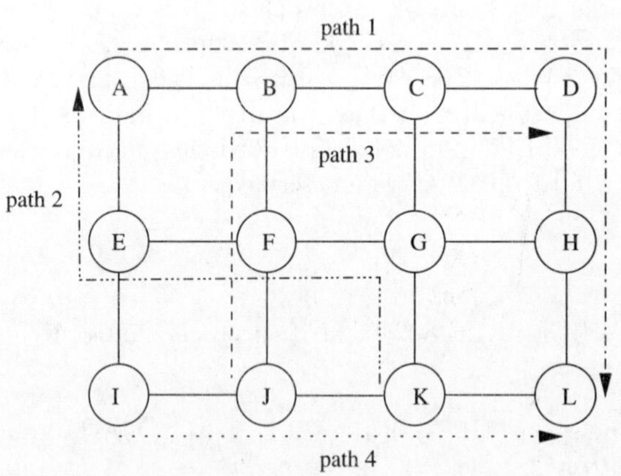

(a)

node ID	free slots
A	{1,2,4,6,7,8}
B	{1,2,5,6,7}
C	{2,4,6,7,8}
D	{1,4,5,6,7,8}
E	{1,3,5,6,7}
F	{1,2,4,6,7,8}
G	{1,3,5,6,7}
H	{2,3,5,6,8}
I	{1,2,4,6,8}
J	{1,5,6,7,8}
K	{2,3,5,6,8}
L	{1,3,4,5,8}

(b)

Figure 10.31. Network topology and slot table for Problem 8.

i. Node A issues a call setup packet having a 2-slots bandwidth requirement. The call setup packet traverses along the PATH1.

ii. Node K issues a call setup packet having a 1-slot bandwidth requirement. The call setup packet traverses along the PATH2.

iii. Node J issues a call setup packet having a 2-slots bandwidth requirement. The call setup packet traverses along the PATH3.

iv. Node I issues a call setup packet having a 2-slots bandwidth requirement. The call setup packet traverses along the PATH4.

9. Consider the network topology shown in Figure 10.32. Construct T and T_{LCF} for path $A \rightarrow B \rightarrow C \rightarrow D \rightarrow E \rightarrow F$. What is the maximum end-to-end bandwidth for that path?

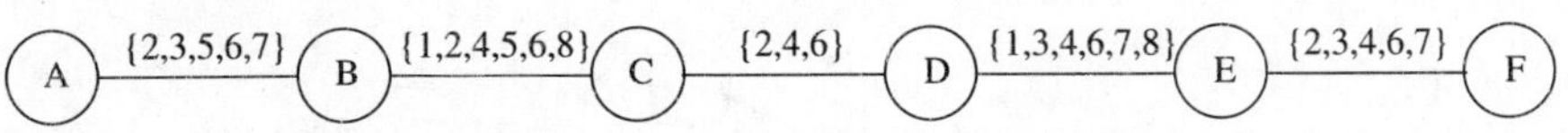

Figure 10.32. Network topology for Problem 9.

10. What are the pros and cons of using the hop-by-hop path bandwidth calculation algorithm proposed in the BR and OQR protocols over the approach used during the unipath discovery operation of the OLMQR protocol for the end-to-end path bandwidth calculation?

11. Write an algorithm for SWAN AIMD rate control mechanism.

12. Compare the admission control mechanisms of INSIGNIA and SWAN frameworks.

13. Assume that a session S requires X units of bandwidth and an intermediate node I has Y $(Y < X)$ units of bandwidth available at that node. Discuss how node I responds after receiving a *RouteRequest* packet for the session S in INSIGNIA and INORA frameworks.

14. Assume that a QoS session is being initiated by the source node A to the destination node I with six units of bandwidth requirement. Let the DAG created by TORA protocol for this QoS session be as shown in Figure 10.33. In this figure, the label on each link specifies the available bandwidth on that link. Discuss how the QoS session is admitted (or rejected) in case of coarse feedback mechanism and class-based fine feedback mechanism for the following two cases:

   (a) $BW_{EH} = BW_{FI} = 3$ and $BW_{HI} = 10$.

   (b) $BW_{FI} = 2$, $BW_{EH} = 3$, and $BW_{HI} = 4$.

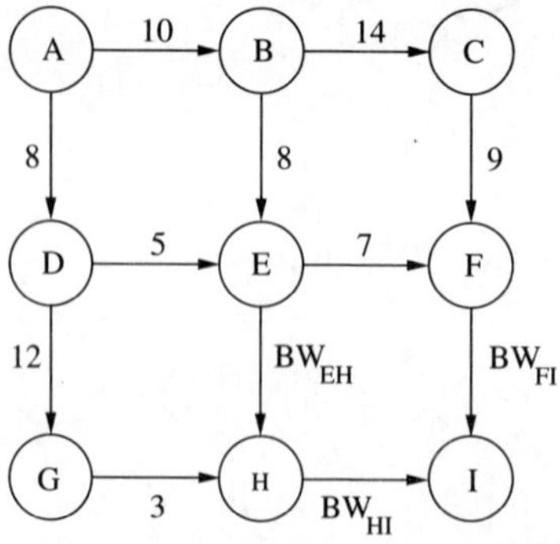

Figure 10.33. Topology for Problem 14.

15. In a military-vehicular ad hoc wireless network using PRTMAC, formed by 500 nodes distributed uniformly in a battlefield area of 1,000 m $\times$ 1,000 m, calculate the number of nodes contending for the data channel and for the control channel. The transmission range of the data channel is 250 m.

16. In Problem 15, find the probability that a beacon gets collided, when the beacons are generated periodically with a period of $P_b = 10$ seconds. Assume the beacon length to be equal to 1 ms.

17. In a PRTMAC system, assume that the beacon transmissions are carried out in the control channel with a period of 5 seconds, with an accurate prediction mechanism. Calculate the probability that an impending collision from another node traveling with a constant velocity of 20 m/s goes undetected. Assume that the transmission range of the data channel is 240 m and the probability of a beacon getting collided is 0.2.

BIBLIOGRAPHY

[1] P. Karn, "MACA: A New Channel Access Method for Packet Radio," *Proceedings of ARRL/CRRL Amateur Radio* 9th *Computer Networking Conference 1990*, pp. 134-140, September 1990.

[2] IEEE Standards Board, "Part 11: Wireless LAN Medium Access Control (MAC) and Physical Layer (PHY) Specifications," *The Institute of Electrical and Electronics Engineers, Inc.*, 1997.

[3] M. Gerla and J. T. C. Tsai, "Multicluster, Mobile, Multimedia Radio Network," *ACM/Baltzer Wireless Networks Journal*, vol. 1, no. 3, pp. 255-265, October 1995.

[4] S. Mangold, S. Choi, P. May, O. Klein, G. Hiertz, and L. Stibor, "IEEE 802.11e Wireless LAN for Quality of Service," *Proceedings of the European Wireless 2002*, vol. 1, pp. 32-39, February 2002.

[5] M. Veeraraghavan, N. Cocker, and T. Moors, "Support of Voice Services in IEEE 802.11 Wireless LANs," *Proceedings of IEEE INFOCOM 2001*, vol. 1, pp. 488-497, April 2001.

[6] M. A. Visser and M. E. Zarki, "Voice and Data Transmission Over an 802.11 Wireless Network," *Proceedings of IEEE PIMRC 1995*, vol. 2, pp. 648-652, September 1995.

[7] IEEE 802.11 TGe, "EDCF Proposed Draft Text," *TR-01/131r1*, March 2001.

[8] IEEE 802.11 TGe, "Hybrid Coordination Function (HCF)–Proposed Updates to Normative Text of D0.1," *TR-01/110r1*, March 2001.

[9] IEEE 802.11 TGe, "HCF Ad Hoc Group Recommendation–Normative Text to EDCF Access Category," *TR-02/241r0*, March 2001.

[10] IEEE 802.11 TGe, "Proposed Normative Text for AIFS–Revisited," *TR-01/270r0*, February 2003.

[11] S. Sheu and T. Sheu, "DBASE: A Distributed Bandwidth Allocation/Sharing/Extension Protocol for Multimedia over IEEE 802.11 Ad Hoc Wireless LAN," *Proceedings of IEEE INFOCOM 2001*, vol. 3, pp. 1558-1567, April 2001.

[12] C. R. Lin and M. Gerla, "Real-Time Support in Multi-hop Wireless Networks," *ACM/Baltzer Wireless Networks Journal*, vol. 5, no. 2, pp. 125-135, March 1999.

[13] B. S. Manoj and C. Siva Ram Murthy, "Real-Time Traffic Support for Ad Hoc Wireless Networks," *Proceedings of IEEE ICON 2002*, pp. 335-340, August 2002.

[14] S. Chen and K. Nahrstedt, "Distributed Quality-of-Service Routing in Ad Hoc Networks," *IEEE Journal on Selected Areas in Communications*, vol. 17, no. 8, pp. 1488-1504, August 1999.

[15] S. H. Shah and K. Nahrstedt, "Predictive Location-Based QoS Routing in Mobile Ad Hoc Networks," *Proceedings of IEEE ICC 2002*, vol. 2, pp. 1022-1027, May 2002.

[16] S. De, S. K. Das, H. Wu, and C. Qiao, "Trigger-Based Distributed QoS Routing in Mobile Ad Hoc Networks," *ACM SIGMOBILE Mobile Computing and Communications Review*, vol. 6, no. 3, pp. 22-35, July 2002.

[17] C. E. Perkins, E. M. Royer, and S. R. Das, "Quality of Service for Ad Hoc On-Demand Distance Vector Routing," *IETF Internet Draft, draft-ietf- manet-aodvqos-00.txt*, July 2000.

[18] C. R. Lin and J. Liu, "QoS Routing in Ad Hoc Wireless Networks," *IEEE Journal on Selected Areas in Communications*, vol. 17, no. 8, pp. 1426-1438, August 1999.

[19] C. E. Perkins and P. Bhagwat, "Highly Dynamic Destination-Sequenced Distance-Vector Routing (DSDV) for Mobile Computers," *Proceedings of ACM SIGCOMM 1994*, pp. 234-244, August 1994.

[20] C. R. Lin, "On-Demand QoS Routing in Multi-Hop Mobile Networks," *Proceedings of IEEE INFOCOM 2001*, vol. 3, pp. 1735-1744, April 2001.

[21] Y. Chen, Y. Tseng, J. Sheu, and P. Kuo, "On-Demand, Link-State, Multipath QoS Routing in a Wireless Mobile Ad Hoc Network," *Proceedings of European Wireless 2002*, pp. 135-141, February 2002.

[22] V. Vidhyashankar, B. S. Manoj, and C. Siva Ram Murthy, "Slot Allocation Schemes for Delay-Sensitive Traffic Support in Asynchronous Wireless Mesh Networks," *Proceedings of HiPC 2003*, LNCS 2913, pp. 333-342, December 2003.

[23] B. S. Manoj, V. Vidhyashankar, and C. Siva Ram Murthy, "Slot Allocation Strategies for Delay-Sensitive Traffic Support in Asynchronous Ad Hoc Wireless Networks," to appear in *Journal of Wireless Communications and Mobile Computing*, 2004.

[24] P. Sinha, R. Sivakumar, and V. Bharghavan, "CEDAR: A Core Extraction Distributed Ad Hoc Routing Algorithm," *IEEE Journal on Selected Areas in Communications*, vol. 17, no. 8, pp. 1454-1466, August 1999.

[25] R. Braden, D. Clark, and S. Shenker, "Integrated Services in the Internet Architecture: An Overview," *IETF RFC1633*, June 1994.

[26] S. Blake, D. Black, M. Carlson, E. Davies, Z. Wang, and W. Weiss, "An Architecture for Differentiated Services," *IETF RFC2475*, December 1998.

[27] R. Braden, L. Zhang, S. Berson, S. Herzog, and S. Jamin, "Resource reSerVation Protocol (RSVP) – Version 1 Functional Specification," *IETF RFC 2205*, September 1997

[28] H. Xiao, K. G. Seah, A. Lo, and K. C. Chua, "A Flexible Quality of Service Model for Mobile Ad Hoc Networks," *Proceedings of IEEE Vehicular Technology Conference*, vol. 1, pp. 445-449, May 2000.

[29] A. K. Talukdar, B. R. Badrinath, and A. Acharya, "MRSVP: A Resource Reservation Protocol for an Integrated Services Network with Mobile Hosts," *ACM/Baltzer Wireless Networks Journal*, vol. 7, no. 1, pp. 5-19, January 2001.

[30] S. B. Lee, A. Gahng-Seop, X. Zhang, and A. T. Campbell, "INSIGNIA: An IP-Based Quality of Service Framework for Mobile Ad Hoc Networks," *Journal of Parallel and Distributed Computing*, vol. 60, no. 4, pp. 374-406, April 2000.

[31] D. Dharmaraju, A. R. Chowdhury, P. Hovareshti, and J. S. Baras, "INORA–A Unified Signalling and Routing Mechanism for QoS Support in Mobile Ad Hoc Networks," *Proceedings of ICPPW 2002*, pp. 86-93, August 2002.

[32] H. Ahn, A. T. Campbell, A. Veres, and L. Sun, "Supporting Service Differentiation for Real-Time and Best-Effort Traffic in Stateless Wireless Ad Hoc Networks," *IEEE Transactions on Mobile Computing*, vol. 1, no. 3, pp. 192-207, September 2002.

[33] T. Sandeep, V. Vivek, B. S. Manoj, and C. Siva Ram Murthy, "PRTMAC: An Enhanced Real-Time Support Mechanism for Tactical Ad Hoc Wireless Networks," *Technical Report*, Department of Computer Science and Engineering, Indian Institute of Technology, Madras, India, June 2001. (A shorter version of this report has been accepted for presentation at *IEEE RTAS 2004*.)

[34] T. Bheemarjuna Reddy, I. Karthigeyan, B. S. Manoj, and C. Siva Ram Murthy, "Quality of Service Provisioning in Ad Hoc Wireless Networks: A Survey of Issues and Solutions," *Technical Report*, Department of Computer Science and Engineering, Indian Institute of Technology, Madras, India, July 2003.

Chapter 11

ENERGY MANAGEMENT IN AD HOC WIRELESS NETWORKS

11.1 INTRODUCTION

The nodes in an ad hoc wireless network are constrained by limited battery power for their operation. Hence, energy management is an important issue in such networks. The use of multi-hop radio relaying requires a sufficient number of relaying nodes to maintain the network connectivity. Hence, battery power is a precious resource that must be used efficiently in order to avoid early termination of any node.

Energy management deals with the process of managing energy resources by means of controlling the battery discharge, adjusting the transmission power, and scheduling of power sources so as to increase the lifetime of the nodes of an ad hoc wireless network. Efficient battery management, transmission power management, and system power management are the three major means of increasing the life of a node. Battery management is concerned with problems that lie in the selection of battery technologies, finding the optimal capacity of the battery, and scheduling of batteries, that increase the battery capacity. Transmission power management techniques attempt to find an optimum power level for the nodes in the ad hoc wireless network. On the other hand, system power management deals mainly with minimizing the power required by hardware peripherals of a node (such as CPU, DRAM, and LCD display) and incorporating low-power strategies into the protocols used in various layers of the protocol stack. This chapter concentrates on the issues involved and the solutions for energy management in ad hoc wireless networks.

11.2 NEED FOR ENERGY MANAGEMENT IN AD HOC WIRELESS NETWORKS

The energy efficiency of a node is defined as the ratio of the amount of data delivered by the node to the total energy expended. Higher energy efficiency implies that a greater number of packets can be transmitted by the node with a given amount of energy reserve. The main reasons for energy management in ad hoc wireless networks are listed below:

- **Limited energy reserve:** The main reason for the development of ad hoc wireless networks is to provide a communication infrastructure in environments where the setting up of a fixed infrastructure is impossible. Ad hoc wireless networks have very limited energy resources. Advances in battery technologies have been negligible as compared to the recent advances that have taken place in the field of mobile computing and communication. The increasing gap between the power consumption requirements and power availability adds to the importance of energy management.

- **Difficulties in replacing the batteries:** Sometimes it becomes very difficult to replace or recharge the batteries. In situations such as battlefields, this is almost impossible. Hence, energy conservation is essential in such scenarios.

- **Lack of central coordination:** The lack of a central coordinator, such as the base station in cellular networks, introduces multi-hop routing and necessitates that some of the intermediate nodes act as relay nodes. If the proportion of relay traffic is large, then it may lead to a faster depletion of the power source for that node. On the other hand, if no relay traffic is allowed through a node, it may lead to partitioning of the network. Hence, unlike other networks, relay traffic plays an important role in ad hoc wireless networks.

- **Constraints on the battery source:** Batteries tend to increase the size and weight of a mobile node. Reducing the size of the battery results in less capacity which, in turn, decreases the active lifespan of the node. Hence, in addition to reducing the size of the battery, energy management techniques are necessary to utilize the battery capacity in the best possible way.

- **Selection of optimal transmission power:** The transmission power selected determines the reachability of the nodes. The consumption of battery charge increases with an increase in the transmission power. An optimal value for the transmission power decreases the interference among nodes, which, in turn, increases the number of simultaneous transmissions.

- **Channel utilization:** A reduction in the transmission power increases frequency reuse, which leads to better channel reuse. Power control becomes very important for CDMA-based systems in which the available bandwidth is shared among all the users. Hence, power control is essential to maintain the required signal to interference ratio (SIR) at the receiver and to increase the channel reusability.

11.3 CLASSIFICATION OF ENERGY MANAGEMENT SCHEMES

The need for energy management in ad hoc wireless networks, discussed in the previous section, points to the fact that energy awareness needs to be adopted by the protocols at all the layers in the protocol stack, and has to be considered as one of the important design objectives for such protocols. Energy conservation can be implemented using the following techniques:

- Battery management schemes
- Transmission power management schemes
- System power management schemes

Maximizing the life of an ad hoc wireless network requires an understanding of the capabilities and the limitations of energy sources of the nodes. A greater battery capacity leads to a longer lifetime of the nodes. Increasing the capacity of the batteries can be achieved by taking into consideration either the internal characteristics of the battery (battery management) or by minimizing the activities that utilize the battery capacity (power management). The system power management approach can be further divided into the following categories:

- Device management schemes
- Processor power management schemes

Figure 11.1 provides an overview of some of the techniques at different layers of the protocol stack that fall into three categories: battery management, transmission power management, and system power management schemes. Though these schemes cannot be strictly classified under the different layers of the OSI protocol stack as they reside in more than one layer, the classification provided in this section is based on the highest layer in the protocol stack used by each of these protocols.

11.4 BATTERY MANAGEMENT SCHEMES

Battery-driven systems are those systems which are designed taking into consideration mainly the battery and its internal characteristics. They try to maximize the amount of energy provided by the power source by exploiting the inherent property of batteries to recover their charge when kept idle. In the following sections, we discuss the ways in which the energy efficiency of mobile wireless communication can be enhanced through the use of improved battery management techniques as described in [1], [2], [3], [4], [5], [6], and [7]. Recent research results in this area have proved that, by varying the manner in which energy is drawn from the batteries, significant improvement can be obtained in the total amount of energy supplied by them. In the section that follows, we also discuss some of the battery characteristics [8] which are used throughout our discussions on battery management.

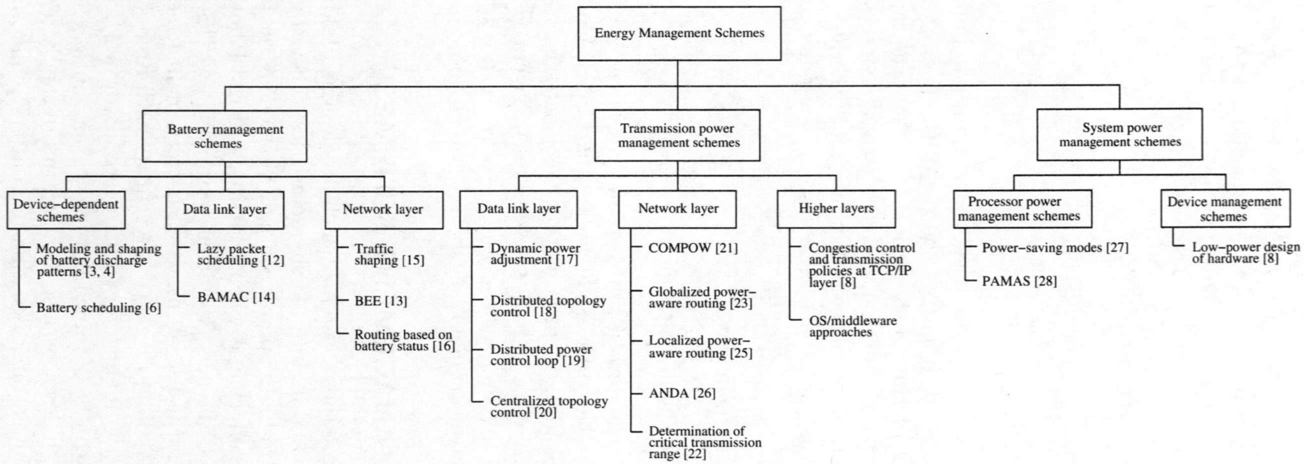

Figure 11.1. Classification of energy management schemes.

11.4.1 Overview of Battery Characteristics

The major components of batteries are illustrated in Figure 11.2. A battery mainly consists of an anode, a cathode, an electrolyte medium, and a case. The anode is often a metal and the cathode a metallic oxide. The electrolyte is a salt solution that promotes the ion flow. The porous separator is used to prevent a short circuit between anode and cathode by keeping them from touching one another. The battery is contained in a structural support (case) that provides dimensional stability and a positive and a negative electrode for discharging (or recharging) the cell. The positive ions move from the anode toward the cathode through the electrolyte medium and the electrons flow through the external circuit. A number of separate electrochemical cells can also be combined within the same case to create a battery.

- **Battery technologies:** The most popular rechargeable battery technologies developed over the last two decades are comprised of nickel-cadmium, lithium ion, nickel metal-hydride, reusable alkaline, and lithium polymer. The main factors considered while designing a battery technology are the energy density (the amount of energy stored per unit weight of the battery), cycle life [the number of (re)charge cycles prior to battery disposal], environmental impact, safety, cost, available supply voltage, and charge/discharge characteristics.

- **Principles of battery discharge:** A battery typically consists of an array of one or more cells. Hence, in the subsequent sections, the terms "battery" and "cell" are used interchangeably. The three main voltages that characterize a cell are: (1) the open circuit voltage (V_{oc}), that is, the initial voltage under a no-load condition of a fully charged cell, (2) the operating voltage (V_i), that is, the voltage under loaded conditions, and (3) the cut-off voltage (V_{cut}) at which the cell is said to be discharged. All the cells are defined by three main capacities:

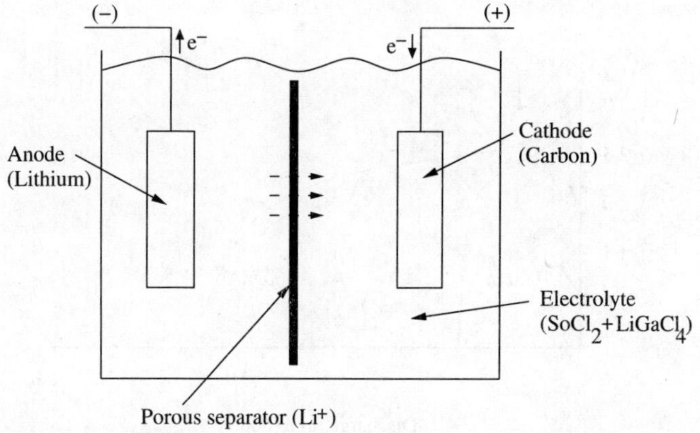

Figure 11.2. Basic structure of a lithium/thionyl chloride battery.

- Theoretical capacity: The amount of active materials (the materials that react chemically to produce electrical energy when the cell is discharged and restored when the cell is charged) contained in the cell refers to its theoretical capacity. A cell cannot exceed its theoretical capacity.
- Nominal (standard) capacity: This corresponds to the capacity actually available when discharged at a specific constant current. It is expressed in ampere-hours.
- Actual capacity: The energy delivered under a given load is said to be the actual capacity of the cell. A cell may exceed the actual capacity but not the theoretical capacity.

The constant current discharge behavior of lithium-manganese dioxide ($LiMnO_2$) cells with $V_{oc} = 3\ V$ and $V_{cut} = 1\ V$ is shown in Figure 11.3 [9]. The discharge curve is flat most of the time and a gradual slope is developed as the voltage reaches the cut-off voltage. The performance of a cell's discharge is measured using the following parameters:

- Discharge time: The time elapsed when a fully charged cell reaches its cut-off voltage and has to be replaced or recharged is called the discharge time of the cell.
- Specific power (energy): This is the power (energy) delivered by a fully charged cell under a specified discharge current. It is expressed in watt-per-kilogram (watt-hour-per-kilogram).
- Discharge current: There are mainly two models of battery discharge: constant current discharge and pulsed current discharge. In pulsed current discharge, the battery switches between short discharge periods and idle periods (rest periods). Chiasserini and Rao in [4] illustrate the performance of the bipolar lead-acid battery subjected to a pulsed discharge

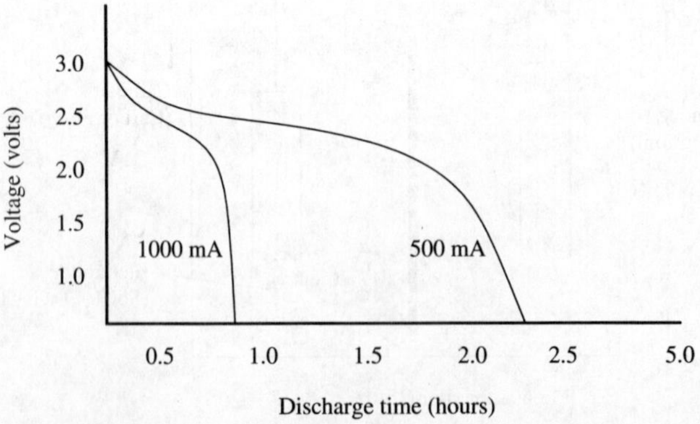

Figure 11.3. Discharge pattern of a cell when $V_{oc} = 3\ V$ and $V_{cut} = 1\ V$.

current of six current pulses. After each discharge, which lasts for 3 ms, the cell was idled for 22 ms during which no recharging was allowed to take place. Figure 11.4 shows the current density and the corresponding cell voltage. The cell is able to recover and revert to its initial open circuit voltage during the first four rest periods. After the fifth current pulse, the rest period of 22 ms turns out to be inadequate for the cell recovery.

- **Impact of discharge characteristics on battery capacity:** The important chemical processes that affect the battery characteristics are given below.
  - Diffusion process: When the battery is actively involved in discharging, that is, at a non-zero current, the active materials move from the electrolyte solution to the electrodes and are consumed at the electrode. If this current is above a threshold value called the *limiting current*, the active materials get depleted very quickly. But as the current decreases, the concentration of the active materials around the electrode drops. By increasing the rest time periods of the battery, longer lifetimes can be achieved due to the recovery capacity effect, which is explained later in this section. In the following discussion, we will concentrate on some of the battery management techniques which increase idle periods for batteries.

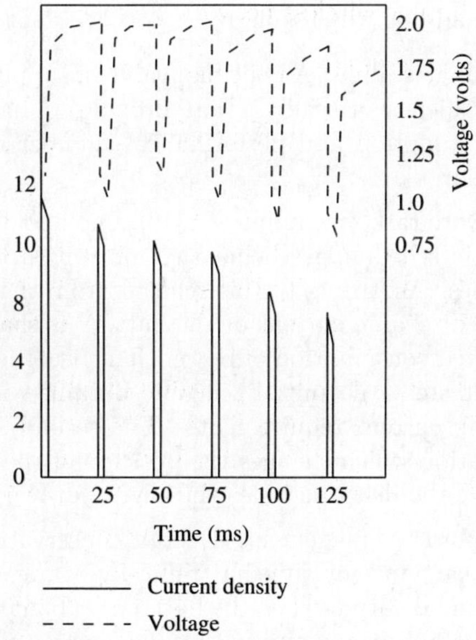

Figure 11.4. Performance of a bipolar lead-acid cell subjected to six current impulses with pulse length = 3 ms and rest period = 22 ms.

- Passivation process: The cell discharge is limited not only by the diffusion process but also by a process called *passivation*, which induces in the cell the precipitation of crystals which are produced by the discharge due to the chemical reactions on the electrode. This phenomenon increases during higher current densities.

Two important effects to be considered for understanding the battery's discharge properties are stated below.

- Rate capacity effect: As the intensity of the discharge current increases, an insoluble component develops between the inner and outer surfaces of the cathode. The inner layer becomes inaccessible as a result of this phenomenon, rendering the cell unusable even while a sizable amount of active materials still exists. This effect depends on the actual capacity of the cell and the discharge current.
- Recovery capacity effect: This effect is concerned with the recovery of charges under idle conditions. By increasing the idle time, one may be able to completely utilize the theoretical capacity of the cell.

- **Battery models:** Battery models depict the characteristics of the batteries used in real life. The pros and cons of following battery models are summarized in [8]: analytical models, stochastic models, electric circuit models, and electrochemical models. Finally, a battery efficient system architecture is proposed and the following approaches are suggested to enable longer life of the nodes of an ad hoc wireless network:

  - Supply voltage scaling: An optimal value of supply voltage (v_{dd}) is maintained, by means of scaling, that provides a balance between battery charge consumption and performance (number of packets transmitted per unit charge).
  - Battery-aware task scheduling: In [10], Luo and Jha proposed a battery-aware static scheduling scheme that optimizes the discharge power of the batteries. According to this scheme, from the knowledge of the task graph, the discharge current of the battery is shaped in order to reduce the unwanted consumption of power. This is done as a two-step process. In the first step, the initial schedule obtained is adjusted in order to reduce peak current requirements. The second step consists of a local transformation which changes the position of the scheduled events so as to minimize the delay and also the energy drawn off the cell.
  - Dynamic power management: Energy conservation can be achieved at the nodes carrying multimedia traffic by a graceful degradation of the quality of audio output when the battery is about to reach the completely discharged state. Many approaches have been suggested to achieve this. In one such policy, the audio device outputs high-quality sound when the remaining battery charge is above a certain threshold value. Once it falls below the threshold value, the audio device tries to degrade the output

sound quality. These policies mainly exploit the recovery capacity effect of the batteries to attain theoretical capacity.

- **Battery scheduling:** The use of multiple batteries in mobile nodes has become very common. The key aspect behind this kind of an architecture is the property of charge recovery by the battery when it remains in idle state. A detailed description of the charge recovery property of the battery can be found in the next section.

- **Smart battery standard (SBS):** This is an emerging technology toward the development of batteries that consume low power. The main aim of SBS is to create standards by which the systems become aware of the batteries and interact with them in order to provide a better performance.

11.4.2 Device-Dependent Schemes

The lifetime of a node is determined by the capacity of its energy source and the energy required by the node. Recent works in [1], [2], [4], [5], [6], and [7] show that the battery life can be improved by introducing techniques which make efficient utilization of the battery power. In this section, some of the device-dependent approaches that increase the battery lifetime by exploiting its internal characteristics are discussed.

Modeling and Shaping of Battery Discharge Patterns

The stochastic model of the discharge pattern of batteries introduced in [5] employs the following two key aspects affecting the battery life: the rate capacity effect and the recovery effect.

Effect of Battery Pulsed Discharge

Recent works such as in [3], [4], and [5] show that pulsed current discharge applied for bursty stochastic transmissions improves the battery lifetime. If pulsed current discharge is applied to a cell, significant improvement in the specific energy delivered is realized. In such an environment, higher specific power can be obtained for a constant specific energy. In [4], a model for battery pulsed discharge with recovery effect is considered. The model proposed consists of a battery with a theoretical capacity of C charge units and an initial battery capacity of N charge units. Battery behavior is considered as a discrete-time Markov process with the initial state equal to N and the fully discharged state 0. Time is divided into slots (frames). Each packet for the node is transmitted in one time slot and the battery enters the previous state by losing a charge unit. If the battery remains idle, it recovers one charge unit and enters the next state. The results suggest that at the most C (theoretical capacity) packets can be transmitted if the battery is given enough time to recover. The passivation (surface modifications of metallic materials which cause an increase in their resistance to the corrosion process) time constant is assumed to be greater than the discharge time to fully drain the theoretical capacity of the cell. Thus, the passivation effects can be neglected. In [4], Chiasserini and Rao studied the battery behavior under two different modes of pulsed discharge.

Binary Pulsed Discharge

In this mode, if there are packets in the queue, transmission of a packet occurs in one time slot; one charge unit is recovered if the queue is empty. The current required for transmission is drained during the entire time frame. The Markov chain for binary pulsed discharge is shown in Figure 11.5. An additional dummy state is added to the Markov chain representing the cell behavior, which represents the start of the discharge. The cell is modeled as a transient process and the packet arrival follows a Bernoulli process. If the probability that a packet arrives in one time frame is stated as $a_1 = q$ and the probability for transmitting a packet in a time slot is given by a_1, then the probability of recovery is given by $a_0 = (1 - q)$. The cell can never cross the charge state of N. The gain obtained in this scheme is given by $G = \frac{m_p}{N}$, where m_p is the total expected number of packets transmitted and N is the amount of charge in a fully charged battery. The gain, however, cannot exceed C/N where C is the theoretical capacity.

Generalized Pulsed Discharge

In a particular time frame, either one or more packets are transmitted or the cell is allowed to recover a charge unit. The quantity of the impulse is equal to the current required to transmit all the packets, and the duration of the impulse is equal to a fraction of the time frame. In the remaining fraction of the time frame, the cell is allowed to recover one unit. In this case, if a_k is the probability that k packets arrive in one burst arrival and M is the maximum number of packets per burst, then the probability of recovery is given by $a_0 = 1 - \sum_{k=1}^{M} a_k$. In each time frame, the cell can move from state z to $z - k + 1$, where $0 < z < N$, with

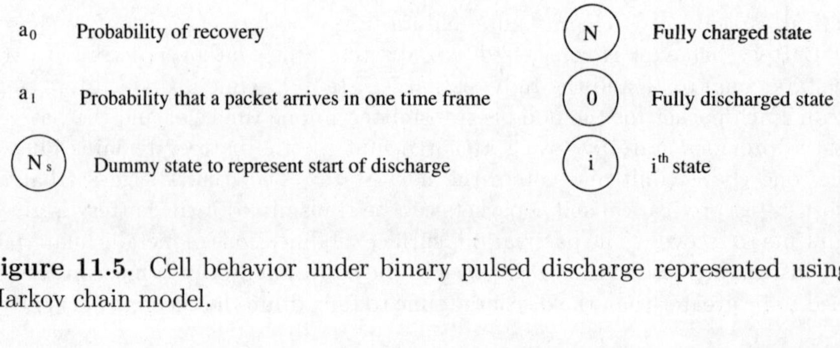

Figure 11.5. Cell behavior under binary pulsed discharge represented using the Markov chain model.

a probability of a_k. Generalized pulsed discharge can be represented as a general Markovian chain, as shown in Figure 11.6. The probability to move from one state to another is specified in Figure 11.6.

An optimal discharge strategy which provides a solution to extend the lifetime of a battery by exploiting the internal battery characteristics is proposed in [1]. The total amount of charge delivered by the battery lies between N and C units. The system is assumed to be a stochastic process with N states $(x_0, \ldots, x_N)$. Each state i is denoted by the tuple (n_i, c_i), where n_i and c_i are the remaining charge and capacity left in the cell, respectively. Thus the initial state is given by (N, C). When the battery which is in state (n_i, c_i) delivers q charge units, it moves to the state $(n_i - q, c_i - q)$. If the battery remains idle for one charge unit, it moves to the state $(n_i + 1, c_i)$. The battery expires if either c_i or n_i becomes 0. When the battery is in state (n_i, c_i), and idles for one unit of time, the probability of recovering one charge unit is given by

$$P_r(n_i, c_i) = \begin{cases} e^{-g}(N - n_i) - \phi(c_i) & : \text{if } \; 1 \le n_i \le N, 1 \le c_i \le C \\ 0 & : \text{otherwise} \end{cases} \tag{11.4.1}$$

Here g is a constant value and $\phi(c_i)$ is a piecewise constant function of the number of charge units delivered which are specific to the cell's chemical properties. Using stochastic dynamic programming, an optimal policy for discharging the cell is proposed in [1]. The cells are then scheduled based on their recovery process. Another battery capacity model is suggested by Simunic *et al.* [7], with which battery lifetime can be estimated accurately. The efficiency of the battery is given by

$$Efficiency = \frac{E_{Cycle}}{E_{Rated}} \tag{11.4.2}$$

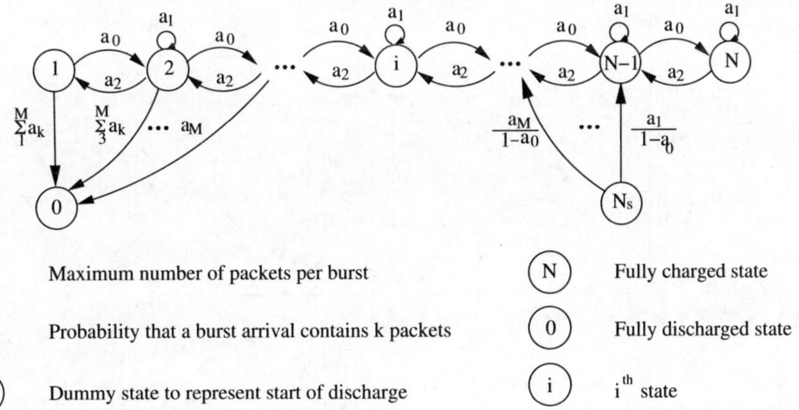

Figure 11.6. Cell behavior under generalized pulsed discharge represented using the Markov chain model.

that is, the ratio of actual capacity (E_{Cycle}) to the rated capacity (E_{Rated}) which is derived from the battery specification. If the battery voltage is nearly a constant, efficiency is given by the ratio of actual to the rated current. Using this model, the battery lifetime estimation can be made as follows:

- The manufacturer specifies the rated capacity, the time constant, and the discharge plot of the battery.
- The discharge current ratio, which is the ratio between the specified rated current (i_{rated}) and the calculated average current (i_{avg}), is computed.
- Efficiency is calculated by the interpolation of points in the discharge plot as seen in Figure 11.7, which shows the variation of efficiency with the current ratio.

Lower efficiency corresponds to a shortened battery lifetime and vice versa.

Battery-Scheduling Techniques

Chiasserini and Rao in [6] proposed battery-scheduling techniques that improve the battery lifetime. In a battery package of L cells, a subset of batteries can be scheduled for transmitting a given packet, leaving other cells to recover their charge. The following approaches are applied to select the subset of cells:

- **Delay-free approaches:** In the above context, *job* is defined as a demand for battery discharge which can be satisfied by the subset of cells. As soon as

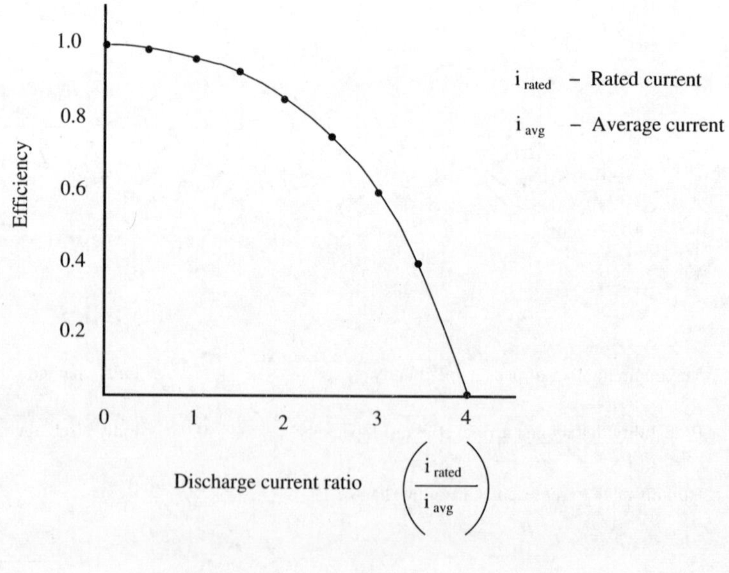

Figure 11.7. Variation of battery efficiency with discharge current ratio.

a job arrives, the battery charge for processing the job will be provided from the cells without any delay. The scheduling scheme for batteries can be any one of the following:

- Joint technique (JN): As soon as a job arrives, the same amount of current is drawn equally from all the cells, which are connected in parallel. If there are L cells, the current discharged from each of them is $\frac{1}{L}$ times the required supply.
- Round robin technique (RR): This scheme selects the battery in round robin fashion and the jobs are directed to the cells by switching from one to the next one, which takes place as shown in Figure 11.8 (a). The job from job queue gets energy from the battery selected by the transmission module based on round robin technique.
- Random technique (RN): In this technique, any one of the cells is chosen at random with a probability of $\frac{1}{L}$. The selected cell provides the total supply required, as shown in Figure 11.8 (b).

• **No delay-free approaches:** In these kinds of approaches, the batteries coordinate among themselves based on their remaining charge. In one such technique, a threshold is defined for the remaining charge of the cell. All the cells which have their remaining charge greater than this threshold value become eligible for providing energy. Delay-free approaches such as round robin scheduling can be applied to these eligible cells. The cells which are not eligible stay in the recovery state. This enables the cells to maximize their capacity. The general battery discharge policy employed in portable devices

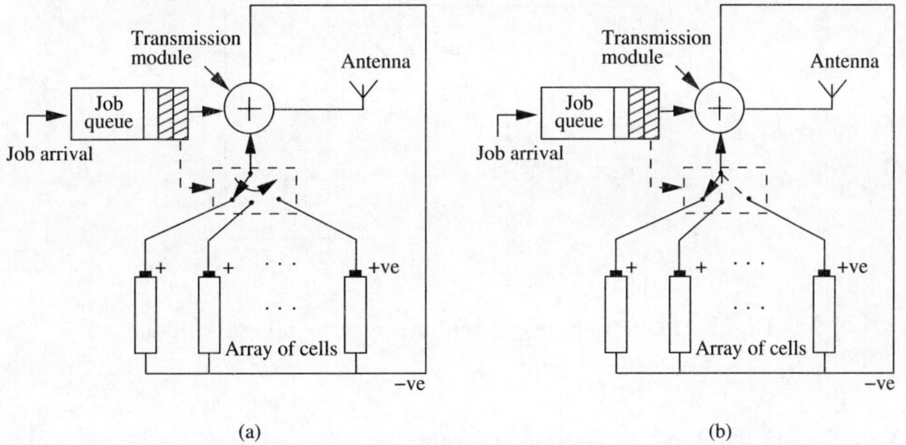

Figure 11.8. Battery-scheduling techniques: (a) round robin technique (b) random technique.

completely drains battery packs one after the other. Bruni *et al.* have shown in [2] that this kind of policy is inefficient.

Using heterogeneous batteries: This section examines a new model suggested by Rong and Pedram in [11] for a battery-powered electronic system, which is based on the continuous time Markovian decision process (CTMDP). It attempts to exploit the two main characteristics of the rechargeable batteries, that is, the recharging capability under no load condition and the rate capacity effects discussed earlier. The main objective is to formulate an optimization problem which tries to minimize the charge delivered by the battery, thereby effectively utilizing the battery capacity. The problem framed is solved using the linear programming approach. The model proposed in [11] correlates the model of the batteries with that of the power-managed portable electronics. The model consists of two power sources which have different discharge characteristics and capacities. The jobs that arrive are serviced using the power from any of the two batteries, where the batteries are scheduled alternatively. The model, which is depicted in Figure 11.9, uses a battery scheduler, job queue, and the job requester. Each of the three components can be modeled using the Markovian model. In this case, the battery scheduler is a simple selector that chooses between the two batteries in an alternating fashion, or it can be a scheduler that uses the round robin scheduling technique. The formulated problem finds an optimal policy that tries to minimize the usage of the batteries that poses a constraint on the number of waiting requests in the job queue.

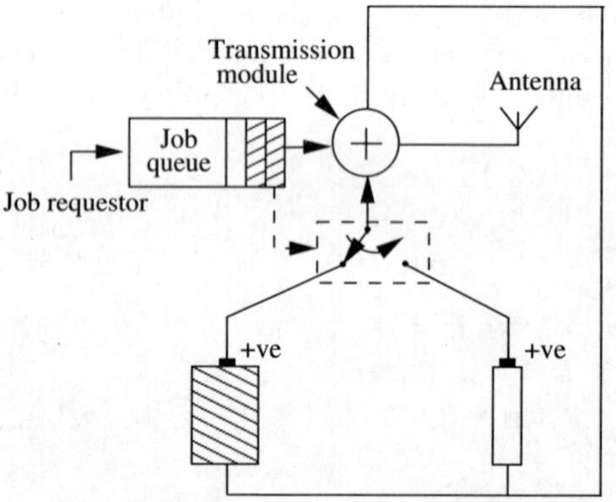

Figure 11.9. Heterogeneous battery-scheduling technique.

11.4.3 Data Link Layer Solutions

The data link layer solutions take into consideration battery characteristics while designing the protocols. Designing a battery-aware scheduling technique and maximizing the number of packets being transmitted are conflicting objectives. The following schemes attempt to find a trade-off between them. Subsequent sections deal with:

- Lazy packet scheduling scheme
- BAMAC protocol

Lazy Packet Scheduling Scheme

The basic principle behind the development of this scheme is that in many of the channel coding schemes for wireless transmission, the energy required to transmit a packet can be reduced significantly by minimizing the transmission power and increasing the duration of transmission. But this may not suit practical wireless environment packets. Hence, a transmission schedule is designed taking into account the delay constraints of the packets. The energy optimal offline and online scheduling algorithms proposed by Prabhakar *et al.* in [12] consider a transmitter-receiver pair. All packets are of equal length. Let $\vec{\tau} = (\tau_1, \ldots, \tau_M)$ be their transmission durations obtained as a schedule where M is the number of packets to be transmitted and $w(\vec{\tau})$ be the energy required to transmit a packet over the duration $\vec{\tau}$. Let d_i, $\{i = 1, \ldots, M\}$ denote the packet inter-arrival times and $k_0 = 0$. Then the following parameters are defined:

$$m_1 = \max_{k\in\{1,\cdots,M\}} \{\frac{1}{k}\sum_{i=1}^{k} d_i\} \tag{11.4.3}$$

$$k_1 = \max\{k : \frac{1}{k}\sum_{i=1}^{k} d_i = m_1\} \tag{11.4.4}$$

For $j \geq 1$,

$$m_{j+1} = \max_{k\in\{1,\ldots,M-k_j\}} \{\frac{1}{k}\sum_{i=1}^{k} d_{k_j+i}\} \tag{11.4.5}$$

$$k_{j+1} = k_j + \max\{k : \frac{\sum_{i=1}^{k} d_{k_j+1}}{k} = m_{j+1}\} \tag{11.4.6}$$

Optimal Offline Schedule

Assuming the arrival times of all the packets (t_i, $\{i = 1, \ldots, M\}$) are known *a priori* and $t_1 = 0$, the problem is to find optimal values for $\tau_i, 1 \leq i \leq M$, so as to minimize $w(\vec{\tau}) = \sum_{i=1}^{M} w(\vec{\tau_i})$. A necessary condition for optimality is

$$\tau_i > \tau_{i+1} \quad \forall\, i \in 1, 2, \ldots, M-1. \tag{11.4.7}$$

The optimal offline schedule suggested in [12] is given by

$$\vec{\tau_i^*} = m_j \qquad if, k_{j-1} < i \leq k_j \tag{11.4.8}$$

such that, $\vec{\tau_i^*}$ is feasible and $\sum_{i=1}^{M} \vec{\tau_i^*} = T$ (time window) and satisfies the necessary condition for optimality stated above. m_j denotes the maximum packet inter-arrival time among all the packets that arrive after the arrival of packet j.

Online Algorithm

Assuming an offline schedule as described above, the time at which packet j starts its transmission is given by

$$T_j^* = \sum_{i=1}^{j-1} \vec{\tau}_i^* \qquad (11.4.9)$$

b_j the backlog when the jth packet starts its transmission is given by

$$b_j = \max(k : \sum_{i=1}^{k-1} D_i < T_j^* - j) \qquad (11.4.10)$$

where D_i is the inter-arrival time of M packets. The time $t < T$ at which a packet j starts its transmission when there is a backlog of b packets can be set equal to the expected value of the random variable $E(\vec{\tau}(b,t))$, which is evaluated numerically.

$$\vec{\tau}(b,t) = \max_{k \in 1,\ldots,M-(j+b_j)} \left(\frac{1}{k+b_j} \sum_{i=1}^{k} D_i\right). \qquad (11.4.11)$$

The lazy packet scheduling scheme combined with battery recovery properties is found to be providing energy saving up to 50% [13].

Battery-Aware MAC Protocol

The battery-aware MAC (BAMAC) protocol [14] is an energy-efficient contention-based node scheduling protocol, which tries to increase the lifetime of the nodes by exploiting the *recovery capacity effect* of battery. As explained earlier in this chapter, when a battery is subjected to constant current discharge, the battery becomes unusable even while there exists a sizable amount of active materials. This is due to the rate capacity effect of the battery. If the battery remains idle for a specified time interval, it becomes possible to extend the lifetime of the battery due to the recovery capacity effect. By increasing the idle time of the battery, the whole of its theoretical capacity can be completely utilized. Also, Equation 11.4.1 clearly shows that this effect will be higher when the battery has higher remaining capacity and decreases with a decrease in the remaining battery capacity. The BAMAC protocol tries to provide enough idle time for the nodes of an ad hoc wireless network by scheduling the nodes in an appropriate manner. It tries to provide uniform discharge of the batteries of the nodes that contend for the common channel. This can be effected by using a round robin scheduling (or fair-share scheduling) of these nodes.

In the BAMAC protocol, each node maintains a battery table which contains information about the remaining battery charge of each of its one-hop neighbor nodes. The entries in the table are arranged in the non-increasing order of the remaining battery charges. The RTS, CTS, Data, and ACK packets carry the remaining battery charge of the node from which they originated. A node, on listening to these packets, make a corresponding entry in its battery table. The objective of the back-off mechanism used in BAMAC protocol is to provide a near round robin scheduling of the nodes. The back-off period is given by

$$back-off = Uniform[0, (2^n \times CW_{min}) - 1] \times rank \times (T_{SIFS} + T_{DIFS} + T_t)$$

where, CW_{min} is the minimum size of the contention window and $rank$ is the position of that entry in the battery table of the node. T_{SIFS} and T_{DIFS} represent the SIFS and DIFS durations. Their values are same as those used in IEEE 802.11. T_t is the is the longest possible time required to transmit a packet successfully, including the RTS-CTS-Data-ACK handshake. The node follows the back-off even for the retransmission of the packets. When this back-off scheme is followed, nodes with lesser $rank$ values back off for smaller time durations compared to those with higher $rank$ values. $Uniform[0, (2^n \times CW_{min}) - 1]$ returns a random number distributed uniformly in the range 0 and $(2^n \times CW_{min} - 1)$, where n is the number of transmission attempts made so far for a packet. Thus the nodes are scheduled based on their remaining battery capacities. The higher the battery capacity, the lower the back-off period. This ensures near round robin scheduling of the nodes. Hence, a uniform rate of battery discharge is guaranteed across all the nodes. This guarantees alternate periods of transmission and idling of the node, which leads to alternate periods of discharge and recovery of the battery, as illustrated in Figure 11.10. In this protocol, whenever a node gains access to the channel, it is allowed

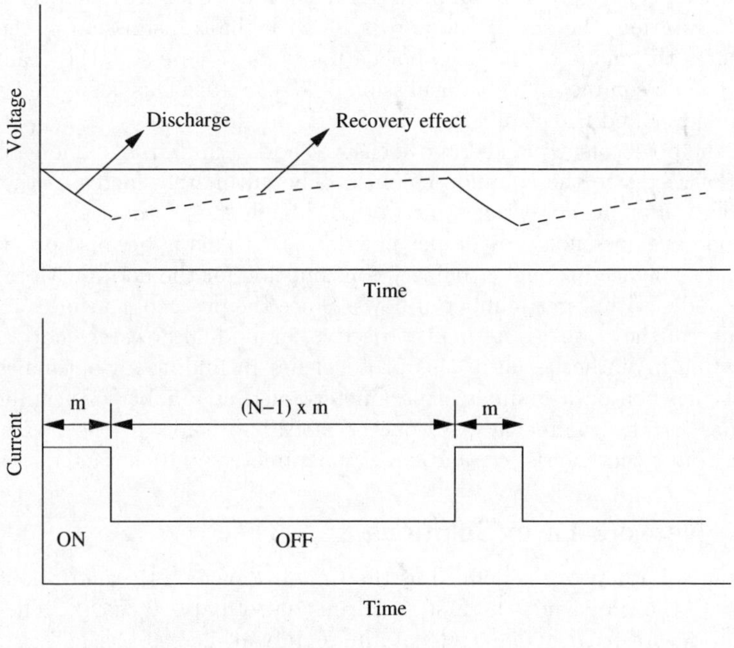

Figure 11.10. Illustration of BAMAC.

to transmit only one packet, giving rise to an average idle time of $(N-1) \times m$, where N is the total number of nodes contending for the common channel and m is the average time taken by a node for transmission of a packet. This improves the lifetime of the battery as it gains more idle time to recover charge because of the recovery capacity effect.

BAMAC(K) Protocol

Unlike the BAMAC protocol wherein the nodes are allowed to transmit only one packet on gaining access to the channel, in the BAMAC(K) protocol proposed in [14], K packets are transmitted consecutively by the node on gaining access to the channel. This provides a discharge time of $K \times m$ and an average recovery time of $(N-1)\times m\times K$ for the nodes. Though a larger value of K results in higher recovery time, it also increases the discharge time of the battery during the transmission of K packets. This increases the rate capacity effect due to faster depletion of the battery charge. A smaller value of K, on the other hand, decreases the recovery time of the battery. Hence, choosing an appropriate value for K is very important for optimum performance of the protocol

In the BAMAC(K) protocol, whenever the node attempts to gain access to the channel, it waits for DIFS time duration before transmitting the first packet. If no other neighbor transmits in this duration, the active node (the node that gains access to the channel) initiates its transmission. For transmitting each of the next $K-1$ packets, it waits only for an SIFS duration; if the channel remains idle during this SIFS duration, the active node proceeds with the transmission of the packet. This ensures that none of the neighboring nodes gains access to the channel until the active node completes the transmission of K packets. This is ensured since the neighbors never find the channel idle for DIFS time duration.

Both the protocols explained above ensure short-term fairness among the nodes in terms of access to the common channel. This ultimately increases the lifetime of the nodes in an ad hoc wireless network. Though the protocol provides fairness to the nodes, it does not provide per flow fairness. This is because providing per flow fairness may lead to higher battery consumption for the nodes with more flows than the nodes with lesser number of flows. Since the protocol considers improving the lifetime of the nodes as its main objective, individual flows are not taken into consideration. Another issue in this protocol lies in finding an optimal value for K, which depends on a number of parameters such as number of neighbor nodes contending for the channel access, packet arrival rate for all the nodes, packet deadline, traffic characteristics, and battery parameters and characteristics.

11.4.4 Network Layer Solutions

The lifetime of a network is defined as the time from which the network starts operating until the time when the first node runs out of battery charge. The network layer solutions for battery management aim mainly at increasing the lifetime of the network. The major solutions provided focus primarily on developing routing protocols that use routing metrics such as low energy cost and remaining battery charge.

Traffic-Shaping Schemes

This section discusses some of the traffic-shaping schemes ([3] and [15]), which are based on the battery discharge characteristics.

The scheme proposed in [3] uses the same basic model for the battery explained in earlier sections and is based on the fact that most of the network traffic is bursty. Introducing some acceptable delays in the battery discharge requests paves the way for the battery to be idle for a few time slots. This allows charge recovery to a certain extent. A proper analysis of the traffic characteristics provides a discharge-shaping technique by introducing battery idle times that trade off energy efficiency and delay. We shall now discuss an algorithm that increases the lifetime of the battery by shaping the network traffic.

Shaping Algorithm

The main goal of the algorithm is to introduce delay slots in the battery discharge process. This is done by defining a threshold which is expressed in terms of the amount of charge. The model used in this algorithm consists of a battery with a nominal capacity of N, a charge request rate of α_N, a theoretical capacity of T, and a threshold (expressed as a state of charge) of B.

Whenever the state of the battery drops below B, it is allowed to recover through idling. The remaining requests that arrive at the system are queued up at the buffer L with a large buffer size to guarantee zero loss probability. As soon as the battery recovers its charge and enters state $B+1$, it starts servicing the queued-up requests. By applying this model of shaping discharge to the cell, the gain obtained is given by $G = \frac{T}{N}$. A large value for M, which is equal to $N - B$, is favorable, since it results in higher service rates and smaller average delays for the discharge requests. Performance improves as the value of M increases. While considering the ON-OFF process, each requiring one charge unit, the ON-OFF times are random variables based on the Pareto distribution,

$$P(x) = \beta k^{\beta} x^{-\beta-1} \qquad k > 0, x \geq k, 0 < \beta < 2 \qquad (11.4.12)$$

Thus with an additional delay in the discharge requests, a significant improvement in the performance of the battery can be achieved.

Strategies for Blocking Relay Traffic

One of the main issues concerned with ad hoc wireless networks is the relay traffic. As mentioned earlier, each node deals with two kinds of traffic: relay traffic and its own traffic. A trade-off is reached between the blocking probability of the relay traffic and the battery efficiency. The intermediate nodes may not wish to transmit the whole of the neighbors' traffic. A method that calculates the optimal fraction of relay traffic, proposed by Srinivasan *et al.* in [15], is discussed in this section. The model used has N number of nodes which are uniformly distributed in an ad hoc wireless network with $R(s, d)$ set of available routes between source s and destination d. If $P(k, r)$ is the power required to transmit from node k to the next node through route r, then the energy cost associated with route r is given by

$$Energycost = \sum_{k \in r, k \neq d} P(k, r) \tag{11.4.13}$$

Whenever a traffic session is generated at the source node s, a route is selected which has minimum energy cost. A relay or an intermediate node can either allow the session traffic by sending an acknowledgment to s or block it by sending a negative acknowledgment to s. If the latter is chosen, on receiving the negative acknowledgment, the source repeats the process for the next best route on the basis of energy cost. If all the routes are blocked, the session is said to be blocked. Each node tries to behave selfishly when there is relay traffic. The amount of selfishness is defined using a quantity called *sympathy*. $sympathy(k, r)$ denotes the sympathy associated with k^{th} node in route r. The value of *sympathy* lies between 0 and 1, which reflects the willingness of the node to accept the relay traffic. The value 0 reflects complete unwillingness and 1 reflects complete willingness of the node to accept relay traffic. It is calculated based on some of the factors affecting transmission such as energy constraints of the node and the node's location in the network. The relay node rejects relay traffic based on the total amount of data the source intends to send to the destination and the strategy used. The following two strategies are considered in order to explore the above discussed trade-off which is based on the sympathy level:

- Random strategy: Assuming a session between source s and destination d, available routes are stored in $R(s, d)$ in the increasing order of the sympathy level. Whenever the kth node in route r receives a session request, it accepts it with a probability $sympathy(k, r)$.

- Pay-for-it strategy: According to this strategy, each node keeps an account of the help that it had received from other nodes relaying its messages, termed *credit,* and the amount of help it has given to others by allowing the relay traffic, that is, *debit.* The node tries to help if it has received more help in the recent past and rejects if its own traffic has been rejected often, that is, the node tries to find a balance between these two parameters.

The number of packets dropped by the relay nodes decreases as the number of selfish users decreases.

Battery Energy-Efficient Routing Protocol

The battery energy-efficient (BEE) routing protocol [13] is an energy-efficient routing protocol that attempts to combine the lazy packet scheduling and the traffic-shaping schemes.

The BEE routing protocol tries to find a balance in the energy consumption among all the nodes in the ad hoc wireless network. In this protocol, a new metric called *energy cost* is introduced. From the available routes, the one which has the lowest energy cost is selected. Even a route with a greater number of hops may be selected provided the route along these nodes has minimum power per link. The algorithm also insists on selecting the path with the higher battery charge in all the nodes in order to allow the recovery effect of the batteries with lesser remaining charge. The network has K nodes with S source nodes and D destination nodes. Any source node $s \in S$ can transmit to the destination node $d \in D$ through the relay nodes. The initial and instantaneous amount of battery charge are B_i and b_i, respectively, for any node i. The transmission range of all the nodes is ρ. Any node i is reachable to any other node j, only if the distance (d_{ij}) between them is less than ρ. The energy required to transmit from node i to node j is given by $e_{ij} = (\frac{d_{ij}}{\rho})^4$, only if i lies within the reach of j and vice versa. Otherwise, the energy required to transmit is infinity.

The energy required to transmit is discretized into few energy levels between e_{min} and e_{max}. The mean energy required for node i to transmit a packet is given by $e_i = \frac{1}{|R_i|}\sum_{j \in R_i} e_{ij}$, where R_i is the set of nodes whose distance from i is less than ρ. Whenever the number of packets transmitted by a node decreases, the energy level of the node increases significantly. This increase in energy level can be expressed as $\Gamma(\lambda_i, e_i)$, which is a function of transmission rate λ_i and the mean energy e_i. The new battery status (state of charge of the battery) at any time instant is given by $b_i + \Gamma(\lambda_i, e_i)$ if node is idle. The energy cost function used in BEE can be defined for k^{th} route r^k_{sd} as follows:

$$F_k = \sum_{l_{ij} \in r^k_{sd}} [\Psi(\lambda_i)e_{ij} + P_{ij}] - min_i \in r^k_{sd} b_i \tag{11.4.14}$$

where s and d are the source and generic destination nodes, respectively; l_{ij} is the link between nodes i and j of route r^k_{sd}; $\Psi(\lambda_i)$ is the weighting function such that $\Psi(\lambda_i) = A.\lambda_i$ with A being a constant; otherwise, $\Psi(\lambda_i) = 1$. P_{ij} denotes the energy penalty that occurs whenever a power level higher than the mean power level is required and is equal to $\max(0, e_{ij} - e_i)$. $min_i \in r^k_{sd} b_i$ is the minimum value of battery status among the nodes of the route r^k_{sd}. The routing protocol can be stated as follows: Whenever source s has data to send to destination d, it calculates the energy cost function of all the routes leading to the destination d. The best route among them is one with minimum cost function, that is, $F_m = \min_{r^k_{sd}} F_k$. One main disadvantage of this scheme is that the complexity of the algorithm depends on the number of routes for which the cost has to be computed. One alternative suggested to this algorithm to reduce the complexity is that the source selects a set of routes c on a random basis from the available list of routes to the destination. The BEE algorithm is then applied to this set alone to find the optimal route.

Energy Conserving Routing Based on Battery Status

The main goal of battery management schemes is to exploit the discharge characteristics of the battery and allow recovery. Energy-efficient routing protocols are designed to take into account this battery information in the selection of the best route. Chang and Tassiulas proposed an energy-efficient routing protocol [16] which tries to maximize the battery lifetime. The algorithm converges to a maximizing flow problem when there is a single power level and provides an optimal lifetime for batteries. In order to maximize the lifetime, the traffic should be routed so as to balance the energy consumption among the nodes rather than trying to reduce the power consumed. Most of the previous work which deals with minimizing the overall energy consumption tries to route the packets through the path that has the minimum energy consumption per unit packet. But these routing techniques remain static, which leads to a more rapid draining of the battery charge through those routes. In [16], Chang and Tassiulas try to find the optimal traffic split (distributing the traffic between multiple routes that exist between the source and the destination) which takes into consideration the remaining charge of the battery. The assumptions made are mentioned below. An ad hoc wireless network with N nodes is considered as a directed graph $G(N, A)$ where A is the set of all directed links (i, j), where $i, j \in N$. S_i denotes the set of all nodes reachable for node i, such that for any existing link (i, j), $j \in S_i$. Initial battery store for node i is given by E_i. $Q_i^{(c)}$ and $q_i^{(c)}$ denote the rates at which information is generated and transmitted, respectively, at node i for commodity $c \in C$, where C denotes the set of all commodities. Each commodity contains a set of flows. e_{ij} is the energy required to transmit an information from node i to node j, $O^{(c)}$ is the set of source nodes, and $D^{(c)}$ is the set of destination nodes. $q_i^{(c)}$ denotes the flow of the commodity, which is the rate at which data is transferred between node i and node j. The optimal flow condition states that the total incoming flows for node n must be equal to the total outgoing flows, as shown in Figure 11.11.

The flow augmentation (FA) algorithm is performed at each node to obtain the flow which, in turn, is used to split the incoming traffic. Each node performs the following at each iteration:

- Each source node $o \in O^{(c)}$ for a commodity c calculates the shortest cost path to its destination $d \in D^{(c)}$.

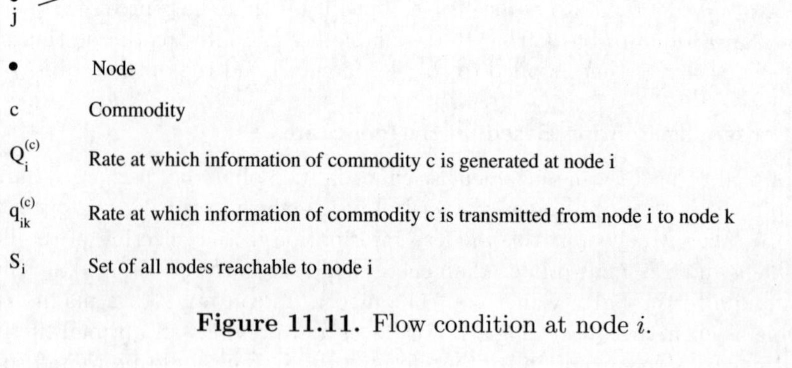

Figure 11.11. Flow condition at node i.

- Then the flow is increased by an amount of $\lambda Q_i^{(c)}$, where λ is the step size.
- The shortest cost path is recalculated and the process is repeated until the first node runs out of its initial energy.

The main objective of the algorithm is to find the link with minimal cost that leads to maximization of the lifetime of the network. The three main factors that influence the cost c_{ij} are the energy required to transmit a packet over the link, e_{ij}, the initial energy E_{ij}, and the instantaneous energy $\underline{E}_{ij}$. A good link of flow augmenting path must consume less energy and should select the path which has more energy remaining. Simultaneous optimization of these factors is not possible. Hence, a balance point is reached between these two. Taking this into account, the cost calculation is given by

$$c_{ij} = e_{ij}^{x_1} \underline{E}_i^{-x_2} E_i^{x_3} \qquad x_1,\ x_2,\ x_3 \geq 0 \tag{11.4.15}$$

$x_1,\ x_2,\ x_3$ are weighting factors for the items e_{ij}, $\underline{E}_i$ and E_i, respectively. The path cost is the sum of the costs of all the links on the path.

The flow redirection (FR) algorithm is an inference of the following observation, proof for which can be obtained in [16]: "In a single source, single destination environment or multiple source, multiple destination environment, without any constraints on the information generation rates, the minimal lifetime of every path remains the same under the optimal flow condition. In case of multiple source and destination, common-source and common-destination nodes are assumed with zero cost link that connects all the sources and destinations, respectively, that is, $e_{d\bar{d}} = 0 \quad \forall\, d \in D^{(c)}$." If there exists a flow from source $o \in O^{(c)}$ to destination $\bar{d}^{(e)}$ which uses the minimum total transmitted energy path with a flow value of $Q_o^{(e)}$, then the steps taken to reroute the flow to a different destination are:

- Determine the paths in which redirection is going to take place.
- Calculate the amount of redirection, that is, the percentage of flows per commodity to be redirected, given by $\epsilon_i^{(e)}$.
- Redirect the flows through certain path to $\bar{d}^{(e)}$ by decrementing an amount $\epsilon_i^{(e)}$ from the outgoing flows and by adding same amount to the flows of the selected path.

11.5 TRANSMISSION POWER MANAGEMENT SCHEMES

The components used in the communication module consume a major portion of the energy in ad hoc wireless networks. In this section, we investigate some of the means of achieving energy conservation through efficient utilization of transmission power such as selection of an optimal power for communication. The variation in transmission power greatly influences the reachability of a node. Increasing the transmission range not only increases coverage, but also the power consumption rate at the transmitter. This section deals with finding a trade-off between the two contradictory issues, that is, increasing the coverage of a node and decreasing its battery consumption.

11.5.1 Data Link Layer Solutions

As stated earlier, transmitter power greatly influences the reachability of the node and thus the range covered by it. Power control can be effected at the data link layer by means of topology control and constructing a power control loop. This section describes different power-based solutions at the data link layer. Recent works in [17], [18], [19], and [20] suggest that a proper selection of power levels for nodes in an ad hoc wireless network may lead to saving of power and unnecessary wastage of energy. Some of the solutions proposed to calculate the optimum transmission range are as follows:

- Dynamic power adjustment policies
- Distributed topology control algorithms
- Constructing distributed power control loop
- Centralized topology control algorithm

Dynamic Power Adjustment Based on the Link Affinity

Ad hoc wireless networks are prone to constant link failures due to node mobility, hence the stability of routes cannot be assured in such situations. But frequent link failures lead to reduced throughput. A protocol that selects a route which has a very low probability of link failures is proposed in [17]. A parameter called *affinity* that decides the stability of a route is defined. Node m samples a set of signals from the node n and calculates the affinity (a_{nm}) as follows:

$$a_{nm} = \begin{cases} high & : \text{if } \delta S_{nm(ave)} > 0 \\ \frac{S_{thresh} - S_{nm(current)}}{\delta S_{nm(ave)}} & : \text{otherwise} \end{cases} \tag{11.5.1}$$

that is, the link is assumed to be disconnected between two nodes n and m if the signal strength $S_{nm(current)}$ is well below the threshold signal strength (S_{thresh}). $\delta S_{nm(ave)}$ is the average of the rate of change of signal strength over the last few samples.

Each node transmits *Hello* packets to its neighbors periodically with constant power. As soon as the receiver hears one such *Hello* packet, it calculates the signal strength of the transmitter ($S_{t,t+\tau}$) using the relation specified below. If the time interval of the arrival of *Hello* packets is represented by τ, then

$$S_{t,t+\tau} = \begin{cases} S_H - \{\frac{S_H - S_{thresh}}{a} * \tau\} & \text{if moving farther and } \tau < a \\ S_H & \text{if moving closer and } \tau < a \\ S_{thresh} & \text{otherwise} \end{cases} \tag{11.5.2}$$

where S_H is the signal strength of the *Hello* packet received, τ is the time period between two successive *Hello* packets, and a is the link affinity between a node and its neighbor. After calculating the signal strength, the node adjusts its transmission power (P_T) accordingly. The new adjusted transmission power ($P_{t,t+\tau}$) is given by

$$P_{t,t+\tau} = P_T * \frac{S_{thresh}}{S_{t,t+\tau}} \tag{11.5.3}$$

Each node transmits with the minimum power that is required to reach the destination. Thus dynamic power adjustment can be made in a distributed way. The authors of [17] show that around 3% to 5% of power saving can be obtained using the above scheme.

Distributed Topology Control Mechanisms

Now we shall discuss the algorithm proposed in [18], which uses a distributed power control mechanism as opposed to the centralized one used in [21] and [22], which is explained in the next section. According to this algorithm, each node of the ad hoc wireless network independently runs a localized algorithm and decides the appropriate power level to be used by that node. A node increases the power directionally until it finds one node in all the directions. Then it tries to increase the lifetime of the nodes to a greater extent by reducing the transmission power and having less coverage of the nodes while guaranteeing the same connectivity as the one achieved when the nodes are maximally powered. The principle behind this algorithm is that the topology of the network can be changed by choosing the appropriate power level. An improper network topology may suffer from poor network utilization, lesser battery life, and higher delays.

The model used in this algorithm uses a cone-based topology on a two-dimensional surface. The model assumes a set V of n nodes in the plane. Each node consists of a power supply unit, memory, and processor for performing simple calculations. Any node n can send a broadcast message with varying powers ranging between $0 \leq p \leq P$. Whenever a node n initiates a broadcast message, all nodes that receive the message (set N) reply with an acknowledgment. Thus, node n becomes aware of the set N. When any two nodes u and v exchange broadcast and acknowledgment messages, they become aware of the directions of each other which are separated by a degree of π. Hence, nodes u and v transmit with a power ρ and $\rho + \pi$, respectively. Techniques such as angle of arrival (AOA), which is used to calculate the direction of the node, are assumed to be available. Now we will look into the algorithm in detail, which consists of two phases.

In the first phase, which is a neighbor discovery process, a distributed algorithm is run on each node so as to form a connected network. This is done as follows. Starting with a small value for the power level, each node sends a broadcast message. Any node receiving it sends an acknowledgment back to the sender. The sender records all the acknowledgments that it received along with the information about the direction. It determines whether there exists at least one neighbor in each cone of degree α. Each node u starts with the initial value for the growing power p. If node u discovers any neighbor v, it adds it to the local neighbors set $N(u)$. The node keeps increasing the power until one node is found in each cone or till the power p reaches the maximum value P. This termination condition can be mathematically formulated using Figure 11.12. That is, each node in the set $N(u)$ covers a cone for any node u. If the union of all the cones forms an angle greater than 2π, the algorithm enters phase 2. The inference made from phase 1 is that, if there is a

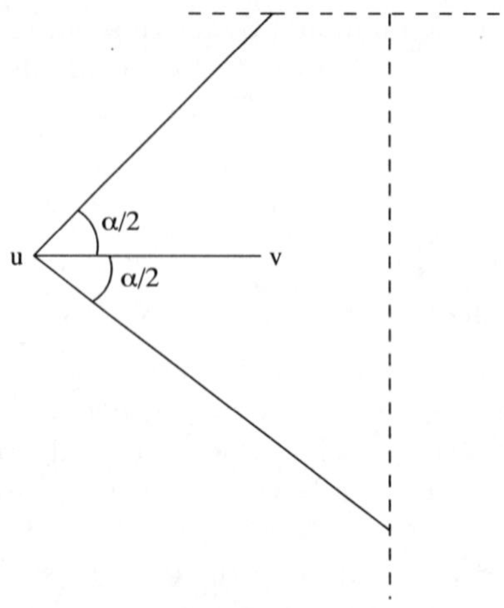

Figure 11.12. Coverage determination.

node v in the cone when sending with a maximum power P, then there is always another node v' in the cone when sending with a minimum power $p(u)$, that is, the algorithm is symmetric.

In the second phase, the redundant edges are removed without affecting the network connectivity of nodes and without removing the minimum power path. If two nodes v and w exist in the same cone for node u, with v, $w \in N(u)$ and $w \in N(v)$, then node w is removed from $N(u)$ if it satisfies the condition

$$P(u, v) + p(v, w) \leq q \times p(u, w) \tag{11.5.4}$$

where $q \geq 1$ and $P(u, w)$ denotes the power required by node u to reach w. For values of P smaller than $\frac{2\pi}{3}$, the algorithm guarantees maximum connected set.

Constructing Distributed Power Control Loop

The following is a distributed approach which tries to attain an optimal power level for the nodes in an ad hoc wireless network. The authors in [19] proposed a power control loop which increases the battery lifetime by 10-15% and the throughput by around 15%. The algorithm is tested on the model that assumes mobility, group communication, and fading due to blockages such as manmade obstacles. The proposed algorithm works at the MAC layer in a distributed fashion. The main objective behind the algorithm is to reduce the energy cost of communication between the nodes and thereby increasing the battery lifetime and the effective bandwidth. The main reasons behind the need for distributed algorithms in ad hoc wireless networks are the mobility and absence of a central arbiter which can inform the nodes about the power levels to be used. The algorithm aims at allowing each node to use different power levels while transmitting to different nodes and at the same time maintaining the connectivity of the network. This is because nodes that are closer require less power for transmission, but on using a common power level as suggested by Kawadia *et al.* in [21], interference may be increased when higher power is selected as a common power level and used for communicating with nearby nodes. The power control algorithm has been incorporated into the IEEE 802.11 MAC protocol. We now discuss the modifications made to the 802.11 MAC protocol.

- Unlike the usual IEEE 802.11 DCF protocol which uses only one common power level, the algorithm suggested by Agarwal *et al.* in [19] uses ten different power levels varying with a step size of one tenth of the maximum power level available.

- The format of the message header is also modified as shown in Figure 11.13. The headers of the CTS and Data frames are modified to include the information of the ratio of the signal strength of the last received message to the minimum acceptable signal strength of the node under consideration. When the receiver receives the RTS signal from the transmitter, it attaches to the CTS the ratio information calculated and sends it back to the sender. Similarly, when the sender gets the CTS packet, it includes the ratio in the Data frame and sends it to the receiver. Thus in a single transmission of RTS-CTS-Data-ACK exchange, both the sender and the receiver learn about the transmit power levels of each other.

- The MAC layer for each node constructs a table which holds information about the transmit power levels of all the neighboring nodes. The information stored in the table consists of the exponential weighted average (EWA) history of the ratio for all the neighbors. This table may be small because of the

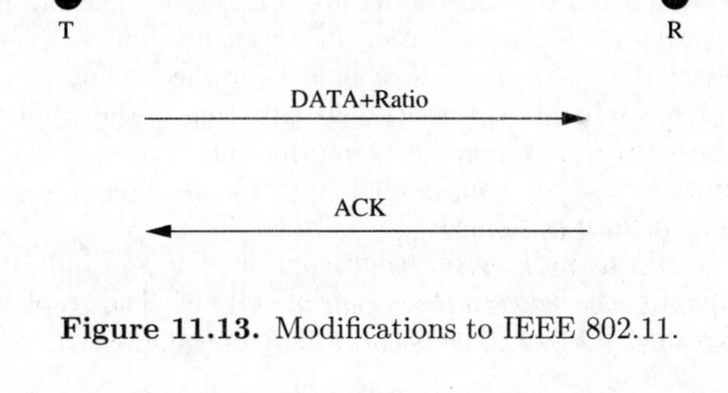

Figure 11.13. Modifications to IEEE 802.11.

fewer number of neighbors in the ad hoc network environment. There exist two situations where the table has to be considered. Whenever a message is received from a node, the receiver looks up into the EWA history. If the node is not present in the EWA table, a new entry is made in the table. The second scenario is, when the receiver is not within the range of the transmitter, that is, when there is no reply for the RTS or the Data signal, the transmitter increases its power by the step size of one tenth of the maximum power available. Similarly, the receiver increments its power level if there is no reply for the CTS sent by it. This process is continued until the node becomes reachable or the maximum power is reached.

Thus each node calculates which level of power is to be used for transmission to each of its neighboring nodes.

The authors in [23] have also proposed a dynamic topology construction algorithm in which nodes, by overhearing the RTS-CTS packets, calculate their transmission powers. The main aim of this power control dual channel (PCDC) protocol is to choose the transmission power in such a way that the connectivity between nodes remains unaffected but at the same time increases the number of simultaneous transmissions in the network, and thereby the throughput. The main difference between this protocol and the centralized topology control algorithm proposed in [20] lies in the use of dual channels, one for transmission of control packets and the other for data packets. The use of a separate channel for control packets avoids to a great extent those collisions caused by the hidden terminal problem. This is effected as follows. Whenever a node M hears transmission of RTS packets in its control channel with power p, destined for another node N, which causes interference to the on-going reception at node M, it sends a special CTS packet which makes the RTS sender withdraw its transmission. The duration of withdrawal is specified by node M based on the time duration for which the on-going transmission of node M is estimated to last. This increases the end-to-end throughput of the system and also reduces the overall power consumption at the nodes.

Centralized Topology Control Algorithm

The algorithm suggested by Ramanathan and Rosales-Hain in [20] is a centralized algorithm which adjusts the power level of the nodes to create the desired topology. The problem is constrained as an optimization problem with power level as the optimization objective and the constraints are connectivity and biconnectivity.

Unlike the conventional representation of an ad hoc wireless network using a graph, as represented in [21] and [18], a model for the ad hoc wireless networks assumed in [20] keeps separate the entities contributing to the ability to communicate. Some of the entities are the mobility information, propagation characteristics, and node parameters such as transmission power and antenna direction. These parameters can be defined as follows.

Any graph is said to be k-vertex/edge connected if and only if there are k-vertex/edge-disjoint paths between every pair of vertices. The graph is connected if $k = 1$ and biconnected if $k = 2$. The network is represented as $M = (N, L)$, where

N is the set of nodes and L is the location information $N \rightarrow (Z_0^+, Z_0^+)$, which is the set of coordinates on the plane. The parameter vector of a node is given by $P = \{f_0, f_1, \ldots, f_n\}$, where $f_i : N \rightarrow R$ is a real value where the parameter vector includes antenna configuration, spreading code, and hardware. In [20] the authors consider only one parameter, that is, power for node u ($p(u)$). Hence, $P = \{fp\}$. The propagation is represented as $\gamma : L \times L \rightarrow Z$ where L represents the set of location coordinates in the plane and $\gamma(l_i, l_j)$ gives the propagation loss at l_j for the packet whose source is l_i. For successful reception

$$p - \gamma(l_i, l_j) \geq S \tag{11.5.5}$$

where S is the receiver sensitivity which is the threshold strength required for reception of signals and is assumed to be known *a priori* and γ is assumed to be a monotonically increasing function of the geographical distance given as

$$\lambda(d) = \gamma(d(l_i, l_j)) + S \tag{11.5.6}$$

where P must be greater than $\lambda(d)$ to achieve successful transmission where $\lambda(d)$ is the *least power function* which specifies the minimum power required to transmit to a node which is at a distance d. Given an ad hoc wireless network represented by $M = (N,\ L)$ and the transmit power function p and the least power function λ, the induced graph can be represented as G=(V, E), where V corresponds to the nodes in N, and E is the set of undirected edges such that $(u,\ v) \in E$ if and only if both $p(u)$ and $p(v)$ are greater than $\lambda(d(u,\ v))$.

The constrained optimization problem can thus be stated as a *connected min-max power (CMP)* problem. The problem is to find a per-node minimal assignment of transmit powers $p : N \rightarrow Z^+$, such that the graph $(M,\ \lambda,\ L)$ that is induced remains connected and the power factor $Max_{u \in N}(p(u))$ has a minimum value. For the biconnected graph, the problem can be stated as *biconnected augmentation min-max power* (BAMP). Given the graph as in the previous definition, the problem is to find a per-node minimal set of power increments ($\delta(u)$) such that the induced graph $(M,\ \lambda,\ p(u)+\delta(u))$ remains biconnected and the power factor $Max_{u \in N}(p(u)+\delta(u))$ has a minimum value. We shall now look into two major types of algorithms that are used to generate connected and biconnected graphs that satisfy the given constraints.

The *connect* algorithm is similar to the minimum cost spanning tree algorithm. The basic idea used in this algorithm is to iteratively merge the connected components until only one component is left. The input to the algorithm is a graph in which the nodes use the minimum power for transmission and hence remain partially connected (M). The following steps are performed in order to carry out this algorithm:

Step 1: First, the connected node pairs are sorted in the increasing order of the mutual distance.

Step 2: If the nodes are in different network components, the power of the nodes are increased so as to reach the other nodes.

Step 3: Step 2 is repeated until the whole network becomes connected.

The *biconnect* algorithm attempts to discover a biconnected graph from the given graph M so as to satisfy the objectives and the constraints. The extension to the biconnected network from the algorithm *connect* can be done as follows:

Step 1: The biconnected components are identified in the graph induced by the algorithm *connect* based on the depth-first search method.

Step 2: The nodes are arranged in non-decreasing order of the connected node pairs as done in the previous algorithm.

Step 3: Nodes which are in different components of the network are connected by adjusting the power appropriately, and this step is repeated until the network becomes biconnected.

Now the graph obtained may not be per-node minimal because adjusting the power to reach the nodes of other components may lead to addition of edges that are not critical. These edges are termed as *side-effect edges*, as shown in Figure 11.14. The numbers of the form $s - p$ denote *step number – power* assigned during the step, $d(s)$ denotes distance d between the corresponding nodes, and the step s during which the edge has formed. Figure 11.14 (a) is per-node minimal and Figure 11.14 (b) is obtained by reducing the power level to 1 but still maintaining the connectivity. To restore the per-node minimal for the graph shown in Figure 11.14 (b), a post-processing phase is carried out after applying the aforementioned algorithms.

Let S be the list of sorted node pairs. In the *post-processing algorithm*, the power of the nodes is reduced to the minimum possible value without affecting the connectivity of the induced graph.

In Figure 11.14 (b), Step 1 connects A and C and Step 2 connects B and D with power level 1. By increasing the power level to 2 in Step 3, A and B get

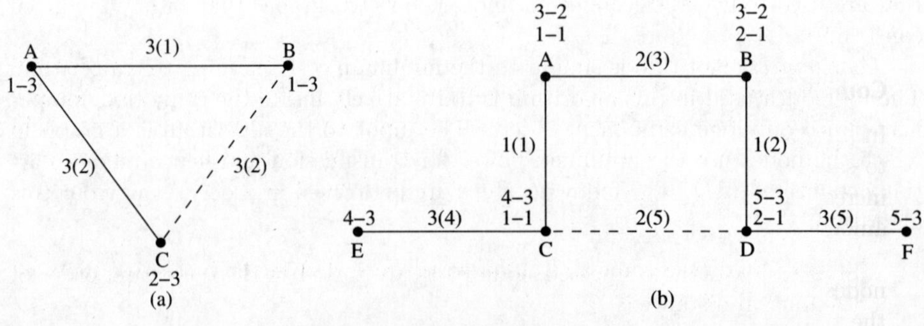

Figure 11.14. Side-effect edges.

connected. In Steps 4 and 5, CE and DF are formed with power level 3. This creates a side-effect edge CD. Hence, the power of A and B can be reduced back to 1, without affecting the graph connectivity.

11.5.2 Network Layer Solutions

The power at the network layer can be conserved by reducing the power consumed for two main operations, namely, communication and computation. The communication-related power consumption is mainly due to the transmit-receive module present in the nodes. Table 11.1 lists the power consumption of the communication module during different modes of operation. Whenever a node remains active, that is, during transmission or reception of a packet, power gets consumed. Even when the node is not actively participating in communication, but is in the listening mode waiting for the packets, the battery keeps discharging. The computation power refers to the power spent in calculations that take place in the nodes during routing and power adjustments. The following section discusses some of the power-efficient routing algorithms. In general, a routing protocol which does not require large tables to be downloaded or greater number of calculations is preferable. Also, reducing the amount of data compression that is done before transmission may decrease the communication power but ultimately increases the number of computation tasks. Hence, a balance must be reached between the number of computation and communication tasks performed by the node, which are contradictory to each other.

Table 11.1. Power consumed by Lucent ORiNOCO wireless LAN PC card in different modes

Mode	Power Spent (Watts)
Transmit	1.4
Receive	0.9
Doze	0.05

Common Power Protocol

In [21], the authors propose a common power protocol (COMPOW) that attempts to satisfy three major objectives: increasing the battery lifetime of all the nodes, increasing the traffic-carrying capacity of the network, and reducing the contention among the nodes.

The main reason behind the need for an optimal transmit power level for the nodes in an ad hoc wireless network is that battery power is saved by reducing the transmission range of the node. This also leads to a connected network with minimum interference. In [21], the authors put forth the following reasons for a common power level in an ad hoc wireless network:

- For the proper functioning of the RTS-CTS mechanism: If there are different power levels for each node, CTS of a lesser-powered node may not be heard by its neighboring nodes. Hence, a neighboring node may start transmitting, which leads to collision.

- For proper functioning of link-level acknowledgments: Whenever the transmitter (T) sends a packet to a receiver (R), the power level at R must be at least equal to that of T so that the acknowledgment sent by node R reaches node T. This implies that the power level of any two neighbors in an ad hoc wireless network must be equal. By transitivity, this is extended to multi-hop neighbors and thus a common power level is necessary for all the nodes in the network. If the common power level selected is a very high value, then this may lead to interference among the nodes as shown in Figure 11.15 (a). On the other hand, if the value is too low, the reachability of nodes may become very weak, which in turn may render the network partitioned as shown in Figure 11.15 (b). Hence, choosing an optimum value is a difficult task. For calculating the common power level, the following solution is proposed in [21]. A network with n nodes is considered for study with a data rate of W bits/sec, in a circular area of A square meters. The common range to be calculated is assumed to be r. A successful transmission from node T to node R requires that there cannot be any other transmission occurring around a distance of $(1 + \Delta)r$ from R, where $\Delta > 0$. Now, let us consider two simultaneous transmissions, one from T to R and another from T' to R' separated by distance of $\bar{r}$ and $\bar{r}'$, respectively, as shown in Figure 11.16. Then the distance between the receivers R and R' is given by

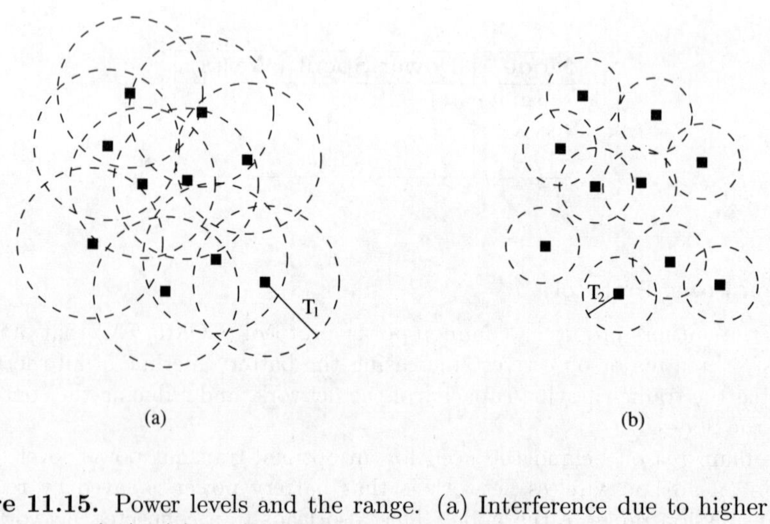

Figure 11.15. Power levels and the range. (a) Interference due to higher transmission range (T_1). (b) Partition of the network due to lower transmission range (T_2).

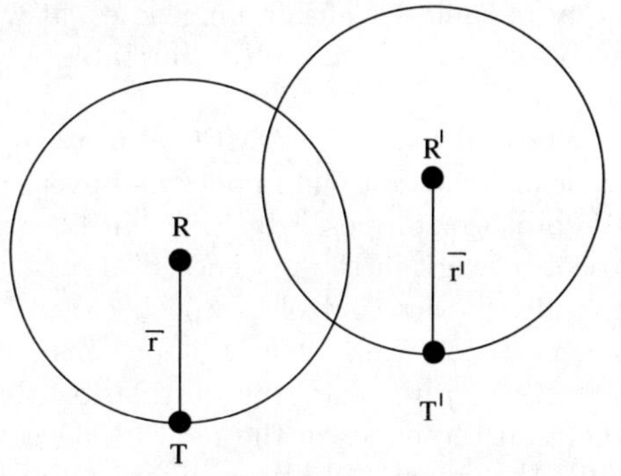

Figure 11.16. Successful simultaneous transmissions.

$$|R - R'| \geq \frac{\Delta\bar{r} + \bar{r'}}{2} \tag{11.5.7}$$

A conclusion drawn from the above discussion is that a circle of radius $\frac{\Delta\bar{r}}{2}$ and a circle of radius $\frac{\Delta\bar{r'}}{2}$ are disjoint. Hence, the distance between any transmitter and receiver must be less than the common range r. If the common range for n nodes is $r(n)$ then the problem can be stated as

$$\lim_{n \to +\infty} Prob(Network\ is\ connected) = 1 \tag{11.5.8}$$

if and only if

$$r(n) = \sqrt{\frac{A(logn + k(n))}{\pi n}}, \quad k(n) \to +\infty \ \ as \ \ n \to +\infty \tag{11.5.9}$$

The maximum throughput that can be supported by the network is given by

$$\lambda(n) = O\left(\frac{1}{\sqrt{nlogn}}\right) bits/sec \tag{11.5.10}$$

In a practical wireless environment, factors such as the number of nodes and the area of the domain may not be known *a priori*. In such cases, rather than to deal with the range factor, it is convenient to deal directly with the power level P. To find the smallest power level that is required to ensure network connectivity, [21] proposes the following network feedback strategy. The power level for a node j is given by P_j. Let $R(P)$ denote the set of nodes which are connected when the common power level ($p_0 = p_1 = \cdots = p_n$) is maintained at a particular value P, and let RP_{max} be the maximal reachable set. By analyzing the feedback and adjusting the power, the smallest power level required can be obtained. That is,

$$P(t) = P_i \text{ such that } R(P_{i-1}) \neq R(P_{max}) \tag{11.5.11}$$

Here, t denotes an instant of some set of sampling times, at which the power levels are changed. $R(P)$ can be obtained from the routing tables and hence the common power level can be calculated in real life.

Kawadia and Kumar proved that the COMPOW protocol works well only in a network with a homogeneous distribution of nodes and exists only as a special case of the CLUSTERPOW protocol proposed by them in [24]. They have extended their COMPOW protocol to work in the presence of non-homogeneous dispersion of the nodes. This extended power control protocol, called CLUSTERPOW, is a power control clustering protocol, in which each node runs a distributed algorithm to choose the minimum power p to reach the destination through multiple hops. Unlike COMPOW, where all the nodes of the network agree on a common power level, in CLUSTERPOW the value of p can be different for different nodes and is proved to be in non-increasing sequence toward the destination. The authors in [24] have provided an architectural design to implement CLUSTERPOW at the network layer. This loop-free power control protocol can work in the presence of any underlying routing protocol.

Globalized Power-Aware Routing Techniques

Minimizing the overall transmission power and distributing the power consumption evenly among the nodes are contradictory to each other. Now we shall look into the routing algorithms suggested in [25] and [26] that attempt to find a balance between these two factors by using new node metrics, as described below, for the route selection process. In [27], the authors have proposed a power optimal scheduling and routing protocol which tries to minimize the total average power in the network, subjected to constraints such as peak transmission power of the nodes and achievable data rate per link.

Minimum Power Consumption Routing

The power required to transmit a packet from node A to node B is inversely proportional to the nth power of the distance (d) between them, that is, $\frac{1}{d^n}$, where n varies from 2 to 4 depending on the distance and terrain between the nodes. A successful transmission from node n_i to node n_j requires the signal to noise ratio (SNR) of the node j to be greater than a specific threshold value Ψ_j. This can be mathematically represented, which shows that for a successful transmission, the SNR at receiver node n_j given by SNR_j must satisfy the condition:

$$SNR_j = \frac{P_i G_{i,j}}{\sum_{k \neq i} P_k G_{k,j} + \eta_j} \Psi_j(BER) \qquad (11.5.12)$$

where P_i is the transmission power of host n_i; $G_{i,j} = 1/d^n_{i,j}$ is the path gain between hosts n_i and n_j; η_j is the thermal noise at the host n_j; and BER is the bit error rate which is based on the threshold Ψ_j.

The total transmission power for route l is the sum of the transmission powers of all nodes in the route. According to minimum power consumption routing (MPCR), the preferred route is the one with minimum total transmission power among all the available routes between the source and the destination. This routing algorithm can be realized by modifying the Dijkstra's shortest path algorithm. But this may select a path with a greater number of hops, which passes through nodes that require less power. Hence it may result in increasing the end-to-end delay of the packets in the network. In addition to this, involving greater number of hops may reduce the stability of the route because of the node mobility which is one of the inherent characteristics of ad hoc wireless networks. Hence, the Bellman Ford algorithm is considered, which takes into account transmission power as a cost metric. The power cost is given by

$$C_{i,j} = P_{transmit}(n_i, n_j) + P_{transceiver}(n_j) + Cost(n_j) \tag{11.5.13}$$

where $P_{transmit}(n_i, n_j)$ is the transmitter power of node i to reach node j and $P_{transceiver}(n_j)$ is the transceiver power of node j, which tries to select the route with the fewer number of hops. The cost function at node n_i is given by

$$Cost(n_i) = \min_{j \epsilon NH_{(i)}} C_{i,j} \tag{11.5.14}$$

where $NH(i) = \{j, n_j \text{ is a neighbor node of } n_i\}$. This algorithm tries to reduce the overall power consumption of the network, but still lacks in the ability to reduce the power consumed at individual nodes. As shown in Figure 11.17, node n may be a common node used by multiple flows simultaneously. This may render node n deprived of all its battery charge faster than other nodes in the network.

Minimum Variance in Node Power Levels

The main motivation behind this metric is to ensure that all the nodes are given equal importance and no node is drained at a faster rate compared to other nodes

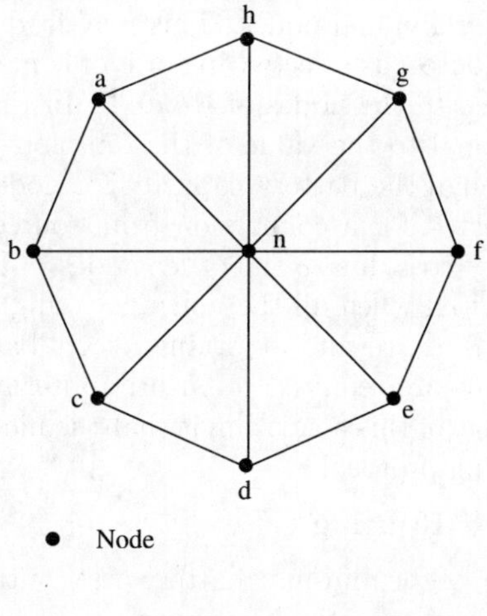

Figure 11.17. An illustration of the disadvantage in using shortest path routing.

in the network. This problem is similar to the load sharing problem in distributed systems, which is an NP-complete problem. Woo *et al.* in [26] suggest a scheme called join the shortest queue (JSQ) that tries to attain the optimal solution. According to this algorithm, for transmitting a packet, a node selects the next-hop node so that it has the least amount of traffic among all neighbors of the node. The objective can also be realized by applying a round robin selection of next-hop neighbors.

Minimum Battery Cost Routing

In the minimum battery cost routing (MBCR) algorithm, individual battery charges are taken into consideration while selecting the route, that is, the path selected must not contain the nodes that have less remaining battery capacity. This may be done in many ways. If c_i^t denotes the battery cost at any time instant t, $f(c_i^t)$ represents the battery cost function of host n_i. Now suppose the function reflects the remaining battery capacity of the node, then

$$f_i(c_i^t) = \frac{1}{c_i^t} \tag{11.5.15}$$

which means that higher the value of the function f_i, the more unwilling the node is to participate in the route selection algorithm. If a route contains N nodes, then the total cost for the route R_i is the sum of the cost functions of all these N nodes. The routing algorithm selects that path with the minimum value of the total cost among all the routes that exist between the source and the destination.

$$R_i = \min(R_j) \quad \forall j \in A \tag{11.5.16}$$

Here A is the set of all routes from source to destination. The major drawback with this scheme is that the use of summation of the remaining energy of the nodes as a metric selects the path which has more remaining energy on an average for all of its nodes rather than for individual nodes. This may lead to a condition as shown in Figure 11.18, where Route 1 is selected in spite of some of its nodes having less battery capacity compared to the nodes of Route 2. In Figure 11.18, "Node x(y)" denotes node x with y equal to the value of the function $f_x(c_x^t)$ for the node x at the time instant t. Although the battery capacity for node 3 is too little, the path containing node 3 is selected for a connection from source to destination because Route 1 has lesser battery cost due to the other nodes in the path.

The algorithm works well when all the nodes have higher battery capacity, but because the network nodes have almost drained their battery charges, some discharge control mechanisms are required to ensure uniform drainage from the batteries. The main advantage of this algorithm is that the metrics used can be directly incorporated in the routing protocol.

Min-Max Battery Cost Routing

The objective function of the min-max battery cost routing (MMBCR) algorithm is to make sure that route selection is done based on the battery capacity of all the individual nodes. Hence, the battery cost is defined as

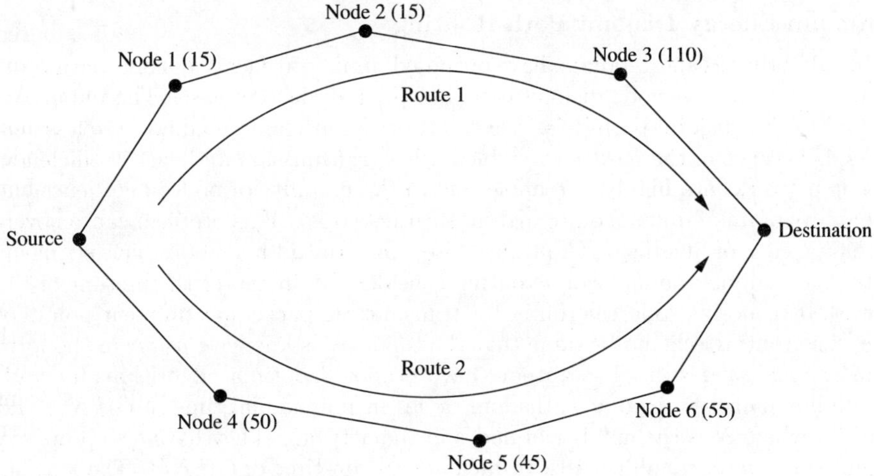

Figure 11.18. Illustrating the disadvantage of minimum-cost routing.

$$R_j = Max_{i \in route_j} f_i(c_i^t) \tag{11.5.17}$$

Therefore, the desired route is given by $R_i = \text{Min}(R_j,\ j \in A)$ where A is the set containing all possible routes. A variant of this routing algorithm minimizes the maximum cost after routing N packets to the destination or after a time period of t seconds. This tries to postpone the first node failure, which ultimately leads to a longer network lifetime. This algorithm ensures uniform discharge from the batteries. A closer look at it reveals that the path chosen does not ensure minimum transmission power and hence rapidly reduces the lifetime of all the nodes.

Conditional Min-Max Battery Cost Routing

In order to solve the contradictory issues that exist in the algorithms previously mentioned, instead of using the battery cost, conditional min-max battery cost routing (CMMBCR) considers the battery capacity directly. A threshold value γ is defined for the remaining battery capacity of the nodes. Only those paths that have sufficiently higher capacity for all their nodes compared to the threshold participate in the routing algorithm. Once the competing routes are decided, the usual MTPR algorithm is applied on them so as to choose the path with minimum total transmission power. The battery capacity of route j R_j^c at time t is

$$R_j^c \;=\; Min_{i \epsilon route_j} C_i^t \tag{11.5.18}$$

Any route j can participate in the routing process only if $R_j^c \geq \gamma$.

Minimum Energy Disjoint Path Routing

In [28], Srinivas and Modiano have proposed minimum energy disjoint path routing for two cases: (a) node disjoint case and (b) link disjoint case. The important need for having disjoint paths in wireless networks, especially in ad hoc wireless networks, is because of the need for reliable packet transmission and energy efficiency. Ad hoc networks are highly unreliable due to the mobility of nodes and hence the probability of link failure is quite high in such networks. This problem can be overcome by means of link disjoint routing. Also, since the ad hoc nodes have stringent battery constraints, node disjoint routing considerably increases the lifetime of the network by choosing different routes for transmitting packets at different points of time. The routing schemes assume that the topology is known *a priori* in the form of an *energy-cost* graph. The authors have proposed optimal algorithms for finding minimum energy disjoint paths and nodes in polynomial time of $O(kN^3)$ and $O(kN^5)$, where N is the number of nodes in the network. They have also proposed a number of sub-optimal heuristics which have a run-time of $O(kN^2)$. The authors have also proposed a distributed version of the optimal algorithms.

Localized Power-Aware Routing Techniques

The aim of this routing protocol is to find the shortest route to the destination so as to increase the net lifetime of the power source (battery), using localized algorithms. Local algorithms are distributed greedy algorithms that try to achieve a global objective based on the information available locally at the node. In this section, we will look into one of the power-cost-aware routing protocols proposed by Stojmenovic and Lin in [29]. This protocol is based on the basic principle that, by placing intermediate nodes at the desired location between two nodes separated by a distance d, the transmission power can be made proportional to the distance d rather than d^α [25], where $\alpha \geq 2$. The protocol tries to find the route that minimizes the total power needed, which increases the battery lifetime. The protocol is designed to satisfy the following main objectives:

- Use of location dependent routing: The distance between the nodes is a vital piece of information that has to be taken into account to minimize the energy required for each routing task. The location of the nodes can be obtained by the methods specified below:
  - Using the global positioning system (GPS), the location of the nodes can be obtained by using the information obtained from the satellite.
  - Receiving control messages from the neighbors at regular intervals and observing the signal strengths obtained at different points of time provides the data about their distance from the node concerned. This process which is explained in [17], is discussed in Section 11.5.1.
- The routing protocol must be loop-free. This is to ensure that the path selected uses minimum power for transmission.

- When shortest path routing is applied to a graph, it may so happen that the same node is involved in several routing tasks. This eventually decreases the lifetime of that node. The protocol must distribute the traffic load so as to avoid this problem.

- The routing protocol must be designed in such a way to reduce the amount of information exchanged among the nodes, since communication incurs loss of energy. Increase in the number of communication tasks also increases the traffic in the network, which results in loss of data, retransmissions, and hence more energy consumption. The number of communication tasks can be decreased by avoiding centralized algorithms and avoiding maintenance of large routing tables.

- Selection of routes must be flexible and should try to avoid memorizing past traffic or the routes, thereby avoiding large storage.

- Another main objective is to achieve maximum packet delivery for dense networks.

- Adaptation of the routing algorithm to the continuous topological changes is important as far as ad hoc wireless networks are concerned.

The algorithm is aimed at selecting a single path for a particular routing task which guarantees delivery of packets. This is due to the mobile nature of the nodes in the ad hoc wireless networks which leads to frequent topological changes. The model suggested by Stojmenovic and Lin in [29] makes use of min-power-graphs which are constructed as follows. The concept used here is similar to that suggested in [22]. Two nodes are said to be connected neighbors if and only if they satisfy the condition given below:

$$d(A, B) < min(t(A), t(B)) \qquad (11.5.19)$$

where $t(x)$ and $d(A,\ B)$ denote the transmission range of the node x and the distance between nodes A and B, respectively. Min-power-graphs are built using this equation. If $t(x)$ is same for all values of x, $x \in$ set of nodes, then the graph is said to be a unit graph. Before going into its details, we will discuss the model used in describing the properties.

- The power needed for transmitting and receiving is assumed to be $u(d) = ad^{\alpha}$ $+ \ c$ where c is a constant dependent on the energy spent on processing for encoding and decoding. The parameter a is adjusted according to the physical environment.

- When sender S sends packets directly to the destination D, let the distance between nodes S and D be given by $|SD| = d$. If the packets traverse through an intermediate node A, let $|SA| = x$ and $|AD| = d - x$. Let the intermediate node be placed at any arbitrary location. The following properties hold for the prescribed model:

Lemma 1: There always exists an intermediate node A between source S and destination D which reduces the energy consumption if packets from S destined to D are routed through it when the condition $d > (c/(a(1-2^{1-\alpha})))^{1/\alpha}$ holds. Maximum power saving is achieved when A is placed exactly in the midpoint of SD.

Lemma 2: If $d > (c/(a(1-2^{1-\alpha})))^{1/\alpha}$, then by dividing SD into n equal intervals, n being the nearest integer to $d(a(\alpha-1)/c^{1/\alpha})$, maximum power saving can be obtained. The minimal power is then given by

$$u(d) = dc(a(\alpha-1)/c)^{1/\alpha} + da(a(\alpha-1)/c)^{(1-\alpha)/\alpha} \qquad (11.5.20)$$

First we discuss a power-saving algorithm, then a cost-saving algorithm, and finally an efficient power-cost saving algorithm derived from the previous two algorithms.

Power-Saving Localized Routing (SP-Power) Algorithm

The centralized version of the above algorithm can be effected by using Dijkstra's single-source shortest weighted path algorithm, where the edge weight is $u(d) = ad^\alpha + c$. This is referred to as the *SP-power algorithm.* Now the corresponding localized algorithm is as follows.

Power Calculation: Now we will calculate the power required to transmit a packet from node B (source or intermediate node) to node D. Let node A be the neighbor of B and let $|AB| = r$, $|BD| = d$, and $|AD| = s$. The power needed to transmit from node B to node A is $u(r) = ar^\alpha + c$. Assuming that the power required for the rest of the transmissions ($v(s)$) in the network is uniformly distributed, by applying the above *Lemma 2* we have

$$v(s) = sc(a(\alpha-1)/c)^{1/\alpha} + sa(a(\alpha-1)/c)^{(1-\alpha)/\alpha} \qquad (11.5.21)$$

The power-saving localized routing algorithm from source S to destination D is given below.

Step 1: Let $A := S$.

Step 2: Let $B := A$.

Step 3: Each node B, which may be a source or intermediate node, will select one of its neighbors A so as to minimize the power $p(B, A) = u(r) + tv(s)$ and sends the message to neighbor node A.

Step 4: Steps 2 and 3 are repeated until the destination is reached, that is, $A = D$, or the delivery has failed, in which case $A = B$.

Cost-Saving Localized Routing (SP-Cost) Algorithm

This algorithm is based on the relation $f(A) = 1/g(A)$, where $f(A)$ is the cost of node A, and $g(A)$ denotes the lifetime of the node A, normalized to be in the interval (0, 1). The localized version of this algorithm under constant power is mentioned below.

Cost Calculation: The total cost $c(A)$ incurred when a packet moves from node B to node D via intermediate node A is the sum of the node's cost $f(A) = 1/g(A)$ and the estimated cost of the route from node A to node D. This cost $f(A)$ of the neighbor node A holding the packet is known to node B. The cost of the nodes in the rest of the path is proportional to the number of hops between nodes A and D, which is in turn proportional to the distance $s = |AD|$ between nodes A and D, and inversely proportional to the transmission radius R. Thus the cost $f(A)$ is ts/R, where t is a constant. Hence, the total cost can be considered to be $c(A) = ts/R + f(A)$ or $c(A) = tsf(A)/R$. The cost-saving localized routing algorithm from source S to destination D is given below.

Step 1: Let $A := S$.

Step 2: Let $B := A$.

Step 3: Let each node B, which may be a source or intermediate node, select the neighbor node A that minimizes the cost $c(A)$. If node D is one of the neighbors, send the message to node D, else send it to node A.

Step 4: Steps 2 and 3 are repeated till the destination is reached, that is, $A = D$, or the delivery has failed, in which case $A = B$.

Power-Cost-Saving Localized Routing Algorithm

In order to arrive at the power-cost algorithm, the power-cost of sending a message from node B to its neighbor node A must be known, which can be power-cost$(B, A) = f(A)u(r)$, where $|AB| = r$, or power-cost$(B, A) = \alpha u(r) + \beta f(A)$, where α and β are constants. Depending on which factor is used, either sum or product, the power-cost-saving localized routing algorithm from source S to destination D can be represented as SP-Power*Cost or SP-Power+Cost algorithm.

Step 1: Let $A := S$.

Step 2: Let $B := A$.

Step 3: Let each node B, which may be a source or intermediate node, select the neighbor node A that minimizes the value pc(B, A) = power-cost(B, A) + $v(s)f(\acute{A})$. Send the message to A.

Step 4: Steps 2 and 3 are repeated till the destination is reached, that is, $A = D$, or the delivery has failed, in which case $A = B$.

Energy-Efficient Ad Hoc Wireless Networks Design Algorithm

In ad hoc wireless networks, cluster formation is done to provide better local coordination, hierarchical addressing, and better scheduling of resources. Such cluster formations involve distributed identification of a cluster-head that has the responsibilities of internal coordination and scheduling. Inter-cluster communication is achieved through cluster gateway nodes. In a power-constrained network, the additional responsibilities deplete the energy reserve of the cluster-head faster than the other nodes in the cluster. This results in the constant change of the cluster-heads and inefficient topology. We will now discuss an energy-efficient network design algorithm called ad hoc network design algorithm (ANDA), which tries to maximize the network lifetime for a static network where cluster-heads are known *a priori*. The basic idea used in this protocol is that the cluster-heads dynamically adjust the power level; and hence, the cluster size varies based on the remaining battery charge. Thus energy is uniformly drained from the cluster-heads, which in turn increases the lifetime of the network.

Let S_C and S_N be the set of cluster-heads and the set of member nodes in the network, respectively. Assuming that the cluster-heads are known *a priori* and fixed, the following functions are proposed by Chiasserini *et al.* in [30]:

- *Covering:* This function tries to find an optimal coverage for all cluster-heads. The algorithm states that all the nodes choose one among the set of cluster-heads that projects maximum lifetime. The resulting network configuration guarantees minimum energy consumption.

- *Reconfiguration:* Finding an appropriate time for the network reconfiguration is a crucial task. It is dependent on the remaining energy of the cluster-heads. Assuming all nodes have the same energy initially and they select their cluster-heads based on the covering algorithm specified above, now after a time period t from the last reconfiguration, the remaining energy at the cluster-heads after a time interval t is given by

$$E_i^{(new)} = E_i^{(old)} - t(\alpha r_i^2 + \beta |n_i|) \tag{11.5.22}$$

where E_i $(i = 1, \ldots, C)$ indicates the remaining energy at the cluster-head i; α and β are constant weighting factors; r_i is the radius of coverage of the cluster-head; and n_i denotes the number of nodes under the cluster-head i. The function covering is performed on the nodes again to assign them to the corresponding cluster-heads. Thus, a load-balancing approach of nodes is carried out in the network to increase the network lifetime.

Determination of Critical Transmission Range

The authors of [22] propose a centralized algorithm that calculates the minimum power level for each node that is required to maintain network connectivity based on the global information from all the nodes. This minimum value of the node's transmission range is termed the critical transmission range. This algorithm aims at finding the transmission range for each of the nodes which acts as a trade-off between increasing the network connectivity, the spatial reusability, and the battery-life extension. The optimal value of the reception range depends on the following factors:

- Mobility: Due to the mobile nature of ad hoc wireless networks, the links are frequently formed and broken. This greatly affects the optimum range to be considered.

- Propagation models: Different fading mechanisms exist, some of which are frequency-dependent. The propagation model that results from these mechanisms has been developed for different network operational conditions such as atmospheric conditions and man-made obstacles. The use of an appropriate propagation model can help in studying the changes required in the physical layer under different environmental conditions.

- Power level: The larger the power level, the larger is the transmission range covered. This leads to faster depletion of the battery charge and hence reduces the lifetime of the node.

- Antenna configuration: Another factor to be considered is how the power is spatially distributed, which actually depends on the antenna model used.

The problem is to find the critical transmission range for each of the nodes so that the network remains connected. But finding nodes within the transmission range is unpredictable due to node mobility. Some mobility model has to be assumed, to simulate the changing topology. This model can be viewed as a set of snapshots which represent the location of the mobile, separated by a small time interval. The authors of [22] prove that most of the results discussed are not sensitive to the mobility pattern under consideration. Since the topology keeps changing, maintaining the same power level for all the nodes as suggested in [21] is not possible. Hence, further investigation is done in this regard.

The network can be represented as a graph $G(v, e)$ with v vertices and e edges. The links of a network are divided into two types: essential links which are necessary to maintain a fully connected network, and the rest of the links that form the set of redundant links. In order to find the essential links a new concept called direct neighbor is introduced by Sanchez *et al.* in [22]. Nodes a and b are said to be direct neighbors of one another, if and only if the only closest node to a is b, and vice versa. Now the problem can be stated as

$$min(l) \in R^+ \mid Conn(G(v, e^*)) = 1 \qquad (11.5.23)$$

where

$$e^* = \{(i, j) \in v^2 \mid |x_i - x_j| \leq l\} \qquad (11.5.24)$$

where x_i denotes the location of node i, l is the transmission range, and $Conn(G)$ is the connectivity of the graph, which is the ratio of the number of pairs of the nodes that are connected to the maximum number of such possible connections. It varies between 0 and 1. A link between two nodes i and j is said to be a critical link (CL) if the removal of the link leaves the network partitioned. It is the link between two *direct neighbors* that determines the critical range. Thus, the *direct neighbors* can communicate without any nodes' intervention, as shown in Figure 11.19. (a,b),

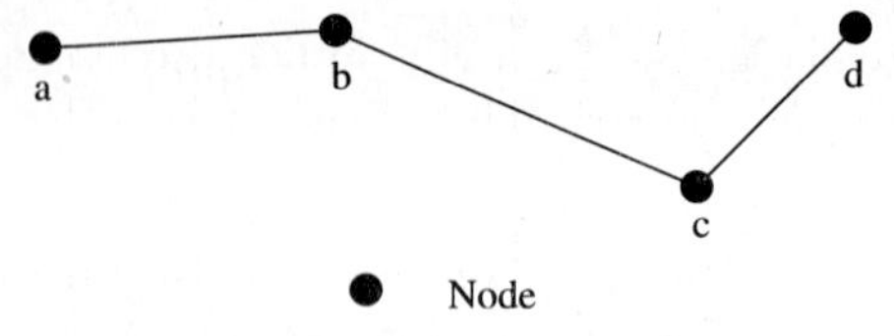

Figure 11.19. A sample network.

(*b*,*c*), and (*c*,*d*) are *direct neighbors* whereas (*b*,*d*) is not, as *c* acts as a link between them. From this, a direct neighbor graph (DNG) G_d is created.

$$G_d = \{G(v, e^d)|\ \forall\ m \in e^* \mid m \text{ is a direct neighbor}\} \tag{11.5.25}$$

Now the problem of finding a critical link is the same as finding the longest link in DNG. On removing the loops in DNG, which can be done by deleting the links in increasing order of their lengths without affecting the network connectivity, CL can be found. The edges of DNG are a subset of e^*. A minimum spanning tree (MST) of the graph G that covers all the network nodes and aims at minimizing some link value can be used to construct DNG. This is given by

$$MST = G(v, e')| \overset{Conn(G)=1}{\min_{(i,j)\in e^t}} |x_i - x_j| \tag{11.5.26}$$

where e^t is the set of edges that provides network connectivity and represents the minimum sum of the graph edge lengths. It is also the subset of the direct neighbor set e^d. Thus, by creating the DNG graph and applying MST algorithm for the graph, the critical link of the graph of the network can be found. This value of CL gives the critical range and hence the optimal power level for the corresponding nodes.

11.5.3 Higher Layer Solutions

This section describes some of the power-aware techniques handled at the TCP/IP and the application layers of the protocol stack. The protocols used at these layers incorporate in them power control and energy conservation.

Congestion Control and Transmission Policies at the TCP/IP Layer

Due to the mobile nature of ad hoc wireless networks and the absence of the central coordinator such as a base station, the links are highly error-prone, which results in a large number of retransmissions, which in turn constantly invokes congestion control mechanisms. This kind of a situation is highly intolerable in the case of ad hoc wireless networks because of the limited energy source. Protocols at the TCP layer have to take into account the energy reserve while allowing retransmissions. Some of the suggestions made in [8] are listed below.

- The number of retransmissions can be considerably reduced by making a few modifications in the error-control protocols at the TCP/IP layer, such as selective retransmissions and explicit loss notification (ELN), which try to differentiate between congestion and other types of losses.

- Error correlation greatly affects the energy conservation of ad hoc wireless networks. TCP layer protocols that take into consideration error bursts and backing off, which result in energy efficiency of the system, can be used.

- If the packets are not received, a probe cycle may be initiated by exchanging some probe messages with the transmitter rather than using congestion control algorithms directly. This achieves higher throughput while consuming less energy due to reduced traffic in the network.

OS/Middleware and Application Layer Solutions

A quadratic relationship exists between power and voltage. Circuit speed reduction can be done which reduces significantly the power consumed at the node. This can be compensated by implementing parallelism and pipelining. Thus, the throughput can be increased for a small value of energy consumption. There exist some sporadic events that are triggered by external events. Hence, the system can be shut down during the period of inactivity.

At the application layer, protocols such as advanced configuration and power interface (ACPI) and power management tools such as power monitor are developed to assist programmers in creating power-efficient applications. Using proxy servers results in traffic redirection or reduction in traffic, such as in multimedia where video quality may be scaled down and audio sent, and in redirecting local traffic when it arrives along with the network traffic. This results in a huge amount of energy conservation. Some energy-conserving metrics can also be used in database design such as energy per transaction. In the context of multimedia, energy conservation can be achieved by reducing the number of bits in the compressed video and by transmitting only selected information. Multimedia traffic consumes a large amount of energy in ad hoc wireless systems.

11.6 SYSTEM POWER MANAGEMENT SCHEMES

This section deals with power control in the peripherals and the processor of nodes in an ad hoc wireless network. Efficient design of the hardware brings about significant reduction in the power consumed. This can be effected by operating some of the peripheral devices in power-saving mode by turning them off under idle conditions. System power consists of the power used by all hardware units of the node. This power can be conserved significantly by applying the following schemes:

- Processor power management schemes

- Device power management schemes

11.6.1 Processor Power Management Schemes

Processor power management schemes deal with techniques that try to reduce the power consumed by the processor, such as reducing the number of calculations performed. In this section, we discuss some of the power management techniques that are applied at the hardware level when there is a request from the higher layers.

Power-Saving Modes

The nodes in an ad hoc wireless network consume a substantial amount of power even when they are in an idle state since they keep listening to the channel, awaiting request packets from the neighbors. In order to avoid this, the nodes are switched off during idle conditions and switched on only when there is an arrival of a request packet. This primarily has two advantages: reducing the wastage in power consumed when the node is in the listen mode, and providing idle time for the batteries of the node to recover charges. Since the arrival of request packets is not known *a priori*, it becomes difficult to calculate the time duration for which the node has to be switched off. One solution to this problem suggested in [31] calculates the node's switch-off time based on the quality of service (QoS) requirements. An assumption made in these systems is that the battery subsystem of the node can be remotely powered on by means of a wake-up signal that is based on RF tag technology. RF tags are used as transponders for remote localization, activation, and identification of objects within a short range. Hard QoS requirements make the node stay active most of the time. This results in high consumption of the battery charge.

In [31], the authors suggest a distributed technique in which a sleep pattern is selected by the power source, that is, the node selects different timeout values (duration of sleep) depending on the traffic delay and the remaining charge of the battery. The model used assumes that $\{1, \ldots, L\}$ are the sleep states, with L corresponding to the fully active state, while 0 corresponds to the deep sleep state. The deeper the sleep state of the node, the longer is the lifetime of the battery, and the larger is the delay encountered in packet transmission or reception. To implement the remote activation of the nodes, a switch called remote activated switch (RAS) is used, as shown in Figure 11.20. As soon as the node enters the idle state, it is switched off by the RAS switch. The receiver of the RAS switch still listens to the channel. It is designed to be either fully passive or powered by the battery. The remote neighbors send the wake-up signal and a sequence. The receiver, on receiving the wake-up signal, detects the sequence. The logic circuit compares it with the standard sequence for the node. It switches on the node only if both the sequences match.

We now discuss a power management scheme using RAS which exploits the aforementioned technique of switching off the nodes. The model used in this scheme assumes the following:

- $\{1, \ldots, L\}$ are the sleep states, where L is the fully active state.

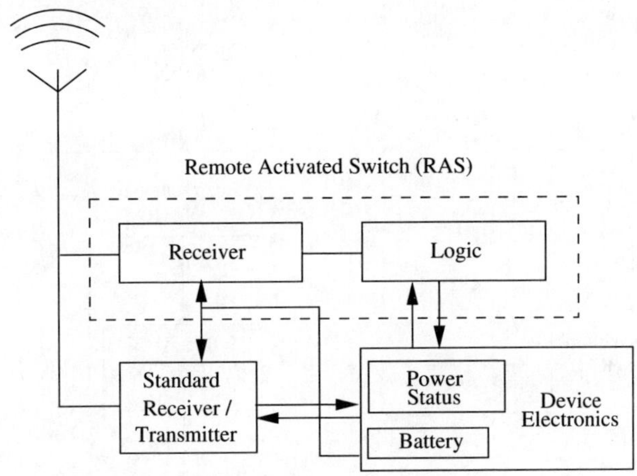

Figure 11.20. Remote activated switch.

- Let the power consumption when the node is in one of the sleep states be denoted by $P_i(i = 1, \ldots, L)$ and the corresponding delay overhead be given by $W_i(i = 1, \ldots, L-1)$, where $P_1 > P_2 > \cdots > P_L$ and $W_1 > W_2 > \cdots > W_{L-1}$.

- Let P_l^t be the power spent on the transition from state l (l=1, ... ,L-1) to state L, and Z_l be the minimum time that a node has to stay in a state in order to achieve a positive energy gain. The value of sleep time Z_l in state l is given by

$$Z_l = \max\{0, \frac{W_i(P_i^t - P_{i+1}) + W_{i+1}(P_{i+1} - P_{i+1}^t)}{P_{i+1} - P_i}\} \tag{11.6.1}$$

- The nodes select the sleep patterns $T^{(q)} = [T_1^{(q)} T_2^{(q)} \ldots T_{L-1}^{(q)}]$ based on the constraint

$$T_{l-1}^{(q)} \geq T_l^{(q)} + Y_l \tag{11.6.2}$$

where $q = \{1, \ldots, Q\}$, Y_l is the system parameter greater than Z_l and the value of Q depends on the battery status and QoS requirement.

- Any node currently in state l wakes up if and only if

$$\tau_j \geq T_l^{(q)} + Y_l \tag{11.6.3}$$

where τ_j is the time spent by the node in state j.

The power management scheme using RAS is shown in Figure 11.21. The figure shows the steps followed by transmitters and receivers that use this scheme.

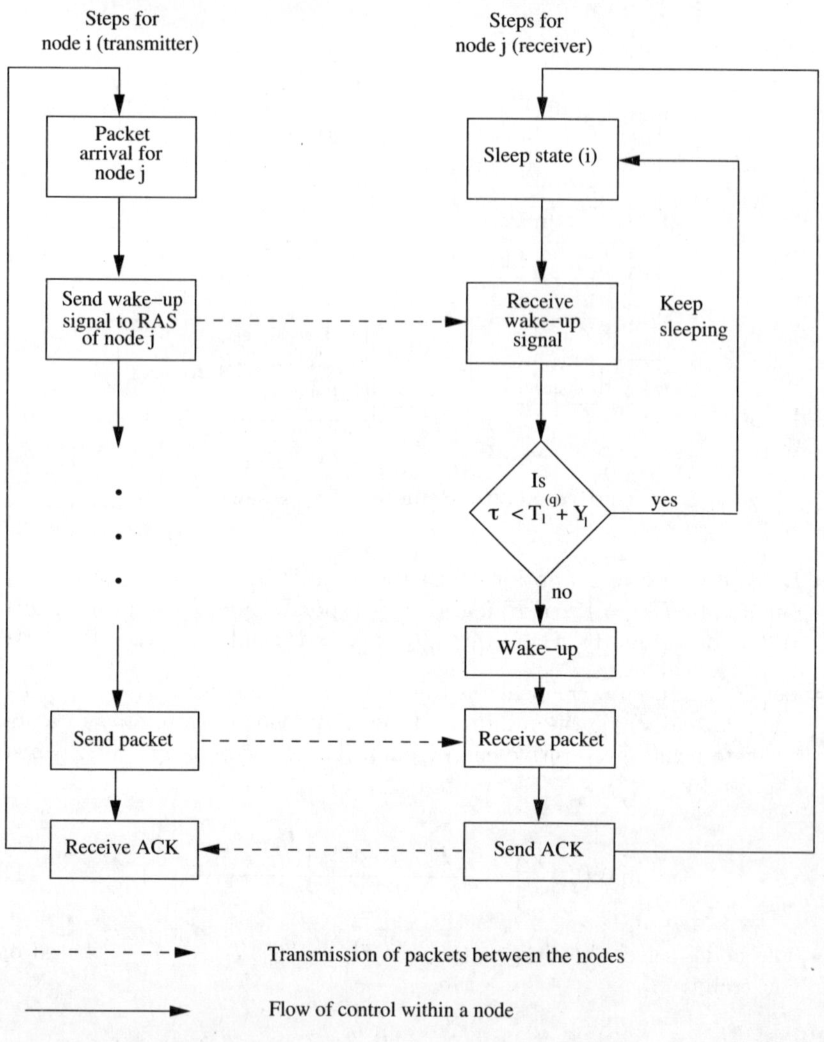

Figure 11.21. Power management scheme using remote activated switch.

Power-Aware Multi-Access Signaling

Power-aware multi-access signaling (PAMAS) [32] is another approach for determining the time duration for which the node should be turned off. This scheme suggests the addition of a separate signaling channel in the MACA protocol [33]. The RTS-CTS signaling takes place in this separate channel, which determines the time period for which the node has to be powered off. The algorithm is divided into two parts:

Addition of separate signaling channel: This can be explained through a state diagram as shown in Figure 11.22. A node can be in any one of the six states represented within the boxes. The algorithm for the MAC layer using PAMAS is described below. Initially, when a node neither transmits nor receives packets, it stays in the *idle* state.

- *Packet transmission:*
  - As soon as the node gets a packet for transmission, it transmits an RTS and enters the *Await CTS* state.
  - If it does not receive the CTS, it enters the binary exponential back-off (*BEB*) state. A node also enters the *BEB* state if it hears a busy tone when a neighboring node which is actively transmitting sends a busy tone in the control channel.
  - After receiving the CTS, it enters the *Transmit packet* state and starts transmitting the packet.
- *Packet reception:*
  - As soon as a node receives an RTS, it sends a CTS back to the sender and enters the *Await packet* state, only if no other neighboring nodes exist in the *Await CTS* or *Transmit packet* state.

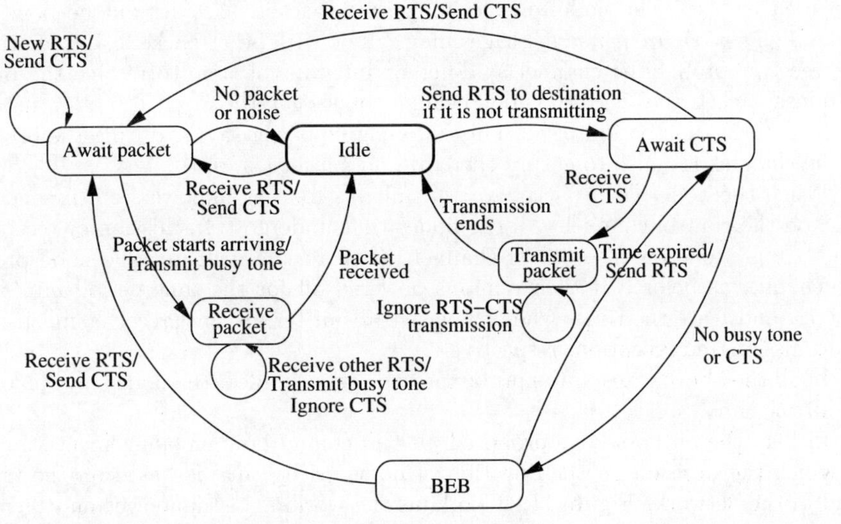

Figure 11.22. PAMAS protocol.

– If packets arrive on time, the node enters the *Receive packet* state and starts receiving the packets.

– If the packet does not arrive on time, the node enters the *idle* state again.

Powering off the radios: We now discuss the conditions under which the node enters the power-off mode:

- Condition 1: The node has no packets for transmission.
- Condition 2: A neighbor node is transmitting or receiving packets, that is, the channel is busy.

The following protocol is carried out by each node to decide on the duration for which it should be powered off. When the node has packets to be transmitted, but the channel is occupied by one of its neighbors, it knows when the neighbor finishes transmission and the channel becomes free for use. If the neighbor takes t_1 seconds to finish its transmission, then the time for which the node should be powered off is t_1 sec.

When the node powers on again and hears on-going transmission started by some other node, it again goes back to the sleep state. The time (t_2) for which it remains so is based on an additional feature that has been added to the protocol which is described as follows. As soon as the node wakes up and has packets to send, it sends a probe packet *t_probe*(l) to all its neighbors on the control channel, where l is the maximum packet length. If transmitters exist whose transmissions are expected to end in the time period $(l/2, l)$, they respond to the probe packet, with a *t_probe_response*(t) packet, where t is the duration for which the transmission is expected to last. The node, on receiving these packets, decides on the power-off time. The *t_probe_response* messages may collide with other packets. In such cases, the receiver probes the channel at different intervals of time to receive the probe response packet, that is, if the node hears a collision in the interval (t_1, t_2), it turns itself off for a time period of t_1. This is to enable power saving during the probing of the channel and also to ensure there are no packet losses.

When the node has a packet to transmit, as soon as it powers on, it sends an RTS on the control channel. All the nodes that undergo transmission or reception send a busy tone on the control channel. If there is a collision, the node probes the channel as before. It then remains powered off for the time period $\min(t, r)$, where t and r are the times when the last transmitter and last receiver finished its transmission and reception, respectively.

In all the above cases, if the probe message gets collided, the node stays powered on all the time.

In [34], the authors have proposed an on-demand power management strategy in which the decision on the duration of nodes' sleep time is based on the traffic load in the network. Figure 11.23 explains the working of this power management strategy. As shown in the figure, every node in the ad hoc wireless network can be in either one of the two modes: power-saving mode (PSM) and active mode (AM). In the AM, the node remains awake and can transmit and receive packets. But in

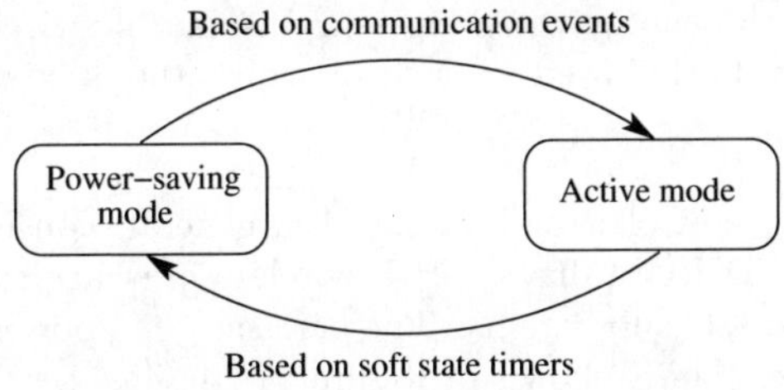

Figure 11.23. Illustration of on-demand power management.

PSM, the node remains in the sleep state and wakes up periodically to check for messages. The transition from PSM to AM is triggered by communication events such as routing messages, and the reverse transition is triggered by the expiration of timers called keep-alive timers. The timer value is calculated based on the type of packet received recently. For example, a route update message does not set the timer value and hence does not trigger a transition of the node to active state. But an RREQ packet is used in setting the timer value. Different packet types set the timers with different values. The PSM of the neighboring nodes, which can be obtained through the *Hello* packet, also influences the timer value of a node. It has been proven that by choosing an appropriate sleep time for nodes based on the timers, a balance in the trade-off between the packet delay, energy consumption, and throughput can be attained.

In [35], Zheng *et al.* have proposed another asynchronous wakeup schedule that proves advantageous in the presence of various traffic patterns. The wakeup schedule is modeled as a block design problem and solved. The wakeup protocol thus obtained has proven to be resilient to packet collision and network dynamics.

11.6.2 Device Power Management Schemes

Some of the major consumers of power in ad hoc wireless networks are the hardware devices present in the nodes. Various schemes have been proposed in the design of hardware that minimize the power consumption.

Low-Power Design of Hardware

Low-power design of hardware results in a significant improvement in the energy conservation. Some of the low-power design suggestions include varying clock speed CPUs, disk spin down, and flash memory. We now look into some of the sources of power consumption in the ad hoc wireless networks and the corresponding solutions to reduce power consumption as suggested in [8].

- Major sources of power consumption in ad hoc wireless networks are the transmitters and receivers of the communication module. The design of transceivers has a significant effect on the power consumption. Hence, great care must be taken while designing them. Switching off various units of the hardware while idling reduces the energy consumption. Instead of switching off fully, different stages may be followed in each of which there exists a different power requirement.

- The main hardware of a mobile node, in general, consists of the LCD display, DRAM, CD ROM drive, CPU, wireless interface card (in the case of a computer), and I/O subsystems. The percentage of power consumed by some of these components is shown in Figure 11.24. The section that follows will give a brief overview of the various means of power consumption and some effective solutions suggested for low-power design of the hardware devices.

CPU Power Consumption

The energy required for the CPU operation depends largely on the clock frequency (F). As the clock rate increases, frequent switching of the logic gates between different voltage levels (V), that is, the ground voltage and the peak voltage, takes place, which leads to higher power consumption. Another effect that significantly influences the CPU power consumption is the chaining of transistors. The larger the capacitance (C) of these transistors, the higher the energy required. Hence, the total power required by the CPU is proportional to CV^2F. The solution suggested is as follows:

- The parameter C can be set during the chip design.
- The values of F and V can be set dynamically at run-time which, along with power-aware CPU scheduling policies, reduces the power consumption significantly.

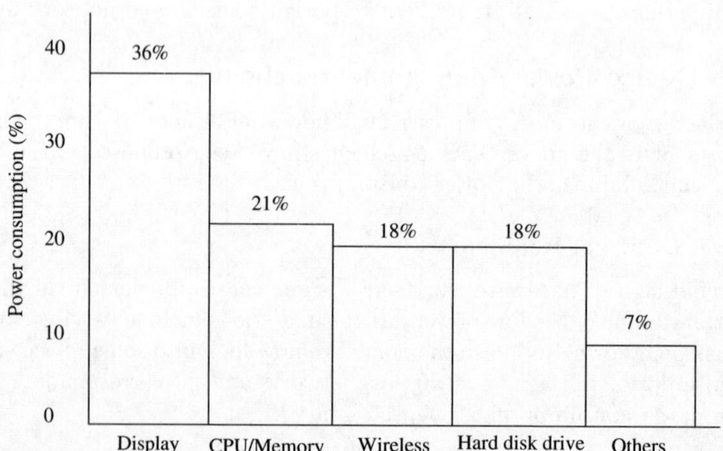

Figure 11.24. Power consumed by various units of hardware.

Power-Aware CPU Scheduling

A small reduction in the value of the voltage V produces a quadratic reduction in the power consumed. This is the motivation behind the power-reduction techniques proposed by Weiser *et al.* in [36]. But the voltage V cannot be reduced until the clock rate is reduced. Some of the approaches to reduce the clock rate suggested in [36] are given below. The CPU utilization can be balanced between peak usage and idle periods. Idle periods or sleep times can be classified into hard or soft sleep times. Hard sleep times are unavoidable and result in compulsory idle times, whereas the soft idle times are those which can be delayed to a maximum extent possible. These two kinds of sleep times must be taken into account while balancing the CPU activities. The following algorithms suggest a few methods to calculate the CPU clock cycle time:

- OPT algorithm: This algorithm works under the ideal condition that future activities and CPU usages are known *a priori* and so the system can adjust its CPU clock rate accordingly and so maximum reduction in power consumption can be obtained by balancing the periods of activities. Though this algorithm remains unrealistic, it can be used as a benchmark for other algorithms.
- FUTURE algorithm: The basic principle behind the FUTURE algorithm is similar to that of OPT, the only difference being the window used for prediction. In the FUTURE algorithm, optimizations are made over a small window. Hence, it is realizable in a practical wireless environment.
- PAST algorithm: Rather than peering into the future, this algorithm looks into the past and decides the clock rate. It is highly unreliable because it sees the past and assumes the future with some probability.

The window size on which these algorithms work acts as a trade-off between power consumption and response times.

Hard Disk Drive (HDD) Power Consumption

As mentioned earlier, the basic source for power consumption in hard disks is the disk spin. Various approaches have been suggested for turning off the drives and to bring down the speed of spinning. We now see how the spin-down can be performed on the disk drives.

- By using historical data: One method suggested is based on the traces of disk usage collected over a long period of time. By analyzing various spin-down thresholds, an optimal value for threshold has to be agreed upon, which acts as a balance between the two contradictory requirements of reducing power consumption and reducing the access delays.
- Spin-up/spin-down policies: Some of the policies used in deciding the time at which the hard disk speed has to be varied are given below.
  - Optimal-optimal policy: According to this policy, by having a complete knowledge of the future, the optimal values for spin-down can be obtained. This tells when the disk has to be spun down to obtain maximum

efficiency, and when the disk has to be spun up to get ready for the next disk operation. This is an unrealistic policy.

- Threshold-demand policy: This algorithm forces spin-down of the disk only after attaining a certain threshold value. But the spin-up takes place only if there exists a request for the disk.
- Predictive-predictive policy: Both the spin-up and spin-down time values are predicted based on the past values.

11.7 SUMMARY

This chapter listed various issues regarding energy management at the different layers of the protocol stack. It mainly concentrated on three major divisions in energy management: battery management, transmission power management, and system power management techniques. A summary of the solutions and the future directions at the device level, data link layer, and the network and higher layers are specified in Tables 11.2, 11.3, and 11.4, respectively.

Developing battery-efficient system architecture that has low cost and complexity remains a crucial issue. Designing smart battery packs that can select appropriate battery discharge policies under different load conditions is a challenging problem. Other issues that exist at the physical layer include efficient battery-scheduling techniques, selection of an optimal transmission power for the nodes,

Table 11.2. Device-dependent energy management schemes

Solution(s) Proposed	Underlying Technique	Future Directions
Modeling and shaping of the battery discharge using binary and generalized pulsed discharge [3],[4].	Developing stochastic model of battery discharge that tracks charge recovery for the batteries.	Shape the actual discharge of the batteries in order to attain the optimal discharge profile.
Battery-scheduling techniques (RR, RN, JN) [6].	Developing scheduling algorithms to select a subset of cells from an array of cells and discharging from these cells to optimize the charge recovery mechanism.	Optimal cost-effective scheduling schemes that provide an efficient performance without significant delay in packet delivery.
Low-power design of hardware [8].	Designing varying clock-speed CPUs, disk spin-down and flash memory.	Predictive power control and clock-speed variation scheme.

Table 11.3. Solutions of energy management at the data link layer

Solution(s) Proposed	Underlying Technique	Future Directions
BAMAC protocol [14].	Increasing the lifetime of the nodes by discharging the batteries of the nodes contending for a common channel in a uniform manner.	Modification of the protocol to suit real-time traffic.
Distributed power control mechanisms [18], [19].	Adding power control to the medium access layer of ad hoc wireless networks that reduces the transmission power level to the minimum required value.	New algorithms that reduce energy cost of communication in ad hoc wireless networks.
Centralized topology control algorithms [20], [22].	Finding the optimal transmission power for each node in the network that increases the network lifetime.	A distributed version of these algorithms that exploit the network transmission parameters.
PAMAS [32].	Exploiting the option of turning off the nodes in the ad hoc wireless network while they are in the idle state.	The trade-off between throughput performance and the delay in transmitting the packet.

and finding the appropriate time duration for switching off the nodes. Further investigation at the data link layer is required in order to address issues regarding relay traffic, such as finding an optimal strategy that decides the amount of allowable relay traffic for a node. Developing battery-aware MAC scheduling algorithms for the nodes that increase the lifetime of the nodes is an important issue. At the network layer, many issues for research remain open such as designing an efficient routing algorithm that increases the network lifetime by selecting an optimal relay node. Energy efficiency at the application layer is becoming an important area of research. It includes development of power management analysis tools and ACPIs which assist in writing power-aware application programs.

Table 11.4. Solutions of energy management at the network and higher layers

Solution(s) Proposed	Underlying Technique	Future Directions
Traffic-shaping schemes [15].	Introducing delay slots in battery discharge and developing strategies for blocking relay traffic.	Shaping the relay traffic under varying node densities and optimal resource sharing in ad hoc wireless networks.
BEE [13].	Finding a balance between the energy consumption of all the nodes by combining lazy packet scheduling with traffic-shaping schemes.	Algorithms that combine BEE with energy-efficient communication algorithms.
COMPOW protocol [21].	Finding an optimal common power that increases the battery lifetime of all the nodes in the network and reduces contention among the nodes.	Efficacy with mobility and reducing the latency while switching between the power levels.
MPCR, MBCR, MMBCR, CMMBCR [25].	Using power-aware metrics in routing algorithm.	Optimal balance between the contradictory issues that exist in these algorithms.
Localized routing algorithms [29].	Developing power, cost, and power-cost routing algorithms that are loop-free.	Study of these algorithms in the mobile scenario with lower degrees of nodes in the network and design of a general formula for power, cost, and power-cost routing algorithms.
ANDA [30].	Maximizing the network lifetime by adjusting the power levels of all the cluster-heads.	Merging ANDA with algorithms related to clustering such as cluster-head rotation.
Transmission policies such as ELN [8].	Reducing the number of retransmissions and implementing error control.	Efficient protocols that reduce the amount of retransmissions.
Developing ACPIs and power monitor tools.	Designing protocols to assist programmers in creating power-efficient applications.	Power-aware design of general-purpose application software.

11.8 PROBLEMS

1. List at least two battery technologies that have the following properties. List your answers in the order of performance.

   (a) Higher energy density

   (b) Flat discharge characteristics

   (c) Low cost

   (d) Fast charging capability

   (e) Higher lifetime

2. Which battery is being commonly used for portable mobile nodes such as laptops? Give a few reasons to support your answer.

3. Which of the three delay-free battery-scheduling techniques suggested in Section 11.4.2 performs better under (a) high-traffic and (b) low-traffic scenarios of an ad hoc wireless network consisting of large number of nodes? Assume that the network uses a MAC protocol that provides short-term fairness among the nodes.

4. Let the maximum nominal and theoretical capacities N and C of a battery be equal to 25 units and 200 units, respectively, and let the current nominal and theoretical capacities of the battery be (a) $n_i = 23$ and $c_i = 190$, respectively. If the battery remains idle for three slots and then is subjected to two current pulses followed by three more idle slots, what will be the remaining capacities of the battery at the end of the idle slots when the binary pulsed discharge pattern is followed? Analyze and summarize your results. Repeat the same with (b) $n_i = 22$ and $c_i = 1$ and (c) $n_i = 2$ and $c_i = 130$. In case (c) assume one idle slot followed by transmission of five packets followed by five idle slots. Assume that the battery always transmits whenever it has a packet queued up for transmission and always gains one unit of charge when it idles for one time slot.

5. Suggest a few metrics that can be associated with battery-aware routing techniques.

6. List the possible steps of the algorithms executed at the source and the intermediate nodes of an ad hoc wireless network that follow the following strategies: (a) random strategy and (b) pay-for-it strategy. Assume a session between source s and destination d. Let $R(s, d)$ be the set containing the available routes between s and d, $sympathy(k, r)$ be the sympathy level of the kth node in route r, and $credit(k, r)$ and $debit(k, r)$ be the credit and debit of kth node in route r, respectively.

7. What are the advantages of distributed power control algorithms in ad hoc wireless networks over the centralized power control algorithms?

8. Let S and D be the source and the destination nodes of an ad hoc wireless network such that the distance between S and D be denoted by $|SD| = d$. If A is the intermediate node between S and D, then prove that the greatest energy saving is obtained when A is in the middle of S and D and $d > (c/(a(1-2^{1-\alpha})))^{1/\alpha}$. The variables c, a, and α are used in the same context as referred to in Section 11.5.2.

9. Consider the ad hoc network as described in the previous problem. If the distance between S and D is divided into n subintervals with $n > 0$, prove that the greatest power saving is obtained when the intermediate nodes are equally spaced and the optimal value of n is approximately equal to $d(a(\alpha-1)/c)^{1/\alpha}$.

10. Prove that the localized power-efficient routing algorithm discussed in Section 11.5.2 is loop-free.

11. In the maximum-battery-cost-routing algorithm (MBCR), the cost of a node is a function of the remaining battery capacity of the node. Express the cost function of a node in terms of some parameters of the node other than battery capacity.

12. What are the disadvantages of clustering in ad hoc wireless networks?

13. What are the pros and cons of adding a separate signaling channel for an ad hoc wireless network? Suggest a few methods for calculating the time for which the nodes of an ad hoc wireless network that uses single channel should be switched off while they remain in the idle state.

14. Describe how an energy-efficient multimedia processing and transmission can be achieved.

BIBLIOGRAPHY

[1] M. Adamou and S. Sarkar, "A Framework for Optimal Battery Management for Wireless Nodes," *Proceedings of IEEE INFOCOM 2002*, pp. 1783-1792, June 2002.

[2] D. Bruni, L. Benini, and B. Ricco, "System Lifetime Extension by Battery Management: An Experimental Work," *Proceedings of International Conference on Compilers, Architecture, and Synthesis for Embedded Systems 2002*, pp. 232-237, October 2002.

[3] C. F. Chiasserini and R. R. Rao, "Improving Battery Performance by Using Traffic-Shaping Techniques," *IEEE Journal on Selected Areas of Communications*, vol. 19, no. 7, pp. 1385-1394, July 2001.

[4] C. F. Chiasserini and R. R. Rao, "Pulsed Battery Discharge in Communication Devices," *Proceedings of MOBICOM 1999*, pp. 88-95, November 1999.

[5] D. Panigrahi, C. F. Chiasserini, S. Dey, and R. R. Rao, "Battery Life Estimation of Mobile Embedded Systems," *Proceedings of IEEE VLSI Design 2001*, pp. 55-63, January 2001.

[6] C. F. Chiasserini and R. R. Rao, "Energy-Efficient Battery Management," *Proceedings of IEEE INFOCOM 2000*, vol. 2, pp. 396-403, March 2000.

[7] T. Simunic, L. Benini, and G. D. Micheli, "Energy-Efficient Design of Battery-Powered Embedded Systems," *IEEE Transactions on VLSI Design*, vol. 9, no. 1, pp. 15-28, May 2001.

[8] K. Lahiri, A. Raghunathan, S. Dey, and D. Panigrahi, "Battery-Driven System Design: A New Frontier in Low-Power Design," *Proceedings of ASP-DAC/VLSI Design 2002*, pp. 261-267, January 2002.

[9] http://www.duracell.com/oem/primary/lithium/performance.asp

[10] J. Luo and N. K. Jha, "Battery-Aware Static Scheduling for Distributed Real-Time Embedded Systems," *Proceedings of IEEE DAC 2001*, pp. 444-449, June 2001.

[11] P. Rong and M. Pedram, "Battery-Aware Power Management Based on Markovian Decision Processes," *Proceedings of ICCAD 2002*, pp. 712-717, November 2002.

[12] B. Prabhakar, E. U. Biyikoglu, and A. E. Gamal, "Energy-Efficient Transmission over a Wireless Link via Lazy Packet Scheduling," *Proceedings of IEEE INFOCOM 2001*, vol. 1, pp. 386-394, April 2001.

[13] C. F. Chiasserini, P. Nuggehalli, and V. Srinivasan, "Energy-Efficient Communication Protocols," *Proceedings of IEEE DAC 2002*, pp. 824-829, June 2002.

[14] S. Jayashree, B. S. Manoj, and C. Siva Ram Murthy, "A Battery-Aware MAC Protocol for Ad Hoc Wireless Networks," *Technical Report*, Department of Computer Science and Engineering, Indian Institute of Technology, Madras, India, October 2003.

[15] V. Srinivasan, P. Nuggehalli, C. F. Chiasserini, and R. R. Rao, "Energy Efficiency of Ad Hoc Wireless Networks with Selfish Users," *Proceedings of EW 2002*, February 2002.

[16] J. H. Chang and L. Tassiulas, "Energy-Conserving Routing in Wireless Ad Hoc Networks," *Proceedings of IEEE INFOCOM 2000*, pp. 22-31, March 2000.

[17] S. Agarwal, A. Ahuja, and J. P. Singh, "Route-Lifetime Assessment-Based Routing (RABR) Protocol for Mobile Ad Hoc Networks," *Proceedings of IEEE ICC 2000*, vol. 3, pp. 1697-1701, June 2000.

[18] R. Wattenhofer, L. Li, P. Bahl, and Y. M. Wang, "Distributed Topology Control for Power-Efficient Operation in Multi-Hop Wireless Ad Hoc Networks," *Proceedings of IEEE INFOCOM 2001*, pp. 1388-1397, April 2001.

[19] S. Agarwal, R. H. Katz, S. V. Krishnamurthy, and S. K. Dao, "Distributed Power Control in Ad Hoc Wireless Networks," *Proceedings of IEEE PIMRC 2001*, vol. 2, pp. 59-66, October 2001.

[20] R. Ramanathan and R. Rosales-Hain, "Topology Control of Multi-Hop Wireless Networks using Transmit Power Adjustment," *Proceedings of IEEE INFOCOM 2000*, pp. 404-413, March 2000.

[21] V. Kawadia, S. Narayanaswamy, R. Rozovsky, R. S. Sreenivas, and P. R. Kumar, "Protocols for Media Access Control and Power Control in Wireless Networks," *Proceedings of IEEE Conference on Decision and Control 2001*, vol. 2, pp. 1935-1940, December 2001.

[22] M. Sanchez, P. Manzoni, and Z. J. Haas, "Determination of Critical Transmission Range in Ad Hoc Networks," *Proceedings of MMT 1999*, October 1999.

[23] A. Muqattash and M. Krunz, "Power-Controlled Dual Channel Medium Access Protocol for Wireless Ad Hoc Networks," *Proceedings of IEEE INFOCOM 2003*, vol. 1, pp. 470-480, April 2003.

[24] V. Kawadia and P. R. Kumar, "Power Control and Clustering in Ad Hoc Networks," *Proceedings of IEEE INFOCOM 2003*, vol. 1, pp. 459-469, April 2003.

[25] C. K. Toh, "Maximum Battery Life Routing to Support Ubiquitous Mobile Computing in Wireless Ad Hoc Networks," *IEEE Communications Magazine*, vol. 39, no. 6, pp. 138-147, June 2001.

[26] M. Woo, S. Singh, and C. S. Raghavendra, "Power-Aware Routing in Mobile Ad Hoc Networks," *Proceedings of IEEE MOBICOM 1998*, pp. 181-190, October 1998.

[27] R. L. Cruz and A. R. Santhanam, "Optimal Routing, Link Scheduling, and Power Control in Multi-Hop Wireless Networks," *Proceedings of INFOCOM 2003*, vol. 1, pp. 702-711, April 2003.

[28] A. Srinivas and E. Modiano, "Minimum Energy Disjoint Path Routing in Wireless Ad Hoc Networks," *Proceedings of MOBICOM 2003*, pp. 122-133, September 2003.

[29] I. Stojmenovic and X. Lin, "Power-Aware Localized Routing in Wireless Networks," *Proceedings of IPDPS 2000*, vol. 2, no. 11, pp. 1122-1133, May 2000.

[30] C. F. Chiasserini, I. Chlamtac, P. Monti, and A. Nucci, "Energy-Efficient Design of Wireless Ad Hoc Networks," *Proceedings of Networking 2002*, pp. 376-386, May 2002.

[31] C. F. Chiasserini and R. R. Rao, "A Distributed Power Management Policy for Wireless Ad Hoc Networks," *Proceedings of IEEE WCNC 2000*, vol. 3, pp. 1209-1213, September 2000.

[32] S. Singh and C. S. Raghavendra, "Power-Aware Multi-Access Protocol with Signaling for Ad Hoc Networks," *ACM Computer Communication Review*, vol. 28, no. 3, pp. 5-26, July 1998.

[33] P. Karn, "MACA — A New Channel Access Method for Packet Radio," *Proceedings of ARRL/CRRL Amateur Radio Computer Networking Conference 1990*, pp. 134-140, September 1990.

[34] R. Zheng and R. Kravets, "On-Demand Power Management for Ad Hoc Networks," *Proceedings of IEEE INFOCOM 2003*, vol. 1, pp. 481-491, April 2003.

[35] R. Zheng, J. C. Hou, and L. Sha, "Asynchronous Wakeup for Ad Hoc Networks," *Proceedings of ACM MOBIHOC 2003*, pp. 35-45, June 2003.

[36] Weiser, M. B. Welch, A. Demers, and S. Shenker, "Scheduling for Reduced CPU Energy," *Proceedings of USENIX Association, OSDI 1994*, pp. 13-23, November 1994.

[37] S. Jayashree, B. S. Manoj, and C. Siva Ram Murthy, "Energy Management in Ad Hoc Wireless Networks: A Survey of Issues and Solutions," *Technical Report*, Department of Computer Science and Engineering, Indian Institute of Technology, Madras, India, March, 2003.

Chapter 12

WIRELESS SENSOR NETWORKS

12.1 INTRODUCTION

Sensor networks are highly distributed networks of small, lightweight wireless nodes, deployed in large numbers to monitor the environment or system by the measurement of physical parameters such as temperature, pressure, or relative humidity. Building sensors has been made possible by the recent advances in micro-electro mechanical systems (MEMS)[1] technology.

Each node of the sensor network consists of three subsystems: the sensor subsystem which senses the environment, the processing subsystem which performs local computations on the sensed data, and the communication subsystem which is responsible for message exchange with neighboring sensor nodes. While individual sensors have limited sensing region, processing power, and energy, networking a large number of sensors gives rise to a robust, reliable, and accurate sensor network covering a wider region. The network is fault-tolerant because many nodes are sensing the same events. Further, the nodes cooperate and collaborate on their data, which leads to accurate sensing of events in the environment. The two most important operations in a sensor network are data dissemination, that is, the propagation of data/queries throughout the network, and data gathering, that is, the collection of observed data from the individual sensor nodes to a sink.

Sensor networks consist of different types of sensors such as seismic, thermal, visual, and infrared, and they monitor a variety of ambient conditions such as temperature, humidity, pressure, and characteristics of objects and their motion. Sensor nodes can be used in military, health, chemical processing, and disaster relief scenarios. Some of the academic and industry-supported research programs on sensor networks include working on Smart Dust at the University of California, Berkeley (UCB), and wireless integrated network sensor (WINS) at the University of California, Los Angeles (UCLA).

The applications of sensor networks are described in the next section, followed by the differences between ad hoc and sensor networks. The major issues and

[1]MEMS devices are miniature structures fabricated on silicon substrates in a similar manner to silicon integrated circuits. However, unlike electronic circuits, these are mechanical devices.

challenges involved in the design of sensor networks are then listed, and the two major forms of sensor network architecture — layered and clustered — are discussed. Various protocols for the major operations of data dissemination and gathering are then described, followed by specialized MAC protocols developed or modified to suit sensor networks. Techniques adopted by sensor nodes to discover their location and the measures to assess the quality of coverage of a sensor network are described. Finally, some sensor-network specific issues such as energy-efficient hardware design, synchronization, transport layer protocols, security, and real-time communication are discussed.

12.1.1 Applications of Sensor Networks

Sensor nodes are used in a variety of applications which require constant monitoring and detection of specific events. The military applications of sensor nodes include battlefield surveillance and monitoring, guidance systems of intelligent missiles, and detection of attack by weapons of mass destruction, such as chemical, biological, or nuclear. Sensors are also used in environmental applications such as forest fire and flood detection, and habitat exploration of animals. Sensors can be extremely useful in patient diagnosis and monitoring. Patients can wear small sensor devices that monitor their physiological data such as heart rate or blood pressure. The data collected can be sent regularly over the network to automated monitoring systems which are designed to alert the concerned doctor on detection of an anomaly. Such systems provide patients a greater freedom of movement instead of their being confined to a hospital. Sensor nodes can also be made sophisticated enough to correctly identify allergies and prevent wrong diagnosis.

Sensors will soon find their way into a host of commercial applications at home and in industries. Smart sensor nodes can be built into appliances at home, such as ovens, refrigerators, and vacuum cleaners, which enable them to interact with each other and be remote-controlled. The home can provide a "smart environment" which adapts itself according to the user's tastes. For instance, the lighting, music, and ambiance in the room can be automatically set according to the user's preferences. Similar control is useful in office buildings too, where the airflow and temperature of different parts of the building can be automatically controlled. Warehouses could improve their inventory control system by installing sensors on the products to track their movement. The applications of sensor networks are endless, limited only by the human imagination.

12.1.2 Comparison with Ad Hoc Wireless Networks

While both ad hoc wireless networks and sensor networks consist of wireless nodes communicating with each other, there are certain challenges posed by sensor networks. The number of nodes in a sensor network can be several orders of magnitude larger than the number of nodes in an ad hoc network. Sensor nodes are more prone to failure and energy drain, and their battery sources are usually not replaceable or rechargeable. Sensor nodes may not have unique global identifiers, so unique addressing is not always feasible in sensor networks.

Sensor networks are data-centric, that is, the queries in sensor networks are addressed to nodes which have data satisfying some conditions. For instance, a query may be addressed to all nodes "in the south-east quadrant," or to all nodes "which have recorded a temperature greater than 30 °C." On the other hand, ad hoc networks are address-centric, with queries addressed to particular nodes specified by their unique address. Hence, sensor networks require a different mechanism for routing and answering queries. Most routing protocols used in ad hoc networks cannot be directly ported to sensor networks because of limitations in memory, power, and processing capabilities in the sensor nodes and the non-scalable nature of the protocols.

An important feature of sensor networks is data fusion/aggregation, whereby the sensor nodes aggregate the local information before relaying. The main goals of data fusion are to reduce bandwidth consumption, media access delay, and power consumption for communication.

12.1.3 Issues and Challenges in Designing a Sensor Network

Sensor networks pose certain design challenges due to the following reasons:

- Sensor nodes are randomly deployed and hence do not fit into any regular topology. Once deployed, they usually do not require any human intervention. Hence, the setup and maintenance of the network should be entirely autonomous.

- Sensor networks are infrastructure-less. Therefore, all routing and maintenance algorithms need to be distributed.

- An important bottleneck in the operation of sensor nodes is the available energy. Sensors usually rely only on their battery for power, which in many cases cannot be recharged or replaced. Hence, the available energy at the nodes should be considered as a major constraint while designing protocols. For instance, it is desirable to give the user an option to trade off network lifetime for fault tolerance or accuracy of results.

- Hardware design for sensor nodes should also consider energy efficiency as a primary requirement. The micro-controller, operating system, and application software should be designed to conserve power.

- Sensor nodes should be able to synchronize with each other in a completely distributed manner, so that TDMA schedules can be imposed and temporal ordering of detected events can be performed without ambiguity.

- A sensor network should also be capable of adapting to changing connectivity due to the failure of nodes, or new nodes powering up. The routing protocols should be able to dynamically include or avoid sensor nodes in their paths.

- Real-time communication over sensor networks must be supported through provision of guarantees on maximum delay, minimum bandwidth, or other QoS parameters.

- Provisions must be made for secure communication over sensor networks, especially for military applications which carry sensitive data.

The protocols which have been designed to address the above issues have been classified in Figure 12.1.

12.2 SENSOR NETWORK ARCHITECTURE

The design of sensor networks is influenced by factors such as scalability, fault tolerance, and power consumption [1]. The two basic kinds of sensor network architecture are layered and clustered.

12.2.1 Layered Architecture

A layered architecture has a single powerful base station (BS), and the layers of sensor nodes around it correspond to the nodes that have the same hop-count to the BS. This is depicted in Figure 12.2.

Layered architectures have been used with in-building wireless backbones, and in military sensor-based infrastructure, such as the multi-hop infrastructure network architecture (MINA) [2]. In the in-building scenario, the BS acts an an access point to a wired network, and small nodes form a wireless backbone to provide wireless connectivity. The users of the network have hand-held devices such as PDAs which communicate via the small nodes to the BS. Similarly, in a military operation, the BS is a data-gathering and processing entity with a communication link to a larger network. A set of wireless sensor nodes is accessed by the hand-held devices of the soldiers. The advantage of a layered architecture is that each node is involved only in short-distance, low-power transmissions to nodes of the neighboring layers.

Unified Network Protocol Framework (UNPF)

UNPF [2] is a set of protocols for complete implementation of a layered architecture for sensor networks. UNPF integrates three operations in its protocol structure: network initialization and maintenance, MAC, and routing protocols.

- **Network Initialization and Maintenance Protocol**

  The network initialization protocol organizes the sensor nodes into different layers, using the broadcast capability of the BS. The BS can reach all nodes in a one-hop communication over a common control channel. The BS broadcasts its identifier (ID) using a known CDMA code on the common control channel. All nodes which hear this broadcast then record the BS ID. They send a beacon signal with their own IDs at their low default power levels. Those nodes which the BS can hear form layer one since they are at a single-hop distance from the BS. The BS now broadcasts a control packet with all layer one node IDs. All nodes send a beacon signal again. The layer one nodes record the IDs which they hear, and these form layer two, since they are one hop away from layer one nodes. In the next round of beacons, the layer one nodes inform the BS of the layer two nodes, which is then broadcast to the entire network. In

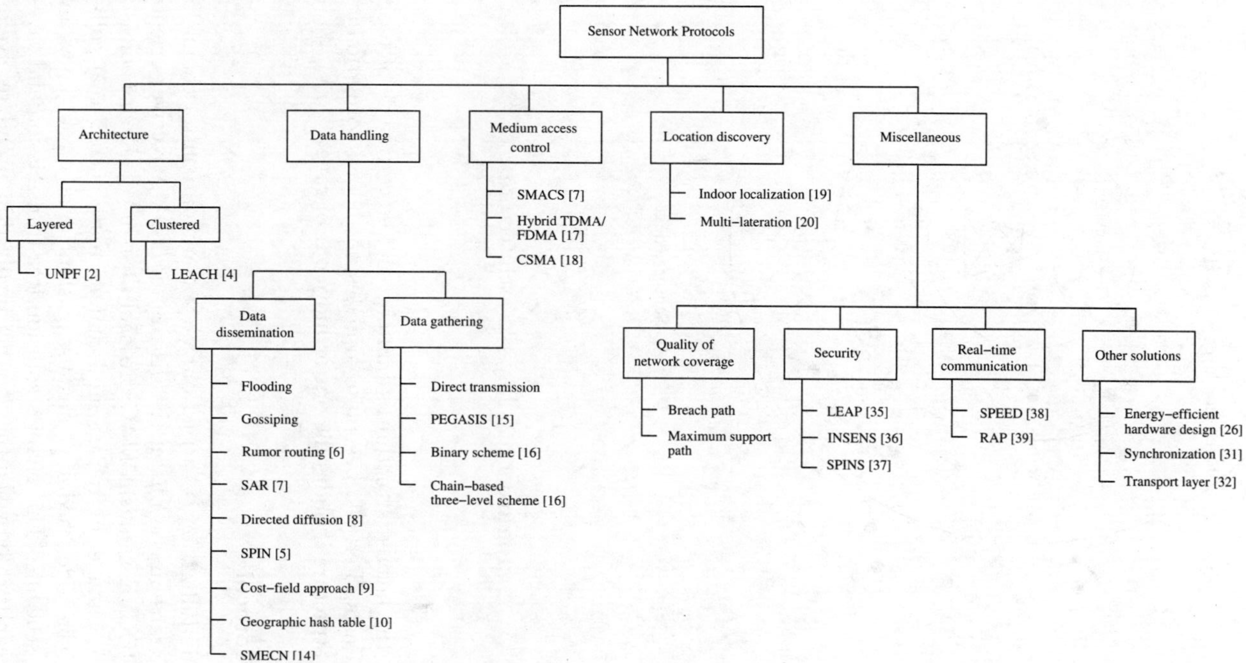

Figure 12.1. Classification of sensor network protocols.

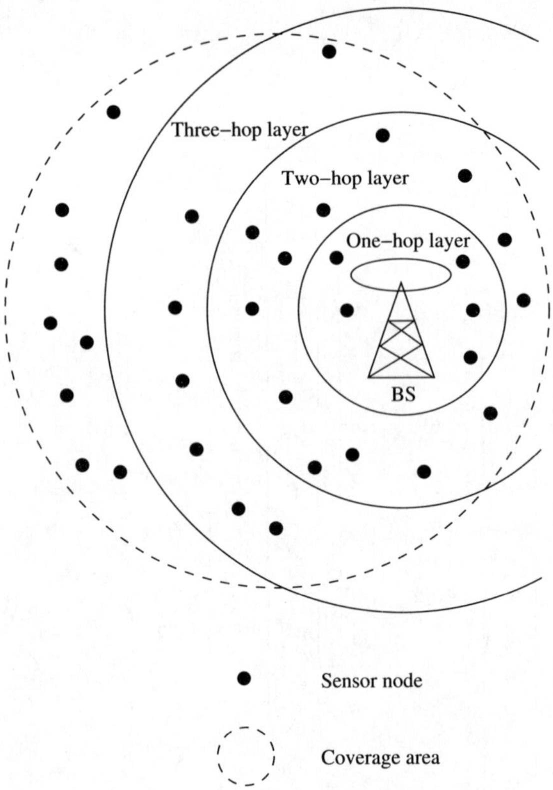

Figure 12.2. Layered architecture.

this way, the layered structure is built by successive rounds of beacons and BS broadcasts. Periodic beaconing updates neighbor information and alters the layer structure if nodes die out or move out of range.

- **MAC Protocol**

Network initialization is carried out on a common control channel. During the data transmission phase, the distributed TDMA receiver oriented channel (DTROC) assignment MAC protocol [3] is used. Each node is assigned a reception channel by the BS, and channel reuse is such that collisions are avoided. The node schedules transmission slots for all its neighbors and broadcasts the schedule. This enables collision-free transmission and saves energy, as nodes can turn off when they are not involved in a send/receive operation. The two steps of DTROC are channel allocation (the assignment of reception channels to the nodes) and channel scheduling (the sharing of the reception channel among the neighbors). DTROC avoids hidden terminal and exposed terminal problems by suitable channel allocation algorithms.

- **Routing Protocol**

  Downlink from the BS is by direct broadcast on the control channel. The layered architecture enables multi-hop data forwarding from the sensor nodes to the BS. The node to which a packet is to be forwarded is selected considering the remaining energy of the nodes. This achieves a higher network lifetime. Existing ad hoc routing protocols can be simplified for the layered architecture, since only nodes of the next layer need to be maintained in the routing table.

A modification to the UNPF protocol set termed the UNPF-R [2] has been proposed. It makes the sensor nodes adaptively vary their transmission range so that network performance can be optimized. While a very small transmission range could cause network partitioning, a very large transmission range will reduce the spatial reuse of frequencies. The optimal range is determined through an algorithm similar to simulated annealing.[2] This is a centralized control algorithm in which the BS evaluates an objective function periodically. For a transmission range R, the objective function is $f(R) = \frac{\epsilon \times d}{n/N}$, where N is the total number of sensors in the system; n is the number of nodes in layer one; ϵ is the energy consumption per packet; and d is the average packet delay. The BS selects a new transmission range R' as follows. If no packet is received by the BS from any sensor node for some interval of time, the transmission range is increased by Δr, a predefined increment. Otherwise, the transmission range is either decreased by Δr with probability $0.5 \times (n/N)$, or increased by Δr with probability $[1 - 0.5 \times (n/N)]$. The objective function is reevaluated with the new transmission range. If $f(R') < f(R)$, then the transmission range R' is adopted. Otherwise, R is modified to R' with probability $e^{\frac{(f(R)-f(R')) \times (n/N)}{T}}$, where T is the temperature parameter, as in simulated annealing. The advantage of the UNPF-R is that it minimizes the energy $\times$ delay metric, and maximizes the number of nodes which can connect to the BS. The minimization of the energy $\times$ delay metric ensures that transmission should occur with minimum delay and with minimum energy consumption. The two conflicting objectives are together optimized by minimizing their product.

12.2.2 Clustered Architecture

A clustered architecture organizes the sensor nodes into clusters, each governed by a cluster-head. The nodes in each cluster are involved in message exchanges with their respective cluster-heads, and these heads send messages to a BS, which is usually an access point connected to a wired network. Figure 12.3 represents a

[2] Simulated annealing algorithm is an optimization heuristic in which an objective function is evaluated for different values of the independent variable. A value which provides an inferior objective value is also accepted with a probability, which is reduced as the algorithm progresses. This is to escape local minima of the objective function. The progress of the heuristic is indicated by the decreasing temperature parameter.

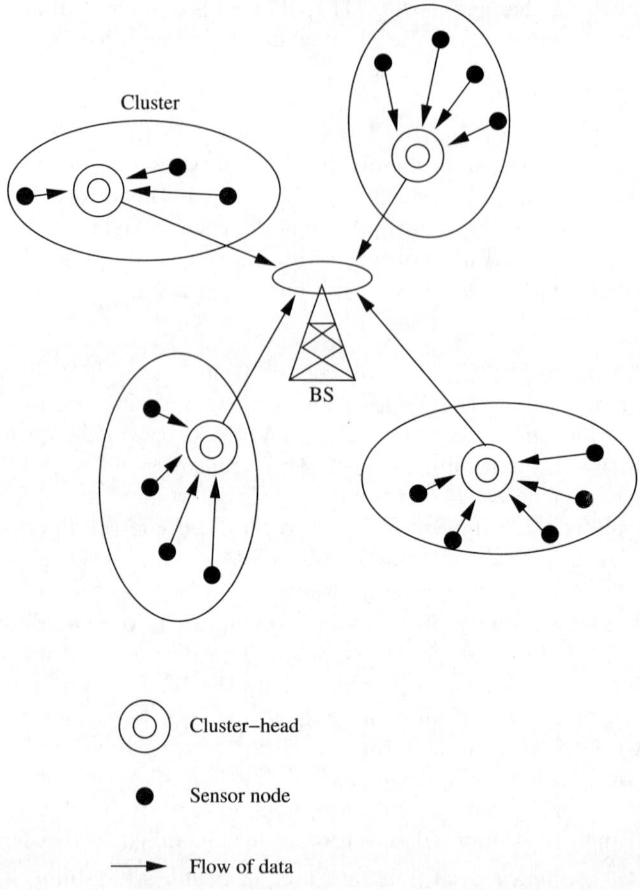

Figure 12.3. Clustered architecture.

clustered architecture where any message can reach the BS in at most two hops. Clustering can be extended to greater depths hierarchically.

Clustered architecture is specially useful for sensor networks because of its inherent suitability for data fusion. The data gathered by all members of the cluster can be fused at the cluster-head, and only the resulting information needs to be communicated to the BS. Sensor networks should be self-organizing, hence the cluster formation and election of cluster-heads must be an autonomous, distributed process. This is achieved through network layer protocols such as the low-energy adaptive clustering hierarchy (LEACH) [4].

Low-Energy Adaptive Clustering Hierarchy (LEACH)

LEACH is a clustering-based protocol that minimizes energy dissipation in sensor networks. LEACH randomly selects nodes as cluster-heads and performs periodic reelection, so that the high-energy dissipation experienced by the cluster-heads in communicating with the BS is spread across all nodes of the network. Each iteration of selection of cluster-heads is called a round. The operation of LEACH is split into two phases: set-up and steady.

During the set-up phase, each sensor node chooses a random number between 0 and 1. If this is lower than the threshold for node n, $T(n)$, the sensor node becomes a cluster-head. The threshold $T(n)$ is calculated as

$$T(n) = \begin{cases} \frac{P}{1-P[r \times mod(1/P)]} & \text{if } n \in G \\ 0 & \text{otherwise,} \end{cases}$$

where P is the desired percentage of nodes which are cluster-heads, r is the current round, and G is the set of nodes that has not been cluster-heads in the past $1/P$ rounds. This ensures that all sensor nodes eventually spend equal energy. After selection, the cluster-heads advertise their selection to all nodes. All nodes choose their nearest cluster-head when they receive advertisements based on the received signal strength. The cluster-heads then assign a TDMA schedule for their cluster members.

The steady phase is of longer duration in order to minimize the overhead of cluster formation. During the steady phase, data transmission takes place based on the TDMA schedule, and the cluster-heads perform data aggregation/fusion through local computation. The BS receives only aggregated data from cluster-heads, leading to energy conservation. After a certain period of time in the steady phase, cluster-heads are selected again through the set-up phase.

12.3 DATA DISSEMINATION

Data dissemination is the process by which queries or data are routed in the sensor network. The data collected by sensor nodes has to be communicated to the BS or to any other node interested in the data. The node that generates data is called a *source* and the information to be reported is called an *event*. A node which is interested in an event and seeks information about it is called a *sink*. Traffic models have been developed for sensor networks such as the data collection and data dissemination (diffusion) models. In the data collection model, the source sends the data it collects to a collection entity such as the BS. This could be periodic or on demand. The data is processed in the central collection entity.

Data diffusion, on the other hand, consists of a two-step process of interest propagation and data propagation. An *interest* is a descriptor for a particular kind of data or event that a node is interested in, such as temperature, intrusion, or presence of bio-agents. For every event that a sink is interested in, it broadcasts its interest to its neighbors and periodically refreshes its interest. The interest is propagated across the network, and every node maintains an interest cache of all events to be reported. This is similar to a multicast tree formation, rooted at the sink. When an event is detected, it is reported to the interested nodes after referring to the interest cache. Intermediate nodes maintain a data cache and can aggregate the data or modify the rate of reporting data. The paths used for data propagation are modified by preferring the shortest paths and deselecting the weaker or longer paths. The basic idea of diffusion is made efficient and intelligent by different algorithms for interest and data routing.

12.3.1 Flooding

In flooding, each node which receives a packet broadcasts it if the maximum hop-count of the packet is not reached and the node itself is not the destination of the packet. This technique does not require complex topology maintenance or route discovery algorithms. But flooding has the following disadvantages [5]:

- Implosion: This is the situation when duplicate messages are sent to the same node. This occurs when a node receives copies of the same message from many of its neighbors.
- Overlap: The same event may be sensed by more than one node due to overlapping regions of coverage. This results in their neighbors receiving duplicate reports of the same event.
- Resource blindness: The flooding protocol does not consider the available energy at the nodes and results in many redundant transmissions. Hence, it reduces the network lifetime.

12.3.2 Gossiping

Gossiping is a modified version of flooding, where the nodes do not broadcast a packet, but send it to a randomly selected neighbor. This avoids the problem of implosion, but it takes a long time for a message to propagate throughout the network. Though gossiping has considerably lower overhead than flooding, it does not guarantee that all nodes of the network will receive the message. It relies on the random neighbor selection to eventually propagate the message throughout the network.

12.3.3 Rumor Routing

Rumor routing is an agent-based path creation algorithm [6]. Agents, or "ants," are long-lived entities created at random by nodes. These are basically packets which are circulated in the network to establish shortest paths to events that they encounter. They can also perform path optimizations at nodes that they visit. When an agent finds a node whose path to an event is longer than its own, it updates the node's routing table. Figure 12.4 illustrates the working of the rumor routing algorithm. In Figure 12.4 (a), the agent has initially recorded a path of distance 2 to event $E1$. Node A's table shows that it is at a distance 3 from event $E1$ and a distance 2 from $E2$. When the agent visits node A, it updates its own

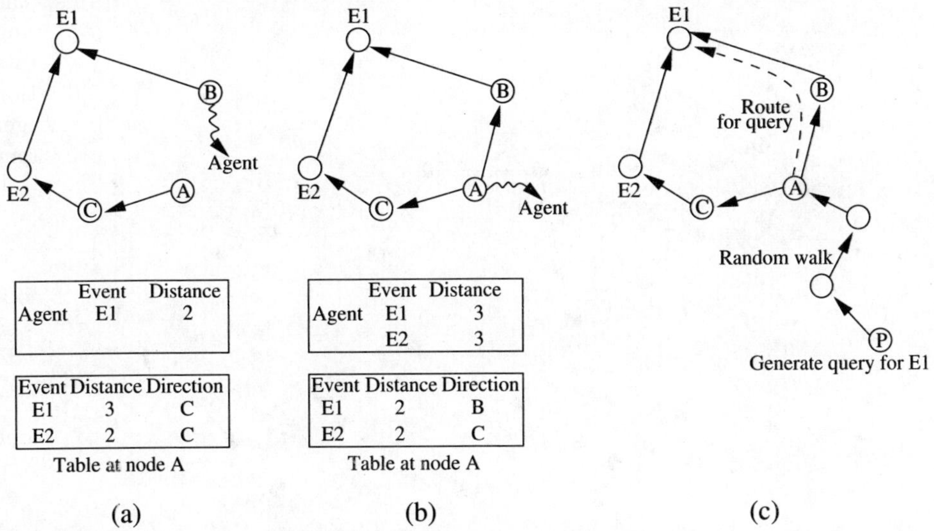

Figure 12.4. Rumor routing.

path state information to include the path to event $E2$. The updating is with one hop greater distance than what it found in A, to account for the hop between any neighbor of A that the agent will visit next, and A. It also optimizes the path to $E1$ recorded at node A to the shorter path through node B. The updated status of the agent and node table is shown in Figure 12.4 (b).

When a query is generated at a sink, it is sent on a random walk with the hope that it will find a path (preestablished by an agent) leading to the required event. This is based on the high probability of two straight lines intersecting on a planar graph, assuming the network topology is like a planar graph, and the paths established can be approximated by straight lines owing to high density of the nodes. If a query does not find an event path, the sink times out and uses flooding as a last resort to propagate the query. For instance, as in Figure 12.4 (c), suppose a query for event $E1$ is generated by node P. Through a random walk, it reaches A, where it finds the previously established path to $E1$. Hence, the query is directed to $E1$ through node B, as indicated by A's table.

12.3.4 Sequential Assignment Routing

In [7], a set of algorithms which performs organization and mobility management in sensor networks is proposed. The sequential assignment routing (SAR) algorithm creates multiple trees, where the root of each tree is a one-hop neighbor of the sink. Each tree grows outward from the sink and avoids nodes with low throughput or high delay. At the end of the procedure, most nodes belong to multiple trees. An instance of tree formation is illustrated in Figure 12.5. The trees rooted at A and

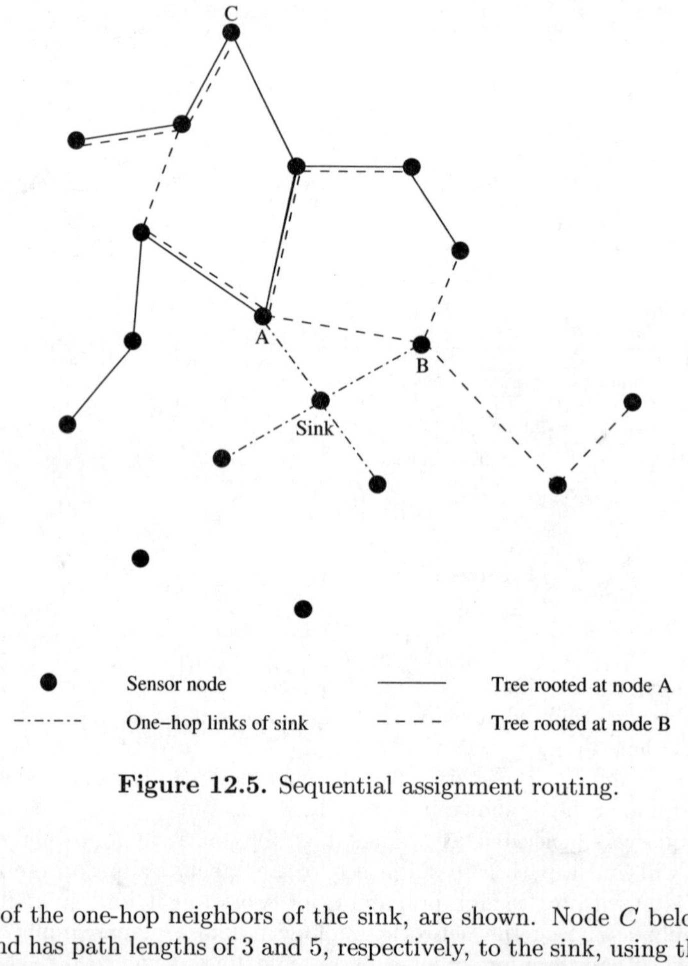

Figure 12.5. Sequential assignment routing.

B, two of the one-hop neighbors of the sink, are shown. Node C belongs to both trees, and has path lengths of 3 and 5, respectively, to the sink, using the two trees. Each sensor node records two parameters about each path through it: the available energy resources on the path and an additive QoS metric such as delay.

This allows a node to choose one path from among many to relay its message to the sink. The SAR algorithm chooses a path with high estimated energy resources, and provisions can be made to accommodate packets of different priorities. A weighted QoS metric is used to handle prioritized packets, which is computed as a product of priority level and delay. The routing ensures that the same weighted QoS metric is maintained. Thus, higher priority packets take lower delay paths, and lower priority packets have to use the paths of greater delay. For example, if node C generates a packet of priority 3, it follows the longer path along tree B, and a packet of priority 5 (higher priority) will follow the shorter path along tree A, so that the priority $\times$ delay QoS metric is maintained. SAR minimizes the average weighted QoS metric over the lifetime of the network. The sink periodically triggers a metric update to reflect the changes in available energy resource after some transmissions.

12.3.5 Directed Diffusion

The directed diffusion protocol is useful in scenarios where the sensor nodes themselves generate requests/queries for data sensed by other nodes, instead of all queries arising only from a BS. Hence, the sink for the query could be a BS or a sensor node. The directed diffusion routing protocol [8] improves on data diffusion using interest gradients. Each sensor node names its data with one or more attributes, and other nodes express their interest depending on these attributes. Attribute-value pairs can be used to describe an interest in intrusion data as follows, where an interest is nothing but a set of descriptors for the data in which the querying node is interested.

```
type = vehicle              /* detect vehicle location */
interval = 1 s              /* report every 1 second */
rect = [0, 0, 600, 800]     /* query addressed to sensors within this
                               rectangle*/
timestamp = 02:30:00        /* when the interest was originated*/
expiresAt = 03:00:00        /* till when the sink retains interest in
                               this data*/
```

The sink has to periodically refresh its interest if it still requires the data to be reported to it. Data is propagated along the reverse path of the interest propagation. Each path is associated with a gradient that is formed at the time of interest propagation. While positive gradients encourage the data flow along the path, negative gradients inhibit the distribution of data along a particular path. The strength of the interest is different toward different neighbors, resulting in source-to-sink paths with different gradients. The gradient corresponding to an interest is derived from the interval/data-rate field specified in the interest. For example, if there are two paths formed with gradients 0.8 and 0.4, the source may send twice as much data along the higher gradient path compared to the lower gradient one. For the interest mentioned earlier, a sensor may send data of the following kind:

```
type = vehicle              /* type of intrusion seen */
instance = car              /* particular instance of the type */
location = [200,250]        /* location of node */
confidence = 0.80           /* confidence of match */
timestamp = 02:45:20        /* time of detection */
```

The diffusion model allows nodes to cache or locally transform (aggregate) data. This increases the scalability of communication and reduces the number of message transmissions required.

The concept of reinforcement is used to update a node's interest along a particular path. For example, suppose the sink wants more frequent updates from the sensors which have detected an event. It reinforces the path by sending an interest with a higher data-rate requirement, in effect increasing the gradient of that path. On the other hand, if the sink needs only fewer updates, it applies negative reinforcement by sending an interest of lower required data-rate.

The directed diffusion model uses data naming by attributes and local data transformations to reflect the data-centric nature of sensor network operations. The local operations of data aggregation are application-specific. Gradients model the network-wide results of local interactions by regulating the flow of data along different paths, depending on the expressed interest.

12.3.6 Sensor Protocols for Information via Negotiation

A family of protocols called sensor protocols for information via negotiation (SPIN) is proposed in [5]. SPIN uses negotiation and resource adaptation to address the deficiencies of flooding. Negotiation reduces overlap and implosion, and a threshold-based resource-aware operation is used to prolong network lifetime. Meta-data, or data describing data, is transmitted instead of raw data. This requires fewer bytes and can be in an application-specific format. SPIN has three types of messages: ADV, REQ, and DATA. A sensor node broadcasts an ADV containing meta-data describing the actual data. If a neighbor is interested in the data, it sends a REQ for the data. Then the sensor node sends the actual DATA to the neighbor. The neighbor again sends ADVs to its neighbors and this process continues to disseminate the data throughout the network. This simple version of SPIN is shown in Figure 12.6.

SPIN is based on data-centric routing, where the nodes advertise the available data through an ADV and wait for requests from interested nodes. SPIN-2 expands

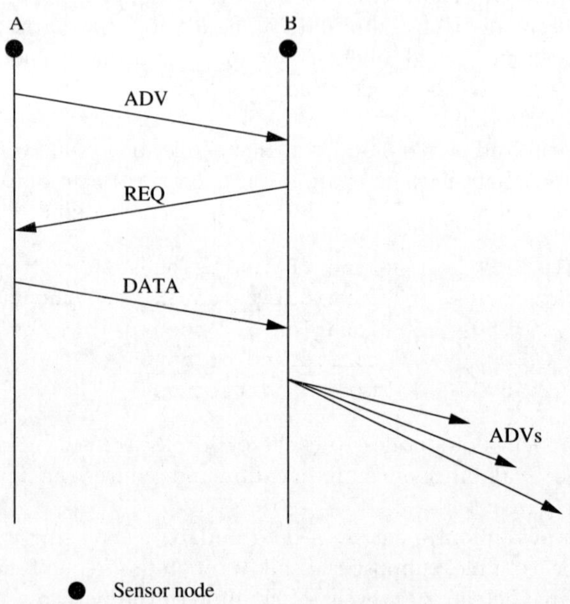

Figure 12.6. SPIN protocol.

on SPIN, using an energy or resource threshold to reduce participation. A node may participate in the ADV-REQ-DATA handshake only if it has sufficient resources above a threshold.

12.3.7 Cost-Field Approach

The cost-field approach [9] considers the problem of setting up paths to a sink. It is a two-phase process, the first phase being to set up the cost field, based on metrics such as delay, at all sensor nodes, and the second being data dissemination using the costs. At each node, the cost is defined as the minimum cost from that node to the sink, which occurs along the optimal path. Explicit path information does not need to be maintained.

Phase 1 sets up a cost field starting from the sink node. A sink broadcasts an ADV packet with its own cost as 0. When a node N hears an ADV message from node M, it sets its own path cost to $min(L_N, L_M + C_{NM})$, where L_N is the total path cost from node N to sink, L_M represents the cost of node M to sink, and C_{NM} is the cost from node N to M. If L_N was updated, the new cost is broadcast through another ADV. This is a flooding-based implementation of the Dijkstra's algorithm. In order to reduce the high communication costs associated with flooding, a back-off-based approach is used. The main reason for overhead is that a node broadcasts its updated cost immediately, whether it was the optimal cost or not. Instead, the back-off modification makes a node defer its ADV instead of immediately broadcasting it. The time to defer is heuristically determined as $\gamma \times C_{MN}$, where γ is a parameter of the algorithm.

The working of the cost-field approach with back-off is illustrated in Figure 12.7. The numbers on the links indicate link costs. The value of γ is assumed to be 10. Initially, nodes N and P did not have a path to the sink and hence had their costs set to ∞. In Figure 12.7 (a), node M broadcasts an ADV, which is received by nodes N and P. They tentatively fix their costs to L_M+2 and L_M+5, respectively, and set their back-off timers to 20 and 50, respectively. Figure 12.7 (b) shows the costs after 20 time units, when node N's back-off timer expires. Node N finalizes its cost to $L_M + 2$ and broadcasts an ADV, which is heard by node P. Since $L_N + 1 < L_M + 5$, node P updates its cost and sets a new back-off timer to 10. The unnecessary ADV of node P's earlier non-optimal cost is avoided by setting the back-off timer. Finally, at 30 time units, node P finalizes its cost to $L_N + 1$ and broadcasts an ADV, as shown in Figure 12.7 (c).

Phase 2 is the data dissemination process. Once the cost field is established, a source sends its message to sink S with cost C_S. The message also contains a cost-so-far field, initially set to 0. Each intermediate node forwards the packet if the cost recorded in the packet plus its own cost equals the original source-to-sink cost. This ensures that the original optimal path is used whenever a packet is routed. While forwarding, the intermediate nodes also update the cost-so-far field.

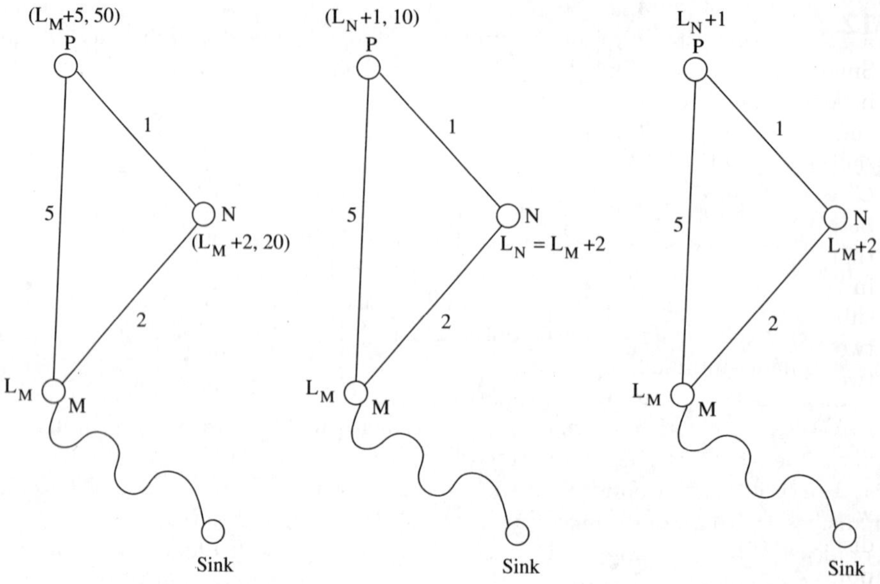

(a) Time T, after M's ADV (b) Time T + 20, after N's ADV (c) Time T + 30, after P's ADV

Figure 12.7. Cost-field approach.

12.3.8 Geographic Hash Table

Geographic hash table (GHT) is a system based on data-centric storage [10], inspired by Internet-scale distributed hash table (DHT) systems such as Chord [11] and Tapestry [12]. GHT hashes keys into geographic coordinates and stores a (key, value) pair at the sensor node nearest to the hash value. The calculated hash value is mapped onto a unique node consistently, so that queries for the data can be routed to the correct node. Stored data is replicated to ensure redundancy in case of node failures, and a consistency protocol is used to maintain the replicated data. The data is distributed among nodes such that it is scalable and the storage load is balanced. The routing protocol used is greedy perimeter stateless routing (GPSR) [13], which again uses geographical information to route the data and queries. GHT is more effective in large sensor networks, where a large number of events are detected but not all are queried. In this case, the data observed is stored in a distributed manner across all nodes, instead of being routed to a central external storage. Queries are routed to the nearest node which contains a copy of the relevant data. This makes the storage and traffic distribution uniform.

12.3.9 Small Minimum Energy Communication Network

Small minimum energy communication network (SMECN) is a protocol proposed in [14] to construct a sub-network from a given communication network. If the entire sensor network is represented by a graph G, the subgraph G' is constructed such that the energy usage of the network is minimized. The number of edges in G' is less than that of G, but all nodes of G are retained in G'. The connectivity between any two nodes is not disrupted by the subgraph. G' is constructed such that the energy required to transmit data from a node to all its neighbors is lower in G' than in G. SMECN also follows the minimum energy (ME) property in its subgraph construction, that is, there exists an ME path in subgraph G' between any two nodes that are connected in G. The power required to transmit data between two nodes u and v is modeled as

$$p(u, v) = t \times d(u, v)^n$$

where t is a constant, n is the path loss exponent indicating the loss of power with distance from the transmitter, and $d(u, v)$ is the distance between u and v. Let the power needed to receive the data be c. Since the transmission power increases exponentially with distance, it would be more economical to transmit data by smaller hops. Suppose the path between u (*i.e.*, u_0) and v (*i.e.*, u_k) is represented by $r = (u_0, u_1, ...u_k)$, such that each (u_i, u_{i+1}) is an edge in the subgraph G', then the total power consumed for the transmission is

$$C(r) = \sum_{i=0}^{k-1}(p(u_i, u_{i+1}) + c)$$

The path r is the ME path if $C(r) \leq C(r')$ for all paths r' between u and v in the graph G. The subgraph G' is said to have the ME property if there exists a path r in G' which is an ME path in G, for all node pairs (u, v). SMECN uses only the ME paths from G' for data transmission, so that the overall energy consumed is minimized.

12.4 DATA GATHERING

The objective of the data-gathering problem is to transmit the sensed data from each sensor node to a BS. One round is defined as the BS collecting data from all the sensor nodes once. The goal of algorithms which implement data gathering is to maximize the number of rounds of communication before the nodes die and the network becomes inoperable. This means minimum energy should be consumed and the transmission should occur with minimum delay, which are conflicting requirements. Hence, the *energy* $\times$ *delay* metric is used to compare algorithms, since this metric measures speedy and energy-efficient data gathering. A few algorithms that implement data gathering are discussed below.

12.4.1 Direct Transmission

All sensor nodes transmit their data directly to the BS. This is extremely expensive in terms of energy consumed, since the BS may be very far away from some nodes. Also, nodes must take turns while transmitting to the BS to avoid collision, so the media access delay is also large. Hence, this scheme performs poorly with respect to the energy $\times$ delay metric.

12.4.2 Power-Efficient Gathering for Sensor Information Systems

Power-efficient gathering for sensor information systems (PEGASIS) [15] is a data-gathering protocol based on the assumption that all sensor nodes know the location of every other node, that is, the topology information is available to all nodes. Also, any node has the required transmission range to reach the BS in one hop, when it is selected as a leader. The goals of PEGASIS are as follows:

- Minimize the distance over which each node transmits
- Minimize the broadcasting overhead
- Minimize the number of messages that need to be sent to the BS
- Distribute the energy consumption equally across all nodes

A greedy algorithm is used to construct a chain of sensor nodes, starting from the node farthest from the BS. At each step, the nearest neighbor which has not been visited is added to the chain. The chain is constructed *a priori*, before data transmission begins, and is reconstructed when nodes die out. At every node, data fusion or aggregation is carried out, so that only one message is passed on from one node to the next. A node which is designated as the leader finally transmits one message to the BS. Leadership is transferred in sequential order, and a token is passed so that the nodes know in which direction to pass messages in order to reach the leader. A possible chain formation is illustrated in Figure 12.8. The delay involved in messages reaching the BS is $O(N)$, where N is the total number of nodes in the network.

12.4.3 Binary Scheme

This is also a chain-based scheme like PEGASIS, which classifies nodes into different levels. All nodes which receive messages at one level rise to the next [16]. The number of nodes is halved from one level to the next. For instance, consider a network with eight nodes labeled $s0$ to $s7$. As Figure 12.9 shows, the aggregated data reaches the BS in four steps, which is $O(log_2 N)$, where N is the number of nodes in the network. This scheme is possible when nodes communicate using CDMA, so that transmissions of each level can take place simultaneously.

12.4.4 Chain-Based Three-Level Scheme

For non-CDMA sensor nodes, a binary scheme is not applicable. The chain-based three-level scheme [16] addresses this situation, where again a chain is constructed as

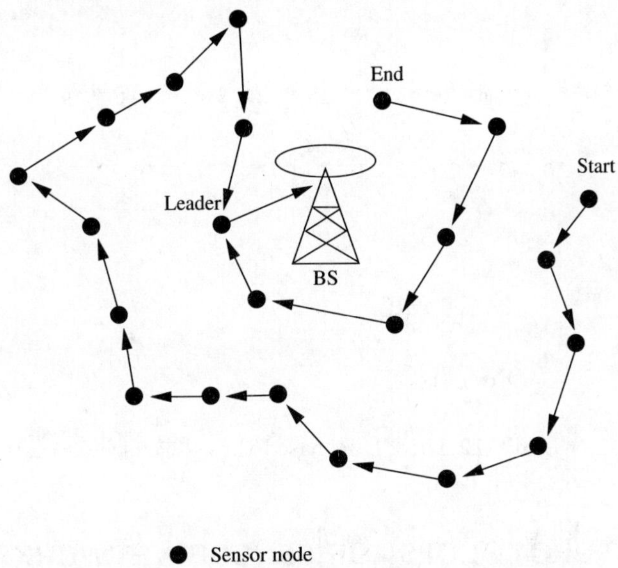

Figure 12.8. Data gathering with PEGASIS.

in PEGASIS. The chain is divided into a number of groups to space out simultaneous transmissions in order to minimize interference. Within a group, nodes transmit one at a time. One node out of each group aggregates data from all group members and rises to the next level. The index of this leader node is decided *a priori.* In the second level, all nodes are divided into two groups, and the third level consists of a message exchange between one node from each group of the second level. Finally, the leader transmits a single message to the BS. The working of this scheme is illustrated in Figure 12.10. The network has 100 nodes, and the group size is ten for the first level and five for the second level. Three levels have been found to give the optimal energy × delay through simulations.

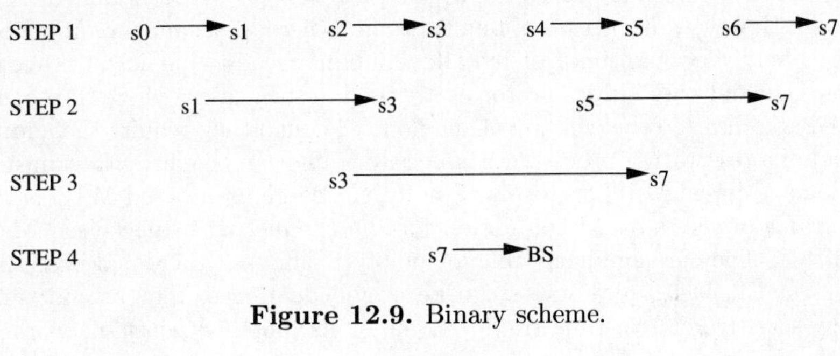

Figure 12.9. Binary scheme.

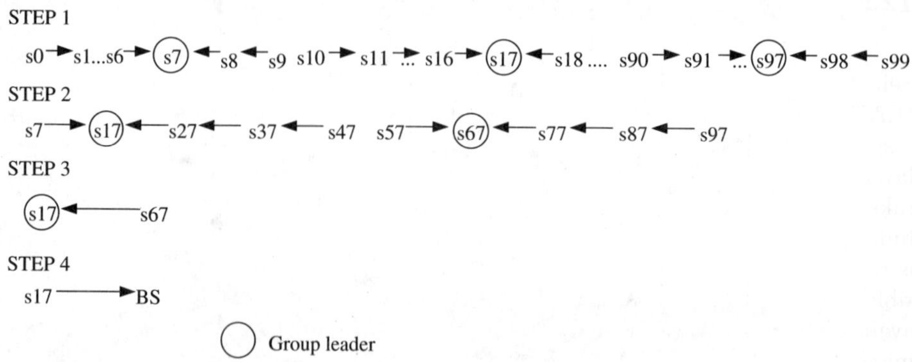

Figure 12.10. Chain-based three-level scheme.

12.5 MAC PROTOCOLS FOR SENSOR NETWORKS

MAC protocols in sensor networks must create a network infrastructure to establish communication links among the thousands of randomly scattered sensors. It must also ensure fair and efficient sharing of communication resources among the nodes, so that the overall lifetime of the network can be maximized. The challenges posed by sensor network MAC protocols make them distinct from other wireless based networks. Unlike infrastructure-based cellular networks, there is no single controlling authority in sensor networks, so global synchronization becomes difficult. Power efficiency is of utmost concern in sensor networks. They also encounter frequent topology changes due to mobility and failure. These factors emphasize the need for MAC protocols specific to sensor networks.

There are three basic kinds of MAC protocols used in sensor networks: fixed-allocation, demand-based, and contention-based. Fixed-allocation MAC protocols share the common medium through a predetermined assignment. They are appropriate for sensor networks that continuously monitor and generate deterministic data traffic, since all nodes which have been allotted the channel can make use of their slot in each round. Fixed-allocation protocols provide a bounded delay for each node. However, in the case of bursty traffic, where the channel requirements of each node may vary over time, a fixed allocation may lead to inefficient usage of the channel. Demand-based MAC protocols are used in such cases, where the channel is allocated according to the demand of the node. Though they require the additional overhead of a reservation process, variable rate traffic can be efficiently transmitted using demand-based MAC protocols. Finally, the contention-based MAC protocols involve random-access-based contention for the channel when packets need to be transmitted. They are again suitable for bursty traffic, but there is a possibility of collisions and no delay guarantees can be provided. Hence, they are not suitable for delay-sensitive or real-time traffic. Some of the popular sensor network MAC protocols have been briefly described in the next section.

12.5.1 Self-Organizing MAC for Sensor Networks and Eavesdrop and Register

Self-organizing MAC for sensor (SMACS) networks and eavesdrop and register (EAR) are two protocols which handle network initialization and mobility support, respectively. SMACS is a distributed protocol for network initialization and link-layer organization [7]. In this protocol, neighbor discovery and channel assignment take place simultaneously in a completely distributed manner. A communication link between two nodes consists of a pair of time slots, at a fixed frequency, which is randomly chosen at the time of establishing the link. Such an assignment is possible in sensor networks without interference from neighboring nodes because the available bandwidth is much larger than the data rate required for a message transmission between two nodes. This scheme requires synchronization only between communicating neighbors, in order to precisely define the slots to be used for their communication. Power is conserved by turning off the transceiver during idle slots, and using a random wake-up schedule during the network start-up phase.

The EAR protocol enables seamless connection of nodes under mobile and stationary conditions. This protocol makes use of certain mobile nodes, besides the existing stationary sensor nodes, to offer service to maintain connections. Mobile nodes eavesdrop on the control signals and maintain neighbor information. The mobile nodes assume full control over connections and can drop connections when they move away. Mobility is hence made transparent to SMACS, since it is independently handled by EAR.

12.5.2 Hybrid TDMA/FDMA

This is a centrally controlled scheme which assumes that nodes communicate directly to a nearby BS. A pure TDMA scheme minimizes the time for which a node has to be kept on, but the associated time synchronization costs are very high. A pure FDMA scheme allots the minimum required bandwidth for each connection. The hybrid TDMA/FDMA scheme, proposed in [17], uses an optimum number of channels, which gives minimum overall power consumption. This is found to depend on the ratio of power consumption of transmitter to receiver. If the transmitter consumes more power, a TDMA scheme is favored, since it can be switched off in idle slots to save power. On the other hand, the scheme favors FDMA when the receiver consumes greater power. This is because, in FDMA, the receiver need not expend power for time synchronization by receiving during the guard band between slots, which becomes essential in a TDMA scheme.

12.5.3 CSMA-Based MAC Protocols

Traditional CSMA-based schemes are more suitable for point-to-point stochastically distributed traffic flows. On the other hand, sensor networks have variable but periodic and correlated traffic. A CSMA-based MAC protocol for sensor networks has been described in [18]. The sensing periods of CSMA are constant for energy efficiency, while the back-off is random to avoid repeated collisions. Binary exponential back-off is used to maintain fairness in the network. An adaptive transmission rate control (ARC) is also used, which balances originating and route-through traffic in nodes. This ensures that nodes closer to the BS are not favored over farther nodes. ARC uses linear increase and multiplicative decrease of originating traffic in a node. The penalty for dropping route-through traffic is higher, since energy has already been invested in making the packets reach until that node. ARC performs phase changes, that is, it staggers the transmission times of different streams so that periodic streams are less likely to collide repeatedly. Hence, CSMA based MAC protocols are contention-based and are designed mainly to increase energy efficiency and maintain fairness.

12.6 LOCATION DISCOVERY

The location information of sensors has to be considered during aggregation of sensed data. This implies each node should know its location and couple its location information with the data in the messages it sends. A low-power, inexpensive, and reasonably accurate mechanism is needed for location discovery. A global positioning system (GPS) is not always feasible because it cannot reach nodes in dense foliage or indoors. It also consumes high power and makes sensor nodes bulkier. Two basic mechanisms of location discovery are now described.

12.6.1 Indoor Localization

Indoor localization techniques [19] use a fixed infrastructure to estimate the location of sensor nodes. Fixed beacon nodes are strategically placed in the field of observation, typically indoors, such as within a building. The randomly distributed sensors receive beacon signals from the beacon nodes and measure the signal strength, angle of arrival, and time difference between the arrival of different beacon signals. Using the measurements from multiple beacons, the nodes estimate their location. Some approaches use simple triangulation methods, while others require *a priori* database creation of signal measurements. The nodes estimate distances by looking up the database instead of performing computations. However, storage of the database may not be possible in each node, so only the BS may carry the database.

12.6.2 Sensor Network Localization

In situations where there is no fixed infrastructure available and prior measurements are not possible, some of the sensor nodes themselves act as beacons. They have their location information, using GPS, and these send periodic beacons to other nodes. In the case of communication using RF signals, the received signal strength indicator (RSSI) can be used to estimate the distance, but this is very sensitive to obstacles and environmental conditions. Alternatively, the time difference between beacon arrivals from different nodes can be used to estimate location, if RF or ultrasound signals are used for communication. This offers a lower range of estimation than RSSI, but is of greater accuracy.

Localization algorithms require techniques for location estimation depending on the beacon nodes' location. These are called multi-lateration (ML) techniques. Some simple ML techniques are described in what follows [20]:

- Atomic ML: If a node receives three beacons, it can determine its position by a mechanism similar to GPS. This is illustrated in Figure 12.11.
- Iterative ML: Some nodes may not be in the direct range of three beacons. Once a node estimates its location, it sends out a beacon, which enables some other nodes to now receive at least three beacons. Iteratively, all nodes in the network can estimate their location. This is shown in Figure 12.12. The drawback of this multi-hop method is that errors are propagated, hence estimation of location may not be accurate.
- Collaborative ML: When two or more nodes cannot receive at least three beacons each, they collaborate with each other. As shown in Figure 12.13, node A and node B have three neighbors each. Of the six participating nodes, four are beacons, whose positions are known. Hence, by solving a set of simultaneous quadratic equations, the positions of A and B can be determined.

A directionality-based localization approach has been explored in [21]. This assumes that beacon nodes have broadcast capability to reach all nodes of the network, and a central controller rotates the beacons with a constant angular velocity ω radians/s. A constant angular separation is maintained between the beacon nodes. Nodes in the network measure the angles of arrival of beacon signals to estimate their location. The errors in this technique occur due to non-zero beam-width from the beacons. The beam is not a straight line as theoretically imagined, but it has a finite width. Hence, the measurement of the angle of the beacon signal will be inaccurate.

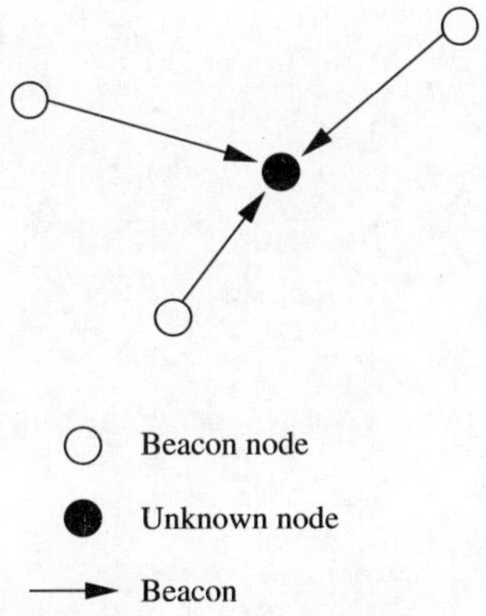

Figure 12.11. Atomic multi-lateration.

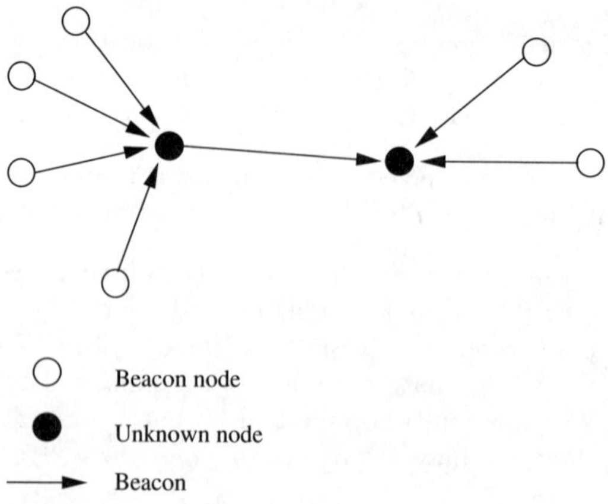

Figure 12.12. Iterative multi-lateration.

The authors of [22] propose an algorithm which derives the location of sensor nodes based mainly on information about connectivity between nodes. The all-pairs shortest paths algorithm is run on the network graph, which has edges indicating connectivity between nodes. Hence, the shortest distance between each pair of nodes is obtained. A mathematical technique called multi-dimensional scaling (MDS), an $O(n^3)$ algorithm (where n is the number of sensors), is used to assign locations to nodes such that the distance constraints are satisfied. The obtained picture of the network could be a rotated or flipped version of the actual network. If the actual positions of any three nodes in the network are known, then the entire network can be normalized (rotated or flipped) to obtain a very accurate localization of all other nodes.

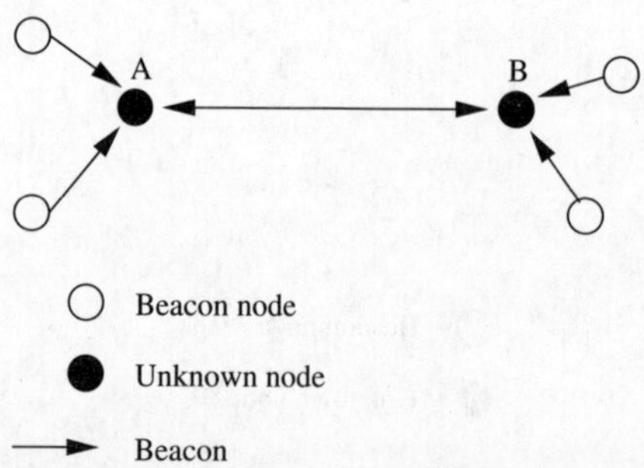

Figure 12.13. Collaborative multi-lateration.

12.7 QUALITY OF A SENSOR NETWORK

The purpose of a sensor network is to monitor and report events or phenomena taking place in a particular area. Hence, the main parameters which define how well the network observes a given area are "coverage" and "exposure." In this section, we shall formally define the coverage and exposure problems, and briefly describe some mathematical techniques to solve them.

12.7.1 Coverage

Coverage is a measure of how well the network can observe or cover an event. Coverage depends upon the range and sensitivity of the sensing nodes, and the location and density of the sensing nodes in the given region. The *worst-case* coverage defines areas of breach, that is, where coverage is the poorest. This can be used to determine if additional sensors need to be deployed to improve the network. The *best-case* coverage, on the other hand, defines the areas of best coverage. A path along the areas of best coverage is called a maximum support path or maximum exposure path.

The coverage problem is formally defined as follows: Given a field A with a set of sensors $S = \{s_1, s_2, \ldots, s_n\}$, where for each sensor s_i in S, its location coordinates (x_i, y_i) are known, based on localization techniques. Areas I and F are the initial and final locations of an intruder traversing the field. The problem is to identify P_B, the maximal breach path starting in I and ending in F. P_B is defined as the locus of points p in the region A, where p is in P_B if the distance from p to the closest sensor is maximized.

A mathematical technique to solve the coverage problem is the Voronoi diagram. It can be proved that the path P_B will be composed of line segments that belong to the Voronoi diagram corresponding to the sensor graph. In two dimensions, the Voronoi diagram of a set of sites is a partitioning of the plane into a set of convex polygons such that all points inside a polygon are closest to the site enclosed by the polygon, and the polygons have edges equidistant from the nearby sites. A Voronoi diagram for a sensor network, and a breach path from I to F, are shown in Figure 12.14.

The algorithm to find the breach path P_B is:

- Generate the Voronoi diagram, with the set of vertices V and the set of edges E. This is done by drawing the perpendicular bisectors of every line segment joining two sites, and using their points of intersection as the vertices of the convex polygons.

- Create a weighted graph with vertices from V and edges from E, such that the weight of each edge in the graph is the minimum distance from all sensors in S. The edge weights represent the distance from the nearest sensor. Smaller edge weights imply better coverage along the edge.

- Determine the maximum cost path from I to F, using breadth-first search.

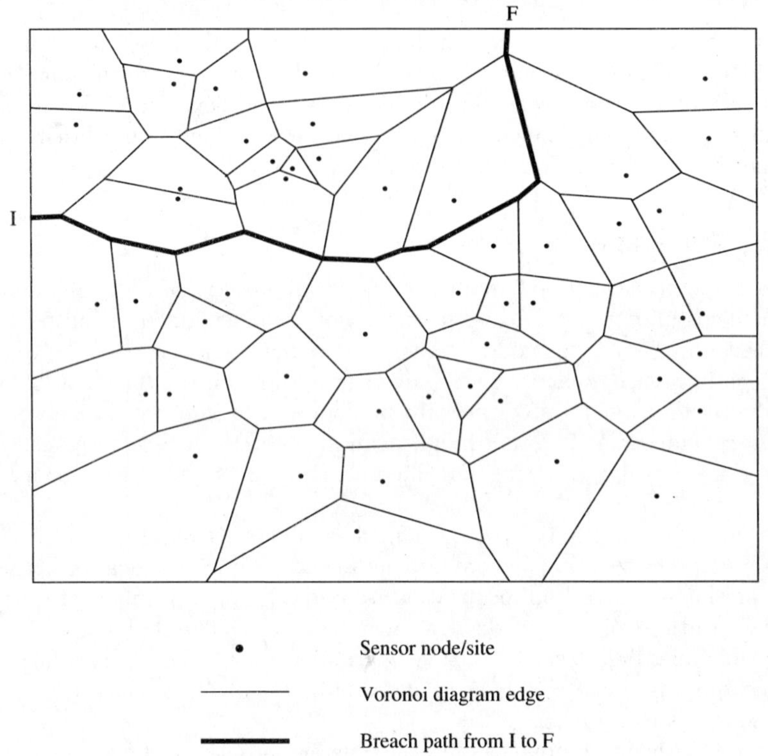

Figure 12.14. Voronoi diagram.

The maximum cost implies least coverage. Hence, the required breach path is along this maximum-cost path determined from the Voronoi diagram. The breach path shows the region of maximum vulnerability in a sensor network, where the coverage provided by the sensors is the weakest.

A related problem is that of finding the best-case coverage. The problem is formally stated as finding the path which offers the maximum coverage, that is, the maximum support path P_S in S, from I to F. The solution is obtained by a mathematical technique called Delaunay triangulation, shown in Figure 12.15. This is obtained from the Voronoi diagram by connecting the sites whose polygons share a common edge. The best path P_S will be a set of line segments from the Delaunay triangulation, connecting some of the sensor nodes. The algorithm is again similar to that used to find the maximum breach path, replacing the Voronoi diagram by the Delaunay triangulation, and defining the edge costs proportional to the line segment lengths. The maximum support path is hence formed by a set of line segments connecting some of the sensor nodes.

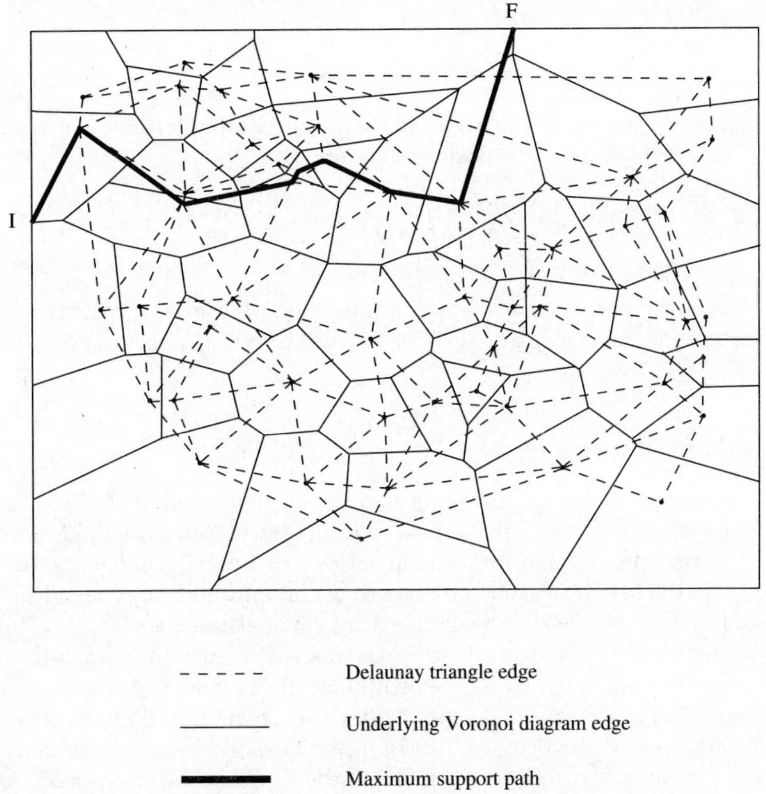

Figure 12.15. Delaunay triangulation.

12.7.2 Exposure

Exposure is defined as the expected ability of observing a target in the sensor field. It is formally defined as the integral of the sensing function on a path from source node P_s to destination node P_d. The sensing power of a node s at point p is usually modeled as

$$S(s,p) = \frac{\lambda}{[d(s,p)]^k}$$

where λ and k are constants, and $d(s,p)$ is the distance of p from s. Consider a network with sensors $s_1, s_2, \ldots, s_n$. The total intensity at point p, called the all-sensor field intensity, is given by

$$I_A(F,p) = \sum_{i=1}^{n} S(s_i, p)$$

The closest-sensor field intensity at p is

$$I_C(F,p) = S(s_{min}, p)$$

where s_{min} is the closest sensor to p. The exposure during travel of an event along a path $p(t)$ is defined by the exposure function

$$E[p(t), t_1, t_2] = \int_{t_1}^{t_2} I_{AorC}(F, p(t)) \left| \frac{dp(t)}{dt} \right| dt$$

where $\frac{dp(t)}{dt}$ is the elemental arc length, and t_1, t_2 are the time instances between which the path is traversed. For conversion from Cartesian coordinates $(x(t), y(t))$,

$$\frac{dp(t)}{dt} = \sqrt{\left(\frac{dx(t)}{dt}\right)^2 + \left(\frac{dy(t)}{dt}\right)^2}$$

In the simplest case of having one sensor node at $(0, 0)$ in a unit field, the breach path or minimum exposure path (MEP) from $(-1, -1)$ to $(1, 1)$ is shown in Figure 12.16.

It can also be proved that for a single sensor s in a polygonal field, with vertices $v_1, v_2, \ldots, v_n$, the MEP between two vertices v_i and v_j can be determined as follows. The edge (v_i, v_{i+1}) is tangent to the inscribed circle at u_i. Then the MEP consists of the line segment from v_i to u_i, part of the inscribed circle from u_i to u_j, and the line segment from u_j to v_j. This is shown in Figure 12.17.

The exposure problem is still unsolved for two points in the same corner, or for points within the inscribed circle. For the generic exposure problem of determining the MEP for randomly placed sensor nodes in the network, the network is tessellated with grid points. An example is shown in Figure 12.18. To construct an $n \times n$ grid of order m, each side of a square is divided into m equal parts, creating $(m + 1)$

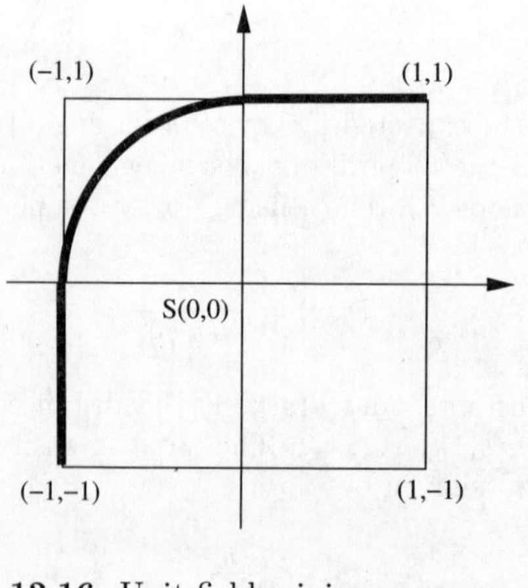

Figure 12.16. Unit field minimum exposure path.

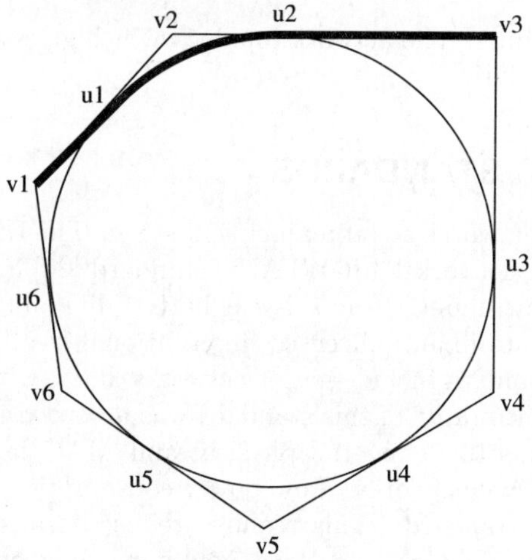

Figure 12.17. Polygon field minimum exposure path.

vertices on the edge. Within each square, all vertices are connected to obtain a grid. Higher order grids have greater accuracy. For each edge in the grid network, the exposure function is used to determine the edge weights, and the MEP is defined as the shortest path, determined by Dijkstra's algorithm.

The mathematical concept of exposure is important for evaluating the target detection capability of a sensor network. Sensors are deployed in a given area to detect events occurring in the field of interest. The nodes collaborate among themselves (perform data fusion) through the exchange of localized information, and reach a decision about the location and movement of a given event or target. In [23], Clouqueur *et al.* discuss a probabilistic protocol for target detection, where the observations made by individual sensors are collaborated, and the presence or movement of a target is probabilistically determined by data fusion, with allowance

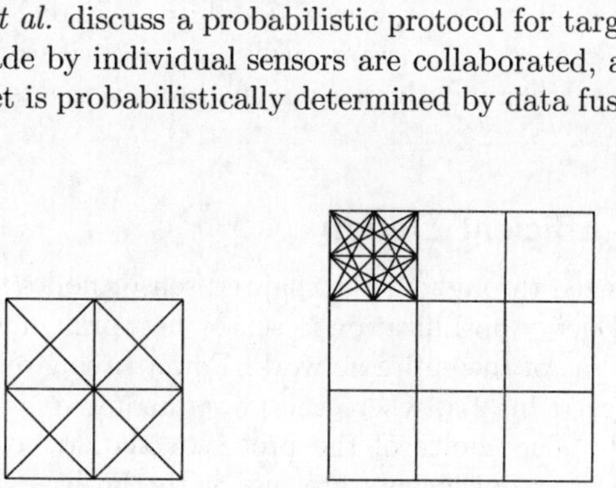

Figure 12.18. Generic minimum exposure path.

for noise in data recording. The network topology which gives a maximum exposure is also determined analytically.

12.8 EVOLVING STANDARDS

Standards for sensor networks are at an incipient stage. The IEEE 802.15.4 low-rate wireless personal area networks (LR-WPANs) standard [24] investigates a low data rate solution with multi-month to multi-year battery life and very low complexity. It is intended to operate in an unlicensed, international frequency band. Potential applications of this standard include sensor networks, home automation, and remote controls. The eighteenth draft of this standard was accepted in May 2003.

This standard aims to define the physical and MAC layer specifications for sensor and other WPAN networks. Low power consumption is an important feature targeted by the standard. This requires reduced transmission rate, power-efficient modulation techniques, and strict power management techniques such as sleep modes. Different network configurations and topologies were compared, and star and mesh networks were found to be favorable. The standard also proposes a generic frame structure whose length can be varied according to the application.

Other standards under development include the SensIT project [25] by the Defense Advanced Research Projects Agency (DARPA) which focuses on large distributed military systems, and the ZigBee Alliance [26], which addresses industrial and vehicular appliances. The IEEE 1451.5 wireless smart transducer interface standard is still under review. It is proposed to include multiple combinations of MAC and physical layers, using the IEEE 802 approach as a model.

12.9 OTHER ISSUES

This section deals with some issues that are recently being explored in sensor networks, such as energy-efficient hardware and architecture, real-time communication on sensor networks, transport layer protocols, and security issues. Because these are mostly in the research stage, there are many improvements to be made on these fronts.

12.9.1 Energy-Efficient Design

As has been emphasized throughout the chapter, sensor nodes have a very stringent energy constraint. Energy optimization in sensor networks must prolong the life of a single node as well as of the entire network. Power saving in the micro-controller unit has been analyzed in [27], where the power required by different processors has been compared. The choice of the processor should be application-specific, such that performance requirements are met with the least power consumption. Computation can be carried out in a power-aware manner using dynamic power management (DPM). One of the basic DPM techniques is to shut down several components of the sensor node when no events take place. The processor has a time-varying computational load, hence the voltage supplied to it can be scaled to meet only the instantaneous processing requirement. This is called dynamic voltage scaling (DVS).

The software used for sensor networks such as the operating system, application software, and network software can also be made energy-aware. The real-time task scheduler should actively support DVS by predicting the computation and communication loads. Sensor applications can use a trade-off between energy and accuracy by performing the most significant operations first, so that premature termination of the computation due to energy constraints does not affect the result by a large margin.

The communications subsystem should also perform energy-aware packet forwarding. The use of intelligent radio hardware enables packets to be forwarded directly from the communication subsystem, without processing it through the micro-controller. Techniques similar to DVS are used for modulation, to transmit data using a simpler modulation scheme, thereby consuming less energy, when the required data transmission rate is lower. This is called modulation scaling.

Besides incorporating energy-efficient algorithms at the node level, there should be a network-wide cooperation among nodes to conserve energy and increase the overall network lifetime. The computation-communication trade-off determines how much local computation is to be performed at each node and what level of aggregated data should be communicated to neighboring nodes or BSs. Traffic distribution and topology management algorithms exploit the redundancy in the number of sensor nodes to use alternate routes so that energy consumption all over the network is nearly uniform.

12.9.2 Synchronization

Synchronization among nodes is essential to support TDMA schemes on multi-hop wireless networks. Also, time synchronization is useful for determining the temporal ordering of messages sent from sensors and the proximity of the sensors. Usually, sensor nodes are dropped into the environment from which data has to be collected, and their exact positions are not fixed before deployment. Hence, synchronization is the only way by which the nodes can determine their relative positions. Further, in order to furnish aggregate data to the monitor node, the sensors must evolve a common timescale using their synchronized clocks, to judge the speed of a moving target or phenomenon. Sensors must be able to recognize duplicate reports of the same event by different nodes and discard them, which means that the node must be able to precisely determine the instant of time at which the event occurred. There are two major kinds of synchronization algorithms: one which achieves long-lasting global synchronization, that is, lasts throughout the network for its entire lifetime, and one which achieves a short-lived or pulse synchronization where the nodes are synchronized only for an instant.

Synchronization protocols typically involve delay measurements of control packets. The delays experienced during a packet transmission can be split into four major components [28]: send time, access time, propagation time, and receive time. The send time is the time spent at the sender to construct the message. The access time is the time taken by the MAC layer to access the medium, which is appreciable in a contention-based MAC protocol. The propagation time reflects the time taken by the bits to be physically transmitted through the medium over the distance separating the sender and receiver. The receive time is the time for processing required in the receiver's network interface to receive the message from the channel and notify the host of its arrival. If the arrival time is time-stamped at a low layer, overheads of context switches and system calls are avoided, and the arrival time-stamp closely reflects the actual arrival time, with the only non-determinism introduced being due to reception of the first bit.

Many existing synchronization algorithms for sensor networks rely on the time information obtained through the GPS to provide coarse time synchronization. The accuracy of time synchronization provided by GPS depends on the number of satellites observed by the GPS receiver. In the worst case, with only one observed satellite, GPS offers an accuracy of 1 μs [29]. However, GPS is not a suitable choice for sensor networks because GPS receivers cannot be used inside large buildings and basements, or underwater, or in other satellite-unreachable environments where sensor networks may have to be deployed.

A low-power synchronization scheme called *post facto* synchronization has been proposed by Elson and Estrin in [30] for wireless sensor networks. In this scheme, the clocks of the nodes are normally unsynchronized. When an event is observed, a synchronization pulse is broadcast by a beacon node, with respect to which all nodes normalize their time-stamps for the observation of the event. This scheme offers short-lived synchronization, creating only an "instant" of synchronization among the nodes which are within transmission range of the beacon node. The propagation delay of the synchronization pulse is assumed to be the same for all nodes.

Yoram Ofek [31] proposed a global synchronization protocol based on exchange of control signals between neighboring nodes. A node becomes a leader when elected by a majority of nodes in the network. A distributed election protocol is used which ensures the presence of a unique leader for the network. The leader then periodically sends synchronization messages to its neighbors. These messages are broadcast in turn to all nodes of the network. The time-difference bounds have been theoretically analyzed, and fault-tolerance techniques have been added to account for errors in the synchronization messages.

A long-lasting synchronization protocol is proposed in [32], which ensures global synchronization of a connected network, or synchronization within connected partitions of a network. Each node in the network maintains its own local clock (real clock) and a virtual clock to keep track of its leader's clock. A unique leader is elected for each partition in the network, and virtual clocks are updated to match the leader's real clock. The leader election process occurs as follows. On power-up, every node makes an attempt to either locate a leader in its partition or claims to be a leader itself. A node decides, with a small probability, to stake a claim for leadership and announces its claim with a random number sent on the claim packet. This *LeaderAnnouncement* packet also contains the transmission power used by the node. A node which receives this claim applies a correction for the propagation delay experienced by the claim packet (calculated based on received power), and updates its virtual clock to the expected value of the leader's real clock at that instant. Time-stamping of claims is performed at the physical layer, to avoid the variable queuing and medium access delays introduced by the MAC layer. The claim is flooded throughout the partition, bounded by a TTL field. In case two nodes within a partition stake a leadership claim, the one whose *LeaderAnnouncement* has a higher random number resynchronizes to the leader whose *LeaderAnnouncement* has the lower random number, and then rebroadcasts the *LeaderAnnouncement* of the node that generated the lower random number. In the highly unlikely case of two leaders generating the same random number, node ID is used for resolution. Periodic beaconing ensures that synchronization is maintained throughout the partition, and nodes which join it later also synchronize their clocks.

Resynchronization is the process of synchronizing different network partitions that are independently synchronized to different clocks to a common clock. In dynamic networks such as sensor networks, frequent changes in topology make resynchronization an important issue. Resynchronization takes place in situations such as the merging of two partitions due to mobility, where all clocks in a partition may need to be updated to match the leader of the other partition, as shown in Figure 12.19.

The typical TDMA superframe structure is shown in Figure 12.20. Presynch frames define the start and end of a superframe, control frames transmit control information, and data frames are the TDMA time slots allotted to the nodes involved in data transfer. A positive shift in resynchronization is defined as the transmission of a data packet at an absolute time later than the slot in the current frame structure. Negative shift is defined as advancing the start of a superframe to transmit the

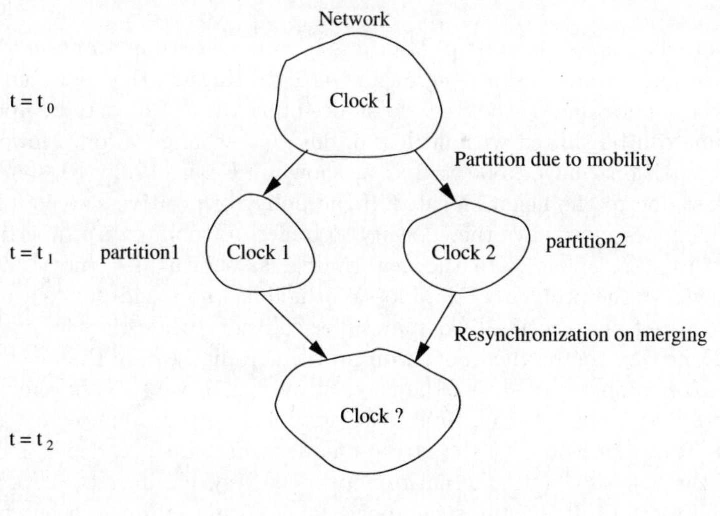

Figure 12.19. Resynchronization.

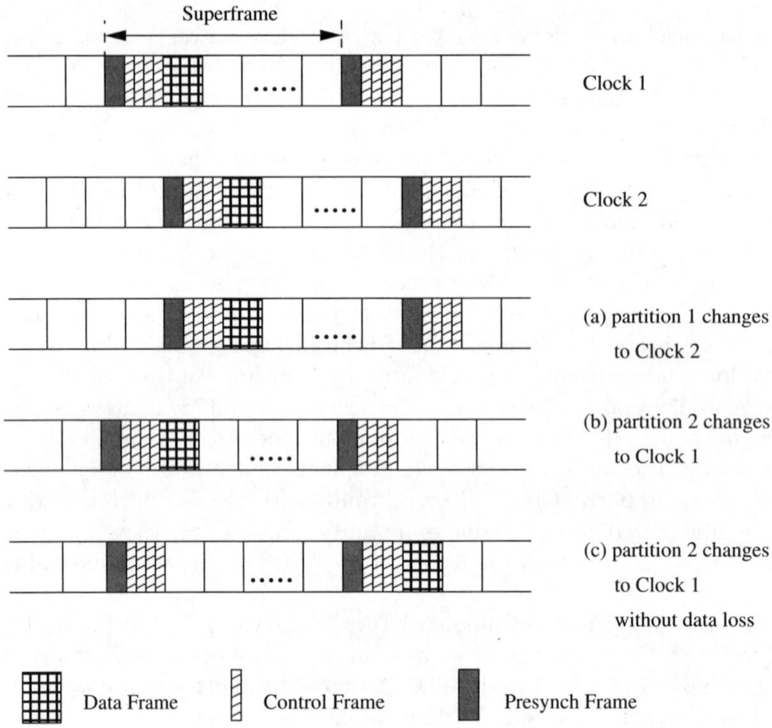

Figure 12.20. Shifting of frames on resynchronization.

data packets earlier than the start of transmission in the current frame structure. Resynchronization maintains slot assignment to routes through the node, but shifts the start of the superframe. If the clocks of nodes of partition 1 have to be updated, the superframe can be shifted without loss of data or reconfiguration. However, if the clocks of partition 2 have to be shifted, as shown in Figure 12.20 (b), some data frames are lost due to the negative shift. If the policy of positive shift is followed uniformly, the nodes must have the capacity to buffer up to an entire superframe's data packets to start afresh with the new timing, as shown in Figure 12.20 (c). Buffering alleviates the problem of data loss on the link whose end-points are being resynchronized, but neighboring links may suffer collisions when they follow different clocks. Hence, as the resynchronization proceeds radially from the new leader, there is data loss along the head of the resynchronization wave. This remains for a time period proportional to the time taken for the *LeaderAnnouncement* packet to propagate from the leader to the farthest node, which in turn depends on the diameter of the network, until the entire network is resynchronized. Also, different methods for transmitting the synchronization information have been studied. Out-of-band synchronization uses a separate control channel for sending claim and beacon packets. Collisions are reduced to a great extent for the control packets. However, the available bandwidth for data transmission is reduced, and the cost of the mobile nodes increases because of the need for an additional radio interface. In in-band synchronization, control information for synchronization shares the same channel with the data packets, as shown in Figure 12.21 (a). This leads to a greater number of collisions, but avoids an additional channel or bandwidth reservation. Piggy-backing can be used to reduce explicit control packets. Control information is piggy-backed onto outgoing data packets, as in Figure 12.21 (b). This involves very low overhead with each packet and leads to considerable bandwidth saving. A control packet carrying the synchronization information is originated only if there are no data packets to be sent from the node. The scheme can also be applied with piggy-backing on the link-level acknowledgments. In sensor networks, data usually flows from all sensors to the monitor, which is a fixed node with greater computing and power resources than the sensors. If the monitor is forced to be the leader, the synchronization information moves in the reverse direction, that is, along the link-level acknowledgments sent by the nodes for each hop of the data packets, as shown in Figure 12.21 (c). Using simulation studies, this has been observed to be the most efficient mechanism.

12.9.3 Transport Layer Issues

The major issue in transport layer protocols for sensor networks is the provision of reliable data delivery. This assumes special significance in the design of general-purpose sensor networks, where groups of nodes may need to be reconfigured or reprogrammed to suit an evolving application. This may require disseminating a

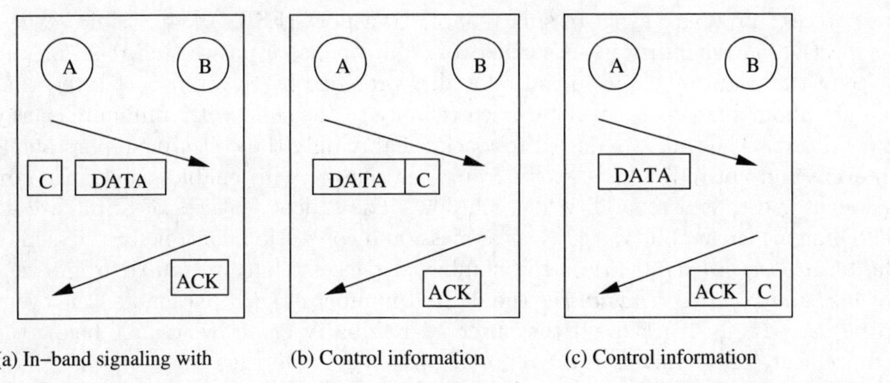

Figure 12.21. In-band signaling.

code segment to some nodes, where loss of even a single line of code would render the retasking operation a failure.

In [33], a reliable, robust, scalable, and customizable transport protocol, pump slowly fetch quickly (PSFQ), is proposed. The key concept behind the protocol is that a source node distributes data at a slow rate (pump slowly), and a receiver node which experiences data loss retrieves the missing data from immediate neighbors quickly (fetch quickly). PSFQ assumes that data loss is due to poor link conditions rather than traffic congestion. It proposes a hop-by-hop error recovery scheme, rather than holding only the destination node responsible for error detection. The overhead of requiring intermediate nodes to keep track of forwarded data is justified in sensor networks, because most transmissions are intended for groups of sensors, so intermediate nodes are also intended receivers.

PSFQ consists of three functions: message relaying (pump), error recovery (fetch), and selective status reporting (report). The pump operation disseminates data to all target nodes, performs flow control, and localizes loss by ensuring caching at intermediate nodes. Hence, the errors on one link are rectified locally without propagating them down the entire path. When a receiver detects gaps in the received sequence numbers, a loss is indicated, and it goes into fetch mode. It requests a retransmission from neighbor nodes. An attempt is made to aggregate losses, that is, many message losses are batched into a single fetch operation, which is especially appropriate for bursty losses. PSFQ supports a report operation to provide feedback on data delivery status to the source. The farthest target node initiates its report on the reverse path of data, and all intermediate nodes append their reports to the same. Hence, PSFQ ensures that data segments are delivered to all intended receivers in a scalable and reliable manner, even in an environment where the radio link quality is poor. It has been observed that the ratio between the fetch and pump rates should be around 5 for maximum effectiveness.

A recent protocol, event-to-sink reliable transport (ESRT) [34], studies a new perspective on reliability in sensor networks. It defines event-to-sink reliability in place of the traditional end-to-end reliability provided by the transport layer, that is, data about the event is to be carried reliably to the sink, with minimum energy expenditure. The sink is required to track reliably only the collective report about an event and not individual reports from each sensor. This enables a relaxation in stringent end-to-end reliability for each flow. The salient features of ESRT are its self-configuring capability, energy awareness, and congestion control. ESRT defines the term *observed reliability* as the number of packets that are routed from event to sink, and *required reliability* as the desired number of such packets for the event to be successfully tracked. If the observed reliability of an event falls below the requirement, ESRT increases the reporting frequency. On the other hand, if the reliability level required has been exceeded, ESRT decreases the reporting frequency in order to conserve energy. The frequency at which sensors must send their reports is conveyed to them through broadcasts from the sink, after appropriate calculations, so that the required reliability is achieved. Congestion control is achieved by monitoring buffer levels at forwarding sensors.

Reliability in the reverse direction, from sink to sensors, is discussed in [35]. The different kinds of reliability required in this direction are listed. On one hand, small queries are sent in a single packet, whereas software that needs to be updated in the sensors may be sent across multiple packets. Accordingly, reliability should be ensured for single or multiple packets, depending on the content. Further classification is based on the intended set of receivers. A message may need to be sent to all sensors of the network, or to all within a sub-area (as in a location-based query), or maybe to a subset of sensors which, among themselves, covers a certain area. In the last case, not all sensors in the area need to receive the message, but only a small subset, the union of whose coverage areas adds up to the required area, needs to reliably receive the message. One of the ways to ensure any of these forms of reliability is to use some nodes as recovery servers, which retransmit the message to sensors which did not receive it.

12.9.4 Security

Sensor networks, based on an inherently broadcast wireless medium, are vulnerable to a variety of attacks. Security is of prime importance in sensor networks because nodes assume a large amount of trust among themselves during data aggregation and event detection. From a set of sensor nodes in a given locality, only one final aggregated message may be sent to the BS, so it is necessary to ensure that communication links are secure for data exchange. The basic kinds of attacks on sensor networks at the network layer level have been listed in [36]. Cryptographic solutions based on symmetric or public key cryptography are not suitable for sensor networks, due to the high processing requirements of the algorithms.

Routing protocols can be affected by spoofing or altering the routing information exchanged between nodes. This can lead to errors in routing, higher latency, or even partitioning of the network. The Sybil attack [37] occurs when a single node presents itself as multiple entities to the network. This can affect the fault tolerance of the network and mislead geographic routing algorithms. Encryption and authentication using a globally shared key can prevent these attacks caused by an outsider trying to corrupt the messages in the network.

A selective forwarding attack is a situation when certain nodes do not forward many of the messages they receive. The sensor networks depend on repeated forwarding by broadcast for messages to propagate throughout the network. Sinkhole attacks are those which make a malicious node seem very favorable to the routing algorithm so that most data is routed through it. This node then performs selective forwarding, or acts as a "sink." Sensor networks are especially vulnerable to sinkhole attacks because most traffic is toward the BS. So, providing a single "favorable" route is likely to influence a large number of nodes to route their data through the malicious node. The wormhole attack lures traffic through a very long path by giving false information to the nodes about the distance between them. This increases latency by avoiding some other possible shorter paths.

Wormhole and sinkhole attacks are difficult to counter because routing information supplied by a node is difficult to verify. However, geographic routing protocols are not affected by these attacks since the routes are considered on demand using the location coordinates, so false distances can be verified.

Hello flood attacks can be caused by a node which broadcasts a *Hello* packet with very high power, so that a large number of nodes even far away in the network choose it as the parent. All messages now need to be routed multi-hop to this parent, which increases delay. This can be avoided by checking the bidirectionality of a link, so that the nodes ensure that they can reach their parent within one hop. The rest of this section deals with some protocols that have been proposed to improve security in sensor networks.

Localized Encryption and Authentication Protocol (LEAP)

Localized encryption and authentication protocol (LEAP) [38] is a key management protocol (a protocol to distribute cryptographic keys) for sensor networks based on symmetric key algorithms, that is, the same key is used by sender and receiver. In a network, requiring every pair of nodes to have a shared key to be used for communication between them is ideal for security, because an attack on any one node does not compromise the security of other nodes. However, in sensor networks, the neighbors of a node may not be known in advance, hence this sharing of keys must take place after the network is deployed, which will cause a high overhead. Also, sensor networks may employ certain processing optimizations such as a node's deciding not to report an event if it overhears its neighbor reporting the same. Such optimizations will be precluded by the usage of a separate key for each neighboring pair. On the other hand, having a common key for all nodes in the network has lower overhead, but compromise of any node affects the entire system.

LEAP uses different keying mechanisms for different packets depending on their security requirements. For instance, routing information, which is usually in broadcast mode, does not require confidentiality, whereas aggregated data sent to the BS must be confidential. Every sensor node maintains four types of keys: an individual key which it shares with the BS; a group key shared with all nodes of the network and the BS; a cluster key shared between a node and its neighbors; and a pairwise shared key with each of its neighbors. The individual key is preloaded into the node before deployment, and is used for transmission of any special information between the BS and the node, such as exclusive instructions to a node, or report from a node to BS about the abnormal behavior of a neighboring node.

It is assumed that the time required to attack a node is greater than the network establishment time, during which a node can detect all its immediate neighbors. A common initial key is loaded into each node before deployment. Each node derives a master key which depends on the common key and its unique identifier. Nodes then exchange *Hello* messages, which are authenticated by the receivers (since the common key and identifier are known, the master key of the neighbor can be computed). The nodes then compute a shared key based on their master keys. The common key is erased in all nodes after the establishment, and by assumption, no node has been compromised up to this point. Since no adversary can get the common key, it is impossible to inject false data or decrypt the earlier exchange messages. Also, no node can later forge the master key of any other node. In this way, pairwise shared keys are established between all immediate neighbors. The cluster key is established by a node after the pairwise key establishment. A node generates a cluster key and sends it encrypted to each neighbor with its pairwise shared key. The group key can be preloaded, but it should be updated once any compromised node is detected. This could be done, in a naive way, by the BS's sending the new group key to each node using its individual key, or on a hop-by-hop basis using cluster keys. Other sophisticated algorithms have been proposed for the same. Further, the authors of [38] have proposed methods for establishing shared keys between multi-hop neighbors.

Intrusion Tolerant Routing in Wireless Sensor Networks (INSENS)

Intrusion tolerant routing in wireless sensor networks (INSENS) [39] adopts a routing-based approach to security in sensor networks. It constructs routing tables at each node, bypassing malicious nodes in the network. The protocol cannot totally rule out attack on nodes, but it minimizes the damage caused to the network. The computation, communication, storage, and bandwidth requirements at nodes are reduced, but at the cost of greater computation and communication at the BS. To prevent DoS attacks, individual nodes are not allowed to broadcast to the entire network. Only the BS is allowed to broadcast, and no individual nodes can masquerade as the BS, since it is authenticated using one-way hash functions (*i.e.*, a hash function whose inverse is not easy to obtain). Control information pertaining to routing must be authenticated by the BS in order to prevent injection of false routing data. The BS computes and disseminates routing tables, since it does not face the computation and energy constraints that the nodes do. Even if an intruder takes over a node and does not forward packets, INSENS uses redundant multipath routing, so that the destination can still be reached without passing through the malicious node.

INSENS has two phases: route discovery and data forwarding. During the route discovery phase, the BS sends a request message to all nodes in the network by multi-hop forwarding (not using its broadcast). Any node receiving a request message records the identity of the sender and sends the message to all its immediate neighbors if it has not already done so. Subsequent request messages are used to identify the senders as neighbors, but repeated flooding is not performed. The nodes respond with their local topology by sending feedback messages. The integrity of the messages is protected using encryption by a shared key mechanism. A malicious node can inflict damage only by not forwarding packets, but the messages are sent through different neighbors, so it is likely that it reaches a node by at least one path. Hence, the effect of malicious nodes is not totally eliminated, but it is restricted to only a few downstream nodes in the worst case. Malicious nodes may also send spurious messages and cause battery drain for a few upstream nodes. Finally, the BS calculates forwarding tables for all nodes, with two independent paths for each node, and sends them to the nodes. The second phase of data forwarding takes place based on the forwarding tables computed by the BS.

Security Protocols for Sensor Networks (SPINS)

Security protocols for sensor networks (SPINS) [40] consists of a suite of security protocols that are optimized for highly resource-constrained sensor networks. SPINS consists of two main modules: sensor network encryption protocol (SNEP) and a micro-version of timed, efficient, streaming, loss-tolerant authentication protocol (μTESLA). SNEP provides data authentication, protection from replay attacks, and semantic security, all with low communication overhead of eight bytes per message. Semantic security means that an adversary cannot get any idea about the plaintext even by seeing multiple encrypted versions of the same plaintext. Encryption of the plaintext uses a shared counter (shared between sender and receiver). Hence, the same message is encrypted differently at different instances in time. Message integrity and confidentiality are maintained using a message authentication code (MAC). This is similar to a checksum derived by applying an authentication scheme with a secret shared key to the message. The message can be decrypted only if the same shared key is present. The message also carries the counter value at the instance of transmission (like a time-stamp), to protect against replay attacks.

μTESLA ensures an authenticated broadcast, that is, nodes which receive a packet can be assured of its sender's identity. It requires a loose time synchronization between BS and nodes, with an upper bound on maximum synchronization error. The MAC keys are derived from a chain of keys, obtained by applying a one-way function F (a one-way function is one whose inverse is not easily computable). All nodes have an initial key K_0, which is some key in the key-chain. The relationship between keys proceeds as $K_0 = F(K_1)$, $K_1 = F(K_2)$, and, in general, $K_i = F(K_{i+1})$. Given $K_0, K_1, ..., K_i$, it is not possible to compute K_{i+1}. The key to be used changes periodically, and since nodes are synchronized to a common time within a bounded error, they can detect which key is to be used to encrypt/decrypt a packet at any time instant. The BS periodically discloses the next verification key to all the nodes and this period is known to all nodes. There is also a specified lag of certain intervals between the usage of a key for encryption and its disclosure to all the receivers. When the BS transmits a packet, it uses a MAC key which is still secret (not yet disclosed). The nodes which receive this packet buffer it until the appropriate verification key is disclosed. But, as soon as a packet is received, the MAC is checked to ensure that the key used in the MAC has not yet been disclosed, which implies that only the BS which knows that yet undisclosed key could have sent the packet. The packets are decrypted once the key-disclosure packet is received from the BS. If one of the key-disclosure packets is missed, the data packets are buffered till the next time interval, and then authenticated. For instance, suppose the disclosure packet of K_j does not reach a node; it waits till it receives K_{j+1}, then computes $K_j = F(K_{j+1})$ and decrypts the packets received in the previous time interval.

12.9.5 Real-Time Communication

Support for real-time communication is often essential in sensor networks which are used for surveillance or safety-critical systems. The communication delay between sensing an intrusion and taking appropriate action greatly affects the quality of tracking provided by a surveillance system. Similarly, in a nuclear power plant, the detection of an abnormality in temperature or pressure must be conveyed in real-time to the control system in order to take immediate action. Hence, delay guarantees on routing would be extremely useful for such systems. Two protocols which support real-time communication in sensor networks — SPEED and RAP — are discussed in this section.

SPEED

A stateless protocol, SPEED, which supports real-time communication in sensor networks, has been proposed in [41]. SPEED is a localized algorithm which provides real-time unicast, real-time area-multicast (multicast to all nodes in a particular region), and real-time anycast support for packet transmission. SPEED has minimal overheads, as it does not require routing tables. It is compatible with best-effort MAC layer, not requiring any special MAC support. It also distributes traffic and load equally across the network using non-deterministic forwarding.

The SPEED protocol requires periodic beacon transmissions between neighbors. It also uses two specific on-demand beacons for delay estimation and congestion (back-pressure) detection. These are used to adapt to changes in the network. The load at a node is approximated using single-hop delay. The measurement is made using data packets which pass by a node, instead of separate probe packets. This minimizes the overhead. Nodes also respond using the delay estimation beacon to inform neighbors of the estimated delay. Routing of packets is performed by stateless non-deterministic geographic forwarding (SNGF). Using geographic information, packets are forwarded only to the nodes which are closer to the destination. Among the eligible closer nodes, the ones which have least estimated delay have a higher probability of being chosen as an intermediate node. If there are no nodes that satisfy the delay constraint, the packet is dropped.

SPEED uses a neighbor feedback loop (NFL) to maintain the estimated delay fairly constant, so that frequent updates of delay estimates are not required. When a packet has to be dropped, that is, there is no path which can meet the delay constraint, the sending rate to the downstream nodes (nodes which are closer to the receiver) is reduced to avoid congestion, thereby maintaining the delay. The NFL issues a back-pressure beacon indicating the average delay. The increased delay is noted, and SNGF accordingly reduces the probability of selecting the congested downstream nodes for routing, until eventually the congestion eases out and delay is reduced. This can continue recursively, propagating the back-pressure from downstream to upstream nodes (nodes which are closer to the sender), to relieve congestion in a hotspot.

Many geographic routing algorithms may encounter a situation when there is no node close to the destination to forward a packet. This is called a "void." SPEED uses a void-avoidance technique by issuing a back-pressure beacon with estimated delay as infinite. This will trigger a search for alternative paths. Hence, if there exists any path to a destination satisfying the delay constraint, it will be detected by SPEED. The protocol provides support for real-time communication over sensor networks by providing guarantees on the maximum delay.

RAP

RAP [42] provides APIs for applications to address their queries. An application layer program in the BS can specify the kind of event information required, the area to which the query is addressed, and the deadline within which information is required. The underlying layers of RAP ensure that the query is sent to all nodes in the specified area, and the results are sent back to the BS. The protocol stack of RAP consists of location addressed protocol (LAP) in the transport layer, velocity monotonic scheduling (VMS) as the geographic routing protocol, and a contention-based MAC scheme that supports prioritization.

LAP is a connectionless transport layer protocol which uses location to address nodes instead of a unique addressing scheme such as IP address. It supports three kinds of communication: unicast, area multicast, and area anycast. VMS is based on the concept of packet-requested velocity, which reflects both the timing and the distance constraints. Hence, requested velocity is a measure of the urgency of the packet. If a packet can travel at its requested velocity, that is, can cover the required distance within a specified time, then it can meet its deadline. VMS gives higher priority to packets which have requested higher velocities. The velocity of a packet is calculated as the ratio of the geographic distance between sender and receiver, to the deadline. Dynamic VMS recalculates the velocity at each intermediate node, so that a packet which has been slower than its requested velocity until then can be given higher priority. This is mapped onto a MAC layer priority, which is handled by the contention-based MAC layer. The protocol hence provides convenient services for application layer programs that require real-time support.

12.10 SUMMARY

Sensor networks realize an all-pervasive distributed network to create an intelligent environment. The possible applications of sensor networks are wide-ranging, from intelligent buildings and sensor-controlled chemical plants, to habitat-monitoring and covert military operations. The direction of research in sensor networks is toward overcoming the challenges of scalability, reliability, robustness, and power-efficiency, so that a variety of applications can be implemented in highly constrained scenarios.

Hardware design of sensor nodes needs to be further miniaturized, and power-efficient hardware and operating systems need to be developed. On the MAC layer, provisions need to be made for mobility of sensor nodes. New ideas for routing, network establishment, and maintenance are still in the development stage. Similarly, a transport layer protocol with limited power consumption and computational costs, and which is capable of interfacing with TCP or UDP, is still on the drawing board. Handling the sensed data, and development of application-specific query languages, would greatly help in fine-tuning the performance of sensor networks for a variety of applications. With open problems at every layer of the sensor network protocol stack, this is a highly promising concept still in its incipient stages.

12.11 PROBLEMS

1. What are the major differences between ad hoc wireless networks and sensor networks?

2. What are the advantages of a clustered architecture over a layered architecture in a sensor network?

3. Consider the third iteration of LEACH protocol. If the desired number of nodes per cluster is ten, what is the threshold calculated for a node during its random number generation?

4. In the implementation of rumor routing, an agent has the table shown in Table 12.1, and arrives at node A with Table 12.2 from node B. What will be the status of the tables after updating?

Table 12.1. Agent table

Event	Distance
E1	4
E2	2

Table 12.2. Node table

Event	Distance	Direction
E1	3	D
E2	3	C
E3	3	D

5. Consider the network topology shown in Figure 12.22. Sequential assignment routing has been applied to form trees rooted at nodes C and D. Node F originates packets of priority 2 and 3. Which routes will they take to reach the sink?

6. Consider the network topology shown in Figure 12.23. The numbers on the edges indicate the cost of the particular hop. Show the step-wise working of the cost-field approach to determine optimal paths, if the sink is D.

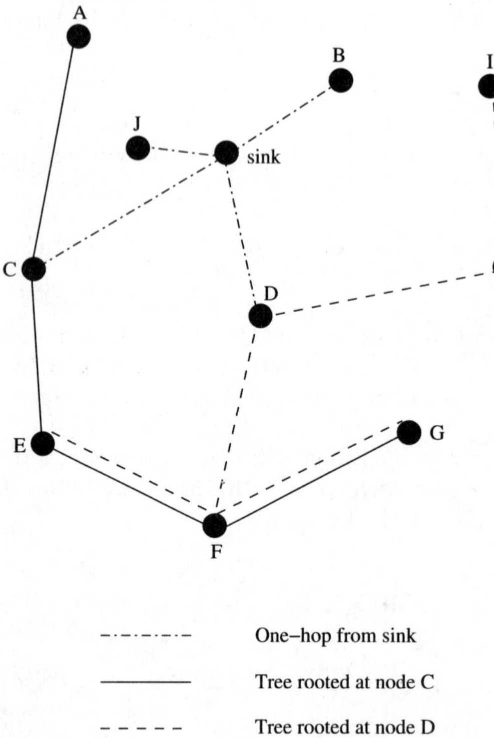

Figure 12.22. Network topology for Problem 5.

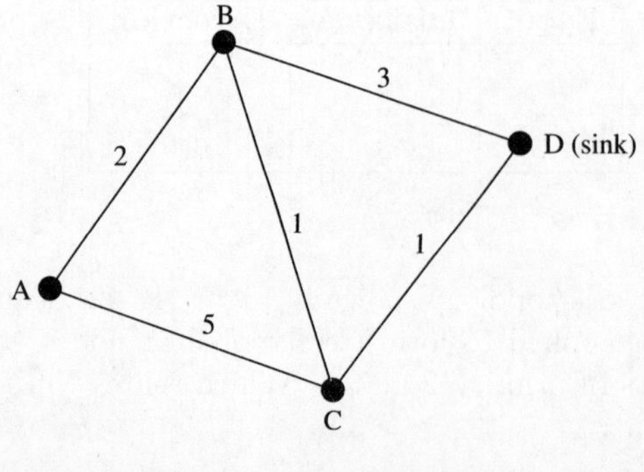

Figure 12.23. Network topology for Problem 6.

7. A network offers three paths between nodes A and B, as shown in Figure 12.24. The numbers on each link indicate (energy, delay) of the link. If the cost-field approach is followed for routing, which path will be chosen if the metric employed is (a) energy, (b) delay, and (c) energy $\times$ delay?

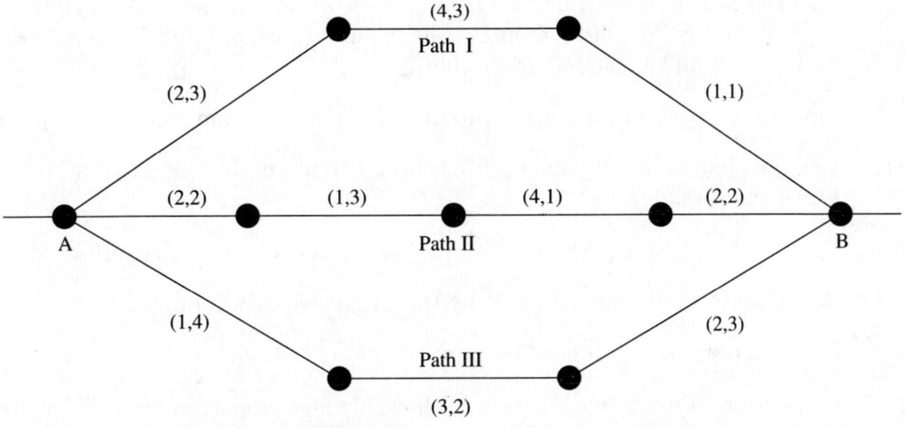

Figure 12.24. Network topology for Problem 7.

8. Consider the network topology graph shown in Figure 12.25. The numbers on the edges indicate the distance between the two end-points. SMECN is used to construct a minimum exposure path between u and v. What is the shortest path between u and v in terms of distance? What is the ME path if (a) $t = 3 \times c$? (b) $t = c$? Assume the pathloss exponent to be 2.

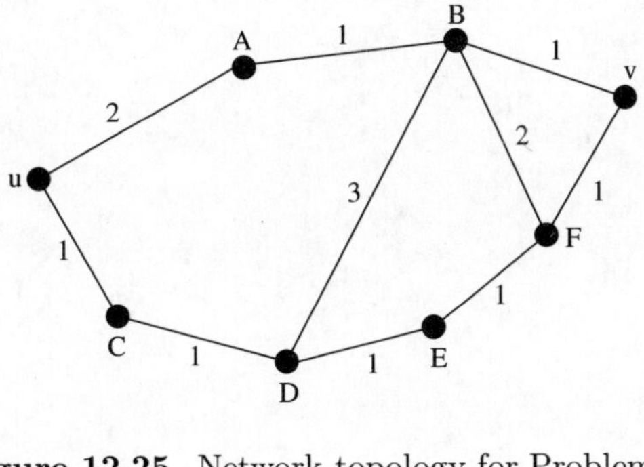

Figure 12.25. Network topology for Problem 8.

9. How does the hybrid usage of TDMA and FDMA minimize energy consumption?

10. A node X receives three beacons from nodes A, B, and C at (0, 0, 0), (2, 6, 0), and (3, 4, 0), respectively. From the received signal strengths, it determines the distances to A, B, and C to be $\sqrt{26}$, $\sqrt{2}$, and $\sqrt{5}$, respectively. Find the coordinates of X.

11. Consider sensors placed at $(3, 4)$, $(2, 5)$, $(-4, 3)$, $(1, 1)$, and $(-3, -2)$. If the parameters λ and k in the sensing power computation are 1 and 2, respectively, what are I_A and I_C at the origin (0,0)?

12. Construct a 2×2 grid of order 2 to find the generic minimum exposure path.

13. Why are public key cryptographic solutions not suitable for the security of sensor networks?

14. Describe the Sybil attack and the sinkhole attack on sensor networks.

15. What are the assumptions made by the LEAP security protocol?

16. How is authenticated broadcast provided by μTESLA?

17. In a sensor network which uses RAP for real-time support, a node X located at $(2, 3)$ receives packets as follows. A packet from a node A located at $(1, 2)$ and destined for node B at $(4, 5)$ with deadline 10 arrives eight time units after origination. Also, a packet from a node $C(0, 1)$ to node $D(5, 13)$ with deadline 10 arrives three time units after origination. If VMS is used for packet scheduling, which packet gets higher priority? If dynamic VMS is used, which packet gets higher priority?

BIBLIOGRAPHY

[1] I. F. Akyildz, W. Su, Y. Sankarasubramaniam, and E. Cayirci, "A Survey on Sensor Networks," *IEEE Communications Magazine*, vol. 40, no. 8, pp. 102-114, August 2002.

[2] J. Ding, "Design and Analysis of an Integrated MAC and Routing Protocol Framework for Large-Scale Multi-Hop Wireless Sensor Networks," *Technical Report*, Department of Computer Science and Electrical Engineering, University of Maryland, Baltimore, July 2002.

[3] "Integrated MAC and Routing Protocol Framework for Large-Scale Multi-Hop Wireless Sensor Networks," Department of Electrical Engineering and Computer Science, Washington State University, http://jaguar.eecs.wsu.edu/~jding1/presentations/poster.ppt

[4] W. Heinzelman, A. Chandrakasan, and H. Balakrishnan, "Energy-Efficient Communication Protocol for Wireless Microsensor Networks," *Proceedings of HICSS 2000*, pp. 4-7, January 2000.

[5] W. R. Heinzelman, J. Kulik, and H. Balakrishnan, "Adaptive Protocols for Information Dissemination in Wireless Sensor Networks," *Proceedings of ACM MOBICOM 1999*, pp. 174-185, August 1999.

[6] D. Braginsky and D. Estrin, "Rumor Routing Algorithm for Sensor Networks," *Proceedings of ACM Workshop on Wireless Sensor Networks and Applications 2002*, pp. 22-31, September 2002.

[7] K. Sohrabi, J. Gao, V. Ailawadhi, and G. J. Pottie, "Protocols for Self-Organization of a Wireless Sensor Network," *IEEE Personal Communications Magazine*, vol. 7, no. 5, pp. 16-27, October 2000.

[8] C. Intanagonwiwat, R. Govindan, and D. Estrin, "Directed Diffusion: A Scalable and Robust Communication Paradigm for Sensor Networks," *Proceedings of ACM MOBICOM 2000*, pp. 56-67, August 2000.

[9] F. Ye, A. Chen, S. Lu, and L. Zhang, "A Scalable Solution to Minimum Cost Forwarding in Large Sensor Networks," *Proceedings of IEEE ICCCN 2001*, pp. 304-309, October 2001.

[10] S. Ratnasamy *et al.*, "GHT: A Geographic Hash Table for Data-Centric Storage," *Proceedings of ACM Workshop on Wireless Sensor Networks and Applications 2002*, pp. 78-87, September 2002.

[11] I. Stoica, R. Morris, D. Liben-Nowell, D. Karger, M. F. Kaashoek, F. Dabek, and H. Balakrishnan, "Chord: A Scalable Peer-to-Peer Lookup Service for Internet Applications," *Proceedings of ACM SIGCOMM 2001*, pp. 149-160, August 2001.

[12] B. Y. Zhao, J. Kubiatowicz, and A. D. Joseph, "Tapestry: An Infrastructure for Fault-Tolerant Wide-Area Location and Routing," *Technical Report UCB/CSD-01-1141*, Computer Science Division, U. C. Berkeley, April 2001.

[13] B. Karp and H. T. Kung, "GPSR: Greedy Perimeter Stateless Routing for Wireless Sensor Networks," *Proceedings of ACM MOBICOM 2000*, pp. 243-254, August 2000.

[14] L. Li and J. Y. Halpern, "Minimum Energy Mobile Wireless Networks Revisited," *Proceedings of IEEE ICC 2001*, pp. 278-283, June 2001.

[15] S. Lindsey and C. S. Raghavendra, "PEGASIS: Power-Efficient Gathering in Sensor Information Systems," *Proceedings of IEEE ICC 2001*, vol. 3, pp. 1125-1130, June 2001.

[16] S. Lindsey, C. S. Raghavendra, and K. M. Sivalingam, "Data-Gathering Algorithms in Sensor Networks Using Energy Metrics," *IEEE Transactions on Parallel and Distributed Systems*, vol. 13, no. 9, pp. 924-935, September 2002.

[17] E. Shih, S. Cho, N. Ickes, R. Min, A. Sinha, A. Wang, and A. Chandrakasan, "Physical Layer Driven Protocol and Algorithm Design for Energy-Efficient Wireless Sensor Networks," *Proceedings of ACM MOBICOM 2001*, pp. 272-286, July 2001.

[18] A. Woo and D. Culler, "A Transmission Control Scheme for Media Access in Sensor Networks," *Proceedings of ACM MOBICOM 2001*, pp. 221-235, July 2001.

[19] K. M. Sivalingam, "Tutorial on Wireless Sensor Network Protocols," *International Conference on High-Performance Computing 2002*, Bangalore, India, December 2002.

[20] A. Savvides, C. Han, and M. B. Srivastava, "Dynamic Fine-Grained Localization in Ad Hoc Networks of Sensors," *Proceedings of ACM MOBICOM 2001*, pp. 166-179, July 2001.

[21] A. Nasipuri and K. Li, "A Directionality-Based Localization Scheme for Wireless Sensor Networks," *Proceedings of ACM Workshop on Wireless Sensor Networks and Applications 2002*, pp. 105-111, September 2002.

[22] Y. Shang, W. Ruml, Y. Zhang, and M. P. J. Fromherz, "Localization from Mere Connectivity," *Proceedings of ACM MOBIHOC 2003*, pp. 201-212, June 2003.

[23] T. Clouqueur, V. Phipatanasuphorn, P. Ramanathan, and K. K. Saluja, "Sensor Deployment Strategy for Target Detection," *Proceedings of ACM Workshop on Wireless Sensor Networks and Applications 2002*, pp. 42-48, September 2002.

[24] IEEE 802.15 Working Group for Wireless Personal Area Networks, www.ieee802.org/15/pub/TG4.html

[25] DARPA SensIT Program Homepage, http://dtsn.darpa.mil/ixo/sensit

[26] The official Web site of the Zigbee Alliance, www.zigbee.org/

[27] V. Raghunathan *et al.*, "Energy-Aware Wireless Microsensor Networks," *IEEE Signal Processing Magazine*, vol. 19, no. 2, pp. 40-50, March 2002.

[28] H. Kopetz and W. Schwabl, "Global Time in Distributed Real-Time Systems," *Technical Report 15/89*, Technische Universitat Wien, 1989.

[29] A. Ebner, H. Rohling, R. Halfmann, and M. Lott, "Synchronization in Ad Hoc Networks Based on UTRA TDD," *Proceedings of IEEE PIMRC 2002*, vol. 4, pp. 1650-1654, September 2002.

[30] J. Elson and D. Estrin, "Time Synchronization for Wireless Sensor Networks," *Proceedings of IEEE IPDPS Workshop on Parallel and Distributed Computing Issues in Wireless Networks and Mobile Computing 2001*, pp. 1965-1970, April 2001.

[31] Y. Ofek, "Generating a Fault-Tolerant Global Clock Using High-Speed Control Signals for the MetaNet Architecture," *IEEE/ACM Transactions on Networking*, vol. 3, no. 2, pp. 169-180, April 1995.

[32] Archana, S., B. S. Manoj, and C. Siva Ram Murthy, "A Novel Solution for Synchronization in Wireless Ad Hoc and Sensor Networks," *Technical Report*, Department of Computer Science and Engineering, Indian Institute of Technology, Madras, India, October 2003.

[33] C. Y. Wan, A. T. Campbell, and L. Krishnamurthy, "PSFQ: A Reliable Transport Protocol for Wireless Sensor Networks," *Proceedings of ACM Workshop on Wireless Sensor Networks and Applications 2002*, pp. 1-11, September 2002.

[34] Y. Sankarasubramaniam, O. B. Akan, and I. F. Akyildz, "ESRT: Event-to-Sink Reliable Transport in Wireless Sensor Networks," *Proceedings of ACM MOBIHOC 2003*, pp. 177-188, June 2003.

[35] S. J. Park and R. Sivakumar, "Sink-to-Sensor Reliability in Sensor Networks," Poster presentation at *ACM MOBIHOC 2003*, June 2003.

[36] C. Karlof and D. Wagner, "Secure Routing in Wireless Sensor Networks: Attacks and Countermeasures," *Proceedings of IEEE Workshop on Sensor Network Protocols and Applications 2003*, pp. 113-127, May 2003.

[37] J. Douceur, "The Sybil Attack," *Proceedings of IPTPS 2002*, March 2002.

[38] S. Zhu, S. Setia, and S. Jajodia, "LEAP: Efficient Security Mechanisms for Large-Scale Distributed Sensor Networks," *Proceedings of ACM Conference on Computer and Communications Security 2003*, pp. 62-72, October 2003.

[39] J. Deng, R. Han, and S. Mishra, "INSENS: Intrusion Tolerant Routing in Wireless Sensor Networks," Poster presentation at *IEEE ICDCS 2003*, May 2003.

[40] A. Perrig, R. Szewczyk, V. Wen, D. E. Culler, and J. D. Tygar, "SPINS: Security Protocols for Sensor Networks," *Proceedings of ACM MOBICOM 2001*, pp. 189-199, July 2001.

[41] T. He, J. A. Stankovic, C. Lu, and T. Abdelzaher, "SPEED: A Stateless Protocol for Real-Time Communication in Sensor Networks," *Proceedings of IEEE ICDCS 2003*, pp. 46-57, May 2003.

[42] C. Lu, B. M. Blum, T. F. Abdelzaher, J. A. Stankovic, and T. He, "RAP: A Real-Time Communication Architecture for Large-Scale Wireless Sensor Networks," *Proceedings of IEEE RTAS 2002*, pp. 55-66, September 2002.

[43] S. Archana and C. Siva Ram Murthy, "A Survey of Protocols and Algorithms for Wireless Sensor Networks," *Technical Report*, Department of Computer Science and Engineering, Indian Institute of Technology, Madras, India, July 2003.

Chapter 13

HYBRID WIRELESS NETWORKS

13.1 INTRODUCTION

Next-generation wireless systems are expected to support a wide range of advanced services which would include many that are currently supported in wired systems but are difficult to achieve in wireless environments because of resource constraints. For many of these services to be realized, an infrastructure with high data rates is necessary. The ability to support both data and voice traffic has been recognized as a desirable characteristic of future wireless systems. The phenomenal growth of the Internet and wireless connectivity has caused an explosive need for higher capacity wireless networks which can efficiently handle a variety of network loads; service highly mobile users with smooth hand-offs; offer connectivity through a variety of access points; manage both best-effort and real-time connections concurrently with QoS support for delay-sensitive applications; and above all, be extendible from the existing infrastructure to form the basis for the next-generation wireless systems. While the first-generation (1G) cellular networks mainly focused on the aspects of frequency reuse, terminal mobility, handoff techniques, and channel assignment, the second-generation (2G) wireless networks focused on improving the technology to make a much better reuse of spectrum, standardizing the technology across the world, and limited data support. The 3G wireless networks focus on efficient support for data traffic, flexible reuse of resources, seamless connection to the Internet, and support for packet-based real-time traffic. The next-generation wireless networks face new challenges in the form of increasing volume of traffic with the increase in the number of users and the average traffic generated by users. The increasing attention of the research community on these issues has resulted in hybrid wireless network architectures that combine multi-hop radio relaying and infrastructure support to provide high-capacity wireless networks. This chapter discusses the recently proposed hybrid wireless network architectures, routing, pricing, power control, and load balancing in such networks.

13.2 NEXT-GENERATION HYBRID WIRELESS ARCHITECTURES

In addition to the development of broadband physical layers, next-generation wireless networks are expected to reuse the spectrum better. In this section, some recent architectures, which exploit multi-hop relaying for efficient spectrum reuse, provide better data rates at the fringes of a cell, and support large user volumes, are described. The recent attempts at throughput enhancement in traditional cellular networks include multi-hop cellular network (MCN) [1], [2], integrated cellular and ad hoc relaying system (iCAR) [3], hybrid wireless network (HWN) architecture [4], self-organizing packet radio networks with overlay (SOPRANO) [5], multi-power architecture for cellular networks (MuPAC) [6], [7], and throughput enhanced wireless in local loop (TWiLL) [8], of which iCAR, SOPRANO, and TWiLL have direct support for real-time traffic. Extensions of MCN and MuPAC which support real-time traffic have been proposed in [2] and [7], respectively. The basic ingredient in these throughput enhancement attempts has been the introduction of ad hoc network characteristics. Reducing the power of transmission (hence, reaching the destination in multiple hops) is one of the well-known techniques to enhance the network throughput, to increase the network service area, and to enable higher data rates due to lesser interference at the receiver node. Though these hybrid architectures [1]–[8] (they are referred to as hybrid architectures because they exhibit properties of both single-hop and multi-hop wireless networks) are still confined to the research arena, they are believed to be potential candidates for the next-generation wireless systems, as mentioned in [9]. A concept group for Third-Generation Partnership Project (3GPP) has included opportunity-driven multiple access (ODMA) [10] which uses multi-hop relaying, in the initial versions of the draft standard [11], although it currently appears to have been excluded in order to clear the concerns of signaling overhead, complexity, and to achieve a finalized standard. Examples of other hybrid architectures include mobile assisted data forwarding (MADF) system [12], ad hoc-GSM (A-GSM) [13], directional throughput-enhanced wireless in local loop (DWiLL) [14], and unified cellular and ad hoc network (UCAN) [15]. The major advantages of such multi-hop architectures are reduced interference, extended coverage, broadband support over extended range, increased reliability, and support for large number of users.

13.2.1 Classification of Hybrid Architectures

Figure 13.1 illustrates a classification of the hybrid architectures. These architectures can be mainly classified as the ones which use dedicated relay stations and the ones which use host-*cum*-relay stations. Here, the dedicated relay stations do not originate data traffic on their own; rather, they assist in forwarding on behalf of the senders. The connection between the dedicated relay stations and the base station (BS) is either by wired link or by using point-to-point wireless links. These architectures can further be divided in to two categories: multi-mode systems and single-mode systems. In multi-mode systems, the mobile hosts (MHs) act either in

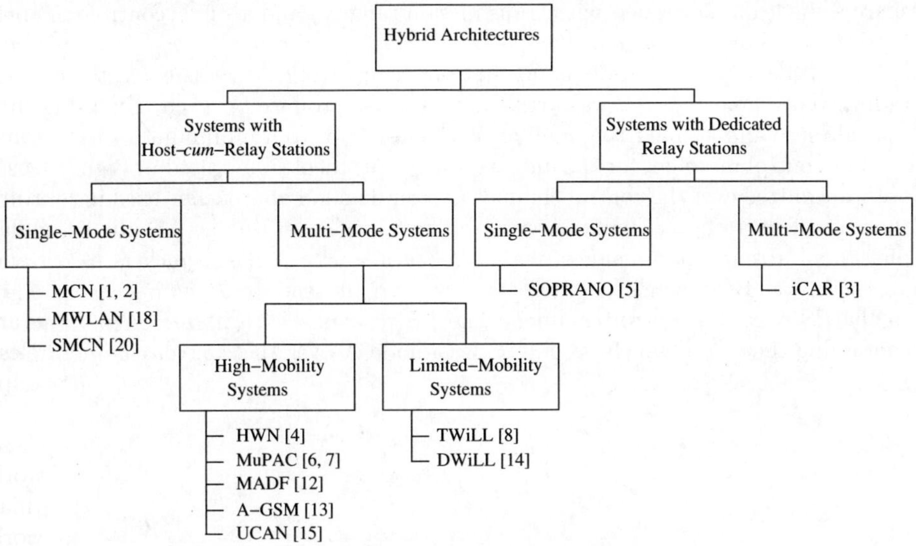

Figure 13.1. Classification of hybrid architectures.

single-hop mode or in multi-hop mode depending on the architecture. In single-mode systems, MHs operate only in multi-hop mode.

13.2.2 The MCN Architecture

Multi-hop cellular networks (MCN) [1], [2] is a novel cellular architecture where a connection between the source and the destination is established over a multi-hop path. Ananthapadmanabha *et al.* proposed a realistic architecture and provided a unicast routing protocol in [2] for best-effort and real-time traffic. MCN suggests that the transmission power of the MHs and the BS over the data channel (r) be reduced to a fraction $\frac{1}{k}$ (where k is referred to as the reuse factor) of the cell radius (R), as shown in Figure 13.2. This means that more than one node (in fact, up to a maximum of k^2 nodes) can transmit simultaneously on the same channel. The node density expected in MCN is fairly high, hence the chances of a network partition within a cell are quite small. The average path length (in terms of number of hops) for a path between the source and the destination increases linearly with k and the number of simultaneous transmissions possible increases as k^2. Hence, the throughput is expected to increase linearly with k. The actual gain will be probably lower because of the overhead of the routing protocol and the possibility of absence of relaying nodes along the straight line path. All MHs in a cell take part in the topology discovery wherein each MH regularly sends to the BS information about the beacon power received from its neighbors. This information is used by the BS to estimate distances between MHs. For best-effort communication, all the cells share a single data channel (with transmission range r) and a single control channel (with transmission range R). That is, the transmission range on the data channel is kept at half of the cell radius and that on the control channel is equal to the cell radius. The value of $k = 2$ was arrived at as a compromise between increasing the spatial reuse and keeping the number of wireless hops to a minimum.

The control messages for the unicast routing protocol (illustrated in Figure 13.2) are transmitted over the control channel. An on-demand approach is used in routing the best-effort traffic through the system. The routing protocol has a route discovery phase and a route maintenance phase. When a source A has a packet to send to a destination B to which a path is not known, it sends a *RouteRequest* packet to the BS over the control channel. The BS responds with a *RouteReply* packet containing the route, which is sent back to node A over the control channel. The

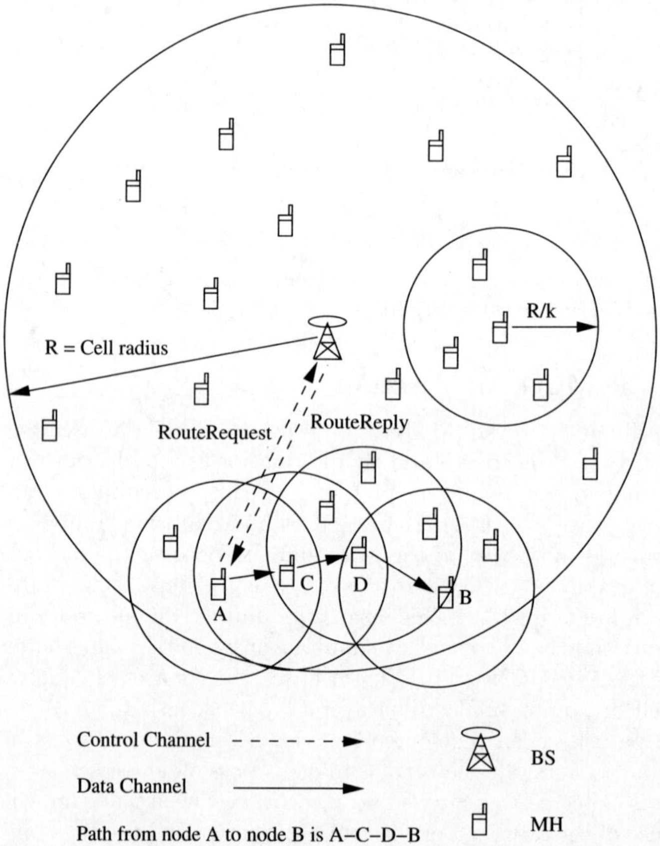

Figure 13.2. MCN best-effort architecture. *Reproduced with permission from [16], © Elsevier, 2004.*

route is computed by Dijkstra's shortest path search algorithm. The source A, upon reception of the *RouteReply* packet, transmits the data packet with the entire route information contained in it, to the next node on the path, which in turn forwards it. This path is also cached in its local route cache (RC). Subsequent packets to the same destination are source-routed using the same path until the RC entry times out or a route break is detected. When a node C detects a break or an interference on the route from node A to node B, node C sends a packet similar to a *RouteRequest* packet to the BS, which sends a new route to node A and node C.

A real-time scheme for MCN is proposed in which the available bandwidth is split into one control channel and several data channels. These data channels are not clustered among the cells and hence can be used throughout the system area. While the transmission range on the data channels is kept at half of the cell radius, that on the control channel is equal to the cell radius. The BS also chooses the data channels to be used along each wireless hop in a route. The BS then broadcasts this information (the path and the channels to be used in every hop) over the control channel. Upon receiving this information, the sender, the intermediate nodes, and the destination become aware of the call setup. If the call cannot be established, the sender is appropriately informed by a unicast packet over the control channel. The channels for each hop in the wireless path obtained are allocated through a first-available-channel allocation policy. For each hop, the channels are checked in a predetermined order and the first channel that satisfies the constraints is used. At some point in the duration of the call, the wireless path may become unusable. This may happen because of either a path break or interference due to a channel collision. The MH that detects this sends a packet, similar to the *RouteRequest* packet, called *RouteError* packet to the BS, which computes a new route and broadcasts the information to all MHs in the cell. In order to reduce the chance of call dropping, some channels are reserved for rerouting calls, that is, of N_{ch} channels, N_{ch}^r are reserved for handling route reconfiguration requests. The greater the value of N_{ch}^r, the less the probability of dropping during reconfiguration.

13.2.3 The MADF Architecture

The mobile assisted data forwarding (MADF) [12] architecture is a hybrid architecture in which a multi-hop radio relaying system is overlaid on the existing cellular networks. The main objective of this system is to dynamically divert the traffic load from a *hot cell* (highly loaded cell) to cooler (lightly loaded) cells in its neighborhood. The MHs use multi-hop relaying to transfer a part of the traffic load from the hot cell to neighboring cells. A small number of designated channels called *forwarding channels* are used to establish multi-hop paths to achieve this purpose. Figure 13.3 shows the operation of this architecture. In the figure, BS X controls the hot cell and the MHs A and P cannot complete their calls through BS X. MADF architecture permits the MH A to set up a call through MH B to BS Y. Similarly, MH P can set up a call relayed over MHs Q and R to reach the wired network through BS Z. The intermediate nodes that forward data are called *forwarding agents*. The

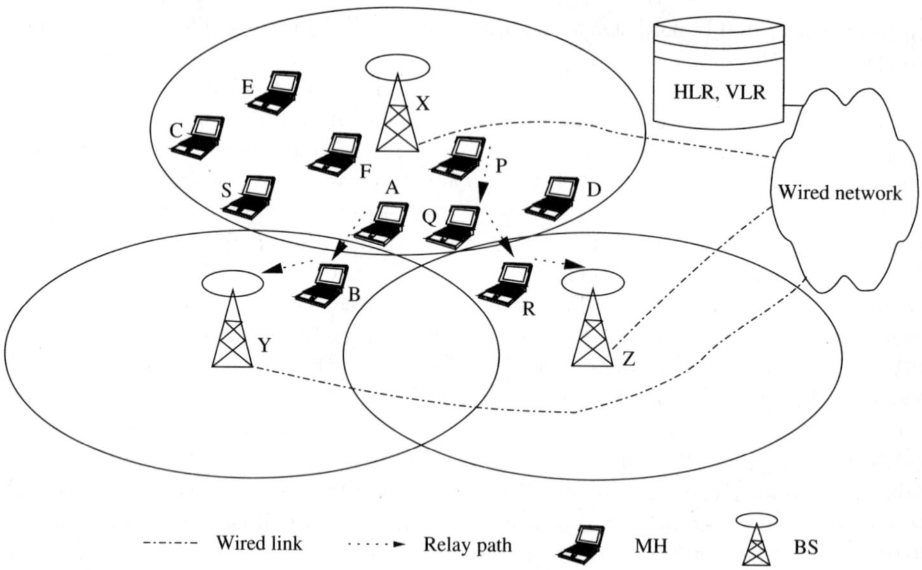

Figure 13.3. Illustration of data forwarding in MADF architecture.

forwarding agents can be fixed, semi-mobile,[1] or even fully mobile nodes. Figure 13.3 shows fully mobile relay stations. Even in a hot cell, the call acceptance rate can be improved by diverting traffic to cooler cells. The channel allocation scheme suggests a set of fixed channels and a set of forwarding channels in every cell. The number of forwarding channels is much lower than that of the fixed channels. The use of forwarding channels is entirely under the control of MHs.

The forwarding agents that are willing to forward data on behalf of other nodes measure the traffic load in the cell using the average packet delay and the number of packets transmitted per second in the cell. If the traffic load in the cell is lower than a particular threshold, then a forwarding agent broadcasts a *free signal* message in the forwarding channels, indicating its willingness to relay data packets. During the forwarding process, the forwarding agent keeps monitoring the traffic in the forwarding channel so that it can avoid admitting more calls or packets which may lead to overloading of the forwarding channels. The MHs in a cell start collecting the free signal messages from potential forwarding agents. When the traffic load in the cell exceeds a certain threshold, the MHs in that cell identify forwarding agents for making further calls. The choice of a forwarding agent is based on the signal quality experienced by the hot cell MH and the traffic in the neighboring cell. A forwarding agent stops accepting fresh relaying packets once the forwarding channels are loaded beyond a threshold. MADF architecture does not specify any

[1]Transceivers that are fixed on limited mobility structures that are generally not mobile but can be moved by some external means.

routing protocol. MHs hold the responsibility for routing the relayed packets and are expected to use one of the existing routing protocols for ad hoc wireless networks.

13.2.4 The iCAR Architecture

The integrated cellular and ad hoc relaying system (iCAR) [3] is a next-generation wireless architecture that can easily evolve from the existing cellular infrastructure. In particular, it enables a cellular network to achieve a throughput closer to its theoretical capacity by dynamically balancing the load among different cells. In a normal single-hop cellular network (SCN), which refers to the traditional cellular network, even if the network load does not reach the network capacity, several calls may be blocked or dropped because of isolated instances of congestion in the system. To counter this, the fundamental idea that iCAR deploys is to use a number of ad hoc relaying stations (ARSs) placed at appropriate locations to relay excess call traffic from a hot cell (heavily loaded cell) to relatively cooler (lightly loaded) cells around it. Thus the excess bandwidth available in some cells can be used to accept new call requests from MHs in a congested cell and to maintain calls involving MHs that are moving into a congested cell.

ARSs are wireless routing devices that can be placed anywhere in the system to route calls from one cell to another. They can communicate with a BS, another ARS, or an MH through appropriate radio interfaces. The placement of ARSs is decided by the network operator and there can be limited mobility as well. The ARS radio interfaces operate on less power than those of BSs, and hence not all MHs can set up relay routes to BSs through an ARS. A control protocol for the call management handles the coordination between MHs, BSs, ARSs, and mobile switching centers (MSCs). iCAR proposes three modes of relaying, namely, primary, secondary, and cascaded relaying, which are illustrated in Figures 13.4 (a), 13.4 (b), and 13.4 (c), respectively.

Primary Relaying

In an SCN, if an MH X in a congested cell A makes a call request and no data channel (henceforth referred to as DCH) is free at BS A, the call will be blocked. Through primary relaying, it can still set up the call by establishing a relay route with an adjacent BS B through one or more ARSs. The interfaces used for communication with BSs are called C interfaces and the ones used with ARSs are called R interfaces. Thus primary relaying involves the switching over of an MH X from the C interface to the R interface in order to communicate with the ARS.

Secondary Relaying

Since the coverage probability of an ARS in a practical environment is not 1, not all MHs will be able to set up relay routes to nearby BSs. Secondary relay capitalizes on the fact that not all MHs with on-going calls in the ARS coverage region are using primary relay. The idea is to free up a DCH from the BS A for use by MH X. Assume that an MH Y in the ARS coverage region has acquired a DCH from BS A to complete a call setup. On BS A's request, MH Y could switch over to the R interface and set up a relay route to a neighboring cool (lightly loaded)

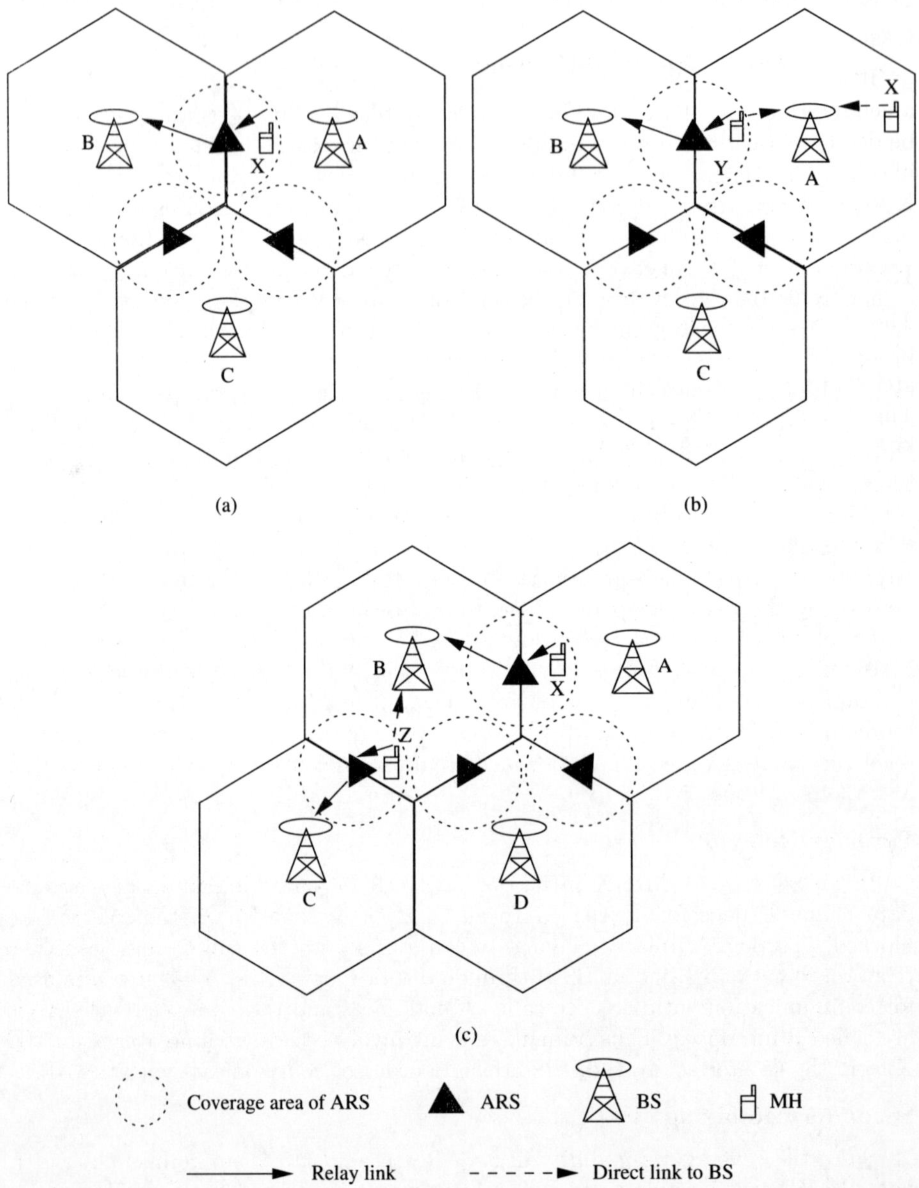

Figure 13.4. iCAR relaying strategies: (a) Primary relay (b) Secondary relay (c) Cascaded relay. *Reproduced with permission from [8], © ACM, 2003.*

cell B. Once the switch-over is complete, the DCH that has been freed can now be used by MH X.

Cascaded Relaying

It is possible that a relay path can be set up between an MH X in cell A with a neighboring BS B which is unfortunately congested. Here secondary relaying can be deployed to free a DCH in cell B by establishing a relay path between an MH Z in cell B and a neighboring cool cell C. The freed DCH can be used to connect MH X to BS B through a relay path.

13.2.5 The HWN Architecture

The hybrid wireless network architecture (HWN) [4], a multi-hop cellular architecture, operates in two modes, namely, ad hoc mode and cellular mode. This architecture has been extended to operate in multiple cells by Jayanth *et al.* [17]. This architecture requires the global positioning system (GPS), as nodes need to know their exact geographical location. In the cellular mode, nodes send packets to the BS which forwards them to the destination (similar to SCN). Figure 13.5 (a) shows the operation of HWN in the cellular mode. In this mode, the node A sends its packets to its BS B, which forwards them to the destination node C. The

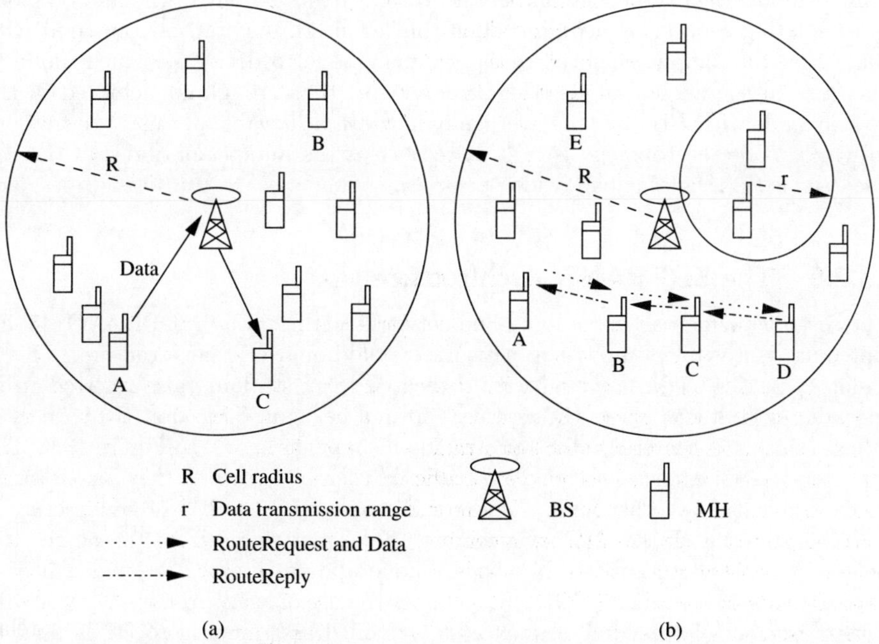

Figure 13.5. The HWN architecture (a) in cellular mode (b) in ad hoc mode.

transmission range of every node is R (the cell radius). In the ad hoc mode, nodes use dynamic source routing protocol (DSR) to discover routes. The operation of the ad hoc mode is shown in Figure 13.5 (b). The transmission range of the nodes is r, where r is chosen by the BS such that all the nodes can reach one another (no partitions will therefore occur). In this mode, the node A uses the DSR protocol to discover routes to the destination D: it floods *RouteRequest* packets which reach the destination D after being relayed by nodes B and C. Node D now sends a *RouteReply* packet which reaches the node A, which then begins to send packets along the newly discovered route. Here, the path from node A to node D does not involve the BS at all. Neither is the BS (node E in this example) necessary for finding the route. The ad hoc mode works well for dense topologies and the cellular mode is better suited to sparse topologies. An algorithm operates at the BS which uses the network topology (GPS-provided location information is sent by the nodes to the BS periodically) to decide in which mode the cell should operate so as to maximize the throughput. This decision is broadcast to all the nodes. In the ad hoc mode, partitions are avoided by having the BS periodically check the topology and broadcasting the minimum power required to keep the network connected. Studies show that HWN architecture performs better than the current generation packet data networks based on the SCN architecture. Apart from throughput, battery power consumed and fairness have also been found to be improved by HWN architecture. The switching algorithm at the BS operates as follows: If the BS is in cellular mode, then the BS estimates the expected throughput in the ad hoc mode by simulating a packet scheduling algorithm. This throughput is compared with the actual throughput achieved in the cellular mode to decide in which mode to operate. In the ad hoc mode, the BS compares the throughput achieved in the ad hoc mode with $BW/2n$ (the average bandwidth achievable per user) to find out to which mode the topology is best suited. (n is the number of nodes in the cell and BW is the available bandwidth per cell).

13.2.6 The SOPRANO Architecture

The self-organizing packet radio ad hoc networks with overlay (SOPRANO) [5] architecture is a wireless multi-hop network overlaid on a cellular structure. This is a slotted packet CDMA system with dedicated relay stations (also referred to as repeaters or routers), where the repeaters form a hexagon (placed at the corners of a hexagon that is centered at BS) or a random shape, as shown in Figure 13.6. The repeaters are not expected to generate traffic on their own; rather, they help forward traffic originated by other MHs. Neighbor discovery in SOPRANO architecture is done on powering up the MH by receiving the carrier signal from the nearest repeater. A registration process by which a node updates its location makes it easy to find a node in the cell. SOPRANO assumes the use of asynchronous CDMA with a large number of spreading sequences. A channel assignment process is used to inform every node about the channel to be used by that node. SOPRANO aims at providing high data rate Internet access by using inexpensive relay stations. Two separate frequency bands are assumed to carry the information, one each for up-

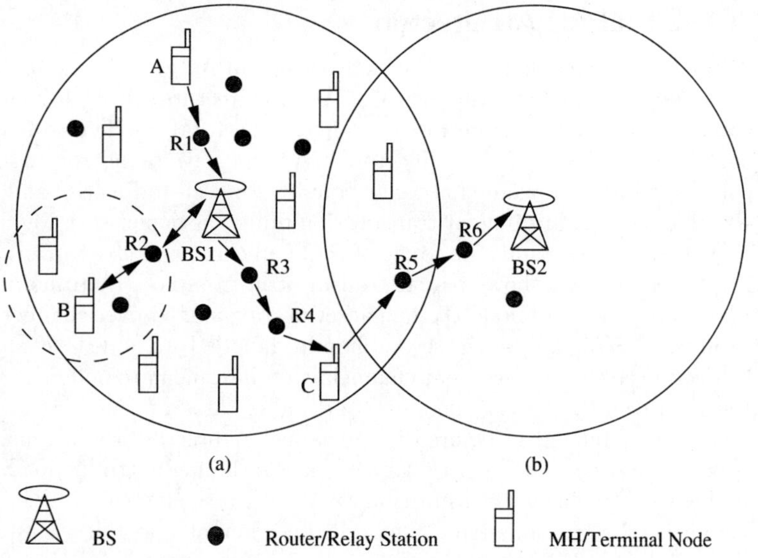

Figure 13.6. The SOPRANO architecture: load balancing through BS2 when BS1 is heavily loaded.

streams and downstreams. The upstream and downstream may include repeaters which operate in a TDD scheme wherein time is divided into slots and every slot is shared by nodes and relay stations by the CDMA scheme. MHs are assigned one channel while the routers are expected to operate in all the channels as per BS's decision. The BS instructs MHs and routers about the channels to be used for a call. An MH or router works either in transmit mode or in the receive mode. The mode of operation is decided by the BS during the call setup/path setup phase. Scheduling of the transmission slots on a path is done such that the system capacity is maximized and at the same time reduction in interference is achieved. Synchronization of the entire network is a must to achieve a collision-free TDD system. Two routing strategies were proposed for this architecture, namely, minimum path loss (MPL) and minimum path loss with forward progress (MFP). In the former case, the packets are forwarded to a receiver with minimum link propagation loss, whereas in the latter case a node sends a packet to a receiver with minimum link propagation loss along with a transmission direction toward the BS. Both the schemes are ensured to be loop-free by keeping track of the traversed path in the packet. SOPRANO can employ dynamic load balancing schemes, as shown in Figure 13.6. The downlink to the node C is BS1-R3-R4-C and if the uplink through the same path gives rise to increased interference, or if it does not have enough slots to accommodate the uplink stream, then the upstream can be completed through BS2. The upstream path now is C-R5-R6-BS2. This makes use of the lightly loaded neighborhood BSs.

13.2.7 The MuPAC Architecture

The multi-power architecture for cellular networks (MuPAC) [6], [7] is a multichannel architecture, wherein an n-channel MuPAC architecture has $n+1$ channels, each operating at a different transmission range. The total bandwidth is divided into a control channel (used for sending topology information and routing-related control packets) using a transmission range equal to the cell radius (R) and n data channels, each operating at different transmission ranges as depicted in Figure 13.7, where channel 1 uses a transmission range of $R/3$ and channel 2 uses a transmission range of $R/2$. Figure 13.8 shows packet transmission from node A to node C. On its first hop from node A to node B, the packet is transmitted over channel 1 with transmission range $R/3$, and for the second hop from node B to node C, channel 2 is used. The decision to use a particular channel at an intermediate node in a path is a local decision taken by the intermediate node and is based on the current load on each channel. The bandwidth required for a channel is directly proportional to the transmission range used. This is necessary to handle the higher traffic load offered to the channel with the higher transmission range. With the increase in the number of data channels, the transmission ranges will be different and the corresponding bandwidths will be used. A topology update mechanism is employed for the BS in order to obtain the topology graph of the network.

The path selection at the BS is done by assigning appropriate edge weights which are directly proportional to the approximate distances between the nodes. This approximate distance is measured using either GPS information or the received signal strength. Edge weight is set to ∞ if a node cannot be reached even with the highest power data channel. Dijkstra's shortest path algorithm is used to obtain a minimum weight path. Thus, the weight assignment scheme is aimed at providing

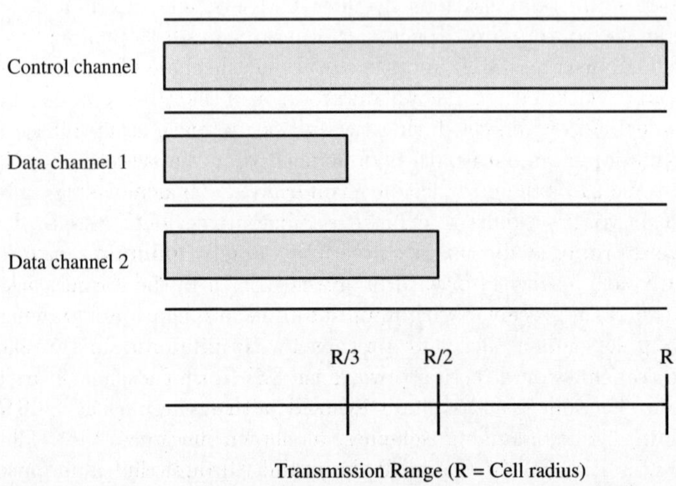

Figure 13.7. A two-channel MuPAC network.

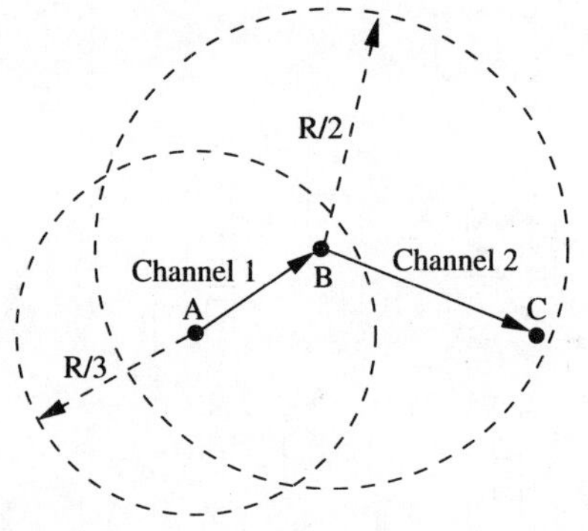

Figure 13.8. Packet transmission in a two-channel MuPAC network.

a good balance between bandwidth reuse and decreasing the total number of hops. MuPAC solves the network partition problem at low node density by using the single-hop control channel for data transmission.

13.2.8 The TWiLL Architecture

The throughput enhanced wireless in local loop (TWiLL) architecture proposed in [8] is a multi-hop architecture for limited mobility systems such as wireless in local loop (WLL), which can be used in high-density traffic environments. Henceforth, MH will also be used to refer to a WLL or TWiLL subscriber, whether it is a stationary fixed subscriber unit (FSU) or an FSU with limited mobility. The bandwidth available is split into one control channel and several data channels which are not clustered between cells. TWiLL solves the problem of network partitions by allotting a channel *ch* in single-hop mode when there is no multi-hop path to the BS. That is, in TWiLL, every channel is designated as a multi-hop channel (MC) or a single-hop channel (SC), as illustrated in Figure 13.9. An MH transmits in the control channel, SCs with a range of R (cell radius), and in the MCs with a range of $r = R/2$, thus keeping the reuse factor $k = 2$ among the MCs. The call establishment process is similar to that in MCN. To establish a call, an MH sends a *RouteRequest* packet to the BS over the control channel. The BS computes a multi-hop path and allocates MCs along the path from the MH to itself, using the same method as in MCN. If such a path cannot be obtained, then the MH is given an SC to communicate directly with the BS. The allocation of channels in single-hop mode reduces the spatial reuse of bandwidth, thus reducing the network throughput, but it will also increase the number of accepted calls when the node density is less, thus increasing the network throughput. The optimal number of SCs can be calculated as a function of the required probability of blocking for the entire system. The TWiLL architecture is illustrated in Figure 13.9. MHs A and B are connected to the BS through multi-hop paths. Node A can reach the BS over one hop while B does so over two hops. MH C is in a partition and cannot reach the

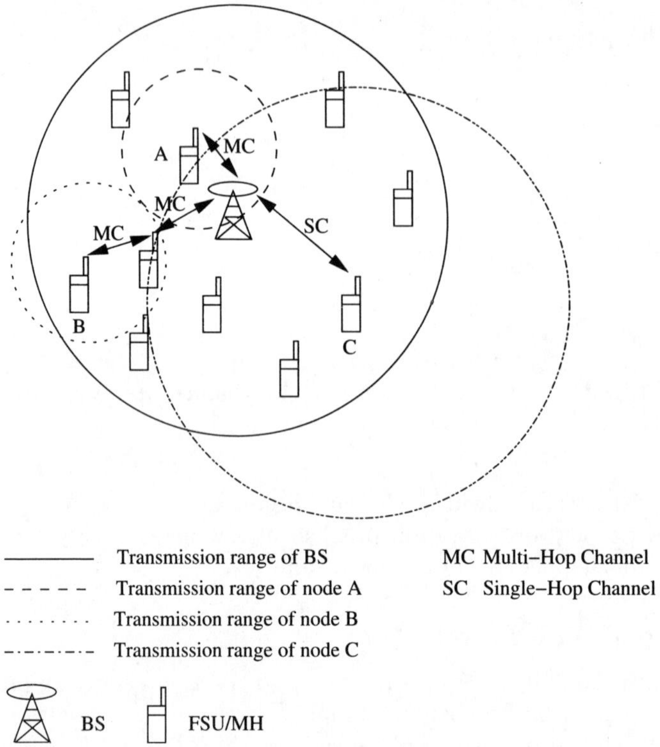

Figure 13.9. The TWiLL architecture. *Reproduced with permission from [8], © ACM, 2003.*

BS through a multi-hop path. Hence, it is allowed to use an SC to communicate with the BS in single-hop mode.

The probability that a call's destination is within the same cell as the call's source is defined as the *locality* of the system. In TWiLL, locality of traffic is used to improve the throughput by a technique called shortcut relaying. Figure 13.10 describes shortcut relaying used in TWiLL. MH A sets up a call to MH E which is present in the same cell. Under a normal WLL-like call setup, a path would be set up between MH A and the BS B and another would be set up between BS B and MH E, as shown in Figure 13.10 (a). However, this would not be very efficient since MH A might be able to reach MH E without going through the BS. This is shown in Figure 13.10 (b), where MH A sets up a two-hop path to MH E. This path setup is coordinated by the BS since MH A does not have knowledge of the network topology. When such a path is computed, the BS is also assumed as a relaying node and thus a path through the BS may also be selected if it is the optimal one. In Figure 13.10 by choosing a two-hop path (Figure 13.10 (b)) over a four-hop path

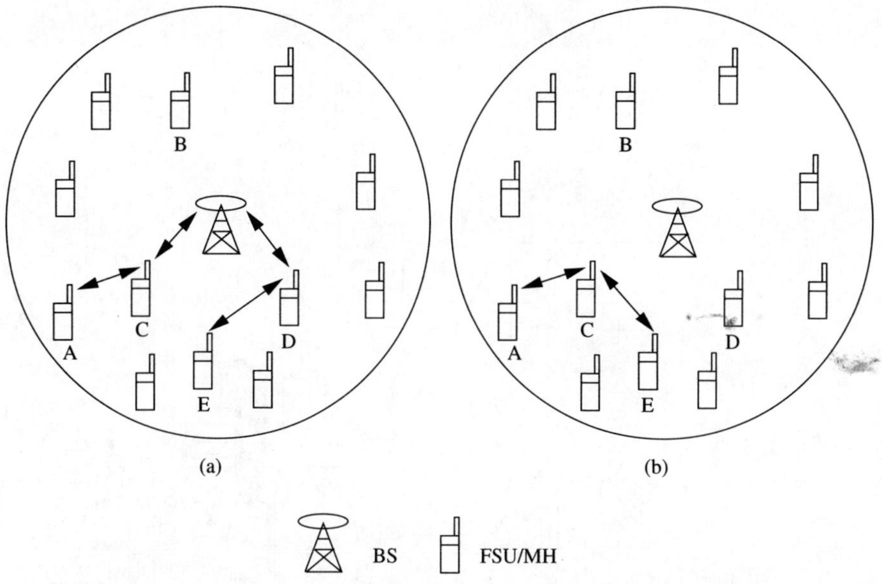

Figure 13.10. Local relaying (a) normal and (b) shortcut.

(Figure 13.10 (a)), bandwidth is conserved by shortcut relaying of local calls. Since in TWiLL, users are mostly stationary or have very limited mobility, the number of path reconfigurations will be fewer than in MCN, thus improving the quality of service. The stationary nature of subscribers also enables the use of directional antennas at the MHs, thus reducing the interference incurred at nodes in TWiLL.

13.2.9 The A-GSM Architecture

The ad hoc GSM (A-GSM) [13] architecture was proposed by Aggelou and Tafazolli as an extension to the GSM cellular architecture for providing extended service coverage to *dead spots*.[2] This is a multi-hop-relaying-enabled extension to the GSM architecture in which service coverage can be provided to the regions that are not covered otherwise. It aims to use the existing GSM system modules and entities with minimal changes for providing compatibility with existing GSM systems.

Figure 13.11 illustrates the use of A-GSM architecture. A node that is not in the coverage area of the BSs P and Q can still receive the services through relay nodes. For example, node B can make and receive calls through node A and node K can avail network services relayed through the path Q → E → F → G. Generally at the fringes of a cell, the signal strength may not be strong enough to give good coverage inside buildings located there. A-GSM can improve the service

[2]These are areas within a cell where providing service coverage is very difficult due to the geographical properties. Examples of dead spots include inside the basements of buildings and subways.

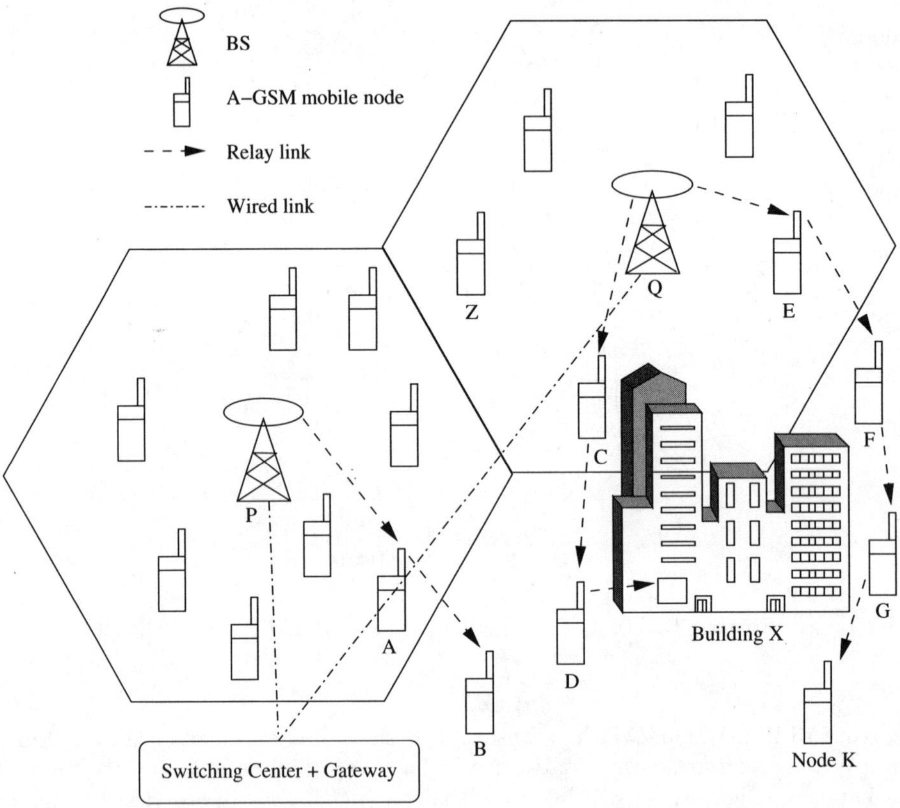

Figure 13.11. A typical scenario of A-GSM relaying.

coverage in such cases as shown in Figure 13.11 where the building X can receive better signals through the relay-enabled A-GSM node (node D). The GSM link layer protocol is modified to handle a beaconing scheme by which every A-GSM-enabled node originates beacons periodically. The important fields carried in the beacon are *Relay*, *RelayCapacity*, *LinkToBTS*, *HopsToBTS*, and *RelayToBTS*. The *LinkToBTS* flag carries information about its reachability to the BS. This flag is set to *ON* if the sender of the beacon has a direct link to the BS. If *LinkToBTS* is set to the *OFF* state, then the beacon carries information about the hop length (in *HopsToBTS*) to reach the BS and the address of the next hop field (in the *RelayToBTS*) in the beacon node to reach the BS. If a sender has no path to the BS, then it sets the *Relay* field in the beacon to –1, indicating that it has neither a direct link nor a multi-hop relay path to the BS. The *RelayCapacity* field carries information about the current level of resources available for the purpose of relaying. Every relayed call requires bandwidth, buffer space, and processing time. Hence, there is a limit to the total number of calls that an A-GSM node can relay. A resource manager module running in an A-GSM node executes a call admission control procedure for accepting or rejecting a call through the node. Once a call is admitted, the necessary resources are reserved. The *RelayCapacity* flag is set to ON to indicate that the sender node has sufficient resources to admit relay calls and is set to OFF when the node is not in a position to accept any more calls. Alternatively, an A-GSM node can avoid transmitting the beacons to indicate that the node is busy, hence implicitly conveying its inability to accept calls for relaying. When an A-GSM node that has only a relay link to the BS requires to set up a call to a destination, it sends a call request to its next-hop node to the BS. The relay node would in turn relay the message to the BS. This relaying process may take place over multiple hops. Once the BS receives the call request, it forwards it to the mobile switching center (MSC), which then sets up the call and sends an acknowledgment to the caller. This is similar to the traditional GSM network (refer to Section 3.5.1 for more details on GSM). All the intermediate nodes relaying the call request and acknowledgment reserve resources for completion of the call. Unlike GSM networks, handoffs can occur within a serving cell itself. Due to mobility and low signal strength, a node connected directly to the BS can be handed off to an A-GSM relay node if no other neighboring BSs are available. This handoff is called GSM to A-GSM handoff. Thus, A-GSM architecture defines A-GSM to A-GSM, GSM to A-GSM, and A-GSM to GSM handoffs within a serving BS. The handoff process in A-GSM requires three steps: link quality measurements, initiation or trigger, and handoff control. The link quality measurements are aimed at measuring the radio link quality between a given node and its neighbors in addition to measuring the quality of the direct link to the BS if any such direct link exists. A handoff initiation is done based on the signal strength measured at the A-GSM nodes. An A-GSM handoff (A-GSM to GSM and GSM to A-GSM handoffs) may be triggered due to failure of the serving BS, or because of change of BS, change of frequency band or time slot due to signal quality degradation, or due to the mobility of nodes or relay stations. An A-GSM to A-GSM handoff is only the change in the relay path for an A-GSM call and is triggered when the averaged signal strength from any BS does not exceed a threshold, when the averaged signal strength of the serving A-GSM node goes below a threshold, and when another neighbor A-GSM node has a signal strength greater than a threshold. A GSM to A-GSM handoff is performed when the signal strength from the serving BS falls below a certain threshold with no neighbor BS with sufficient signal strength and when the averaged signal strength from a neighbor A-GSM node is higher than the threshold. A hybrid handoff control scheme is employed for the handoff control phase. For handoffs from A-GSM call (a call that is currently being relayed over multiple hops) to GSM call (call with direct link to BS) and GSM call to GSM call (handoff between two BSs), a mobile assisted handoff scheme is employed. For a GSM call to A-GSM call or an A-GSM call to A-GSM call handoff, a mobile controlled handoff is used where the A-GSM nodes make radio measurements and handoff decisions. Simulation studies carried out in [13] using A-GSM architecture have shown that A-GSM provides better throughput in cellular environments with a high number of *dead spots*.

13.2.10 The DWiLL Architecture

Directional throughput-enhanced wireless in local loop (DWiLL) [14] is a high-performance architecture for wireless in local loop (WLL) systems. DWiLL uses the dual throughput enhancement strategies of multi-hop relaying and directional antennas. The major advantages of DWiLL include a reduction in the energy expenditure at the fixed subscriber units (FSUs)[3] and the ability to provide enhanced throughput when the number of subscribers becomes large. The WLL systems are explained in detail in Section 3.7.

The system architecture of DWiLL is similar to the TWiLL [8] architecture. The spectrum is divided into a number of channels. The key difference between TWiLL and DWiLL is the use of directional relaying by the FSUs in DWiLL. DWiLL assumes that the directional antenna at the FSU is oriented in the direction of the BS. Since there is no significant requirement for the directionality to be changed, there is no need for a sophisticated electronically and dynamically steerable antenna. The FSU uses the directional antenna to transmit control information, beacon signals, and the data messages. Also, due to directionality, the wireless link-level connectivity between two nodes is not symmetric. The system works by building the topology information at the BS as in BAAR protocol (discussed in Section 13.3.1). Each node reports the set of nodes from which it receives a beacon, along with the received power to the BS. DWiLL designates the data channels into two categories, namely, multi-hop channels (MCs) and single-hop channels (SCs). The SCs are further divided into uplink channels (ULCs) and downlink channels (DLCs). MCs operate over a transmission range of r meters, where r is a fraction of the cell radius R ($r = \frac{R}{k}$ where $k = 2$). The SCs operate over a transmission range of R meters. ULCs are assigned to those nodes that do not find intermediate relaying station to use MCs for setting up data paths to the BS. The DLCs are used by the BS for the downlink transmissions to the FSUs. Figure 13.12 shows the call setup process in DWiLL. Figure 13.12 (a) shows a unidirectional call from FSU A to FSU E. It uses multi-hop relaying from FSU A to the BS and a DLC channel on the downlink from BS to FSU E. Another example is shown in Figures 13.12 (b)(i) and 13.12 (b)(ii), where a duplex path is set up between FSU A and FSU B. On the unidirectional path $A \rightarrow B$, a single MC is allotted and the unidirectional path $B \rightarrow A$ is obtained through $B \rightarrow BS \rightarrow A$ (an MC is assigned between FSU B and BS, and a DLC is assigned between BS and FSU A).

Figures 13.13 (a)–(d) illustrate the advantages of DWiLL compared to the traditional single-hop WLL systems. Figure 13.13 (a) shows the channel allocation between FSUs B and D in a given cell. It requires two channels (one uplink channel and one downlink channel) for the $B \rightarrow BS$ and another two channels for the $BS \rightarrow D$ link. These four channels cannot be reused in the cell. In such a scenario, DWiLL can complete the call with four MCs as shown in Figure 13.13 (b). The advantage of the DWiLL system is that these MCs operate over shorter transmission ranges, making them reusable within the same cell.

[3]This refers to the equipment that resides at the premises of a subscriber and is part of fixed WLL services. In certain WLL systems, limited mobility is provided for FSUs, making the power consumption of mobile nodes an important issue as in cellular networks.

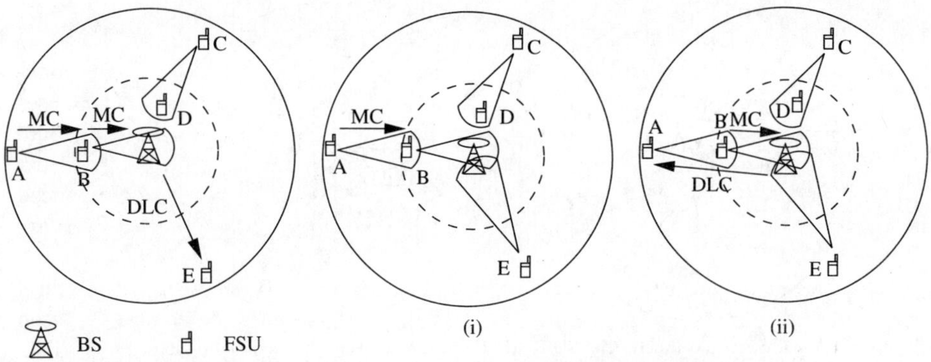

(a) Channel allocation for a unidirectional call from FSU A to FSU E

(b) A call between FSU A and FSU B
(i) A–>B is assigned an MC
(ii) B–>A requires one MC between FSU B and BS and a DLC between BS and FSU A

Figure 13.12. Illustration of call setup in DWiLL.

Figure 13.13 (c) shows another situation in the traditional WLL system, where the call between FSUs A and B requires four channels for duplex communication. For the same situation, DWiLL requires only three channels: one MC for the unidirectional link A $\rightarrow$ B, and one MC and one DLC for the B $\rightarrow$ A unidirectional path obtained through B $\rightarrow$ BS $\rightarrow$ A. Even in this case, the two MCs used can be reused within the cell.

The BS chooses the appropriate transmission mode (multi-hop/single-hop) to the FSU by means of channel selection. DWiLL assumes that the BS uses omnidirectional relaying, though in practice the BS too may be equipped with multiple directional antennas (this can reduce interference significantly and hence improve the throughput).

13.2.11 The UCAN Architecture

Unified cellular and ad hoc network (UCAN) [15] architecture is a hybrid wireless network architecture that combines wireless WANs and 802.11b-based ad hoc wireless networks. Similar to MCN [2], iCAR [3], and MuPAC [6], [7], UCAN also requires every node to have multiple radio interfaces, narrow bandwidth, high transmission range interface to directly communicate with BS and the high bandwidth, low transmission range interface for multi-hop communication. Unlike any of the previous architectures, UCAN is a technology-specific architecture, where the WAN part is a CDMA-based 1xEVDO-HDR (1xEvolution-Data Only-High Data Rate)[4] network and the ad hoc communication is based on the popular IEEE 802.11b standard. Similar to other hybrid wireless network architectures, UCAN also aims at

[4]This is one of the members of the CDMA2000 family of 3G standards.

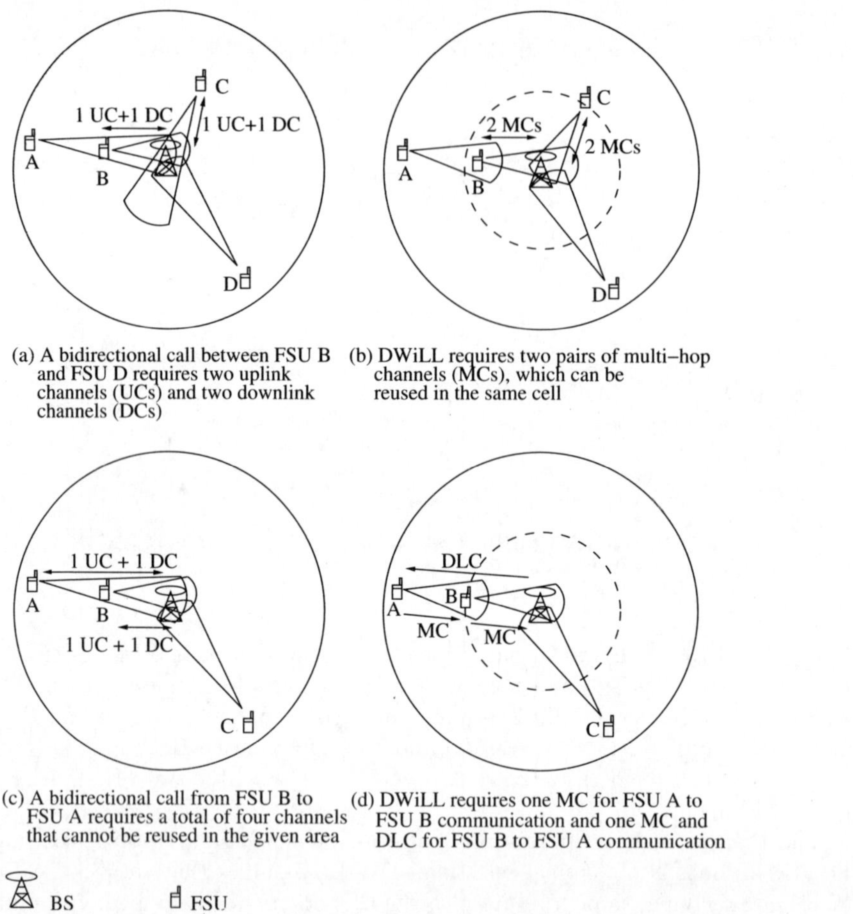

(a) A bidirectional call between FSU B and FSU D requires two uplink channels (UCs) and two downlink channels (DCs)

(b) DWiLL requires two pairs of multi–hop channels (MCs), which can be reused in the same cell

(c) A bidirectional call from FSU B to FSU A requires a total of four channels that cannot be reused in the given area

(d) DWiLL requires one MC for FSU A to FSU B communication and one MC and DLC for FSU B to FSU A communication

Figure 13.13. A comparison of DWiLL with traditional WLL systems.

enhancement of the capacity of network and improvement of the throughput for the end-user. An illustration of UCAN architecture is shown in Figure 13.14.

The BSs shown in the figure follow the 1xEVDO-HDR standard as discussed in [15]. The data rates obtained by MHs in this network vary with distance from the BS and the link quality. The bandwidth can vary between 38 Kbps and 2.4 Mbps (with early 1xEVDO-HDR networks) depending upon the link quality. MH F in Figure 13.14 is so close to the BS X that it has a strong wireless link to the BS, hence MH F achieves high throughput. The MH D is near the boundary of the cell covered by the BS Y, and hence it has a weak wireless link leading to reduction in throughput achieved. The nodes that have weak wireless links over the WAN interface (MHs B, R, and D in Figure 13.14) can make use of their 802.11b interfaces,

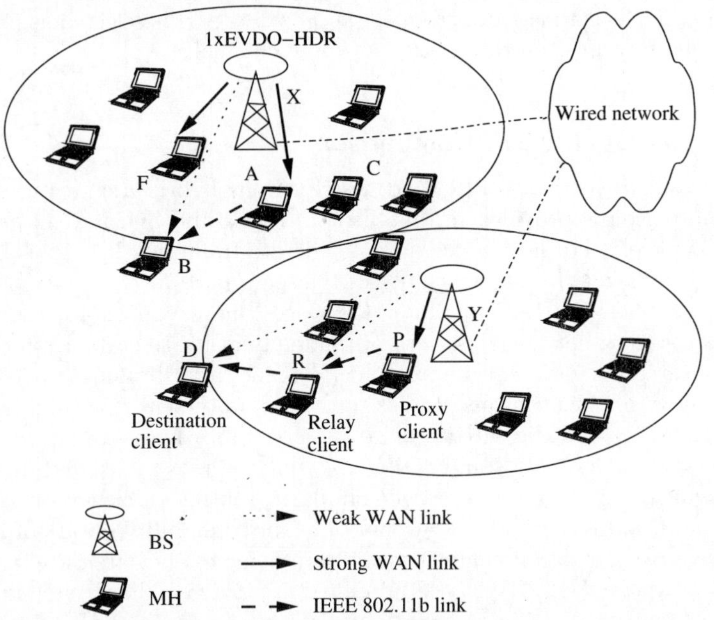

Figure 13.14. The UCAN architecture.

to improve throughput. For example, when a *destination client* (an MH that receives packets from the BS) experiences low data rates over its WAN interface, it originates *RouteRequest* packets through its 802.11b interface, in order to detect a node with a strong link to the BS (referred to as *proxy client*) by using a proxy discovery protocol. The proxy discovery protocol is a routing protocol that maintains routing state information at every intermediate node in the path so that packets to and from the BS follow the same path. Once a proxy client is identified, it registers with the BS its status as proxy client for a given destination. In Figure 13.14, the MH A acts as proxy client for MH B, and MH P acts as proxy client for MH D. Once the proxy client is registered, the BS sends further packets addressed to the destination to the proxy client and the proxy client, using the routing state information for the destination client, forwards the packets over its 802.11b interface. The intermediate nodes that relay data packets and *RouteRequest* packets are referred to as *relay clients*. The selection of proxy client is made based on the received signal quality at the WAN interface. Two approaches are used for discovery of proxy clients: a greedy routing approach and an on-demand routing approach. The greedy routing approach uses a proactive routing strategy in which every node maintains the signal quality information of all its neighbor nodes. The on-demand approach uses a reactive mechanism in which a *RouteRequest* is propagated through the network. Among the number of MHs that contend for becoming a proxy client by sending registration requests to the BS, the most appropriate MH is selected as proxy client based on the transmission rate of the downlink channel.

13.2.12 A Qualitative Comparison

Having classified (in Figure 13.1) and described the hybrid architectures, a qualitative comparison is provided in this section. iCAR and SOPRANO proposed to use dedicated relay stations which do not generate traffic on their own. iCAR uses a seed-growing algorithm [3] which places the seed ARSs at the boundary between two cells, whereas SOPRANO architecture places the relay stations either in layers forming a cocentric hexagonal pattern or randomly inside the cell. Even though the use of dedicated relay stations improves performance and simplifies the routing and resource allocation mechanisms, it adds to the cost of the entire system because the relaying hardware and the wired or wireless interconnections are expensive. Routing efficiency is high in iCAR and SOPRANO due to the presence of dedicated relay stations which do not originate traffic. Similarly, routing efficiency is high in both TWiLL and DWiLL due to the stationary or limited mobility nature of MHs.

MuPAC uses a dynamic channel switching scheme to alleviate the effects of path breaks. HWN, MCN, MuPAC, and UCAN require increased hardware requirements in terms of multiple interfaces operating at different powers and associated power sources. The cost increases with miniaturization of hardware and hence a proportional cost increase does not occur for an MH in TWiLL and DWiLL. Network partitions refer to the chances of an MH becoming isolated from its neighbors and hence unable to obtain the services of the system. iCAR, HWN, MuPAC, TWiLL, DWiLL, and UCAN handle network partitions. In iCAR, any node can either connect to the BS directly or through the nearby ARS, hence connectivity does not depend on the presence or absence of neighbor MHs. SOPRANO assumes the density of relay stations to be high enough to avoid partitions, but placing a relay station in an ideal location may not be feasible in practice. In HWN, all the nodes use transmission power that is sufficient enough to maintain the network connected in the ad hoc mode, but chances of transient partitions still exist. MCN does not handle the situation caused by the occurrence of network partitions. MuPAC handles partitions by using single-hop control channels when nodes get isolated. TWiLL and DWiLL use a novel approach of classifying channels as single-hop channels (SCs) and multi-hop channels (MCs). The use of WAN and LAN interfaces avoids network partitions in UCAN architecture. Isolated nodes are assigned a single-hop channel by the BS as and when it is required. Generally, the chances of partitions are less in TWiLL as it is a limited mobility system. Performance at high mobility is better with systems having dedicated relay stations. A novel use of channels with higher transmission range in MuPAC does not degrade the performance as much as in MCNs. iCAR, SOPRANO, TWiLL, and DWiLL envision connection-oriented systems using TDMA, FDMA, or CDMA, whereas MCNs, HWN, MuPAC, MADF, and UCAN are mainly proposed for packet-based data cellular systems. Architectural extensions using connection-oriented systems have been proposed for MCN, MuPAC, and MADF. HWN architecture proposed to use polling mechanism in its cellular mode of operation, hence it can support real-time traffic in that mode, whereas in the ad hoc mode it does not provide any such support.

The control overhead is higher for systems which use relaying by MHs (except in the case of TWiLL and DWiLL), as the instability of MHs could increase the number of control packets such as route reconfiguration packets. TWiLL is a limited mobility system which does not suffer from this, though it uses MHs for relaying.

As pricing is another important requirement for any system to survive, ease of implementation of pricing schemes is an issue in these architectures. Systems which do not use MHs for relaying do not require new pricing schemes as none of the MHs expends power for any other MHs. In cases where MHs are used for relaying traffic on behalf of other nodes, new pricing schemes are required. It is difficult to implement pricing schemes in MADF and HWN when it is operating in the relaying or ad hoc mode. It is easier to implement multi-hop pricing schemes in architectures such as MCN, MuPAC, A-GSM, UCAN, TWiLL, and DWiLL where the BS handles the path-finding and the route reconfiguration process. Provisioning of multicast routing support is easier in systems such as MADF, iCAR, UCAN, TWiLL, and DWiLL since certain single-hop channels can be designated as multicast channels. Other architectures may require extensions to ad hoc multicast routing protocols adapted to these architectures. Table 13.1 provides a qualitative comparison of all the hybrid architectures described in this section.

13.2.13 Open Issues in the Next-Generation Hybrid Architectures

This section discusses the major challenges that the next-generation hybrid architecture must address in order to meet the expectations of the emerging applications and the growing number of users.

Flexibility

Flexibility is the keyword to the next-generation wireless networks where users are provided ubiquitous access to information and data, irrespective of their physical location. Transparent handoff to various heterogeneous networks is necessary to effect this. User preferences may influence the handoff parameters and decisions. For example, while a particular user prefers to use networks based on cost of access, others might prefer to get minimum handoffs at a higher cost of operation. In the context of next-generation hybrid wireless networks, flexibility refers to the freedom provided to a user to work either in the infrastructure mode or in ad hoc mode, in the presence of infrastructure. A user can choose to work in the unlicensed band to communicate with another nearby node when he finds that he can reach the node without any support from fixed infrastructure. Since these nodes can use the unlicensed band of the radio spectrum, the cost of communication is reduced. When a user finds that the intended recipient of his information is not reachable directly, he can opt for using the support of available infrastructure by registering with one or more available service providers. Even in the presence of the service providers, one can opt for connection to the service provider either over a single wireless hop or over multiple wireless hops.

Table 13.1. Comparison of hybrid wireless architectures

Issue	MCN	iCAR	HWN	SOPRANO	MuPAC	TWiLL	A-GSM	DWiLL	MADF	UCAN
Dedicated Relay Stations	No	Yes	No	Yes	No	No	No	No	No	No
Routing Efficiency	High	High	Low	High	High	High	High	High	Low	Low
Cost of MH	High	Low	High	Low	High	Low	Low	Low	Low	High
Routing Complexity	Low	Low	High	Low	Low	Low	Low	Low	High	High
Network Partitions Handled	No	Yes	Yes	No	Yes	Yes	Yes	Yes	Yes	Yes
Expected Performance at High Mobility	Low	Good	Low	Good	Good	Not applicable	Low	Not applicable	High	Low
Connection or Packet-Based	Both	Connection	Packet	Connection	Both	Connection	Connection	Connection	Both	Packet
Real-Time Traffic Support	Yes	Yes	Cellular mode	Yes	Yes	Yes	Yes	Yes	Yes	No
Multiple Interfaces	Yes	Yes	Yes	No	Yes	Yes	Yes	No	No	Yes
Control Overhead	High	Low	High	Low	High	Low	Low	Low	High	High
Relay by MH	Yes	No	Ad hoc mode	No	Yes	Yes	Yes	Yes	Yes	Yes
New Pricing Schemes Required	Yes	No	Yes	No	Yes	Yes	Yes	Yes	Yes	Yes
Ease of Implementation of Pricing Schemes	Easy	Easy	Difficult	Easy	Easy	Easy	Easy	Easy	Difficult	Easy
Technology-Dependent	No	No	No	No	No	No	Yes	No	No	Yes

Pricing Schemes

In the traditional cellular networks, pricing is a simple problem for which straightforward solutions exist. In iCAR [3] and SOPRANO [5], the relaying is done by dedicated relay stations which are under the direct control of the BSs. Hence, existing pricing schemes can be used without major changes. In the hybrid wireless network architectures such as in [1], [2], [6], the relaying is done by the intermediate nodes which expend their own resources such as the battery charge and buffers, and hence they should be paid for the services rendered. Architectures such as HWN [4], which operate both in cellular mode and ad hoc mode, require new pricing schemes when they operate in ad hoc mode but can make use of existing schemes while operating in cellular mode. New multi-hop pricing schemes such as the one proposed in [8] are required for these architectures.

Multimedia Support

Support for multimedia traffic in hybrid wireless network architectures is still in a very early stage of research, especially for packet-based systems. Better multi-hop scheduling with or without assistance from the BS is required for this purpose. This problem is more complex in HWN architecture when it is operating in ad hoc mode.

QoS at High Mobility

Due to the smaller transmission range and frequent path breaks due to the instability of relaying nodes, further mechanisms are required to achieve good performance at high mobility. Dynamic prediction-based approaches can be used to take proactive steps to reroute calls before the actual break of the path. The path selection process can be modified to select a path which is more resilient to mobility.

Resource Management

Resource allocation is one of the major issues which takes multiple dimensions ranging from path finding to the use of a particular mode of operation. The BS, with its resources and availability of location and state information of the system, can effectively be used for this task. The following are some of the major resources which require dynamic allocation strategies:

- Channel allocation (time slot/code/frequency to be used at every hop)
- Power control (the transmission power to be used at every hop)
- Mode switching and selection (whether to work in ad hoc mode or infrastructure mode)
- Packet scheduling (reordering packets to introduce the property of fairness or priority)

Load Balancing

In the hybrid architectures, the locality of traffic (the fraction of calls originated in a cell that terminate in the same cell) influences the capacity that can be achieved by the system. The system capacity increases with increasing locality of traffic.

With low values of locality, the BS gets saturated early, affecting the maximum achievable throughput of the system. Load balancing might increase the throughput by rerouting the calls originated by the nodes at the boundary of cells through the neighborhood BSs. Also, intelligent routing can be used to avoid going through the BS even when the call is to be terminated outside the cell, in a neighboring cell. In this case, the call can be completed through multiple hops through the boundary of the cell. Load balancing is one of the primary objectives of the iCAR architecture.

Network Partitions

In TWiLL, network partitions are handled by means of allocating SC channels for those nodes in the partitions. A temporary solution in the form of making use of the single-hop control channel is provided in the MuPAC system. The use of the control channel for data transmission causes several problems, including the control channel becoming the bottleneck and loss of critical control packets which could cause severe performance degradation. Better solutions are required for packet data architectures to handle the problem of partitions.

Power Control

Power control in hybrid networks is important and more complex than in cellular networks. This is because of the multi-hop nature of the former and is essential in reducing the interference and improving efficiency of CDMA systems. New efficient power control algorithms are required for these types of networks.

Routing Protocols

Efficiency of the routing protocol is key to the performance of these hybrid systems. New routing protocols with different routing constraints, including the SNR, path diversity, etc., are required which can make use of partial or full topological information available at the BS.

Efficient Dissemination of Control Information

Registration and location information, neighbor topology updates, path maintenance packets, etc., need to be made available quickly and efficiently to the router/BS which is responsible for handling them. Aggregation of these control packets would reduce the control overhead. The assumption of a single-hop control channel simplifies this problem but demands more resources in terms of battery charge and computing power. Hence, efficient control protocols are required for the architectures in which the single-hop control channel is not used.

Path Reconfiguration Mechanisms

In order to improve the performance of the system, efficient path reconfiguration mechanisms are required. End-to-end reconfiguration as used in [2] or local reconfiguration can be used. In those architectures where dedicated relay stations are used, proactive reconfiguration mechanisms can be employed.

Support for Multicast Routing

An important requirement of next-generation wireless networks is the support for multicast routing protocols. Since these architectures use multiple short-range hops, efficient multicast routing protocols are required to support applications such as broadcast audio/video, multimedia multicast groups, and video conferencing.

13.3 ROUTING IN HYBRID WIRELESS NETWORKS

Routing in the traditional cellular networks is fairly simple and does not extend to hybrid wireless networks. The routing protocols available for ad hoc wireless networks also fail to provide good solutions for hybrid wireless networks for the following reasons:

- **Presence of BSs:** Hybrid wireless networks have extremely reliable and fixed BSs. Current ad hoc routing protocols do not exploit the BSs or fixed infrastructures whenever they exist.

- **Presence of a high-speed wired backbone:** The presence of the wired network connecting the BSs has to be considered in the design of the routing protocol. Communication among BSs is possible at high speeds and low costs. This holds for exchange of routing information also.

- **High routing overhead:** The fully distributed nature and highly dynamic topology in ad hoc wireless networks result in significant routing overhead. This affects the achieved throughput when a particular routing protocol is used. Hence, routing protocols designed for ad hoc wireless networks are not suitable in hybrid wireless networks where throughput enhancement is the primary objective. Routing in hybrid wireless networks can make use of BSs such that the routing overhead can be significantly reduced.

- **Scalability:** The current routing protocols for ad hoc wireless networks have high routing overhead and hence their performance degrades when used in a large network. In hybrid wireless networks such as MCNs, one has to provide continuous connectivity to a large number of users as in traditional cellular networks. Consequently, the requirements for MCNs are not met by these ad hoc wireless network routing protocols.

This section discusses the existing routing protocols for hybrid wireless networks, which effectively utilize ad hoc radio relaying in the presence of fixed infrastructure for enhanced network capacity.

13.3.1 Base-Assisted Ad Hoc Routing

The base-assisted ad hoc routing (BAAR) [18] protocol was proposed for the MCN architecture. It can be used in other similar architectures. The main feature of the BAAR protocol is that it efficiently makes use of the BS for routing. The BAAR protocol tries to assign as much responsibility as possible to the BSs, for reasons mentioned earlier. The BSs are a natural choice for maintaining all kinds of databases, including the location database and the topology information database. Mobile hosts (MHs) obtain information from these databases maintained at the BSs by exchanging control packets over the control channel. The computation of routes is also done at the BSs. Routing between the BS and the MS uses source routes (the overhead in the header is not a significant issue since the number of wireless hops is small).

Role of MSs and BSs

Every node is required to participate in the neighbor discovery process. Each node maintains a neighbor table based on the *hello beacons* it receives periodically every T seconds from the node in its capture area. These entries time out with a period T_n ($> T$), where T is the beacon interval. The other entries in the table include the current received power (Rx_p) and the received power last notified to the BS (Rx_{np}). Whenever there is an appreciable difference between Rx_p and Rx_{np}, new entries are added to the neighbor table or old ones removed from it. The node sends incremental updates to the BS over the control channel. Thus, the BS has up-to-date information about all the links in its cell. When a node A has packets to be sent to node B, it sends a *RouteRequest* packet for a route to the BS it is registered to. The BS then computes a suitable route using the information it has and sends back a *RouteReply* to the requesting node. The BS has to maintain the topology graph of the nodes in its service area, in a suitable data structure. It is also required to exchange this information with the BSs in its neighbor cells. This enables computation of purely wireless routes between MHs that may belong to different cells, but lie close to each other at the boundary of the cells. When a BS receives a *RouteRequest* from a node, it uses BAAR protocol to compute the route as described in the following section.

Operation of BAAR Protocol

Here $BS(x)$ denotes the BS to which node x is registered. (x, y) denotes a link between nodes x and y, and $w(x, y)$ the edge weight of the link (x, y). The operator "." stands for concatenation. Here node S is the source node and node D is the destination node.

1. *If* $BS(S) = BS(D)$,
   - *Run the shortest path algorithm over all the wireless links in the common cell.*
   - *Return the path obtained.*
2. *Else if* $BS(S)$ *and* $BS(D)$ *are BSs in adjacent cells,*
   - *Get the state of links in the adjacent cell to which node D belongs.*
   - *Run the shortest path algorithm over all the wireless links in the two adjacent cells, including the link* $(BS(S),\ BS(D))$ *with* $w(BS(S), BS(D)) = 0$.
   - *Return the path obtained.*

3. *Else*

- *Run the shortest path algorithm over all the links in node S's cell and obtain the shortest path* p1 *from source node* S *to* $BS(S)$.
- *Similarly, obtain the shortest path* p2 *from* $BS(D)$ *to* D.
- *Return* $p1.(BS(S), BS(D)).p2$.

Figures 13.15, 13.16, and 13.17 illustrate the operation of the BAAR protocol. The first case where the source S and destination D belong to the same cell is illustrated in Figure 13.15. The source node S originates a *RouteRequest* packet over the control channel to the $BS(S)$ requesting a path to the destination node D. In this case, the $BS(S)$ runs a shortest path algorithm within the nodes in the cell and returns the path to node S. Figure 13.16 shows the second case, where the source node S and destination node D belong to neighboring cells. In this case, the $BS(S)$ merges the topology graph of nodes in both the source cell and the destination cell, taking the cost of the wired path between $BS(S)$ and $BS(D)$ to be zero. This can result in two path choices: (i) path I, through the boundary of the two cells (for example path $S \rightarrow A \rightarrow B \rightarrow D$ in Figure 13.16) and (ii) path II, through the wired network (for example path $S \rightarrow H \rightarrow BS(S) \rightarrow BS(D) \rightarrow R \rightarrow D$ in Figure 13.16).

Figure 13.17 shows the third case, where the source node and destination node belong to two different non-neighboring cells. In this case, the only way to obtain

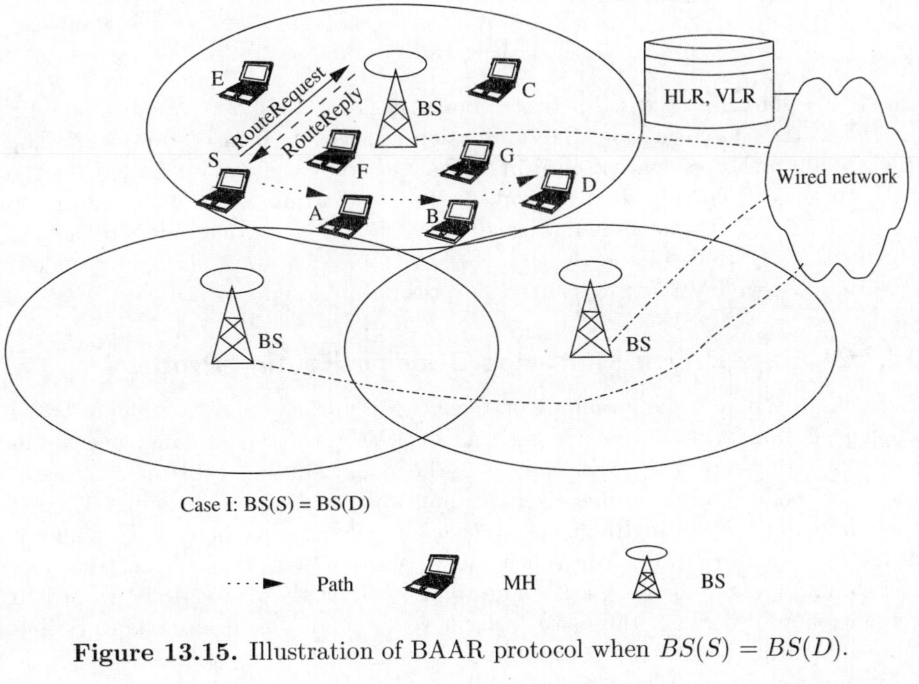

Figure 13.15. Illustration of BAAR protocol when $BS(S) = BS(D)$.

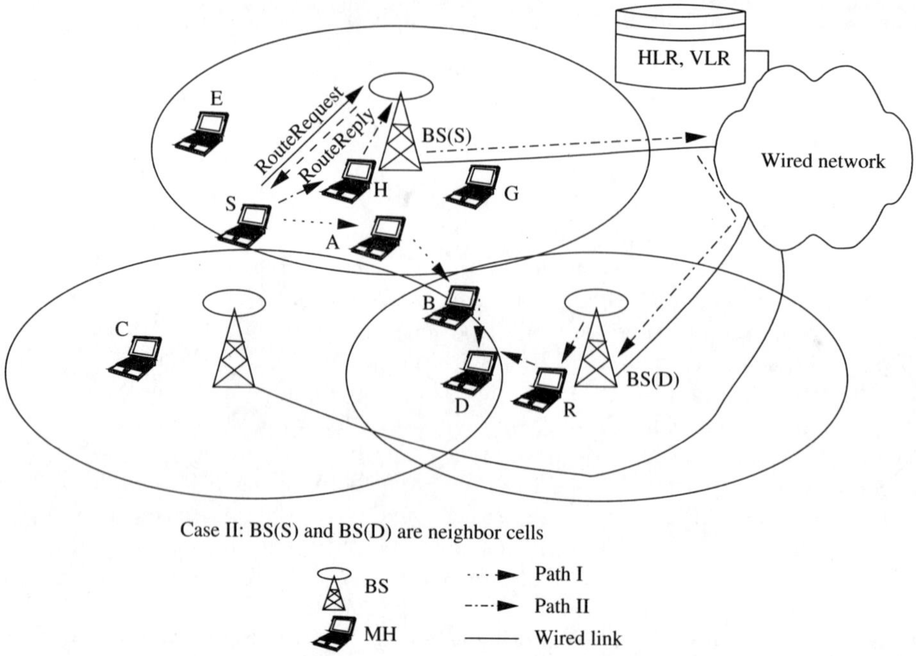

Figure 13.16. BAAR protocol routing over neighboring cells.

the path is through the wired network connecting both the BSs. A multi-hop path is used in the source cell and the destination cell as shown in Figure 13.17. The $BS(S)$ runs a shortest path algorithm within the source cell and concatenates the path obtained from the wired network for the destination cell and the multi-hop path from $BS(D)$ to the node D. Hence, the path obtained from node S to node D is $S \rightarrow H \rightarrow BS(S) \rightarrow WiredNetwork \rightarrow BS(D) \rightarrow P \rightarrow D$. The source node S uses this route information to source route the data packets.

13.3.2 Base-Driven Multi-Hop Bridging Routing Protocol

The base-driven multi-hop bridging protocol (BMBP) was proposed for a prototype implementation for multi-hop wireless LANs (MWLANs) [19]. It is implemented in such a way that a part of BMBP resides in the mobile stations (MSs) and a part in the access points (APs) in order to enable multi-hop routing and roaming. The AP that computes the routing table called *bridging table* for a particular MS is known as the *associated-AP* of the MS. When a new packet arrives at a node and the node has a routing entry for the packet's destination, then the packet is forwarded to the next hop; otherwise, the packet is sent to the node's associated-AP, possibly through multiple hops. The protocol works by building bridging tables at each node (MS and AP). In the bridge table, the destination sequence number, as proposed

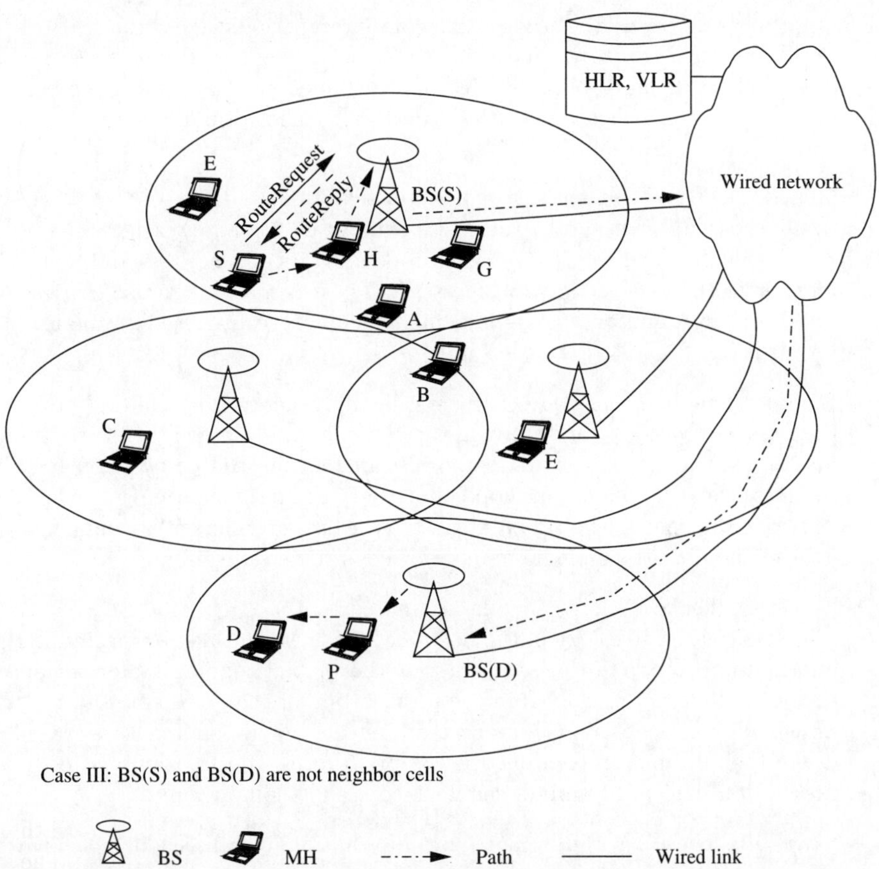

Figure 13.17. BAAR protocol routing over distant cells.

in DSDV protocol [20], is used to prevent loops in routing. The MS additionally time-stamps the entry to avoid stale entries. For each destination in the bridge table, the node records the next-hop node and the hop-count on the corresponding path to the destination node.

Messages Used in BMBP

The protocol uses four message types: *Beacon*, *Hello*, *Bridge*, and *Care-of*. The APs and MSs periodically generate the *Beacon* and *Hello* messages, respectively, to let their neighbors know of their presence. The *Beacon* is flooded to help the MSs identify their associated-APs. The *Beacon* packets are forwarded by MSs which receive them. The *Hello* messages are also broadcast messages that are relayed toward the AP through other MSs that append their own information (the address and a sequence number). The bridges are computed periodically by the AP using information obtained from the *Hello* messages and are sent to the respective MSs to update their bridging tables. Each node increments its sequence number on transmitting either a *Beacon* or a *Hello* message. The format and use of each message type are described below.

- **Beacon:** The *Beacon* message is periodically originated by an AP that wishes to advertise its presence to the MSs in its cell. An MS which receives the *Beacon* will forward the *Beacon* after appending its own address and sequence number to the received *Beacon* message. The *Beacon* packet also carries the hop-count information that is incremented by every intermediate node that forwards it.

- **Hello:** The *Hello* messages are periodically generated by the MSs toward their associated-APs. On receiving the *Hello*, each MS appends its own information as it forwards the *Hello* and also increments the *entry count* field in the message. The maximum hop-count that the *Hello* message traverses can be limited to reduce the probability of path breaks. The entry count field is used for keeping track of the number of hops traversed so far.

- **Bridge:** The *Bridge* message is originated by the AP periodically. It contains the routing topology of the network, which was formed by gathering the information from *Hello* messages about the partial topology information of the nodes that were involved in the propagation of that *Hello* message. Each *Bridge* message consists of one or more entries corresponding to every MS, where the information contained is comprised of the destination address, the destination sequence number, the next-hop, and the hop-count.

- **Care-of:** The *Care-of* message is for information interchange over the wired network that interconnects the APs. It contains the list of MSs that are currently associated with the AP and that originates the *Care-of* message.

All the messages except the *Bridge* message are transmitted in the broadcast mode. Figure 13.18 shows the direction of the different message types in BMBP.

BMBP Procedures

As part of the BMBP, specific procedures must be executed at the MSs and the APs in order to achieve multi-hop routing. Each MS keeps track of the set of APs from which it has received a *Beacon* message in an array called *data.* If the hop-count specified by the new *Beacon* is less than the hop-count to the MS's current associated-AP, then the node accepts the new AP as its associated-AP. In case the *Beacon* is from its associated-AP and the sequence number on the *Beacon* indicates a newer packet, then it updates the hop-count in its *data* array with the hop-count of the received *Beacon.* An MS will further forward a received *Hello* message only if the associated-AP field in the *Hello* message corresponds to its own associated-AP; otherwise, the MS just discards the *Hello.* The message is dropped when the hop-count field of the received *Hello* message has reached the threshold for the maximum

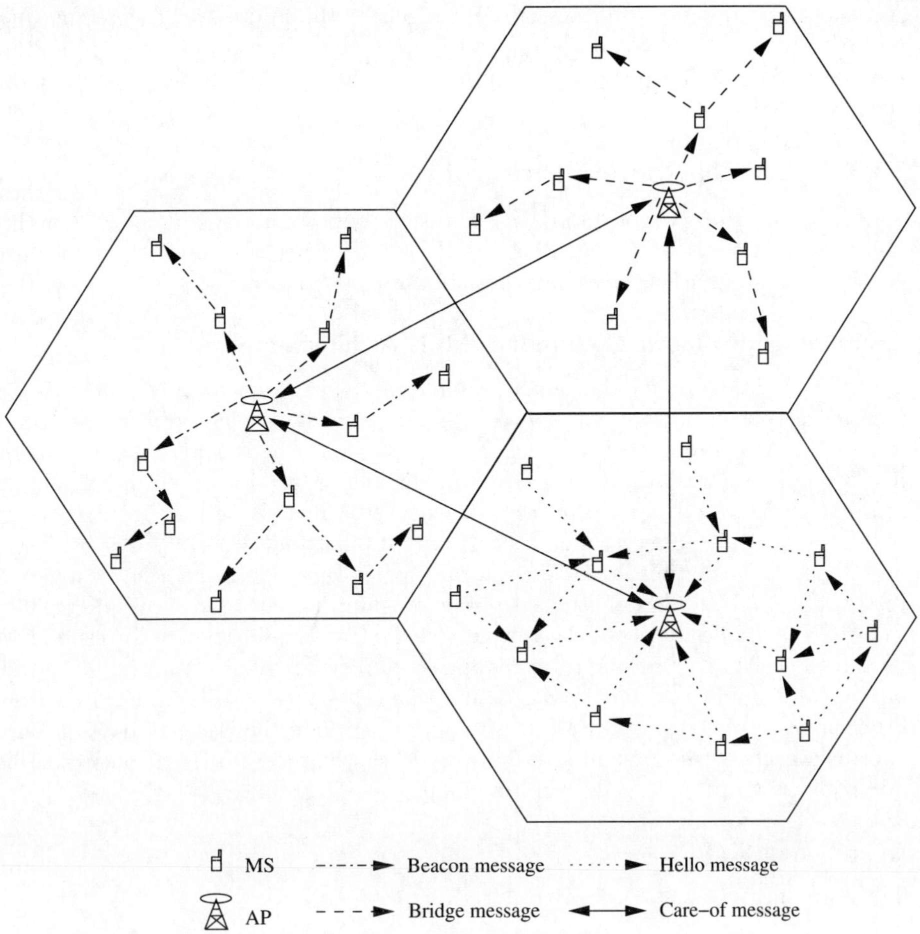

Figure 13.18. Messages used in BMBP.

number of wireless hops. When a *Hello* is received at an AP, the AP does a sanity check to verify if the *Hello* is indeed intended for it and, if so, it then proceeds to process the *Hello* message. The path taken by the *Hello* message gives the AP a partial bridge table which it computes and stores in the form of a structure called *BSTable*. If the hop-count of the newer *Hello* indicates a shorter path to the MS that originated the *Hello*, then the AP sends a *Bridge* message to the MS. When an MS receives a *Bridge* message, it checks if it is the intended destination of the message. If it is the destination, it just replaces its bridging table entries with those present in the message. Otherwise, it looks up its local bridge table to find the next hop to the destination and forwards the *Bridge* if an entry exists. The

AP, on receiving a *Care-of* message, just records the address of the remote MS (the MS that is registered to the AP which originated the *Care-of* message) and its associated-AP. An MS uses the bridging table to forward the packets it receives to the associated-AP.

13.3.3 SMCN Routing Protocol

The SMCN routing protocol (SMRP) has been proposed for single-interface multi-hop cellular network (SMCN) [21]. SMRP provides mechanisms for routing both control and data packets, possibly through multiple hops.

Issues in Routing for Single Interface MCN Architecture

The fundamental problem that is encountered in single interface MCN systems is the transfer of control information between the MSs and BSs. MCN routing protocols such as BAAR assume that all control information can be reliably sent in a single hop from or to the BS through the control channel. Also, since the control packets are sent on a separate interface, the data packets can now be transmitted with a higher success rate. Even though the success rate in a single interface system is not very high, the reduction in the success rate is not very significant when compared to the added advantages of reduced power consumption and lower cost of devices. In SMCN, all nodes transmit data using a single transmission power corresponding to half the cell radius. Using a lower range would typically result in the formation of a greater number of partitions, a situation wherein an MS gets isolated from other nodes in the network. SMRP attempts to address the issues of how to route control packets efficiently and also how an MS can find the BS that is nearest to it and register with that BS over multiple hops.

Operation of SMRP

The main messages used in SMRP are

- *Registration Request* (*RegReq*)
- *Registration Acknowledgment* (*RegAck*)
- *RouteRequest*
- *RouteReply*
- *NeighborMessage*/*Beacon*
- *NeighborUpdate* (*NeighUpdt*)

Each node (both BS and MS) periodically generates the *Beacon* messages. This message contains information regarding the set of BSs that the node has a route to and also the hop-count to each such BS. For example, when the BS generates the *Beacon* message, it will send as part of the message its own address and the hop-count to be 0. Every MS that forwards the *Beacon* increments the hop-count.

When a *Beacon* reaches another MS, it has to process the message and find the BS nearest to it on the basis of the hop-count metric. Each node will keep track of the data such as the list of BSs that are accessible to it, the hop-counts to each BS, and the next-hop nodes to reach the BS. When the *Beacon* arrives at a node, if the *Beacon* has come from an MS which is the current next-hop node to its registered BS, it will simply update the contents of its local routing data with the new routing data. Then the MS proceeds to compute the new BS to register to by finding the BS with the smallest hop-count. It also keeps track of the current next-hop node to its nearest BS. An MS registers with a new BS if it has shorter distance (in terms of the number of hops) compared to the BS with which the MS currently registered. In order to reduce the vulnerability of the path, nodes will not register to a particular BS if the hop-count exceeds a particular threshold. The MS then sends a *RegReq* to the nearest BS computed by originating the *RegReq* to its current next-hop node to that BS. Each MS has the additional responsibility of forwarding such control packets on behalf of other nodes. Further, as the *RegReq* is being propagated toward the BS, each node will append its address into the *Path* field of the *RegReq* packet to facilitate routing of the *RegAck* packets. When the *RegReq* has reached the intended BS, the BS then generates a *RegAck* to be sent to the MS that originated the *RegReq*. *RegReq* follows the path specified in the *RegReq* packet. The *RegAck* proceeds in the reverse route of the *RegReq*, with the path being copied into the *RegAck* so that each MS knows where to forward the *RegAck*. An MS is said to have completed the registration when it receives the *RegAck*; only then can the MS participate in routing and data transmission.

Each MS will receive the *Beacon* from its neighbors and will record the received power of the *Beacon* in a neighbor table. If the difference between the received power of the newly received *Beacon* and the previously recorded power exceeds a particular threshold, the MS will have to send a *NeighUpdt* message to the BS informing it of the changes in the neighborhood topology. Each MS also checks periodically to see if a *NeighUpdt* message needs to be sent to the BS. Along with the received power, the MS also time-stamps the entry with its local system clock. If it does not receive another *Beacon* from the same neighbor within a prescribed time interval, then it concludes that the neighbor has moved too far away from it, and indicates this to the BS with some large negative value in the *NeighUpdt*. During the periodic neighbor table check that an MS performs, it also checks if any of the nodes that it has recorded as a next-hop node to the BS have moved away from it, that is, either the received power is negligibly small or no *Beacon* has been received. In such a case, it has to update the local routing table to indicate the new status. The *NeighUpdt* is forwarded to the BS in much the same way as the *RegReq*, in the sense that each MS will forward the *NeighUpdt* to the next hop toward the BS. The BS uses the neighbor information to determine the current network topology. It uses an *incidence matrix* to keep track of the received powers between neighboring nodes, and this is used as an approximate distance estimate for the routing process. Figure 13.19 shows the routing of control packets over multiple hops in SMRP. Under a high-mobility scenario, the next-hop information that each MS maintains may become obsolete, but the situation is expected to stabilize within

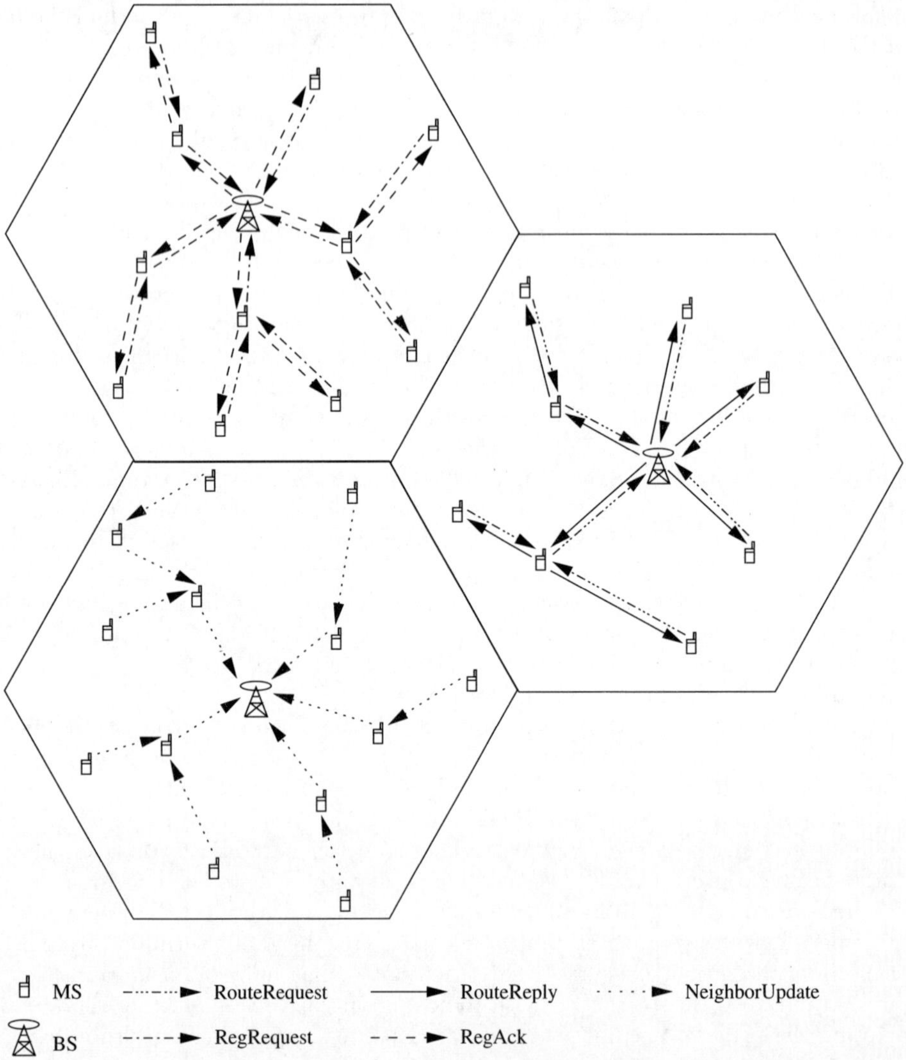

Figure 13.19. Illustration of messages used in SMCN routing protocol (SMRP).

a time interval equal to the period of the *Beacon* message, as the MS will receive the current topology information through the *Beacons* from its neighbors. An MS does not have any routing information locally available except the address of the next-hop node to the registered BS for transmitting control information.

Whenever a packet arrives from the higher layers, the MS will send a *RouteRequest* to its BS requesting the BS for a path to the destination. Based on the topology information it has, the BS finds out the shortest path to the destination and sends it to the requesting node through a *RouteReply* packet. The route for a destination MS, residing within the cell, may take multiple hops without passing through the BS, similar to the BAAR protocol. This can lead to a reduction in load at the BS. A BS is assumed to know the set of MSs that are registered with other BSs; this information can be transmitted over the wired network that interconnects the BSs. The *RouteRequest* and *RouteReply* propagation mechanism is identical to the registration mechanism. Once the *RouteReply* arrives at the source MS, it will start sending packets to the destination, using the source routing mechanism. The destination can optimize the routing efficiency to a certain extent by copying the route through which it received the packet, to reduce the number of *RouteRequest* packets that it may have to generate later. If, at an intermediate node, it is discovered that the route has become stale, then a new *RouteRequest* is generated to perform the route reconfiguration. The frequency of such route updates is dependent on the degree of mobility; at low mobility, the number of route reconfigurations will be minimal. The reconfiguration can be performed either locally by the node that discovered the path break, in which case the sender is unaware of the route change, or it can be performed on an end-to-end basis with the *RouteReply* reaching the source of the data packet.

13.3.4 DWiLL Routing Protocol

DWiLL routing protocol (DRP) [14] was proposed for the DWiLL architecture discussed in Section 13.2.10. It is different from the routing protocols used in traditional WLL systems and the TWiLL architecture since DWiLL architecture uses multi-hop relaying with directional antennas. Due to the directionality, the wireless link-level connectivity between two nodes is not symmetric. DRP works by building the topology information at the BS as in BAAR protocol. Each node reports the set of nodes from which it receives a *Beacon*, along with the received power of the *Beacon* to the BS. The *Beacon* packets are received with a certain power high enough to cause interference at nodes that are within the directional sector of radius $r(1+\alpha)$ and angle $\theta(1+\beta)$, where r is the multi-hop channels' transmission range (equal to half the cell radius), and θ is the azimuth of the directional antenna (the angular extent of transmission measured with respect to the radial line from the BS to the FSU), and α and β are the interference parameters as illustrated in Figure 13.20. The BS builds up two topology matrices: the multi-hop connectivity graph (MCG) g and the single-hop connectivity graph (SCG) G. The BS also builds two interference matrices: *r-interference matrix* for multi-hop channel interference and the *R-interference matrix* for single-hop channel interference. These interference matrices contain the interference information that every channel in its category (multi-hop channel or single-hop channel) experiences at every node with a significant level of interference such that the channel cannot be used at that node. The r-interference matrix and the R-interference matrix contain information regarding the interference on all the channels used in MCG and SCG, respectively. For an omnidirectional antenna system, the incidence matrix can be assumed to be symmetric, but this is not the case in the DWiLL architecture. The region in which a signal can be successfully received with a multi-hop channel is r meters in the direction of the receiver node and at an angle θ from the center line (refer to Figure 13.20). The region affected by electromagnetic interference from a multi-hop channel is extended to a distance of $r(1 + \alpha)$ radially and at an angular deviation of $\theta(1 + \beta)$, as shown in Figure 13.20. Any channel that is free of interference at a given link is considered usable in that link. Figure 13.21 (a) shows the sample topology in a cell. The MCG g and SCG G are shown in Figures 13.13 (b) and 13.13 (c), respectively for the topology in Figure 13.13 (a).

Whenever a source node needs to set up a call session with another node, it requests the BS by sending a *RouteRequest* packet over the single-hop control channel. The BS in turn uses the DRP to obtain a path to the destination node (this path may or may not include the BS), marks the necessary channels at every intermediate link, and replies with a *RouteReply* packet to the sender node. In DRP, a duplex path from a source node S to any destination node D is split into two simplex paths, that is, from node S to node D, the *forward simplex path* and from node D to node S, the *reverse simplex path.* Each simplex path is further divided into two parts, that is, the forward simplex path is divided into (i) uplink from S

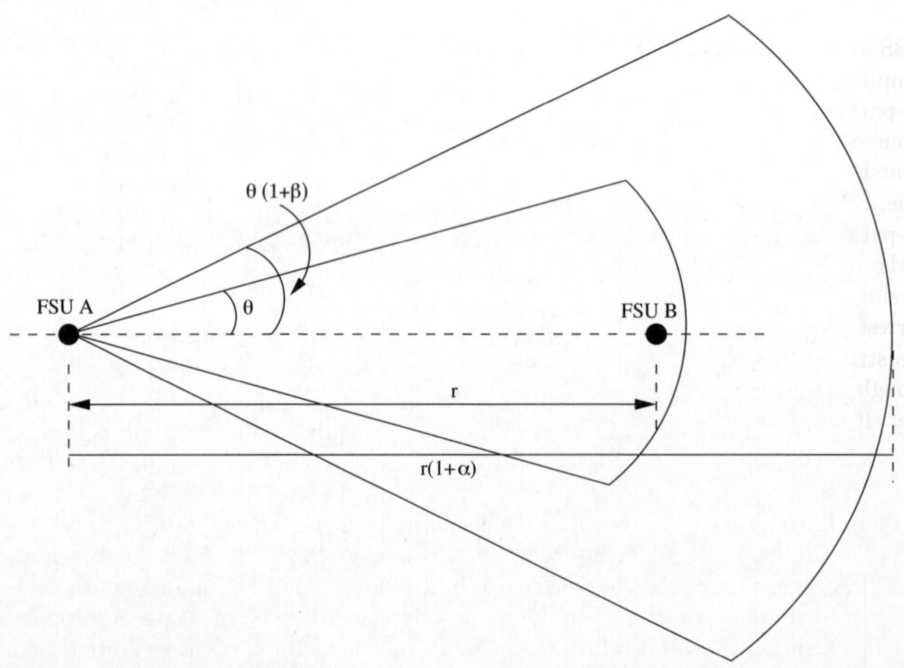

Figure 13.20. A simplified view of the interference regions in a directional transmission system.

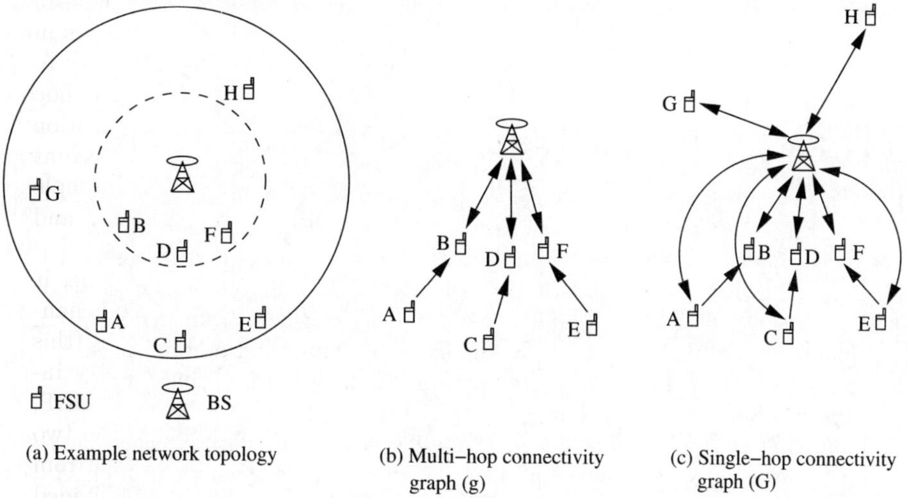

Figure 13.21. The connectivity graphs used for routing in DWiLL.

to BS and (ii) downlink from BS to D, and the reverse simplex path is divided into (i) uplink from D to BS and (ii) downlink from BS to S. These paths are called sub-paths and are listed in Table 13.2. The sub-paths shown in Table 13.2 apply to a source-destination pair residing in the same cell. In cases where the destination is outside the cell, for an outgoing call, only sub-path 1 and sub-path 4 are applicable. Similarly, for an incoming call from a node that resides outside the cell, only sub-path 2 and sub-path 3 are applicable. The BS runs a shortest path algorithm on the MCG data structure for a simplex forward path from node S to node D. If the simplex forward shortest path from node S to node D is not available, then a shortest path algorithm is executed for the sub-path 1 and sub-path 2 in the MCG data structure. If the shortest path for sub-path 1 and sub-path 2 cannot be found through MCG, then a shortest path scheme is attempted in the SCG data structure. If the forward path-finding process is successful, then similar path finding

Table 13.2. Sub-paths used in DWiLL routing protocol

Sub-paths for a Duplex Path from node S to node D	From	To
Sub-path 1	S	BS
Sub-path 2	BS	D
Sub-path 3	D	BS
Sub-path 4	BS	S

is attempted for the reverse path from node D to node S; otherwise, the call is rejected. If the call establishment is successful, a *RouteReply* packet is sent to the node S, carrying the complete path to the destination.

13.4 PRICING IN MULTI-HOP WIRELESS NETWORKS

Multi-hop wireless relaying is used in a wide variety of domains and applications. Ad hoc wireless networks were intended for military, tactical, and emergency search-and-rescue applications, where the need for an instant, infrastructure-less, and reliable communication network that could be deployed with very low cost and low time-investment was felt. Gradually, the applications have extended to civilian domains, for providing low-cost network access to homes and mobile users. Wireless mesh networks present an interesting use of multi-hop relaying for wide area networks with predominantly domestic users. The hybrid wireless network architectures aimed at throughput enhancement, as discussed earlier in this chapter, raise the issue of pricing, which is of utmost importance to both the user and the service provider.

13.4.1 Issues in Pricing in Multi-Hop Wireless Networks

- **Reimbursement:** Some suggestions for pricing in multi-hop wireless networks, such as the ones proposed for packet-based multi-hop cellular networks [21] and throughput enhanced wireless in local loop (TWiLL) [8], have been directed toward the concept of reimbursement, that is, every intermediate node that forwards a packet or participates in voice traffic on behalf of another node is reimbursed for the resources that it has expended. The extent of reimbursement is a significant factor that will determine the willingness of the intermediate nodes to expend some power for others.

- **Fairness:** One of the major concerns of pricing in MCNs and ad hoc wireless networks is fairness, the idea that every node pays an equal amount per unit of traffic that it has generated. Fairness can be viewed in multiple dimensions, such as on a per flow basis or a per byte basis. However, fairness considerations on a per byte basis with respect to data traffic and per second for voice traffic seem more reasonable, as they are usage-based pricing approaches. The issue of reimbursement leading to a profit in incentive-based (when the extent of reimbursement exceeds the resource expenditure) schemes is unavoidable in multi-hop relaying, as there can exist nodes that forward more packets than they originate, and hence the received payments exceed the total expenditure.

- **Service provider's revenue:** The service provider's revenue assumes significance in infrastructured networks such as MCNs and wireless mesh networks, where a service provider is responsible for providing network access and hence the profit earned by the provider is a prerequisite for the network to function. All the pricing schemes must ensure that the revenue earned by the service provider is above a certain minimum amount, below which it is infeasible for the provider to sustain the network. This is especially true when incentive-based schemes are to be considered where it may happen that the service provider's income is meager.

- **Aggregation of pricing information:** In traditional cellular networks, accounting is usually done at the base stations (BSs), that have complete information regarding the usage patterns with respect to both voice and data traffic. One issue that is to be addressed, especially in the multi-hop domain, is how the pricing information is aggregated. Clearly it may not be possible for the BSs to be a part of all the traffic that is generated in the network due to the use of multi-hop relaying paths that exclude the BSs, and as a result the BSs may be unaware of the amount of traffic that has been sent. The pricing information such as the total call time or total data sent, if aggregated at the nodes themselves, presents another dimension to the pricing problem: trust. It may happen that the person operating the device can tamper with the information either to make a profit for himself or ensure a loss for others.

- **Secure transfer of accounting information:** One should also consider feasible means of transmitting the pricing information securely to a trusted accounting station (typically the service provider). Security over wireless networks is a major issue and this may necessitate a dedicated secure channel between the accounting stations and the mobile hosts (MHs) for transmitting the pricing information.

- **Translation of resource expended into cost:** The power expended by a node in forwarding packets for other nodes has to be suitably converted into a monetary entity. This may have to take into account the battery power consumed for transmitting a unit amount of data, the life of the battery, and the extent to which it affects the node's status to transmit its own data.

13.4.2 Pricing in Military Ad Hoc Wireless Networks

In deployment scenarios of ad hoc wireless networks such as military applications, search-and-rescue operations, and tactical applications, the kinds of mobile nodes involved are largely homogeneous, dedicated nodes whose sole motive is to play an active role in the communication. Such scenarios do not involve *opportunistic*[5] nodes, and hence pricing as a means of stimulation for participation is redundant, as the service availability is guaranteed. However, pricing frameworks can be used for power optimization in such networks. This is applicable when there are different kinds of mobile hosts, those that have a good battery backup (*e.g.*, vehicle-mounted-mobile devices) and those that have low battery backup (such as handheld devices). Pricing can play an important role in routing the packets predominantly through nodes that have greater battery supply, and pricing mechanisms can be used for a certain degree of power/throughput optimization as suggested in [22].

[5]These are nodes that originate their own traffic, but are unwilling to relay other nodes' traffic.

13.4.3 Pricing in Multi-Hop Wireless WANs

In metropolitan areas, providing low-cost, high-bandwidth Internet access with very low deployment cost and setup time may become a reality with the emergence of wide area ad hoc wireless networks, where some of the relay nodes act as *gateways* to the wired network (Internet). Another network model is to use the multi-hop wireless mesh network for local communication for domestic use in a community of houses. In the former case, the gateway nodes can perform the required authentication and accounting mechanism to charge the users for the network access. This can be typically like the charging mechanisms employed by Internet service providers (ISPs), either on a time usage or a byte usage basis. In the other case, when the network is intended for the use of a local community, the presence of a central coordinator cannot be assumed, and many of the problems that are faced in ad hoc networks, discussed in the following section, need to be addressed. Since wireless mesh networks are intended for commercial deployment, the pricing should be as realistic as possible, with the charging units having a realistic correspondence to actual monetary units. Also the network operator's revenue is to be guaranteed to be above a minimum threshold, so that the maintenance of the network is feasible.

13.4.4 Pricing in Ad Hoc Wireless Networks

The primary motive in pricing for ad hoc wireless networks is to stimulate participation, that is, to ensure that mobile nodes act as routers in forwarding data and control messages on behalf of other MHs. Such a pricing framework should be decentralized, due to the absence of any perceivable supporting infrastructure.

The Terminodes Approach

Several pricing models have been presented as part of the terminodes project [23] to enforce service availability[6] in ad hoc wireless networks. The typical application scenarios of such networks are in civilian deployment of wireless wide area networks, without any infrastructure. The *packet purse* and *packet trade* models [24] introduce the concept of a virtual currency called *nuggets,*[7] that is used to stimulate cooperation. The models assume a tamper-proof security module in each node, a public key infrastructure, greediness of nodes, and absence of infrastructure. The major objectives of packet purse and packet trade models [24] are:

- Packet replay should be detected and such packets should be dropped in order to ensure that the nugget counters are maintained correctly.
- A node should be prevented from overloading the network.
- An intermediary node should not be able to feign information such as packets sent and the number of nuggets.

---

[6]Service availability refers to the availability of packet forwarding service from the nodes in an ad hoc wireless network.

[7]These are also referred to as *beans*, *nuglets*, and *virtual currency*.

- The users should not be able to tamper with the hardware/software in order to get undue benefits.

- Packets that are lost cause a net loss of nuggets in the network over time. Hence, there needs to be a mechanism for periodically replenishing the system with nuggets.

Packet Purse Model

In this model, the source node of a packet loads the packet with a certain number of nuggets. At every intermediate node, the packet is forwarded if there is a non-zero nugget count in the *pricing header* of the packet, else the packet is dropped. Each forwarding node reduces the nugget count in the packet by one, and also increases its nugget count[8] by one. Since the hop-count to the destination is unpredictable (due to node mobility), it is likely that if the source loads fewer nuggets the packet is dropped, and if it loads an excess, then either the destination acquires them or the nuggets are dropped. Typically, if the destination is an application server, this can be viewed as some service cost paid by the source node. This model prevents a node from overloading the network, as it would mean a loss of nuggets.

Packet Trade Model

The packet trade model does not suffer from the problems of the packet purse model where packets can get dropped due to insufficient nugget count in the packet, resulting in a loss of nuggets for the source node. These problems are solved in the packet trade model by introducing a buy-and-sell transfer model. Each intermediate node at a distance k hops from the source will sell the packet to the next hop for exactly k nuggets, that is, the next hop reduces the nugget count by k and the current node increases the nugget count by k. Thus it can be seen that each intermediate node will get exactly one nugget for the service rendered by it. The destination of the data flow is the ultimate buyer of the packet. This may lead to indiscriminate flooding by certain malicious sources, as the sources do not require any nuggets to transmit data. However, nodes can exert their discretion and avoid such users that are known to overload the network. The solution to the security problem is achieved by the security module that will insert the pricing header (typically between the link and network layers). The use of symmetric counters between neighbors overcomes the problem of packet replay. The model also introduces a concept of a fine, in case any node misbehaves (*i.e.*, it has claimed a nugget for forwarding a packet but has not actually forwarded it), and the security module at the misbehaving node will reduce its nugget count the next time it forwards a packet containing the fine counter.

[8]Every node keeps track of its stock of nuggets. A destination node that receives the packet increases its stock of nuggets and an originating node decreases its stock of nuggets on origination of a packet.

A Counter-Based Approach to Stimulate Cooperation

Buttyan and Hubaux proposed in [25] a simple counter-based mechanism at each node in the ad hoc wireless network in order to stimulate cooperation. The model is studied both analytically and with the use of simulations. The counter-based approach assumes that the nodes are deployed for civilian applications, and each node is under the complete control of the "selfish" user. The selfishness is defined later in this section. The selfishness of nodes is measured in terms of their originated traffic. The problem of such users can be nullified with the use of a tamper-proof security module at each node and cryptographic protection of packets. The counter-based mechanism works as follows. Each node maintains a *credit count*[9] that is indicative of the level of service it can expect from the network. When a node wants to send a packet over n hops to a destination, it can do so only if its credit count exceeds n, and whenever a node forwards a packet, its credit count is incremented by one unit. Each node is modeled as having two incoming flows, IN_f (the packets that arrive for forwarding) and IN_o (the packets that the node has originated), and two outgoing flows, OUT (the sent packets) and DRP (those packets that are dropped). Both the outflows can be considered as a combination of the forwarded (out_f and drp_f) and originated (out_o and drp_o) packets. The node's state is described by the state variables b and c. The former is essentially the number of packets that can be sent by the node under the battery power constraint and the latter is nothing but the *credit count.* The model can be formulated as one with linear constraints, and the selfishness can be represented as the maximization of out_o. It can be shown that the optimal value of out_o is $\frac{B+C}{N+1}$, where B and C denote the initial values of b and c, and N is the expected number of intermediate nodes to the destination. In case the node runs out of credits before reaching the optimum value, it can buffer its own packets and can forward packets for other nodes to enhance its *credit count.* An extended model is also presented in [25], that considers a uniform traffic flow r_o for the originated packets and r_f for the packets that need to be forwarded. Depending on the ratio between the two quantities, the node can reach the optimum value with or without dropping any of its own packets. Further, a decision mechanism is provided that will decide which packets will be forwarded and which ones will be dropped. Four forwarding strategies (rules) are used which depend on the current value of f, the number of packets forwarded so far:

1. if $f < \frac{NB-C}{N+1}$ then forward, else drop

2. if $f < \frac{NB-C}{N+1}$ then
    if $c \leq C$, then forward
    else forward with probability $\frac{C}{c}$ or drop with probability $1 - \frac{C}{c}$
  else drop

3. if $f < \frac{NB-C}{N+1}$ then
    if $c \leq C$, then forward

[9]This is also referred to as *nugget count* or *nuglet count.*

else drop
else drop

4. if $f < \frac{NB-C}{N+1}$ then
if $c \leq C$, then forward with probability $1 - \frac{c}{C}$
else drop with probability $\frac{c}{C}$
else drop

In all four rules, packets are dropped after the threshold $f = \frac{NB-C}{N+1}$ has been reached. Once the required value of f is reached, the node has enough credits to drain its battery out by sending only its own packets. The four rules differ in what happens before this threshold is reached. In Rule 1, packets are always forwarded (hence it is most cooperative), while in the other rules, the forwarding decision depends on the current value c of the credit count. Rules 2, 3, and 4 are less cooperative, in this order. As discussed earlier in this section, there is a need for a tamper-proof security module within which the important functionalities are contained, and tampering with the other modules cannot provide any incentive for the malicious nodes. The packet forwarding protocol described in [25] uses sending and receiving counters at the neighboring nodes (to detect packet replay), a credit count that keeps track of the number of credits a node owes its neighbor, and a symmetric key for the session. The system should perform periodic credit synchronization in order to actually earn credits for forwarding, as the credit counter is not increased immediately, rather updated by the security module in the next-hop node. The value of the period is crucial, as it may so happen that a node does not earn credits even though it was a forwarder, as its neighbor may have moved out of range. One of the key concerns is that the protocols and synchronization should consume resources (bandwidth and power) as little as possible.

A Microeconomic Framework for Enabling Forwarding

A microeconomic framework is proposed by Ileri *et al.* in [26] to design a system that encourages forwarding through a reimbursement mechanism. The metric used for optimizing the user's utility in the network is the average amount of data received per unit of energy expended. The network revenue is simply the net difference between the charges paid by all the users and the reimbursements paid out to intermediate routing nodes. The charge paid by every user is simply λT_i, where T_i represents the throughput of user i, and λ is fixed for a given service class. If a forwarding node j allots a fraction k_j of its throughput T_j for forwarding, then the network revenue is given by $revenue = \sum_i \lambda T_i - \sum_j \mu k_j T_j$, where μ is the unit reimbursement paid to the forwarding nodes. [26] studies the system proposed in a small network consisting of two nodes and one access point. One major concern is that in larger networks, routing acquires significance in the pricing/ reimbursement strategy. Self-organization can be essentially viewed as a joint optimization problem where the access point tries to maximize its net revenue while the individual users try to maximize their net utilities (the net utility of a user is defined as utility minus payment). The designed pricing mechanism successfully induces forwarding, when forwarding can enhance the perceived utilities. [26] shows that the forwarding scheme is more beneficial for both the access point and the nodes when compared to the non-forwarding scheme.

Sprite — A Cheat-Proof, Credit-Based System

Zhong *et al.* proposed in [27] a credit-based system to provide forwarding incentive for "selfish" users that does not require tamper-proof hardware as in the packet purse model or the packet trade model [24]. The essential idea is to make the nodes keep a receipt for every message received and have a credit clearance service (CCS) to which a node reports receipts of messages that it has received/forwarded. The two key aspects are (i) the system does not assume tamper-proof hardware (hence selfish users can maximize their utilization) and (ii) an incentive scheme by which a node receives sufficient credit for forwarding. Sprite addressed two key issues in the context of pricing: who is to be charged (source or destination) and who is to reimbursed (a forwarder who attempted or a forwarder that actually succeeded in transmitting to the next hop). Charging the source is reasonable from the context of overloading, denial of service (DoS) attacks on the destination, and colluding. Also, the CCS assumes a node has forwarded a message only if a successor of that node reports a valid receipt of the message. The objectives of the scheme are the prevention of cheating actions and providing incentive for forwarding. In particular, the scheme does not attempt a balanced payment (in which the total charge to sender equals the total reimbursement). The three forms of selfish actions that a node may exhibit are recording a receipt but not forwarding the message, not reporting a receipt, and making false claims of having received a message. A natural way of motivating a node to forward is to give greater reimbursement for forwarding successfully. This means that for a given message, the last node to have received the message gets a reimbursement less than that received by its predecessors. For example, if a message is sent along the path 1-2-3-4-5-6, where node 1 is the source, node 6 is the destination, and node 4 is the last node to have received the message (*i.e.*, nodes 5 and 6 do not report any receipts for the message), then the sender node 1 has to pay an amount α each to nodes 2 and 3, and an amount β to node 4, such that $\beta < \alpha$. There could be collusion between the sender and some of the forwarding nodes, such that the last forwarding nodes do not report their receipts, and then the colluding group stands to gain $\alpha - \beta$. Sprite [27] avoids it by charging the sender an extra amount in case the destination does not receive the message, thereby preventing the colluding group from obtaining any gain. There is also a mechanism to prevent nodes from making false claims regarding receipt of messages. If the destination colludes with the forwarding nodes (*i.e.*, the intermediaries forward only the receipt instead of the entire message), then it requires a higher layer confirmation mechanism between the source and destination to validate the receipt of the entire message. However, if the destination does not collude with the intermediaries, then it will not report any partial receipt, and hence the CCS can accordingly reduce the reimbursement for the intermediaries by an appropriate factor γ. Thus, in effect, the payment that the source has to make to a node i is given by

$$P_i = \begin{cases} \alpha & \text{if } i < e = d \\ \beta & \text{if } i = e = d \\ \gamma\alpha & \text{if } i < e < d \\ \gamma\beta & \text{if } i = e < d \end{cases} \tag{13.4.1}$$

where e is the last node that reports receipt of the message and d is the destination of the message. Refer to [27] for the details of the required protocol procedures for sending and receiving messages and a game-theoretic analysis of the receipt-submission game. Even though the system prevents cheating without the use of tamper-proof hardware, the assumptions of fast and reliable access to the CCS, and the requirement of the nodes to report all receipts, are unrealistic, and these may not be valid in a practical ad hoc wireless network.

13.4.5 Pricing in Hybrid Wireless Networks

In the traditional cellular network model, pricing is not a major issue, as the base station acting as a representative of the service provider has the complete information on the traffic being generated within a cell. Furthermore, every byte of data has to be forwarded through the BS directly without involving other mobile hosts. The above scenario is not applicable for MCNs where the routing is far more complex, involves relaying by intermediate nodes, and not all traffic passes through the BS. The situation becomes far more complex if mobility is considered. Mobility may lead to multiple route reconfigurations, and the BS may be unaware of the path through which the packet reached the destination if local reconfiguration is performed.

A Micropayment Scheme for MCNs

Micropayments are useful for stimulating cooperation among rational (selfish) participants, and can be used to ensure routing of data and voice packets. Traditional schemes for micropayment in the physical world presuppose the knowledge of the number of parties to be paid and the amount to be paid. This assumption may not be valid in a mobile environment. The scheme proposed by Jakobsson *et al.* in [28] has four major components.

1. A strategy for users to determine how packets should be routed. Each node has a threshold reward (or payment) level that needs to be satisfied in order for it to forward, and each source accords each originated packet a reward level proportional to the packet's importance.

2. A verification module at the BSs to check for valid payments.

3. An aggregation technique that works on the principle of probabilistic selection of payment tokens (or winning tickets). This system probabilistically selects intermediate nodes eligible for payment for their forwarding service.

4. An auditing technique that allows for the detection of cheating on the part of the nodes. The audit technique uses the information gathered from the winning tickets (also called payment tokens or payment claims) and packet transmission information from the bases in order to detect cheating or colluding among nodes.

In this scheme, all source nodes that originate a packet attach a payment token to each packet, and all intermediate nodes on the path to destination verify whether the token corresponds to a winning ticket (similar to a lottery system). Winning tickets are reported to the nearby BSs periodically. The BSs receive payment claims and the packet with tokens. After removing and verifying the authenticity of the tokens, the packets are forwarded through the wired network to their destinations. The winning tickets and payment tokens are sent to accounting stations for further processing. The scheme advocates one token per packet rather than one per payee, allowing all nodes that transmitted the packet successfully to claim a winning ticket. Winning tickets are communicated to the nearby BSs that forward the packet along the backbone after confirming the validity of the payment tokens. An interesting strategy used is that the intermediaries profit not only from their own winning tokens but also from that of their immediate neighbors in the path. This scheme has the following advantages:

- The encouragement for both transmitting the packet and reporting the information to the authority is provided.

- The system ensures that fewer tickets need to be deposited, as there are multiple rewards per deposited ticket.

- It allows for an efficient cheat-managing system.

The packet-originating node is charged per packet originated whereas the intermediate nodes relaying the packet are paid per winning ticket. It is more important to note that the intermediate nodes are not paid for each packet they handle, rather for the winning tickets they and their neighbors along the path collect. The payments are finalized by a central accounting authority. Some of the common forms of cheating or colluding that could occur in a multi-hop relaying environment include the following:

1. **Selective acceptance:** An intermediate node receives and forwards packets with winning tickets, but not those without winning tickets.

2. **Packet dropping:** An intermediate node drops packets after receiving them (without forwarding them further), irrespective of whether or not a winning claim is made.

3. **Packet sniffing:** A node can claim to have forwarded a packet with a winning ticket by merely snooping on the wireless channel.

4. **Greedy collection:** A node can collect large number of tickets.

5. **Information tampering:** Tampering with claims and reward levels.

For each of the described cheating strategies, [28] presents a detection mechanism within the proposed framework, thus providing a way to nullify the presence of malicious nodes.

Pricing for Voice Traffic

In traditional cellular networks, voice traffic predominates over data traffic, and it is essential to study certain pricing issues for such traffic. The use of multi-hop relaying is identified as a useful throughput enhancement alternative to traditional cellular networks. Since intermediate nodes are involved in call relaying, it is imperative that some form of cost reimbursement be provided. The three pricing schemes for call relaying in a multi-hop relaying architecture for enhancing throughput in wireless in local loop system are discussed below.

Position-Independent Cost-Reimbursement (PICR)

PICR [8] introduces the reimbursement factor α. In particular, αC_p is the amount per second that gets paid by the call originator to every relaying MH in its optimal path to the BS, where C_p is the transmission cost per second that an MH incurs. In this scheme, the amount that gets reimbursed to a relaying MH is independent of its position. The term $N(d)\alpha C_p$ refers to the cost-reimbursement done to the subscriber by the $N(d)$ (average relay load at a distance d measured in terms of hop-count) MHs that have optimal paths passing through the subscriber's MH, and $R(d)$ is the number of hops that needs to be reimbursed at distance d from the BS. The average cost of operation per unit time at a distance d from the BS is $A(d) = S + R(d)\alpha C_p + N(d)(1-\alpha)C_p$, where S is the bill-rate. The revenue obtained by the network operator is given by $N_R = St_{av}D$, where t_{av} is the average call time, and D is the average node density in a cell. To satisfy the $N_{R_{min}}$ minimum revenue limit, the bill-rate S should be $= \frac{N_{R_{min}}}{t_{av}D}$.

Position-Independent Cost-Reimbursement, Position-Dependent Bill-Rate (PICR-PDBR)

The factor α was introduced in the PICR scheme in order to pay back some part of the cost incurred at a relaying MH. It is found that irrespective of the value of α, the simulation studies did not exhibit a high degree of fairness; this is the essential motivation for the subsequent PICR-PDBR scheme, hence the notion of a position-dependent bill-rate $S(d)$, which represents the bill-rate at a distance d from the BS. Now the average cost becomes $A(d) = S(d) + R(d)\alpha C_p + N(d)(1-\alpha)C_p$. To ensure fairness (in this context, fairness refers to the uniform call cost per node per unit time), $S(d)$ is calculated such that $\forall d \ A(d) = A$ where A is the uniform cost incurred by all MHs. The revenue obtained by the network operator N_R is given by

$$N_R = \sum_{0 \le i < R/\delta} S(i\delta) t_{av} (\frac{D}{\pi R^2}) \times 2\pi i \delta^2 \tag{13.4.2}$$

$$\Rightarrow N_R = \frac{2D\delta^2 t_{av}}{R^2} \sum_{0 \le i < R/\delta} S(i\delta) i \tag{13.4.3}$$

where δ is an elementary step that can be made as small as desired and R is the radius of the cell. The $S(d)$ values are calculated for satisfying the fairness and minimum revenue requirements. The conditions that must be satisfied for meeting the fairness and minimum revenue requirements are given below.

$$N_R = \frac{2D\delta^2 t_{av}}{R^2} \sum_{0 \le i < R/\delta} S(i\delta) i \ge N_{R_{min}} \tag{13.4.4}$$

$$S(d) = A - R(d)\alpha C_p - N(d)(1-\alpha)C_p \tag{13.4.5}$$

This calculation can be easily made by assuming an arbitrary value for A and using the second equation first and then shifting the obtained $S(d)$ values appropriately to satisfy the first inequality. It is observed in [8] that PICR-PDBR maintains a fairer system than PICR. In fact, the system is absolutely fair at $\alpha = 1.0$, which satisfies the $\alpha \ge 1$ condition required for relaying the MH incentive.

Position-Dependent Cost-Reimbursement, Position-Dependent Bill-Rate (PDCR-PDBR)

It is observed in [8] that in the PICR pricing scheme, if α, the reimbursement factor, is the same for all routing nodes, fairness cannot be achieved in the system. Also in the PICR-PDBR scheme, the system is fair when $\alpha = 1.0$, which is just sufficient for compensating a node's forwarding expenditure. It is more encouraging for the nodes to participate in the forwarding process when they are reimbursed more than the expenditure (leading to an incentive). Hence in the PDCR-PDBR scheme, $\alpha(d)$ is defined as the reimbursement factor at distance d from the BS. This is the essence of the PDCR-PDBR scheme. To ensure fairness, $\alpha(d)$ needs to be computed such that $\forall d \; A(d) = A$, where A is the uniform cost incurred by all MHs. To ensure that a relaying MH has an incentive to relay calls, $\alpha \ge 1$ must hold. Thus, minimum $\alpha(d)$ must be calculated using the following two constraints:

$$\alpha(d) = \frac{S + \sum_{1 \le i \le R(d)} \alpha(d_i) C_p - A}{N(d) C_p} + 1 \tag{13.4.6}$$

and $\alpha(d) \ge 1$. Note that in spite of the above precautions, fairness may not be achieved because of the presence of non-relaying MHs, that is, MHs for which the

Hop-by-Hop and Retransmit Attempts Scheme for Reimbursement (HRSR)

The HRSR scheme incorporates the features of the HHSR scheme (link-level success) and the concept of partial reimbursement for the retransmit attempts made by the intermediate nodes. In this scheme too, the pricing information has to be aggregated at all the nodes in the network, each keeping track of the number of bytes it has successfully forwarded to the next hop, and the total number of attempts it has made for each source for which it functions as an intermediary. This is the most general scheme that takes into account the entire traffic that a node forwards, but it requires a tamper-proof pricing and security module at the MHs as in the HHSR scheme. $F_{s_{ij}}$ represents the total number of bytes that have been successfully forwarded to the next hop. In addition to this, each intermediate node gets partial reimbursement for the retransmitted bytes $F_{r_{ij}}$ in order to give greater incentive for forwarding. The total reimbursement is given by $I_{repay_i} = \sum_{j=1}^{N} F_{s_{ij}} \alpha C_p + \sum_{j=1}^{N} F_{r_{ij}} R_{repay} C_p$, where R_{repay} is the reimbursement factor for the retransmitted bytes.

13.4.6 Open Issues in Pricing for Multi-Hop Wireless Networks

- Pricing obviously needs some currency units and in any commercial system, the pricing needs to be done in absolute real-world monetary units rather than in a virtual nugget-like currency. The translation of the pricing parameters into absolute monetary units requires a deeper understanding of the issues involved.

- Most pricing systems will need to dynamically decide on the pricing parameters such as the reimbursement factors and the type of reimbursement, in order to take into account the network size and the offered load.

- Typically, there may be multiple service providers arranged in a hierarchy in order to provide the Internet access. In such scenarios, the interoperability of such hierarchical systems, through a pricing and/or an authentication mechanism, needs to be studied.

- There has been a proliferation of Wi-Fi *hotspots* in areas of commercial interest such as cafes and restaurant chains that aim to provide easy Internet access to customers. Such hotspots use an access point (AP) to which users who have a wireless device can connect using IEEE 802.11b technology. Pricing in hotspots and across hotspots, and pricing issues in the integration of GSM and multi-hop wireless LANs (MWLANs) [19], need to be addressed.

- Security is a key aspect of pricing, as both nodes and the network operator are likely to suffer if there are malicious users in the network. This can be either in the form of tamper-proof hardware/software modules on the mobile devices, or cryptographic/authentication systems that can function even under worst case considerations. Security in ad hoc and other multi-hop wireless networks needs further exploration, as these networks are bound by the limited storage and processing capabilities of the mobile devices.

- Since most schemes involve cost-reimbursement, pricing becomes closely linked with routing, as the determination of the forwarders and the distance to the destination (hop-count) are necessary. The dependence between the pricing framework and the routing protocol is another key issue that is unaddressed so far.

13.5 POWER CONTROL SCHEMES IN HYBRID WIRELESS NETWORKS

Power control schemes which assume importance in hybrid wireless networks as such networks are designed to serve large number of power-constrained mobile nodes. The HWN architecture utilizes power control in ad hoc mode to maximize the achievable system throughput. Based on the topology of the network formed by the mobile nodes, a BS estimates the minimum necessary power to achieve connectivity. The maximum achievable throughput is a function of the transmission power used. The higher the transmission power, the lower the throughput achieved. The BS switches to cellular mode if the power being used for connectivity in the ad hoc mode results in lower throughput than what could be achieved in the cellular mode. In this section, a power control scheme that can be used with networks such as MCN and MuPAC is discussed.

The MCN and MuPAC architectures provide limited flexibility in terms of the transmission ranges. The transmission power must be chosen from a limited set of available values in the case of MuPAC and is fixed in case of MCN. Though MuPAC does not restrict the number of data channels that can be used, increasing the number of data channels does not increase the flexibility because the available bandwidth is divided into many different channels, thus increasing the transmission time. Also, the increase in the number of data channels increases the number of interfaces, which may add to the complexity of the devices. Further, at low loads this is not advisable because most of the data channels will remain idle. MCN, on the other hand, forces the use of a single data channel for all nodes.

13.5.1 Issues in Using Variable Power in IEEE 802.11

The IEEE 802.11 [29] MAC protocol uses the request-to-send/clear-to-send (RTS-CTS) mechanism for gaining access to the channel, as explained in detail in Section 2.3. In this mechanism, the sender node, on sensing the channel to be idle, sends an RTS message to the receiver. The receiver replies with a CTS message. These messages are heard by the neighbor nodes, which then avoid transmissions for a time interval specified by the network allocation vector (NAV) so as to avoid collisions with on-going transmissions. Thus, the probability of hidden terminals affecting the transmission is reduced. However, when variable transmission powers are used in 802.11, this effect will be nullified. This is illustrated with an example in Figure 13.22. Node A wishes to send data to node B. Node A finds the channel idle and hence sends an RTS message to node B. Since node B is close enough, node A uses a lesser transmission power sufficient for node B to hear its packets. Node B replies

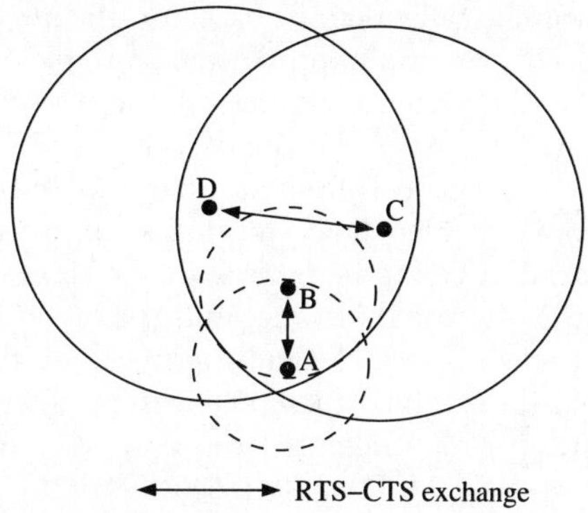

Figure 13.22. Problem in using variable power with MAC protocols such as IEEE 802.11.

with a CTS with the same power. Neither the RTS nor the CTS is heard by node C and node D, which are relatively further apart. Thus, node A is ready to send data to node B.

When the transmission between nodes A and B is going on, node C wishes to send data to node D and finds the channel idle. Since node D is relatively further apart from node C, node C uses a higher transmission power. This transmission reaches node B and hence a collision occurs at node B. Thus, the origination of a new packet transmission causes interference with an existing packet transmission. This problem would not have occurred if node A and node B were to use the same transmission power as node C. In this case, the RTS or the CTS of nodes A and B would be heard at nodes C or D or both, and hence collision would be avoided. Therefore, variable transmission power cannot be directly used in IEEE 802.11 MAC protocol-based networks.

13.5.2 Power Optimization Scheme for Hybrid Wireless Networks

A power optimization scheme for hybrid wireless networks is proposed in [30]. In this scheme, the available bandwidth is considered to be divided into n channels, each of which can operate at a different transmission ranges similar to the MuPAC architecture. Using the periodic *Hello* packets used in MCN and MuPAC architectures and received power at every node, an approximate distance to the sender is calculated. The MuPAC uses α as a safety factor to calculate this distance. In addition to this, a factor called the *mobility margin* as the mobility safety factor is also introduced in this power optimization scheme. This is necessary because the factor α only compensates for the error in the calculation of the distance. However, mobility may cause two nodes to move further apart before the next *Hello* message is transmitted. This is accounted for by the mobility margin. The mobility margin may be transmitted by the BS to every node at the time of registration. The value of the mobility margin may be defined based on the degree of power saving (a smaller value for the mobility margin improves the power saving) and the resilience required against mobility (the higher the value for the mobility margin, the better the resilience against mobility). Also, the value of the mobility margin depends on the frequency of neighbor updates. In case of high-update frequency, a low value of the mobility margin would suffice; however, in case of low-update frequency, the value of the mobility margin needs to be increased. The transmission power is estimated from the sum of the estimated distance and the mobility margin value. This is done as follows. Let the estimated distance between the two nodes be x meters and s be the mobility margin. Thus, the transmission power needed to transmit up to a distance of radius $x+s$ is calculated to be t_x. Let r_i be the transmission range of the channel selected by MuPAC for transmission. If t_x is greater than the transmission power to be used for r_i then the transmission takes place, using r_i as the transmission range. Otherwise, the RTS and CTS packets are exchanged over the transmission range r_i but the data transmission takes place with the reduced transmission power t_x. The acknowledgment is also sent over the reduced transmission range. Though the transmission range for sending DATA and ACK packets is changed, the transmission range of the RTS and CTS messages is unchanged. Hence, in this case a higher throughput is achieved due to reduced interference in the data channel. An alternative study carried out by Jung and Vaidya in [31] increased the transmission range of RTS and CTS to the maximum possible value and found a reduction in throughput achieved. This is because the number of nodes that hear the RTS or the CTS is greater than the number of nodes that hear the actual transmission.

In the scheme proposed by Bhaya *et al.* in [30], the transmission power of RTS-CTS is kept constant and at the same time, the transmission power for DATA and ACK is reduced to the sufficient value required for communication. The effect of this scheme is as follows. The number of nodes affected by transmission are the same as that in the non-power-optimized system. It is important that these nodes are prevented from transmitting data in order to prevent a collision with the on-going transmission as per the collision avoidance paradigm. Therefore, in the example shown in Figure 13.22, nodes C and D hear the RTS sent by node A or the CTS sent by node B, though they may not hear the actual DATA and ACK transmissions. Thus, node C and node D are prevented from transmitting data on the same data channel as node A and node B for the transmission duration specified by RTS-CTS. Hence, the chances of collision during the data transmission interval are reduced. This is illustrated in Figure 13.23.

The above method of preventing certain nodes from transmission even though they are not directly affected by the on-going data transmission is necessary in the interest of the nodes that are currently involved in transmission. The performance of this scheme depends on the estimated distance between the nodes, which is used for the calculation of the reduced transmission range. It requires a nearly accurate estimation of the distance between two nodes. This can be done in many ways. If the GPS information is available at the nodes, the receiver can convey its GPS coordinates to the sender using the CTS. Alternatively, the sender can translate the

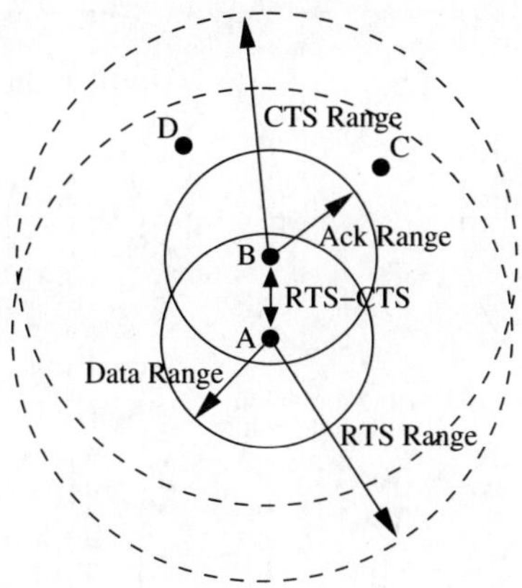

Figure 13.23. Illustration of the power optimization scheme for hybrid wireless networks.

signal strength information of the *Hello* messages received from the neighbors to approximate distance information. In this case, the transmission power information may be included in all packets transmitted by the nodes. Further, to overcome errors due to varying signal strengths, a node may average the received power for the last few receptions from the same sender.

13.6 LOAD BALANCING IN HYBRID WIRELESS NETWORKS

Load balancing refers to the distribution of relay traffic load uniformly throughout the network so that no region in the network is particularly overloaded. The need for load balancing arises from the fact that the amount of relay traffic (traffic relayed by a node) in a static multi-hop wireless network is dependent on the position of the nodes in the network and the node density in the region. As a result, the nodes close the center of the network need to relay more traffic than the nodes away from the center when the shortest path routing is used. De Couto *et al.* studied this problem in [32]. This problem is not very severe in highly mobile ad hoc wireless networks. But it assumes significance in static ad hoc wireless networks, wireless mesh networks, wireless sensor networks, and hybrid wireless networks where nodes are not highly mobile. This is illustrated in Figure 13.24 (a). The relay traffic is highest at the center of the network, and decreases with increasing distance from the center of the network. This may lead to rapid draining of battery charge for the nodes at the center of the network, which can result in the formation of a ring in which the coverage at the center of the network is affected. This is shown in Figure 13.24 (b). This problem is referred to as the *ring formation* problem and is the motivation for load balancing schemes in multi-hop wireless networks. This position-dependent traffic load causes the nodes in a high traffic area to drain their battery charge faster, thus reducing their normal life span. Also, the overall

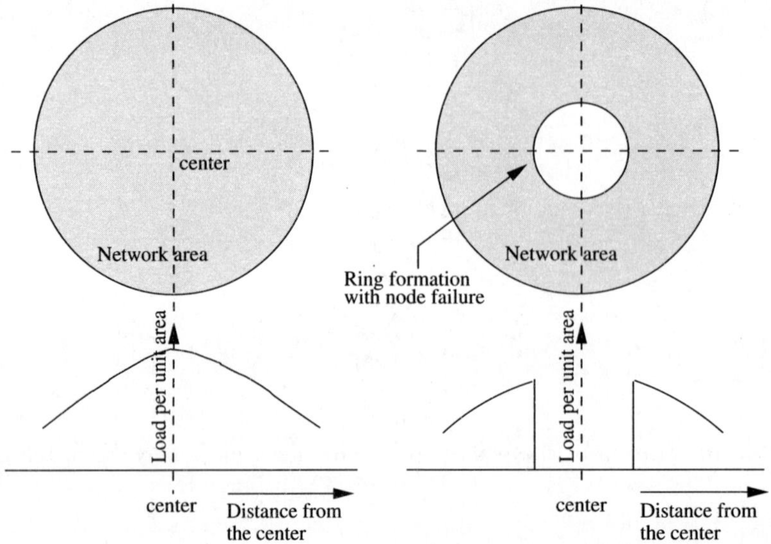

(a) Load density varies with distance from the center of the network

(b) Formation of ring due to the failure of nodes at the center of the network caused by excessive relay traffic

Figure 13.24. Effect of imbalance in load distribution.

network throughput achieved can be significantly reduced with high load imbalance in the network. Therefore, load-balancing mechanisms can also be considered as throughput enhancement schemes. In case of high mobility, problem arising out of the ring formation may not be very serious because a node always keeps changing position between regions of high load and regions of low load. The reduction in life span can be fatal in two ways. First, it may cause a partition in the network due to the absence of intermediate relay nodes. Second, it may lead to dropping of the on-going calls in the network, thus reducing the throughput achieved in the network. Load balancing can be used to solve this problem. Instead of using the traditional shortest path routing which introduces load imbalance in some regions of the network, a longer path may redistribute the load and hence improve the battery life of the nodes. However, the shortcomings of this method are a longer path length which may lead to an overall increase in battery power consumption, an increase in the end-to-end delay in packet delivery, and a reduction in throughput. Though an increase in delay and a reduction in throughput seem to be a direct consequence of the increase in the path length of a wireless network, this may not always be the case in multi-hop wireless networks. In multi-hop wireless networks, a decrease in the average distance between the neighbor nodes in a path can improve the achievable physical layer data rate between them. For example, using IEEE 802.11b standard, two nodes can engage in data transfer at a maximum rate of 11 Mbps if they are sufficiently closer to each other. As the distance between the sender and receiver increases, the data rate is reduced to different lower rates by a mechanism called automatic rate fallback (ARF) to 5 Mbps, 2 Mbps, and 1 Mbps. In addition to the higher transmission rates at the wireless links, the load-balancing schemes utilize the unutilized bandwidth in the lightly loaded regions of the network. Load-balancing can also lead to a reduction in queuing delay and collisions.

The load balancing is important in hybrid wireless networks such as MCNs, when the traffic locality is low (traffic locality is defined as the fraction of originated calls that gets terminated in the same cell). Traffic locality varies between 0 and 1, where locality = 0 refers to the case where the source and destination are in different cells and locality = 1 refers to a situation where all the calls get terminated within the cell. With low values of traffic locality, the probability that the BS will become saturated is high. Load balancing can improve performance in such situations. Mesh networks are a special category of ad hoc wireless networks in which the multi-hop radio relaying mechanism is used but with all nodes fixed, with some nodes acting as gateways to the wired Internet. These networks mainly aim at providing high-speed Internet access to residential areas. Traffic is routed using multiple hops to a gateway which is connected to the Internet. In a given network, there may be multiple gateways. A mesh network can provide fast and low-cost access to the Internet for the mobile users also. Because of the presence of multiple gateways, choosing the correct gateway and routing to the gateways becomes an important issue. Regions close to the gateways may become hotspots[10] in the network, resulting in larger packet loss due to congestion. Preferred ring-based routing and load-balancing schemes proposed in [33] can be used for load distribution in MCNs, ad hoc wireless networks, and wireless mesh networks.

13.6.1 Preferred Ring-Based Routing Schemes

The routing protocol used needs to convey the global topology to all the nodes in the network. Routing protocols such as DSDV [20] or fisheye state routing [34] (discussed in detail in Chapter 7) can be used for this purpose. Using the global topology, each node builds the connectivity matrix for the network and using Warshall's algorithm, the *Center* (the BS in the case of an MCN or the center node in the case of an ad hoc wireless network) of the network is determined in $O(n^3)$ time, where n is the number of nodes in the network. Ties are resolved in favor of the node with lesser node ID in order to maintain uniformity across all the nodes in the network. Once the *Center* is determined, the *Rings* to which the node belongs can be determined. A *Ring* is an imaginary division of the network into concentric rings about the *Center* of the network. Figure 13.25 shows an example of such division of network into *Rings* based on physical distance from the *Center* of the network. Determination of the *Center* is fairly straightforward in case of the hybrid wireless networks such as MCN because the base station can serve as the ideal *Center* of the network. It also lies geographically at the center of the cell. The base station can use the received power over the control channel or the GPS information to determine the approximate distance of the node from itself and thus

[10]Hotspots refer to heavily loaded network nodes.

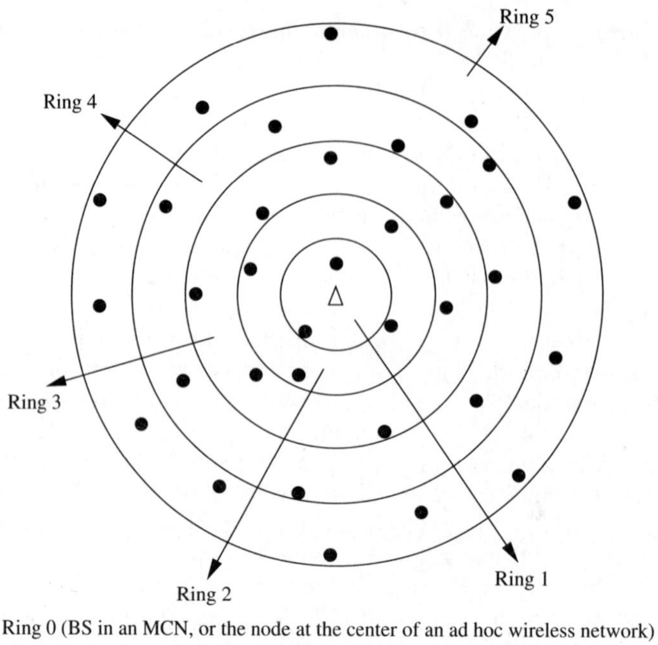

Figure 13.25. Example for determining *Rings*.

divide the nodes into various *Rings*. The choice of number of *Rings* is decided by the base station. Distance from *Center* refers to the hop-count from the *Center*. In ad hoc wireless networks, the *Center* is a node which is determined periodically from the topology information obtained using the routing protocols. Thus, distance from *Center* in this case refers to the hop-count from the *Center*. It is not necessary that all nodes in the network use these approaches for load redistribution. Hence, in this scheme, nodes that participate in load redistribution and those that do not participate can coexist in the same network without any compatibility problems.

The preferred ring-based load-balancing schemes can be used in wireless mesh networks by finding a *Center* which is closest to the gateway nodes.

The three schemes proposed for load distribution and throughput improvement are:

1. Preferred inner ring routing scheme (PIRS)
2. Preferred outer ring routing scheme (PORS)
3. Preferred destination ring routing scheme (PDRS)

A variation of the third scheme is called the preferred source ring routing scheme (PSRS). The key idea behind all these schemes is that any traffic generated by a node in *Ring* i for a node in *Ring* j must not be allowed to go beyond the *Rings*

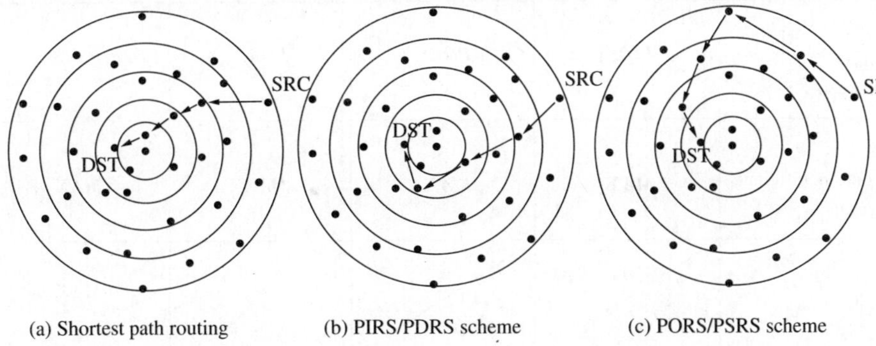

Figure 13.26. Example for routing using different schemes.

enclosed by *Rings i* and *j*. Figure 13.26 (a) shows an example of shortest path routing between source (SRC) and destination (DST) nodes.

Preferred Inner-Ring Routing Scheme (PIRS)

The traffic generated by *Ring i* for *Ring j* must not go beyond the *Rings* enclosed by *Rings i* and *j*. Further, a packet must be preferably routed through the **inner** of the two *Rings*. For the nodes belonging to the same *Ring*, the packet must be transmitted in the same *Ring*. For nodes belonging to different *Rings*, the packet must go across the *Rings* in radial direction[11] and all angular transmissions (transmission within a given *Ring*, that is, sender and receiver of transmission belong to the same *Ring*) must take place in the inner of the two *Rings* (of the source or the destination). For example, in Figure 13.26 (b), the source lies in *Ring 5* while the destination lies in *Ring 2*. Hence, no transmission must pass through *Ring 0* or *Ring 1* unless unavoidable. Further, in the path traced, transmission reaches *Ring 2* from *Ring 5* before the angular transmission begins (refer to Figure 13.26 (b)). Tracing the route in this scheme only requires changing the weights of the edges in the adjacency matrix of the graph before running Dijkstra's shortest path algorithm. Figure 13.27 presents the weights that can be used. The overall complexity of finding the route is the same as in the shortest path routing.

Preferred Outer-Ring Routing Scheme (PORS)

This is the exact reverse of the previous scheme. The packet remains for the maximum time in the outer of the *Rings* of the source and destination.

The traffic generated by *Ring i* for *Ring j* must not go beyond the *Rings* enclosed by the *Rings i* and *j*. Further, the packet must be preferably routed through the **outer** of the two *Rings*. In other words, for the nodes belonging to the same *Ring*, the packet must be transmitted in the same *Ring*. For nodes belonging to different *Rings*, the packet must go across the *Rings* in radial direction and all angular

[11]To reduce the number of hops, it does not go along the radial direction, but the condition forces the use of minimum number of nodes for transmission in the intermediate *Rings*.

Edge between \ Scheme	*PIRS*	*PORS*	*PDRS*	*PSRS*
Nodes in different *Ring*	*MAX*	*MAX*	*MAX*	*MAX*
Nodes in destination *Ring*	*MAX* / *1*	*1* / *MAX*	*1*	*MAX*
Nodes in source *Ring*	*1* / *MAX*	*MAX* / *1*	*MAX*	*1*
Nodes in other *Ring*	*MAX*	*MAX*	*MAX*	*MAX*

Figure 13.27. Edge weight to be used in Dijkstra's algorithm for finding route. MAX represents some large number larger than maximum number of hops possible but less than unconnected. Some entries are divided into 2. The upper half represents the case of destination *Ring* larger than the source *Ring* and the lower half for the case where the destination *Ring* is smaller than the source *Ring*. *Reproduced with permission from [33], © Springer-Verlag, 2003.*

transmissions must take place in the outer of the two *Rings* (of the source and the destination). In Figure 13.26 (c) the source and the destination *Rings* are *Ring 5* and *Ring 2*, respectively, and the angular transmission takes place in *Ring 5*.

Preferred Destination/Source Ring Routing Schemes

The two schemes presented earlier make a trade-off between hop-count and load distribution. The *PORS* affects the hop-count more than the *PIRS*. But the main goal is to move the load from the *Center* toward the periphery of the network. This is better accomplished by *PORS*. *PDRS* and *PSRS* attempt to strike a balance between the two. In *PDRS*, the angular transmission takes place in the *Ring* of the destination node, while in *PSRS*, this transmission takes place in the source node's *Ring*. These two schemes are a hybrid of the *PIRS* and *PORS*, that is, they represent *PORS* in some cases and *PIRS* in other cases. Also the average hop-count in *PDRS* and *PSRS* lies between that of *PIRS* and *PORS*.

The rule followed by PDRS (PSRS) is described in what follows. The traffic generated by *Ring* i for *Ring* j must not go beyond these *Rings* and, the *Rings* enclosed by these *Rings*. Further, the packet must be preferably transmitted in the *Ring* of destination (source) node. For the nodes belonging to the same *Ring*, the packet must be transmitted in the same *Ring*. For nodes belonging to different *Rings*, the packet must go across the *Rings* in radial direction and all angular transmissions must take place in the *Ring* of the destination (source) node. In the examples used to demonstrate the *PIRS* and *PORS*, *PDRS* traces the same path as shown in Figure 13.26 (b), while *PSRS* traces the path shown in Figure 13.26 (c).

The *PSRS* and the *PDRS* which appear very similar may give varied performance in case of wireless mesh networks where the destination may be the nearest gateway node. Here the assumption that the heavy load region is the *Center* of the network may not hold as there can be multiple gateway nodes in the same network. Hence, *PDRS* may end up overloading a few *Rings* containing the destination and thus leading to imbalance. Thus, *PSRS* is preferred over *PDRS* in such networks. In mesh networks, *PSRS* and *PDRS* may vary in the average hop-count as well.

An example case of wireless mesh networks with three gateways placed randomly over the entire network is illustrated in Figure 13.28. Preferred ring-based schemes use the following three heuristics in order to balance the load on the gateways:

- **Least Loaded Gateway (LLG):** In this heuristic, the nodes direct their traffic to the least loaded gateway. The load on the gateway is measured by the number of data packets relayed by it to the wired network. The load information of the gateway is communicated by the routing protocol. The

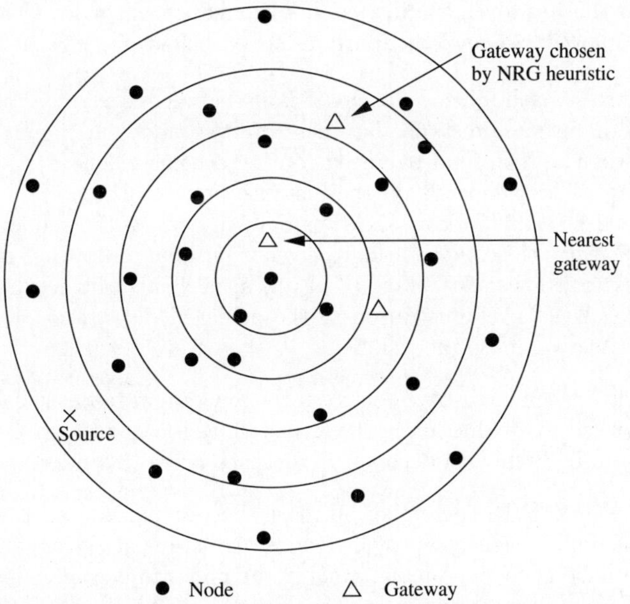

Figure 13.28. Illustration of the mechanisms for the choice of gateway.

motivation behind this heuristic is to load all gateways to the same extent and hence attain load balancing across the gateways.

- **Nearest Ring Gateway (NRG):** In this heuristic, the source node prefers a gateway whose *Ring* is closer to its own *Ring*. This need not be the nearest gateway in terms of number of hops (refer to Figure 13.28). The motivation behind using this heuristic is to involve the least number of *Rings* for a transmission of a data packet.

- **Hybrid Approach Gateway (HAG):** In this heuristic, a source node identifies the gateways based on both of the above alternatives and chooses the desired gateway in a uniformly distributed fashion. This essentially selects either LLG or NRG scheme with equal probability.

Normally, an increase in hop length results in decrease in throughput. However, in preferred ring-based routing schemes, it is found that an average increase of hop length of about 4% to 8% comes along with an increase in throughput of 25% to 30%. The reasons behind the improved throughput are discussed below.

Why System Throughput Improves When Load Balancing Schemes Are Used

In addition to improving the battery life of the nodes and avoiding partitions in ad hoc wireless networks, load balancing schemes are also used for improving system throughput as discussed earlier in this chapter. There are several factors that contribute to the improvement in system throughput when load-balancing techniques are employed. Some of them are discussed below.

1. The shortest path scheme has fewer hops between a source-destination pair. Hence, the average distance between any two nodes in the shortest path is higher than that in load-balanced routing schemes. The data rate achieved in a wireless link depends on the distance between the nodes. For example, IEEE 802.11b can work at 11 Mbps data rate when the sender and receiver are sufficiently close. When the receiver is far apart from the sender, 802.11b switches to a lower data rate by using a mechanism called automatic rate fallback (ARF). An illustration of the achieved data rate for 802.11b with distance is given in Figure 13.29.

2. The shorter average distance between the sender and receiver of a wireless link can also lead to a reduction in the error rate as the signal to noise ratio (SNR) is likely to be high with decrease in distance between sender and receiver.

3. The region near the center lies on more shortest paths than the region away from the center. Hence, when the load is high, any packet loss and subsequent retransmissions near the center cause more nodes to become affected than the packet loss in any other region of the cell.

4. As a result of many paths being set up via the region near the center, the queue length at the nodes near the center grows longer than that at the nodes

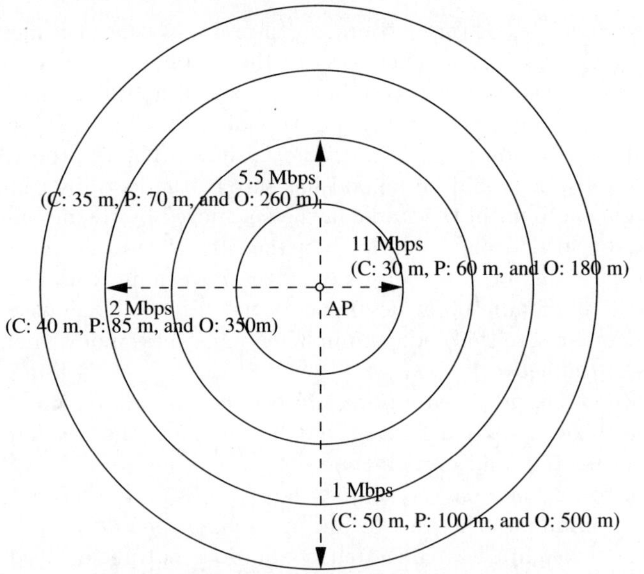

Figure 13.29. Variation of achieved data rate with distance in IEEE 802.11b.

near the periphery of the network. If the number of nodes in the network increases or the rate of data generation at every node increases, more packets will be accumulated in these hotspots, resulting in queue overflow.

5. The number of collisions also increases as the number of nodes contending for the channel increases. Though the number of nodes in the system is the same, the number of nodes in the hot area which have packets for transmission is larger. The number of collisions observed from the studies was found to be less by 4% in the case of *PORS* as compared to the shortest path.

6. Bandwidth near the periphery is unused in shortest path schemes because the number of routes set up via the periphery is much smaller.

13.6.2 Load Adaptive Routing Schemes for Throughput Enhancement and Load Balancing

A measure of load balancing is *traffic fairness* and this can be achieved for every node in the network when the number of packets relayed by a particular node is proportional to its originated traffic. However, defining traffic fairness as above may not lead to an efficient utilization of network resources. Hence, Bhaya *et al.* [35] defined traffic fairness from a load balancing point of view. According to [35], the *fairness index*, c, of the network, is the ratio of the maximum relay load (experienced by any node in the network) to the average relay load over the entire network. Ideally, when the load is uniform across all nodes in the network, the fairness index is 1.0. The lower the value of the fairness index of the network, the better the load distribution and hence traffic fairness. Though the standard deviation of the relay traffic would be a good measure of the load distribution, it may not provide a true account of the traffic fairness achieved by the mobile nodes, as some nodes may be loaded more lightly than the rest. For achieving fairness, reduction of the relay traffic at the node relaying the maximum traffic (the node which experiences the maximum unfairness) is essential. Since all the other nodes relay less traffic as compared to this node, minimizing the fairness index will improve the overall fairness of the network.

Hassanein and Zhou [36] proposed a novel protocol called load-balanced ad hoc routing (LBAR) protocol. LBAR defines a new metric called the *degree of nodal activity* for representing the load on a mobile node. The main objective of this scheme is to find a route between the source and the destination such that the end-to-end delay is minimum. The idea behind minimizing the delay is based on the fact that a path through a less congested region does not get delayed due to the contention for the channel by intermediate nodes. A setup message is sent by the source in order to locate a path, and the destination responds with an ACK containing the selected path. However, fairness in LBAR becomes a secondary objective.

Zhang *et al.* proposed a multipath routing scheme in [37] and analyzed the effect of distribution of load among multiple paths using a queuing model that incorporates the traffic among these paths. They also consider load balancing as an optimization problem. The use of delay as considered by Zhang *et al.* as a criterion for load distribution may work well in case of real-time traffic where the nodes have preassigned slots to access the channel. However, in the case of best-effort traffic, using delay as a criterion may lead to unexpected results because of the unbounded delay at times of high contention.

The dynamic load-balancing schemes consider the traffic load at every intermediate node in a path to make routing or load-balancing decisions. The traffic load at the nodes can be approximated by means of queue length at the nodes, number of packets transmitted in the channel within a time frame, number of collisions experienced, or the number of traffic flows passing through the node. The load-balancing and throughput enhancement schemes described in this section can function independent of the parameter considered for traffic load measurement.

Some of the heuristics for dynamic load balancing and traffic fairness among the mobile nodes proposed by Bhaya *et al.* in [35] are the following:

- Preferred hybrid routing scheme (PHRS)
- Base station supported reconfiguration (BSR)
- Multiple exclusive shortest path routing (MSPR)
- Multipath dispersion heuristic (MPD)

These heuristics are primarily based on the trade-off between the complexity of computation, load-balancing capability, and the increase in path length.

Preferred Hybrid Routing Scheme (PHRS)

This scheme is a hybrid version of the basic schemes such as the *PORS, PIRS, PDRS,* and the shortest path routing schemes explained in the preceding section. In case of *PHRS*, a node evaluates the path using the *PORS*, *PIRS*, and the shortest path scheme and, based on the maximum relay load along these paths, chooses the least loaded path. An alternative to this approach may be to use the average load along the computed paths to decide the least loaded path. Using the average load may address the issue of load balancing, but it may not provide the fairest distribution of the load. Further, choosing a path based on the lowest average load may not lead to an effective increase in throughput because of the chances of any highly loaded bottle-neck nodes that may present. Since the complexity of computing a path using *PORS* and *PIRS* is the same as that of Dijkstra's shortest path algorithm, the overall time complexity of *PHRS* increases only by a constant factor as compared to that of Dijkstra's shortest path algorithm.

Base Station Supported Reconfiguration (BSR)

This routing scheme is applicable only in the case of the MCNs where the BSs help the mobile nodes in path finding during routing. In the case of the MCNs, all *RouteRequests* are sent to the BS, which responds with a route from the source to the destination. All nodes update the BS periodically with their neighbor nodes information. The BS may use any approach to route the traffic between the source and the destination. It may use a simple shortest path routing based on hop-count or other routing strategies, such as *PORS* or least loaded path, to distribute the load on the network effectively. In BSR, the BS reconfigures the paths for the nodes experiencing low effective throughputs. The idea behind this scheme is the fact that the low throughput in a dense network is due to collisions and due to the increase in queuing delays at nodes. The probability of collision is higher in a region where load density is higher. The reconfiguration of routes where the throughput is low tends to redistribute the load from hotter regions (heavily loaded regions) of the network to cooler regions (lightly loaded regions). In the BSR scheme, along with the neighbor information, each node updates the BS with information about the number of packets sent to a particular destination and the packets it has received from a particular destination within a specific time period. The use of a time window enables measuring of current throughput rather that the throughput attained over a long period of time. Throughput computation at the BS is done in order to calculate the throughput obtained by all the nodes in the network.

In BSR, the BS calculates the average throughput of the network and determines the nodes that have low throughput. It then broadcasts a *REQUEST-RECONF* message over the network which contains a list of K source nodes (K is fixed by the BS *a priori* depending upon the size of the network) which are currently experiencing low throughput. The value of K must be chosen judiciously because a high value of

K may result in increased control overhead, while a low value of *K* may not improve the performance of the network as the number of routes reconfigured will be small. These nodes can then decide whether they need reconfiguration and if required, they will request the BS for reconfiguration of the route to the destination using the *RECONF* message. The BS may now reroute the traffic for the requesting node along some lightly loaded paths in the network. Thus, the BSR scheme is a dynamic scheme which tries to achieve load balancing by rerouting calls dynamically, thereby increasing the throughput.

Multiple Exclusive Shortest Path Routing (MSPR)

The routing schemes that are based on the maximum load as the edge weight function in the shortest path computation, increase the path length dramatically, which is not advisable as the overall power consumption of the network will be high. *PORS* and *PIRS* attempt to tackle this problem by striking a balance between the two. *PORS* and *PIRS* are static routing schemes and they cannot balance the load well when traffic generation is not uniformly distributed.

The MSPR attempts to reduce the average hop length of the selected paths along with traffic fairness by searching for multiple paths in a region close to the shortest path. Finding the first k-shortest paths may be computationally complex and may not lead to sufficient benefits as well, because the paths may not be sufficiently node-disjoint.

A *k-MSPR* heuristic determines the path as follows. Dijkstra's shortest path algorithm is used to find the shortest path in the network. This is the first of the k paths obtained by the MSPR heuristic. Next, all edge weights of the links in the shortest path are increased to a high value (a high value which is not infinity prevents selection of such nodes as far as possible and at the same time does not prevent the selection of the node if no other path exists). Dijkstra's shortest path algorithm is run again to find the next shortest path. This process is continued to obtain k maximally disjoint paths. Among these k paths, the one which is least loaded is chosen. Figure 13.30 shows the *k-MSPR* algorithm in brief.

Multipath Dispersion (MPD) Heuristic

This heuristic borrows an idea from the primary-backup approach used for reliability of connections [38]. Here, one primary path and multiple backup paths are selected for the purpose of routing. Since best-effort traffic does not make any reservations along the paths chosen, it does not lead to blocking of unutilized bandwidth. A node may use one or more disjoint paths to transmit its data. The main objective of using multiple paths here is to distribute the load evenly among the paths. The intuition behind using this scheme is that the load on a path will get distributed over multiple paths and hence a region handling heavy traffic will be able to distribute its load over a wider region. This may be particularly useful in military applications such as battlefields where the traffic between a few select points is much higher than that between other points in the network. The presence of multiple routes also provides connection reliability. The characteristics of the MPD scheme are as follows:

```
The k–MSPR Heuristic
HIGH_VALUE = a high value <> INFINITY

PathSearch(Source, Destination){
    // Dijkstra's shortest path search (with edge weights = 1)
}

k–MSPR(Source, Destination, k){
    for i:=1 to k do{
        PathSearch(Source, Destination);
        for all nodes in the returned path
    }
    change all edge weights to HIGH_VALUE
    evaluate max(load on each node) for k paths
    return the path with the least value of
        max(load on each nodes of k paths)
}
```

Figure 13.30. The *k-MSPR* algorithm.

- The load along the paths may be split in a random manner either uniformly or based on the current load on each of the paths.
- In case of TCP-based traffic where the data received is acknowledged, the distribution of load may be based on the throughput achieved along each of the paths. This cannot be applied for UDP traffic since the UDP throughput achieved cannot be easily determined at the source node.

Table 13.3 provides a comparison of the various load balancing schemes discussed in this chapter.

13.7 SUMMARY

The popularity of wireless networks has given rise to many additional requirements of the system, especially the requirements of greater bandwidth and demand for high-quality multimedia services. In addition to the broadband physical layer, frequency reuse provided by different architectures also assumes importance. Several potential next-generation wireless network architectures utilize the multi-hop relaying for increased frequency reuse. A number of such wireless architectures, called hybrid wireless architectures in this chapter, were described in detail. Several issues in these architectures, such as routing, pricing, power control, and load balancing were also described in this chapter. Future networks include a variety of heterogeneous networks with widely varying capabilities, from Bluetooth scatternets to satellite networks. Seamless interoperation of such networks is an important requirement for next-generation wireless networks.

Table 13.3. Comparison of various load-balancing schemes

Feature	PIRS	PORS	Shortest Path	PHRS	BSR	MSPR	MPD
Application	All*	All	All	All	MCN only	All	All
Dynamicity	Static	Static	Static	Dynamic	Highly Dynamic	Depends on k	Less Dynamic
Algorithmic Complexity	$O(N^2)$	$O(N^2)$	$O(N^2)$	$3 \times O(N^2)$	$O(N^2)$	$k \times O(N^2)$	$2 \times O(N^2)$
Average Hop-Count	$\approx 1.18x$	$\approx 1.9x$	x	$\leq 1.9x$	Unknown	$\propto k$	$\leq (x+1)^{\dagger}$
Interoperability‡	Yes	Yes	Yes	Yes	No	Yes	Yes
Changes at Mobile Nodes in Hybrid Wireless Networks	Nil	Nil	Nil	Minor (only updates)	Large (protocol)	Minor (only updates)	Minor (handle multiple paths)
Usage Without Topology Discovery	No	No	Yes	No	N/A	Yes	Yes

* Here "All" refers to hybrid wireless networks such as MCNs and ad hoc wireless networks.

† This assumes the network is reasonably dense so that the hop-count of the next shortest path is at most one more than that of the shortest path.

‡ Interoperability refers to the ability of the network where the nodes with and without load balancing capability can coexist.

13.8 PROBLEMS

1. Find the number of nodes affected by a single call from a node to its BS for (a) a packet-based SCN network, and (b) a packet-based MCN network. Use the following data: number of nodes in the network = N, SCN transmission range = R meters, MCN control transmission range = R meters, MCN data transmission range = r = R/2 meters, and the number of hops for the node to reach the BS = 3. Consider only a single cell with radius R meters.
2. How does the value of reuse factor (k) influence the performance of MCNs?
3. Calculate the value of reuse factor (k) required for an average neighbor density of 6. The number of nodes in the cell is 100 and the cell radius (R) is 500 meters.
4. Calculate the average number of neighbors in an MCN, assuming uniformly distributed nodes with the cell radius $R = 500$ m, the data transmission range $r = 250$ m, and the number of nodes in the network N = 100.
5. Calculate the average bandwidth that can be obtained in the cellular mode in an HWN system with a total system bandwidth of 2 Mbps and 500 nodes in the cell when (a) all nodes have active sessions and (b) 25% of the nodes have active sessions.
6. Calculate the time complexity of BAAR protocol.
7. Assuming the BS in an MCN using BAAR protocol has to service n *RouteRequests* in unit time, calculate the computational burden at the BS.
8. A particular service provider has decided to incorporate the PICR scheme for pricing in a multi-hop voice wireless network. The service provider covers a subscriber base of 20,000 subscribers with ten BSs with one BS per cell. Calculate the minimum bill-rate (currency units per minute) to be charged for meeting a minimum revenue ($N_{R_{min}}$) of 25,000 currency units. The average monthly call duration per subscriber is 100 minutes.
9. Assuming a uniform relay load throughout the network with the PICR scheme in use, calculate the average cost of operation for (a) a node at four hops away from BS and (b) a node at one hop away from the BS. $C_p = 0.1$ currency units, $S = 0.25$ currency units, $\alpha = 0.8$, and relay load = 10.
10. In the above problem, what would be the average costs of operation if the relay loads at four hops away from BS and one hop away from the BS are 2 and 10, respectively?
11. Assuming the PICR-PDBR scheme, in a voice multi-hop network, calculate the average cost of operation for (a) a node at four hops away from BS and (b) a node which is one hop away from the BS. $C_p = 0.1$ currency units, $\alpha = 0.8$, and *relay load* = 10. Assume that bill-rate (S) for a node at one hop away from BS is 0.3, and at four hops away from BS is 0.1.

12. In the above problem, what would be the average costs if the relay loads at four hops away from BS and one hop away from the BS are 2 and 10, respectively?

13. Discuss the factors to be considered for deciding an appropriate cost-reimbursement factor for networks with and without power constraint nodes.

14. Assume that Metro-MCN-Networks LLC Inc. hired you for a modification of their country-wide multi-hop cellular network that runs BAAR protocol. Their requirements include a change in routing protocol that enables the users to dynamically enable and disable their willingness to relay others' traffic. Can you help them tinker the BAAR protocol for incorporating this feature?

15. Having successfully completed the BAAR protocol modification at the Metro-MCN-Networks LLC Inc., they have requested that you solve the control overhead problem in their campus-wide MWLAN. One colleague suggested the aggregation of control packets at the intermediate nodes to reduce the control overhead. Can you list the pros and cons of such a technique?

16. Discuss the pros and cons of increasing the mobility margin in the power control scheme for hybrid wireless networks.

17. In the power control scheme for hybrid wireless networks, what are the two important factors that should be considered to decide the value of the *mobility margin*?

18. Calculate the average path length for (a) shortest path scheme (b) PIRS (c) PORS and (d) PDRS in a circular-shaped network with radius R m.

19. For the PORS scheme, consider r as the mean distance of *Ring* i from the *Center* and α as the thickness of *Ring* i, where $r = \frac{r_i + r_{i+1}}{2}$, $\alpha = r_{i+1} - r_i$, $r_{i+1} = r + \alpha/2$, and $r_i = r - \alpha/2$. For any given *Ring* i, assuming a uniform node distribution with locality = 1.0, find the following:
    (a) Traffic load offered by nodes in *Ring* i to *Ring* i.
    (b) Traffic offered by nodes in all the *Rings* k to *Ring* i and vice versa, where $0 \leq k < i$.
    (c) Traffic offered by nodes in all the *Rings* m to *Ring* i and vice versa, where $i < m \leq R$, where R is the radius of the network.
    (d) Traffic crossing *Ring* i.
    (e) Total traffic in *Ring* i.

20. Search the literature and discuss how auto rate fallback (ARF) in IEEE 802.11b works.

BIBLIOGRAPHY

[1] Y. D. Lin and Y. C. Hsu, "Multi-Hop Cellular: A New Architecture for Wireless Communications," *Proceedings of IEEE INFOCOM 2000*, pp. 1273-1282, March 2000.

[2] R. Ananthapadmanabha, B. S. Manoj, and C. Siva Ram Murthy, "Multi-Hop Cellular Networks: The Architecture and Routing Protocol," *Proceedings of IEEE PIMRC 2001*, vol. 2, pp. 78-82, October 2001.

[3] H. Wu, C. Qiao, S. De, and O. Tonguz, "Integrated Cellular and Ad Hoc Relaying Systems: iCAR," *IEEE Journal on Selected Areas in Communications*, vol. 19, no. 10, pp. 2105-2115, October 2001.

[4] H. Y. Hsieh and R. Sivakumar, "Performance Comparison of Cellular and Multi-Hop Wireless Networks: A Quantitative Study," *Proceedings of ACM SIGMETRICS 2001*, pp. 113-122, June 2001.

[5] A. N. Zadeh, B. Jabbari, R. Pickholtz, and B. Vojcic, "Self-Organizing Packet Radio Ad Hoc Networks with Overlay," *IEEE Communications Magazine*, vol. 40, no. 6, pp. 140-157, June 2002.

[6] K. J. Kumar, B. S. Manoj, and C. Siva Ram Murthy, "MuPAC: Multi-Power Architecture for Packet Data Cellular Networks," *Proceedings of IEEE PIMRC 2002*, vol. 4, pp. 1670-1674, September 2002.

[7] K. J. Kumar, B. S. Manoj, and C. Siva Ram Murthy, "RT-MuPAC: Multi-Power Architecture for Voice Cellular Networks," *Proceedings of HiPC 2002*, LNCS 2552, pp. 377-387, December 2002.

[8] B. S. Manoj, D. C. Frank, and C. Siva Ram Murthy, "Throughput Enhanced Wireless in Local Loop (TWiLL) — The Architecture, Protocols, and Pricing Schemes," *ACM Mobile Computing and Communications Review*, vol. 7, no. 1, pp. 95-116, January 2003.

[9] M. Frodigh, S. Parkvall, C. Roobol, P. Johansson, and P. Larsson, "Future-Generation Wireless Networks," *IEEE Personal Communications Magazine*, vol. 8, no. 5, pp. 10-17, October 2001.

[10] 3GPP TR 25.924 V 1.0.0, "Opportunity Driven Multiple Access (ODMA)," December 1999.

[11] 3GPP TR 25.833 V 1.1.0, "Physical Layer Items Not for Inclusion in Release '99," April 2000.

[12] X. Wu, S. H. G. Chan, and B. Mukherjee, "MADF: A Novel Approach to Add an Ad Hoc Overlay on a Fixed Cellular Infrastructure," *Proceedings of IEEE WCNC 2000*, vol. 2, pp. 23-28, September 2000.

[13] G. N. Aggelou and R. Tafazolli, "On The Relaying Capability of Next-Generation GSM Cellular Networks," *IEEE Personal Communications Magazine*, vol. 8, no. 1, pp. 40-47, February 2001.

[14] B. S. Manoj and C. Siva Ram Murthy, "A High-Performance Wireless Local Loop Architecture Utilizing Directional Multi-Hop Relaying," *Technical Report*, Department of Computer Science and Engineering, Indian Institute of Technology, Madras, India, June 2002.

[15] H. Luo, R. Ramjee, P. Sinha, L. Li, and S. Lu, "UCAN: A Unified Cellular and Ad Hoc Network Architecture," *Proceedings of ACM MOBIHOC 2003*, pp. 353-367, September 2003.

[16] B. S. Manoj, R. Ananthapadmanabha, and C. Siva Ram Murthy, "Multi-Hop Cellular Networks: The Architecture and Routing Protocols for Best-Effort and Real-Time Communications," revised version submitted to *Journal of Parallel and Distributed Computing*, 2004.

[17] K. J. Kumar, B. S. Manoj, and C. Siva Ram Murthy, "On the Use of Multiple Hops in Next-Generation Cellular Architectures," *Proceedings of IEEE ICON 2002*, pp. 283-288, August 2002.

[18] B. S. Manoj, R. Ananthapadmanabha, and C. Siva Ram Murthy, "Multi-Hop Cellular Networks: The Architecture and Routing Protocol for Best-Effort and Real-Time Communication," *Proceedings of IRISS 2002*, Bangalore, India, March 2002.

[19] Y. D. Lin, Y. C. Hsu, K. W. Oyang, T. C. Tsai, and D. S. Yang, "Multi-Hop Wireless IEEE 802.11 LANs: A Prototype Implementation," *Journal of Communications and Networks*, vol. 2, no. 4, December 2000.

[20] C. E. Perkins and P. Bhagwat, "Highly Dynamic Destination-Sequenced Distance-Vector Routing (DSDV) for Mobile Computers," *Proceedings of ACM SIGCOMM 1994*, pp. 234-244, September 1994.

[21] V. Sekar, B. S. Manoj, and C. Siva Ram Murthy, "Routing for a Single Interface MCN Architecture and Pricing Schemes for Data Traffic in Multi-Hop Cellular Networks," *Proceedings of IEEE ICC 2003,* vol. 2, pp. 969-973, May 2003.

[22] N. Feng, S. C. Mau, and N. B. Mandayam, "Joint Network-Centric and User-Centric Pricing and Power Control in a Multi-Cell Wireless Data Network," *Proceedings of CISS 2002*, March 2002.

[23] http://www.terminodes.org

[24] L. Buttyan and J. P. Hubaux, "Enforcing Service Availability in Mobile Ad Hoc WANs," *Proceedings of ACM MOBIHOC 2000*, pp. 87-96, August 2000.

[25] L. Buttyan and J. P. Hubaux, "Stimulating Cooperation in Self-Organizing Mobile Ad Hoc Networks," *ACM/Kluwer Journal for Mobile Networks (MONET), Special Issue on Mobile Ad Hoc Networks*, vol. 8, no. 5, October 2003.

[26] O. Ileri, S. C. Mau, and N. Mandayam, "Pricing for Enabling Forwarding in Self-Configuring Ad Hoc Networks of Autonomous Users," *Technical Report*, WINLAB, Rutgers University, October 2003.

[27] S. Zhong, J. Chen, and Y. R. Yang, "Sprite: A Simple, Cheat-Proof, Credit-Based System for Mobile Ad Hoc Networks," *Proceedings of IEEE INFOCOM 2003*, vol. 3, pp. 1987-1997, March 2003.

[28] M. Jakobsson, J. P. Hubaux, and L. Buttyan, "A Micro-Payment Scheme Encouraging Collaboration in Multi-Hop Cellular Networks," *Proceedings of Financial Cryptography 2003*, January 2003.

[29] IEEE Standards Board, "Part 11: Wireless LAN Medium Access Control (MAC) and Physical Layer (PHY) Specifications," *The Institute of Electrical and Electronics Engineers, Inc.*, 1997.

[30] G. Bhaya, B. S. Manoj, and C. Siva Ram Murthy, "Power Control Scheme for Multi-Hop Cellular Networks," *Technical Report*, Department of Computer Science and Engineering, Indian Institute of Technology, Madras, India, December 2002.

[31] E. S. Jung and N. H. Vaidya, "A Power Control MAC Protocol for Ad Hoc Networks," *Proceedings of ACM MOBICOM 2002*, pp. 36-47, September 2002.

[32] D. S. J. De Couto, D. Aguayo, B. A. Chambers, and R. Morris, "Performance of Multi-Hop Wireless Networks: Shortest Path Is Not Enough," *Proceedings of the First Workshop on Hot Topics in Networks, ACM SIGCOMM 2002*, October 2002.

[33] G. Bhaya, B. S. Manoj, and C. Siva Ram Murthy, "Ring-Based Routing Schemes for Load Distribution and Throughput Improvement in Multi Hop Cellular, Ad Hoc, and Mesh Networks," *Proceedings of HiPC 2003*, LNCS 2913, pp. 152-161, December 2003.

[34] A. Iwata, C. C. Chiang, G. Pei, M. Gerla, and T. W. Chen, "Scalable Routing Strategies for Ad Hoc Wireless Networks," *IEEE Journal on Selected Areas in Communications*, vol. 17, no. 8, pp. 1369-1379, August 1999.

[35] G. Bhaya, B. S. Manoj, and C. Siva Ram Murthy, "Load Adaptive Routing Schemes for Fairness in Relay Traffic in Multi-Hop Cellular and Ad Hoc Wireless Networks," *Technical Report*, Department of Computer Science and Engineering, Indian Institute of Technology, Madras, India, December 2002.

[36] H. Hassanein and A. Zhou, "Routing with Load Balancing in Wireless Ad Hoc Networks," *Proceedings of ACM International Workshop on Modeling, Analysis, and Simulation of Wireless and Mobile Systems*, pp. 89-96, July 2001.

[37] L. Zhang, Z. Zhao, Y. Shu, and O. W. W. Yang, "Load Balancing of Multipath Source Routing in Ad Hoc Networks," *Proceedings of IEEE ICC 2002*, pp. 3197-3201, April 2002.

[38] G. Phani Krishna, M. Jnana Pradeep, and C. Siva Ram Murthy, "An Efficient Primary-Segmented Backup Scheme for Dependable Real-time Communication in Multihop Networks," *IEEE/ACM Transactions on Networking*, vol. 11, no. 1, pp. 81-94, 2003.

[39] D. B. Johnson and D. A. Maltz, "Dynamic Source Routing in Ad Hoc Wireless Networks," in *Mobile Computing*, edited by T. Imielinski and H. Korth, pp. 153-181, Kluwer, 1996.

[40] B. S. Manoj, D. C. Frank, and C. Siva Ram Murthy, "Performance Evaluation of Throughput Enhancement Architectures for Next-Generation Wireless Systems," *ACM Mobile Computing and Communications Review*, vol. 6, no. 4, pp. 77-90, October 2002.

[41] B. S. Manoj and C. Siva Ram Murthy, "On Using Multi-Hop Relaying in Next-Generation Wireless Networks," *Technical Report*, Department of Computer Science and Engineering, Indian Institute of Technology, Madras, India, August 2003.

Chapter 14

RECENT ADVANCES IN WIRELESS NETWORKS

14.1 INTRODUCTION

It is expected that by the end of this decade, the predominant mode of Internet access will be over wireless networks. In order to support high data rates on short ranges, new promising technologies such as the ultra-wide band (UWB) transmission scheme and optical wireless wavelength division multiplexing (WDM) networks are currently under research. In addition to the demands on increasing the performance of the lower layers of the system, pressure is on to efficiently reuse the existing frequencies, which also demands allocation of newer bands for commercial wireless communications. The additional areas of interest in next-generation wireless networks include freedom in selection of a particular network from a set of heterogeneous wireless networks, seamless handoff across such networks, support for multimedia traffic, and all the above services at an affordable cost. The first step in this is the development of wireless fidelity (Wi-Fi) systems that integrate the high-speed wireless LANs with the wide-area packet cellular infrastructure. The demand for high bandwidth for high-quality multimedia traffic has also resulted in utilizing the WDM technology in wireless communication. This chapter discusses the recent advances in the area of wireless networking such as UWB technology, wireless fidelity, optical wireless communication, and IEEE 802.11a/b/g.

14.2 ULTRA-WIDE-BAND RADIO COMMUNICATION

The major differences between the ultra-wide band (UWB) technology and the existing narrow-band and wide-band technologies are the following: (i) The bandwidth of UWB systems, as defined by the Federal Communications Commission (FCC), is more than 25% of the center frequency or a bandwidth greater than 500 MHz. (ii) The narrow-band and wide-band technologies make use of a radio frequency (RF) carrier to shift the base band signal to the center of the carrier frequency, whereas the UWB systems are implemented in a carrier-less fashion in which the modulation scheme can directly modulate base band signals into an impulse with very sharp rise and fall time, thus resulting in a waveform ranging several GHz of bandwidth [1]. These impulses have a very low duty cycle.[1] A simple example of UWB transmission is any radio frequency (RF) transmission with a bandwidth of 500 MHz at a center frequency of 2 GHz. The IEEE 802.11b is a narrow-band system with a 22 MHz bandwidth with center frequencies ranging from 2.412 GHz to 2.462 GHz. One of the major approaches for generating a UWB waveform is the impulse-based approach which is explored later in this section. The principles behind the operation of UWB technology had been applied in radar applications since the 1980s. With proper emission restrictions (restrictions on the upper limit on the transmission power) in place, the UWB spectrum can overlay the existing narrow-band spectrum, resulting in much more efficient use of the existing radio spectrum.

Figure 14.1 illustrates the upper limits of transmission power permitted for the UWB system in comparison with the IEEE 802.11a. The 802.11a standard has three bands, each with 100 MHz bandwidth with an upper limit of effective isotropically radiated power (EIRP),[2] as follows: 16 dBm for the band 5.15-5.25 GHz, 24 dBm for the band 5.25-5.35 GHz, and 30 dBm for the upper band with frequency range 5.725-5.825 GHz. This results in a power spectral density of -3 dBm/MHz,

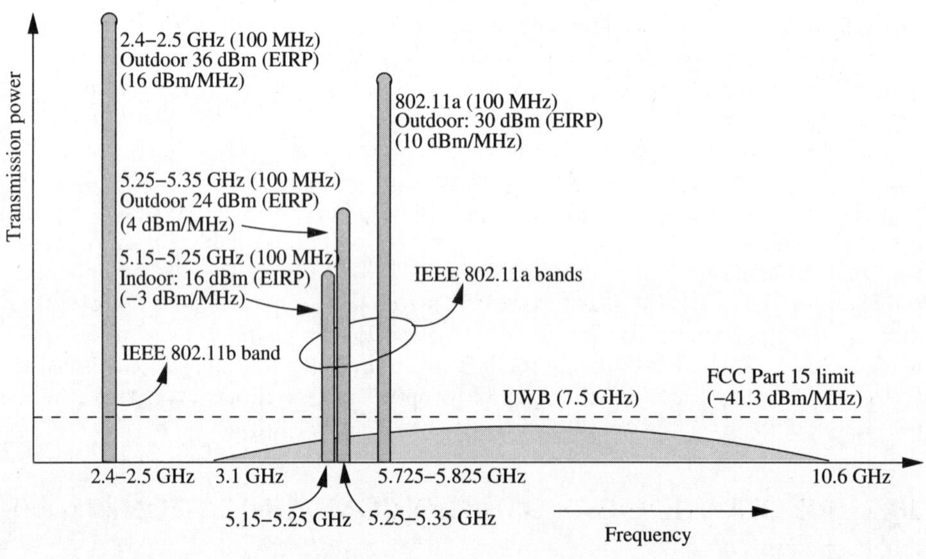

Figure 14.1. Spectrum of UWB systems compared with IEEE 802.11b and 802.11a.

---

[1]The duty cycle of a pulse is defined as the ratio of the duration of the pulse to the sum of the pulse duration and the period between two successive pulses. Lower values of duty cycle refer to shorter pulse durations compared to the period.

[2]EIRP of a transmission system in a given direction is the transmission power that would be needed with an isotropic (otherwise called omnidirectional) radiator to produce the same power density in the given direction.

4 dBm/MHz, and 10 dBm/MHz, respectively. In contrast to these, FCC has restricted the maximum power emitted by the UWB systems to be a power spectral density of -41.3 dBm/MHz. As depicted in Figure 14.1, the UWB signals appear as noise to the legacy systems that utilize the same spectrum.

14.2.1 Operation of UWB Systems

The operation of UWB systems is based on transmission of ultra-short pulses (that yields a wide-band bandwidth signal [1]) which are also called monocycles. Each monocycle is similar to a single cycle of an ultra-high-frequency sine wave and is a single ultra-short pulse. The pulse widths of monocycles range from 0.10 to 1.6 ns with pulse-to-pulse intervals of between 25 and 1,000 ns. A single monocycle of width 0.5 ns is shown in Figure 14.2 (a). Figure 14.2 (b) shows a pulse train involving a sequence of monocycles generated at regular intervals. Monocycles are inherently wide bandwidth signals. The time domain function of a Gaussian monocycle (any sharp pulse whose time domain behavior can be represented as the first derivative of a Gaussian distribution) can be represented as $g(t) = \frac{t}{\tau} \times e^{-(\frac{t}{\tau})^2}$, where t is time and τ is the time-decaying factor that decides the duration of the monocycle. It can be represented as $g(f) = -j \times f\tau^2 e^{f\tau^2}$ in the frequency domain, where the center frequency (f_c) is proportional to $\frac{1}{\tau}$. This means that the shorter the monopulse duration, the higher the center frequency. For a monopulse similar to the one shown in Figure 14.2 with a pulse width of 0.2 ns, the center frequency is around 5 GHz. The frequency domain function of an ultra-short pulse shows that the frequency components spread across (bandwidth) a wide range greater than 110% that of the center frequency. For a monopulse with 0.2 ns duration, the bandwidth is approximately around 5 GHz, as illustrated in Figure 14.3.

In the pulse train of monocycles, the information is modulated either by using pulse position modulation (PPM) or similar modulation techniques [2], [3]. The effect of modulation with PPM on a train of monocycles is illustrated in

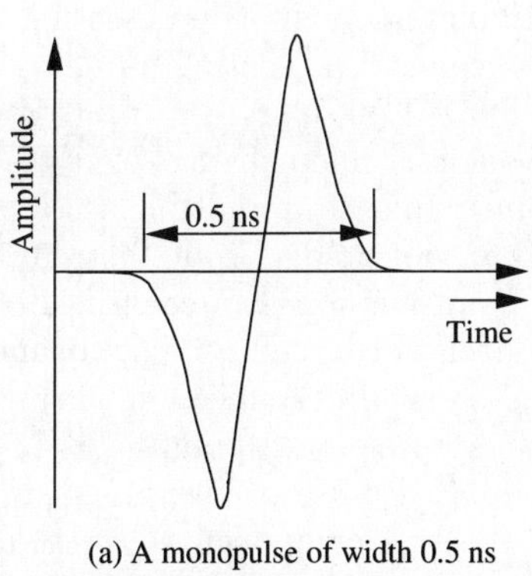

(a) A monopulse of width 0.5 ns

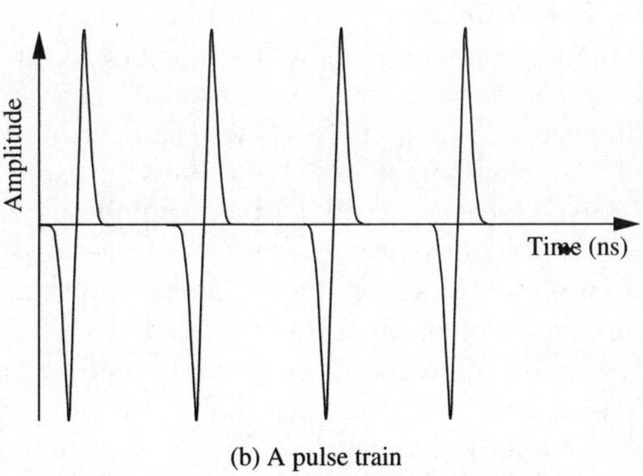

(b) A pulse train

Figure 14.2. An illustration of a monocycle and pulse train.

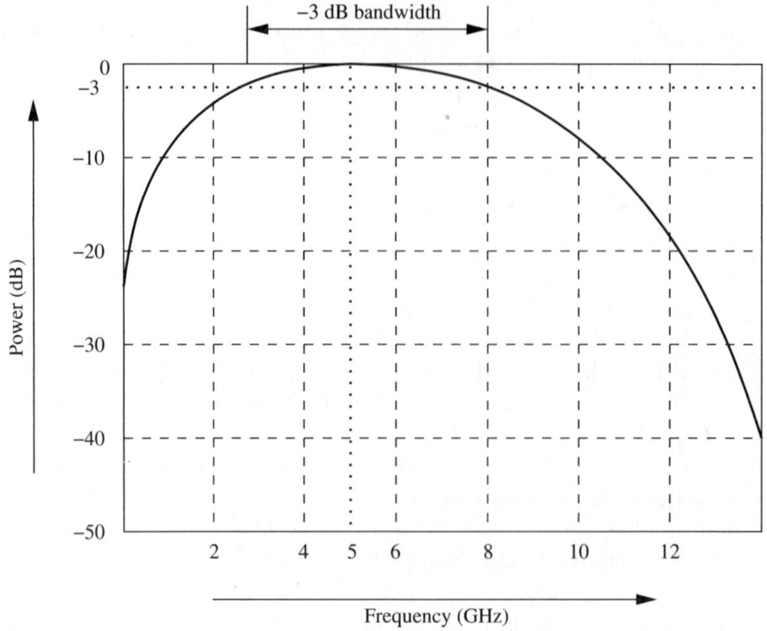

Figure 14.3. Bandwidth of a monocycle.

Figure 14.4 (a), in which the precise timing of the monocycle varies with respect to its nominal position. For example, in a 100 Mpps (million pulses per second) system, monocycles are transmitted every 10 ns (refer to Figure 14.4). In this case, a PPM modulation may advance the monocycle for 10 picoseconds for representing a digital bit "1" and delay the monocycle for 10 picoseconds representing a digital "0". An encoding can be performed over the unmodulated pulse train to provide distinct time-hopping codes for channelizing the pulse train (channelizing refers to provisioning of multiple channels in the same spectrum). In such a multiple access system, each user would have a unique pseudo-random noise (PN) codes for encoding the pulse trains so that multiple simultaneous transmissions can coexist [4]. Figure 14.4 (b) shows the PN coded pulse train without data content. The nominal pulse position in the train is decided by the PN code as shown in Figure 14.4, where the pulses appear with different time differences from the reference points Tn+1, Tn+2, and Tn+3. This encoding process on the modulated pulse train makes the signal appear like white noise (noise containing all spectral components). A receiver cannot detect the transmission and the corresponding data without having the unique pseudo-random (time-hopping) code with which it was coded, even when the receiver is located at very close proximity to the transmitter.

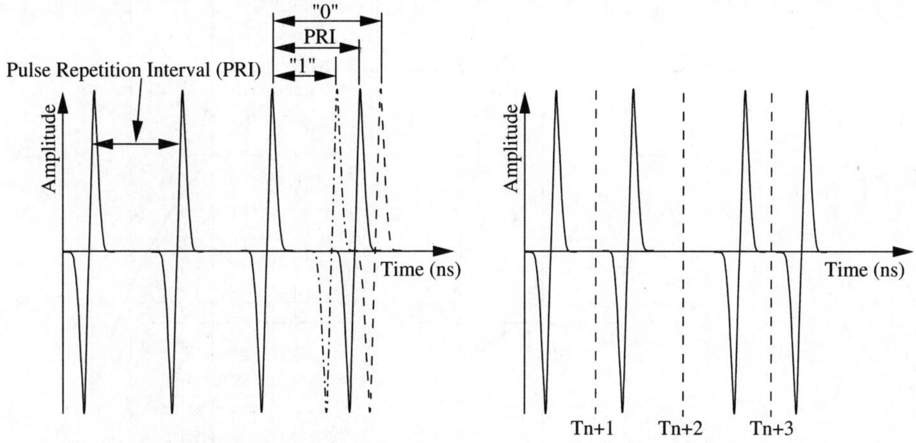

(a) A pulse position modulated pulse train (b) A pseudo–random coded pulse train

Figure 14.4. The PPM coded pulse train.

14.2.2 A Comparison of UWB with Other Technologies

The trends that drive the short-range wireless technologies, including UWB technology, are the following: (i) The growing demand for data and multimedia capability in portable devices at high data rate but at low cost and power consumption. (ii) Increasing pressure on the wireless spectrum demanding higher reuse. (iii) Decreasing semiconductor cost and availability of low-power devices. Some of the competing technologies for such wireless access scenario are IEEE 802.11b, IEEE 802.11a, and Bluetooth.

The bandwidth, transmission range, and capacity comparison in bps/m^2 among 802.11b, Bluetooth, 802.11a, and UWB is illustrated in Figures 14.5, 14.6, and 14.7, respectively. In North America, the 802.11b spectrum ranges from 2,400 MHz to 2,483 MHz and is divided up into 11 channels from 2,412 MHz to 2,462 MHz, spaced 5 MHz apart, as illustrated in Figure 14.8. However, each channel is 22 MHz wide, resulting in great overlap. For example, the spectrum of channel 1 centered at 2,412 MHz overlaps with neighboring channels 2, 3, 4, and 5. This leads to a situation where, at any given point, only three channels (1, 6, and 11) can be simultaneously used. In Europe, of the 13 permitted channels, four channels can be used simultaneously.

Hence in North America, three channels among the 11 channels can operate simultaneously, providing a total bandwidth of 33 Mbps. The capacity of the system with respect to the area of coverage can be 33 Mbps over an area of a circle with a 100 m radius. Hence, 802.11b has a capacity of 1 Kbps/m^2.

Bluetooth has a transmission range of 10 m in its low-power mode and a peak bandwidth of 1 Mbps. There can be around 10 simultaneous piconets operating in

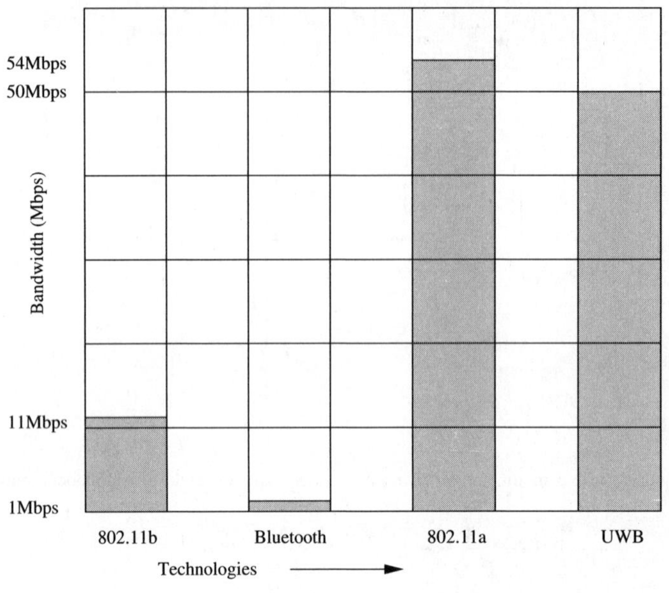

Figure 14.5. Bandwidth comparison of UWB with other competing technologies.

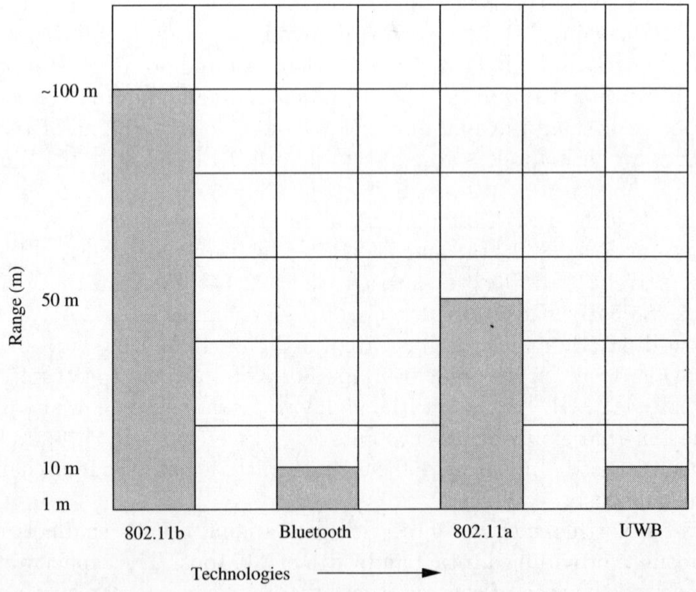

Figure 14.6. Range comparison of UWB with other competing technologies.

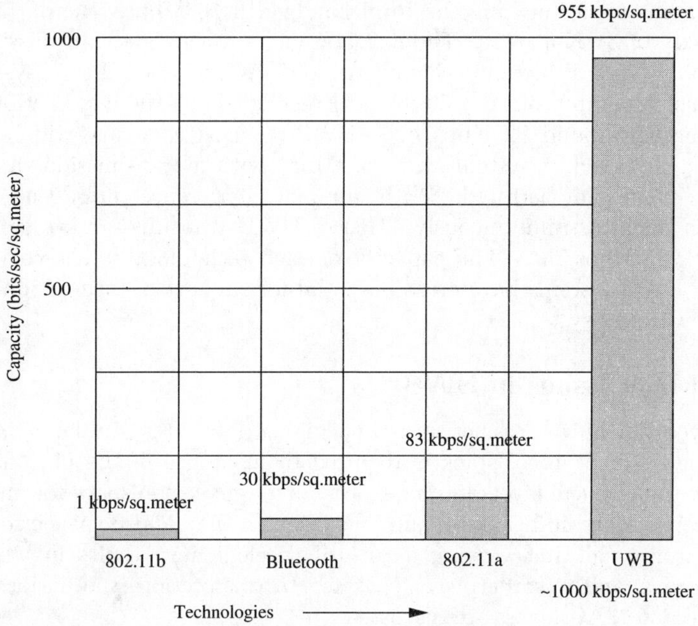

Figure 14.7. Capacity comparison of UWB with other competing technologies.

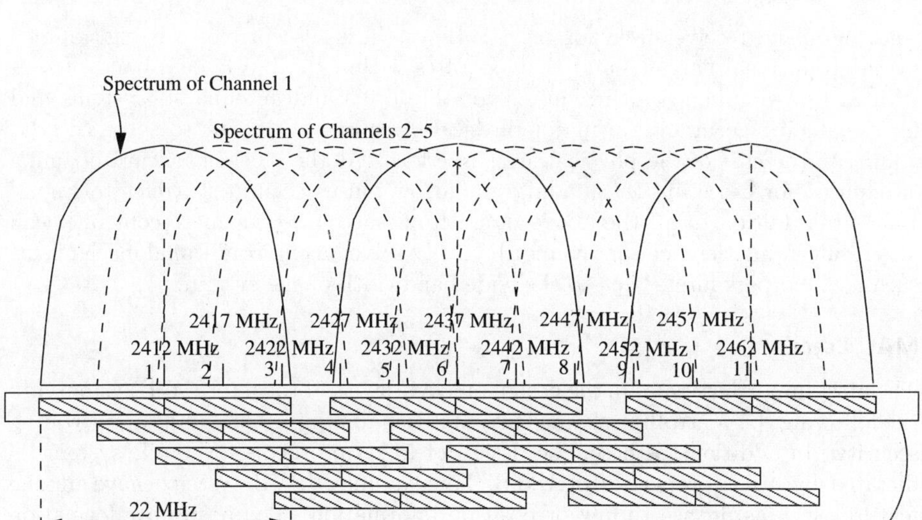

Figure 14.8. An illustration of 802.11b channels.

the 10 m radius circle, making the total bandwidth 10 Mbps. Hence, Bluetooth has a capacity of 31 Kbps/m^2. The 802.11a has an operating transmission range of around 50 m with a maximum bandwidth of 54 Mbps, and 12 such systems can simultaneously operate within a 50 m radius circle. Thus the capacity of 802.11a is approximately around 82 Kbps/m^2. UWB systems can vary widely in their projected capacities, but systems with 50 Mbps with a transmission range of 10 meters have been demonstrated. With minimal interference, more than six such systems can operate simultaneously. Hence, UWB transmission systems have a capacity of 955 Kbps/m^2. The capacity of each technology is shown in Figure 14.7. Thus, UWB systems have great potential for support of future high-capacity short-range wireless systems.

14.2.3 Major Issues in UWB

With the evolution of wireless networks with higher data rates, similar advancement in all the layers of protocol stack is required to make use of the high data rates provided by the physical layers [5]. The increasing application space for multimedia communications puts additional requirements on latency and delay performances. The error-prone and time-varying nature of wireless link results in large delay, delay variations, and misordering of packets. In this section, the major issues at the physical and MAC layers are discussed.

Physical Layer

The wide-band receiver used for the UWB system is susceptible to being jammed by the traditional narrow band that operates within the UWB pass band. Issues such as the wide bandwidth required for the filters and antenna subsystems and filter-matching accuracy are major physical-layer issues difficult to solve without adding to the cost of the physical-layer interface. At the receiver, accurate timing is required for detecting the modulated narrow pulses accurately (refer to Figure 14.4). In addition to all the above, noise from the on-board micro-controller can also result in interference that cannot be easily solved using traditional mechanisms such as band-pass filters, due to the wide bandwidth of the system.

MAC Layer

The most important issues in the design of MAC-layer protocol for UWB systems are the following: (i) controlling channel access, (ii) maintaining QoS, and (iii) providing security. In addition to these, the design of MAC protocols for UWB systems is dictated by other properties of UWB. For example, UWB systems have unique features such as precise timing or position information. The position information can be obtained by the following feature. A 10 GHz UWB system can distinguish signal echos (signal echos refer to the multipath signals that arrive at the receiver through different paths with corresponding delays) that differ by 100 picoseconds. By analyzing the different signal echos that take different delays, a nearly exact position and time information can be obtained. For a 10 GHz system, the delay differences could be of the order of 100 ps and hence an accuracy of 3 cm in distance estimation can be achieved. Utilization of this feature at the MAC layer can improve performance of multimedia unicast and multicast communication. MAC protocols can also benefit from the flexibility of trading off the throughput with respect to the transmission range. The peak amplitude of pulses and the pulse repetition frequency (PRF) can be varied to obtain constant average power. Using this mechanism, MAC protocol can provide different data rates and transmission ranges on a per-link or per-packet basis. UWB systems can be implemented using spread spectrum technology in order to provide better coexistence with the existing systems. Above all, another major aspect of the design of MAC protocol for UWB systems is the compatibility and coexistence of UWB systems with existing WPANs and WLANs.

14.2.4 Advantages and Disadvantages of UWB

Promising applications of UWB in communication include cable-free audio/video devices, broadband WPANs, and high-speed wireless links. UWB systems are potential candidates for high data rate, low power, and short- to medium-range communication applications. UWB systems have a wide range of applications other than communications, some of which are automobile collision-detection devices, medical imaging similar to x-rays and ultrasound scans, through-wall imaging for detecting people and objects in law-enforcement applications, and ground-penetrating radars in construction applications. The major advantages of UWB systems include simplicity of implementation, high data rate, inherent robustness to multipath fading, flexibility of operation, low power consumption, and low cost of implementation. The disadvantages of UWB systems include stringent design requirements of communication subsystems and chances of interference from the existing technologies.

14.3 WIRELESS FIDELITY SYSTEMS

Wireless fidelity (Wi-Fi) system is the high-speed wireless LAN that was originally intended to extend the wired Ethernet in offices to wireless clients. The coverage area and ability to support high bit rates are the two major reasons behind the name Wi-Fi. Though the popular wireless LAN standards IEEE 802.11b and 802.11a are considered as the standard Wi-Fi candidates, conceptually any high-speed wireless LAN protocol such as HiperLAN can be used. The integration of *Wi-Fi hotspots* (wireless LAN access points) with wide area wireless networking technologies such as GSM and GPRS provides an added advantage for the mobile nodes. Such an integrated system provides secure, reliable, and high-speed wireless connectivity. A Wi-Fi network can be used to connect computers to each other, to the Internet, and to wired networks. Wi-Fi networks operate in the unlicensed 2.4 GHz and 5 GHz radio bands, with an 802.11b or 802.11a, or with products that contain both bands (dual band), so that they can provide an enriched network experience. Wi-Fi systems are potential candidates for provisioning high-speed multimedia content delivery in areas such as indoor offices, airport lounges, and shopping malls. The Wi-Fi Alliance [6] is a non-profit international association formed in 1999 to certify interoperability of IEEE 802.11-based products in order to make the objectives of Wi-Fi a reality. The advantages of Wi-Fi systems are ease-of-use, high-speed Internet access, low cost of operation, and flexibility of reconfiguration.

14.3.1 The Service Provider Models for Wi-Fi Systems

- **The Wi-Fi micro carrier model:** In this model, small business operators can set up their own access points (APs) and maintain customer relations and billing with subscribers. An example of this category is a restaurant operating a small Wi-Fi system with a set of APs on its premises.

- **The franchisee-franchisor model:** This model for Wi-Fi systems is that a franchisor company making an agreement with a franchisee (*e.g.*, a restaurant which has an inbuilt Wi-Fi system for its internal purposes) for providing Wi-Fi connectivity on a revenue-sharing basis. The external communication, access network costs, and back-office softwares may be supplied and maintained by the franchisor. Hence, the franchisor company can extend its services to the public.

- **The Wi-Fi carrier model:** In this model, a particular company referred to as Wi-Fi carrier can own, deploy, and operate a number of Wi-Fi system-enabled APs at public places. The subscribers can utilize the designated carrier's network services in their coverage area based on acceptable billing models.

- **The Wi-Fi aggregator model:** This model refers to an abstract service provider which strikes wholesale partnerships with Wi-Fi operators. Such aggregators mainly focus on two major things: (i) reselling of the services provided by the Wi-Fi operators and (ii) giving their subscribers access to a large number of networks. The advantages of this model are easy scale-up of network coverage as the aggregator does not own the infrastructure and hence, by having more partnerships, a service provider can increase the coverage area and the customer base.

- **The extended service provider model:** The synergistic operation of Wi-Fi systems with existing cellular systems, especially with the 3G systems, can increase profits for cellular operators. This can even lead to reduction in the deployment cost of 3G systems. The widespread deployment of Wi-Fi systems can be considered as complementing the 3G systems. Also, the availability of wireless devices equipped with Wi-Fi and cellular interfaces encourages the possibility of switching to the Wi-Fi systems whenever an AP is detected. *Vertical handoff* (the handoff performed across two networks which are operating at widely varying coverage regions, for example, the handoff performed between a wireless LAN and wireless WAN) can be used to switch back to the wide area cellular networks as and when necessary in such cases. Thus, the extended service provider model envisions the provisioning of Wi-Fi services as an extension to the existing service provided by the cellular network operators. An illustration of such a model is shown in Figure 14.9, where the

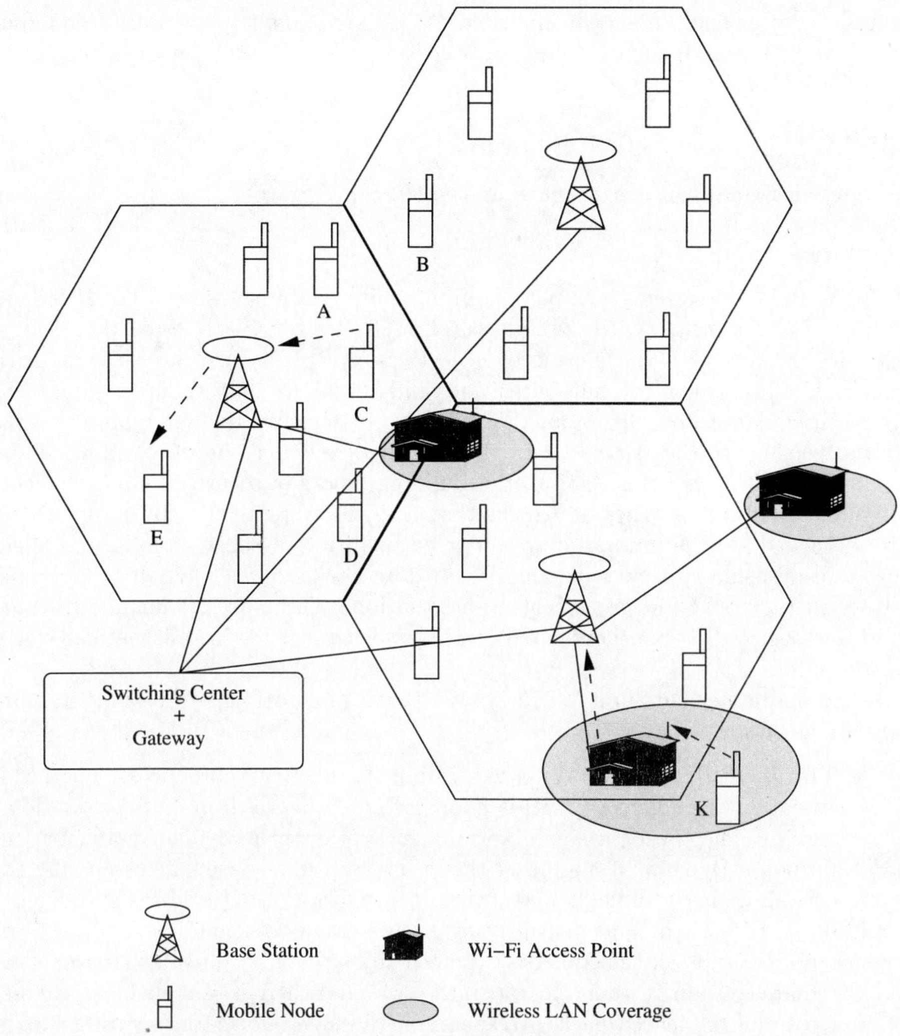

Figure 14.9. A typical scenario of a Wi-Fi system.

users can access information and data through either the Wi-Fi AP or through the 3G BS. Referring to Figure 14.9, the node C can communicate directly with the BS in order to make a call to the destination node E, and node K can communicate with the BS through the nearest Wi-Fi AP. This choice of selection can be based on network availability, cost of access, bandwidth availability, and specific user requirements. Such a system requires placement of APs at places such as crowded traffic junctions, public places, and railway stations. The presence of multiple service providers trying to use this model would result in managed utilization of spectrum in the bands assigned for 802.11.

14.3.2 Issues in Wi-Fi Systems

The major issues in Wi-Fi systems are the following:

Security

Security in Wi-Fi systems is important and requires viable solutions. The IEEE has proposed a long-term security architecture for 802.11, which is referred to as robust security network (RSN). The major objectives of RSN, which is based on IEEE 802.1x, are access control, authentication, and key management. Even though the research work done by Mishra and Arbaugh in [7] indicates the vulnerability of the initial versions of the IEEE 802.1x, IEEE 802.1x remains the only viable option as of this writing. In this section, the building blocks of security, major elements of threats and major types of attacks that a Wi-Fi system could be subjected to, are discussed. In the infrastructure mode of operation of 802.11 standard, a client has to establish a relation with an AP called an *association*. There are three main states in the transition process of an association. These are (i) unauthenticated and unassociated, (ii) authenticated and unassociated, and (iii) authenticated and associated.

The major elements in a Wi-Fi system that can contribute to its vulnerability are discussed below.

- **The eavesdropper:** Wireless medium is inherently broadcast, making it trivial for the adversary with a good receiver to eavesdrop on other stations' traffic. The use of a single key for encryption proved inadequate for this purpose. Dynamic changing of the encryption key is a must for making the eavesdropping a difficult task. Since the majority of the Wi-Fi systems are not WEP (wired equivalent privacy – any privacy solution in which the effort required to break the cipher is expected to be roughly equivalent to the effort required by an intruder to tap into a wired Ethernet)-enabled, it is easier for an inexperienced hacker to snoop on valuable informations such as credit card number, PIN numbers, or social security numbers from a public hot spot Wi-Fi system.

- **The rogue client:** The authenticated client that has intentions of acting against the organization is another problem aggravated by the wireless environment. A Wi-Fi client running a routing protocol and an additional wired interface can easily use its authentication information on its corporate wireless network infrastructure to transfer valuable information. This problem is extremely difficult to trace and prevent as the loss incurred cannot easily be detected.

- **The rogue AP:** The rogue AP is the unauthorized AP attached to an organization's network, either on its premises or in its authorized employee's home. An AP without the necessary security provisions, or in situations where the security is disabled temporarily, can act as a rogue AP. Even with security features such as WEP, organizations that use virtual private network (VPN) to secure their wireless LANs may be compromising valuable data. A VPN is a communication network between two or more machines or networks, built for the private use of an organization over a shared public infrastructure such as the Internet. The mechanisms to identify rogue APs include sniffing, pinging, and wide spectrum channel monitoring around the premises of the organization.

The two major types that make a Wi-Fi network particularly vulnerable are session hijacking and man-in-the-middle attack. Both exploit the authentication process.

- **Session hijack attack:** In this case, a hijacker node waits until the successful completion of a node's authentication process with an AP. On completion of the authentication, the hijacker sends a forged disassociate message that appears to be originated by the AP. On reception of the message, the original node gets disassociated. The AP continues to consider the node as part of the network. Now the hijacker, using the originally authenticated node's address, can utilize network services until the session expires. The initial versions of IEEE 802.1x security framework for 802.11 protocols are vulnerable to this attack, as indicated in [7].

- **Man-in-the-middle attack:** When an adversary spoofs messages to represent itself as AP to a node under attack, it can obtain the victim's authentication information. Using this information, the adversary authenticates with another AP masquerading as the victim node. This attack is known as man-in-the-middle attack. This attack is particularly effective when the victim node is not present within the transmission range of the AP.

Authentication

The IEEE 802.1x standard is proposed for providing authentication and controlling traffic to a protected network, as well as for dynamically changing encryption keys. The use of the extendible authentication protocol (EAP) as a framework for wired and wireless network provides a mechanism for multiple authentication methods such as certificate-based authentication, Kerberos[3] authentication method, token cards, one-time passwords, and public key authentication. The interactions between specific entities in 802.1x is discussed below. The beginning of a series of messages that mark the authentication process is when a client (referred to as the supplicant in 802.1x terminology) sends an EAP-start message to the AP (authenticator). The authenticator responds with an EAP-request identity message. The client replies with an EAP-response packet that contains the identity of authentication server. This authentication server uses a specific authentication algorithm to verify the credentials of the client by performing verification based on any of

[3]Kerberos is a network authentication protocol, designed to provide strong authentication for client/server applications using secret key cryptography.

the multiple authentication mechanisms mentioned above. Upon verification of the client's information, the AP originates an EAP-success or EAP-failure packet to the client. Once the authentication is successful, the AP permits data traffic between the client and the network.

Quality of Service (QoS)

Provisioning of QoS is important in supporting time-sensitive traffic such as voice and video. The IEEE 802.11e standard which is under consideration is aiming at providing enhanced QoS for Wi-Fi systems.

Economics of Wi-Fi

- **Billing schemes:** Billing for Wi-Fi systems assumes importance as the commercial viability is a major factor for the existence of the Wi-Fi systems. The possible billing schemes include the following. The entity that has the responsibility of customer care and billing can be different from the one which actually provides network access service. Such an agency is referred to as a *billing agency*. The billing agency can employ different methods to accept payments from the customers. The major billing approaches that can be employed in the case of Wi-Fi systems are flat-rate schemes and volume-based pricing. In flat-rate schemes, the user is permitted to utilize network services for a specified amount of time without restricting the bandwidth. The volume-based approach charges the customer based on the amount of data transacted over the network. In addition to both these schemes, business establishments can provide Wi-Fi services as a value addition to the customers visiting the premises for the core business activity. In such cases, the billing is not considered as the bandwidth provided belongs to the organization's unutilized bandwidth; it can even be considered without additional charging.

- **Revenue sharing model:** In the franchisee-franchisor model and the aggregator model, the sharing of revenue is important as there exist multiple business entities in the process of customer relationship, billing, and providing service. Different revenue-sharing models that include a fixed fraction-sharing model (in which the amount shared among the parties involved is prefixed) and a variable-fraction volume-based sharing model (in which the percentage of revenue that goes to different parties involved varies with the volume of data transferred) can be employed. Such systems can consider a constant-fraction for the user per bit of data transferred, and in the high-traffic-density environments, a variable rate per bit of data transferred can be used. In the variable-fraction per bit of data transferred, application level mechanisms which communicate the cost of communication at any particular location is essential.

Spectrum Issues

The issues related to spectrum management are important as Wi-Fi becomes popular with its increasing use as a critical business communication infrastructure. The current allocation of free ISM band in the 2.4 GHz band for 802.11b raises several questions of interference. The source of interferences can be either naturally generated by other devices that are designated to operate at the same band or artificially generated by a rogue interference generator node. Since ISM band is unlicensed, any user with an 802.11 interface can disrupt communication at a specific location without inviting prosecution. The major interference sources for Wi-Fi systems are the following: (i) interference from cordless phones, microwave ovens, and Bluetooth-enabled devices and (ii) interference from jammers. The first issue can be reduced to some extent by the following ways: (i) use of different frequency-hopping patterns for communication-technology-related interference as used in Bluetooth standard and (ii) proper electromagnetic shielding in devices using microwave band for non-communication-related purposes such as microwave ovens.

14.3.3 Interoperability of Wi-Fi Systems and WWANs

The wireless wide area networks (WWANs) use base stations (BSs) that cover a few tens of kilometers in radius. Traditionally, the mobile nodes (MNs) communicated only through these BSs. However, WWANs provide only low bandwidth (in terms of tens of Kbps) compared to the broadband LAN services (in terms of tens of Mbps) offered by the Wi-Fi networks. There are many situations where a WWAN cannot serve the requirements well, some of which are the following: (a) interiors of buildings, basements of buildings, subways, and where the signal-to-noise ratio (SNR) may not be sufficient to provide a high-quality service (b) heavily loaded BSs (cells), where the call blocking ratio is high due to the high offered traffic and the limited spectrum of WWAN. The Wi-Fi operators can leave the responsibilities such as billing, brand-building, and advertising to the WAN SP and hence, the Wi-Fi SPs are likely to derive a sustained revenue at a low cost. Hence, the interoperability between heterogeneous networks, especially between the existing WWANs and Wi-Fi networks, is mutually beneficial. The growing deployment of Wi-Fi hotspots[4] also underlines the importance of using them as a complementary access system to the existing WWANs. The concentration of hotspots is especially high in places such as commercial complexes, business districts, airports, and educational institutions. In such scenarios, an MN currently registered with a WWAN BS may enter the coverage regions of Wi-Fi APs very frequently. The MN may then choose to relinquish its connection to the WWAN BS and instead use the Wi-Fi AP for communication. This leads to a reduction in the load on the cellular network, thus enabling more MNs to be supported. The interoperability of Wi-Fi systems and cellular networks is advantageous to both the network users and the network service providers. This interoperability is the basis of the extended service provider model described in Section 14.3.1.

[4] Wi-Fi hotspots are implemented using WLAN access points (APs) and hence, hotspots and APs are used interchangeably.

Network Selection and Wi-Fi Interoperability

The current-generation MNs such as laptop and palmtop computers can support a wide variety of functions such as voice calls, streaming multimedia, Web browsing, and e-mail, in addition to other computing functions. Many of them have multiple network interfaces that enable them to communicate with different networks. Active research is underway to develop a commercially viable unified network interface that can operate across several heterogeneous networks (multimode 802.11 interface, discussed in Section 14.5, is a beginning in this direction). These varied functions place different requirements on the communication networks supporting them. For example, an MN involved in streaming multimedia traffic requires a much higher bandwidth than an MN involved in downloading a Web page. A *user-profile* defines the abstract behavior of the network user or an MN derived from the resource requirements and the user preferences. All nodes that follow a particular user-profile can be considered to belong to a particular class. The user-profile [8] of an MN determines the overall network access behavior of the MN, including whether it should connect to the Wi-Fi AP or to the WWAN BS when it is in the coverage regions of both. Switching between Wi-Fi APs and WWAN BSs on the basis of user-profiles also balances the different network resources among the different MNs with varied requirements. Three distinct user-profiles for MNs are given below:

- **Bandwidth-conscious user-profile**: This is the profile of a user or a class of users in which the user always chooses to connect to a network with the highest free bandwidth. An estimate of the free bandwidth available at an AP or a BS is sent along with beacon messages periodically transmitted by the respective AP or BS. A bandwidth-conscious MN, on receiving such a beacon of sufficient signal strength, will switch to the advertising BS or AP if the advertised bandwidth is greater than the free bandwidth estimate at the BS or the AP it is currently registered to. In order to avoid frequent switching between two different networks, a bandwidth threshold can be used. In such a bandwidth-threshold-based network-switching process, a handoff decision is made only when the new network has a bandwidth difference that exceeds the bandwidth-threshold. The MNs with high bandwidth requirements (like those engaged in multimedia data transfers) can possess this type of user profile.

- **Cost-conscious user-profile**: This user-profile represents the profile of a user or a class of users in which the user always prefers to be connected to the network that offers the lowest per-byte transmission cost among the available choices of networks. Each BS or AP advertises its associated per-byte transmission cost in the periodic beacons sent by it. A cost-conscious MN will switch to a different AP or BS only if the advertised transmission cost is less than that of the AP or BS with which it is currently registered. APs or BSs belonging to the same network service provider may advertise different transmission costs depending on their current load or their geographic location. An MN engaged in non-real-time file transfer can possess a cost-conscious user-profile.

- **Glitch-conscious user-profile**: A *glitch* is defined as an interruption in the transmission or connectivity which occurs when an MN switches to a new AP or BS. Thus an MN with a glitch-conscious user-profile tries to minimize the number of both vertical[5] and horizontal[6] handoffs it undergoes in order to achieve the smoothest possible transmission. One strategy to achieve this goal is to remain connected with the BSs of the cellular network as long as possible. The larger coverage regions of the cellular BSs result in fewer horizontal handoffs than in the case of the Wi-Fi APs.

In all the three different user-profiles, maintaining connectivity is of utmost importance to the MN. This may result in an MN registering with an AP or a BS whose parameters go against the MN's user-profile. For example, when a cost-conscious MN finds the signal strength from its registered AP falling below a specified threshold value, it will switch to a different AP or BS in its coverage region, even if the per-byte transmission cost associated with the new AP or BS is higher than that of its current AP. The user-profile of an MN does not remain constant with time. For example, an MN involved in low-bandwidth non-real-time file transfers may switch to a bandwidth-intensive multimedia application. Moreover, an MN can possess multiple user-profiles at the same time – for example, MNs can be both bandwidth- and glitch-conscious. In such cases, the decision to switch between the APs and the BSs is more complex.

The user-profile of an MN determines its behavior and resource consumption as it moves across the terrain, encountering different APs and BSs on the way. The distinct behavior of nodes belonging to different user-profiles is described through Figure 14.10. An MN moves from point A to point E along the dotted line shown in the figure. The maximum bandwidth available at BS1 is 100 Kbps, while those available at the two APs, AP1 and AP2, are 11 Mbps each. In the scenario depicted here, it is assumed that the free bandwidth available at AP1 is much less than that at either BS1 or AP2 due to a very large number of MNs currently registered with AP1. It is also assumed that the free bandwidth available at AP2 is greater than that at BS1. The per-byte transmission cost associated with the BS is higher than that associated with the APs. The behavior of each class of MNs that holds the various user-profiles is described below:

Bandwidth-Conscious MNs

A bandwidth-conscious MN always tries to register with the BS or AP offering the maximum free bandwidth. It can be seen from Figure 14.10 that the MN registers with the sole AP (AP1) accessible to it at the beginning of its journey (point A). At point B, it comes under the transmission range of BS1 also. Since BS1 has more free bandwidth than AP1, the MN will switch over to BS1 and will remain registered with it until it reaches the point D. On entering the range of AP2 at point D, the MN switches over to AP2, although it is still in the range of the

[5] A handoff that takes place across different networks, for example, across a WAN and a LAN, is called a vertical handoff.

[6] The handoff that takes place between two network access entities of the same type, for example, between two APs or between two BSs, is called a horizontal handoff.

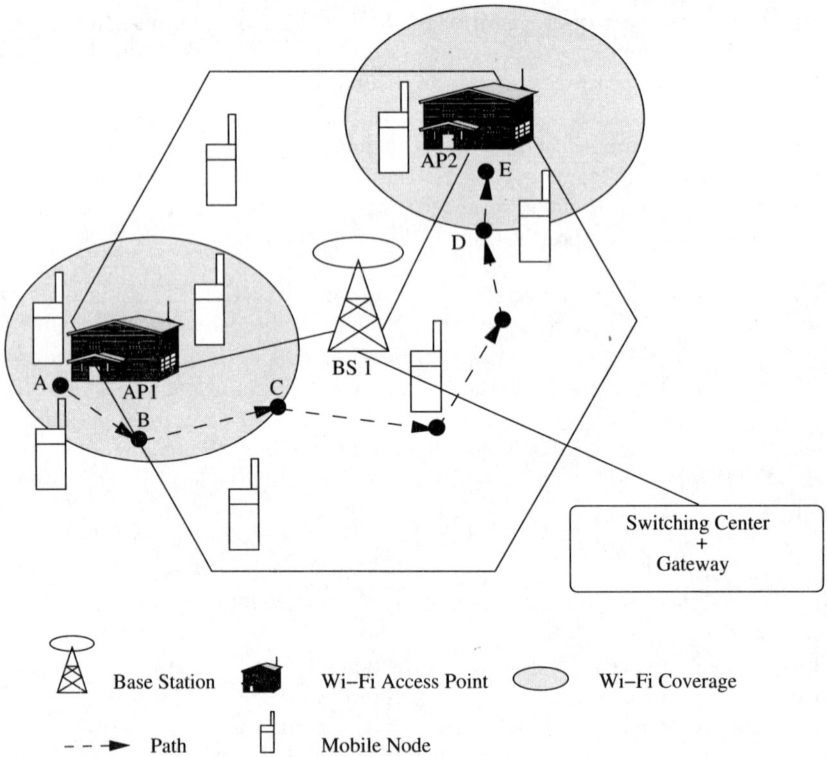

Figure 14.10. Behavior of MNs with different user-profiles as they move across the terrain.

BS. This is because AP2 advertises a higher amount of free bandwidth than BS1. The MN remains with AP2 until the end of its journey (point E).

Cost-Conscious MNs

A cost-conscious MN tries to register with the least-cost AP or BS at all times. After starting its journey from point A, the MN remains registered with AP1 until point C is reached. It must be noted here that the MN does not switch over to BS1 after it enters BS1's transmission range at point B. This is because the transmission cost associated with BS1 is higher than that associated with AP1. At point C, however, the MN goes out of the range of AP1 and is thus forced to register with the higher cost BS1 in order to maintain connectivity. On reaching point D, the MN registers with the lower cost AP2 and remains with it until the end of its journey (point E).

Glitch-Conscious MNs

A glitch-conscious MN tries to minimize the number of glitches in its connection by registering with the BS that has a much larger coverage. The MN remains registered with AP1 between points A and B. However, once it enters the range of BS1 at point B, it switches over to the BS and remains with it until the end of the journey at point E.

Table 14.1 shows the points at which network switching takes place as the MN at point A moves across the terrain shown in Figure 14.10.

Table 14.1. Handoff behavior of the MNs that have different user-profiles

User-Profile	Switching at		
	Point B	Point C	Point D
Bandwidth-Conscious	Yes	No	Yes
Cost-Conscious	No	Yes	Yes
Glitch-Conscious	No	Yes	No

14.3.4 Pricing/Billing Issues in Wi-Fi Systems

Pricing/billing schemes for a Wi-Fi systems assume significance as the commercial viability is very important for the survival of Wi-Fi systems. In the case of Wi-Fi systems, unlike the traditional wired or wireless networks, the entity that has the responsibility of customer care and billing can be different from the one which actually provides network access service. Henceforth, the agency that is responsible for providing customer care, billing, and revenue-sharing is referred to as the *billing agency* (BA). The BA can employ different methods to accept payments from the network users. The major billing approaches that can be employed in the case of Wi-Fi systems are *flat-rate pricing* and *volume-based pricing.* In the case of flat-rate pricing, a user is permitted to utilize the network services for a specified period of time without any restrictions on the bandwidth consumed. The volume-based pricing approach charges a user based on the amount of data which he/she transacted over the network. In addition to both these schemes, business establishments can provide Wi-Fi network services as a value addition to customers visiting their premises for core business activities. In such cases, the billing is not an issue as the bandwidth provided belongs to the organization's (unutilized) bandwidth, hence it can be provided without additional charging.

In the franchisee-franchisor and the aggregator models discussed in Section 14.3.1, the sharing of revenue is important because there exist multiple business entities in the process of customer relationship, billing, and providing the actual network access service. Two major revenue-sharing models are: the *fixed-fraction sharing model* (in which the revenue shared among the service providers involved is fixed *a priori*) and the *volume-based variable-fraction sharing model* (in which the percentage of revenue that goes to the different parties involved varies with the volume of data transferred). Service providers can consider a constant rate for the user per-bit of data transferred when the traffic load is light, and in the high-traffic-density environments, a variable rate per-bit of data transferred can be used. In the variable rate scheme, application layer mechanisms which periodically communicate the cost of communication are essential.

Billing Agency Models

The two main aspects in any network access are the actual network service provisioning and billing. Traditionally, these two were carried out by the same agency. However, in Wi-Fi systems, there can be different models for the BA and the network service providers (SPs). As illustrated in Figure 14.11, the basic BA models are the following: (a) billing-free service model, (b) associated BA model, (c) revenue-sharing-based BA model, (d) integrated BA model, and (e) independent BA model.

- **Billing-free service model:** As shown in Figure 14.11 (a), this is the simplest model and is in existence today in most places. Here, the WAN BA charges MNs for their network access whereas the Wi-Fi access is provided as a free service to the users, or as a value addition to the Wi-Fi SP's existing core business service.

- **Associated BA model:** In this case, Wi-Fi systems and WANs have independent BAs, each charging the MNs for their respective network services provided. Figure 14.11 (b) illustrates this model.

- **Revenue-sharing-based BA model:** This model provides a single-point billing for the MNs, and at the same time enables the MNs to avail network services from WAN or Wi-Fi networks. A revenue-sharing agreement must be in place between the WAN BA and the Wi-Fi SPs. Since the WAN SPs cover a larger geographical area, it is more appropriate to have the BA closely associated with the WAN SP as shown in Figure 14.11 (c).

- **Integrated BA model:** This model is the one that is associated with the extended service providers who operate a range of different networks including WANs and Wi-Fi APs. This is illustrated in Figure 14.11 (d).

- **Independent BA model:** In this case, the BA is an independent entity that has billing tie-ups with several WAN SPs and Wi-Fi SPs. This is illustrated in Figure 14.11 (e).

14.3.5 Pricing/Billing Schemes for Wi-Fi Systems

Volume-based pricing schemes and directions for pricing in the various Wi-Fi service models are discussed in this section. The two key aspects of the pricing schemes are the revenue-sharing approach and a reimbursement scheme for link-level successful delivery (similar to the pricing model discussed in [9]). Since the traffic considered in a Wi-Fi system includes at least a single AP, the AP can act as an efficient

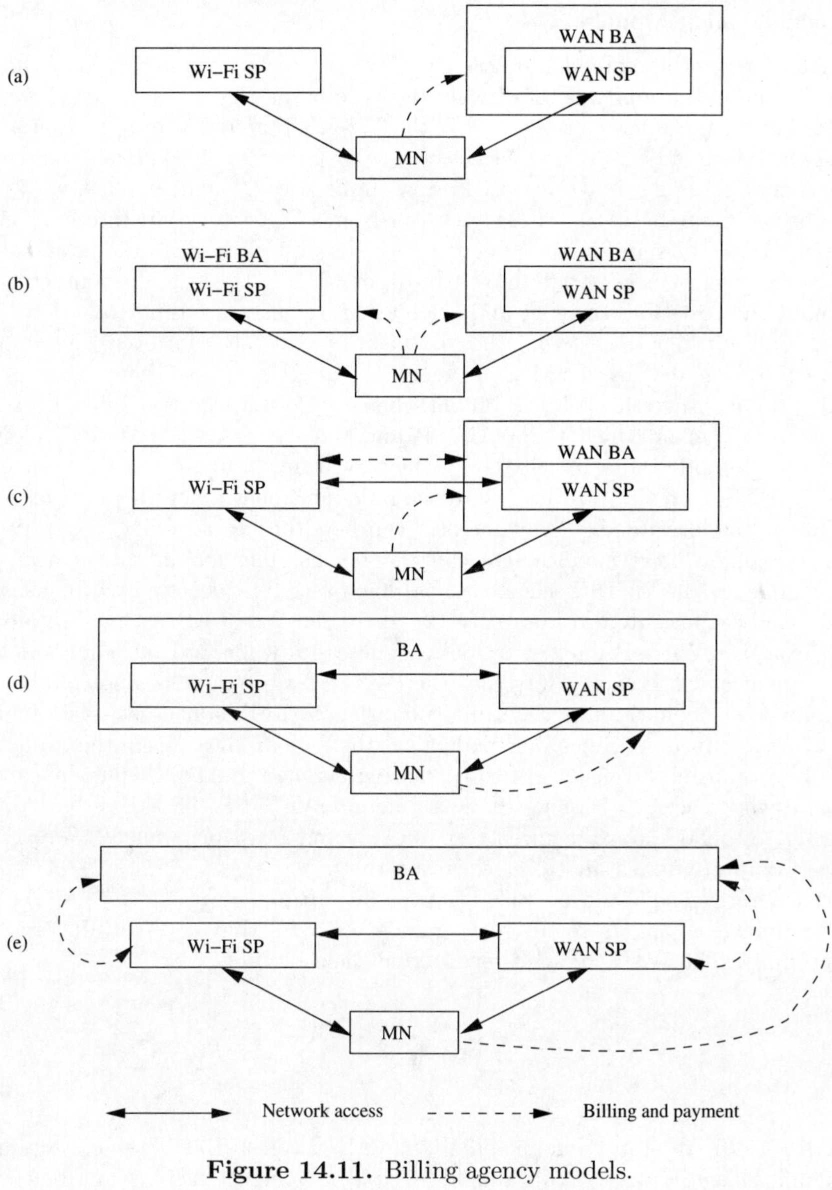

Figure 14.11. Billing agency models.

and reliable accounting station with the responsibility of gathering and delivering the pricing information to the BA. Also, due to the nature of the traffic, it can be observed that there is no need to aggregate any packet information at the MNs.

Revenue-Sharing Models

Revenue-sharing-based models describe the way in which the money paid by the user is split among the multiple SPs involved in providing the network access service. Generally, in the micro-carrier model, the Wi-Fi SP may lease a fraction of the bandwidth from an existing wired network service provider and resell it among its customers either on a flat-rate-based or a volume-based pricing scheme.

The number of bytes that an MN receives from a Web site or any Internet content provider would be significantly larger than the request packets it originates, leading to an asymmetry in the traffic flow. In order to take the traffic asymmetry into account, the costs for receiving and originating a packet are different. Here R_{cost} (O_{cost}) is the per-byte cost that the MN has to pay the BA for every byte of data received (originated) by the MN.

We can envisage the simplest billing scheme (referred to as the *simple revenue-sharing scheme*) in which the WAN SP and the BA treat the Wi-Fi AP as an intermediate node that participates in the forwarding process and reimburses the Wi-Fi AP through the Wi-Fi SP with the reimbursement amount β for every byte of data it has forwarded. The value of β can be either an absolute monetary unit or a fraction of the revenue the BA charges the customer for the same traffic unit. The simple revenue-sharing scheme is a volume-based fixed-fraction revenue-sharing model and is illustrated in Figure 14.12, where the Wi-Fi AP receives an amount of β from the WAN BA for every packet successfully forwarded. The value of β has to be decided in such a way that the WAN SP's revenue does not fall below a minimum threshold, and at the same time, the Wi-Fi SPs also get a fair share of the generated income. Since this model pays the APs on the basis of the traffic that they have transmitted instead of equal sharing among the APs, this model can also be used when each AP is operated by a different Wi-Fi SP. The MNs that connect directly to the WAN are required to pay O_{cost} and R_{cost} per unit of traffic or per byte of data originated and received, respectively.

The net revenue of the Wi-Fi SP is given by $NR_{WiFi} = \sum_{j=1}^{num\_nodes} F_{WiFi_j} \times \beta$, where the F_{WiFi_j} refers to the total number of bytes that a Wi-Fi SP forwarded successfully. The WAN SP's net revenue will then become

$$NR_{WAN} = \sum_{i=1}^{num\_nodes} Paid_i - \sum_{j=1}^{num\_APs} NR_{WiFi_j} \tag{14.3.1}$$

where $Paid_i$ is the total amount paid by an MN i to the BA. The SP's revenue is a significant entity, as the SPs should be able to generate a minimum revenue in order to sustain the network services. This minimum revenue can be achieved by altering the values of β, R_{cost}, and O_{cost} depending on the node density and the traffic distribution in the network.

The volume-based variable-fraction revenue-sharing model uses different values of β_i depending on the volume of traffic transacted through a particular Wi-Fi SP. For example, a three-level volume-based variable-fraction revenue-sharing model may reimburse β_1 for $0 < t < T_1$, β_2 for $T_1 < t < T_2$, and β_3 for $T_2 < t$, where t

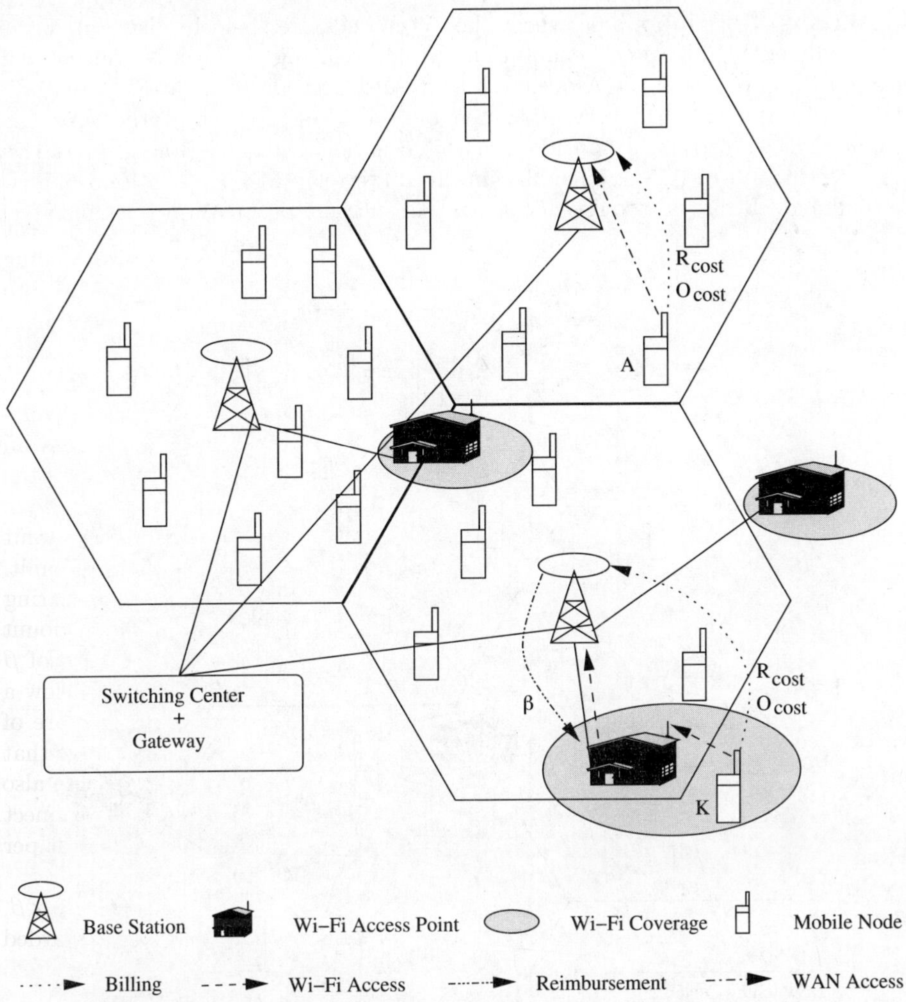

Figure 14.12. Illustration of simple revenue-sharing scheme.

is the traffic-volume transacted through a particular Wi-Fi SP, and T_1, T_2, and T_3 are the different traffic-volume thresholds.

Another scenario in which a Wi-Fi system can operate is the *multi-hop Wi-Fi system*[7] where the MNs can also act as the forwarders on behalf of other nodes which are unable to communicate directly to the Wi-Fi AP. Examples of such systems are the single interface MCN (SMCN) architecture [9] and multi-hop WLANs

[7] Multi-hop Wi-Fi systems are also called ad hoc Wi-Fi systems as they use ad hoc radio relaying in the Wi-Fi environment.

(MWLANs) [10]. In an area where the Wi-Fi APs are sparsely distributed with a broadband Wi-Fi system, enabling multi-hop relaying at the MNs can actually extend the coverage of the APs as well as raise the revenue-generation potential of the Wi-Fi SPs. The catch here is to enable the MNs to forward others' packets. Figure 14.13 illustrates a multi-hop Wi-Fi scenario where MN B accesses the services of the Wi-Fi AP through the intermediate relay MN K. The following two pricing models discuss the revenue-sharing in such multi-hop Wi-Fi systems.

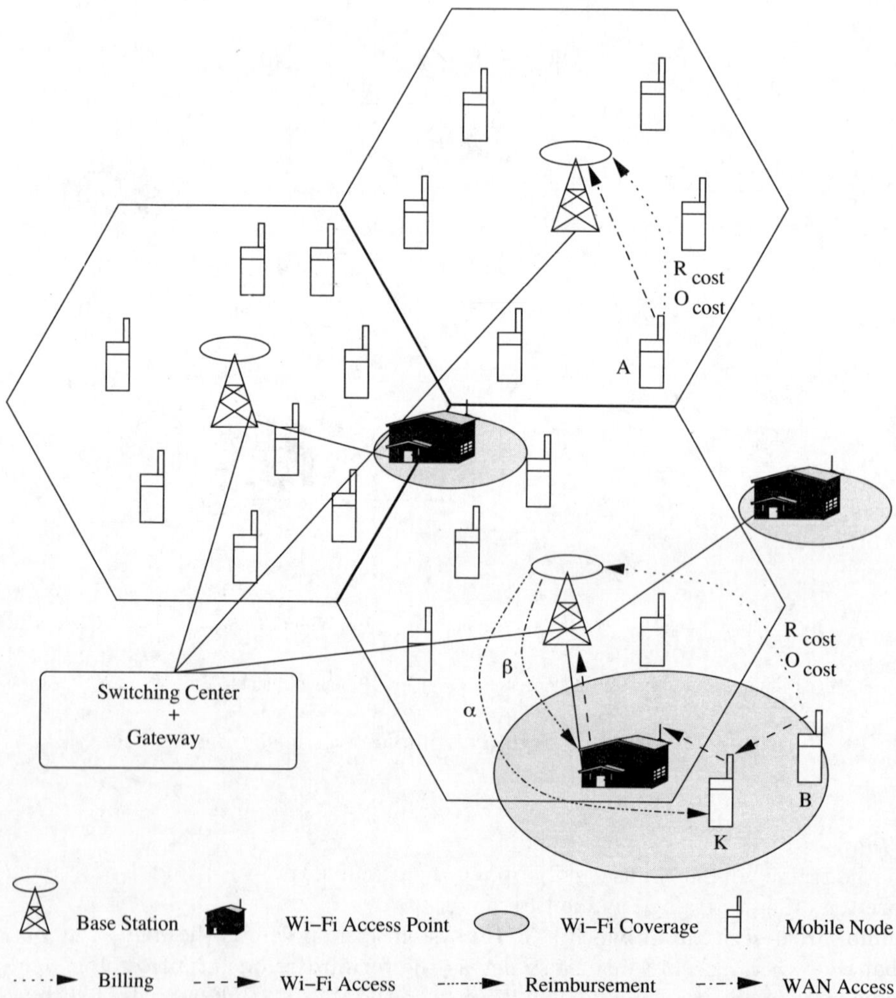

Figure 14.13. Illustration of multi-hop Wi-Fi revenue-sharing scheme.

The payment scenarios in a multi-hop Wi-Fi network with a fixed-fraction revenue-sharing model (referred to as *multi-hop Wi-Fi revenue-sharing scheme*) is depicted schematically in Figure 14.13. In this case, for every successfully delivered packet, the Wi-Fi AP receives an amount of β per packet and an intermediate MN that forwards the packet on behalf of the original sender receives an amount of $\alpha \times C_p$, where α is the reimbursement factor and C_p is the cost incurred at the intermediate node per byte of data forwarded by MNs.

The net revenue of Wi-Fi SP is given by $NR_{WiFi} = \sum_{j=1}^{num\_nodes} F_{WiFi_j} \times \beta$. Equation 14.3.2 represents the total reimbursement ($Repay_i$) that an MN i receives from the BA, where $F_{s_{ij}}$ is the total number of bytes that a node i has successfully forwarded for node j.

$$Repay_i = \sum_{j=1, j\neq i}^{num\_nodes} F_{s_{ij}} \times \alpha \times C_p \tag{14.3.2}$$

The reimbursement factor α is applicable only for those MNs that are acting as forwarders for other MNs. The reimbursement becomes an incentive when $\alpha > 1$.

The WAN SP's net revenue then becomes

$$NR_{WAN} = \sum_{i=1}^{num\_nodes} Paid_i - \sum_{i=1}^{num\_nodes} Repay_i - \sum_{j=1}^{num\_APs} NR_{WiFij} \tag{14.3.3}$$

Another possible revenue-sharing scenario is where the billing parameter R_{cost} is split into separate billing parameters R_{costAP} and R_{costBS} for the Wi-Fi SP and WAN SP, respectively. Similarly, O_{cost} is split into O_{costAP} and O_{costBS}. As per this scheme, those MNs which access the WAN SP through a Wi-Fi SP need to pay a differentiated billing rate O_{costBS} and R_{costBS} for the unit traffic originated or received by them, respectively. Such MNs are required to pay the Wi-Fi SPs O_{costAP} and R_{costAP} for their packet forwarding services. Lowering the O_{costBS} and R_{costBS} compared to O_{cost} and R_{cost} can help the WAN SPs to reduce the traffic load by encouraging more MNs to use the Wi-Fi services. Due to the differentiation in the billing rates, this scheme is called *differentiated multi-hop revenue-sharing scheme.* This scheme requires the MNs to individually pay the Wi-Fi SP and the WAN SP as shown in Figure 14.14. In this case, the amount to be paid by a given MN ($Paid_i$) is obtained as follows:

$$Paid_i = T_{o_i} \times O_{costAP} + T_{r_i} \times R_{costAP} + T_{o_i} \times O_{costBS} + T_{r_i} \times R_{costBS} \tag{14.3.4}$$

where T_{o_i} is the total number of bytes originated by a node and T_{r_i} is the total number of bytes received by a node. The Wi-Fi SP's revenue is simply the amount that is paid by the MNs that have been in its domain, and no further sharing needs to be done with the WAN SP. O_{costBS} and R_{costBS} provide bandwidth or additional service provided by the WAN. In this scenario, one can think of a number of Wi-Fi SPs each offering the service to a customer who can choose the SP (if available) which provides the most economical service. The registration mechanism by which

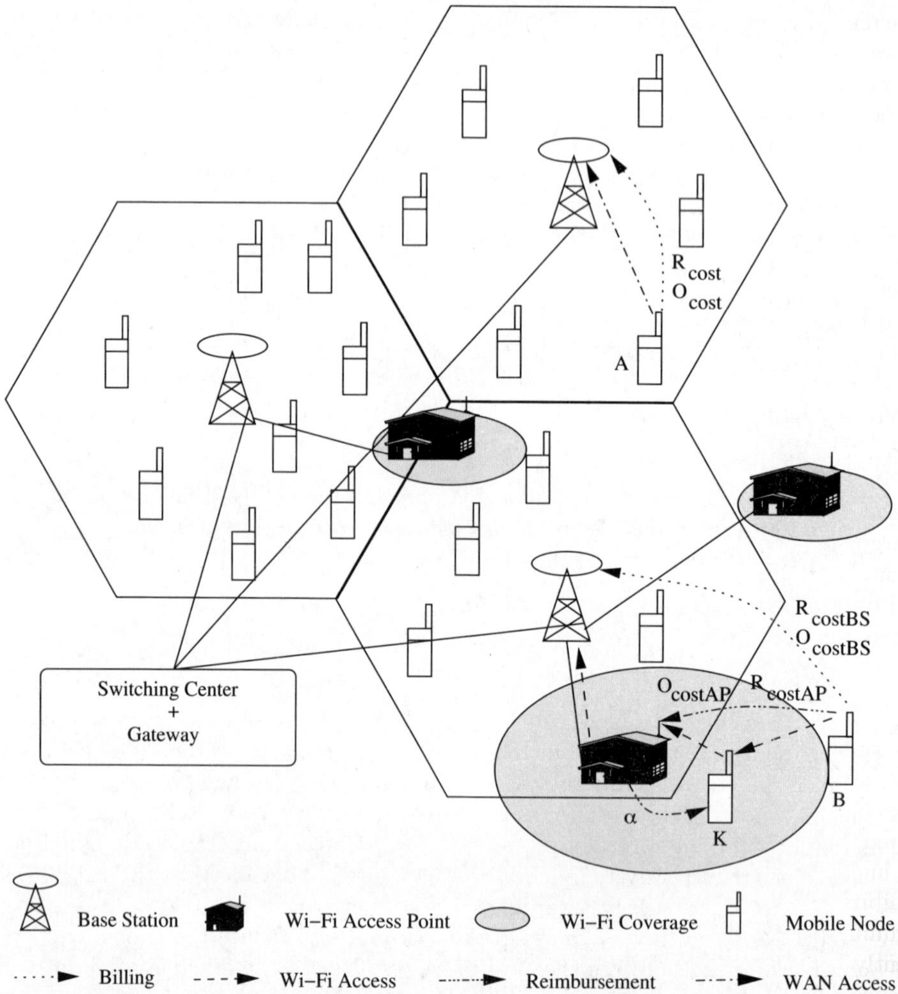

Figure 14.14. Illustration of differentiated multi-hop revenue-sharing scheme.

an MN joins a particular network may also have to take into account the cost parameters such as O_{costAP} and R_{costAP}.

Similar to the previous scheme, the following equation represents the total reimbursement that an MN receives from the BA.

$$Repay_i = \sum_{j=1, j \neq i}^{num\_nodes} F_{s_{ij}} \times \alpha \times C_p \qquad (14.3.5)$$

The reimbursement comes into the picture only for the MNs that act as intermediate forwarding nodes for other MNs.

Equation 14.3.6 gives the total expenditure that an MN incurs from forwarding on behalf of other MNs.

$$P_i = \sum_{j=1,j\neq i}^{num\_nodes} F_{s_{ij}} \times C_p + \sum_{j=1,j\neq i}^{num\_nodes} F_{r_{ij}} \times C_p \tag{14.3.6}$$

where $F_{r_{ij}}$ is the total number of retransmitted bytes that a node i has forwarded on behalf of node j. The following equation shows the total amount that an MN has to pay the BA:

$$Paid_i = T_{o_i} \times O_{cost} + T_{r_i} \times R_{cost} \tag{14.3.7}$$

In this case, the WAN SP may distinguish between those MNs connected directly to them from those MNs that connect through a Wi-Fi AP. In such a differentiated charging mechanism, in order to encourage people to spare the low bandwidth WAN resources, the O_{cost} and R_{cost} for those MNs which access the WAN SP directly will be greater than O_{costBS} and R_{costBS} which are applied for those MNs which access the WAN through Wi-Fi APs.

14.4 OPTICAL WIRELESS NETWORKS

The discussion of wireless networks so far was restricted to the communication based on radio waves. Optical wireless communication enables communication using infrared rays and light waves operating at frequencies well beyond the visible spectrum for high data rate local communication. Optical wireless communication technology exhibits a number of properties that makes it a suitable alternative to indoor RF communication. The advantages of optical wireless communication include significantly less interference due to its lack of penetration through walls, positioning of spectrum at a completely unregulated and unlicensed band, increased security, and high data rate. Optical wireless technology promises broadband data delivery at short ranges in point-to-multipoint LANs and point-to-point medium-distance optical links. Optical wireless transmission can be classified into short-range communication and long-range communication systems. A comparison of these two types of optical wireless transmission schemes is given in Table 14.2. Long-range communication systems are mainly used for outdoor point-to-point optical links and short-range systems are used in indoor and outdoor applications. Unlike the long-haul networks [12] in fiber-based optical networks, the long-range optical wireless systems can operate over a distance of hundreds of meters only. The short-range systems operate over a distance of few meters. With the ever-growing demand for broadband wireless connectivity, the utilization of RF spectrum is a bottleneck due to the spectrum congestion, licensing requirements, and unsuitability of certain bands for broadband applications.

Table 14.2. A comparison of optical wireless technologies

Issue	Short-Range	Long-Range
Distance	< 10 m	<1,000 m
Data Rate	9600 bps to 4 Mbps	<10 Gbps
Source Power	Low	High
Preferred Transmitter	LED	Laser
Preferred Receiver	PIN Diode	Avalanche Diode
Mode of Propagation	Line of Sight (LoS) and Diffused	LoS
Effect of Atmospheric Conditions	Limited	Significant
Cost of Equipment	Low	High

14.4.1 Short-Range Infrared Communication

The use of infrared radiation for wireless communication was first proposed in the late 1970s. In 1993, the Infrared Data Association (IrDA), a non-profit organization, was founded by major hardware, software, and communications equipment manufactures for establishing and promoting an infrared standard that provides convenient cordless connectivity and fosters application interoperability over a broad range of platforms and devices. As a result of the activities of IrDA, short-range infrared communication was in widespread use in the last decade, with IrDA interfaces built into several hundred million electronic devices including desktop, notebook, palm PCs, printers, digital cameras, public phones/kiosks, cellular phones, pagers, PDAs, electronic books, electronic wallets, toys, watches, and other mobile devices. IrDA has developed standards that work on widely ranging data transfer rates (9.6 Kbps – 4 Mbps).

The indoor short-range communication can be classified into (i) directed transmission and (ii) diffusion-based transmission. In the directed transmission, the transmitter and receiver are required to be pointed to each other and there should exist a line of sight (LoS) transmission link between them. In diffusion-based transmission, the transmitter and receiver need not have a LoS for communication. Table 14.3 summarizes the differences between directed and diffused transmission systems.

14.4.2 Optical Wireless WDM

The use of wavelength division multiplexing (WDM) [12], [13], which is a method of sending many light beams of different wavelengths simultaneously down the core of an optical fiber, has been successful in utilizing the tremendous bandwidth offered by the optical fibers. Operating on the basis of the WDM technology, an optical wireless network utilizes different wavelengths between a point-to-point wireless link. This enables carrying large number of simultaneous sessions across a sender-

Table 14.3. A comparison of two short-range transmission schemes

Issue	Directed	Diffused
Line of Sight	Required	Not required
Vulnerability to Link Blockage	High	Low
Multipath Degradation	Low	High
Path Loss	Low	High
Data Rate	High	Lower
Power Efficiency	High	Low
Coverage	Low	High
Support for Mobility and Roaming	Difficult	Easier
Suitability for LANs	Less suitable	More suitable
Suitability for Point-to-Point Long-Range Links	Highly suitable	No

receiver pair. The system operates at a spectrum centered around 1,330 nm or 1,550 nm in order to be compatible with the wavelengths used for traditional fiber-based WDM systems.

First-generation point-to-point wireless WDM systems with four wavelengths, each carrying a data rate of 2.5 Gbps, make a total transfer rate of 10 Gbps. An illustration of such a system is shown in Figure 14.15. Optical wireless WDM technology is expected to increase the bandwidth to a greater extent over the wireless technologies. However, it has several disadvantages, some of which are (i) LoS link

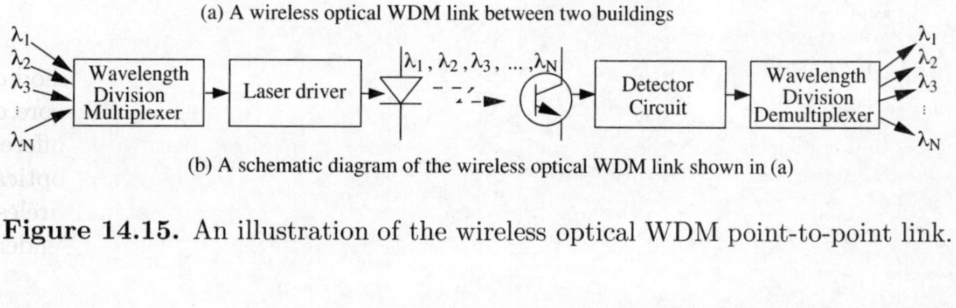

Figure 14.15. An illustration of the wireless optical WDM point-to-point link.

break can lead to the loss of a tremendous amount of data and (ii) dense smoke, rain, birds, kites, and other atmospheric changes can lead to link breaks. Reliability is one of the major design objectives for designing wireless optical WDM networks. Figure 14.16 illustrates an optical wireless metro area ring network formed by several optical wireless point-to-point links. In cities where laying of fibers is difficult or expensive, optical wireless rings are suitable alternatives.

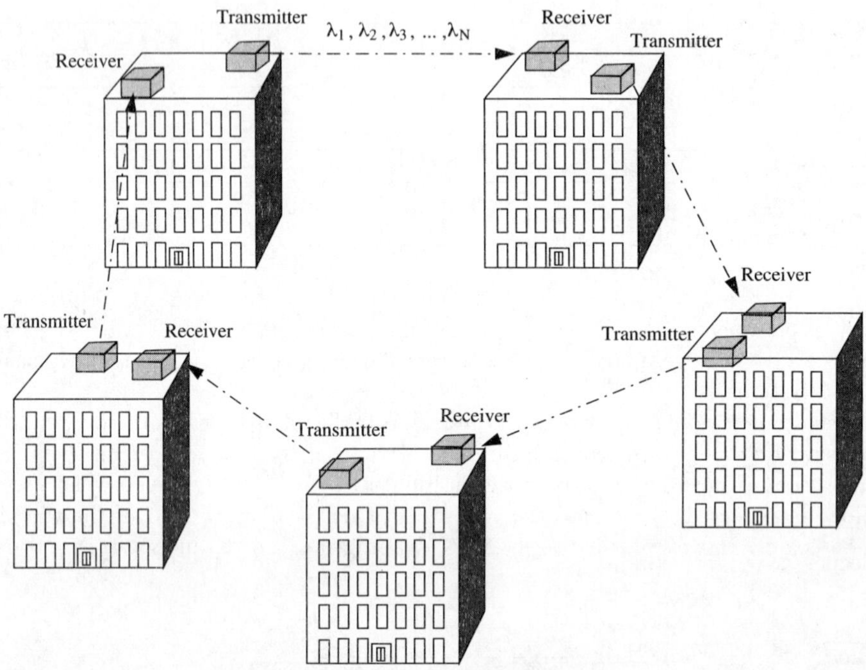

(a) A wireless optical WDM ring formed among many buildings.

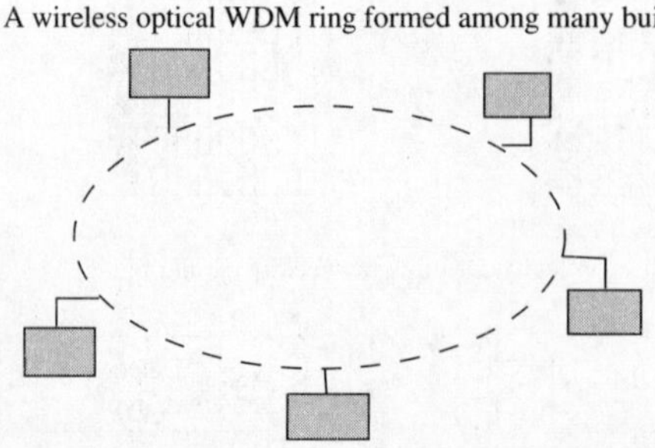

(b) A schematic diagram of the optical wireless WDM ring formed by the links in (a).

Figure 14.16. An illustration of the wireless optical WDM ring network.

14.4.3 Optical Wireless WDM LAN

The infrared wireless LANs fail to exploit the bandwidth available in the 1,330 nm and 1,550 nm optical spectrum windows to the fullest. The multi-party communication in a wireless LAN environment using the radio spectrum can be replaced with the optical wireless WDM (OWWDM) system that can provide a much higher bandwidth with reduced interference and high wavelength reusability. Table 14.4 compares the RF wireless LANs and OWWDM wireless LANs. An illustration of the operation of OWWDM LAN is shown in Figure 14.17. Here the stations can connect to the OWWDM AP (access point) using the control wavelength (λ_C) and obtain any data wavelength (λ_i where $i = 1, 2, \ldots, N$) for data transfer. This necessitates wavelength tuning capabilities at the stations and the ability to operate over multiple wavelengths at the APs.

Table 14.4. RF wireless LANs versus OWWDM wireless LANs

Issue	RF	OWWDM
Spectrum	Permitted bands can be used without license indoors. Outdoor applications require license in several countries.	Unlicensed
Bandwidth	High (of the order of Mbps)	Huge (of the order of Gbps)
Coverage	Large	Limited
Signal Fading	High	Medium
Security	Vulnerable to eavesdropping	Eavesdropping is a concern only in diffusion-based outdoor systems.
Blockage	No	Communication link can be blocked by opaque bodies
Eye Hazard	No	Yes
Support for Mobility	Yes	Difficult
Cost of Equipment	High	Low
Carrier Frequency	Lower	High
Effect of Multipath	High	Low
Required Receiver Sensitivity	Low	High
Interference	From other RF sources	Only from light sources
Receiver Cost	High	Low
Transmitter-Receiver Alignment	Not required	Required in point-to-point systems

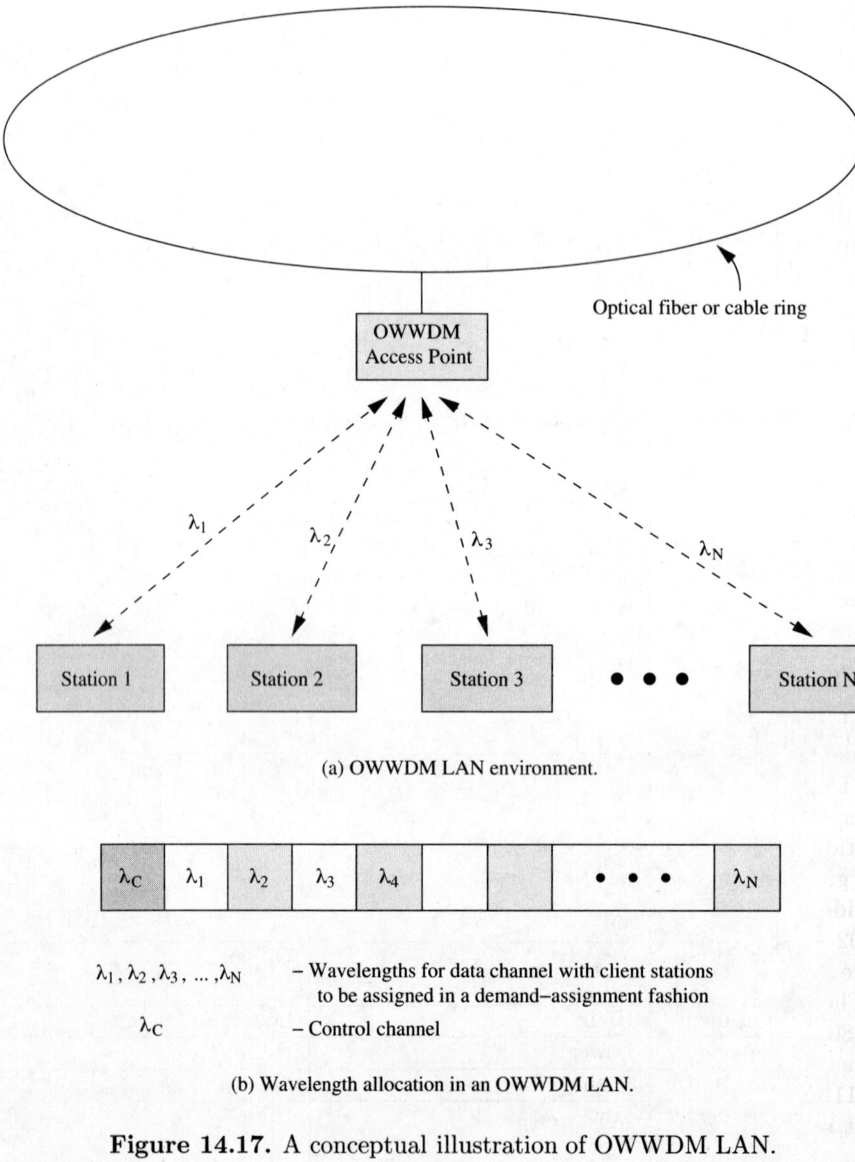

Figure 14.17. A conceptual illustration of OWWDM LAN.

Issues in OWWDM LANs

The major issues in a practical OWWDM LAN include the choice of the protocols and subsystems. MAC protocols for optical fiber LANs that belong to the category of broadcast-and-select[8] can be extended to work in OWWDMs also. The choice of receivers includes PIN diodes or Avalanche diodes with the former less expensive and sensitive and the latter more sensitive. The choices for a transmitter include Laser diodes or inexpensive LEDs, with the former able to provide higher data rate with more wavelengths at a higher cost. Additional capabilities such as wavelength conversion and tuning are required for the operation of OWWDM. The factors affecting the performance of such a system are path loss, interference, receiver sensitivity, and protocol efficiency. In addition to all other technical issues, safety issue assumes significance in the design due to the fact that high optical energy in the specified optical bands can damage human eyes. This necessitates an upper limit on the maximum power of the transmitter used in the system.

14.5 THE MULTIMODE 802.11 — IEEE 802.11a/b/g

The different physical layer specifications for IEEE 802.11 operating at different frequency bands led to another dimension of the operation of WLANs, that of interoperability of different IEEE 802.11 [14] variants such as 802.11b, 802.11a, and 802.11g. In addition to these physical layer differences, enhancements to the basic medium access mechanism such as 802.11e for providing QoS support add to this interoperability issue. Thus was born the *multimode 802.11* (*i.e.*, IEEE 802.11a/b/g) which can operate in all of the three major variants of IEEE 802.11 standards to date. Multimode 802.11 is the WLAN client interface implementation that can seamlessly work with APs operating according to any of the three IEEE 802.11 standards. At any given time, an IEEE 802.11a/b/g interface works with only one AP. The major advantage of multimode 802.11 is the backward compatibility of the newer 802.11a and 802.11g with the millions of existing 802.11b installations. In addition to the interoperability issue, the use of 802.11a and 802.11b simultaneously at a given service area can actually increase the available bandwidth. Table 14.5 provides the total bandwidth available at any given service area by the combination of 802.11a, 802.11b, and 802.11g. The entire bandwidth may not be available for a single station as it requires several network interface cards and APs.

The multimode WLAN clients that are 802.11a/b/g compatible can connect to any 802.11 AP transparently, leading to enhanced roaming services across different networks. An 802.11a/b/g client that detects carriers from different APs using 802.11b, 802.11a, and 802.11g is illustrated in Figure 14.18 (a). The 802.11a/b/g client searches for carrier in all the bands of operation and selects the most ap-

[8] A broadcast-and-select network consists of a passive star coupler connecting the nodes in the network. Each node is equipped with one or more fixed-tuned or tunable optical transmitters and one or more fixed-tuned or tunable optical receivers. Different nodes transmit messages on different wavelengths simultaneously. The star coupler combines all these messages and then broadcasts the combined message to all the nodes. A node selects a desired wavelength to receive the desired message by tuning its receiver to that wavelength.

Table 14.5. Comparison of bandwidth in multimode 802.11 systems

System	Number of Simultaneous Channels		Total Bandwidth*	
	USA	Europe	USA	Europe
802.11b	3	4	33 Mbps	44 Mbps
802.11a	13†	19	702 Mbps	1,026 Mbps
802.11g	3	4	162 Mbps	216 Mbps
802.11b/g	3	4	162 Mbps	216 Mbps
802.11a/b/g/	16†	23	864 Mbps	1,242 Mbps

*Total physical layer bandwidth (at a sufficiently close distance to an AP) that can be made available to all nodes in a given service area (which may require several APs).

†Expected to add another 11 channels in North America.

propriate one. Hence, the new generation of wireless LAN clients are expected to be capable of operating in all these different modes, requiring a wide frequency tuning range (2 GHz to 6 GHz), multiple MAC implementations, and multiple baseband processors. Early solutions for multimode 802.11 were designed to have an integrated chipset that combined the 2.4 GHz chipset, 5 GHz chipset, and the baseband chipset for PCI,[9] mini-PCI, or CardBus interfaces. A multimode LAN client is designed to automatically select the strongest detected carrier. A schematic diagram of the 802.11a/b/g client interface is shown in Figure 14.18 (b). The major difference between this interface card and the traditional single-mode client interface is the presence of multiple chipsets that can process the physical layer information belonging to different 802.11 variants. It is necessary to have at least two such physical layer chipsets — one for the 2.4 GHz frequency range and the other for the 5 GHz frequency range to handle 802.11a/b/g physical layers. The control signals from the baseband controller are used to switch to one particular physical layer chipset depending on the availability and signal strength from the APs. In addition to the switching decisions, the baseband module includes the MAC implementations and the interfaces to the PC bus standards such as PCI, mini-PCI, and CardBus.

Figure 14.19 shows a time line of multimode 802.11 in wireless LANs. One can expect a k-mode operation which can cover almost every existing standard ranging from wireless PANs and LANs to satellite networks with a single handset by using programmable interfaces by 2010. It is extremely difficult to predict the evolution of wireless networks beyond a certain point, considering the revolutionary changes that wireless communication technology has undergone every decade. The existing solutions for 802.11a/b/g based on embedding different chipsets handling different physical layer standards are not scalable for a flexible k-mode operation

[9]PCI stands for peripheral component interconnect — a widely accepted local bus standard developed by Intel Corporation in 1993. mini-PCI is a PCI-based industry-wide standard for modems. CardBus is a 32-bit extension of PCMCIA (Personal Computer Memory Card International Association), PC card standard and has operation speeds up to 33 MHz.

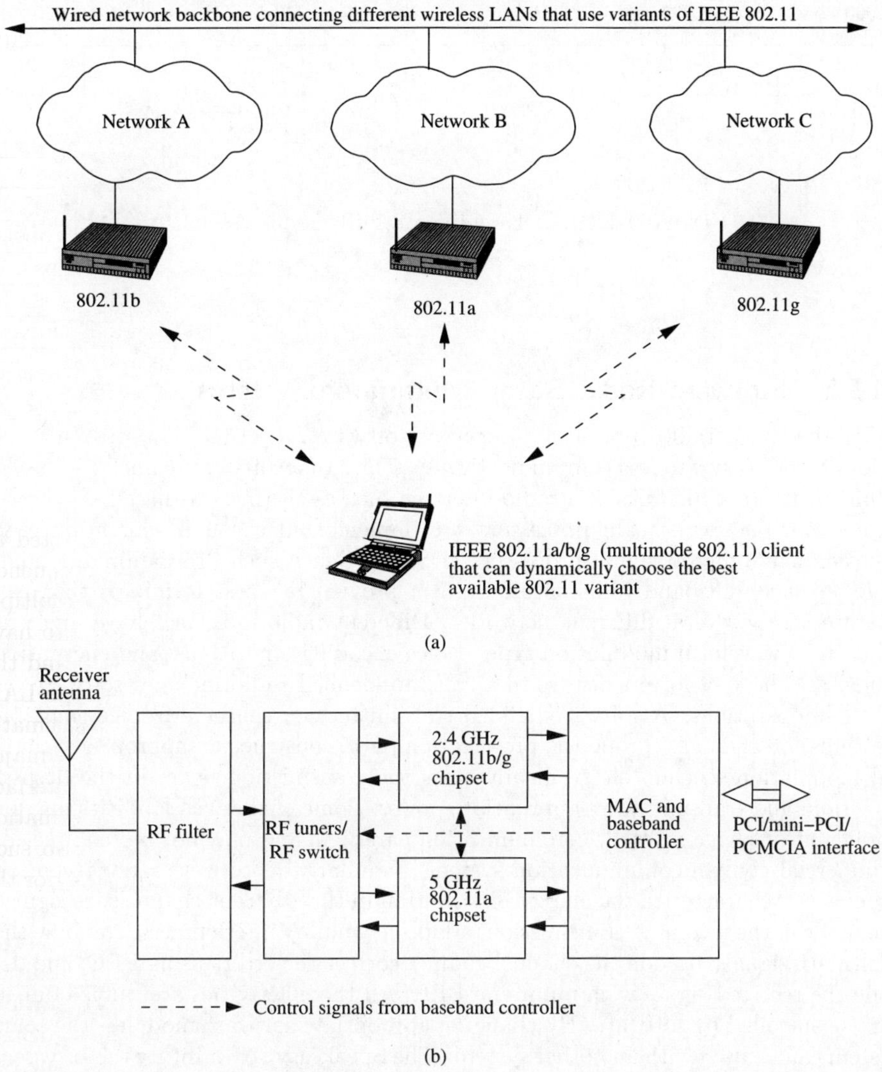

Figure 14.18. Multimode 802.11. (a) Operation of the multimode client. (b) Schematic diagram of the multimode 802.11 client interface.

where interoperability among many networks would be required. Also, the cost of such multimode interfaces increases linearly with the number of modes of operation. Programmable radio interfaces (popularly known as *software-defined radio* or *software radio* in short) point the way toward achieving a low cost k-mode wireless communication interface for future wireless communication networks.

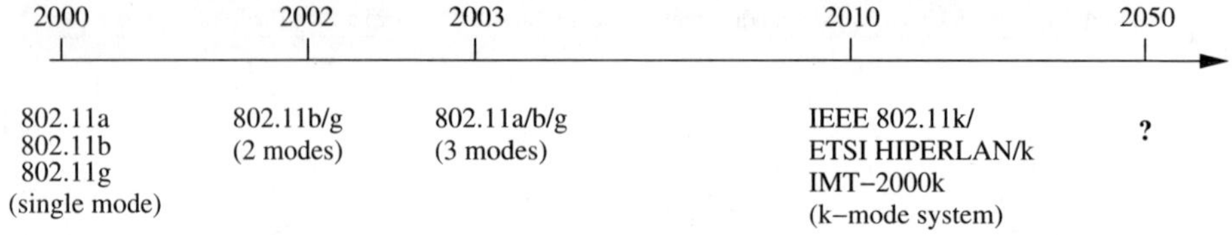

Figure 14.19. A time line of multimode systems.

14.5.1 Software Radio-Based Multimode Systems

Software-defined radio (popularly known as software radio) [15] is a proven technology in military wireless communication systems. In a software-defined radio system, the different modules of a radio interface such as IF (intermediate frequency), base band, and bit stream processors are implemented through general-purpose programmable digital signal processors (DSPs). Hence such programmable, multimode, and multiband radio interfaces can provide low-cost terminals that can operate across several different networks. Different radio interfaces have different RF bands, waveform modulation types, voice-encoding algorithms, and encryption schemes. The waveform software in a software-defined radio includes software for all these operations. A software-defined transmitter can characterize the available transmission channels, probe the propagation path, construct an appropriate channel modulation scheme, electronically steer the transmission beam in the desired direction, and transmit at an appropriate power. Some of these capabilities may be applicable only in a military communication paradigm and may not be required for commercial civilian communication systems. Similarly, a software radio reception process can characterize the energy distribution in the adjacent channels, recognize the mode of the incoming transmission, adaptively nullify interferences, combine the multipath signals, decode the channel coding, correct the errors using FEC, and decode the received signal with minimum BER. All these activities are carried out in DSPs controlled by software. By changing appropriate software modules, the same system can be used with any other system. The Speakeasy [16] military radio system was designed to emulate more than 15 military radio systems. The waveform software required for a particular network can be preloaded, taken from a predefined standard set, or downloaded through over-the-air (OTA) data interfaces. Figure 14.20 shows an illustration of a software-defined multimode and multiband wireless network interface operating across a GSM network and a W-CDMA network, by downloading network specific waveform software. Every network is assumed to be using a control protocol or a control channel with the necessary information about the waveform software. In the presence of a new network, the multimode interface detects the network over the control channel and downloads the waveform software and the necessary protocol software. Once the new waveform and associated protocol software are loaded onto the radio interface, the mobile terminal becomes

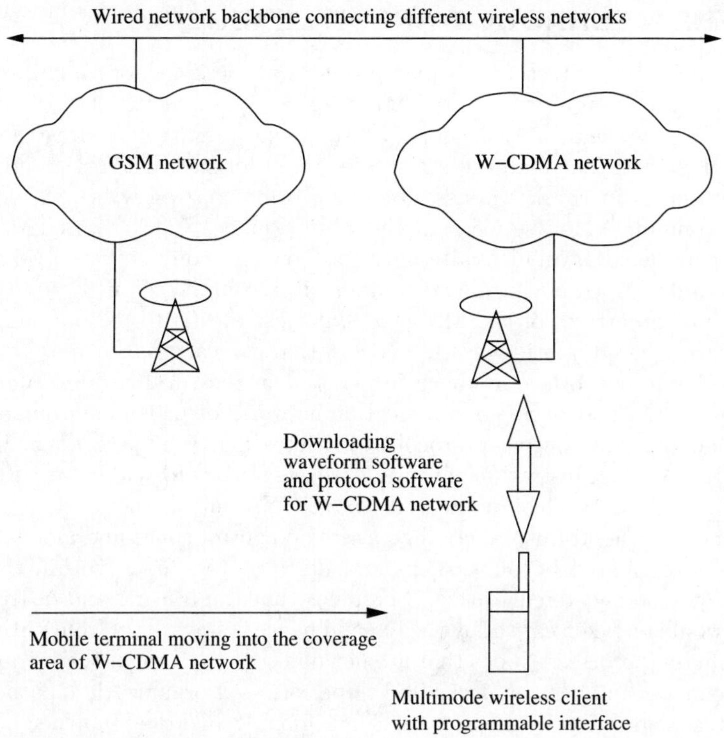

Figure 14.20. An illustration of operation of software-defined multimode wireless client.

ready to communicate with the network similar to any other single-mode interface designed for that network.

Even though software radio is currently used at the BSs only, recent developments are aimed at using it at the terminal side also. The major constraints for achieving this are processing power and power consumption. Experimental software radio-based terminals that operate in dual mode between GSM and W-CDMA exist today. They consume higher amounts of power for operation compared to the existing single-mode terminals. With the development of high-speed processors and low-power VLSI devices, multimode terminals based on software radio may soon become a reality.

14.6 THE MEGHADOOT ARCHITECTURE[10]

The Meghadoot[11] architecture is a packet-based wireless network architecture for low-cost rural community networks. Traditional wireless networks for rural telephony, such as wireless in local loop (WLL), require extensive infrastructure for service deployment. The high investments required for such networks, and the low revenue prospects in rural regions, discourage commercial service providers from providing communication services in the rural regions. Packet-based radio networks are considered as an ideal alternative for low-cost community networks, both in the urban developed environments and also in the rural regions.

The major objectives of the Meghadoot project are (i) to develop a fully distributed packet-based hybrid wireless network that can carry voice and data traffic, (ii) to provide a low-cost communication system in the rural regions, and (iii) to provide an alternate low-cost communication network for urban environments.

Meghadoot uses a routing protocol called infrastructure-based ad hoc routing protocol (IBAR). An illustration of routing process in the Meghadoot architecture is shown in Figure 14.21. The infrastructure node (IN) controls the routing process in its k-hop neighborhood (also referred to as k-hop control zone) and aids in routing for the calls originated beyond k-hops and destined to a node within the k-hop region. Any node registered to the IN assumes that the routing and other control activities would be taken care of by the IN, and hence it stays away from initiating its own path-finding process. Nodes that are not under the control of the IN operate in the ad hoc mode, and hence are required to perform self-organization, path-finding, and path reconfiguration by themselves. The region beyond the k-hop neighborhood from an IN is called the ad hoc routing zone. Meghadoot requires the gateway nodes (GNs) to hold the additional responsibility of interfacing the nodes in the ad hoc routing zone (operating in the ad hoc mode) to the IN in order to enable such nodes to find routes efficiently to the nodes inside the control zone of the IN.

Nodes in the ad hoc routing zone broadcast *RouteRequest* packets in order to find a path to the destination. Every intermediate node that receives the packet forwards it further until the packet reaches the destination. When the destination node receives the packet, it responds by sending back a *RouteReply* packet. This mechanism is similar to that used in the dynamic source routing (DSR) [18] protocol. The disadvantage of using DSR protocol is the high control overhead generated by the broadcast packets used for path-finding. In Meghadoot, this routing overhead is reduced whenever INs are present. Whenever a GN receives a *RouteRequest* packet, instead of flooding, it forwards the packet to the IN. If the destination node lies in its control zone, the IN returns the path information to the GN using a *RouteReply* packet. The GN receives the *RouteReply* packet and forwards to the original sender node. Using the IBAR protocol, the IN maintains the approximate

[10]This is a prototype development project [17] currently being carried out at the High Performance Computing and Networking Laboratory, Department of Computer Science and Engineering, Indian Institute of Technology, Madras, India.

[11]This is a Sanskrit word meaning *Cloud Messenger*, derived from the epic love story written by legendary Indian poet Kalidasa. The theme of this story is a *message* sent by an exiled Yaksha in Central India to his beloved wife in the Himalayas through a *cloud*.

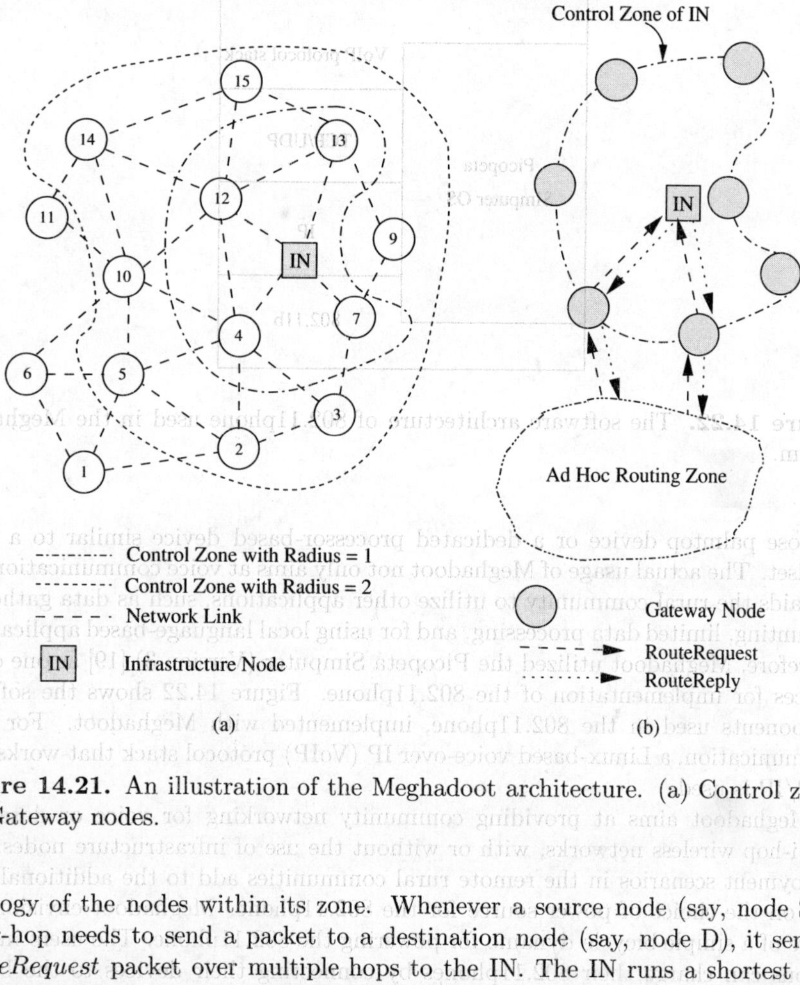

Figure 14.21. An illustration of the Meghadoot architecture. (a) Control zones. (b) Gateway nodes.

topology of the nodes within its zone. Whenever a source node (say, node S) in the k-hop needs to send a packet to a destination node (say, node D), it sends a *RouteRequest* packet over multiple hops to the IN. The IN runs a shortest path algorithm to find a path to node D and returns the path found to node S. Node S can now start using this path provided by IN. When a path break is detected by the sender node, it sends a new *RouteRequest* packet to the IN for reconfiguring the broken path. If an intermediate node detects a path break, it sends a *RouteError* packet to the IN, upon reception of which the IN obtains a new path and informs the sender.

14.6.1 The 802.11phone

The end user equipment in Meghadoot is an IEEE 802.11 enabled device, either a laptop computer with an 802.11 adapter, or a small handheld device with an 802.11 interface. Meghadoot is aimed at deployment in rural areas, where other communication infrastructure is not available, using an 802.11phone (an inexpensive handheld device with an 802.11b card). The 802.11phone could either be a general-

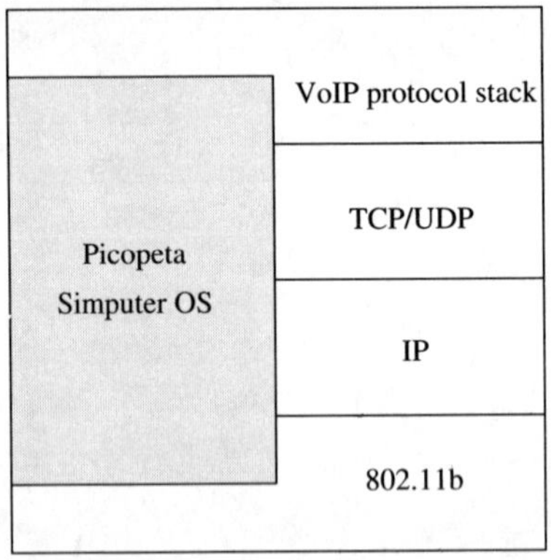

Figure 14.22. The software architecture of 802.11phone used in the Meghadoot system.

purpose palmtop device or a dedicated processor-based device similar to a GSM handset. The actual usage of Meghadoot not only aims at voice communication, but also aids the rural community to utilize other applications, such as data gathering, accounting, limited data processing, and for using local language-based applications. Therefore, Meghadoot utilized the Picopeta Simputer (Version 3) [19] as one of the devices for implementation of the 802.11phone. Figure 14.22 shows the software components used in the 802.11phone, implemented with Meghadoot. For voice communication, a Linux-based voice-over IP (VoIP) protocol stack that works with TCP/IP is used.

Meghadoot aims at providing community networking for voice or data over multi-hop wireless networks, with or without the use of infrastructure nodes. The deployment scenarios in the remote rural communities add to the additional pressure on the choice of power source for the 802.11phone. Meghadoot envisions the usage of a simple bicycle dynamo for powering the 802.11phone. The users in rural regions can charge their 802.11phones by connecting their devices to the bicycle dynamo through a cycle-based 802.11phone (CB8) charger.

Meghadoot provides an affordable communication system for rural communities where service providers may not be willing to establish infrastructure-based communication networks because of their return-on-investment constraints. This is typically the case in many developing nations. Meghadoot provides an ideal alternative in such situations.

14.7 SUMMARY

The evolution of wireless networking has had tremendous influence on human life and the ways of communication, from the first-generation (1G) cellular networks to the current third-generation (3G) wireless networks. The recent popularity of wireless networks has given rise to many additional requirements of the system, especially the requirement of greater bandwidth and demand for high-quality multimedia services. The ultra-wide band technology is providing a new direction to utilize the radio spectrum for the broadband short-range radio communication and hence offering a new direction to the development of wireless networks. The multi-mode IEEE 802.11 systems enable the end users to roam seamlessly across different WLANs — IEEE 802.11a, IEEE 802.11b, and IEEE 802.11g. Wireless fidelity (Wi-Fi) systems are aimed at integrating the high-speed wireless LANs and the wide-area packet cellular networks. Another recent trend in wireless networks is the use of optical wireless WDM networks for high bandwidth support. The Meghadoot wireless network architecture described in this chapter is a low-cost solution for rural community networks.

14.8 PROBLEMS

1. Compare and contrast UWB communication with conventional wide-band communication techniques based on spread spectrum techniques.

2. What is the single most important feature of the UWB system that makes it suitable for RADAR (radio detection and ranging) applications?

3. What are the important features that make the UWB systems applicable for high-speed mobile communications?

4. Calculate the duty cycle and average power to peak power ratio of a UWB system in which the monocycle duration is 0.5 ns, pulse repetition interval is 25 ns, and pulse amplitude is 0.1 mW.

5. Which of the service provider models require revenue-sharing schemes for pricing/billing schemes? Explain why.

6. Discuss the adaptability of the PICR-PDBR scheme (proposed for hybrid wireless networks; discussed in the previous chapter) as a Wi-Fi pricing/billing scheme.

7. A micro service provider installs two APs in his premises. What is the maximum data rate he could provide in his service area for the following cases: (a) Both 802.11b APs, (b) An 802.11b AP and an 802.11a AP, (c) One 802.11b AP and an 802.11g AP, and (d) Both 802.11a APs?

8. A coffee shop owner decided to provide Internet access to his customers. From the Internet Web site www.buy-here-or-get-lost.com, he found the best deals for the 802.11b, 802.11a, and 802.11g APs as $20, $80, and $40, respectively. Advise him on the best possible bandwidth in his shop with a budget limit of $100, given the fact that his neighbor, an ice-cream shop owner, is already using channel 6 among the available 802.11b channels.

9. Consider the simple revenue-sharing scheme. Calculate the revenue generated by a particular Wi-Fi SP for a per-month traffic of (i) 85 traffic units and (ii)

400 traffic units for the following cases: (a) if a fixed β value of 0.02 cents per traffic unit is reimbursed and (b) if a variable β with $\beta_1 = 0.03$ cents, $\beta_2 = 0.01$ cents with a traffic threshold of 100 traffic units is used.

10. Consider the simple revenue-sharing scheme with a volume-based variable-fraction revenue-sharing model where the reimbursement factor β has two levels, that is, β_1 for traffic up to 100 units, β_2 for traffic above 100 units. Assume that the operating cost for a Wi-Fi service provider with a single AP is \$10 per-month. Calculate the corresponding β values for achieving break-even for (a) 85 traffic units per month and (b) 500 traffic units per month.

11. Calculate the revenue generated by a Wi-Fi SP if β is a fraction of the billing rate charged by the BA for the Wi-Fi traffic rate of (a) 85 units per month and (b) 400 units per month. What are your comments on the profitability? Assume that the BA charges a billing rate of 0.1 cents per traffic unit and β is 0.5.

12. Wireless optical WDM rings provide high data rate networks in metropolitan areas. Discuss possible solutions and factors to be considered for providing reliability for a wireless optical WDM ring network.

13. What are the major factors to be considered while selecting a light source for designing a mobile network using optical wireless technology?

14. What is the preferred light source for point-to-point metro area optical wireless links? Give reasons.

15. Give two advantages and disadvantages in using laser diodes as light sources for optical wireless networks.

16. Discuss the advantages and disadvantages of a hardware-based multimode 802.11 terminal.

17. Discuss the advantages and disadvantages of software radio-based multimode terminals.

18. Discuss the effects of increasing the control zone radius in Meghadoot architecture in terms of the resource requirements at the IN and the routing efficiency.

BIBLIOGRAPHY

[1] M. Z. Win and R. A. Scholtz,"Impulse Radio: How It Works," *IEEE Communications Letters*, vol. 2, no. 2, pp. 36-38, February 1998.

[2] F. Ramirez-Mireles and R. A. Scholtz, "Multiple-Access Performance Limits with Time-Hopping and Pulse Position Modulation," *Proceedings of IEEE MILCOM 1998*, vol. 2, pp. 529-533, October 1998.

[3] M. Z. Win and R. A. Scholtz, "Ultra-Wide Bandwidth Time-Hopping Spread-Spectrum Impulse Radio for Wireless Multiple-Access Communications," *IEEE Transactions on Communications*, vol. 48, no. 4, pp. 679-689, April 2000.

[4] R. A. Scholtz, R. Weaver, E. Homier, J. Lee, P. Hilmes, A. Taha, and R. Wilson, "UWB Radio Deployment Challenges," *Proceedings of IEEE PIMRC 2000*, vol. 1, pp. 620-625, September 2000.

[5] J. Foerster, E. Green, S. Somayazulu, and D. Leeper, "Ultra-Wideband Technology for Short- or Medium-Range Wireless Communications," *Intel Technology Journal,* Q2, 2001.

[6] http://www.weca.net

[7] A. Mishra and W. Arbaugh, "An Initial Security Analysis of the IEEE 802.1x Security Standard," *Technical Report*, Computer Science Department, University of Maryland, USA, February 6, 2001. URL: http://www.cs.umd.edu/~waa/1x.pdf.

[8] D. A. Joseph, B. S. Manoj, and C. Siva Ram Murthy, "On The Impact of User-Profiles on the Interoperability of Wi-Fi Systems and Cellular Networks," *Technical Report*, Department of Computer Science and Engineering, Indian Institute of Technology, Madras, India, October 2003.

[9] V. Sekar, B. S. Manoj, and C. Siva Ram Murthy, "Routing for a Single Interface MCN Architecture and Pricing Schemes for Data Traffic in Multi-Hop Cellular Networks," *Proceedings of IEEE ICC 2003*, vol. 2, pp. 967-973, May 2003.

[10] Y. D. Lin, Y. C. Hsu, K. W. Oyang, T. C. Tsai, and D. S. Yang, "Multi-Hop Wireless IEEE 802.11 LANs: A Prototype Implementation," *Journal of Communications and Networks*, vol. 2, no. 4, December 2000.

[11] V. Sekar, B. S. Manoj, and C. Siva Ram Murthy, "A Framework for Interoperability of Wi-Fi Hotspots and Wide Area Packet Cellular Networks," *Technical Report*, Department of Computer Science and Engineering, Indian Institute of Technology, Madras, India, May 2003.

[12] R. Ramaswami and K. Sivarajan, *Optical Networks: A Practical Perspective*, Second Edition, Morgan Kaufmann, 2001.

[13] C. Siva Ram Murthy and G. Mohan, *WDM Optical Networks: Concepts, Design, and Algorithms*, Prentice Hall PTR, New Jersey, 2001.

[14] IEEE Standards Board, "Part 11: Wireless LAN Medium Access Control (MAC) and Physical Layer (PHY) Specifications," in *The Institute of Electrical and Electronics Engineers, Inc.*, 1997.

[15] J. Mitola, "The Software Radio Architecture," *IEEE Communications Magazine*, vol. 33, no. 5, pp. 26-38, May 1995.

[16] R. I. Lackey and D. W. Upmal, "Speakeasy: The Military Software Radio," vol. 33, no. 5, pp. 56-61, May 1995.

[17] B. S. Manoj, B. Ranjan, S. S. Doshi, I. Karthigeyan, and C. Siva Ram Murthy, "Meghadoot: A Packet Radio Network Architecture for Rural Communities," *Proceedings of NCC 2004: 10th National Conference on Communications*, pp. 205-209, January 2004.

[18] D. B. Johnson and D. A. Maltz, "Dynamic Source Routing in Ad Hoc Wireless Networks," *Mobile Computing*, Kluwer Academic Publishers, vol. 353, pp. 153-181, 1996.

[19] http://www.picopeta.com

ABBREVIATIONS

ABAM	Associativity-Based Ad hoc Multicast Routing
ABR	Associativity-Based Routing
ACK	Acknowledgment
ACPI	Advanced Configuration and Power Interface
ACTP	Application Controlled Transport Protocol
A-GSM	Ad hoc GSM
AIFS	Arbitration IFS
AM	Amplitude Modulation
AMPS	Advanced Mobile Phone System
AMRIS	Ad hoc Multicast Routing Protocol Utilizing Increasing Id-numbers
AMRoute	Ad hoc Multicast Routing
ANDA	Ad hoc Network Design Algorithm
AODV	Ad hoc On-demand Distance Vector Routing
AP	Access Point
AQR	Asynchronous QoS Routing
ARAN	Authenticated Routing for Ad hoc Networks
ARC	Adaptive Transmission Rate Control
ARF	Automatic Rate Fallback
ARPANET	Advanced Research Project Agency Network
ARS	Ad hoc Relaying Station
ASK	Amplitude Shift Keying
ATCP	Ad hoc TCP
ATM	Asynchronous Transfer Mode
ATP	Ad hoc Transport Protocol
BA	Billing Agency
BAAR	Base-Assisted Ad hoc Routing
BAMAC	Battery-Aware MAC
BAMP	Biconnected Augmentation Min-max Power
BEB	Binary Exponential Back-off
BEE	Battery Energy-Efficient Routing
BEMRP	Bandwidth-Efficient Multicast Routing Protocol

BER	Bit Error Rate
BFSK	Binary FSK
BIP	Broadcast Incremental Power
BLIMST	Broadcast Link-based Minimum Spanning Tree
BLU	Broadcast Least Unicast
BMBP	Base-driven Multi-hop Bridging Protocol
BPSK	Binary PSK
BR	Bandwidth Routing Protocol
BRAN	Broadband Radio Access Network
BS	Base Station
BSA	Basic Service Area
BSR	Base Station Supported Reconfiguration
BSS	Basic Service Set
BTMA	Busy Tone Multiple Access
BTS	Base Transceiver Station
CA	Certifying Authority
CAM	Core-to-group Address Mapping
CAMP	Core Assisted Mesh Protocol
CATA	Collision Avoidance Time Allocation
CB8	Cycle-Based 802.11phone
CBR	Constant Bit Rate
CCA	Clear Channel Assessment
CCK	Complementary Code Keying
CCS	Credit Clearance Service
CDMA	Code Division Multiple Access
CDPD	Cellular Digital Packet Data
CEDAR	Core Extraction Distributed Ad hoc Routing
CFS	Cost Function Switcher
CGSR	Cluster-head Gateway Switch Routing
CMMBCR	Conditional Min-Max Battery Cost Routing
CMP	Connected Min-max Power
COA	Care of Address
COMPOW	Common Power Protocol
CRC	Cyclic Redundancy Check
CSMA	Carrier Sense Multiple Access
CSMA/CA	CSMA with Collision Avoidance
CSMA/CD	CSMA with Collision Detection
CTMDP	Continuous Time Markovian Decision Process
CTS	Clear-To-Send
CW	Contention Window
DAG	Directed Acyclic Graph
DAMPS	Digital-AMPS
DBASE	Distributed Bandwidth Allocation/Sharing/Extension

DBPSK	Differential BPSK
DBTMA	Dual Busy Tone Multiple Access
DCF	Distributed Coordination Function
DCMP	Dynamic Core-based Multicast Routing Protocol
DDM	Differential Destination Multicast Routing
DECT	Digital Enhanced Cordless Telecommunications
DHT	Distributed Hash Table
DIFS	DCF IFS
DLL	Data Link Layer
DLPS	Distributed Laxity-based Priority Scheduling
D-MAC	Directional MAC
DoS	Denial of Service
DPM	Dynamic Power Management
D-PRMA	Distributed Packet Reservation Multiple Access
DPS	Distributed Priority Scheduling
DPSK	Differential PSK
DQPSK	Differential QPSK
DRP	DWiLL Routing Protocol
DS	Distribution System
DSDV	Destination Sequenced Distance Vector Routing
DSL	Digital Subscriber Loop
DSP	Digital Signal Processor
DSR	Dynamic Source Routing
DSSS	Direct Sequence Spread Spectrum
DTROC	Distributed TDMA Receiver Oriented Channel Assignment
DVS	Dynamic Voltage Scaling
DWiLL	Directional Throughput-Enhanced Wireless in Local Loop
DWOP	Distributed Wireless Ordering Protocol
E^2MRP	Energy-Efficient Multicast Routing Protocol
EAP	Extendible Authentication Protocol
EAR	Eavesdrop And Register
ECN	Explicit Congestion Notification
EDCF	Extended DCF
EDGE	Enhanced Data Rates for GSM Evolution
EESR	End-to-End Successful Transmission Reimbursement Scheme
EIFS	Extended IFS
EIRP	Effective Isotropically Radiated Power
ELN	Explicit Loss Notification
E-NAV	Extended NAV
ERSR	End-to-End and Retransmit Attempts Scheme of Reimbursement

ESRT	Event-to-Sink Reliable Transport
ESS	Extended Service Set
ETSI	European Telecommunications Standards Institute
EY-NPMA	Elimination Yield Non-Preemptive Multiple Access
FA	Foreign Agent
FAMA	Floor Acquisition Multiple Access
FAMA-NTR	FAMA-Non-persistent Transmit Request
FCC	Federal Communications Commission
FDD	Frequency Division Duplex
FDDI	Fiber Distributed Data Interface
FDMA	Frequency Division Multiple Access
FEC	Forward Error Correction
FGMP	Forwarding Group Multicast Routing Protocol
FGMP-RA	FGMP-Receiver Advertising
FHSS	Frequency-Hopping Spread Spectrum
FIFO	First-In-First-Out
FM	Frequency Modulation
FORP	Flow Oriented Routing Protocol
FPRP	Five-Phase Reservation Protocol
FSK	Frequency Shift Keying
FSR	Fisheye State Routing
FSU	Fixed Subscriber Unit
FTP	File Transfer Protocol
FWA	Fixed Wireless Access
GFSK	Gaussian FSK
GHT	Geographic Hash Table
GMSK	Gaussian MSK
GN	Gateway Node
GPRS	General Packet Radio Service
GPS	Global Positioning System
GPSR	Greedy Perimeter Stateless Routing
GSM	Global System for Mobile Communications
HA	Home Agent
HAG	Hybrid Approach Gateway
HAWAII	Handoff Aware Wireless Access Internet Infrastructure
HCF	Hybrid Coordination Function
HHSR	Hop-by-Hop Successful Delivery Reimbursement
HIPERLAN	High-Performance Radio LAN
HLR	Home Location Register
HRMA	Hop Reservation Multiple Access
HRSR	Hop-by-Hop and Retransmit Attempts Scheme for Reimbursement

HSCSD	High-Speed Circuit-Switched Data
HSR	Hierarchical State Routing
HTML	Hypertext Markup Language
HTTP	Hypertext Transfer Protocol
HWN	Hybrid Wireless Network
IARP	Intra-zone Routing Protocol
IBAR	Infrastructure-Based Ad hoc Routing
IBSS	Independent Basic Service Set
iCAR	Integrated Cellular Ad hoc Relaying System
ICMP	Internet Control Message Protocol
ICSMA	Interleaved Carrier-Sense Multiple Access
IDU	Interface Data Unit
IEEE	Institute of Electrical and Electronics Engineers
IERP	Inter-zone Routing Protocol
IETF	Internet Engineering Task Force
IF	Intermediate Frequency
IFS	Inter-Frame Spacing
IMT	International Mobile Telecommunications
IN	Infrastructure Node
INSENS	Intrusion Tolerant Routing in Wireless Sensor Networks
IP	Internet Protocol
IrDA	Infrared Data Association
IS	Interim Standard
ISDN	Integrated Services Digital Network
ISM	Industrial, Scientific, Medical
ISO	International Organization for Standardization
ISP	Internet Service Provider
ITCP	Indirect TCP
ITU	International Telecommunications Union
L2CAP	Logical Link Control and Adaptation Protocol
LAN	Local Area Network
LAP	Location Addressed Protocol
LAR	Location Aided Routing
LBAR	Load Balanced Ad hoc Routing
LCC	Least Cluster Change
LCT	Link Cost Table
LEACH	Low-Energy Adaptive Clustering Hierarchy
LEAP	Localized Encryption and Authentication Protocol
LET	Link Expiration Time
LIFS	Long IFS
LLG	Least Loaded Gateway
LMDS	Local Multipoint Distribution Service

LORA	Least Overhead Routing Approach
LR-WPAN	Low-Rate Wireless PAN
MAC	Medium Access Control
MACA	Multiple Access Collision Avoidance
MACA-BI	MACA-By Invitation
MACA/PR	MACA with Piggy-backed Reservation
MADF	Mobile Assisted Data Forwarding
MAN	Metropolitan Area Network
MANET	Mobile Ad hoc Network
MAODV	Multicast Ad hoc On-demand Distance Vector Routing
MARCH	Media Access with Reduced Handshake
MBCR	Minimum Battery Cost Routing
MC	Multi-hop Channel
MCEDAR	Multicast Core-Extraction Distributed Ad hoc Routing
MCG	Multi-hop Connectivity Graph
MCN	Multi-hop Cellular Network
MCSMA	Multichannel CSMA
MDS	Multi-Dimensional Scaling
MECP	Minimum Energy Consumed Per Packet
MEMS	Micro-Electro Mechanical Systems
MEP	Minimum Exposure Path
MFP	Minimum Path Loss with Forward Progress
MH	Mobile Host
MINA	Multi-hop Infrastructure Network Architecture
ML	Multi-Lateration
MMAC	Multichannel MAC
MMBCR	Min-Max Battery Cost Routing
MMDS	Multichannel Multipoint Distribution Service
MMNC	Minimum Maximum Node Cost
MN	Mobile Node
MPCR	Minimum Power Consumption Routing
MPD	Multipath Dispersion Heuristic
MPDU	MAC Protocol Data Unit
MPL	Minimum Path Loss
MPSP	Multicast Priority Scheduling Protocol
MRL	Message Routing Layer
MRT	Multicast Routing Table
MS	Mobile Station
MSC	Mobile Switching Center
MSDU	MAC Service Data Unit
MSK	Minimum Shift Keying
MSPR	Multiple Exclusive Shortest Path Routing
MST	Minimum Spanning Tree
MT	Mobile Terminal

M-TCP	Mobile TCP
MuPAC	Multi-Power Architecture for Cellular Networks
MWLAN	Multi-hop Wireless LAN
MZRP	Multicast Zone Routing Protocol
NAV	Network Allocation Vector
NDPL	Neighbor Degree-based Preferred Link Algorithm
NFL	Neighbor Feedback Loop
NRG	Nearest Ring Gateway
NSMP	Neighbor Supporting Ad hoc Multicast Routing Protocol
NTS	Not-To-Send
ODMA	Opportunity-Driven Multiple Access
ODMRP	On-Demand Multicast Routing Protocol
OFDM	Orthogonal Frequency Division Multiplexing
OLMQR	On-demand Link-state Multipath QoS Routing
OLSR	Optimized Link State Routing
OQR	On-demand QoS Routing
ORA	Optimum Routing Approach
OSI	Open Systems Interconnection
OTA	Over-The-Air
OWWDM	Optical Wireless WDM
PACS	Personal Access Communication System
PAM	Pulse Amplitude Modulation
PAN	Personal Area Network
PC	Point Coordinator
PCF	Point Coordination Function
PCM	Pulse Code Modulation
PDA	Personal Digital Assistant
PDC	Personal Digital Cellular
PDCR-PDBR	Position-Dependent Cost-Reimbursement, Position-Dependent Bill-Rate
PDRS	Preferred Destination-Ring Routing Scheme
PDU	Protocol Data Unit
PEGASIS	Power-Efficient Gathering for Sensor Information Systems
PHRS	Preferred Hybrid Routing Scheme
PHS	Personal Handyphone System
PICR	Position-Independent Cost-Reimbursement
PICR-PDBR	PICR-Position-Dependent Bill-Rate
PIFS	PCF IFS
PIRS	Preferred Inner-Ring Routing Scheme
PKI	Public Key Infrastructure
PLBA	Preferred Link-Based Algorithm
PLBM	Preferred Link-Based Multicast Routing

PLBQR	Predictive Location-Based QoS Routing
PLBR	Preferred Link-Based Routing
PLBU	Preferred Link-Based Unified Routing
PM	Phase Modulation
PN	Pseudo-random Noise
PORS	Preferred Outer-Ring Routing Scheme
PPM	Pulse Position Modulation
PRF	Pulse Repetition Frequency
PRMA	Packet Reservation Multiple Access
PRTMAC	Proactive RTMAC
PSFQ	Pump Slowly Fetch Quickly
PSK	Phase Shift Keying
PSRS	Preferred Source-Ring Routing Scheme
PSTN	Public Switched Telephone Network
QAM	Quadrature Amplitude Modulation
QBSS	QoS Basic Service Set
QoS	Quality of Service
QPSK	Quadrature PSK
RBAR	Receiver-Based Autorate Protocol
RC	Route Cache
RET	Route Expiry Time
RF	Radio Frequency
RI-BTMA	Receiver-Initiated BTMA
RLL	Radio in Local Loop
RP	Radio Port
RSSI	Received Signal Strength Indicator
RSU	Radio Subscriber Unit
RSVP	Resource Reservation Protocol
RT	Routing Table
RTMAC	Real-Time MAC
RTS	Request-To-Send
RTU	Radio Transceiver Unit
SAP	Service Access Point
SAR	Security-aware Ad hoc Routing
SBS	Smart Battery Standard
SC	Single-hop Channel
SCG	Single-hop Connectivity Graph
SCN	Single-hop Cellular Network
SDMA	Space Division Multiple Access
SDU	Service Data Unit
SEAD	Secure Efficient Ad hoc Distance Vector Routing
SIFS	Short IFS

SIM	Subscriber Identity Module
SMACS	Self-organizing MAC for Sensor Networks
SMCN	Single-Interface MCN
SMECN	Small Minimum Energy Communication Network
SMRP	SMCN Routing Protocol
SMS	Short Messaging Service
SMTP	Simple Mail Transfer Protocol
SNEP	Sensor Network Encryption Protocol
SNGF	Stateless Non-Deterministic Geographic Forwarding
SNR	Signal-to-Noise Ratio
SOPRANO	Self-Organizing Packet Radio Networks with Overlay
SPIN	Sensor Protocols for Information via Negotiation
SPINS	Security Protocols for Sensor Networks
SRMA/PA	Soft Reservation Multiple Access with Priority Assignment
SS7	Signaling System 7
SSA	Signal Stability-based Adaptive Routing
STA	Station
STAR	Source-Tree Adaptive Routing
STP	Signaling Transfer Point
SWAN	Stateless Wireless Ad hoc Network
TCP	Transmission Control Protocol
TCP-BuS	TCP with Buffering Capability and Sequence Information
TCP-ELFN	TCP-Explicit Link Failure Notification
TCP-F	TCP Feedback
TCP/IP	Transmission Control Protocol/Internet Protocol
TCP SACK	TCP with Selective ACK
TDD	Time Division Duplex
TDMA	Time Division Multiple Access
TDR	Trigger-based Distributed QoS Routing
TELNET	Terminal Network
TGe	Task Group e
TIMIP	Terminal Independent Mobility for IP
TORA	Temporally Ordered Routing Algorithm
TTCP	Transaction-oriented TCP
TTL	Time To Live
TTP	Trusted Third Party
TWiLL	Throughput Enhanced Wireless in Local Loop
UCAN	Unified Cellular and Ad hoc Network
UDP	User Datagram Protocol
UMTS	Universal Mobile Telecommunications System
UNPF	Unified Network Protocol Framework

UWB	Ultra-Wide Band
UWC	Universal Wireless Communications
VBR	Variable Bit Rate
VDCF	Virtual DCF
VLR	Visitor Location Register
VMS	Velocity Monotonic Scheduling
VoIP	Voice over IP
VPN	Virtual Private Network
WAE	Wireless Application Environment
WAN	Wide Area Network
WAP	Wireless Application Protocol
WARM	Wireless Ad hoc Real-time Multicasting
WATM	Wireless ATM
WBM	Weight-Based Multicast Routing
WBPL	Weight-Based Preferred Link Algorithm
W-CDMA	Wideband CDMA
WDM	Wavelength Division Multiplexing
WDP	Wireless Datagram Protocol
WEP	Wired Equivalent Privacy
Wi-Fi	Wireless Fidelity
WiLL	Wireless in Local Loop
WINS	Wireless Integrated Network Sensor
WLAN	Wireless LAN
WLL	Wireless in Local Loop
WMAN	Wireless MAN
WML	Wireless Markup Language
WPAN	Wireless PAN
WRP	Wireless Routing Protocol
WSP	Wireless Session Protocol
WTLS	Wireless Transport Layer Security
WTP	Wireless Transaction Protocol
WWAN	Wireless WAN
ZHLS	Zone-based Hierarchical Link State Routing
ZRP	Zone Routing Protocol

INDEX

C

D

M

R

T

U

V

ABOUT THE AUTHORS

C. SIVA RAM MURTHY received the B.Tech. degree in Electronics and Communications Engineering from Regional Engineering College (now National Institute of Technology), Warangal, India, in 1982, the M.Tech. degree in Computer Engineering from the Indian Institute of Technology (IIT), Kharagpur, India, in 1984, and the Ph.D. degree in Computer Science from the Indian Institute of Science, Bangalore, India, in 1988.

He joined the Department of Computer Science and Engineering, IIT, Madras, as a Lecturer in September 1988 and became an Assistant Professor in August 1989 and an Associate Professor in May 1995. He has been a Professor with the same department since September 2000. He has held visiting positions at the German National Research Centre for Information Technology (GMD), Bonn, Germany, the University of Stuttgart, Germany, the University of Freiburg, Germany, the Swiss Federal Institute of Technology (EPFL), Switzerland, and the University of Washington, Seattle, USA.

Dr. Murthy has to his credit over 100 research papers in international journals and over 75 international conference publications. He is the co-author of the textbooks *Parallel Computers: Architecture and Programming* (Prentice-Hall of India, New Delhi, India), *New Parallel Algorithms for Direct Solution of Linear Equations* (John Wiley & Sons, Inc., New York, USA), *Resource Management in Real-time Systems and Networks* (MIT Press, Cambridge, Massachusetts, USA), and *WDM Optical Networks: Concepts, Design, and Algorithms* (Prentice Hall PTR, Upper Saddle River, New Jersey, USA; reprinted by Prentice-Hall of India, New Delhi, India). His research interests include parallel and distributed computing, real-time systems, lightwave networks, and wireless networks.

Dr. Murthy is a recipient of the Sheshgiri Kaikini Medal for the Best Ph.D. Thesis from the Indian Institute of Science and the Indian National Science Academy (INSA) Medal for Young Scientists. He is a co-recipient of Best Paper Awards from the 1st Inter Research Institute Student Seminar (IRISS) in Computer Science, the 5th IEEE International Workshop on Parallel and Distributed Real-Time Systems (WPDRTS), and the 6th International Conference on High Performance Computing (HiPC). He is a Fellow of the Indian National Academy of Engineering.

B. S. MANOJ completed his undergraduate work in 1995 and received the M.Tech. degree in 1998 in Electronics and Communication Engineering from the Institution of Engineers (India) and Pondicherry Central University, Pondicherry, India, respectively. From 1998 to 2000, he worked as a Senior Engineer with Banyan Networks Pvt. Ltd., Chennai, India, where his primary responsibility was design and development of protocols for real-time traffic support in data networks. He currently is an Infosys doctoral student in the Department of Computer Science and Engineering at the Indian Institute of Technology, Madras, where he focuses on the development of architectures and protocols for ad hoc wireless networks and on next-generation hybrid wireless network architectures. He is a recipient of the Indian Science Congress Association Young Scientist Award for the Year 2003. His current research interests include ad hoc wireless networks, next-generation wireless architectures, and wireless sensor networks.